# LA COTTE DE ST. BRELADE

*The rock arch at La Cotte de St. Brelade, seen from the south ravine*

*A century ago the Pleistocene deposits reached the level of the large horizontal joint halfway up the face (the ladder at the bottom of the site provides a scale).*
*Drawn to scale by Joanna Richards, from photographs.*

# LA COTTE DE ST. BRELADE
## 1961–1978

## EXCAVATIONS BY C. B. M. McBURNEY

Edited by

## P. Callow and J. M. Cornford

G. Brochet
P. Callow
C. Cartwright
J. Chaline
J. M. Coles
J. M. Cornford
M. N. Le Coustumer
A. P. Currant
H. Duroy
H. Frame

P. Giresse
F. Hivernel
J. C. C. Hutcheson
J. Huxtable
P. R. Jones
R. L. Jones
D. H. Keen
J. P. Lautridou
M. N. Levant

A. E. Mourant
J. T. Renouf
P. J. Rose
K. Scott
C. A. Shell
C. B. Stringer
B. J. Szabo
A. S. Vincent
B. van Vliet-Lanoë
D. Walton

GEO BOOKS
NORWICH

Published by Geo Books, 34 Duke St., Norwich NR3 3AP.

© Geo Books 1986.

Text typeset in 10 and 8 pt Lasercomp Times Roman at Oxford University Computing Service, after preparation at Cambridge University Computing Service.

Printed in Great Britain at the University Press, Cambridge.

*British Library Cataloguing in Publication Data*

La Cotte de St. Brelade 1961–1978: excavations by C. B. M. McBurney.

1. Excavations (Archaeology) — Channel Islands — Jersey
2. Palaeolithic period — Channel Islands — Jersey
3. Cotte de Saint-Brelade site (Jersey)
4. Jersey — Antiquities

I. Callow, Paul
II. Cornford, Jean M.

936.2'34101    GN772.22.G7

ISBN 0–86094–207–4

# CONTENTS

# FIGURES

# TABLES

# MICROFICHE

## APPENDICES IN MICROFICHE

# CONTRIBUTORS

Dr G. BROCHET, Institut des Sciences de la Terre, 6 Bvd. Gabriel, 21100 Dijon, France.

Dr P. CALLOW, University Computer Laboratory, Corn Exchange St., Cambridge CB2 3QG, UK.

Ms C. CARTWRIGHT, Institute of Archaeology, 31–34 Gordon Sq., London WC1H 0PY, UK.

Dr J. CHALINE, Institut des Sciences de la Terre, 6 Bvd. Gabriel, 21100 Dijon, France.

Prof. J.M. COLES, FBA, Dept. of Archaeology, Downing St., Cambridge CB2 3DZ, UK.

Mrs J.M. CORNFORD, c/o Dept. of Archaeology, Downing St., Cambridge CB2 3DZ, UK.

Dr M.N. LE COUSTUMER, Centre de Géomorphologie du CNRS, Rue des Tilleuls, 14000 Caen, France.

Dr A.P. CURRANT, Dept. of Palaeontology, British Museum (Natural History), Cromwell Rd., London SW7 5BD, UK.

Dr H. DUROY, Centre de Géomorphologie du CNRS, Rue des Tilleuls, 14000 Caen, France.

Ms. H. FRAME, Donald Baden-Powell Quaternary Research Centre, 60 Banbury Rd., Oxford OX2 6PN, UK.

Prof. P. GIRESSE, Centre de Recherches de Sédimentologie Marine, Université de Perpignan, Ave. de Villeneuve, 66025 Perpignan, France.

Dr F. HIVERNEL, c/o Dept. of Archaeology, Downing St., Cambridge CB2 3DZ, UK.

Mr J.C.C. HUTCHESON, c/o Dept. of Archaeology, Downing St., Cambridge CB2 3DZ, UK.

Mrs J. HUXTABLE, Research Laboratory for Archaeology and the History of Art, 6 Keble Rd., Oxford OX1 3QJ, UK.

Mr P.R. JONES, P.O. Box 49, Arusha, Tanzania.

Dr R.L. JONES, Dept. of Geography, Lanchester Polytechnic, Priory St., Coventry CV1 5EZ, UK.

Dr D.H. KEEN, Dept. of Geography, Lanchester Polytechnic, Priory St., Coventry CV1 5EZ, UK.

Dr J.P. LAUTRIDOU, Centre de Géomorphologie du CNRS, Rue des Tilleuls, 14000 Caen, France.

Dr M.N. LEVANT, Centre de Géomorphologie du CNRS, Rue des Tilleuls, 14000 Caen, France.

Dr A.E. MOURANT, FRS, The Dower House, Maison de Haut, Longueville, St. Saviour, Jersey, Channel Islands.

Dr J.T. RENOUF, Education Dept., P.O. Box 142, Highlands, St. Saviour, Jersey, Channel Islands.

Ms P.J. ROSE, Dept. of Archaeology, Downing St., Cambridge CB2 3DZ, UK.

Dr K. SCOTT, Dept. of Archaeology, Downing St., Cambridge CB2 3DZ, UK.

Dr C.A. SHELL, Dept. of Archaeology, Downing St., Cambridge CB2 3DZ, UK.

Dr C.B. STRINGER, Dept. of Palaeontology, British Museum (Natural History), Cromwell Rd., London SW7 5BD, UK.

Dr B.J. SZABO, Branch of Isotope Geology, US Geological Survey, Box 25046 MS 963, Denver Federal Center, Denver, Colorado 80225, USA.

Ms A.S. VINCENT, P.O. Box 49, Arusha, Tanzania.

Dr B. VAN VLIET-LANOE, Centre de Géomorphologie du CNRS, Rue des Tilleuls, 14000 Caen, France.

Dr D. WALTON, Dept. of Physics, McMaster University, Hamilton, Ontario L8S 4L8, Canada.

# FOREWORD

The site of La Cotte de St. Brelade has been the subject of excavations and investigations for one hundred years, from its discovery in 1881 to the great work of Professor Charles McBurney in 1961–62 and 1966–78. It is the purpose of this monograph to set out some of the most salient features of the site as they have been deduced from the latest excavations.

Charles McBurney was my first instructor at Cambridge and in early 1979 it was an honour to accept from him an invitation to see to completion the report on his work at La Cotte, should he be unable to finish the project himself. By late 1979 the work of analysis and interpretation was already under way, and the nucleus of a fine post-excavation team had been assembled by Charles in the persons of Paul Callow, Jean Cornford and Kate Scott. In December 1979 Charles McBurney died and I assumed overall responsibility for the project.

No one will appreciate more than I do that the preparation of a report on such a complex site would be a difficult task for the excavator himself, and to someone unfamiliar with the day-to-day progress of the work, and the gradual development of thought processes about it, the problems are compounded. Yet the report now set out with its many analytical and interpretative elements would, I think and hope, have pleased the excavator.

In manuscript and in several interim reports there are preserved short statements by Charles McBurney in which he set out his aims for the research project at La Cotte. One of these was "to identify the nature and where possible the sources of raw materials ... made use of by the Pleistocene inhabitants of the cave". Another was the establishment of a chronological framework "to clarify and amplify further the geological and palaeontological observations bearing on the age of the settlements". A third was the understanding of the agencies which created and affected the huge bone heaps so spectacularly recovered from the site and now conserved for display and study; "in the long run we hope to have a detailed record which may add a completely new feature to our knowledge of the habits of ancient hunters at this time". A fourth aim, derived from his own archaeological philosophy, was the development of scientific techniques for the study of such intractable material as the lithic and bone assemblages represented; "like all products of human activity which survive from the prehistoric past they combine with a measure of mutual similarity a substantial spectrum of variation. It is part of my purpose to distinguish and define these ... and to examine such associated circumstances as may help to explain them".

There can be no doubt that the results presented here do fulfil many of these aims. And in particular what I would see as the dominant theme for Charles McBurney has been largely realised by the development of the project over the years. This was to comprehend the place of La Cotte de St. Brelade in the whole sequence of events in the later Pleistocene; for him, La Cotte was not the problem nor the sole subject for enquiry, but represented "one more element in the mosaic of changes which link the Lower to the Middle Palaeolithic throughout Western Europe".

The major task in early 1980 was to provide sufficient financial stability to allow the work to proceed, and here I record my personal gratitude to all those museums, colleges, societies, foundations and trusts which responded readily and generously to our applications, and to those individuals who advised and contributed to the project. Every application to every agency was met with sympathy and generosity, and I take this as nothing more than a direct tribute to the memory of Charles McBurney in his great vision and endeavour at La Cotte. The list of acknowledgements printed here in no way can repay those who responded so well and who ensured the continuation of the work. In all of this, the owners of the site, La Société Jersiaise, generously supported the project from its inception.

Since late 1979 the day-to-day work of analysis and interpretation has been the responsibility of Paul Callow, Jean Cornford and Kate Scott, and my thanks go to all three for their dedication and persistence. In particular, I thank Paul Callow who has initiated or undertaken the majority of the investigations into the site structure and the lithic industries, ably assisted by a small army of helpers. And I thank Jean Cornford for a variety of tasks undertaken, among them the literary editing of the many contributions. To Mrs. Anne McBurney I owe an immeasurable debt of gratitude for her constant support and encouragement, and her tireless work in helping the professional team to prepare the volume.

It must be emphasised that this cannot be the book which Charles McBurney himself would have written, though it incorporates and develops many of his ideas. Instead, it is a book written by numerous participants who have made their own original contributions and who have taken the lines of attack most suitable to their own concepts of archaeology in the 1980s, a development of which I am sure Charles would approve.

John Coles

# EDITOR'S PREFACE

When Charles McBurney directed a very limited excavation in the north ravine at La Cotte de St. Brelade in 1961 no one could have foreseen that from this small beginning would develop the largest research project so far undertaken by a predominantly British-based (if international) team into the Old Stone Age of Europe; indeed, until the final years of fieldwork the budget was tiny and the manpower was chiefly supplied by small groups of students. Undaunted by the size and complexity of the site the team carried out the removal of enormous quantities of archaeologically sterile rubble, before being rewarded with the recovery of extraordinarily rich stone industries and the fragile remains of mammoth, woolly rhinoceros and other animals. The justification for such a considerable effort lies in the extreme importance of La Cotte, not only for Palaeolithic archaeology but for Pleistocene stratigraphy. We have here a unique combination: a very long and strongly varying sequence of human occupation, extending back some quarter of a million years and associated with good dating and environmental evidence, and (because of the site's coastal location) an opportunity to study the relationship between marine and terrestrial responses to climatic change during the Middle and Upper Pleistocene.

Charles McBurney's connections with the site began in the 1950s, when he was consulted on several occasions by the then excavator, Father Christian Burdo. Although La Cotte dominated the work of his later years, his interests were always far broader. His investigations in north Africa during and after the second World War, culminating in the publication in 1967 of his great monograph on the Haua Fteah cave, are universally recognised as an immense achievement. A doctoral thesis on the stone tools of Neanderthal man, excavations at Palaeolithic sites in Britain and the Middle East, and research into such diverse fields as microscopic wear traces, Palaeolithic art, and the Old Stone Age of the USSR ... these are eloquent testimony to the extraordinary development of his experience over the years, and to the technical, logistical and intellectual mastery of his subject which was invaluable in the face of the challenge presented by La Cotte. (For a more extensive account of his career the reader's attention is drawn to that by Clark and Wilkinson 1986). Nor should it be forgotten that, in functioning as a training excavation for the Department of Archaeology at Cambridge between 1961 and 1978, La Cotte formed an integral part of his teaching — that other side to his contribution to Stone Age archaeology. In the foreword to a volume dedicated to the memory of Charles McBurney, Professor Grahame Clark referred to the dispersal of his pupils to academic positions throughout the world, writing that "nothing would have given him greater pleasure than the knowledge that for years to come undergraduates in four continents would be grounded in the knowledge of Palaeolithic man to which he had devoted his own life" (in Bailey and Callow 1986, xiv). It may be no exaggeration, therefore, to suggest that the La Cotte 'experience' will continue to exercise, for many years to come, a subtle influence over anglophone thinking about prehistory.

The reader may care to know something of the editors' early association with the project to explain how it came about that this responsibility fell to them. JC was Charles McBurney's research assistant from 1966 onwards, working at first on the typological and statistical analysis of material from several sites and, from 1972, almost entirely on the processing and classification of the artefacts from La Cotte; she took part in the 1969 and 1970 field seasons. PC dug at the site as a student between 1968 and 1973. He was a co-author of the preliminary report which appeared in the 1971 issue of the *Proceedings of the Prehistoric Society*, and spent some time collecting comparative data from French sites with a view to writing a doctoral thesis centred on La Cotte; this was later abandoned in favour of a study on the application of statistical and computer techniques to British and French handaxe series. In 1979, while a Lecturer at the University of Khartoum, he was invited by Charles McBurney (by then seriously ill) to return to Cambridge and fill the need for "an understudy who could carry on to completion the final report for publication on La Cotte, if it should prove necessary (that is to say if [McBurney's] health was inadequate) for someone else to take over" (as McBurney wrote to Professor Grahame Clark). This was facilitated by his appointment later that year to a post created with the aid of a grant from the Science Research Council to investigate industrial variation at the site. With Charles McBurney's death at the end of 1979 the roles of grantholder and Director passed to Dr. (later Professor) John Coles, who despite numerous other commitments took on general responsibility for the project, which of course imposed many demands beyond those of research and publication.

Some statistics will give an idea of the scale of La Cotte and of the operations. The depth of deposit was about 30 m at the turn of the century (the cumulative thickness was substantially greater); of this, some 11 m remained to be explored from 1961 onwards. The Cambridge excavations produced about 130,000 artefacts in all, a figure greatly exceeding the total number of Lower and Middle Palaeolithic finds listed for the British mainland in the *Gazetteer* published in 1968 by the Council for British Archaeology; in fact the overall total for La Cotte is probably nearer 200,000 if earlier finds are included. In the years 1976–78 alone some 80,000 pieces were recovered. To study the evidence produced by this work — a far more expensive process than the excavations themselves — it has been necessary to gather together a large team of specialists from several countries. At the time of Charles McBurney's death the record of his ideas about the post-1970 discoveries and his thoughts about them was necessarily incomplete. Washing, marking and preliminary classification of the finds was still in progress, and most of the specialist investigations remained to be done. Resolution of basic stratigraphic questions and the definition of artefact assemblages for detailed study were completed in early 1981, while the last of the vitally important sedimentological evidence was not available until 1983. Because of the timing of these events, and because the employment of all Cambridge-employed staff on fixed-term contracts imposed a time limit, it was not possible to conduct the research in text-book order; as events proved, moreover, considerable modifications to the original interpretation of the site were required.

Emphasis has been placed on providing as full a report as possible, perhaps at the cost of retaining some structural and editorial imperfections. The entire text was typed on the computer, as being the best way to provide the flexibility needed to cope with frequent changes of interpretation. Then, once arrangements had been finalised with the publisher, it was possible after further processing to use the Joint Academic Network to transfer the text electronically to Oxford, where it was set on Oxford University Computing Service's Lasercomp phototypesetter; most of the tables were set from magnetic tape on the publisher's own equipment.

The 1966–78 finds are in the possession of the owners of the site, La Société Jersiaise, through whose generosity those from 1961 and 1962 have been supplied to, respectively, the Museum of Archaeology and Ethnography (University of Cambridge) and the British Museum.

Finally, we should like to dedicate this book to the memory of past excavators of La Cotte, and to the people of Jersey.

## ACKNOWLEDGEMENTS

Financial support from the following individuals and grant-giving bodies is gratefully recognised:

*EXCAVATION*

1961–78 The Crowther-Beynon Fund; 1976 The British Academy; 1976–78 The British Museum; 1982 The Jersey Heritage Trust.

*POST-EXCAVATION*:

HRH The Prince of Wales; The Boise Fund; The British Academy; The British Museum; Professor J. M. Coles; Corpus Christi College, Cambridge; Corpus Christi College, The Master's Fund; The Crowther-Beynon Fund; The Department of Archaeology, University of Cambridge; The Jersey Heritage Trust; The L. S. B. Leakey Trust; The Leverhulme Trust for a Leverhulme Research Fellowship held by Dr. P. Callow during 1979–82; The Manpower Services Commission for the provision of staff; The Meaker Trust; The Mourant Trust; Miss P. J. Rose; The Science Research Council (now the Science and Engineering Research Council) for research grants GR/A 1711.8, GR/A 58357 and GR/B 21072 awarded to the late Professor C. B. M. McBurney; La Société Jersiaise; The Society of Antiquaries of London.

*PUBLICATION SUBVENTIONS*:

Deputy Michael Bonn, Jersey (for the colour plate); The Jersey Heritage Trust; La Société Jersiaise.

It is appropriate to mention here the particular gratitude of one of the editors (PC) to his wife Margaret for financial support during the academic year 1983–84, so making possible the completion of his contributions to the book.

It is not possible to mention by name all those who in one way or another have contributed directly to the excavations and subsequent work — numbering respectively perhaps 150 and 70 people. But our special gratitude is due to Deputy Michael Bonn, Mrs. Margaret Finlaison and the Mourant Trust committee for the part they played in raising urgently-needed funds from Jersey, to the late Sir Edward Boulton for his generosity in allowing the excavators to camp on his land from 1973 onwards, and to Mr. George Drew, Acting Curator of the Museum of La Société Jersiaise, for his cooperation and practical assistance on innumerable occasions over the years. Without the support of these and other members of the Société the research and publication would have been impossible.

Our thanks are also due to Cambridge University's Department of Archaeology for providing work areas and other facilities for many years, to the University Museum of Archaeology and Ethnography for storage space, technical and practical help, and to the University Computer Laboratory for both resources and expert advice. Kate Scott led the work of bone conservation, ably assisted by Jill Foley, Ralph Weyment and Veronica Scott, while Lawrence Smith prepared the microtine fauna for study. Pamela Rose, Richard Burnell and Simon Newham all gave freely of their time in the later stages of the project. Photographic preparation of the figures was carried out by Gwil Owen. The work of numerous illustrators is gratefully acknowledged. Phil Dean drew the quartz artefacts; Pam Lucas and Ian Bennett inked the principal section drawings (Figs. 6.3 to 6.7) after compilation by Pamela Rose; Joanna Richards drew the frontispiece. Particular thanks are due to Hazel Martingell for undertaking the considerable task of drawing the flint and 'stone' artefacts.

Our debt to our co-authors and their parent institutions, and to those others who are mentioned in the text as having contributed expert assistance, will be apparent. Nevertheless, Dr. Jean-Pierre Lautridou must be singled out for gaining the collaboration of so many colleagues in the study of the complex geology of the site; their task was an exceptionally difficult one because many of the sections had been walled up. The characteristic generosity of the late Professor François Bordes, in agreeing to the use of his unpublished data for some of the French industries, is gratefully acknowledged; without this the multivariate analysis described in chapter 32 would have been impossible. And the perceptive reader is likely to recognise the influence of Professor Henry de Lumley's account of the

Hortus cave, as a model for publication that we have attempted to follow.

Special recognition is due to Professor John Coles for his support and advice over the years. Without his determined efforts to guarantee the continuing financial security of the project the work could never have reached completion. And we must pay tribute to the dedication of Mrs. Anne McBurney, expressed in hard work and encouragement, to the goal of bringing to fruition this the eventual product of her husband's labours.

We thank the authors, publishers and editors who have kindly agreed to our use of their copyright material for quotation and illustration as acknowledged in the text, captions and bibliography. Copyright holders not immediately identifiable from the standard format for bibliographic references, and whose illustrations have with permission formed the basis of (or been incorporated in) our own figures, include: Alan R. Liss, Inc. (Fig. 3.1), Elsevier Scientific Publishing Co. (Fig. 3.3), The Geologists' Association (Fig. 4.1), The British Geological Survey (for Her Majesty's Stationery Office, Fig. 5.4), and The Controller, Her Majesty's Stationery Office, and The Hydrographer of the Navy (Figs. 22.3 and 31.2). It must be stressed that in no case does citation of material in this volume alter the ownership or nature of any reserved rights; reproduction of any such material, in any form, is therefore not permitted without the agreement of the holder(s) of those rights.

# PART I

## INTRODUCTION AND BACKGROUND

# INTRODUCTION

P. Callow

Jersey, the most southerly of the Channel Islands, very roughly approximates a 15 by 10 km rectangle aligned east/west, with an area of 11,655 ha (Fig. 1.1). Its permanent population (76,050 in the April 1981 census) is augmented each year by some half million visitors, drawn by attractive scenery, low prices and a climate a degree or so warmer than that of the English south coast (Table 1.1).

The island, whose position on the continental shelf plays an important part in the events described in this book, consists in essence of a plateau about 100 m above sea-level, dipping somewhat towards the south. A considerable proportion of the interior is mantled by loess;

this is a fine silt laid down during the ice ages by winds blowing from areas where a thin cover of vegetation was insufficient to hold down the soil. It was during the colder periods also that banks of loess and rock rubble, known as 'head', built up at the foot of many of the steeper slopes as a result of soil movement accelerated by the alternation of freeze and thaw conditions. Such deposits are particularly evident along the coastal cliffs, wherever they have not been completely removed by marine erosion, and they must have been even more extensive formerly. The loess and head are important features of the Jersey landscape today; also when taken together with buried soils and ancient beach deposits, they

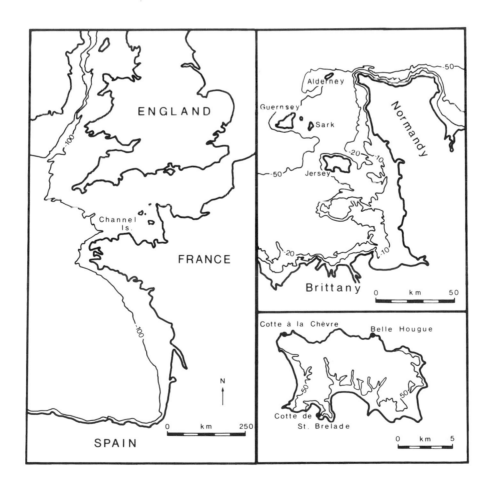

**Fig. 1.1** Location of La Cotte de St. Brelade and the topography of the surrounding area.

| MAJOR DIVISIONS | ERA | PERIOD | EPOCH | | EVENTS IN ARMORICA AND THE ENGLISH CHANNEL |
|---|---|---|---|---|---|
| PHANEROZOIC | Cenozoic | Quaternary | Holocene | | Eustatic sea-level changes (in the range −100 m to c.40 m O.D.) as response to ice sheet formation and destruction. Alternating submergence and exposure of the Channel floor. |
| | | | Pleistocene | 0.01 | |
| | | | | 2 | |
| | | Tertiary — Neogene | Pliocene | 5 | Tertiary sea-level changes (known range from below O.D. to c.130 m O.D.) resulting from tectonic movements. Cutting of significant marine platforms notable below 8 m O.D. and above c.90 m O.D. |
| | | | Miocene | 25 | |
| | | Tertiary — Palaeogene | Oligocene | | |
| | | | Eocene | | Eocene submergence of the Channel floor with limestone deposition in embayments between the islands. Proto Channel Islands probably recognisable. |
| | | | Palaeocene | 65 | |
| | Mesozoic | Cretaceous | | | Cretaceous submergence of the Channel floor with Chalk deposition. Relief distinction between Armorica (high) and Channel floor (low). |
| | | Jurassic | | 144 | |
| | | Triassic | | 213 | Atlantic Ocean begins to open. Triassic downfaulting of Western Approaches initiates the separation of Cornubia from Armorica. Paris and Hampshire/Dorset basins form. |
| | | | | 248 | |
| | Palaeozoic | Permian | | 286 | Hercynian orogeny. Imposition of new structural lines in Armorica and the enhancement of older (Cadomian) ones. |
| | | Carboniferous | | 360 | |
| | | Devonian | | 408 | Shallow seas over Armorica. Grès Armoricain deposited during the Lower Ordovician. |
| | | Silurian | | 438 | |
| | | Ordovician | | 505 | Main period of granitic intrusion in northern Armorica. |
| | | Cambrian | | 590 | |
| PROTEROZOIC (PRE-CAMBRIAN) | | | | | Cadomian orogeny (c.700 to 500 Ma ago). Imposition of major structural grain of northern Armorica. |
| | | Isotopic dates in millions of years Ma | | | Brioverian sediments deposited in open seas and oceans bordering continental margins of uncertain position (1000 to 700 Ma). |
| | | | | | 1000 to 2000 Ma – largely unrepresented. |
| | | | | | Pentevrian (3000 to 2000 Ma) – sedimentation and intrusion followed by metamorphism. |
| | | 4500 | | | |

Fig. 1.2 The evolution of the western English Channel area.

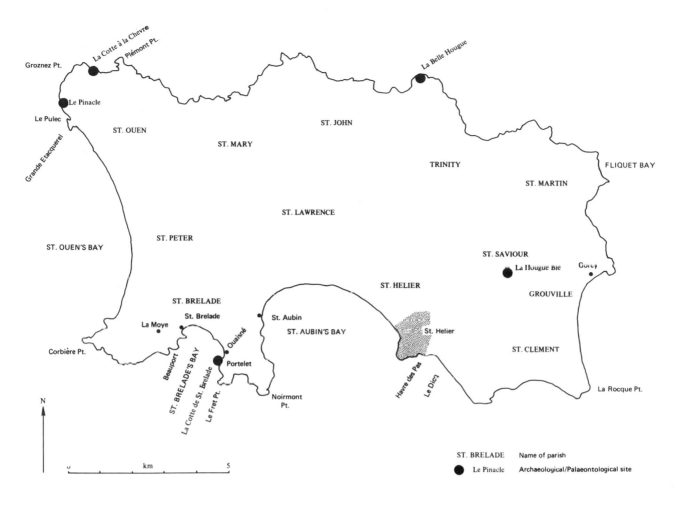

**Fig. 1.3** Jersey: location map for place names mentioned in this and other chapters.

provide the geological evidence required for a reconstruction of the evolution of that landscape during the Quaternary period, in the later stages of which man first visited the area and left traces of his sojourn.

The impact of man himself upon the environment, leaving aside the more obvious effects of urban development and transport systems, is particularly apparent in the vegetation. Much of the island is given over to agriculture (which plays an important though now much diminished role in its economy following the development of tourism and finance), and few areas remain more or less unaffected. Some of the promontories and the head deposits mantling the cliffs carry a covering of heather, gorse and bracken; most of the low-lying marshland has been drained.

Striking evidence of longer-term natural change is provided by 'submerged forests' and peat containing fossil pollen. An example of the former is visible at spring tides in St. Helier harbour, where submerged peats also occur and were studied by H. and M. E. Godwin (1952), who demonstrated that alder (implying damp conditions) was the dominant feature until the low-lying areas were flooded by the rising sea during the post-glacial period.

This relatively small marine transgression was, of course, only the final expression of the rise in sea-level since the coldest part of the last ice age, when the ice-sheets held a greater proportion of the world's free water than is the case today. At such times Jersey became part of the European mainland, as did the British Isles.

The story of the development of the English Channel area is of still wider significance when we come to interpret the evidence of man's earliest visits. The geology determines the general form of the sea-bed and the islands. It accounts for specific features of the landscape which attracted man or preserve the evidence of his former presence; it also controls the availability of raw materials suitable for the manufacture of stone tools. Fig. 1.2 summarises the geological evolution of the region, therefore, and sets the scene for more detailed discussion in later chapters; Fig. 1.3 is a location map, giving place names mentioned in this and other chapters.

To the modern visitor, Jersey's various coasts present startling differences in appearance. The north is characterised by precipitous cliffs and numerous small coves. To the east, reefs are bared over a distance of some 2 km seawards at low tide, while to the west a single

**Fig. 1.4** Aerial view of La Cotte de St. Brelade from the northwest. The north and west entrances are visible in the nearer headland. Note the break of slope marking the line of the dead cliff in the bay beyond, where the mantle of 'head' is being eroded by the modern sea (photo J. T. Renouf).

sweep of sand backed by high dunes runs for 6 km along St. Ouen's Bay. There are three main stretches of beach on the southern coast of the island, separated by rocky headlands; the most easterly runs from La Rocque to Jersey's principal town, St. Helier. St. Aubin's Bay, in the middle of the coast, has St. Helier at its eastern end, while to the west is a projecting stretch of the plateau fringed by cliffs and extending from Noirmont Point to Le Fret Point. West of the latter is St. Brelade's Bay, with the village of that name on its western side. Over the past few years this bay has been developed for tourism, with the building of hotels, to take advantage of the sheltered and picturesque situation and of the very good bathing.

From the village the east side of the bay is seen to be dominated by two reddish granite promontories. The most southerly land in view is Le Fret Point, but the sweep of the shore between this and the end of the sands, at Ouaisné, is broken by the massive protrusion of La Cotte Point, running up to the top of Portelet Common at about 70 m above mean sea-level (Figs. 1.3, 1.4–1.6). As the tide goes down (the range is very great, around 10 m) it becomes possible to scramble around these headlands. In making the circuit of La Cotte Point, the visitor who is not in too much of a hurry to get into the next bay may glance inland, and be startled to see a great fissure, 15 m wide, running deep into the granite and terminating in a precipitous cliff (Fig. 1.7). Though

**Table 1.1** Climatic Data for Jersey (averaged over 82 years)[a]

| | |
|---|---|
| Temperature | (°C) |
| Mean | 11.4 |
| Yearly maximum | 24.3 |
| Yearly minimum | -4.1 |
| Seawater mean | 12.3 |
| Relative humidity (%) | 79.4 |
| Rainfall (mm per annum) | 836.4 |
| Sunshine (hours per annum) | 1949.3 |
| (% of possible) | 43.8 |
| Mean wind speed (mph) | 9.7 |
| Predominant winds | WSW & ENE |
| | *days* |
| Fine | 60 |
| Light cloud | 160 |
| Overcast | 140 |
| Rain | 191 |
| Frost | 5 |
| Ground frost | 45 |
| Fog | 34 |
| Snowfall | 11 |
| Storm | 13 |

[a]From Rey 1980.

**Fig. 1.5** La Cotte Point from the west at low tide (the shadow in the west ravine is cast by the south pinnacle).

he is likely to be impressed by the spectacle of the damage which the power of the sea has wrought upon the solid rock, the full drama of the place may escape him. For he stands at the entrance to La Cotte de St. Brelade, with its record of a quarter of a million years of human prehistory, and the geological events which accompanied this.

At the further end of the ravine, on the left, a dark opening appears to lead into a cave, though closer inspection reveals that in reality a massive rock arch here bridges a second ravine at right angles to the first. The northern extremity of this chasm is visible as an insignificant fissure from the St. Brelade side, while south of the main entrance it continues right through to the next bay. The system of ravines thus forms a letter T, with the stem to the west (1–4 in Fig. 1.8). Though the fissures were originally filled with rubble to about 40 m above sea-level, much of this has been removed during a century of excavation.

A great volume of deposit, with its archaeological contents, was dug out before the Second World War (Fig. 1.9); this consisted almost entirely of material dat-ing to the last ice age. In 1950 however, the late Father Christian Burdo, S.J., (who had already done much work on the upper layers) began excavation at a lower level and demonstrated the existence of much earlier layers with plentiful evidence of human occupation. Forced by the infirmity of old age to cease major operations in 1956, he was keen that work should continue at the site, and that this should be conducted to the highest possible standard. Therefore, between 1961 and 1978 a further programme of excavations was undertaken by a team from the University of Cambridge led by the late Professor Charles McBurney, and provided the greater part of the material finds described in this book (though the more poorly recorded, but very important, collections arising from the earlier work are also considered where appropriate).

If the discovery of Neanderthal remains (teeth from the upper and lower jaw and, subsequently, a small fragment of a child's skull) was the aspect of the site which gave La Cotte its importance before the 1914–1918 war, it is the long archaeological and environmental sequence and its interest for both early prehistory and quaternary

**Fig. 1.6** La Cotte Point from the southeast, showing the two pinnacles and the entrance to the small south ravine.

**Fig. 1.7** The west ravine and the rock arch in 1970. The rock floor in the foreground is awash at high tide. The shelf of rock on the left is probably a relict of the last interglacial.

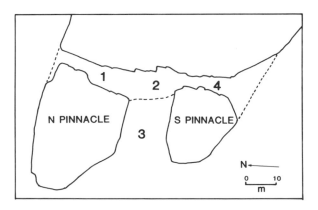

**Fig. 1.8** La Cotte Point, showing its principal features. Key: 1 The north ravine (the site of much of the early work as well as that of Burdo and McBurney); 2 the main area of excavation in 1936–1940; 3 the west ravine (entered by the sea at high tide); 4 the south ravine (partly excavated in 1936–1940).

**Fig. 1.9** (A) The cave entrance in the late 19th century, at the beginning of the excavations, with deposits almost reaching the rock arch. (B) The west ravine in 1910 (compare with Fig. 1.7). In particular, note the head banked up on the right; this was partly removed between 1936 and 1940 (photo E. F. Guiton).

stratigraphy which are the features most sharply brought into focus by the recent work. Between 1961 and 1978 well over 100,000 stone artefacts were recovered from deposits antedating the last interglacial period. In addition to these, and to the expected dense masses of food refuse, there were found the remains of two (or possibly three) major accumulations of mammoth and woolly rhinoceros bones, arguably the result of a specialised hunting technique. Moreover, the re-investigation of a depositional succession with a former total thickness of about 30 m has demonstrated that the record extends back through at least two complex interglacial periods before our own (Fig. 1.10). A more detailed history of the excavations is given in the next chapter.

It must be stressed that the bottom of the sequence at the site has not yet been reached. The 'deep sounding' which was cut in the floor of the north ravine in 1961 and extended in 1962 was reopened by the writer in 1982. Although the results were important, a number of very large blocks of granite impeded progress and it proved impossible to deepen the cuttings to bedrock (which could indeed be far below the lowest level reached, which was about 9 m above mean sea-level). Thus we do not know at what date the filling of the ravines, and perhaps the human occupation, began (the two events need not have been contemporary). For the same reason the timing and detailed mechanism of the opening of the fissures is uncertain, though it is possible to reach some general conclusions about their development (chapter 4).

One further point should be brought to the reader's attention before the excavations are discussed at length. The archaeological finds have all been made within the ravines (the earliest occupation deposits survive only in the northern one). The sea has destroyed everything outside; above, the surface of the headland is almost bare of soil. The area in which excavations have taken place is therefore very restricted, but this is a product of geological processes. If we are to see La Cotte through the eyes of prehistoric man we must take a different view, one which encompasses the whole of the headland and its immediate surroundings. Visitors to the excavations in the north ravine often commented that it was a bleak and inhospitable place (a view heartily endorsed by generations of students) and that it seemed incomprehensible that it should once have been occupied to the extent that its richness would suggest. In fact, conditions are not always so unpleasant, and the degree of discomfort at any time is greatly affected by the strength and direction of the wind (when this is from the west it is funnelled straight through the site). On the other hand, the southern entrance and the modern beach on the same side are sunny and sheltered from the wind, though they offer no protection against rain. Also, the top of the headland, though exposed, provides the best view over the area (Fig. 1.11). So La Cotte is best thought of as a site *complex*, different areas of which might be exploited according to circumstances and of which there remains direct evidence of only a very small part.

**Fig. 1.10** Inside the north ravine at the end of the 1978 season, looking northwards. The step in the cutting is at about the top of layer C (the scale is 1 m long).

**Fig. 1.11**   The view from the top of La Cotte Point, looking towards Beauport.

# 2

# THE HISTORY OF INVESTIGATIONS AT LA COTTE

A. E. Mourant and P. Callow

The history of research at La Cotte de St. Brelade is long (the centenary was celebrated in 1981) and complicated. Numerous amateur and professional investigators have been concerned in the work; both their excavation strategy and their interpretation of their observations have evolved in company with the development of prehistoric archaeology and Pleistocene geology from the early days of both subjects. For the modern investigator, reading the numerous published accounts of this work provides a fascinating, if sometimes frustrating, insight into his own intellectual antecedents: fascinating because of the interplay of preconception and discovery, frustrating because so much of the site's contents have gone for ever. As will be seen later, the order of events at La Cotte ensured that most of the more recent layers (recent in the sense that they represent perhaps the latest third or so of the site's known history) were stripped away before the 1939–1945 war, with rather uneven recording of the discoveries. In recompense, however, this drastic process made it possible to explore much older deposits whose very existence had been discounted by early workers.

## 2.1 THE DISCOVERY

The most generally conspicuous evidence for man's presence in Jersey in prehistoric times is provided by the numerous burial mounds ('hougues') and megaliths of neolithic age, many of which play a part in the rich folklore of the island. In 1685 the historian Jean Poingdestre, attempting a more rational explanation, suggested that the largest such monument (La Hougue Bie) had been built for defensive purposes against the Danes in order to obtain early warning of approaching ships. In subsequent centuries a certain amount of excavation (mostly mere pillaging) took place as the graves' significance was realised, though of course it was not until much later that an idea of their date was obtained. Similarly, in the early years of the 19th century a number of raised beaches and 'submerged forests' were known in Jersey, but the association of some of these with the presence of man (so important in our modern reconstruction of the island's past) was at that time not appreciated.

In 1861 two boys, both aged 16, Samuel Dancaster and Joseph Sinel, were scrambling around the steep granite cliffs of the northwest corner of Jersey when they found flint chippings in an old sea cave 18 m above modern sea-level, known as La Cotte à la Chèvre. This was the first recorded find of evidence of the presence of palaeolithic man in Jersey, but it went unnoticed at the time.

Sinel wrote later (1914, 52) that it was the presence of flint which struck the boys "as something unusual in a flintless island like ours. But like Peter Bell with his primrose, we took no lesson from them, and 'Palaeolithic man' was in those days a doubtful quantity. In fact I do not think we had yet heard of him." Few investigators anywhere in Europe had yet realised that human artefacts were present in the raised beaches and river terraces of the continent, and that they were tens of thousands of years old. It was after all only two years since the publication of Darwin's *The Origin of Species*, and the announcements at the Royal Society and the British Association by several distinguished British scientists (notably John Evans and Joseph Prestwich) that they accepted that Boucher de Perthes' finds of flint implements in the Somme valley were indeed from the terrace gravels, and therefore of considerable antiquity. The remains of Neanderthal man had come to light in Germany in 1857, but were far from being generally accepted as ancient.

In 1881 Dancaster was exploring the cliffs of the southwest of the island in the company of T. Saunders when they found chipped flints at the mouth of a high-level cave known as La Cotte, now known (to distinguish it from the other cave) as La Cotte de St. Brelade. It is pure coincidence that has given similar names, otherwise unique in the island, to these two palaeolithic sites, for they are so named on the Godfray map of 1849, long before the artefacts were discovered.

By the time La Cotte de St. Brelade was discovered there was a general interest in the occurrence of palaeolithic man in Europe, and Dancaster and Sinel returned to La Cotte à la Chèvre and began excavations there. These early excavations were carried out mainly by Sinel and Dancaster alone, and without the official participation of the Société Jersiaise. Subsequent excavators, in addition to Sinel, were Father C. Noury, the geologist, Captain (later Major) N. V. L. Rybot, Mr. J. Sinel, Junior and, in 1911, Dr. R. R. Marett.

## 2.2 THE ROLE OF THE SOCIETE JERSIAISE

Reference has been made above to the Société Jersiaise; it will be helpful to digress at this point in order to explain the nature of this body and the role played by it and its members in the development of Jersey archae-

**Fig. 2.1** (A) Participants in the excavations at La Cotte de St. Brelade, about 1914; seated (left to right) are R. R. Marett, E. T. Nicolle and J. Sinel (photo R. Mollet). (B) Christian Burdo (left) and Charles McBurney in Burdo's laboratory in 1959 (photo Jersey Evening Post).

ology, with special reference to La Cotte.

The society, as the title page of its Bulletin proclaims, was founded in 1873 and exists 'for the study of the History, the Language, the Geology, the Natural History, and the antiquities of the Island and their preservation; also the publication of historical documents'. This reflects a degree of expansion of activity since it was first constituted; however, it is worth noting that of its original three subcommittees, one was archaeological in nature. Regular reports from the various activity sections, together with articles on a wide range of subjects connected with the island, appear in the annual Bulletin, which has missed only four issues since it was first produced in 1875 (a centenary edition published in 1973 summarises the work of the society to that date).

Until comparatively recently, the greater part of the archaeological efforts of the society were concerned with the two palaeolithic caves and the numerous neolithic monuments (of which the best known is La Hougue Bie), though important work was also done at later and multi-period sites — Le Pinacle (neolithic to 2nd century A.D.) and Grosnez Castle, for instance. After the Second World War there was a decline in activity until 1970 (apart from the continuing investigations at La Cotte itself), but the section is flourishing once again and many new excavations have been carried out recently, either under local direction or in conjunction with British universities.

By 1878 a small museum had been formed, consisting mainly of 'curiosities', though subsequently the collections were limited to material of local interest. In 1893 the present building was acquired and has since served both as headquarters and principal museum. Since 1977 the archaeological collections have been housed in a complex of buildings at La Hougue Bie. To this day the museum remains an organ of the Société Jersiaise, and has not been taken over by the States of Jersey.

The work at La Cotte has thus to be seen in the context of a flourishing local learned society which rapidly attracted members — often distinguished — from the mainland as well as the island itself. Equally, funding for the research has variously come from private individuals, from the Société Jersiaise, and from many different trusts and scientific bodies. The existence of the Society has facilitated the safeguarding of a number of important archaeological sites, mostly by deed of gift, though the two most important, La Cotte and La Hougue Bie, were acquired by purchase.

## 2.3    SOME BIOGRAPHICAL NOTES

As with many important archaeological sites, the excavations have been conducted by amateurs and professionals from extremely diverse backgrounds (Fig. 2.1). A complete list of participants in the research at La Cotte would be too long to include here, but the following instances may throw some light on their general character.

Among the earlier investigators of La Cotte de St. Brelade mention has already been made of Samuel Harry DANCASTER (1845–1937). He ran a farm 200 yards from the shore towards the northern end of St. Ouen's Bay. With Sinel he was the joint discoverer in 1861 of prehistoric human implements at La Cotte à la Chèvre, and in the 1880s they were the first to carry out systematic excavations there. In 1881 Dancaster and Saunders jointly discovered prehistoric implements at La Cotte de St. Brelade.

In 1902, when Sinel was spending a weekend at the farm, as he often did, Dancaster summoned his guest one morning to witness a great exposure of the submerged peat and forest beds in the bay, from which the thick layers of superincumbent sand had been washed away overnight. The scene was the subject of dramatic, and now classic, photographs taken by Sinel. Despite his friendship with Sinel and his obvious scientific interests, Dancaster does not appear to have taken part in any excavations other than at La Cotte à la Chèvre, and he was not well known to Société Jersiaise archaeologists.

His life-long friend Joseph SINEL (1844–1929), at first mainly concerned with La Cotte à la Chèvre, later took a very large part in the work at La Cotte de St. Brelade. Many years later, in 1924, the first author got to know him well during the excavation of Jersey's principal neolithic monument, the passage grave of La Hougue Bie.

Sinel left school at 15 and entered the still existing firm of Voisins, where by 1868 he had become manager of the furniture department. But his heart was not in shopkeeping, and he then took the plunge and became a self-employed professional naturalist, primarily a taxidermist, but also a lecturer and writer of successful books on natural history. In 1881 or thereabouts he set up 'The Aquarium' at Havre des Pas, which for many years supplied the needs of institutions in Britain for living and preserved marine biological specimens. In 1907 he became curator of the museum of the Société Jersiaise, and he, more than any other man, is its creator. The splendid marine biology room is still almost as he left it. Both before and after he joined the Museum he took a large part in several archaeological excavations besides those mentioned. He continued to work at La Cotte until the end of the early excavations, in 1917 or 1918.

In the 1870s and 1880s a number of members of the Société took part in excavating several minor megalithic monuments, but they do not appear to have taken much interest in the palaeolithic sites; little archaeological work of any kind was done in the 1890s. In 1901, however, Edmund Toulmin NICOLLE became joint Honorary Secretary. Though he was a barrister and held the office of Vicomte, his chief interest in life now became the Société Jersiaise. He did much to recruit members to the society and to encourage them to take part in its activities. It was he who drew the writer into working with the Archaeological Section. He and Sinel, with others to be mentioned, initiated a new period of active excavation of known archaeological sites and the discovery of new ones, culminating in the justification of Nicolle's intuition and courage in purchasing and excavating the mound of La Hougue Bie.

The early years of this century saw also the entry into archaeological work of Major N. V. L. RYBOT, at first only on his leaves in Jersey. However, after his traumatic experiences in the First World War he was invalided out of the Army and took a very active part in archaeological and historical research. He was a skilled artist and draughtsman and for a great many years was almost the sole archaeological draughtsman of the Société.

Mr. E. F. GUITON too was recruited at this time and, as a highly skilled photographer, made very large contributions to the Société's archaeological work.

Dr. R. R. MARETT, Reader in Social Anthropology at the University of Oxford and Rector of Exeter College, spent much time excavating in Jersey during academic vacations. With an international reputation in the world of anthropology and archaeology, he was a valuable ambassador of the Société and played a large part in securing financial support for archaeological work in Jersey. He was also able to recruit archaeologists at Oxford to take part in excavations on a larger scale than could have been managed by the Société alone. Outside the island La Cotte became known as 'Dr. Marett's Cave'.

Another early investigator was Dr. Andrew DUNLOP. He was a medical practitioner who was also a highly competent geologist and the first man to apply geological principles to the Pleistocene deposits of Jersey.

The owner of the cave, the Seigneur of Noirmont, Mr. G. F. B. DE GRUCHY, was a keen archaeologist who not only encouraged the excavations but also took a considerable part in them. He generously presented to the Société all the objects found on the site, and after his death his heirs sold the site to the Société for a nominal sum.

The Rev. Father Christian BURDO (1881–1961) was a member of the Society of Jesus (the Jesuits), a Chevalier of the French Légion d'Honneur, a Doctor of Philosophy, and ultimately a Membre d'Honneur of the Société Jersiaise (Fig. 2.1B). He first came to Jersey in 1901, and studied at Maison St. Louis. There he was a fellow student of Father Pierre Teilhard de Chardin, who became a life-long friend. After qualifying at Maison St. Louis he studied and taught in England, Paris and the Middle East. In 1924 he returned to Maison St. Louis as Professor of Philosophy and Anthropology. At first his main local scientific interest was geology (to which he had probably been introduced by Teilhard de Chardin), but subsequently his attention shifted to archaeology and, as described below, he directed important excavations at the Pinnacle and at La Cotte de St. Brelade. He was a man of broad interests with a great capacity for friendship, and became a much loved and active member of the Société Jersiaise.

## 2.4 EXCAVATIONS PRIOR TO THE 1939–1945 WAR

Some idea of the scope of the successive phases of excavation up to the present time may be obtained from the simplified diagram of Fig. 2.2.

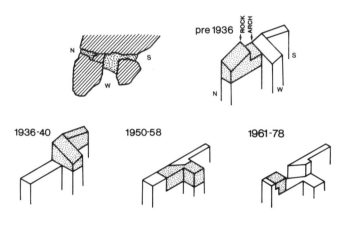

**Fig. 2.2** Stages in the removal of the deposits at La Cotte de St. Brelade.

In the early years of the century several beginnings were made to excavate La Cotte de St. Brelade, but each ended when the danger of overhanging talus on the site became too great. It was only in 1910 that the Société decided to employ quarrymen to clear the overburden and render excavation relatively safe. No plans or diagrams were published with the report of the first season's work. The deposits were described as layers of black soil, which proved to be a combination of ashes, carbonised wood and clay, mixed with whitish masses of bone detritus, compacted into a breccia. It was on 25 August 1910 that the most important find of all was made, that of nine human teeth. The original description reads: "In one part of the most coherent bone mass had been the right half of a human lower jaw, nine teeth being ranged side by side in original position but unfortunately no trace of the once supporting bone was apparent" (Nicolle and Sinel 1911, 72). The excavators present on this occasion were Dr. Paul Chappuis, Mr. A. H. Barreau, Sinel, Dunlop, Nicolle and Guiton, together with Mr. E. Daghorn (a quarryman) and a labourer whose name has not been preserved.

It is appropriate to anticipate and to quote here the account of the discovery of four further teeth in 1911: "On examining some small portions of the clay that had adhered to the rock on the left of the cave, at the spot where human teeth had been found during the previous exploration, four more of these teeth were discovered, bringing the total up to thirteen" (Nicolle and Sinel 1912, 214). The teeth were shown by Dr. (later Sir Arthur) Keith to belong to the Neanderthal type of man (Keith and Knowles 1912a).

To return to the 1910 excavations, numerous flint implements were found, and identified as of Mousterian type, which was consistent with the identification of the human remains. Bones, horns and teeth of rhinoceros, reindeer, horse and bovidae were also found. As on many previous occasions, dangers from falling stones caused the excavations to be brought to a close, but after attention by quarrymen they were resumed in 1911.

By the end of the 1911 digging season the occupation horizons, some 3–4 feet thick, had been excavated over an area some 11 feet square. Once again the dangers of overhanging sterile head stopped further work. A small excavation was carried out in 1912 on the south side of the gully, facing the main cave. In 1913 the British Association for the Advancement of Science, on the initiative of Marett, formed a committee to sponsor further work, and in the years 1914 and 1915 the area excavated was extended to a total of 1200 square feet. Gradually the number of flint fragments grew to a total of over 5000, and the number of animal and bird species represented by bones increased.

Early in the Société Jersiaise excavations it had been realised that the rock structure was not a true cave but a rock arch open to north and south, and that most of the contents had probably entered through a sort of funnel on the north side. It was however a complete surprise when on 3 September 1915 the overburden amounting to some 500 tons suddenly collapsed, and the excavators then at work were fortunate to escape with their lives and indeed without injury. This fall of course put an end to the season's excavations, and a further 200 tons of rubble fell during the winter. Nevertheless, a partial clearance of the fallen material was made in 1916 and excavations were continued on a limited scale in the western part of the deposits, which were found to slope upwards considerably when pursued northwards. Numerous identifiable animal bones were found, and well above the main occupation horizons, some 20–30 feet above the floor level and towards the western wall, was a rich bed of rodent remains.

Work was continued in 1917, without professional labour, but with the participation of a number of boys from Victoria College working under the late Mr. A. J. Robinson. They included Cyril Le Marquand who, before his recent much regretted death, was of great help to the Société in planning for the completion of the La Cotte work in which he was proud to have participated. It was thought in 1918 that little remained to be done in order to complete the excavation of the main cave down to the lowest level of human occupation.

A further rodent bed higher up near the eastern wall was explored in 1918 and 1919. In the latter year, the excavations were visited by Teilhard de Chardin, who recorded his impressions in a letter dated 5 September (Teilhard de Chardin 1965, 305–6); perhaps surprisingly, he seems not to have been to the site when he was in the island before the war, despite his interest in archaeology and palaeontology.

Most unfortunately, and indeed tragically, when the material from these very fruitful years of excavation from 1910 to 1919 came to be examined it was concluded that there was virtually no difference in culture or in fauna between the different levels within the main 3–4 feet of occupation levels. Therefore all the material, both cultural and faunal, was, with a very few exceptions, merged together so that we are now deprived of vital information on what were almost certainly important variations from one bed to the next. Thus the fruits of a vast quantity of devoted work have to a large extent lost their value.

It was about 1921 that Robinson introduced A. E. Mourant to the site. The top rodent bed was then exposed, and Mourant and his brother, in all innocence, did a fair amount of private collecting from it. This bed remained exposed until 1936, when it was finally excavated by Burdo. No official excavations were carried out between 1919 and 1936.

In a paper published in 1935, Mourant had pointed out that there were still very extensive deposits on the site, which could however only be reached if the overlying (and presumably sterile) deposits of head above them in the west ravine were cleared (Mourant 1935). At a meeting of the Executive Committee on 10 January 1936, Marett referred to this and on his proposal the Committee agreed to such clearance, and made a grant of £50 towards the expenses. On 26 March inspection of the site was carried out by Marett, Burdo, Rybot, Lt.-Col. G. A. Beazeley, Major A. D. B. Godfray, Guiton, Lomax and Mourant, and plans were made for the forthcoming excavations, which were to be, essentially, the examination of the rodent bed, which was still exposed, and the removal of all the deposits in the west ravine, working from a succession of horizontal terraces. After the transport of equipment from the Pinnacle site, and elaborate preparations for the considerable engineering operations necessary, excavations began on 13 May.

In the outcome, most of the planned work on the main gully deposits was carried out by Burdo, Lomax and Godfray, who had previously been working together for some years on the complicated stratified Neolithic to Gallo-Roman site at the Pinnacle in northwest Jersey. The recent limited re-examination of this site has shown how very thorough and accurate this work was, and there can be no doubt that the experience acquired there was of great value in the work at La Cotte. When Burdo first came to La Cotte after his excavations at the Pinnacle he realised from the outset the defects of previous work, particularly on the stratigraphical side, and he was determined to record the precise place where every object was found. Lomax's experience as an engineer, first

applied to archaeological work at the Pinnacle, was also of very great value in the mechanically more difficult work at La Cotte.

These three workers, with some considerable help from others, were able to clear from the west ravine the whole of the archaeologically sterile overburden (all of course carefully examined) and to carry out much work not only on the rodent bed but on the archaeologically rich strata beneath it. Meanwhile Marett worked in the south ravine from time to time. To make work in both places safe it was necessary to quarry away much overhanging rock at the northeast corner of the great south pinnacle, which is bounded by these two fissures. The work in the west ravine added very considerably to knowledge of beds equivalent to those excavated in the main cave before 1918, though not to the point of recovering the lost stratigraphy of those deposits. The key to them may still be buried in the ravine.

**Fig. 2.3** The interior of the north ravine in 1914, showing the trench cut by Captain Coltart into the 'peats' (layer 9.3) underlying the 'Mousterian' levels (photo E. F. Guiton?).

17

**Fig. 2.4** (A) Detail of the fossil cliff, just after its discovery in 1953 (photo C. Burdo?).
(B) The fossil cliff as seen in the east face in 1961 (as left by Burdo).

However, the main value of the great task of clearance was to make possible Burdo's own later investigations both on the lower deposits in the gully and, more important still, his (and McBurney's) excavations in the main cave after the Second World War. His pre-war work can also be regarded, in a more personal sense, as a preparation for the much more important work he

carried out later. The programme planned in 1936 was almost complete when it was brought to a premature conclusion by the German occupation of Jersey in 1940. In 1938 Professor F. E. Zeuner, the eminent Pleistocene geologist and archaeologist, visited Jersey and made a critical examination of the Pleistocene cave sites of the two Cottes and Belle Hougue. At La Cotte de St. Brelade the peaty soil (layer 9.3), which underlay the only archaeological deposits then known, was already recognised in 1910, and up to 1940 was tacitly regarded as the base of the whole cultural sequence (Fig. 2.3), probably with a marine raised beach only a short distance below. However, Zeuner (1940) probed deeper into these deposits and showed that there were two distinct 'peat' layers. He advised further excavation of these, and it was almost certainly he who advised Burdo to look for further archaeological deposits below them.

A. E. M.

## 2.5 BURDO'S SECOND CAMPAIGN OF EXCAVATIONS (1950–1956)

After the hiatus caused by the German occupation, Burdo once more began work at La Cotte. He started in the main ravine, but this time at a lower level and against its north wall. At first he encountered the residue of the 'Mousterian' deposits left by earlier workers and the 'peats' on which they had also commented. These, like the underlying loessic layer in this part of the site, exhibited a very steep dip towards the sea. Initially, very few artefacts were found below the 'peats'; lower down, the concentration of tools increased dramatically and moreover appeared to be of quite a different industry from that previously recovered. Among the discoveries were handaxes of characteristic Acheulian type. These finds occurred in a very stony loessic 'head' with, as the excavation progressed into the north ravine, occasional patches of ashy material. One of the more puzzling aspects of the stratigraphy was that these layers, also, maintained a strong dip towards the sea and yet bore very many traces of human habitation.

An important turning point came in 1953, when the yellowish loessic layer vanished and Burdo found that he was now advancing into a high bank of solid ash. Unfortunately he had not preserved a section on the west side of the cutting, as the cave wall itself had been a convenient datum for measuring the position of his finds. However, in the east section a clear boundary, almost vertical in places, was apparent between these two contrasting deposits (Figs. 2.4A and B). North of this interface, the stratification was essentially horizontal, whereas to the south there was a dip of about 15°. He considered a number of possible explanations for this phenomenon. One, that it was the result of human excavation in prehistoric times in order to construct a trap for wild animals, was speedily rejected; alternatively, he thought, it might be an erosional feature caused by run-off from the headland channelled into the west ravine by the ancient topography. The latter interpretation came closer to the truth, but it was left to Burdo's successor at the site to discover the true reason for this 'fossil cliff'. Its most important implication for Burdo's artefact finds was that those discovered in the steeply dipping loess were in a later stratigraphic context than those further north. Indeed they are not even in place, being derived from the top of the 'cliff'. However, be-

tween 1953 and 1956 (when he closed the excavations) he succeeded in recovering large amounts of *in situ* material. Besides his annual reports on the site in the Bulletin of the Société Jersiaise, he published a short monograph on his work at La Cotte (Burdo 1960).

## 2.6 THE McBURNEY EXCAVATIONS (1961–1978)

When Burdo's health was already failing, he had had discussions about the site and its industries with Dr. (later Professor) C. B. M. McBurney (Fig12.1B), to whom the Société Jersiaise in due course entrusted the continuation of the excavation programme. The principal stages through which the site subsequently developed are sketched in Fig. 2.5.

McBurney was struck by the apparent resemblance between the vertical 'cliff', which had so puzzled Burdo, and the one cut by the sea in the deposits of the Grotte du Prince in Monaco during the last interglacial. If the parallel could be confirmed and a raised beach be shown to exist at the foot of the 'cliff', a valuable chronological datum would thereby be established. In 1961 and 1962, therefore, he concentrated on trenching at a lower level than Burdo had reached (the 'deep sounding' seen in Fig. 2.6A), and on recovering fresh samples of artefacts from a more precise stratigraphic context than previous techniques had permitted. As an aid to studying the finer stratigraphy of the ashy layers and as a permanent record of the cliff, he also took a series of latex 'peels' of the sections (see Fig. 2.6B). In the event, the presence of a marine beach at the foot of the cliff was indeed demonstrated, and has, more than any other feature, proved to be the key to the La Cotte sequence.

Fieldwork by McBurney in the Middle East prevented further work at La Cotte de St. Brelade until 1966, apart from a short visit in 1964 while he was excavating at La Cotte à la Chèvre. Plans were put into effect to open up larger areas than had been possible in 1961–1962 (when Burdo's north section had been straightened). In both 1966 and 1968 a considerable amount of largely sterile overburden (of early Weichselian age) was removed so that the final Saalian loesses could be excavated. In 1966 a number of bones of large mammals were discovered in these layers.

In 1968, when the cutting had been extended both northwards and westwards, it became apparent that the mammoth and woolly rhinoceros bones at the base of layer 6.1 formed a deliberate accumulation, of a kind previously unknown in cave deposits of this antiquity, giving La Cotte a new importance (Fig. 2.7A). The fragility and size of the bones, in decalcified loess, meant that special techniques had to be devised for their safe recovery — a particularly difficult operation because very often they were interlocking. The help of conservation staff from the British Museum (Natural History) was of great value in developing the necessary expertise.

In subsequent work until 1977 the site had therefore to be divided into two parts: in the west, progress was greatly slowed by the need to conserve the faunal remains, whereas in the more exposed area to the east the

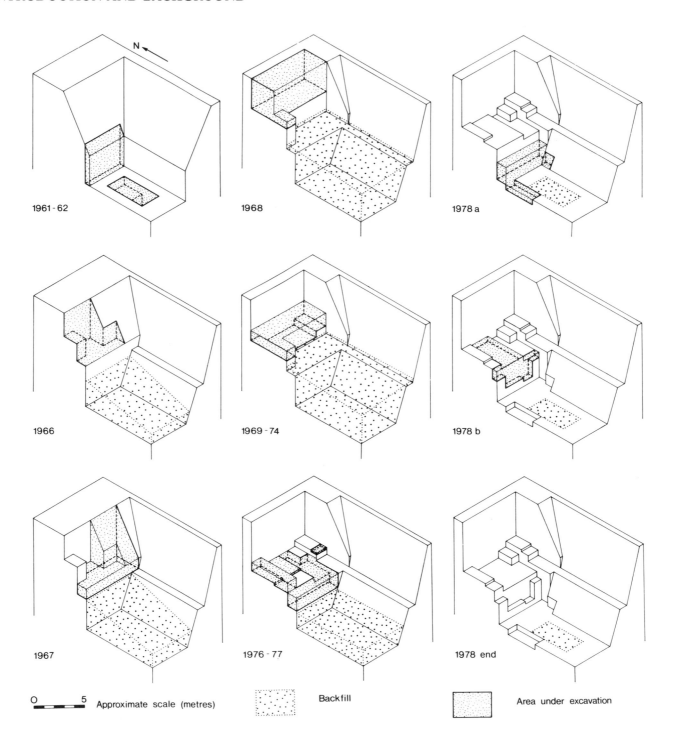

N

1961-62

1968

1978 a

1966

1969-74

1978 b

1967

1976-77

1978 end

O  5  Approximate scale (metres)

Backfill

Area under excavation

**Fig. 2.5** Stages in the removal of the deposits during the McBurney excavations.

bones not only seemed to have been scarcer originally, but what few there were had virtually disappeared as a result of leaching and could not be recovered. The considerable quantities of bone in layer 6 had been completely cleared by 1973 and, after removal of a layer of slabs that had fallen from the west wall in antiquity (see Fig. 2.7B), work commenced on another 'bone heap' directly below the first in an earlier loess, layer 3. By 1976 most of the excavated area had been taken down to the archaeologically rich ashy loess of layer A, and a trench could be sunk in the working floor in an attempt to relocate the underlying layers B and C (the old 1962

**Fig. 2.6** (A) Looking down from the north face of the excavations towards the deep soundings in 1961. (B) Preparation of latex 'peels' of the north and east faces in 1962; the principal deep sounding, enlarged by that time, is in the foreground (the smaller one, close to the rock wall at the entrance to the north ravine, had been abandoned because none of the archaeological material was *in situ*).

**Fig. 2.7**  (A) The upper bone heap (layer 6.1), just after its discovery, with two rhino skulls on top. (B) Granite slabs (part of a single rock fall) between the bone heaps.

**Fig. 2.8** Detail of pebbles in the fossil beach, as re-exposed in 1982.

section had been completely covered up in 1966, during backfilling of Burdo's cutting in the interests of the safety of excavators and site alike).

In the final two years of the excavations under Professor McBurney's direction, the large bones at last no longer impeded progress. With funding (by the British Museum from 1976 onwards) on a larger scale than ever before, the team concentrated on clearing layer A over as large an area as possible and on obtaining further samples from the underlying layers.

After removal of the backfill, the old 1962 north face was cut back in two stages for a further distance of about 140 cm. Because of numerous large boulders, the work proved slower than expected, and hopes of re-opening excavations in the floor of the cave (that left by Burdo, and penetrated only by the limited 'deep soundings' of 1961–1962) were only partly realised before the end of the last digging season in 1978. By that time, though, enormous numbers of artefacts had been recovered; the finds from 1978 exceed the total for 1966–1977. Even from the lowest major stratigraphic unit the sample was quite large enough for analytical purposes. Immediately after the close of work the principal sections were walled in, to protect them from erosion by natural and human agencies, and in the expectation that no further excavations would be undertaken at the site for many years (to allow time for significant advances in scientific techniques to be made).

## 2.7 FURTHER INVESTIGATIONS AT THE SITE (1980–1982)

In September 1980 Callow visited Jersey with two aims. Firstly, it was necessary to undertake a preliminary sorting of the Marett and other collections of stone artefacts which were in store at La Hougue Bie, with a view to

their study in Cambridge over the next few months; secondly, it was already clear that the stratigraphy of the site presented complications calling for further work, so fresh soil and pollen samples were collected. On this occasion, the site was carefully examined by Dr. J. P. Lautridou of Caen, whose invaluable collaboration had been recruited when members of a Quaternary Research Association field meeting visited La Cotte during the previous April.

In the course of the next year the laboratory work on these and other samples, together with an exhaustive analysis of the excavation records and the first (highly tentative) age estimates based on thermoluminescence, made apparent the need for a drastic revision of the environmental sequence and dating, coupled with a modified depositional model. Therefore in August 1981 fresh work was carried out at La Cotte. Such sections as were still accessible were cleaned up and new soil samples taken (notably for micromorphological study by B. van Vliet-Lanoë); two small cuttings were made in the floor of the site, south of the retaining wall, in order to expose the upper part of layer H.

The final stage of field investigations in preparation for this publication was carried out in October–November 1982, when a team under Callow's direction, and with the aid of a grant from the Jersey Heritage Trust, re-opened the 1961–1962 'deep sounding'. An effort was made to minimise disturbance of *in situ* archaeological deposits, as the goals on this occasion were entirely geological: in particular, to obtain fresh confirmation of the 1961 observation of the raised beach (Fig. 2.8), which had never been sectioned, and of its relationship to the fossil cliff. At the same time, all accessible standing sections were cleaned once more, and re-examined with the help of Lautridou and B. van Vliet-Lanoë.

It must be acknowledged that the planning of this brief field season was not without anxiety, because if

McBurney's interpretation of the stratigraphy had been seriously challenged by the new findings the consequences for the future of the project would have been serious indeed. As it turned out, the results were most satisfactory. The beach deposits proved not only to be related to the other deposits as anticipated, but to provide considerable information about sea-level trends at the end of the last interglacial. A previously unsuspected soil was discovered in the remnants of the Weichselian layers. And laboratory examination of samples of sediment from the fossil cliff led eventually to the resolution of uncertainties as to the nature of layer H, the earliest so far discovered, and to an explanation of the conflicts in the environmental data that it had yielded over the years. Evidence accumulating from mid-1979 onwards, to the effect that many of the richest layers were laid down under interglacial conditions and that the flint for tool-making was obtained as beach pebbles, invalidated the hypothesis linking sea-level and raw material that had been proposed in the preliminary report (McBurney and Callow 1971, 200). Because the 1982 soil samples showed that the 'loessic' sediments of layer H with their flint-rich industry had been redeposited during an early temperate phase (and that their granulometry and content of injected later periglacial material were highly misleading) it was possible once more to think in terms of a causal relationship, though one diametrically opposed to the earlier model (see chapter 22). This short final excavation therefore epitomises what is perhaps the dominant theme of the present volume; that is, the inseparability of the archaeology of La Cotte from a broader base of Quaternary research.

P. C.

# PLEISTOCENE CHRONOSTRATIGRAPHY

P. Callow

## 3.1 PLEISTOCENE CHRONOSTRATIGRAPHY TODAY

Until the 1960s, when the impact of potassium argon dating became apparent, the attribution of ages in years (as opposed to relative dating) to deposits older than a few tens of millenia was a matter of guesswork. Early estimates had been based on the proposed rates of geomorphological and other processes and were extremely unreliable. The publication in 1924 of the Milankovitch theory of solar radiation had linked major climatic change to astronomical calculations (available in English translation as Milankovitch 1941), permitting calibration of the Pleistocene climatic sequence. It was never fully accepted, however, and was almost universally rejected once some early radiocarbon dates failed to agree with the theory's predictions.

Thus by the 1950s absolute dates played a limited part in the thinking of researchers who required a sound theoretical model into which to fit their field observations. Instead, almost all research effort was directed towards an understanding of environmental changes in the course of the Pleistocene, in the hope of establishing a reliable system of criteria which could be used to locate any fresh evidence within a detailed sequence of events.

In little more than two decades there has been a significant shift of attitudes towards the search for a time-calibrated scale for the Pleistocene; this has been brought about by two developments in particular:

1. the advent of several dating techniques which are suitable for use in Pleistocene contexts;
2. a realisation that the sediments of the ocean floor provide a more continuous record of world climate than do terrestrial deposits, and that these show that many of the assumptions behind land-based schemes of Pleistocene subdivision are suspect.

On land, also, importance has been attached to finding environmental records of very great duration rather than to piecing together short sequences whose correlation is open to doubt.

## 3.2 TECHNICAL DEVELOPMENTS

### 3.2.1 Chronometric Dating

The development in the late 1940s of the radiocarbon dating method (Libby 1952) provided for the first time a means of obtaining reliable age estimates anywhere in the world — subject to the availability of suitable organic material — as far back as about 40,000 years ago (attempts to extend its application to 70,000 have had limited success). Though one early consequence was an eclipse of the Milankovitch theory, which seemed not to fit some of the new observations, the new technique possessed too limited a range to meet the needs of Pleistocene research: the half-life of $^{14}C$ is simply too short. A considerable amount of effort has therefore been directed towards alternative approaches to chronometric dating; many of these, also, depend upon radioactivity to run the 'clock'.

In 1962, potassium-argon (K/A) dating was used to show that at Olduvai Gorge, in Tanzania, beds contemporary with stone tool-making hominids were about 1.8 million years old — far older than had been expected (Curtis and Evernden 1962; Evernden and Curtis 1965). More recently, Hays and Berggren (1971) demonstrated that the Pliocene-Pleistocene boundary, as officially defined by the first appearance of 'cold water' marine species in a sedimentary sequence in Calabria in southern Italy, was of a similar age, and thus nearly three times older than Penck and Brückner's estimate of 650,000 years. Hays and Berggren's argument rests on the recognition of reversals of the earth's magnetic polarity which match those at Olduvai.

K/A dating is limited in its applicability because it requires volcanic rocks, and because of the very long half-life of $^{40}K$, which makes it unsuitable for ages of less than about a million years (indeed, diorites from southeast Jersey have been dated to 580 million years by this technique). On the other hand, methods based on the decay series of uranium (Fig. 3.1) have been applied to materials as diverse as bone, shell, coral and stalagmite up to about a third of a million years old (Lamb 1977, 68–70; Lalou and Hoang 1979; Schwarcz 1980). They have been particularly valuable for dating marine terraces, thereby throwing light on the relationship between isotopic variation in deep-sea sediments and sea-level changes (see below). Ages for sites of archaeological importance have been obtained on travertines and calcite formations in numerous caves; a number of examples are mentioned elsewhere in this chapter.

Uranium-series dating of fossil bone has had very mixed success, however, because of variation in the time taken for the initial assimilation of uranium and because in some instances the uranium/daughter products system is open rather than closed, so that there may be gain or loss of one or more of the elements employed for dating (see also chapter 16). Even so, it is of considerable interest for non-calcareous deposits if it can be shown that the system remained closed, or if suitable corrections can be made to obtain an open-system date. A technique of particular interest to the archaeologist is thermoluminescence (TL) dating, which was originally developed for fired clay but has proved successful when carried out on burnt flint (see chapter 15) and has much the same age range as the uranium-series methods. More recently, TL has been successfully applied to sediments, including loess (Wintle and Huntley 1982; Wintle 1982); it is therefore likely to become of very great value in non-archaeological contexts.

The last two methods to be mentioned in this brief review do not provide dates as such, but when calibrated by means of dated material have provided age estimates. The first of these, and by far the more important for Pleistocene chronology, is palaeomagnetic reversal chronology — referred to above in the context of the dating of the Pliocene-Pleistocene boundary. The direction of the earth's magnetic field generally varies by a few degrees of inclination and declination, but on very rare occasions undergoes a 180° shift. Apart from a few short-lived events of reversal, the polarity has remained 'normal' since the beginning of the Brunhes epoch, dated to 750 kya; prior to that, and extending as far back as about 2.43 mya, is the Matuyama epoch of reversed polarity. Dating of these changes is by means of K/A and other techniques on volcanic rocks in which evidence of former polarity is preserved (Cox 1969). Identification of polarity changes in sites lacking directly datable materials enables age estimates to be made with confidence, provided that other information is available to limit the possible time range.

Finally, racemisation of amino acids in shell and bone provide a means of correlating younger deposits (the limit is around a quarter of a million years, depending on the particular amino acid studied). The racemisation rate is a function of temperature so the method is restricted to sites with a similar environmental history. An interesting example of its use is the demonstration by Davies (1983) that the D-alloisoleucine to L-isoleucine ratio for shells from fossil beach deposits of the Gower Peninsula, Wales, exhibit two modes and therefore represent two transgressive events separated by a considerable interval; the younger beach yielded values very similar to that for the raised beach in La Belle Hougue cave, Jersey, which has a uranium-series date of c. 121 kya (Keen, Harmon and Andrews 1981).

### 3.2.2 Deep-Sea Cores

The *H.M.S. Challenger* expedition of 1872–1875 showed that in the deeper parts of the ocean the floor covering often consists of fine-grained 'oozes' formed largely from the skeletons of micro-organisms. It was later realised that these sediments preserve a document of past climatic changes expressed as variations in the fossil record. In this respect they may be considered comparable to

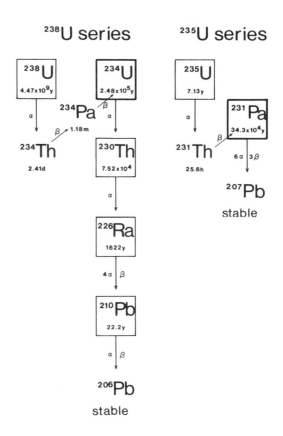

**Fig. 3.1** Decay series of uranium isotopes $^{238}$U and $^{235}$U. Boxes with thick outlines indicate isotopes useful for dating purposes (after Cook *et al.* 1982, Fig. 2).

terrestrial deposits, but with an important difference: they are not exposed to the atmospheric conditions responsible for so many gaps in terrestrial sequences, and therefore offer much better possibilities of continuity. Recent work has shown that occasional erosion is possible as the result of high-speed 'turbidity currents', but these are rare in comparison to erosive events on land.

The invention in 1947 of the Kullenberg coring device (capable of collecting 10-15 m long samples of ocean sediment) made it possible to obtain 'sections' going back hundreds of thousands, and even millions, of years. It became apparent almost immediately that the chemical variations within these cores were as interesting as the contained fossil record; the carbonate content, for example, exhibits cyclic fluctuations which are related to climatic variation through changes in ocean currents. A significant advance was made in 1955, with the development by Emiliani of a technique for measuring the ratio of the oxygen isotopes $^{16}$O and $^{18}$O in the skeletons of fossil foraminifera from the cores (Emiliani 1955). Changes in this ratio were regarded by him as reflecting ocean surface temperature. Subsequent work by Shackleton and Opdyke (1973) has shown that the contribution of temperature is smaller than that of sea-water composition. Since the proportion of the heavier $^{18}$O atoms is lower in water vapour than in the water from

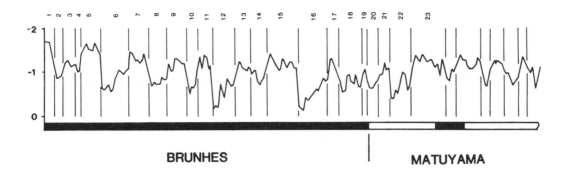

**Fig. 3.2** The oxygen isotope record and magnetostratigraphy of Pacific core V28-239 (after Berggren *et al.* 1980, Fig. 14).

which it has evaporated, the ratio is a function of ocean volume, and hence of the amount of water locked up in the ice-sheets. It therefore provides an indicator of the extent of glaciation — a critical variable in Pleistocene stratigraphy (direct measurement of the ratio in ice from the Greenland and Antarctic sheets yields similar results, but over a shorter time-scale). Most of the temperature component of isotopic variation concerns the water at the bottom of the oceans, and is a product of polar ice development rather than a simple function of insolation intensity.

Emiliani gave numbers to the more conspicuous peaks and troughs in his isotope curves; in his scheme, which has been universally adopted, odd numbers represent periods of low $^{18}$O content in the oceans (i.e. low ice-volume and warmer conditions), beginning with the present interglacial as 1, and numbering backwards through time (see Fig. 3.2).

Identification of magnetostratigraphic boundaries (especially that between the Matuyama and Brunhes epochs), and the calibration (by uranium-series techniques) of the last 200,000 years in certain cores, leaves a gap in the dating of Middle to Upper Pleistocene events which until recently has had to be dealt with by interpolation; with the increasing number of cores available from which to check sedimentation rates, though, the estimates seem to have been reliable. The reasoning behind recent attempts to construct a Quaternary time-scale from magnetostratigraphic, isotopic, chemical and faunal variations in deep-sea cores is set out in detail by Berggren *et al.* (1980).

Publication of Pacific core V28-238 (Shackleton and Opdyke 1973) and others in which the Matuyama-Brunhes boundary could be identified had important implications for students of the sequence on land. They showed very clearly that the number of ice-volume fluctuations was very much greater than was suggested by the evidence previously available (Fig. 3.2). In particular, they made all too clear the incompleteness of most terrestrial records and the unreliability of the schemes of Pleistocene subdivisions that had been so laboriously developed.

Isotope stage 5 was identified by Emiliani as corresponding to the last interglacial of the terrestrial record.

It was divided into five substages (5a–5e, in order of increasing antiquity). Shackleton (1969) proposed that the earliest of these was in fact the interglacial, at about 120 kya (actually a very slight underestimate), and that it was followed by two interstadials (5c and 5a) with cooler periods between.

The presence in several Equatorial Pacific cores of a layer of volcanic ash dated to about 230 kya provides valuable evidence for the dating of isotope stage 7 (Ninkovich and Shackleton 1975). Of particular interest is that this stage of low ice-volume (low $^{18}$O) is subdivided by a short period during which the amount of ice greatly increased. This substage 7b, which began around 230 kya, is suggested by Ruddiman and McIntyre (1982; see also Andrews 1983) to have represented a cooling on a nearly global scale, at least, and to have been of glacial intensity (on the evidence of oxygen isotope and carbonate data, of the percentage of polar foraminifera, and of the amount of ice-rafted sand in north Atlantic cores). Very great uncertainties attach to the percentage of the isotope signal that is due to cooling of bottom waters as opposed to ice-volume change, however, and the probable consequences for terrestrial sites are not easily assessed (though see below).

An important breakthrough occurred with the rehabilitation of the Milankovitch theory (Broecker *et al.* 1968; Hays, Imbrie and Shackleton 1976); it is now widely accepted that, when allowance is made for the lag imposed between insolation and ice-volume changes by the mechanisms responsible for glaciation and deglaciation, there is a very close fit between orbital perturbations and the fluctuations in the ocean sediment record. Because the astronomical calculations of the earth's orbit are extremely reliable they provide the most accurate calibration yet available for the Pleistocene (Imbrie *et al.* 1984). The time-scale of the core which has been employed for comparison with the evidence from La Cotte, V19-30, has been obtained in this way.

### 3.2.3 Continuous or Near-Continuous Records on Land

Although the classic terrestrial glacial/interglacial frameworks have been compiled from several relatively short depositional sequences, two more complete types of re-

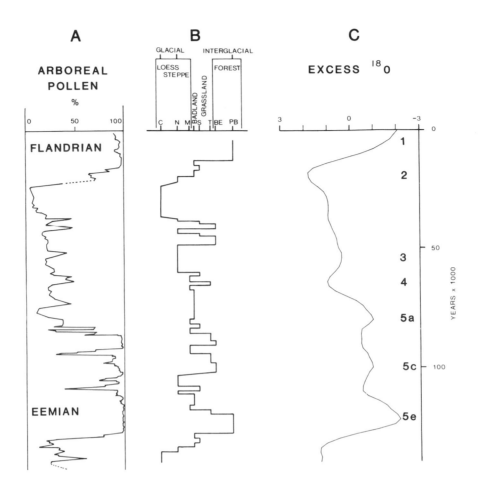

**Fig. 3.3** Some detailed Upper Pleistocene records. (A) La Grande Pile, northeast France: percentage of arboreal and oak-mixed forest pollen (after Woillard 1980). (B) Environment and magnetostratigraphy of the area around Prague and Brno, Czechoslovakia (after Kukla 1977), based on lithology, soils and snails; the reversed-polarity Blake Event occurred at about 110 kya. (C) The stacked, smoothed oxygen isotope record as a function of age on the SPECMAP timescale (based on data in Imbrie *et al.* 1984).

cord are now available for Europe; these are based on (1) palynology and (2) loess stratigraphy (coupled with environmental data from molluscan remains). They yield long and detailed records which can be correlated with the marine data, and also provide a yardstick against which to assess the reliability of the older schemes.

An increasing number of Old World sites have yielded long pollen profiles. The first examples published could not be linked directly to the development of ice-sheets in the classic study areas: they included Lake Zeribar, in western Iran (van Zeist and Bottema 1977 and earlier works), Tenaghi Philippon, in Greece (van der Hammen *et al.* 1971) and Padul, in Spain (Florschütz *et al.* 1971). At Tenaghi Philippon and Padul the last cold complex is preceded by a tripartite temperate maximum. Within the glaciated area of northern Europe, La Grande Pile peat bog (in northeast France) has a succession of organic muds which overlie a Saalian till

and are interstratified with clays representing colder periods (Woillard 1975; 1978; 1980). The base of the profile once again exhibits three important peaks of temperate forest development, of which the first is the most pronounced (Fig. 3.3). Woillard concluded that they correspond to the temperate subdivisions of isotope stage 5, a view supported by a long series of radiocarbon dates for the upper part of the diagram (Woillard and Mook 1982). The significance of this record is that it confirms that these low ice-volume substages are matched by vegetational changes in temperate latitudes.

Exceptionally detailed and complete sedimentary and pedological data is available from the enormously thick loess deposits of central Europe outside the glaciated areas; moreover, calibration has been possible by means of palaeomagnetism. Kukla (1975) showed that it is possible to identify cyclic patterns of soils and lithologic units in the loesses of Czechoslovakia. At least seventeen first-order (glacial) cycles are recorded during the past 1.7 million years (Fink and Kukla 1977), as well as a large number of second-order (stadial) subcycles. The succession therefore matches that of the deep-sea sediments in its complexity; the two are also in fairly good agreement as to the probable duration and magnitude of climatic events.

Of particular interest in view of the concern of the present volume with the later Middle and Upper Pleistocene is that Kukla's cycle B (A is the Holocene) begins with a brownearth (deciduous forest) soil and is followed by two temperate steppe soils. Loess deposition occurred between the pedogenetic events. The loess overlying the forest soil has reversed magnetic polarity corresponding to the short-lived Blake event at about 110 kya (Kukla and Kocí 1972); the soil itself is therefore that of substage 5e, and those that follow are likely to represent 5c and 5a. That aeolian deposition took place in the intervening period is of some importance for the evidence that it provides of the effects on land of the cooler substages.

The beginning of Kukla's cycle C is marked by a double forest soil (with badland conditions intervening) whose age, estimated by interpolation, is very close to that of the twin-peaked isotope stage 7 (Kukla 1977, 351).

Thus the palynological and central European lithological/pedological records indicate that events on land in the course of the Pleistocene reflected very closely those recorded in the ocean sediments. It is also apparent from them that any system of subdivision of the same period which does not allow a similar degree of complexity is bound to be in error; indeed, one of the conclusions reached by Kukla (1977, 366) was that "after many years of widespread miscorrelation, the classical north European terminology can not be saved. The sooner it is abandoned in inter-regional correlations the better".

## 3.3 CONVENTIONAL TERRESTRIAL SCHEMES OF RELEVANCE TO LA COTTE

In view of the comments made in the preceding section it may seem surprising that it has been thought worthwhile to devote space to classic frameworks. Nevertheless, these terminologies remain of importance because of the use made of them, however misguidedly, as labels for more soundly based regional schemes; also there are few instances in which chronometric dates or other means of correlation with the isotope records are available.

### 3.3.1 The Alps and Adjacent Areas

The model proposed by Penck and Bruckner (1909), on the evidence of Bavarian fluvioglacial outwash terraces, described four glacial periods (successively Gunz, Mindel, Riss and Würm); subdivisions of these were later defined by Eberl (1930). But as Kukla (1975) has pointed out, although the original scheme was based only on morphostratigraphic data it has been employed (impro-

perly) as a chronostratigraphic framework and applied far from the stratotypes, with correlations obtained by a counting-back approach. Moreover, current attempts to produce new lithostratigraphic definitions in the type area (Chaline and Jerz 1983) may well add to the confusion by attempting to perpetuate the old terminology.

For our purposes it is sufficient to consider the scheme commonly employed in southern France, as it is in this area that lie many of the archaeological sites to be compared with La Cotte. In Aquitaine very detailed subdivisions have been proposed using the Alpine nomenclature but with local stratotypes (Laville 1976; Laville et al. 1980; Bordes et al. 1980). Here a very large number of minor events have been grouped into three Riss and four Würm stadials; the Würm II/III interstadial coincides with the Middle/Upper Palaeolithic transition and is considered sufficiently important to justify grouping the stadials into Würm ancien (I + II) and Würm récent (III + IV). A similar partition of the last two glaciations is employed in Mediterranean France (Miskovsky 1976). The deposits of the later part of Würm have been dated by radiocarbon and are securely correlated, but linking of the various regional schemes for earlier periods is less sure, and it is not impossible that the same name has been given to different events, particularly in the Middle Pleistocene.

Correlation between the French 'Alpine' stages and the deep-sea core record is far from straightforward. There now seems to be fairly wide agreement that the Riss-Würm interglacial (marked, variously, by pedogenesis, erosion or calcite formation in caves) is equivalent to substage 5e, but labelling of what follows is disputed. Leroi-Gourhan (1983) sees Würm as beginning around 70 kya (isotope stage 4), preceded by two warmer periods (the Amersfoort-Brörup and Odderade 'interstadials' of northwest Europe). But Laville et al. (1984) prefer to retain the designation Würm ancien as embracing the whole period from 5d to the latter part of 3, on the grounds that the phases corresponding to 5c and 5a were not so much temperate as humid in Aquitaine. On both interpretations, sites traditionally dated to Würm I should be viewed as falling between about 115 and 70 kya.

Uranium series dates show that speleothems developed during isotope stage 7 in the Abri Suard, Charente (Blackwell et al. 1983) and Pech de l'Azé II, Dordogne (Schwarcz and Blackwell 1983); at the former site this was the last period of fully temperate climate before the Riss-Würm interglacial. However, there is disagreement about the dating of earlier stages. De Lumley (1976b) favours the equivalence Riss I/II/III = isotope stages 8/7b/6 (and Mindel-Riss = 9), but Duchadeau-Kervazo and Kervazo (1983) argue that Riss II was more important than this would imply, identifying isotope stages 10, 8 and 6 with the three Riss glacial maxima. In fact, because of the lacunae which are evident in some of the key sites (Riss III deposits are lacking at Pech de l'Azé II and in the Ardèche, for example) age estimates given for Riss and earlier deposits must be treated with circumspection until far more chronometric information is available.

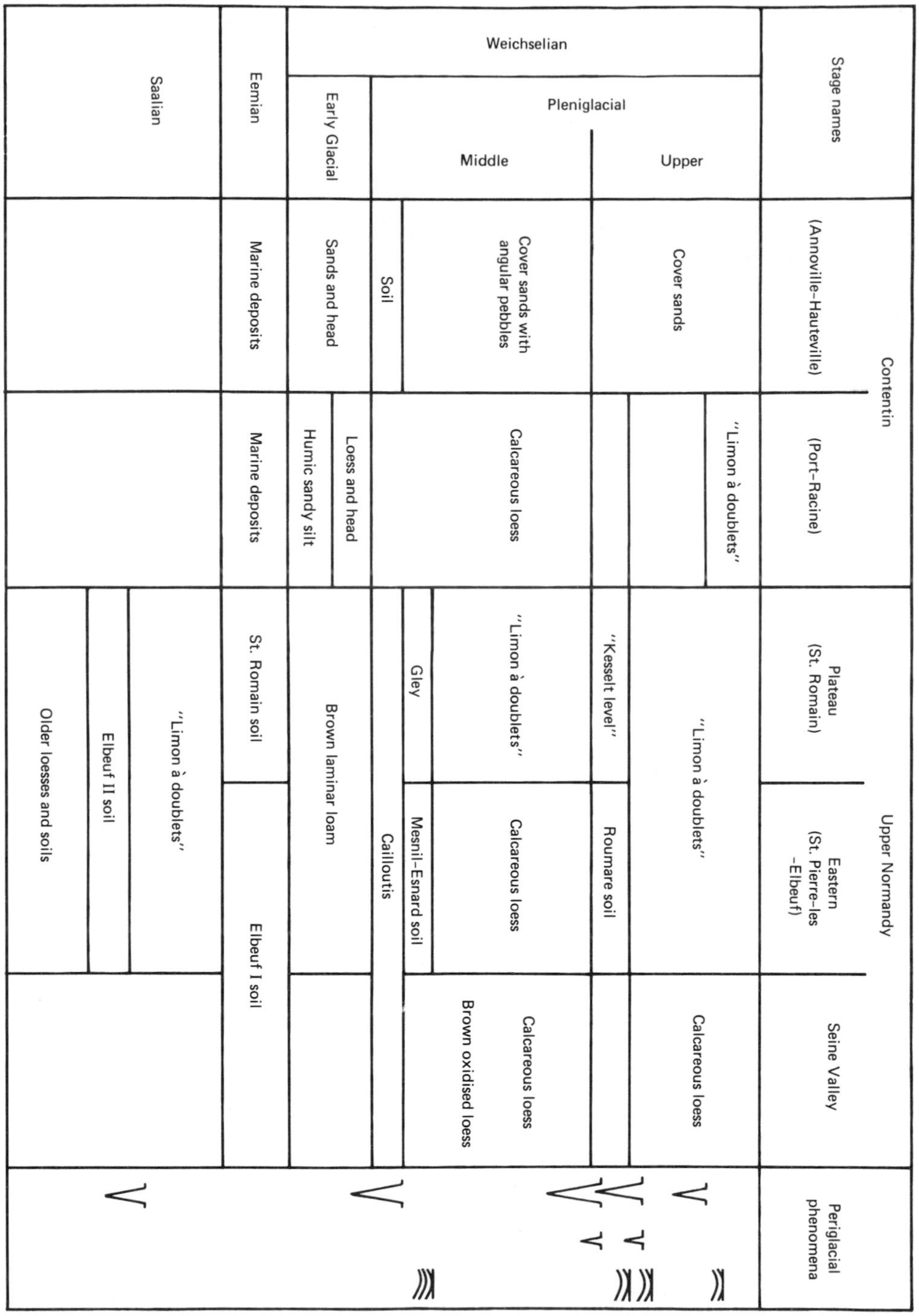

| Stage names | Saalian | Eemian | Early Glacial | Weichselian — Pleniglacial Middle | Weichselian — Pleniglacial Upper |
|---|---|---|---|---|---|
| Contentin (Annoville–Hauteville) | | Marine deposits | Sands and head | Soil; Cover sands with angular pebbles | Cover sands |
| Contentin (Port-Racine) | | Marine deposits | Humic sandy silt; Loess and head | Calcareous loess | "Limon à doublets" |
| Upper Normandy — Plateau (St. Romain) | Older loesses and soils; Elbeuf II soil; "Limon à doublets" | St. Romain soil | Brown laminar loam | Gley; "Limon à doublets"; "Kesselt level" | "Limon à doublets" |
| Upper Normandy — Eastern (St. Pierre-les-Elbeuf) | | | | Mesnil-Esnard soil; Cailloutis; Calcareous loess | Roumare soil; Calcareous loess |
| Seine Valley | Elbeuf I soil | | | Brown oxidised loess; Calcareous loess | Calcareous loess |
| Periglacial phenomena | | | | | |

Fig. 3.4 A very simplified representation of the late Middle and Upper Pleistocene sequence in Normandy, based on data in Lautridou *et al.* (1982) and other sources.

## 3.3.2 Northern Europe

The Scandinavian ice-sheets were far more substantial than those of the classical Alpine area, and extended over Britain and north Germany. In the latter area a complex series of moraines (tills) exists, interstratified with marine deposits in low-lying areas, as also are the tills, glaciofluvial sands and periglacial coversands of the Low Countries. In order of increasing antiquity, the principal tills are named Weichsel, Warthe, Saale and Elster. Between Weichselian sands and the Saale Till in the Netherlands there occur marine clays of the last Pleistocene transgression, the Eemian (Zagwijn 1975). Another transgression, the Holsteinian, corresponds to the Elster-Saale interglacial; this appears to be bipartite, divided into Holsteinian *sensu stricto* and Dömnitzian by the Fühne cold period. The status and position of Warthe have sometimes been questioned, but it now seems likely that the age of the Saale-Warthe warm period, possibly of interglacial status, is given by the uranium-series dates on travertines of c. 220 kya from Ehringsdorf (Kukla 1977) and also perhaps that of 228 kya from Bilzingsleben (Harmon *et al.* 1980), though there are geological grounds for suspecting that the latter date is too young and that the Bilzingsleben deposits should be linked to a still earlier episode. For a discussion of alternative proposals for subdividing the Middle Pleistocene in this area see Nilsson (1983).

In Belgium and north France the Saalian is usually taken as covering the entire period between the Holsteinian and the Eemian, including the Warthe stage. The scheme is based on correlation between a large number of sites in coversands and loess. These exhibit soil and gelifluction horizons which have been claimed to provide the means for matching sections over a large area (Paepe and Sommé 1970; Sommé, Paepe and Lautridou 1980); marine and estuarine deposits in Belgium and Normandy have permitted the incorporation of transgressive events into the succession.

Two provinces of loess may be defined in northwest Europe according to distribution and origins. In the north, deflation of the area close to the Scandinavian ice-sheet, in particular the North Sea basin, has long been known to have been responsible. On the other hand, in the Normandy province the exposed bed of the English Channel has been regarded as the source. But Lautridou (1984, 387) considers that here the loess originated in estuarine deposits formed when the sea was perhaps as little as 20–30 m below its present level, and that a strong case can be made for the rejuvenation of these estuaries (and hence the potential for loess deposition further inland) not only under Early Glacial conditions but also during interstadials, a point to which we shall return briefly in chapter 31.

Some of the loessic formations are very ancient. That at La Londe near Rouen in the Seine valley has reversed magnetic polarity, and antedates the Matuyama-Brunhes boundary (Biquand and Lautridou 1979); so too do loessic sands of the very high terrace of the Somme at Grâce (Bourdier *et al.* 1974, and articles by other authors in the same volume). For our purposes, though,

only the late Middle and Upper Pleistocene need to be considered.

Facies differences are apparent in the Normandy loess. In particular there is a tendency further west for it to have suffered decalcification and formation of fine banding due to segregation of the finer fractions by ice (*limon à doublets*). This implies more humid conditions than for the calcareous loess, so an east-west trend is to be expected, but the position of the boundary between the facies varies considerably through time (Fig. 3.4). In the Cotentin Peninsula and in Brittany to the west (Monnier 1980b) loessic or sandy head developed on slopes and at the foot of sea-cliffs. Correlation between the Weichselian deposits of this area and those in the thicker exposures further east has been facilitated by their evolving heavy mineralogy (Coutard *et al.* 1973), but the dating of events within some of the longer head sequences poses considerable difficulties.

At Grandcamp-les-Bains (Calvados) the loess overlies marine deposits with a maximum altitude of + 5 m above Ordnance Datum (a little above present high tide) which are assigned to the Eemian (Coutard, Lautridou *et al.* 1979). Eemian beaches at a similar level, covered by head, occur widely in the northern part of the Cotentin. At Port Pignot, older and more heavily weathered marine deposits rest on a separate bench at c. + 10 m; covered by Saalian slope deposits, they are attributed to an earlier Saalian or Holsteinian transgression — still higher beaches have been reported elsewhere (Coutard, Giresse *et al.* 1981).

As already mentioned, soils have been used as markers for stratigraphic correlation. A well-developed forest soil, the St. Romain Soil, whose stratotype is situated on the plateau north of the Seine estuary, is attributed to the Eemian, as are the Elbeuf I Soil near Rouen and the Rocourt Soil of Belgium (Sommé, Paepe and Lautridou 1980). In the case of St. Romain the attribution is supported by thermoluminescent dating to 115 ± 10 kya (Wintle, Shackleton and Lautridou 1984). The Weichselian itself is subdivided by means of several marker horizons. The most important of these is a cryoturbated soil, the *Niveau de Kesselt* (named after a site in Belgium and dated to 28.2 kya); this separates the Upper from the Lower Pleniglacial. At the base of the Lower Pleniglacial deposits there is a surface from which deep ice-wedges penetrate the underlying Early Glacial layers, formed under less extreme conditions.

At both St. Romain and St. Pierre-lès-Elbeuf there are long sequences of more ancient loesses and soils. The Elbeuf IV palaeosol, which is overlain by a rich tufa, is thought to be of Holsteinian age; if so, at least two forest soils or soil complexes formed in the course of the Saalian.

## 3.3.3 Britain

The subdivisions of the Quaternary in Great Britain, like those of the Alpine sequence, are of little stratigraphic relevance here, but will concern us when a broader context is required for the archaeological evidence from La Cotte. The basis of a long-accepted scheme for the Brit-

**Table 3.1** Classification of the Middle and Upper Pleistocene in Britain, and Suggested Correlations with Northwest Europe[a]

| Period | | Britain | Northwest Europe | Palynological characteristics |
|---|---|---|---|---|
| HOLOCENE | IGl | Flandrian | Flandrian | High frequencies of *Corylus* pollen in zone FII. |
| UPPER PLEISTOCENE | Gl | Devensian | Weichselian | — |
| | IGl | Ipswichian | Eemian | High frequencies of *Corylus* pollen in zone IpII, and a well-marked *Carpinus* zone (IpIII). |
| MIDDLE PLEISTOCENE | Gl | Wolstonian | Saalian | — |
| | IGl | Hoxnian | Holsteinian | High frequencies of *Hippophaë* pollen in the late glacial (late Anglian). Considerable *Tilia* frequencies in late zone HoII. *Corylus* maximum later in zone HoII. *Abies*, *Picea*, *Carpinus* and *Pterocarya* in zone HoIII. |
| | Gl | Anglian | Elsterian | — |
| | IGl | Cromerian | 'Cromerian complex' | Low frequencies of *Corylus* pollen. High frequencies of *Ulmus* in zone CrII, but *Tilia* frequencies low. *Abies*, *Picea* and *Carpinus* occur in zone CrIII. |
| | Gl | Beestonian | 'Cromerian complex' | — |
| | IGl | Pastonian | 'Cromerian complex' | *Carpinus* in zones PaII and III. *Tsuga* in zone PaIII. No or very low *Abies* in zone PaIII. |

[a]The numbering of the interglacial pollen zones runs from I (pre-Temperate) to IV (post-Temperate). After Sparks and West (1972, 177) and West (1977, 268).

ish Isles has been described by Sparks and West (1972) and West (1977). Detailed correlations were published by Mitchell, Penny *et al.* (1973). The principal types of information employed included: (1) geological evidence of glaciation, or of periglacial action; (2) the pollen record (especially for interglacials); (3) faunal evidence.

The most incontrovertible argument for identifying a glacial episode is the existence of morainic material *in situ*. However, this has not proved straightforward in practice, as is shown by the abandonment of an earlier terminology based entirely on East Anglian deposits in favour of type sites which are more widely dispersed.

Palynology was a critical element in the construction and application of the 1970s classification, as it was a fundamental assumption of the British school of pollen analysts that each interglacial was characterised by a unique vegetational history, providing a 'signature' by which the age of palynological sequences could be established. Faunal remains played a more limited role in the discussion, but it was observed that the assemblages reflected the contrast between glacial and interglacial episodes by the presence or absence of certain species. In addition, variation in the representation of particular taxa in different interglacials is quite striking, and once again may assist in locating a deposit within the Pleistocene (Stuart 1982, 171). Of special interest is the apparent absence of *Hippopotamus amphibius* in Hoxnian deposits, though it is present in the Cromerian and Ipswichian interglacials.

The later part of the sequence and proposed correlations with European stages are given in Table 3.1. The simplicity of this model compared with the evidence from the deep-sea cores, coupled with the incompleteness or ambiguity of many of the most important sites, has caused it to be called into question in recent years (e.g.

by Sutcliffe 1976). The most serious difficulty arose from the identification of the 730 kya Matuyama-Brunhes palaeomagnetic boundary (see above) in deposits in the Netherlands belonging to the first interglacial of the three-interglacial 'Cromerian complex' (van Montfrans 1971; Zagwijn *et al.* 1971), that is, at the end of the British Lower Pleistocene. In the cores, the boundary has been found in stage 19 (stage 5e represents the Ipswichian). The scheme just described allows for only three interglacials in an interval containing six stages of low ice-volume.

A working group set up by the International Quaternary Association subcommittee of the Royal Society has recently attempted to resolve the problem of the number of post-Hoxnian interglacials recorded in Britain — a matter of particular significance for the interpretation of La Cotte de St. Brelade in the context of British, as opposed to continental, archaeological discoveries. A short account of its results is given by Shotton *et al.* (1983). It concluded that at a number of sites (e.g. Stanton Harcourt, in Oxfordshire, and the terraces of the Warwickshire Avon) there are indications of a post-Hoxnian interglacial older than the Ipswichian *sensu stricto*, as defined by the stratotype at Bobbitshole (Ipswich, Suffolk). Apart from the stratigraphic evidence, it was noted that this episode was characterised by *Mammuthus* and *Equus*, as opposed to the *Palaeoloxodon* and *Hippopotamus* of the most clearly dated Ipswichian sites. Moreover, certain localities hitherto attributed to the Ipswichian for rather insubstantial reasons also possess a *Mammuthus/Equus* fauna. Coope, in the same report, points out that the beetle *Anotylus gibbulus* (now found in the Caucasus) was absent in the Ipswichian *sensu stricto* but occurs at some of the sites proposed as falling between the Hoxnian and the Ipswichian.

Thus although the British Pleistocene is now in some disarray, it seems that the faunas may help to discriminate between temperate phases with rather similar pollen records, and that if this is so there are rather strong grounds for believing that at least one additional post-Hoxnian interglacial is present in the British record.

That the Ipswichian is equivalent to isotope substage 5e has been confirmed by dating of stalagmite associated with a *Hippopotamus* fauna in Victoria Cave, Settle, in Yorkshire (Gascoyne *et al.* 1981); ages were obtained in the range 135 ± 8 to 114 ± 5 kya, and the cave earth with fauna probably formed around or before 120 ± 6 kya. It is likely that the Hoxnian, during which there occurred the marine transgression preceding the Ipswichian, corresponds to part of oxygen isotope stage 11, though stage 9 is also possible. The former is estimated by Shackleton (1975) from interpolation within core V28-238 as lasting from 440 to 367 kya; uranium-series dates on bone which suggest a stage 7 age for the Hoxnian (Szabo and Collins 1975) are certainly too young. A stalagmitic floor in Pontnewydd cave in north Wales has been dated to $224^{+43}_{-31}$ kya (Ivanovich *et al.* 1984) and was therefore formed in isotope stage 7 (such speleothems are characteristic of temperate conditions). As yet it cannot be stratigraphically related with any precision to events elsewhere during the Hoxnian-Ipswichian interval.

## 3.4   SEA-LEVELS IN THE MIDDLE AND UPPER PLEISTOCENE

Although the oxygen isotope variations are a function of ice-volume, they do not allow very accurate estimation of sea-level. More direct evidence must therefore be sought from dated marine terraces.

There has been little difficulty in identifying beaches and reefs of substage 5e; estimates of the corresponding sea-level maximum are subject to some uncertainty, but there is universal agreement that it was some metres above the present level. Identification of features corresponding to 5c and 5a is more difficult because (1) in neither case was sea-level as high, so they are usually submerged, and (2) their elevations were apparently very similar. Consequently the most detailed records are for coastlines which have undergone uplift. Reefs dated radiometrically to these substages have been reported above sea-level in such situations, for example in Barbados (Broecker *et al.* 1968; R. K. Matthews 1973), Haiti (Dodge *et al.* 1983) and New Guinea (Chappell 1983). In each case the altitudes have been corrected on the assumption that uplift occurred at a uniform rate; this seems justified by a fair measure of consistency in the results, but inaccuracies of several metres are possible and the exact figures have sometimes been queried (e.g. by Stearns 1976). If an altitude of +6 m is taken for substage 5e, typical values for 5c and 5a are -10 to -13 m and -13 to -19 m.

Oxygen isotope data indicate that the ice-volume in stage 7 was probably somewhat greater than during substage 5e; the longest of the calibrated reef sequences,

that for the Huon Peninsula in New Guinea, also suggests that during 7c the sea-level was well below today's, while that of 7a was of the order of -10 m. On the other hand, the beginning of stage 9 at around 330 kya was marked by a rise to a level comparable to that of the present time (Chappell 1983). In the deep-sea cores the isotopic peak for stage 11 is closer to that of 9 than of 7. This appears to contradict the observation that in many areas marine terraces which must lie within this time range are found well above sea-level: of particular concern for the present discussion are those of Morocco, the Mediterranean and the Atlantic seaboard of Europe (examples occur on Jersey at around +18 m and +30 m as described in chapter 5). Such altitudes cannot be explained in terms of glacio-eustatic variation of the amplitude indicated by the isotopic curves, which evidently do not tell the full story of sea-level (as opposed to ice-volume) change. In fact, there has been a general fall in world sea-level of about 100 m over the last 0.8 my alone. Various explanations have been put forward for this (see Lamb 1977, 113 ff.; Bowen 1978, 158 ff.):

1. an increase in the amount of water absorbed by terrestrial deposits;
2. lowering of the ocean floor associated with continental drift and mountain building;
3. variation in the geoid (the shape of the ocean surfaces) due to gravity;
4. glacio-isostasy (crustal deformation due to the weight of ice-sheets);
5. hydro-isostasy (crustal deformation due to the weight of the oceans).

Although all of these may have been responsible for sea-level variation of local or global significance their relative importance cannot yet be assessed. Glacio-isostasy is the only one of the listed mechanisms whose effects have been documented in detail (e.g. for northern Europe since the last glacial maximum). In some areas it is clear that other factors have produced dramatic results: thus Calabrian marine beds occur at up to 1400 m asl because Italy is at the junction between two continental plates.

In the light of the above, it would appear advisable to treat attempts to correlate terraces according to their altitude with caution, except perhaps locally (and even then subject to the caveat that sea-level stillstands may have occurred more than once at the same level; see also chapter 5). Nevertheless it is certain, for the western English Channel and the Western Approaches, that

1. since isotope stage 6, if not earlier, periods of maximum ice-sheet volume have corresponded to a sea-level of around -100 m or lower (Hamilton and Smith 1972);
2. substage 5e (the Eemian) resulted in a sea-level a few metres above today's (with widespread deposition of beach material), while during 5c and 5a it was possibly a little below;
3. at least one and possibly several marine formations and platforms at +8 m or above probably date to

isotope stages 7 or 9. Thus estuarine deposits were laid down at around + 10 m (5 m above the modern estuary) at Tancarville; at the top is a palaeosol, and a second occurs in the succeeding deposits before the Eemian beach at the same locality (Lautridou *et al.* 1982).

## 3.5 BASIS OF A MODEL FOR INTERPRETING LA COTTE DE ST. BRELADE

The nature of the evidence from La Cotte is such that it is appropriate to draw upon several different kinds of evidence.

1.  The local (Channel Islands) and regional (northern France) stratigraphy is built on the loesses, heads, soils and raised beaches of the area. The terminology employed is that described in section 3.3.2 — that is, the modified northern European scheme of Sommé, Paepe and Lautridou (1980).
2.  The palynological record from La Grande Pile (Woillard 1978) is the best currently available for the Upper Pleistocene of northwest Europe (see section 3.2.3); it will be assumed that the three deciduous forest phases at its base are of interglacial status, that the first and most important is the Eemian, and that comparable vegetational changes occurred in our own area.
3.  The oxygen isotope variations in deep-sea cores are assumed to reflect global climatic changes (expressed, though not strictly synchronously, in terms of ice-volume and also of sea-level). In particular, substage 5e is taken to correspond to the last period of sea-level higher than today's (in the Eemian); the triple peaks of 5 are regarded as equivalent to the three interglacials at the base of La Grande Pile. It is also considered that the twin peaks of stage 7 would have been reflected in events on land. The corresponding sea-level maxima are not known.

To anticipate the more detailed discussion in the second part of this volume, three chronometric techniques have so far been employed at La Cotte: TL (on burnt flint), uranium-series (on bone) and radiocarbon. Of these, TL is regarded as the most reliable. Uranium-series dating on bone, as opposed to calcite, tends to give ages which are too young; also the archaeological deposits at the site are too old for $^{14}$C to yield meaningful results.

34

# THE GEOLOGICAL SETTING AND ORIGIN OF LA COTTE DE ST. BRELADE

J. T. Renouf

It is the intention in this chapter to describe those features of the geology of southwestern Jersey, and of the La Cotte area in particular, that are relevant to understanding the origin of the cave shelter and its subsequent development.

## 4.1  THE GENERAL SETTING

The southwestern corner of Jersey has as bedrock a Cambrian granite complex nearly 600 million years old. It is part of a series of granites and allied plutonic rocks that occur elsewhere in Jersey, on the other islands and on the adjacent continent in France. These granites are generally more resistant to erosion than the surrounding sedimentary or metamorphic rocks and tend to form strong headlands. This is true also of northwestern Jersey, where the other Palaeolithic cave, La Cotte à la Chèvre, is situated.

Within the granite complex of southwestern Jersey, internal structural weaknesses causing differential erosion have been brought about by a number of factors:

1.  differences in strength of the compositional variants of the granite complex;
2.  the original cooling fractures — the joints;
3.  later fractures and intrusions: (a) effects on pre-existing joints of later compressional and tensional stresses; (b) initiation of new fractures as a result of the later compressional and tensional stresses; (c) intrusion of other igneous material, mostly dolerites and lamprophyres, in the form of more or less narrow dykes.

These factors are related to the geological history of the much wider region of Armorica, the western Channel and southwest England. The purpose here is not to describe the events of this history in detail but to make it clear that the resulting fundamental structures were already built into the granite complex when the agents of erosion began to attack it, possibly as early as the Eocene stage of the Tertiary, some 60 mya, or perhaps even earlier. Only since then have the internal structural weaknesses begun to play their part in forming the landscape of Jersey as we see it today. The factors in the final shaping were

4.  changes in land/sea relationships — mainly tectonic;

5.  changes in land/sea relationships — mainly climatic;
6.  chemical and mechanical weathering.

In the present state of knowledge it is not possible to reconstruct the precise sequence of interaction between these factors that led to the formation of the La Cotte headland and cave. All we can do is to give as clear as possible an outline of the evidence as it stands, paying particular attention to the geology of its later stages, the interpretation of which is crucial to the understanding of the archaeological sequence.

## 4.2  INFLUENCES ON THE TOPOGRAPHY OF THE GRANITE AREA

In the following sections the first five factors affecting the southwestern granite complex are examined in broad outline. The period of man's occupation of the cave is discussed in greater detail (4.3). A concluding section (4.4) summarises the evolution of the headland and its cave. Weathering — the sixth factor — is discussed in subsequent chapters.

### 4.2.1  Compositional Variations

Henson (1947) described the broad features of the complex. He identified a number of compositional variants of which three have a particular relevance to La Cotte. The first granitic intrusion was coarse-grained and typified by that found at La Moye and Noirmont. A second intrusion invaded the central area of the coarse-grained granite to form a finer grained variety found on either side of Beauport. Between them there was an intermediate zone where the development of large porphyritic crystals of potassium feldspars gave the rock a distinctive appearance. The width of the intermediate zone and other factors suggest that the intrusion of the Beauport granite followed soon after that of La Moye type, before the latter finally cooled.

The geological sketch map (Fig. 4.1) shows that the La Cotte area lies astride the contact of the two granites and the intermediate zone of mixing. The compositional variation is not the most significant factor in the position of the headland; rather it is the difference in joint styles developed in the intermediate zone of porphyritic crystals. The rock of the cave itself is largely of the porphyritic mixed type.

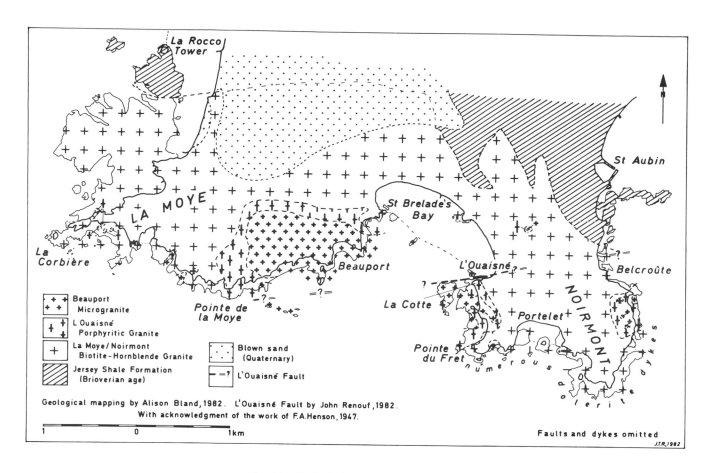

**Fig. 4.1** Geological map of southwest Jersey.

## 4.2.2 The Jointing in the Granites

Tensions within the southwestern granites, as they cooled from the molten state, produced three major sets of joints. The pattern of planes of La Cotte headland is derived from the technique of the contoured stereogram Fig. (4.2) and from orthographic projections Figs. (4.3 and 4.4). Two sets of joints vary about the vertical and lie more or less at right angles to one another. One trends a little west of north, on average; the other varies between E/W and ENE/WSW. Members of the third set dip at lesser angles, generally below 40°, and are variable in strike, though with a pronounced concentration at an easterly dip of some 13°. The effect of this regular arrangement of jointing has led to erosion producing more or less vertical rectangular blocks bounded above and below by somewhat sloping surfaces which are nonetheless commonly close to the horizontal.

Density of the vertical jointing varies. There are more joints in the Beauport variety of granite than in the La Moye type, and the mixed zone between is variable. Within all three types of granite there are concentrations of closely spaced joints separated by zones in which they are much more widely spaced. It is possible that the concentrations of closely spaced planes of weakness may be a feature imposed in part by the effects of subsequent faulting.

## 4.2.3 Later Compressional and Tensional Stresses

Under this heading are included the effects on pre-existing joints as well as the initiation of new planes of weakness. The intrusion of dykes is also considered.

Since the cooling of the granites during the Cambrian hundreds of millions of years ago the southwestern granites have suffered many stresses and strains, the effects of which can be seen today. The La Cotte headland reveals the existence of a number of fault-related features. The map (Fig. 4.5) shows the position of a major zone of faulting, the Ouaisné fault, identified by its reconstituted rock following crushing during faulting. This crush zone bounds the northern cliff of the headland along its base. There is some indication in the distribution of the Beauport and La Moye granite outcrops at Ouaisné of a dextral displacement along the fault. The strike of the fault is c. 085 TN (i.e. slightly north of east), which is consistent with the known lines of weakness in northeastern Armorica along which lateral displacements are both known and suspected. It probably represents a very old line which has been reactivated from time to time since its inception several hundred million years ago. The whole cliff line from La Cotte Point to the col at M.R. 604476, and beyond to Belcroute, probably relates to erosion along the line of weakness provided by this fault zone and has thus had a

36

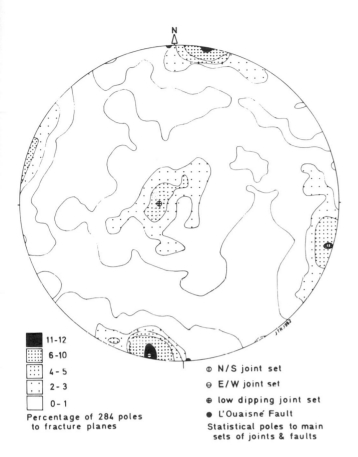

**Fig. 4.2** A stereographic representation of planes of weakness (joints and faults) affecting the granite bedrock at La Cotte. The contours indicate the density of the plotted values, allowing concentrations with similar dip and strike to be identified. These sets are noted in the diagram and are more visually represented in Figs. 4.3 and 4.4.

| | |
|---|---|
| ■ | 11-12 |
| ▦ | 6-10 |
| ⣿ | 4-5 |
| ⠄ | 2-3 |
| □ | 0-1 |

Percentage of 284 poles
to fracture planes

⊕ N/S joint set
⊖ E/W joint set
⊕ low dipping joint set
● L'Ouaisné Fault
Statistical poles to main
sets of joints & faults

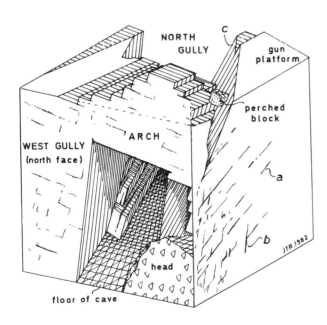

**Fig. 4.3** Orthographic projection of the statistically identified sets of planes of weakness from Fig. 4.2: (A) the three major planes; (B) the planes which affect the rock wall in the vicinity of the bone heaps, and which eventually gave rise to the fall of slabs between the deposition of the two heaps (B').

profound effect on the shaping of the Noirmont/Portelet promontory.

A number of fault planes have been identified within the granites of La Cotte headland and its immediate vicinity. Because of the lack of reference horizons within the granite it is not possible to prove the amount of movement along such faults, but the presence of narrow, lens-like sheets of mylonitic rock, often less than a centimetre thick, between planes of weakness is indicative of intense strain. Movement was very small. This feature is present in vertical, more or less E/W planes in the west ravine and one or two further to the south, notably those bounding the northern margin of the stack at M.R. 593473. More or less N/S trending vertical planes are affected by the so-called fissure-filling locality on the northwestern cliff of La Cotte headland, where the rock has been shattered locally by even stronger fault strains; yet there appears to be virtually no movement along such fault lines.

Only a few faults have been demonstrated by the presence of mylonitic fills, but they affect both major directions of near vertical planar weaknesses. The majority of planar weaknesses undoubtedly originated in joints, and it is likely that in subsequent periods, characterised by compressional rather than tensional stresses, the stress was taken up preferentially on the existing joints. It is possible also that some of the planes dipping at less than 45° are related to the relief of the compressional stresses and are therefore technically reverse faults or thrusts, but this has not been proven.

To summarise, the granitic structure at La Cotte is

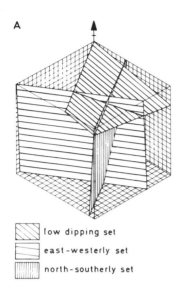

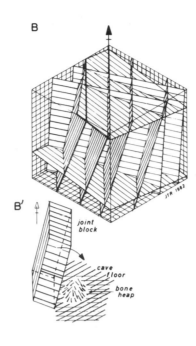

low dipping set

east-westerly set

north-southerly set

**Fig. 4.4** Modified orthographic projection to illustrate how the form of La Cotte cave resulted from erosional dissection of the major planes of weakness. The moment represented is an idealised stage in the archaeological excavations during the late 1960s, when the bone heaps were being recovered.

still dominated by the original pattern of cooling strains developed some 600 million years ago. Later compressional and tensional stresses have largely achieved release by following the joint planes. The only significant exception to this is the major fault zone, the Ouaisné fault, which runs parallel to the northern scarp of La Cotte headland and extends eastward towards Belcroute.

At some stage in the geological history of the island, when tensional stresses predominated, a swarm of more or less E/W trending dykes of doleritic composition were intruded into the granites and diorites that occur along the south coast. Their alignment and attitude suggest that they too tended to exploit the weaknesses of the existing E/W joint set. In the immediate area of La Cotte such dykes are of minor importance, though where they occur they have usually provided significant lines of attack for the processes of erosion and weathering, taking the form of gullies (e.g. north of the Ouaisné fault on the beach outcrops).

### 4.2.4 Changes in Land-Sea Relationships: Mainly Tectonic

By the time of the early Tertiary, when Armorica had taken its general form, the forces of marine and terrestrial erosion and deposition could begin to shape the surface that we see today. A number of more or less level areas at different heights above and below present sea-level have been identified in the general region of Armorica/southwest England and the western Channel. Everard (1977, 55–59) has made a useful summary of these and of the possible order of events with special reference to Cornubia, but his account has wider implications. Although most of these platforms are not of

particular relevance to La Cotte, they are important in the context of the overall shaping of Armorica. Among them, mention may be made of the very widespread level at some 130 m (a remnant is probably preserved at Les Platons in Jersey) and the relatively flat floor of the Channel. The development of such platforms separated by several hundred metres in height is related to changes in sea-level caused by earth movements affecting Armorica and the surrounding region either directly, or indirectly as a result of such movements elsewhere (e.g. building of the Alpine mountain system and the continuing opening of the Atlantic Ocean).

Little by little the succession of events during the Tertiary produced a topography which, in spite of its lack of detail, would be recognisable as that of Armorica as we know it. Jersey, Guernsey, Alderney and Sark would be identifiable either as islands, or, at times of low sea-level, as flat-topped hills rising from the surrounding plain. The basic elements of the river system that patterns Armorica and the western Channel sea floor were all present (cf. Fig. 1.1).

Of particular relevance to southwestern Jersey and the eventual forming of La Cotte cave were significant plateau levels, one at some 65 m O.D. and the other spanning a range of height from somewhat below -5 m O.D. to a little above +8 m O.D. The two were linked by an important cliff feature. The lower of the two plateaux is represented by the present shore platform off St. Brelade's Bay, while the higher of the two is best preserved in the Noirmont/Portelet-La Moye plateau. At the time the higher plateau would have been considerably more extensive than at present and much of the shore platform between Ouaisné and St. Brelade's Bay still unformed. It is unlikely that La Cotte cave existed

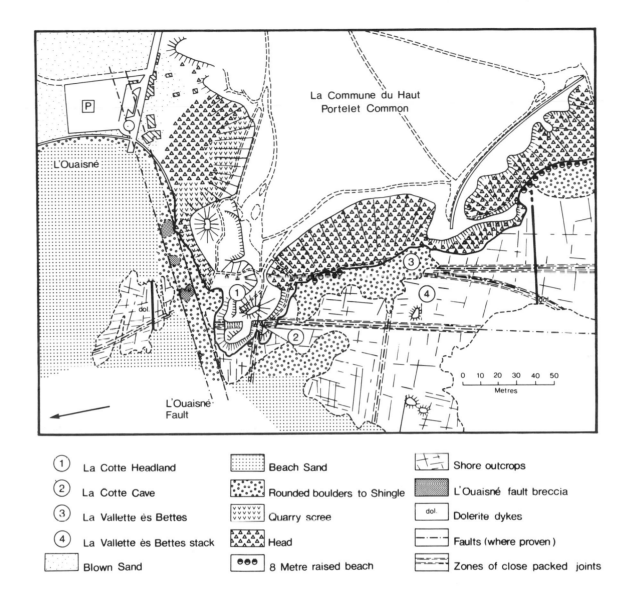

**Fig. 4.5** The immediate vicinity of La Cotte: geology and other features.

① La Cotte Headland
② La Cotte Cave
③ La Vallette ès Bettes
④ La Vallette ès Bettes stack

Blown Sand

Beach Sand
Rounded boulders to Shingle
Quarry scree
Head
8 Metre raised beach

Shore outcrops
L'Ouaisné fault breccia
dol. Dolerite dykes
Faults (where proven)
Zones of close packed joints

even in an incipient form, while the Noirmont/Portelet-La Moye headlands were probably separated from each other by some early precursor of St. Brelade's Bay (Fig. 4.6).

### 4.2.5 Changes in Land-Sea Relationships: Mainly Climatic

The above account gives a picture of the configuration of the land at the end of the Tertiary period, before the onset of the major alternations of cold and warmth that mark the Quaternary. With these began the rise and fall of sea-level related to climate rather than to earth movements, which continued to occur, but with effects less dramatic than those produced by the successive capture and release of huge masses of water by the ice-caps and ice-sheets.

In its later history the Noirmont/Portelet-La Moye plateau was eaten back over some two and a half million years by the combined effects of marine erosion at times of interglacial high sea-levels and of weathering during glacials. The cumulative effect of these processes was:

1. to reduce the overall level of the 65 m plateau;
2. to produce a relief on its surface, notably of small valleys, but including also the scarp along the Ouaisné fault;
3. to bring about cliff recession and in particular the relatively rapid excavation of St. Brelade's Bay.

The processes exploited the weaknesses in the rocks of which the Ouaisné fault, combined with the jointing of the Beauport granites, was of the first importance. Important sea-levels can be identified at 18 m O.D. and

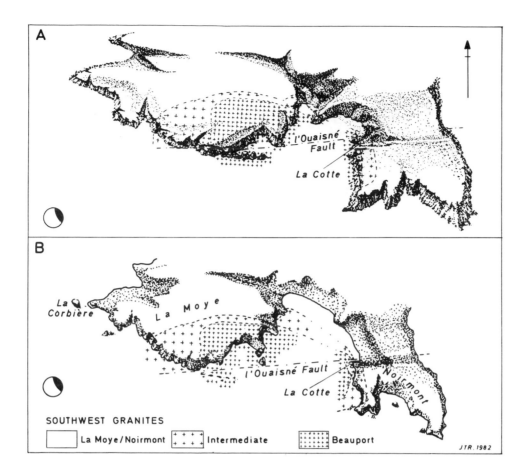

**Fig. 4.6** Representation in oblique aerial view of the development of the modern topography of southwest Jersey (B) from a hypothetical situation prior to 500 kya (A). Erosion was guided by many lines of weakness in the granites, notably joints, faults and dykes. A particularly significant factor in the larger landscape is thought to have been the Ouaisné fault; in the La Cotte headland itself, faults in the west ravine and a north-south fracture zone provided the basic pattern of weaknesses exploited by marine erosion. The eye symbol represents the direction of aerial view; not to scale.

about 30–40 m O.D., besides continuous intermittent recutting of the shore platform proper. Undoubtedly the height of the sea operating at or about the shore platform was not always the same, but it was close enough to continue the process of erosion at the same general level. Many small breaks across the surface suggest these differences in level at different times. It nonetheless remains true to say that the shore platform over a range of some 15 m has been a persistent feature of the topography around Armorica for several million years, just as the earlier 65 m and 130 m heights mark long or recurrent stands of the sea.

The 18 and 30–40 m levels are mostly identified around the Channel Islands by notches and raised beaches along the present cliff lines that mark the descent from the higher plateaux to the shore platform (Mourant 1933 and 1935; Keen 1978). This indicates that they are comparatively recent in age, as the rate of erosion of the cliffs is such that evidence of their presence would otherwise have been totally destroyed.

Recent work on their age and association with the succession of sea-levels and marine deposits suggests that both the 18 and 30–40 m episodes may date back to something in the order of half a million years ago. It is likely that a cliff line separating shore platform from plateau recognisably close to the one we see at present has been in existence for at least as long. It is at this stage that it becomes possible to discuss the La Cotte cave in a little more detail, particularly in relation to its occupation by man.

### 4.3 GEOLOGY AND ARCHAEOLOGY

The lowest recorded stratigraphical unit in La Cotte cave — layer H — lies in a gully cut into the rock at 8 m O.D., and probably even lower at the northern end (to judge from the excavational evidence and from observations made in autumn 1979). Further south, near the junction with the west ravine, it is undercut by a later raised beach at about the same level (McBurney and Callow 1971). The beach deposit, discovered by McBurney in 1961, appears to be related to an undercutting of the lowest stratigraphical unit in a manner closely analogous to the present undercutting of the Weichselian head in the gullies of the Pinnacle (on Jersey's northwest coast) by the present high sea-level. There is, however, no reason for associating the cutting of the platform on which the beach presumably rests with the episode that produced the beach deposits. On the contrary, the fact that this stratigraphical unit itself rests directly on the platform deeper in the cave (further to the north, that is) suggests that it was cut during an earlier episode of erosive activity at or just above the 8 m level.

From this evidence, a strong case can be made for proposing that the essential T shape, or at least the L part of it, of which the stem forms the west ravine, had already been created before man's first known occupation of the cave. It follows also that the cliff line of the headland could not have been far beyond its present western edge. In this context, the maximum known depth (of horizontal penetration) of any cave system cut into the Noirmont/Portelet-La Moye promontory is 170 m, at La Moye Point. The La Cotte cave system is some 120 m in depth, and happens to be L shaped, while the outer, exposed portion of the system is a gully some 50 m long. This suggests that the western cliff line of La Cotte headland at the time of the first 8 m cutting of the main N/S ravine may not have been more than some tens of metres west of its present position.

No direct evidence of the 18 m sea-level, in the form of notches or raised beaches, has been found at La Cotte. Had such remnants existed when the lowest stratigraphical unit was laid down, it seems likely that they would have been protected by it from subsequent destruction. This suggests that traces of the 18 m sea-level were destroyed by erosion after the first cutting of the cave at the 8 m level, before the cooler period of man's first occupation as represented by layer H. A further argument that the 18 m sea-level preceded this is the improbability that these soft occupational deposits would have survived inundation by the 18 m sea. Surely at least some erosional or chemical evidence would have survived as witness to the event. The alternative is to suppose that the west ravine was closed by a massive rockfall, but it seems unlikely that this would have been impregnable against the direct force of Atlantic gales. The history of the cave following the deposition of the lowest stratigraphical unit is marked by at least one rise of the sea to the 8 m level. This event was responsible for the erosion of the basal units to form a cliff, and for the beach on which it partly rests (see chapter 6).

Infilling of the cave by the spalling off of large and small rock fragments from the roof and walls, and from the heights above, followed the lines predictable from the joint and fault pattern. Above about 20 m O.D. the cave system is dominated by larger joints than the zone below, but to east and west the N/S ravine is bounded by nearly vertical planes of weakness. The larger joint blocks at the higher levels would have served as keystones, allowing the seas of earlier times to tunnel into the cliff below without causing the roof to collapse. By the beginning of the period for which the deposits are preserved, however, widening of the main cave had already seriously reduced the strength of the roof, and in the course of subsequent weathering much of it collapsed into the cave, to become embedded in the finer sediments and occupational debris accumulating below. Post-glacial removal of the head by marine erosion and, even more, by this century's archaeological excavations, have exposed the remnant of the cave roof (the present arch) to the forces of subaerial erosion; even in the temperate climate of today the lifespan of the arch is likely to be short, perhaps to be measured in hundreds rather than thousands of years.

## 4.4  SUMMARY OF EVENTS

The original cooling of the southwestern granite complex produced a strong pattern of jointing. Additional variability in this feature was caused by the position of La Cotte headland astride the contact between the two main southwestern granites, those of Beauport and Noirmont/ Portelet-La Moye. Subsequent events over several hundreds of millions of years imposed the additional planar weaknesses of a fault system. The most important single fault was that now marking the scarp south of Ouaisné. Other faults followed to a large extent the main directions of the initial N/S and E/W vertical joint sets. Specifically at La Cotte headland two parallel N/S parallel planes of weakness define the main cave gully, while several parallel fault planes mark the line of the west ravine.

Erosion at the generalised level of the present shore platform eventually reached to within 100 m or less of the present La Cotte headland. Seas at levels of 18 and 30–40 m O.D. notched the cliff separating the shore platform from a plateau level at about 65 m O.D. around the islands and, without much doubt, began the cutting of La Cotte cave at these levels. There then followed a crucial event when a high sea-level at about 8 m O.D. worked back along a fault line of the west ravine and, reaching the N/S planes of weakness, excavated well into the northern arm, the present main cave. The work of erosion of the main cave was facilitated by the fortuitous occurrence of close-set jointing between 8 m and some 20 m O.D. overlain by more massive jointing which produced a keystone effect at higher levels.

The subsequent history of the cave is one of further exploitation of the various lines and planes of weakness, by successive levels of the sea during interglacial periods and by subaerial erosion during the intervening cold phases. The preservation of the material evidence of man's use of the cave is due to the fact that no subsequent rise in sea-level was of sufficient height or duration to remove the deposits that had accumulated during the colder phases. However, weathering and erosion continued slowly but surely to reduce the strength of the keystone jointing in the roof of the system, causing its breakdown block by block.

# 5

# THE QUATERNARY DEPOSITS OF THE CHANNEL ISLANDS

D. H. Keen

The intention in this chapter is to identify and examine the range of features associated in the creation of the landscapes that affected the lives of Palaeolithic man at La Cotte de St. Brelade and elsewhere in the islands. Some of the features relate to the remote geological past when the principal hard rocks of the islands and adjacent areas were formed, but it is with the sequence of events that led to the present geomorphology that this account is concerned. The starting point for this review is in the late Mesozoic, where the first evidence for the origin of the islands occurs, but the bulk of the chapter is concerned with the events of the Middle and Upper Pleistocene with their fluctuating climates and sea-levels, which were the major controls on man's activities in the islands, and of which aspects particularly relating to La Cotte are discussed in chapters 7 and 31.

The terminology used in this chapter is that proposed by Mitchell, Penny *et al.* (1973) for the British Isles. For greater comprehensibility, the north European glacial/interglacial stage names used by French authors have been translated in terms of the British scheme (see Table 3.3).

## 5.1 THE GENERAL SETTING: MESOZOIC AND EARLY TERTIARY

The feature of the Channel Islands that gives them particular interest in the geological history of the last 100 million years is their situation on a shelf of the Armorican massif (Fig. 5.1) lying at a critical height with respect to changes in sea-level over this great period of time. To the northwest of the Alderney/Ushant line the Channel floor has tended to subside ever lower over this period, allowing the accumulation of considerable thicknesses of mostly marine sediments of Mesozoic and Tertiary age. By contrast, the Armorican massif has tended either to remain stable or to rise. The islands on their

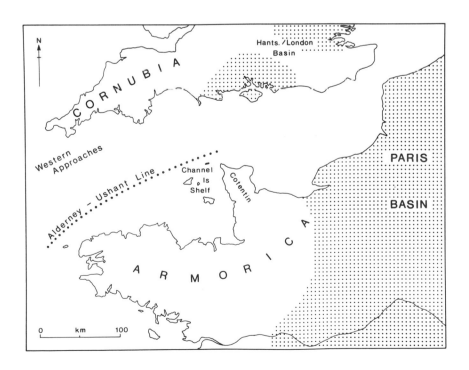

**Fig. 5.1** The Channel Islands shelf and geological provinces in the western Channel.

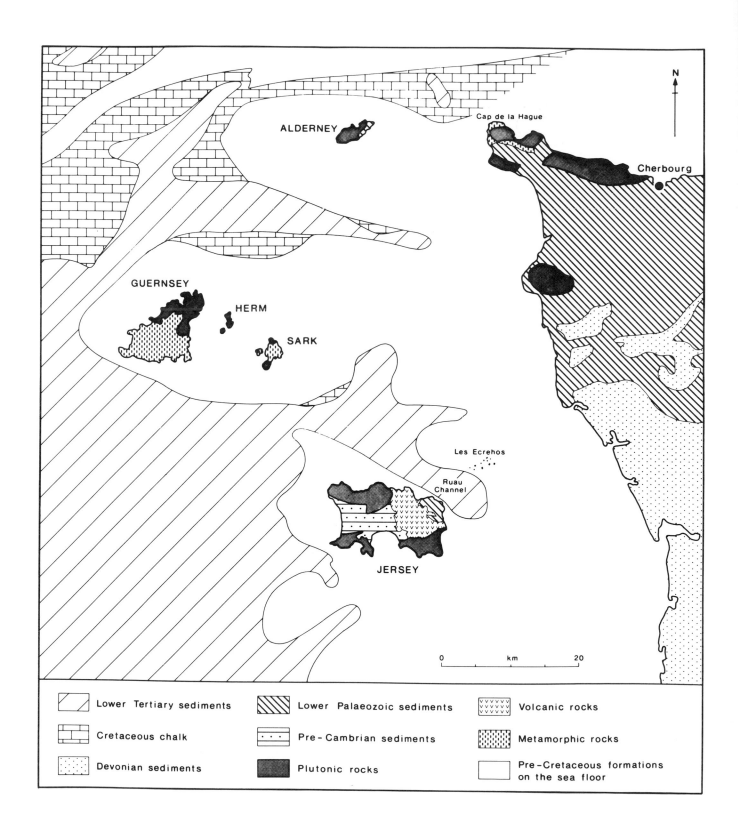

**Fig. 5.2** The solid geology of the Channel floor around the Channel Islands.

shelf at the junction of these two zones have always been an integral part of the massif rather than the Channel, and have been very sensitive even to quite small changes in the level of the sea, whether these were due to movement of the land (tectonic) or of the sea (eustatic). The principal terrains are located on Fig. 5.1. Geological evidence of shoreline and near shoreline marine sediments at certain stages of the Mesozoic enables the existence of the Cornubian and Armorican massifs to be deduced. Deeper water sediments of the same ages characterise the depressed areas of the London/Hampshire and Paris basins and the Channel. The general absence of Mesozoic sediments from the massifs, including the Channel Islands shelf, makes it difficult to make any meaningful statement about the possible distribution of land and sea over these areas.

Towards the end of the Mesozoic era there was an almost world-wide marine transgression over low-lying land areas. In these seas, over substantial areas of northern France, the Channel and Britain, were formed the calcareous oozes and muds that were to become consolidated as the Chalk. The climatic conditions were considerably warmer than those prevailing at present and set, as it were, a high point from which temperatures were to decline steadily during the Tertiary period to culminate in the Quaternary ice ages. The Chalk has a dual interest for the islands and La Cotte. In the warm seas of the time countless billions of sponges lived and died. Their siliceous skeletons supplied the bulk of the silica that went to the formation of the flint for which the rock is famous and which was to be so much prized by Palaeolithic man. For this reason it is not only of academic interest to determine the extent of Chalk outcrops in the area. Mapping of the Channel floor over the past few decades has shown that Chalk oozes and muds were deposited over parts of the Channel Islands shelf (Fig. 5.2). The main Chalk outcrop is to the west of Guernsey, but isolated occurrences are also known south and southeast of Alderney, as well as near the Minquiers. This would seem to indicate that the two islands mentioned may have already existed in outline during the Cretaceous, although by Upper Chalk times they, like the rest of western Europe, were probably submerged under the Chalk sea. However, it is certain that, when man was inhabiting La Cotte cave at times when the Channel floor was partly exposed, the nearest Chalk outcrops (and thus the nearest source of flint *in situ*) were those now beneath the waters of the Channel near the Minquiers and north of Jersey and, further afield, those off Guernsey. Whether man used these outcrops or turned to other sources is discussed in the archaeological sections of this volume.

Profound changes marked the end of the Mesozoic era. The extensive Chalk seas over northwest Europe contracted and many areas became land. The deposits of the Tertiary period, which are the first known after those of the Chalk, are of Eocene age and their distribution in the Channel Islands area is far more extensive than that of the Chalk, though the same problem of interpretation remains as to how far the present outcrops reflect their real and original extent of Eocene times. The deposits

are often rubbly limestones such as the samples taken from the sea floor in the Ruau channel between northern Jersey and the Ecréhous (Renouf 1971). The most that can be said at the time of writing is that there is a wealth of Tertiary evidence indicating the presence of significant land areas within the Armorican and Cornubian massifs and that it therefore seems likely that the present distribution of the Eocene rocks reflects, to a considerable extent, the topography of the sea floor at the time of deposition. If this is the case, then the Channel Islands can be said to have their first positive existence as early as this stage of the Tertiary.

## 5.2 LATE TERTIARY AND EARLY PLEISTOCENE EVENTS

The distribution of deposits of early and middle Tertiary age shows that the sea occupied large areas of the London/Hampshire and Paris basins, the Channel and parts of the Channel Islands shelf at this time. Thus it is clear that subsidence of these areas relative to the massifs occurred progressively. It is otherwise for the later Tertiary and early Pleistocene. In spite of the continuation of relative vertical movements between land and sea, the major differences in response between the massifs and the basins (caused by orogenic and related activities outside the area in the Alps and elsewhere, including the continuing opening of the Atlantic Ocean) had lessened; the massifs and the basins were behaving more or less as one, a state of affairs which has continued up to the present. An examination of the relief of the land, from the Seine to Ushant and from Sussex to Land's End, reveals the presence of a number of recurrent, more or less dissected, plateaux and platforms at different heights. It is these that mark important stages in the development of the whole area and that have influenced the form of the landscape as it appears today. However, neither the plateaux and platforms nor the deposits occasionally associated with them have proved particularly amenable to precise dating or explanation. It is against this background that the whole of the following discussion must be viewed. The platforms, plateaux and associated deposits which have most bearing on the present account lie below about 150 m (all heights are quoted in metres above mean sea-level). Landscape elements above this are not considered here.

Changes in sea-level during the Tertiary period can generally be ascribed to the effects of tectonic movements, whether in the immediate area or elsewhere. As the climate cooled world-wide, there began the series of ice ages which culminated in the vast ice-sheets of Pleistocene times. The appearance of the earliest (the Antarctic ice-cap) probably lies more than 30 million years back in the Tertiary. A vital aspect of the locking up of water in ice-caps, ice-sheets and glaciers was the effect on sea-level. Anticipating information to be given later, the sea fell by about 100 m in the western Channel at the maximum of glaciation and certainly rose to 18 m, if not substantially higher, at times of minimum glaciation during interglacials. Much of the difficulty of accounting for

**Fig. 5.3** Quaternary sites on the Channel coasts.

the various plateaux, platforms and associated deposits lies in the complicated interplay between sea-level movements of an eustatic nature (linked to the waxing and waning of the ice-sheets) and the movements due to tectonic activity.

At some time in the late Tertiary, a major planation surface was produced by a sea-level at some 130 m. This surface is widespread in northern France and southern Britain, although Jersey is the only Channel Island which rises high enough to preserve a remnant (at Les Platons, above the north coast). This surface, and others down to another particularly prominent platform at about 66 m, are most often assigned a Plio/Pleistocene origin — that is, late Tertiary to early Quaternary. The 66 m planation surface is of particular importance in the islands, where it forms extensive plateau levels on Jersey, Guernsey, Alderney and Sark. On Jersey, the small but

important remnant plateau of Ouaisné Common above La Cotte de St. Brelade is at this level, and provides a minimum age for the origin of the landscape features in the area.

Fragmentary deposits of Pliocene age are known from a few sites in southern England (Fig. 5.3), notably from St. Erth in Cornwall (Mitchell et al. 1973a) and Lenham in Kent (Wooldridge 1927), and around Redon in the southeastern part of Brittany. A few Lower Pleistocene sites are known, including Guindy in the Côtes-du-Nord (Morzadec-Kerfourn 1974), St. Sauveur and St. Nicolas-de-Pierrepoint in Manche (Brebion et al. 1973), Le Londe in the Eure (Kuntz et al. 1979) and, north of the Channel, Netley in Surrey (Dines and Edmunds 1929). No deposits are known in the islands. The deposits are found within the height range of the planation surfaces thus suggesting linkage in some way, but

46

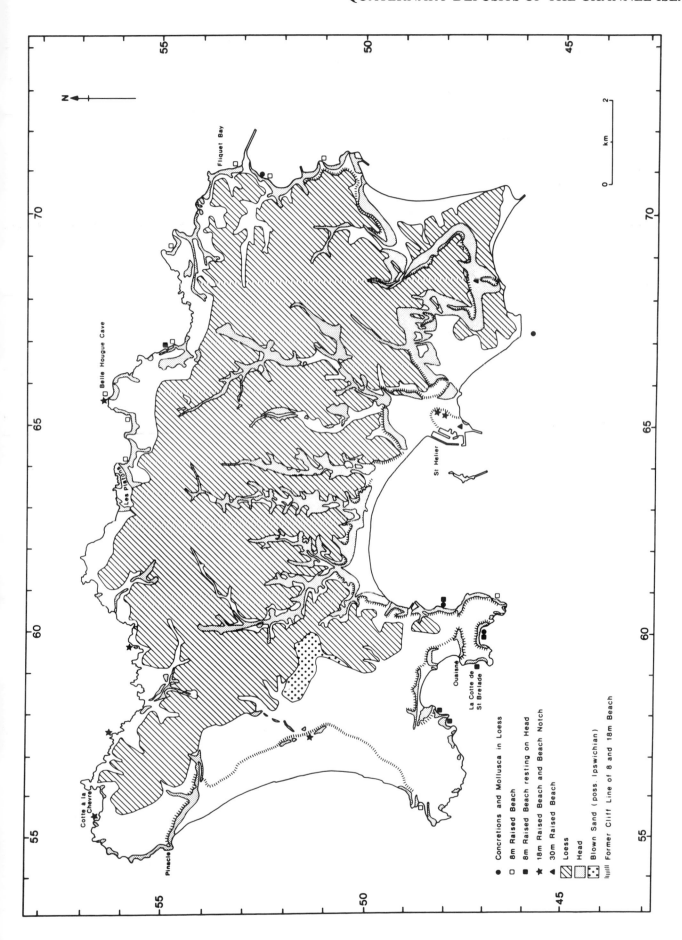

**Fig. 5.4** The Quaternary deposits of Jersey (from Keen 1978).

no more precision can be offered at present. Thus in spite of their considerable extent on both sides of the Channel, and their significance as features of the landscape, their age and origin are obscure.

## 5.3 PLEISTOCENE DEPOSITS AND SEA-LEVELS

### 5.3.1 Surfaces and Deposits between 30 and 40 m

Apart from these Lower Pleistocene deposits, sediments of the Lower and early Middle Pleistocene are rare in the area. The earliest widespread deposits which occur are those relating to sea-levels between 30 and 40 m. Occurrences are known in both Jersey and Guernsey (Keen 1978), on the south coast of England at Slindon in Sussex (Dalrymple 1957), and in the Anse du Brick at Cherbourg in Normandy (Elhai 1963). The age of the deposits is uncertain but similar sea-levels, indicated by the estuarine sections of the terraces of the Seine (Lautridou et al. 1974) and of the Thames (Ovey 1964; Kerney 1971), suggest an age in the Hoxnian interglacial. The fauna and the archaeology of these terrace deposits also suggest this interpretation, but the complexity of such terrace sediments and the lack of any geochronometric means of dating them does not allow this 30–40 m sea-level to be accurately dated.

Within the islands the effects of the 30–40 m sea-level, apart from locally more extensive planation particularly notable in western Guernsey, are confined to modification or notching of the major cliff profile that descends from the 66 m level down to the present composite shore platform. This relationship implies for the first time the existence of planation at or near the present sea-level, and thus has a bearing on interpretation of the 8 m levels to be discussed later. The same sea-level probably played a part in modifying the cliff into which La Cotte cave is cut, though the extreme susceptibility to destruction of the strongly jointed granite precludes any more specific statement, as direct evidence is not preserved. The formation of the system of fissures in the headland is discussed in the previous chapter.

### 5.3.2 Surfaces and Deposits at about 18 to 20 m

Beaches, notches and platforms between 18 and 20 m occur widely throughout the Channel Islands (Fig. 5.4) and adjacent areas of France (Keen 1978), though such features have seldom been identified elsewhere in the western Channel (Mitchell 1977). Considerable differences of opinion exist over the interpretation of the deposits and associated erosional notches and platforms. In Guernsey, deposits at these heights occupy considerable areas in the centre of the island but, generally, the principal effect of the sea which formed them was similar to that of the 30–40 m sea — that is, the cutting of a notch or small platform into the main cliff profile, on which beach deposits may or may not have been preserved. On the French side, Elhai denies the existence of any beaches which are not part of the 'bas-normannien' succession (the now obsolete French term which includes all the features related to the so-called 8 m stage). The point at issue is not the evidence but its interpretation, Elhai considering any beach or notch between the present upper parts of the shoreline and 20 m to belong to a single series of events not ascribable to distinctive, well separated sea-levels. It is certainly true that a number of important deposits of marine origin not only occur between 8 and nearly 18 m (see Moulin Huet below) but are more or less continuous over the height range represented. On the English side of the Channel such deposits reaching above the height of the 18 m beach are known from Plymouth and Portland (Zeuner 1959; Davies and Keen 1985).

If the 8 and 18 m deposits are continuous — and this is by no means proven nor agreed by most workers in the field — then the 18 m level may represent a long stillstand during the descent from the 30 to the 8 m level. This case is argued by Mitchell (1977). However, the locally extensive nature of the platform and associated deposits and the evidence from cave sequences in the islands offer a strong case for the existence of a discrete and important sea-level in its own right. In Moulin Huet cave in Guernsey (Collenette 1916; George 1973; Keen 1978 and 1980), a fill of marine deposits appears to have been continuous from 1 to at least 12 m. The interpretation of such a continuous succession is best explained by a rising sea at the end of a glaciation rather than by a descending one at the end of an interglacial. It can thus be concluded that some time after the 30 m level was formed the sea descended as low as 1 m before rising well above 8 m, and perhaps as high as 18 m. As a consequence of this, the 18 m sea-level must be considered separate from both the 8 and the 30 m sea-levels.

The possibility of a range of marine levels at and below the general 18–20 m series is also suggested by the varied terraces and estuarine deposits of the lower Seine which occupy this height range (Lautridou et al. 1974; Lautridou and Puisségur 1977).

Interglacial conditions, which are probably necessary for an 'extra' raised beach between the possible 30–40 m Hoxnian and 8 m Ipswichian levels could perhaps be accommodated within the fluctuations of the climate indicated by the loesses of the lower Seine. At St Pierre-lès-Elbeuf three post-Hoxnian palaeosols are known, which would allow two further interglacials to be added to the Ipswichian generally recognised (Lautridou et al. 1974).

There are no occurrences of the 18 m level in the immediate vicinity of La Cotte cave and, as noted for the 30–40 m level, any notches formed are likely to have been subsequently destroyed by collapse and erosion of the cliff. However, the cave of La Cotte à la Chèvre on Jersey's northwestern cliffs lies at 18 m and contains marine deposits (see Appendix C). It is thus assumed to have been cut at this time.

### 5.3.3 Surfaces and Deposits at about 8 m

The most widely developed marine deposits in the Channel Islands, and indeed along the Channel coasts as a whole, are those of the 8 m level (Fig. 5.4). In the

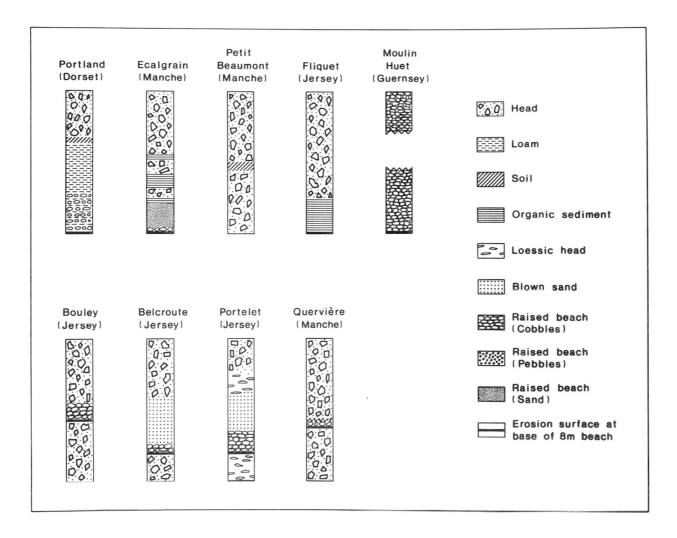

**Fig. 5.5**   The stratigraphy of selected Pleistocene sites in the western Channel.

islands the height variation is from 3 to 11 m; this range is similar for the Eemian beaches of the French coastline and the low raised beaches of southern England, south Wales and Ireland. Stratigraphically, the beach is generally found below a thick layer of periglacial head and rests on a platform cut either in the solid rock or more rarely in head (Fig. 5.5). The erosional platforms over the same height range form one of the most significant and extensively developed planation surfaces in the islands, for example the intertidal expanses of southeastern Jersey, Guernsey's west coast, the Ecréhous and the Minquiers reefs. It is generally accepted that these platforms are of composite origin. They have played an important part in determining the landscapes of the islands at times of lowered sea-level during the later Pleistocene.

Major problems of dating and correlating features of this 3–11 m geomorphological entity, as Bowen (1973) rightly identifies it, result from the paucity of useful faunal data, material susceptible to radio-isotope work and of most of the usual stratigraphical means of comparison. Locally, the detailed stratigraphy of deposits

reveals something of the suspected complexity of events that prevents any simplistic interpretation, such as the notion that everything can be fitted to the Ipswichian interglacial.

In the islands, at least ten sites (Keen 1975) exist where a layer of periglacial material separates the 8 m beach from the underlying erosional platform cut into the solid rock. Two marine phases are needed to explain this evidence, one during which the platform was cut and one during which the beach was deposited, the two being separated by an intervening periglacial episode. This episode is of unknown length and cannot itself be dated.

Similar sites are known from Normandy. The deposits around Grandcamp-les-Bains, Manche (Coutard, Lautridou *et al.* 1979) provide evidence bearing on the possible length of the periglacial episode; it must have been long enough to allow the partial decalcification of a beach gravel resting on Jurassic limestone before the deposition of new marine sediments at the same level. At the Baie de Quervière, west of Cherbourg, the succession platform/head/beach/head is the same as the examples in the islands.

The interpretations of deposits and platforms at the 3–11 m level in southern and southwestern Britain, in south Wales and in southern Ireland are contradictory. In the Gower caves (south Wales) two marine episodes separated by terrestrial deposits are known (Sutcliffe 1980). Sutcliffe holds that the time interval between the two beaches may have been of short duration, with the possibility that the two beaches can be encompassed within the one interglacial. In southwestern Britain and in southern Ireland the stratigraphy of the glacial tills overlying the beaches has been used to assign a possible age to the beaches themselves. Mitchell (1970) and Stephens (1970) have both suggested a Hoxnian age on these grounds. In spite of this, the Gower marine episodes and those further east in southern England have usually been assigned to the Ipswichian, though, it must be stressed, often on less than sound evidence. The doubts raised by such views are reinforced by consideration of the evidence emerging from detailed stratigraphical examination of raised beach sites, the information provided by estuarine deposits, and by isotope and amino acid investigations of the sparse faunas that do exist (Andrews, Bowen and Kidson 1979; Keen, Harmon and Andrews 1981).

In the islands the 8 m beach in Belle Hougue cave on Jersey's north coast has yielded one of the rare faunas from beaches at this level. A date of 121 kya (Keen *et al.* 1981) has been derived from the calcite cementing the top of the beach. The interpretation of the faunal elements, notably the occurrence of the mollusc *Astralium rugosum* (Linné), which has its modern northern limit at the Gironde, has suggested a warmer sea in the Channel than at the present. The date of 121 kya matches the oceanic stage 5e of Shackleton and Opdyke (1973), from deep-sea cores, and offers a reasonable assignment of this beach, at least, to the last interglacial.

The problem of the sites where two marine stages are known is not directly helped by this evidence from Belle Hougue cave, because at this site there is only one beach at the 8 m level. Information beginning to emerge from study of the Portland raised beaches and associated deposits (Davies and Keen 1985) goes a long way to confirming the existence of two very distinct sea-level episodes. The succession here begins with a beach gravel which rises to 15 m and is overlain by a complex of terrestrial deposits which includes horizons of decalcification and a palaeosol which may be of interglacial rank. A molluscan fauna obtained from the beach deposits, though similar to that from Belle Hougue, possesses features which Baden-Powell (1931) held were suggestive of somewhat cooler conditions. Re-examination of this fauna and evidence from microfossils suggest that part of the raised beach is indicative of warm conditions and must be equated with the Ipswichian as at the Belle Hougue cave. Thus the Portland succession, too, shows evidence of two sea-levels: one of Ipswichian age and the other, rising to 15 m, of an earlier date.

The suggestion that some beaches were formed at the 8 m level in somewhat colder climate than that proposed for the Ipswichian interglacial has been used in interpreting sequences in southern Ireland, where some beaches are interbedded with head near their tops (Bowen 1973). However, the available evidence is insufficient to exclude the possibility that the main body of any such beach was laid down under warmer conditions than were its concluding stages. Even if the whole marine episode represented by these deposits corresponds to a beach such as underlies the Portland sequence, it still leaves open the question of the proven Ipswichian 8 m beaches.

A possible interpretation of the extensive platforms of the 8 m feature, cut between 3 and 11 m, has been advanced by a number of authors; it is based on the premise that erosion by the sea would have been much faster under the cold conditions of the glaciations than under the warmer climates of the interglacials (see Dawson 1980). Mitchell (1970) has suggested that the beaches with indications of formation under cooler conditions than are normal for the Ipswichian may have formed under cold climate as the sea fell from the 30 m level. Whether the linking of the model to the 30 m level is acceptable or not, the possibility of 3–11 m sea-levels unrelated to known interglacials should be given serious consideration, particularly in view of the increasing amount of evidence pushing back the age of some elements of the 8 m complex of deposits and erosional levels. The application of such principles might also facilitate correlation of beaches and levels with the sea-level curves based upon the oxygen isotope ratios and isotope dating of deep-sea core successions (e.g. Shackleton and Opdyke 1973). As it stands, the current interpretation of the deep-sea cores gives no explanation for a high sea-level of 30 m within the time span of the Upper Pleistocene, but does identify comparatively numerous episodes when low raised beaches might have been formed.

In summary, the evidence indicates the existence of at least two marine episodes contributing deposits to the 3–11 m complex of surfaces and beaches. The younger of these may be of Ipswichian age with a date of some 120 kya and so correlate with the oceanic stage 5e. The earlier beach and the platforms associated with the complex of levels may be of any age in the Pleistocene. This having been said, the freshness of many of the low beach deposits compared to those at 18 and 30 m tends to encourage the interpretation that most 8 m beach deposits are genuinely of last interglacial age.

### 5.3.4 Periglacial Deposits: Head

The 8 m beaches in the islands and on both sides of the Channel are usually overlain by a layer of head. This ranges from a maximum thickness of 40–50 m below the high cliffs of the Cotentin (Watson and Watson 1970) and Devon (Mottershead 1971) to 3–5 m under the low cliffs of northern Guernsey (Keen 1978). The main control over the thickness of the head is the height of the local solid rock cliff from which the material that makes it up is derived. The higher the rock cliff the thicker the head, to give a normal ratio of 4 or 5 to 1 for cliff height to head thickness (though it should be noted that the head thins both upward and downward to zero, so that the cross section is that of a double sigmoidal curve).

The outcrops of head are not confined to the coastal cliffs but occur anywhere where the slopes are steep enough. The maps of the islands (Keen 1978) show veneers of head draping the sides of all the major valleys and plateau slopes.

As with the 8 m beach on which it rests, there is no general agreement as to the age of the head. Most authorities (see Bowen 1973) believe that the bulk of the head is post-Ipswichian in age and belongs to the Devensian glaciation, but a school of thought represented by Stephens (1970) and Coutard, Helluin et al. (1979) prefers to divide the successions into lower, pre-Ipswichian and upper, post-Ipswichian units on the basis of significant palaeosols which separate them. Apart from the possible example from Portland (Keen 1985), such palaeosols have not been proven north of the Channel. In Normandy the horizons are clearer, though the straight division into pre- and post-Ipswichian units is still open to debate (see Coutard, Helluin et al 1979).

At a few localities near Cap de la Hague, Manche, and including the Baie d'Ecalgrain (Coutard, Helluin et al. 1979; Coope et al. 1986), at Asnelles-Belle-Plage/St Côme de Fresnes, Calvados (West and Sparks 1960), at Fliquet, Jersey (Coope, Jones and Keen 1980) and at Port-Lazo, Finistère (Morzadec-Kerfourn 1974), the base of the head contains organic muds and peaty silts. The Ecalgrain and Fliquet sediments have yielded minimum dates of 44.5 and 25.5 kya respectively (Shotton and Williams 1971 — Birm. 211 and 955). If the bulk of the overlying heads at these localities follow on directly from the muds and silts, any palaeosols within the succession would have to be accomodated within the Devensian, and probably before the maximum cold after 25 kya (Mitchell, Penny et al. 1973).

The pollen and beetles obtained from these deposits show the rapid onset of cold, and the development of open habitat conditions presumed to mark the transition between the warmth of the interglacial indicated by the marine 8 m beaches (where present) and the full periglacial conditions represented by the head. It would seem at least likely that, in some of the cases cited, the underlying 8 m raised beach would be of Ipswichian age, though in no instance is the age proven. In terms of detail the Port-Lazo and Fliquet sites differ from the others in two respects. Neither is seen to rest upon an 8 m marine horizon, and both have yielded either faunal and floral evidence that is indicative of a more rigorous climate, although one in which marine influences are still strong, especially in the lower levels at Fliquet. Study of the beetles at Fliquet suggests maximum summer temperatures of 11 to 12°C and thus a climate of northern Scandinavian severity in immediately pre-head times. In the absence of underlying marine deposits, it is possible that the Port-Lazo and Fliquet deposits may represent the breakdown of the climate at the close of one of the early Devensian interstadials.

One effect of the head has been the filling of many of the gullies and caves cut during earlier periods of high sea-level. This can be clearly seen at a number of localities where erosion is only now revealing the presence of former caves and gullies as at The Pinnacle, in northwest Jersey, and at La Cotte itself, where the huge cave and gully complex was concealed beneath a considerable thickness of head until excavation removed it. It is always possible that other formerly inhabited caves still lie unsuspected beneath the abundant head that mantles cliffs in the islands and on both sides of the Channel.

### 5.3.5  Periglacial Deposits: Loess

Coupled with the head as a major product of the Devensian is the loess. This deposit is widely developed in the islands and on the French and English mainlands. Pre-Devensian loess is also widespread in northern France (Lautridou 1968; Coutard et al. 1970) though rare in England where only late Devensian loess is recognised for certain (Catt 1977).

Wherever the loess occurs it is the usual fine, structureless silt as described by Coutard et al. (1970). It originated under cold, dry conditions such as probably coincided with the maximum intensity of cold during glacial periods. The molluscan faunas recovered from a few sites in the islands and on the French mainland (Lautridou and Puisségur 1977, and Keen 1982a) are all tolerant of the most restricted habitats, thus tending to confirm the harsh climatic regime.

Correlation of the loess sequences with those of the head is difficult. Exposures of the head and the loess are close together in Normandy but there is as yet no direct connection between the two. The head sequences in Normandy may span much of the Devensian, but subdivision remains uncertain. Coutard, Helluin et al. (1979) have proposed an outline stratigraphy for the Baie d'Ecalgrain, but their accompanying discussion highlights the problems of dating that need resolving before the proposed succession can be confirmed. In England no attempt has been made to produce a similar subdivision. Thus the main stratigraphy for the Devensian in the area is best based on the loess, although, as this deposit is largely absent from England, cross-channel correlations are difficult.

### 5.4  SUMMARY AND CONCLUSIONS

Differentiation of the Channel Islands shelf within the Armorican province was probably well under way by the early Tertiary. During the later Tertiary and early Pleistocene stage of the Quaternary, a number of high level surfaces of marine erosion were cut and remnants of these form significant features of the landscape today. Perhaps as early as the late Middle Tertiary, water abstraction from the world's oceans, which nourished the build-up of ice as far afield as Antarctica, complicated the interplay between changing sea-levels and the land movements which had predominated in the area until then. By Pleistocene times, certainly, the succession of glacial stages caused major and rapid changes in sea-level maxima between that of today and heights up to at

**Table 5.1**  Pleistocene Deposits in Jersey[a]

| Stratigraphical terminology | Surfaces and deposits | Representative site in Jersey |
| --- | --- | --- |
| Upper Devensian | Loess | La Motte |
| Devensian | Head | Bonne Nuit Bay |
| Early Devensian (either interstadial or Devensian/Ipswichian transition) | Fliquet peat | Fliquet |
| Ipswichian | Some 8 m raised beaches | Belle Hougue Cave |
| Wolstonian (?) | Lower head of Belcroute, Bouley Bay and Portelet | Portelet Bay |
| Intra-Wolstonian (?) interglacial | 18 m beach | Mont du Jubilee |
| Intra-Wolstonian (?) interglacial | some 8 m beaches | ? |
| Late Hoxnian (?) | some 8 m beaches | ? |
| Hoxnian | 30–40 m beaches | South Hill |
| Tertiary and Lower Pleistocene | Erosion surfaces to 130 m | Island plateaux and surfaces |
| Early Tertiary | First shaping of the islands | Main island |

[a]Stratigraphic terminology after Mitchell, Penny *et al.* (1973).

least 30–40 m. The Channel Islands were particularly susceptible to sea-level changes because of the shallow waters that cover the shelf from which they rise. By the time that man is known to have visited the area, the sea had cut caves in a number of places between present low tide and the platform at some 66 m. These were visited by early man where they proved suitable. In Jersey the two La Cotte caves were both used, and a record of the occupations was left behind in the form of anthropogenic materials which became interbedded with deposits that were the consequence of natural accumulation — the head, the loess and the raised beaches. The plains over which early man hunted had been fashioned over long periods of the Tertiary and, probably, by much increased planation during the Pleistocene.

In spite of the extensiveness of Pleistocene deposits, and a considerable amount of fieldwork in recent years, it remains extremely difficult to establish a reliable succession, let alone the ages of the various events. The only deposit that has so far been fixed with some certainty is the Belle Hougue raised beach at 8 m, which is dated to 121 kya. Table 5.1 is a provisional attempt to arrange the events in the islands into a chronological sequence.

# PART II

## STRATIGRAPHY, ENVIRONMENT AND DATING

# NOTE

The scope of this report is such that it has not been considered practicable to supply a full glossary. The following explanations of terms employed in Part II may be found useful, however.

*Colluviation.* Downslope displacement and accumulation of soft material under wash (rill or sheet), splash or other non-alluvial conditions.

*Diagenesis.* Evolution of a sediment or a rock during burial. It includes mineralogical transformation, hardening and compaction.

*Epigenesis.* Replacement of one mineral by another (mineralogy). Phenomenon occurring after another (wider meaning).

*Gelifraction.* Mechanical weathering of rocks or other cohesive material under the action of frost, generally as a result of the growth of ice in the pores. It results in a finer granulometry.

*Grus.* A granular disintegration product in deep weathering profiles on granite. In humid temperate regions with mildly acid soil conditions, weathering of biotite to hydrobiotite-vermiculite assemblages and 2/1 Al-hydroxy minerals, with accompanying expansion, results in the formation of microfractures (Bustin and Mathews 1979). *In situ* grus retains its original structure but falls apart easily when disturbed by the hand (Isherwood and Street 1976).

*Hydromorphy.* Temporary or permanent water-logging in a sediment or rock, with resulting expression of oxy-reduction features — bleaching or iron hydroxide ('rust') precipitation. Acid soil conditions and the presence of organic matter may cause this phenomenon to be emphasised.

*Illuviation.* Vertical or lateral migration of particles, organic compounds, or salts (carbonates, silica etc.) associated with their relative enrichment in a lower horizon. The term is most usually applied to clay- or silt-size particles, or to organo-mineral complexes. (See also section 9.3.2).

*Loess.* A yellow, calcareous, porous, wind blown silt, characteristic of Pleistocene sequences all over the world, and mostly laid down during glacial periods. In primary context it has a fine syngenetic network of secondary calcium carbonate which is believed to be the result of arid weathering (Kukla 1977); decalcification may occur in the course of subsequent pedogenesis. The source of the dust may be any poorly vegetated area upwind of the deposition site.

*Ranker.* A raw humus accumulation, generally acid, resulting from a low rate of mineralisation of the litter (as a result of a reduced biological activity related to thermal conditions). (See also section 9.3.2).

*Spodosol* (or podsol). Acid soil developing on a porous, and generally poor, substratum (usually sand) under rather humid climatic conditions. It is characterised by a low humification rate, intense weathering of minerals (especially clays) and leaching of organo-mineral complexes to some depth. Occurs widely, from tropical to high arctic regions.

# THE STRATIGRAPHIC SEQUENCE: DESCRIPTION AND PROBLEMS

P. Callow

## 6.1 INTRODUCTION

In chapter 2 an account was given of the long programme of excavation which has taken place at La Cotte de St. Brelade. Of particular relevance here is the contrast between the excavations prior to 1940, which were concerned only with the deposits of the last glaciation, and those from 1950 onwards, which were devoted almost exclusively to the previously unsuspected layers underlying the 'peats' exposed in Captain Coltart's trench (Fig. 2.3). As well as the differences in stratigraphic emphasis, there is one of extent: the earlier work ranged right along the north ravine, the eastern part of the west ravine, and even the small south ravine. On the other hand, Burdo's operations in the 1950s concerned the west ravine only along its north wall, and the north ravine, while McBurney concentrated on the latter only. As a result, our information is at its most detailed for the north ravine, and notably for the lower layers; it is far less exact elsewhere. In this volume, therefore, the later deposits are dealt with somewhat summarily (the sequence described is certainly incomplete). There is however the consolation that those that remain are comparatively accessible to further excavation and study (likely to be more profitable geologically than archaeologically).

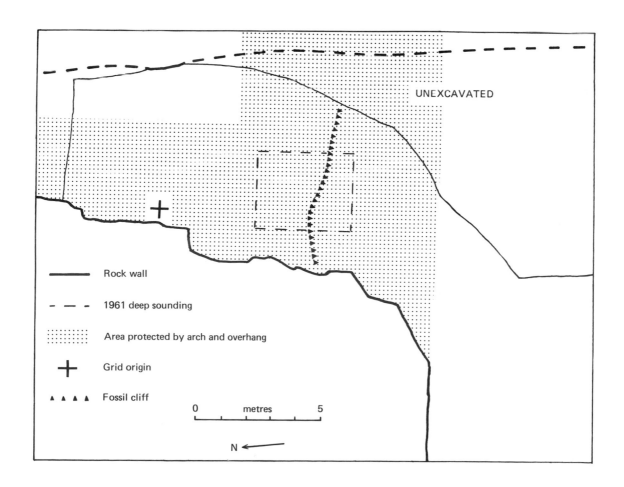

**Fig. 6.1** Schematic plan of the north ravine.

**Table 6.1** Depositional Agencies

| Origin | Context | Transport | Nature |
|---|---|---|---|
| Distant | Primary | Wind | Silt (loess) and sand (cover- or dune-)<br>Pollen, small marine organisms etc. |
| Plateau/Headland | Secondary and tertiary | Soil movement and water | Rolled erratic and local beach and stream pebbles<br>Silt and sand (aeolian and fluviatile)<br>Fine granite weathering products (grus)<br>Shattered granite<br>Archaeological material<br>Biological material (including pollen) |
| Foreshore | Secondary and tertiary | Marine | Rolled erratic and local pebbles<br>Sand (marine)[a]<br>Marine organisms[a]<br>Reworked cave sediments |
| Ravine | Primary | Gravity | Fine granite weathering products<br>Shattered granite |
|  |  | Human | Technological material |
|  |  | Human and animal | Bones (including food remains)<br>Shellfish, birds' eggs[a] etc. |
|  |  | Autochthonous | Vegetable remains, some rodents etc.<br>Freshwater organisms |

[a]Not so far identified at La Cotte.

This chapter is intended to introduce the reader to the broad outlines of the sequence, and to some of the processes which have contributed to it, before the more detailed discussion that follows in chapter 7. Before proceeding further, however, attention is drawn once again to an aspect of the site that is of the greatest significance, and to which some reference has been made in previous chapters. Though most of the deposits in the west ravine were destroyed during the Flandrian transgression, those that remain, and also those in the entrance of the north ravine (see Fig. 6.1) consist largely of head above an ancient raised beach at the lowest level yet reached at the site. But further north there exist quite different and much older deposits, never directly affected by the sea. The interface between the two sets of strata takes the form of a vertical cliff formed when the raised beaches were deposited.

Comparison with the preliminary report (McBurney and Callow 1971) will show that there have been changes in the naming of some of the layers. Because the existence of the steep dip (particularly in the proximity of the east wall) was not known, it was not realised that the lower of the two rich archaeological horizons encountered in the 1960s (then termed layer 2) is in fact the same as the layer A of 1961–1962; similarly the loess underlying it (formerly layer 1) is layer B. For historical reasons, therefore, the numbering runs from bottom to top in the sequence H–A, 3–14. There have also been modifications between 6.2 (formerly $7_1$) and 9 (which is now subdivided only in the southern part of the area excavated). But the revised nomenclature was drawn up before the final laboratory results were available, and is still defective inasmuch as 6.2 is now known to *post*-date the raised beach, 7.1 (see below, and subsequent chapters). Issues of particular importance in arriving at a reading of the stratigraphy are considered in section 6.6.

## 6.2 THE CONTRIBUTING AGENCIES

The sequence is marked by numerous episodes of deposition, pedogenesis and erosion. Since the site is not a true cave, but a fissure with non-vertical sides (and, moreover, in close proximity to the sea), many of the factors which govern the accumulation of sediments in a karstic limestone situation do not apply here (see, for example, Schmid 1969, and many of the references listed by Collcutt 1979). Because of the comparative insolubility of the granite and, in particular, the presence of a (possibly intermittent) reservoir of sediment on the headland itself, the occurrence of a locally derived component in the sediments may stem from many causes; even the material which was brought to the area by aeolian transport may have had a complicated history before reaching

**Table 6.2**  Post-Depositional Agencies

| Agency | Result |
|---|---|
| **Surface or shallow subsurface:** | |
| Wind | Deflation (removal of fine fraction) |
| Rain | 'Splash and spread'; flotation and redistribution of charcoal etc. Rills (channel formation) Sheetwash |
| Frost | Mechanical disaggregation |
| Vegetation | Humus formation and production of acids; root mat stabilising the surface |
| Human or animal | Trampling (compaction and redistribution of deposits); digging (construction of hearths etc.) |
| **Subsurface:** | |
| Percolating water | Eluviation of fine fraction (silts and clays); illuviation of fine fraction at lower levels, lining the pores and coating soil particles Solution of chemical salts (leaching), dissolution of bones etc.; reprecipitation of chemical salts, secondary phosphate formation (nodules, cements etc.) |
| Water saturation and freeze-thaw on sloping ground | Soil creep (several mm per year) Solifluction (several cm per year) Mud flow, slumping (rapid) |
| Water saturation and freeze-thaw on more level ground | Platy structures Frost heaving Ice wedge formation |
| Gravity | Compaction of unconsolidated deposit (distortion and faulting may occur e.g. in the vicinity of large boulders) |

its present resting-place. Table 6.1, though by no means exhaustive, gives some idea of the possibilities. Further complication is introduced by the likelihood, under suitable conditions, of reworking of the deposits by both natural and human agencies; the depositional status may thus be further downgraded (primary to secondary etc.). The most important components of the sediments, quantitatively, are loess, granitic sand (derived from grus) and granite blocks; in certain layers, also, the contribution of phosphatic minerals is extraordinarily high.

Possible mechanisms for erosion and redeposition at La Cotte are listed in Table 6.2. To date, ancient human interference by digging has not been convincingly demonstrated; minor disturbance may be inferred from the presence of double-patinated artefacts, though the

availability of previously worked pieces could also be due, for example, to their having been brought to the surface by freeze-thaw processes. The most spectacular erosional feature on the site is marine in origin, being the vertical cliff cut during the Eemian transgression by the sea entering from the west ravine (see chapter 7). The topography of the site will have controlled to a great extent the effectiveness of runoff from rain or snow. Even today, the overhang of the rock wall gives the west side considerable protection from direct rainfall. Here, water can only enter the site as precipitation driven by strong winds, by draining off more exposed areas, or by running off the dripline or down the west wall (an everyday analogy is provided by the spout of a badly designed jug). On the east side north of the rock arch, in contrast, a considerable amount of water is bound to have entered during periods of high precipitation as a result of the large catchment area provided by the sloping rock face. But against this must be set the correspondingly greater likelihood of debris entering the site from the headland by the same route.

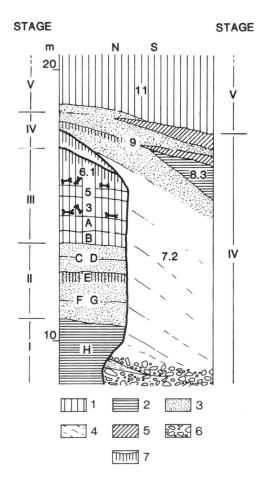

**Fig. 6.2**  Simplified north-south section, showing grouped stratigraphic units. Key: 1 Loess/loessic head; 2 water-laid silt; 3 granitic sand; 4 talus resulting from collapse of fossil cliff; 5 humiferous deposits (ranker soils); 6 marine gravel; 7 truncated forest soil.

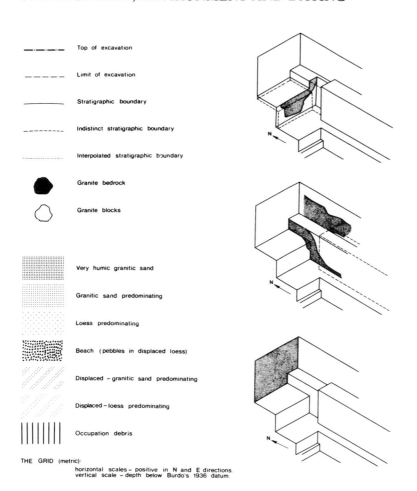

Top of excavation

Limit of excavation

Stratigraphic boundary

Indistinct stratigraphic boundary

Interpolated stratigraphic boundary

Granite bedrock

Granite blocks

Very humic granitic sand

Granitic sand predominating

Loess predominating

Beach (pebbles in displaced loess)

Displaced – granitic sand predominating

Displaced – loess predominating

Occupation debris

THE GRID (metric):
    horizontal scales – positive in N and E directions.
    vertical scale – depth below Burdo's 1936 datum.

**Fig. 6.3** Key to conventions used in the sections, and location on Fig. 6.4 of the principal illustrated sections (Figs. 6.5–7).

A third and very important factor contributing to the geological sequence is pedogenesis. Soil formation can result not only in the formation of humic layers but also in deep chemical changes and the movement of clay particles (see chapter 9). In the particular case of La Cotte, further post-depositional chemical changes appear to have been particularly marked in certain areas, as a result of locally more intense percolation of water and the presence of high concentrations of phosphates from guano and animal bones. In consequence, the sections are not always easy to read. Considerable differences in colour may bear no relation to depositional features, and reliable stratigraphic interpretation depends heavily upon observation of the orientation of stones, bone fragments etc. A particular hazard was encountered on the eastern side of the north ravine, where leaching by water percolating close to the rock wall altered the fine-grained deposits to a uniform very pale brown which completely obscured the layer boundaries; in this instance the initial field interpretation had to be modified in the light of the density of artefacts and the results of laboratory tests on the sediments (see section 8.2.6).

The phosphatic minerals in the La Cotte sediments are of particular interest, and are described in detail in chapter 10. Several stages of the fixing of phosphorus from bone are recorded, and also a distinctive mineral (taranakite) which is derived from guano. These phosphates can yield valuable environmental information, but they are the result of downward migration; they relate not to the layers in which they are found but to leaching from higher up in the sequence. At Cueva Morin, in Spain, phosphorus peaks occurred a metre or so below the occupation horizons which were their source (Butzer 1982, 82). In the case of La Cotte some of the phosphorus appears to have migrated even farther downwards.

It will be seen from the above that field observations are but the first step in arriving at an understanding of the geological events at such a site, and that much depends on detailed examination of material brought back to the laboratory. This specialist work is discussed at length from chapter 8 onwards.

### 6.3 MAJOR STRATIGRAPHIC FEATURES

Before considering the sequence in order of deposition, it may be helpful to identify some of the more immediately obvious groupings of layers. The following points should be considered in relation to Fig. 6.2, to the somewhat simplified isometric section (Fig. 6.4), and to the larger-scale sections which follow (Figs. 6.5–6.7); Fig. 6.3 is the

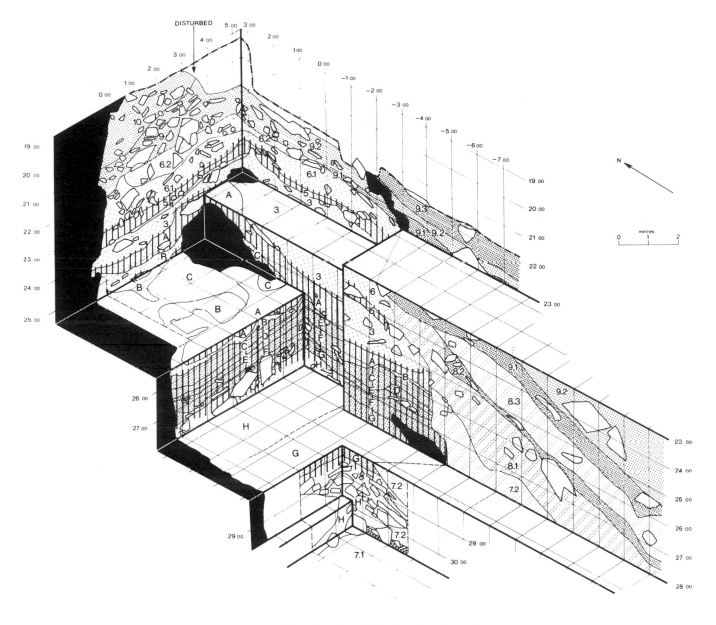

**Fig. 6.4** Isometric section through the deposits in the north ravine.

key to the conventions used.

1. Three principal and distinct loess accumulations are in evidence: (a) that which contributed loessic sediments to layer H; (b) layers B to 6.2, though loessic material in the latter is now known to be merely reworked; and (c) layer 11. Of these, the lowest is not represented *in situ* and can only be inferred; the others appear to span more than one period of deposition. Differences in the heavy minerals suggest some variation in the origin of the aeolian material. As a rule, also, these layers contain numerous sharp-edged granite fragments.

   The second loess accumulation is notable for two rich archaeological layers (A and 5). The occupations giving rise to these appear to have ended quite suddenly, with the abandonment of large quantities of animal remains which were subsequently buried in the loess of layers 3 and 6 (see chapter 18). Similar archaeological layers (though probably without bone heaps of this kind) evidently occurred in the third loessic complex, which was excavated in the early years of this century.

2. Between the lowest and middle loess accumulations, from G to C, the deposits are of rather different character. The *limon* fraction (see chapter 8) is relatively unimportant, there being instead a preponderance of sand derived from the granite. The most

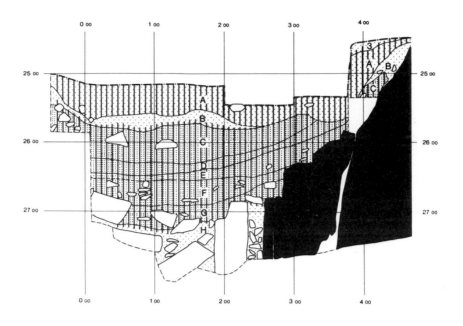

**Fig. 6.5** North face at 0.0 N in 1978.

conspicuous component of the sediments is anthropogenic, with numerous stone tools and manufacturing debris, bone fragments, ash etc. Sometimes there are bands consisting largely of small pieces of partly burned bone extending over large areas of the site.

3. Between the deposition of the second and third loess accumulations there is a major unconformity, most easily visible as the vertical cliff cut into the earlier deposits at about 5 m south (Fig. 6.4). At the foot of this feature, and seen only in the 'deep sounding' cut in 1961–1962 and re-excavated in 1982, is a layer of rounded pebbles (Fig. 2.8). This deposit (layer 7.1), with a matrix apparently derived from earlier but higher layers, was interpreted by McBurney as the storm beach left by a period of high sea-level. The sea has actually undercut the deposits of layer H, which are supported by a network of boulders, and at the northern end of the 'deep sounding' the beach deposits give way to the resulting slumped material. As discussed in section 7.3, both the altitude and the stratigraphic position of this event are entirely consistent with the Eemian transgression.

The layers following this beach in the sequence are extremely varied, but begin with loessic deposits which certainly derive in part from the second accumulation (and also incorporate material from layers H–C), which would have slid down the face of the cliff as the latter degraded. Another contribution may well have been from the surface of the headland, entering in particular via the east wall of the west ravine; this is hard to prove, however, as such material would almost certainly be very similar to the non-archaeological component of the *in situ*

layers of the north ravine. That the initial stages of this process were fairly rapid is suggested by the good state of preservation of the fossil cliff, which stands almost vertical to a height of nearly 5 m (despite the generally uncemented nature of the material out of which it is cut). The layers dip sharply to the south, in contrast to the nearly horizontal stratification of the earlier deposits. At a higher level, the sequence includes gyttja and ranker soils, sometimes referred to as peats in the earlier literature of the site, before the onset of the third loess accumulation (see chapter 9).

Precise delineation of the upper part of the fossil cliff is extremely difficult on the evidence of the available sections because of the leaching already referred to of the deposits close to the east wall (unfortunately all but a thin skin of material adhering to the rock was removed in the 1950s). The truncation of the archaeological layer 5 at about 5 m south of the grid origin, and the running together of the humic layers 8.2 and 9.1 in the same area, suggest that the unconformity corresponds very approximately to the line indicated in Fig. 6.4.

## 6.4 THE STRATIGRAPHIC SUCCESSION

### 6.4.1 Lithostratigraphic Subdivisions

Study of the deposits at La Cotte, and their contents, reveals an extremely complex history, despite the probability that there are many more lacunae resulting from erosion than have yet been recognised, and that the degree of complexity has therefore been underestimated.

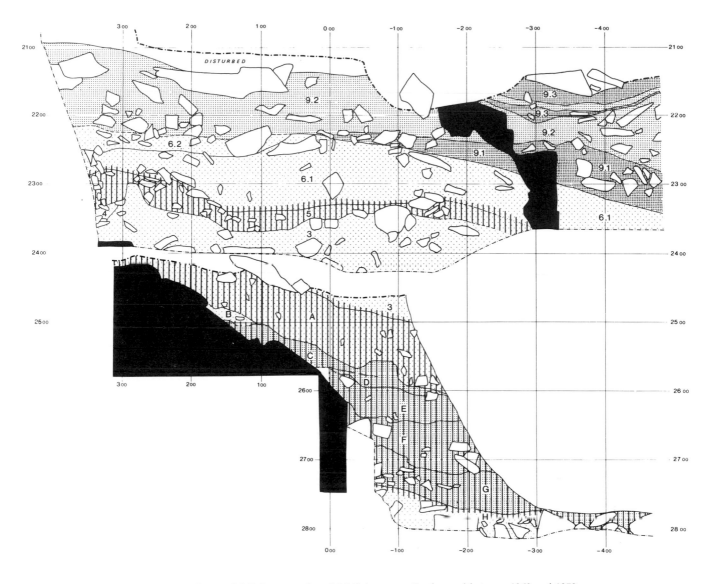

**Fig. 6.6** East face at 3.8 E (lower part) and 5.0 E (upper part), observed between 1968 and 1978.

The classification of deposits cannot be formalised to the degree required by the scheme laid down in the International Stratigraphic Guide (Hedberg 1976), because detailed sedimentological analysis was initiated only after the close of excavations and walling-up (or destruction by erosion) of the most important sections; it is based on such samples as were available without the possibility of checking conclusions in the field. Apart from the likelihood that minor or even major erosional events have not been recognised, the boundaries between the principal layer groupings are not represented in such sections as remain accessible.

Subject to the above reservations, the La Cotte de St. Brelade Formation can be informally divided into four members based on gross lithology:

| Name | Layers |
|---|---|
| Upper Loessic Member | 11–14 (?) |
| Upper Granitic Sand Member | 8.1–10 |
| Lower Loess Member | B–7.2 |
| Lower Granitic Sand Member | H–C |

The topography of the site is such that a great deal of local reworking has occurred; facies differences have arisen because of the various sources of redeposited material. Fine silts occur in both Granitic Sand Members as a result of redeposition of loess in limnic beds under temperate conditions (layers H and 8.2–8.3). Also the collapse of the cliff referred to above, which occurred for the most part under conditions cold enough to produce freeze-thaw structures, caused an accumulation of

61

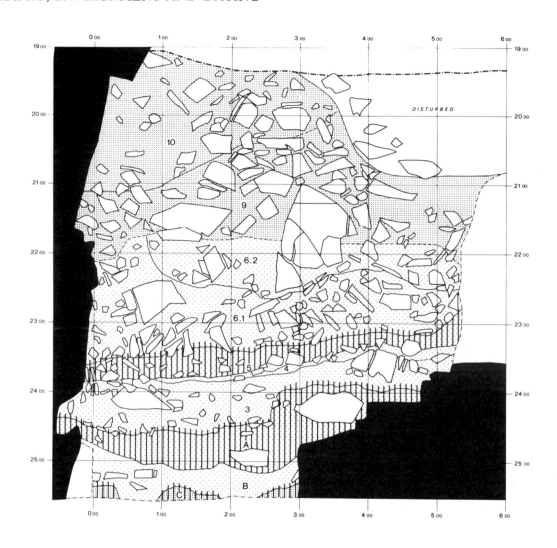

**Fig. 6.7** North face at 3.3 N, observed between 1968 and 1978.

loessic sediments directly above the beach (there is no evidence of aeolian deposition); gelifluction also caused loessic head to be redeposited above the Bt horizon of the forest soil contemporary with the marine incursion (see chapter 9). The rather peculiar topographic situation therefore means that over-reliance on lithology produces a partition of the sequence which largely obscures the principal elements in the site's development.

Because of the problems just outlined, it has been necessary to remodel the lithostratigraphic subdivision in accordance with other data (pedological, biological and morphological) as detailed in the next chapter. It must be stressed that the resulting scheme is interpretative, and remains informal, being an unavoidable departure from the procedures recommended by the International Stratigraphic Guide.

### 6.4.2  The Composite Scheme

The revised major subdivisions will be termed *Stages* (not to be confused with isotopic *s*tages). *Episodes* are

the finest units of resolution of events (depositional, erosional, pedogenetic etc.) which it was considered useful to label formally. Whereas a *layer* (as recorded by the excavators) is a unit of accumulation only, an *episode* may just as easily represent an erosional event or the alteration of already existing deposits — thus a single layer may preserve evidence of more than one episode. Because the scheme retains a strong lithological bias, these Stages do not correlate exactly with climate; in particular, there is a lag at the end of each temperate or cold period while old deposits are reworked and before new sources of sediments become operative.

Events at the site are listed episode by episode in chapter 7. A simplified version, based on Stages and layers only, is given here as an introduction to the interpretation of the succession (Table 6.3 and Fig. 6.2). Colours are described according to the Munsell system.

**Stage 0** concerns the formation of the ravine system during one or more periods of high sea-level (chapter 5). It is likely that this would have resulted in deposition of beach pebbles and rounded boulders at the base of the cave, as well as rounding of the walls. However, the

**Table 6.3**  Subdivision of the La Cotte Succession

| Stage | Episodes | Layers | Principal events |
|---|---|---|---|
| 0 | 0 | none | Formation of ravine system |
| I | 1–2 | (H) | Loess deposition |
| II | 3–12 | H–C | Ponding and rock falls (?); then — few granite blocks; granitic sandy matrix; important interglacial-type *lessivé* soil (layer E); extensive colluviation |
| III | 13–21 | B–6.1 | Increasing proportion of granite blocks; loessic matrix |
| IV | 22–35 | 6.2–9 | Numerous granite blocks and granitic sandy matrix except where older loessic material has been reworked; important interglacial-type soil complex (affecting underlying layers) and a compound marine transgression that resulted in erosion of much of the earlier sediments, followed by intermittent immature slope-soil formation; intermittent ponding |
| V | 36–48 | 10–13 | Numerous granitic blocks in loess (loessic head) passing into sandy head and then pure aeolian sand |
| VI | 49 | 14 | Recent soil |

**Fig. 6.8**  Detail of the 1966 north face. The floor of the cutting is in the top of layer 3 (marked by the presence of large blocks). The scale rests on the base of layer 5, with its top in layer 6. The lower part of 6 is a pure stoneless loess, but solifluction structures are apparent higher up, with renewed rock-falls.

excavations have not reached a deep enough level to reveal the former, and subaerial erosion would have removed evidence of the latter (at least from those parts of the site now visible). This Stage therefore remains presumptive, inasmuch as there is no direct field evidence apart from the existence of the cave itself.

**Stage I** (the earliest period proved by excavation) is marked by a leached, light grey or pale brown loessic matrix (10YR 7/2 or 6/3), insufficient to fill all the cavities between the boulders. Though treated as a single layer (H) it is likely that this corresponds to a complex series of events: in particular, the roof fall (blocks) need not have been strictly synchronous with the aeolian transport of loess, or with the contained archaeological material. Thin section analysis indicates that loess was washed into a pond in the ravine system during a temperate period. Stage I has therefore been taken as the period of deposition of the loess in the region, and is therefore only indirectly represented in the site itself.

**Stage II** (layers H–C) is first known from the 'pond' in which layer H formed, and by roof falls. There is then a progressive reduction in the number of large blocks and a much diminished loess content, the latter being replaced by granitic sand (the result of granular weathering of the bedrock). That this reflects a more temperate climate is suggested both by a similar granulometry during the later Stage IV (see below) and by micromorphological evidence of soil development in layer E. Except in the case of E, colluviation has destroyed the original structure of the rich archaeological occupation layers, but their ashy component, in particular, has caused a general darkening of the deposit, in some cases to 10YR 2/1 (black) or 3/2 (very dark greyish-brown). There are zones of lighter coloured material, for example 10YR 7/2 (pale grey) or 7/3 (very pale brown). Layers F and G are sufficiently complex to have been subdivided during excavation, though with confidence on the east side of the site only. Some of the colour changes are evidently post-depositional (section 6.2). The deposits of this Stage are archaeologically extremely rich.

**Stage III** (layers B–6.1) is distinguished from the preceding one by a predominantly loessic matrix and (as a general rule) large numbers of coarse angular granite fragments ranging from small chips to large boulders; solifluction and other freeze-thaw phenomena are much in evidence, and in the units for which microfauna is preserved this is indicative of extremely cold conditions (chapter 14). The transition from Stage II is not entirely clear-cut; a climatic deterioration was probably already under way in layers D and C. Lenses of pure pale yellow loess (2.5YR 7/4) are first evident in layer B, albeit intercalated with ashy archaeological material (Fig. 6.8); in its present form this probably represents a reworking of two different deposits. The overlying layer A is a very rich breccia of bone fragments, ash and stone artefacts in a loessic matrix, with numerous phosphatic concretions in some areas. Its colour is therefore rather variable: 2.5Y 4/4 (olive brown), 10YR 3/3 (dark brown) or 5/4 (yellowish brown) in the mass. Occasionally, to the west, it is so well cemented that it can only be broken up with the aid of a chisel.

Layer A is overlain by the pale yellow to light olive brown loessic head of layer 3 (2.5Y 7/4 to 5/4) in the lower part of which occurs the first of the two 'bone heaps'; the upper part of 3 seems to have been entirely sterile (see Fig. 6.13). At the top of layer 3 there is a thin and discontinuous slightly darker band (layer 4), followed by the very pale brown (10YR 7/4) layer 5 with artefacts and bone fragments occurring — in less profusion than was the case in A — in a loessic matrix. The overlying layer 6.1 (loessic head) is very similar in colour and granulometry; at its base was a second 'bone heap', resting directly on the surface of 5. A section cut in 1966–1967 at about 1.0 m N suggests that the initial accumulation was of a pure, almost stoneless, loess, with which granite fragments were mixed during a subsequent solifluction episode which almost entirely destroyed the parent deposit (Fig. 6.8).

**Stage IV** (layers 6.2 to 9) has as its most important stratigraphic feature the marine erosion during which the deposits in the southern part of the ravine were entirely removed, being replaced by collapsed material from the upper part of the sequence as it then existed (i.e. chiefly from Stage III). The subsequent deposition was largely a matter of adjustment to the new topography. A marked change is apparent in

the granulometry of some of the layers, however. Whilst the supply of granite blocks persists, there is frequently also an important component of coarse granitic sand in the matrix (cf. Stage III). Though in two layers (7.2 and 8.3) the granulometry is loessic, this is the result of the reworking of earlier layers.

The beach deposit (7.1) consists of rounded and subangular granite pebbles in a matrix derived from the older deposits in the cave (Stage II); the mechanics of the formation of this layer, and of the cliff itself, are discussed in section 6.6.4. Layer 7.2 is essentially loessic, typically light yellowish brown (10YR 6/4) for the most part, and results from the rapid erosion of the upper part of the sea-cut cliff, incorporating occasional lenses of the lower, darker layers (rich in flints, bone fragments and — sometimes — phosphatic concretions).

Layer 8.1 is a granitic sand (5YR in colour) with numerous granite blocks. It is overlain by the second of the loessic deposits of this Stage (8.3, with an organic mud or gyttja, 8.2, at its base); this is bleached and relatively stoneless, in striking contrast to the layers above and below. It is in fact a reworked loess, washed into a pond inside the ravine system. Its lower part is a little sandier (which may explain the heavy mineral variation referred to in section 8.2.7, as the sand fraction — reflecting the regime prevailing throughout this stage — may be penecontemporary with deposition, in contrast to the loess).

Layer 9 — which might better be described as a complex, with its multiple dark yellowish brown (10YR 4/3) ranker soils developed on a slope deposit of light yellowish brown (10YR 6/4) granitic sand and blocks — is comparable in many respects to layers 8.1 and 8.2. In fact, 8.3 is the result of renewed erosion of the Stage III loesses, an aberration in an otherwise consistent regime lasting from layers 8.1 (or earlier, in deposits destroyed with the top of Stage III further north) to 10. By the time the last deposits of this stage were laid down, the loess was no longer vulnerable, and the granitic sand extends over the whole area excavated in the north ravine.

Perhaps one of the most striking aspects of Stages IV and V is the extreme southerly dip of their component layers, in contrast to the more horizontal (if irregular) bedding of Stages II and III. This is clearly due primarily to the drastic topographical deformation caused by marine erosion, resulting in a cliff at least 9 m high within the north ravine (and presumably an equivalent feature at the buried junction of the main and south ravines).

**Stage V** is only poorly understood, as the best exposures of the deposits were removed during the early excavations. All that is now left is the lower part of the sequence in the main ravine, the undug deposits remaining in the south ravine (for which there are no visible sections), and a section on one side of the entrance overlooking St. Brelade's Bay — at the top of, and slightly stepped back from, the main north face of the excavations (it is not shown in Figs. 6.4 and 6.7). This last is at the extreme edge of the fissure system; in view of its inconsiderable thickness and the steep dip of the stratification, it must be presumed to give only a poor representation of deposition during the last glaciation. In this section, four major units attributable to Stage V are recognised. Layer 10 is a granitic sand with blocks, rather like 9 but with no ranker soils and with even less silt in its matrix; near the west wall of the ravine, the lower part of the ravine is virtually a loose scree, with no fines at all. Layer 11 is a loessic head, with an abrupt transition to layer 12, a head with sandy matrix. This is superseded by an entirely stoneless ferruginous sand (layer 13), well cemented and with slight iron panning. Nothing comparable to layers 12 and 13 is referred to by Burdo in his account of the sequence in the main ravine (1960, 16–21), so these are something of a puzzle. Layer 13, in particular, is very distinctive and could hardly fail to have been noticed by him during the 1936–1940 excavations. The most likely explanation is that it was deposited very late in the glaciation, and failed to resist erosion in the main ravine, where the dip was very steep towards the west. If this were so, one would be forced to conclude that the whole of Burdo's layers [B] to [L], with a maximum thickness of over 20 m, is subsumed in our layer 11, which (some 30 m north of his section) is reduced to a thickness of less than two metres. While this cannot be proved, some evidence of the complexity of layer 11 is provided by the discovery, in 1982, of a soil horizon within it. Pedogenesis apart, there is no reason to doubt that the greater part of the Stage V deposits were laid down essentially under conditions of extreme cold.

**Stage VI** comprises the recent soil (layer 14). This is now completely dug away, but Burdo (1960, 17) describes it as 60 cm thick, with artefacts including a 19th century golf ball and a neolithic barbed and tanged arrowhead.

## 6.5　A CYCLICAL MODEL FOR SEDIMENTATION AT LA COTTE

The discussion so far has concentrated upon the details of the succession. But the repeated alternation of loessic and sandy deposits suggests the existence of an underlying and relatively simple mechanism whose workings are obscured by the superimposition of other factors. A key is suggested by consideration of sources of sediment and corresponding climatic conditions.

1.　Aeolian loess deposition generally implies a cold dry (continental) climate.
2.　Deep weathering of the granite, to produce grus, requires temperate humid (oceanic) conditions.

Either type of material could be reworked in the presence of sufficient moisture; the exact mechanism would depend also on temperature. Thus under temperate conditions runoff was responsible for washing loess into freshwater pools, and under colder conditions various freeze-thaw processes played a more important part. As for the granitic sand, freeze-thaw must have contributed to the further breakdown of the weathered granite after grussification, and was certainly active in its redeposition in the ravines (see van Vliet-Lanoë and Valadas 1983 for a discussion of such phenomena); the character of layers G–E, with no traces of frost action, indicates that colluviation could also be effective (chapter 9).

Sedimentation processes at La Cotte may therefore be viewed in terms of a cycle as shown in Fig. 6.9, with subcycles based upon only one of the extremes (grussification or aeolian loess deposition) followed by reworking. Persistence or reintroduction of either sediment type can occur in spite of environmental changes provided that a reservoir of suitable material is available. A full cycle corresponds to two of the Stages referred to above.

## 6.6　SPECIAL ELEMENTS IN THE GEOMORPHOLOGICAL RECONSTRUCTION

### 6.6.1　Layer Topography

At the beginning of the chapter reference was made to the wide range of factors affecting deposition and erosion at La Cotte. While laboratory analysis of the sediments (particularly in thin section) can give important information in this respect, the orientation of stones and other objects and also the morphology of the layer boundaries themselves provide many clues to the processes operating at any one time.

Evidence of the complexity of the surfaces in the lower part of the second block of loessic sediments is provided in the isometric view by the horizontal section at a depth of 25.5 m below datum, as well as the upper part of the more southerly of the two vertical sections (Fig. 6.4). Analysis of the topography shows that layer A in particular fills a number of gullies which in places cut right through B into C. These features, which are drainage-related, show in the east side of the site a tendency to run away from the rock wall in a southwesterly

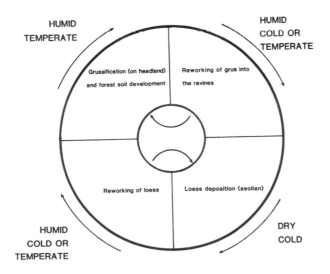

**Fig. 6.9** Simplified model of sedimentation at La Cotte, showing both the basic cycle and the most common subcycles. Note, however, that the prior existence of loessic material in the ravines permitted deviation from the usual sequence in layers 7.2 and 8.3 (loess reworking without renewal of the supply).

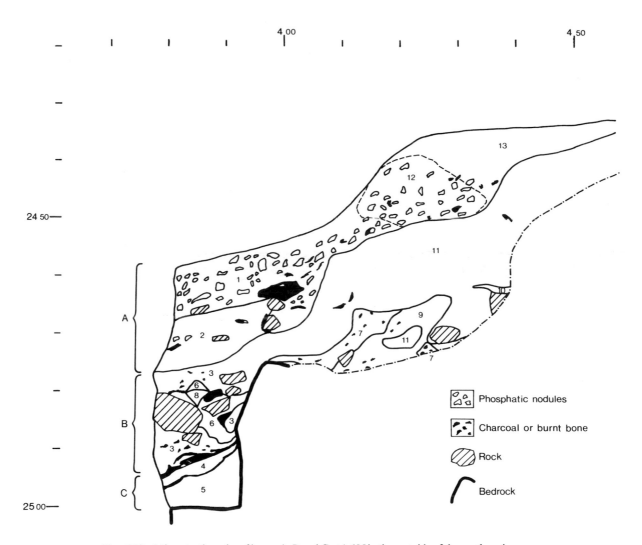

**Fig. 6.10**  Microstratigraphy of layers A, B and C at 1.6 N in the east side of the north ravine, as observed in 1981. (1) Yellowish brown (10YR 5/4), with dense guano and some charcoal; (2) Greyish brown (10YR 5/2), with a little guano and charcoal; (3) Dirty yellowish brown loess (matrix 10YR 5/6); (4) Pale brown (10YR 6/3) with charcoal; (5) Light brownish grey (10YR 6/2) with charcoal; (6) Brownish lenses within layer B; (7) Dark greyish brown (10YR 4/2) with charcoal (in layer B); (8) Dark brown lens in layer B; (9) Dark brown (10YR 4/3), loessic, with some charcoal; (10) Lens of charcoal; (11) Relatively clean yellowish brown loess (10YR 5/6); (12) Zone of pinkish phosphate nodules; (13) As (1), but disturbed during construction of retaining wall.

direction, and are no doubt a consequence of the enhanced runoff caused by the sloping rock face. A further indication of the importance of this feature of the site is the way that the eastern ends of many of the layers are banked up against it (Fig. 6.6). Fig. 6.10 gives some idea of the difficulties encountered during attempts to resolve the detailed stratigraphy of this area. Alternating regimes of deposition and erosion led to initially puzzling features such as that shown in Fig. 6.11, where the dark layer C has been partially eroded by surface water, giving a free-standing pinnacle subsequently buried under loess (B).

Very detailed plotting of the layer interfaces on the basis of closely spaced direct stratigraphic observations was impractical given the other requirements of field-

work. A useful approximation is however given by using the coordinates of the numerous artefacts in the case of the richer layers. Fig. 6.12 shows the result of doing this for layer A. The deposits are seen to be banked up against the walls of the ravine (especially in the east) and the gully running down towards the south near the west wall is clearly defined (many of the larger bones of the lower 'bone heap' in layer 3 were resting in this channel, as mentioned above).

### 6.6.2  Vertical Distribution of Archaeological Material in the North Ravine

Figs. 6.13 and 6.14 show that the density of the lithic artefacts in the north ravine is far from uniform. After

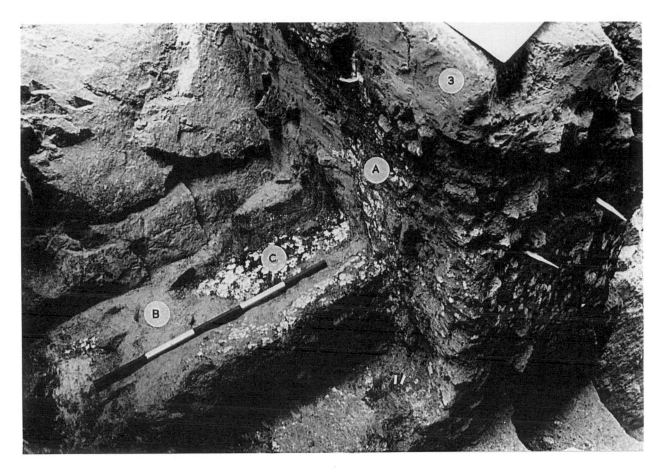

**Fig. 6.11** Steeply dipping layers C–3 lying against the sloping rock face on the east side of the cave. Loessic material (B) has infiltrated behind a pinnacle of layer C (dark, with phosphatic concretions) before being covered by layers A (similar in appearance to C) and 3 (loess).

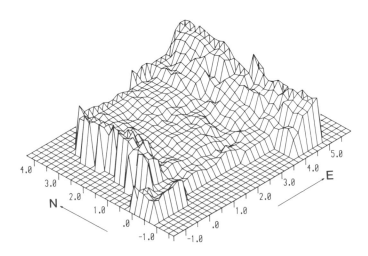

**Fig. 6.12** Isometric topographic drawing of the base of layer A, using artefact provenance data. Note that the surface is shown slightly too high along the west wall north of the grid origin, because layer A had not been completely excavated here (the general trend is correct). Also the northern edge to the east of 3.0 E appears too high here, because the positions of the A/B/C interfaces could not be ascertained from the site records and finds had to be excluded from analysis.

the richness of layers H–A, there is a clear gap in the distribution (corresponding to the greater part of 3 and 4) and then a band of finds which corresponds to 5 and the base of 6. Above this the artefacts fall into two groups: (1) a diffuse spread of pieces throughout the eastern and central part of layers 6.1 and 6.2, and (2) a concentration against the west wall, in layer 9. Neither of these appears to display the pattern one might expect for *in situ* occupation material. Moreover, the former occurs in solifluucted deposits; this, coupled with the easterly distribution, makes it probable that the relatively few artefacts slid into the ravine from a campsite on the headland together with other debris (the advantages of this promontory as a look-out station for game hunting are apparent in Fig. 1.10). As for the material in 9, this lies in a gully running along the west wall, where it was probably concentrated by water action during denudation of the top of layer 6.2. If this reading is correct, the raw material frequencies for these assemblages, which contain about 40% quartz pieces, would support the view that the putative occupation on the top of La Cotte Point was of the same age as layer 5. Such artefacts as occur in the lower part of 3 may also be explained as derived from the top of A, which has a very irregular surface as demonstrated above. During the early stages of deposition of 3, local reworking of minor eminences of archaeological deposit may be expected to have incorporated artefacts in the base of this loess. There is thus no conclusive argument for occupation of the site during either of the periods during which loess was covering the bone heaps; abandonment of the site helps to explain the good state of preservation of the latter (see also chapters 18 and 31).

This raises the further issue of why layer B, which in some repects closely resembles layers 3 and 6, should

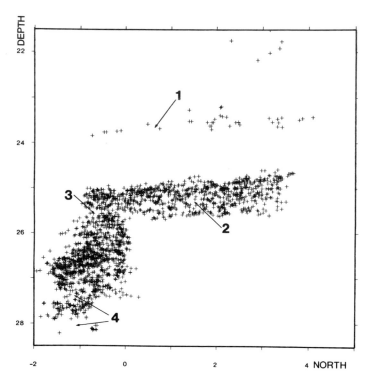

**Fig. 6.13** Vertical projection of artefact provenance data over a 10 cm width, from top to bottom of the sequence, showing variations in density; the projection is located on a N–S line at 2.0 m E.

contain a considerable amount of lithic material and other occupation debris. This occurs as numerous thin lenses of limited extent in the pale yellow loess, giving a streaky appearance. Moreover, the thinness of the layer is in marked contrast to the other, much better developed, loesses within the same complex. It therefore seems that many or all of the artefacts are not *in situ*, but were caught up from the rich layer C when this was being reworked together with the original layer B loess; others may be a consequence of the reworking of B into A, and of confusion over the precise position of the boundary between these layers. The complicated system of gullies at this stage of the site's development has already been mentioned.

It is likely that the site was abandoned for the period beginning with layer 6 and ending with the first of the occupations discovered by Marett and his co-workers, apart from a minor occupation between the two marine transgressions (layer 'H₂', described in chapter 9). The numerous finds made by Burdo in the steeply dipping layers 7.2 and 8.1 have to be regarded as in a derived context. The same is true of the very much sparser pieces in the overlying layers to the south of the cliff. Indeed, anything found in these deposits at a depth lower than about 24 m below datum could originate in the pre-raised beach archaeological layers in the ravine itself, even leaving aside the possible site already referred to on the summit of the headland.

### 6.6.3 Estimating the Position of the Dripline

One of the first questions that springs to the mind of a visitor to La Cotte is whether the site was ever a 'true cave' (that is, effectively roofed over) at any time during the Old Stone Age occupation. Today, what shelter there is in the north ravine is provided by the rock arch at the ravine entrance, and by the overhang of the west wall further to the north. During the excavations it was uncomfortably apparent that much of the cave floor is exposed to rain (on a really windy day, in fact, very little protection is available). Was early man obliged to put up with the same conditions, or was the shelter more extensive in the past? This question also has implications

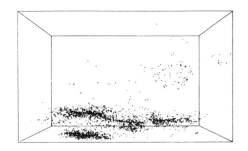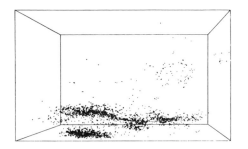

**Fig. 6.14** Stereoscopic pair of scattergrams of artefact provenance, from the top of layer A to layer 9, viewed from the north. The principal scatter corresponds to layer 5; the top of A is just visible. The scatter of points on the right occupies a gully cut into layer 6.2.

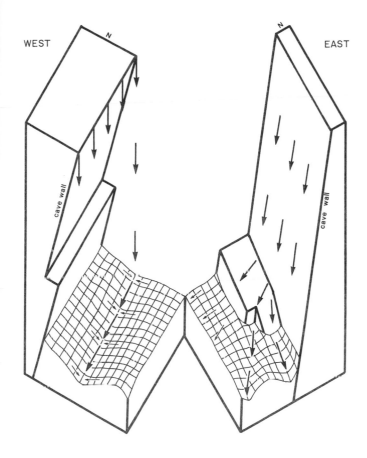

**Fig. 6.15** Major contributions to runoff in the north ravine. In the (schematic) drawing the upper part of this very deep fissure is not depicted. The true position of the top of the west wall (which does not in fact create an effective dripline) corresponds roughly to the division between the two halves of the drawing; to the east the deposits are directly exposed to precipitation.

for microenvironmental conditions affecting erosion, post-depositional soil changes caused by subsurface water, and the conservation of bone and other materials.

The southern face of the rock arch is in the same plane as the entire north face of the west ravine, corresponding to a single major joint plane (chapter 4). Rounding (probably marine) of the rocks at the base extends to the wall itself, suggesting that the latter existed in its present form during the last interglacial. The rock face is now quite stable, and indeed had evidently resisted subaerial erosion during the last glaciation. Since there is no indication of a still earlier marine incursion into the fissure system during the human occupation, it is quite likely that the southern side of the rock arch has remained effectively unchanged throughout this period.

The north face of the arch, and the overhang of the west wall, are of much greater interest inasmuch as their position bears on the exposure to the elements, early in the history of the site, of areas which were certainly occupied. The former is now at about 6 m south of datum, and the lip of the latter 2 m east of datum. However, the topography of the west wall is complex, and

there is a second subsidiary (but very important) dripline at about 0.5 m east, caused by water running in under the lip. This now shows up very clearly on the concrete retaining wall built to protect the north section at about 3 m north. The extent to which these west wall features are recognisable in the Pleistocene deposits is critical in establishing the evolution of the upper parts of the site.

The east side of the cave lacks a dripline, as there is no overhang; instead, water runs down the rock face into the floor of the site; irregularities in the surface of the wall concentrate the water at certain points, causing more intensive erosion locally.

The effects of all these features (Fig. 6.15) are apparent in deposits pre-dating the last glaciation:

1. The gully cut into layers A and B near the west wall, and referred to in section 6.6.1, immediately underlies the subsidiary dripline mentioned above; it certainly extended south of the grid origin, though its southern limit can no longer be determined.
2. Today, water running down the east face at about 1 m north flows into a part of the cave where clear erosional features were detected at the junction of layers B and C (Fig. 6.11).
3. Layers at least as low as F dip steeply close to the east wall as far south as the fossil cliff, apparently as a result of runoff.
4. As further confirmation of observations 1 and 2, there is considerable lateral variation in the colours of layers C–G in the lower north face (Fig. 6.16). This matches the zones currently affected by (a) direct rainfall on the eastern half of the section, and (b) the subsidiary dripline, and must reflect the varying intensity of percolation of water during the last interglacial (at the latest).

Of the above points, 1, 2 and 4 indicate that the rock arch cannot have extended further north than the grid origin, while 3 strongly suggests that it was in about its present position. On the other hand, none of the evidence precludes the possibility that the west wall may have overhung further than today, giving a narrower opening to the top of the ravine (and a little more shelter below); the subsidiary dripline would probably have been at least partly active under such circumstances.

### 6.6.4 The Marine Transgressions

The double transgression which culminated in the cutting of the fossil cliff and deposition of the beach (layer 7.1) is an important feature of the La Cotte succession. As with so many other aspects of the site, however, it poses more problems than are immediately apparent.

1. On an open shore, such a beach could reasonably be interpreted as giving the height of the sea-level maximum. In a deep fissure system containing unconsolidated deposits, however, the processes determining the altitude of the beach deposits are less straightforward; moreover the great tidal range in this part of

**Fig. 6.16** The lower part of the north face in 1978: (A) in August, at 0.6 m south of grid origin; (B) at the end of the excavations, after the section had been cut back to the base line. Note the shelving bedrock on the right.

the Channel and the site's exposure to Atlantic gales must be taken into account. The formation of substantial storm bars, perhaps 2 m higher than high tide level, is to be expected from modern analogies.

2. The nearly horizontal stratification and considerable thickness of the older deposits suggest that the contents of the ravine system were already substantial when the rising sea began its incursion. Consequently, a considerable period of time must have been required for erosion to reach the interior of the north ravine.

3. The energy available for erosion will have been a function of the submerged cross-section of the west ravine. As this is roughly parallel-sided, the effective area of section (and hence the hydraulic energy) is essentially governed by the depth of water. However, an exception must be made in the case where the water level is low enough to be influenced by the topography of the floor. The latter slopes down to the south, so a small change in sea-level results in a considerable difference in cross-sectional area. The lowest point visible is a little above high tide level today; therefore it would require a substantial transgression, in Pleistocene terms, to affect the deposits within the fissures. A rise of about 4 m relative to modern sea-level is necessary for the whole width of the ravine to be filled. Furthermore, the floor shelves upward as the system is entered; although it is partly buried by the talus at the junction of the fissures, the highest point (the threshold of the north ravine) probably lies at about 2 m above the modern high tides (i.e. 7 m asl).

4. By analogy with other sea-caves on Jersey, there may be a shallow depression in the floor inside the entrance, carved out by the undertow. A pothole of this type, filled with beach pebbles, was discovered by McBurney during excavations at Cotte à la Chèvre (Fig. C.2).

5. Once the erosion reached the junction of the ravines, the wave energy would have been distributed over a wider front (bedrock and unconsolidated fill alike). The forces attacking each unit area of the surfaces, and hence the average rate of advance of the sea, will have been correspondingly reduced. Nevertheless, all unconsolidated material would have been removed in due course had the sea not begun to fall.

6. The fossil beach and cliff preserve evidence of conditions not at the peak of the transgression, but at the point when the post-maximum fall in sea-level had proceeded to the extent that the hydraulic energy was no longer sufficient to effect erosion. The equilibrium point, and the height of the fossil beach, thus reflect the altitude and configuration of the rock threshold, and give only a minimum value for the altitude of the transgression. This view is supported by the existence, on the northern side of the west ravine, of a sloping shelf of apparently water-smoothed bedrock (presumably protected from sub-aerial erosion for much of the last glaciation by a rapid build-up of talus over it); this is about 2 m higher than the fossil beach — well above the reach of modern storms — and thus gives a very approximate *minimum* figure for the transgressive peak of 3–4 m above the present.

# INTERPRETING THE LA COTTE SEQUENCE

P. Callow

## 7.1  THE ENVIRONMENTAL SEQUENCE

In order to reconstruct the changing conditions at La Cotte it is necessary to make use of several lines of evidence, no one of which provides a complete record in itself. Moreover, the activities of prehistoric man and the geomorphological complexities of the site add to the problems of interpretation. The account which follows (summarised in Table 7.1) draws heavily upon the work of other contributors to part II of this book. The writer acknowledges a special debt of gratitude to B. van Vliet-Lanoë, whose suggestions about the significance of the La Cotte deposits and knowledge of sites in northern France have played an important part in developing this scheme. Nevertheless, final responsibility for its soundness rests with him alone.

### 7.1.1  Biological Evidence

Faunal remains are somewhat restricted in their stratigraphic occurrence (chapters 13 and 14). The microfauna is really informative for the loesses B, 3 and 6.1 only; the samples from 5 and A may appear explicit insofar as they contain the collared lemming, *Dicrostonyx*, but the very low density of rodent remains in these two occupation layers leaves open the possibility that they are derived from the underlying loesses in view of the extent of reworking which is known to have occurred.

Though the large mammals also are limited in stratigraphic distribution, and the sample sizes are rather small, they provide at least an indication of conditions in some of the industrially rich occupation layers (D, C, A and 5) as well as in the intervening loesses. The species present, if mostly rather catholic in their habits, when taken collectively suggest a fairly open and rather cool environment for the occupation layers, and for the mixed loess B. Reindeer is present in layers C–6.1. Environmental interpretation of the more restricted range of species present in the 'bone heaps' at the base of layers 3 and 6 is more difficult. These are evidently even less representative of the contemporary fauna than is usually the case for archaeological occurrences, but reflect conditions which were either similar to those of the underlying archaeological layers, or else becoming somewhat more steppic. The occurrence of a rich, very 'cold' microfauna tends support to the latter view.

The scarcity of identifiable fauna of any kind in the lower, sandier layers may be attributable to any or all of three separate causes:

1. In the absence of loess the *initial* pH may have been relatively low, rendering the mineral component of the bones liable to early dissolution facilitated by the greater porosity of the matrix.
2. Very intensive human activity, perhaps coupled with a lower sedimentation rate. Trampling and burning resulted, especially in layer E, in a mass of unidentifiable fragments.
3. Chemical weathering during soil formation particularly affected unburnt bone in the more exposed parts of the site, and took the form of the migration of bone phosphate (chapter 10). Layer E, and those below, evidently suffered severely in this respect.

The abundance or otherwise of identifiable bone reflects, in a complex manner, changes in depositional and post-depositional circumstances, and emphasises the contrast between layers laid down in cooler and in more temperate conditions.

Evidence for contemporary vegetation rests on the pollen (chapter 11) and charcoal (chapter 12), neither being very abundant. Apart from the hollow casts found by early excavators in the layers post-dating the fossil beach, no macrofossils are recorded.

From the top of layer H to layer C the pollen suggests that the climate was usually temperate (with mixed woodland, grassland, heath and marsh, with maritime elements), apart from an episode spanning the G/F interface, when more boreal conditions prevailed. This is not contradicted, at least, by the charcoal in G and C (hazel and oak). Charcoal also indicates that during A oak was growing somewhere in the region (possibly well to the south if the wood was brought to the site as an artefact), but no pollen evidence is available. From layer 3 to layer 6 a more open landscape, and very much colder conditions, are demonstrated by Campbell's pollen samples; there may have been some slight climatic amelioration in layers 4 and/or 5, however, and again in the lower part of 6.2. It is impossible to be certain whether this last represents the onset of an interstadial (a further succession of colder deposits having been truncated by erosion) or the first indications of the interglacial complex indicated by the raised beach deposits and by the pollen of Stage IV.

Stage IV is essentially temperate in character, though a cooling is indicated in layer 9.1, and again at the top of 9.3. The large gaps in the palynological record are partly filled by environmental information from other sources.

### 7.1.2 Sedimentological Evidence

As mentioned in the previous chapter, the granulometry of the deposits suggests a number of natural groupings of units. The justification for using these groups as a basis for the definition of the principal stages in the site's evolution is provided by the other sedimentological observations, as well as by the implications of the biological data.

Phenomena explicable by frost action are apparent from the thin sections, as well as from macroscopic observations (platy structures, festooning etc.) in the following layers: H, E, D, C, B, A, 3, 4, 5, 6.1, 6.2, 7.2, 9 and 11. However, of these layers it seems that E, at least, is merely penetrated by frost action from above at a later date. Also, observations made during the 1982 excavations, and micromorphological studies, suggest that freeze-thaw effects seen in layer H are likely to result from the very much later exposure to the effects of deteriorating climate at the end of the main Stage IV marine transgression. There is thus a lack of freeze-thaw activity probably lasting from the deposition of layer H to the erosion of the top of E. This culminates in an important soil, evidence of which is provided by reddish-brown clay illuviation in E.

A second major soil complex, apparently beginning some time after the deposition of 6.1, resulted in several Bt horizons (see chapter 9) which affected layers from the base of 9 down to 4. This period of pedogenesis includes the marine transgression responsible for cutting the fossil cliff and depositing the beach; however, it also included episodes of freeze-thaw. The whole complex has been assigned to Stage IV.

Evidence of the wider significance of this latter series of soils is provided by the differences in mineralogy between the loessic layers 3–6.1 (Stage III) and layer 11, corresponding to the contrast between the Older and Younger Loesses in Normandy.

Of the chemical analyses, those for organic matter and phosphates seem largely to reflect human activity at the site (except in the case of the strongly humiferous ranker soils).

### 7.1.3 Environmental Changes in the La Cotte Record

La Cotte contains evidence of accumulation of deposits under both temperate and very cold conditions. In this section, the former are considered as a group, followed by the latter.

#### TEMPERATE STAGES

There are at least two interglacial complexes represented at La Cotte prior to the Holocene. The first (Stage II) probably lasted from layer H (if not before) to layer C, by which time climatic deterioration is quite marked, and has two temperate phases separated by a more boreal episode, the later being more important in terms of soil development. The second complex (Stage IV) is much more conspicuous in the stratigraphy because of erosion during the accompanying marine transgressions

and the very great thickness of its deposits in the southern end of the north ravine. The resulting topographic changes within the cave caused the immediately preceding deposits, apart from deep weathering products, to be largely displaced or destroyed between periods of stabilisation of the surface. When climatic fluctuations are also brought into consideration, this makes for a very complicated record of deposition and pedogenesis.

The clay illuviation affecting layer E is particularly important, as it implies not only stabilisation of the surface but also conditions sufficiently temperate and long-lived for a soil of forest type to develop. But the apparent lack of a marine incursion into the cave during Stage II, and the low degree of phosphate alteration in the lower layers, suggest that this was not as important an interglacial complex as Stage IV, during which the sea reached an altitude several metres above that of the present day, and phosphate mineralisation proceeded to the production of crandallite. Nevertheless, in Stage II a degree of proximity to the sea is suggested by the occurrence of taranakite deriving from guano (chapter 10), presumably due to the nesting of sea-birds on ledges in the ravine. It is also possible that the sea entered the cave during the earlier of the two temperate phases before the accumulation of the earliest deposits so far reached. It is clear from pedological and stratigraphic evidence that erosion has been responsible for gaps in the sequence at a minimum of two points during this stage (i.e. above and below E).

At first sight the palynological and sedimentological records for Stage IV are contradictory and confusing, as the deposits include material of loessic origin (layer 8.3) and sandy head (9.2 and 8.1) with temperate pollen. The difficulty lies in establishing whether it is the pollen or the bulk of the sediment that has been reworked, or both. In the case of 8.3, a strong case may be made for the latter on the evidence obtained during thin section analysis. Moreover, the episodes of pedogenesis clearly identified in the deposits north of the fossil cliff are not easily tied to events further south.

A particularly critical question concerns conditions following the maximum of the marine transgression. Earlier interpretations (e.g. McBurney and Callow 1971) assumed that the beach found in 1961 represented the highest altitude reached by the sea. However, the 1982 excavations showed that the sea-level must already have fallen somewhat, and that gelifluction occurred between two distinct lenses of storm beach. This brings La Cotte into line with other sites inasmuch as it indicates that the regression was not very substantial until late in the interglacial.

If the first Stage IV pedogenesis (affecting layer 6.1 and below) corresponds to the Eemian transgression, as seems probable, no such straightforward equivalent for the second (affecting layer 6.2, and separated from the first by a phase of gelifluction) is apparent in the deposits overlying the beach. Either the preceding or the subsequent gelifluction in the north must be equivalent to freeze-thaw structures in the deposits banked against the cliff (layer 7.2). At this point the picture is confused by erosion and displacement of material.

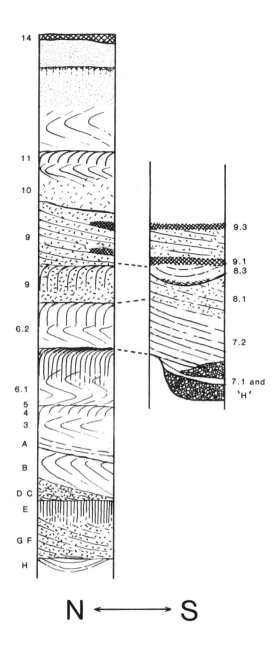

**Fig. 7.1** Composite profile of La Cotte de St. Brelade, showing episodes of pedogenesis and the dynamics of deposition of the sediments north and south of the fossil cliff.

(in which case granular weathering of the granite must already have been producing sediment).

2.  It was laid down under *cold* conditions, during the gelifluction episode between the 6.2 and 9 soils, but includes the topsoil of the former (itself probably sandy). The sloping ground surface may have been too unstable for the 6.2 pedogenesis to have been very effective there, though fine clay illuviation was seen in some samples of sediments overlying the lower unit of the beach (including 7.2) — see chapter 9.

3.  Layer 8.1 corresponds to the period when pedogenesis was taking place in the lower part of layer 9 north of the cliff (the only trace of the earlier, 6.2 soil further south is the fine clay illuviation just mentioned).

Against (1) is the apparent absence from 8.2 and 8.3 of frost structures corresponding to the gelifluction phase which followed the illuviation of 6.2; this leaves the third of the *sols lessivés* without an equivalent further south. Against (3) is that the likelihood of incorporation of large numbers of artefacts from the Stage III (and perhaps II) deposits must have been greatly reduced by the time the layer 9 pedogenesis occurred. In compiling Figs. 7.1–7.4 layer 8.1 has therefore been treated as containing the geliflucted topsoil of the second *sol lessivé* in Stage IV.

The process by which formation of the pond represented by 8.3 became possible is unclear. A localised rock fall at the western entrance (perhaps due to a minor seismic event) is one possibility; another is the development of a storm bar in the same area following renewed marine transgression, though the altitude of the deposits makes this extremely unlikely. If the interpretation suggested above is correct, the pond was formed when the ground surface further north was stabilised by vegetation at the time of the pedogenesis of the base of layer 9; this would explain the virtual absence of coarse sand in the lacustrine sediments.

Rather open vegetation (though some thermophilous pollen is present) points to a deterioration of climate at the time of the ranker 9.1. The subsequent layer of sandy head (9.2) contains oak and alder pollen, but the same reservations apply as for layer 8.1 — reworking of older topsoil with the aid of frost action cannot be ruled out in view of the nature and position of the deposit. In any event, the more temperate character of the conditions under which ranker 9.3 was formed is altogether more convincingly demonstrated by a succession of pollen spectra, though they cannot be regarded as interglacial.

Thus in Stage IV we have three phases of *lessivé* soil formation north of the fossil cliff at the top of layer 6.1 and 6.2 and in the lower part of 9, and at least three (probably four) temperate episodes in the more rapidly accumulating deposits further south (the last of them being less important, and represented by ranker formation). Moreover, there is direct evidence of marine transgression during the first temperate phase. The significance of these observations is discussed in section 7.2.

The pollen content of 8.1 certainly suggests an interglacial, and the sediment appears (from Burdo's description) to have been sandy, resembling that of layer 9. Burdo's excavations show it to be quite rich archaeologically (particularly in comparison to layers 6.2 and 9); this indicates either that there was renewed occupation or that some of the Stage III deposits were still undergoing erosion. Because 8.1 could not be traced in later excavations it has not been possible to establish whether it was affected by freeze-thaw. Therefore several interpretations are possible.

1.  It corresponds to the upper part of the soil affecting 6.2, displaced downslope under temperate conditions

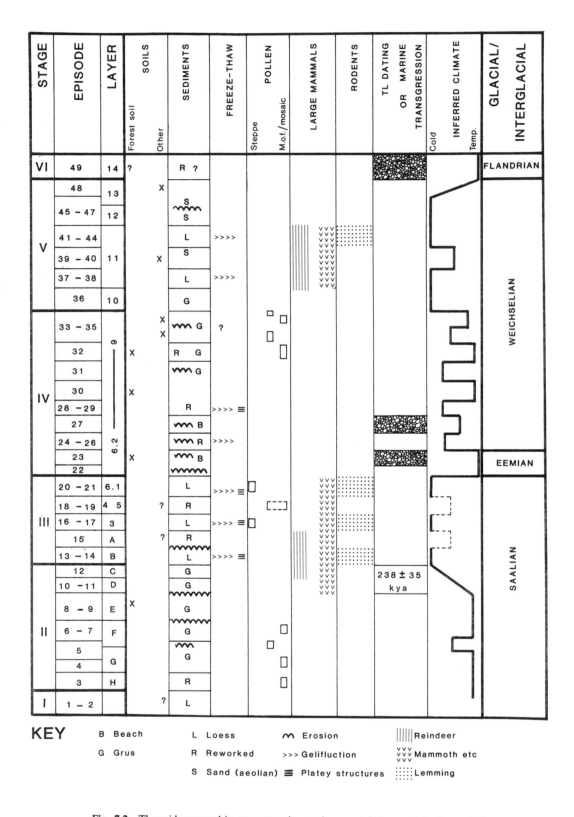

**Fig. 7.2** The evidence used in reconstructing environmental change at La Cotte de St. Brelade.

## COLD STAGES

The general character of Stage I, the first cold phase, for which only indirect evidence survives, is attested by the loessic composition of layer H. This fine component seems not to be in place, however, but to have been washed into a pond in the cave during the temperate Stage II.

The essentially loessic head deposits of Stage III possess some interesting features. There is no doubt that conditions were extremely cold for much of the time that they were being laid down. Also, during loess deposition the climate was probably relatively dry, though the fact that the loess is of *limon à doublets* facies indicates that there was more moisture present than in the easterly facies found in the Paris basin (Lautridou and Giresse 1981). Nevertheless, solifluction in layer 6.1 appears to take place late in its formation and must reflect the onset of wetter conditions.

The human occupation deposits within this stage (essentially layers A and 5, though also including the base of 3 and 6.1 respectively) pose a number of important questions. Are they the result of natural reworking of pre-existing archaeological layers? Or do they provide evidence of the presence of man at times of cool, if not cold, climate? Or, again, do they mark interstadials, more favourable to human existence in the region?

The answer to the first question is provided, at least in part, by a comparison of the lithic industries (chapter 24). They are sufficiently different from the others at the site for it to be clear that there were occupations subsequent to layer C with their own peculiar industrial features. Especially important are the differences between the assemblages from layers A and 5, showing that 5 is not merely reworked A. The 'bone heaps', also, indicate that man was indeed a visitor to the site at the end of the period during which the underlying — and archaeologically richer — layers were laid in their present position. Even if the deposits A and 5 are not strictly in place, therefore, there is no reason to suppose that the position of these units in the stratigraphic sequence are unsound. As to climate, the presence of oak charcoal and of organic matter of ranker type are (admittedly) slight indications that the cold was not extreme in A; such organic material has rather equivocal significance on occupation sites. Again, in 5 there is juniper charcoal, which is unlikely to occur in glacial-maximum conditions, and more importantly the pollen in the underlying layer 4 is temperate in type. These arguments are not as conclusive as one could wish, but tip the balance in favour of the episodes of human occupation having taken place in periods of climatic amelioration which were, nevertheless, too slight or brief to initiate a major change of depositional regime, or to effect a replacement of the mammalian fauna.

The boundary between Stages II and III is poorly defined, in that layers C and D already show a cooling trend, while the first loess of Stage III (layer B) is poorly preserved over much of the site and is mixed with charcoal and other anthropogenic debris. Nevertheless, B is of significance as a true periglacial deposit with a microfauna characteristic of northern Siberia today, and is therefore in marked contrast to C. Though B was not represented among the soil samples submitted to Giresse by Lautridou for heavy mineral analysis (chapter 8), the overlying layer A was included, and differs strongly from the samples of 3 and 6.1 (which follow on within our Stage III). A high percentage of garnet links it to the mineralogy of the analysed layers below (H and F) — also, incidentally, to the Stage V loess, layer 11 (of course this has no direct climatic implications). A possible source for loess at the beginning of Stage III, before other and mineralogically distinct supplies became available, is thus the earlier deposits remaining from Stage I (by then without sufficient plant cover to protect them from deflation).

The evidence from Stage III therefore points to a single glaciation marked by two or more interstadials. Erosion and reworking at the junction of layers C and B could have obscured a long and oscillating decline towards periglacial conditions as at the beginning of the last glaciation. When we can first identify this stage at La Cotte, it is already sufficiently established for an important rodent migration to have taken place.

Stage V is poorly represented because many of its deposits were so effectively quarried away by the early excavators; nothing now links the surviving sections, though it is clear that they include loess and aeolian sands comparable to those of the Cotentin and the Bay of Mont-St-Michel (with the same facies and same mineralogy). Accounts of the upper layers are sketchy, and sometimes appear inconsistent (though erosion may be partly responsible for this); also the descriptions of some of the layers are incomplete or even misleading — for instance in respect of the granulometry. The description of the sequence as published by Burdo (1960) is based principally upon the two sections seen by him in the west ravine in 1936–1940 and 1950 onwards (a direct connection between them seems never to have been achieved) as well as upon the earlier work done in the north ravine. Because of the disappearance during the Second World War of all soil samples and field drawings for the 1936–1940 excavations, the interpretation given below is tentative and certainly incomplete — the considerable dip characteristic of these deposits makes correlation particularly hazardous. The logic behind the new reading of the stratigraphy is set out in Appendix E (microfiche).

At least one episode of soil formation occurs within the loessic heads, with Mousterian occupation both before and after it. Anthropogenic material is totally absent in the upper part of the sequence, above a *cailloutis* of unknown significance, until the topsoil.

## 7.2   DATING THE LA COTTE SEQUENCE

A radiocarbon date of 47,000 ± 1500 B.P. (GrN 2649) was obtained on a sample of uncertain provenance collected by Burdo in December 1957 and has since ac-

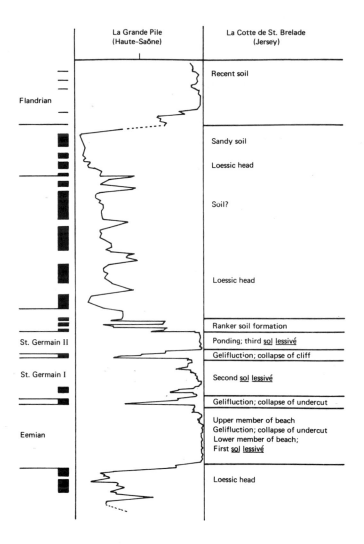

**Fig. 7.3** Comparison of the upper part of the La Cotte sequence with the pollen record from La Grande Pile in northeast France (after Woillard 1980).

3. absolute dating of layers C and D to 238 ± 35 thousand years ago, by thermoluminescence measurements on burnt flint (Chapter 15).

Of these, (1) and (3) give a firm age for the appropriate deposits, but (2) requires more elaborate reasoning, as it does not of itself provide an immediately obvious date expressible in years. Instead, it depends upon the fact that this is the last marine transgression seen to affect the cave deposits prior to the Flandrian (i.e. the present interglacial), and is separated from it by alternating cold and temperate layers followed by a great thickness of loessic head laid down under periglacial conditions. As there is no evidence elsewhere for a comparable marine episode in the course of the Weichselian glaciation, this strongly suggests a Last Interglacial complex date for the fossil beaches. A possible alternative, that the beach deposits were laid down in their present position in an even earlier interglacial, can be dismissed with some confidence. Quite apart from the consequence that the Eemian transgression would then be unaccounted for, the next available candidate for the marine event would fall (according to the evidence of oxygen isotope records from deep-sea cores) within the time range suggested by the dates from layers C and D (i.e. isotope stage 7); the interglacial which has left its mark in the layer E pedogenesis is evidently a more reasonable equivalent for this stage, however. This leaves isotope substage 5e (equated with the last interglacial) as the best choice for the La Cotte Stage IV transgression.

As mentioned in chapter 3, isotope substage 5e is now known from a large number of ocean cores, and is reliably dated to c. 124 kya (Shackleton 1975); on Jersey itself a good match for this date is that for the '8 m' beach at La Belle Hougue cave (chapter 5; Keen, Harmon and Andrews 1981). The series of three temperate soils (*sols lessivés*) identified by van Vliet-Lanoë north of the fossil cliff at La Cotte can only be interpreted as equivalent, successively, to substages 5e, 5c and 5a (significantly, the best developed soil is the earliest). Their palynological equivalents in the La Grande Pile sequence (Woillard 1980) would be the Eem, St. Germain I and St. Germain II interglacials (Fig. 7.3). Correlation between the last two of these and the succession south of the fossil beach cliff is less certain, though several temperate episodes are apparent (see above). The 'pond deposit', layer 8.3, is a strong candidate for St. Germain II (section 9.4), while the pollen in 8.1 may date originally to St. Germain I. The overlying rankers 9.1 and 9.3 fall at the end of isotope substage 5a and the beginning of stage 4, before the Lower Pleniglacial; parallels for this (certainly incomplete) part of the sequence are also to be found at La Grande Pile.

The dual character of the raised beach as established in 1982 is of great interest. Though the Eemian transgression occurred under temperate conditions, the subsequent regression appears from palynological evidence at other sites to have become effective quite late in the interglacial. At La Cotte the reworked hearth in layer 'H2' may well indicate that man was able to reach the island at this time, and that the sea-level had fallen to 15 m

quired currency (e.g. J. G. D. Clark 1977, 35). The laboratory itself had reservations about this determination, however, pointing out that the sample had been treated only with dilute nitric acid and might include recent humic material (Vogel and Waterbolk 1963, 165). It is to be hoped that the date will not be quoted in future.

Uranium-series dating on bone (which if successful would have provided estimates for several layers) has also proved unreliable here, apparently because of strong post-depositional leaching (chapter 16). Bone samples have been submitted for amino acid racemisation determinations, but no results are available as yet.

Calibration of the depositional succession (and its environmental and archaeological evidence) is thus available at three points. *From the top downwards*, these are:

1. the most recent soil, with its late prehistoric and modern artefacts (Stage VI);
2. the episode of high sea-level (Stage IV); and

or more below its present level. The subsequent storm beach cannot have been deposited under such conditions, and can only have two explanations: (1) that it corresponds to substage 5c; or (2) that it represents a minor and usually unrecognised fluctuation of sea-level at the end of Eem, after an initial cooling.

Though possible on the basis of the evidence from La Cotte itself, the first of these is unlikely to be correct, inasmuch as sea-level estimates from marine terraces at Barbados and elsewhere indicate that the transgressive maximum in 5c was about -13 m (Stearns 1976), which is well below that required at La Cotte. The alternative may seem surprising, as such a fluctuation has not generally been recognised. Nevertheless, the most detailed of the cores covering this period, V19-30, shows a well-defined oscillation of suitable amplitude at about 118 kya, on the 5e/5d boundary (see Fig. 7.4). The altitude reached by the *mean* sea-level cannot be reliably estimated from the La Cotte evidence because of the possibility that the second marine gravel was laid down during a quite exceptional storm.

The La Cotte Stage III deposits (and probably also layer C) are equivalent to isotope stage 6. The isotopic record does in fact possess several minor peaks of amplitude similar to that of stage 3 (the interstadial complex during the last glaciation); the human occupations may be associated with these.

Turning now to the earlier deposits, it has already been suggested that the TL dates and the pedogenesis following deposition of layer E point to a general equivalence between La Cotte Stage IV and oxygen isotope stage 7. However, stage 7 (like stage 5) appears to have been multimodal, with two very clear major peaks at about 240 and 210 kya separated by a period of ice advance (probably more important than those of substages 5d and 5b). There was also a third, minor, ice retreat just after 200 kya.

Several interpretations are possible in arriving at an equation between La Cotte Stage II and isotope stage 7; the standard deviation of the dates is too great for these to assist the detailed correlation. Layer C, though included within Stage II, probably belongs to the beginning of isotope stage 6 (presence of reindeer etc.).

1. Layers H–G = c. 210 kya; F–D = c. 200 kya.

2. Layers H–G = c. 240 kya; F–D = c. 210 kya.

3. Layers H–G = isotope stage 9; F–D (only) = stage 7, with lacunae due to erosion.

Of these, (2) and (3) imply that the more boreal character of the pollen at the G/F interface is linked to an important unconformity which was unrecognised in the field; in fact photographs of the lower point of the north face of the 1978 excavations do suggest an unconformity of some kind at about the right level, but this cannot now be verified. But isotope stage 8 would surely have been marked by loess deposition; this is not apparent. In the case of the 230 kya ice readvance, it is much less certain that conditions for loess transport would have

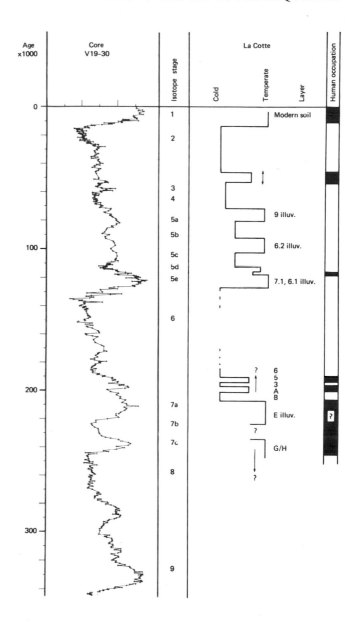

**Fig. 7.4** Comparison of the La Cotte sequence with the oxygen isotope record of core V19-30, time-calibrated by means of Milankovitch orbital fluctuations (courtesy of N. J. Shackleton).

been satisfied. On the other hand, (1) probably gives more emphasis to the c. 200 kya peak than can be justified at present. With some reservations, therefore, the second of the proposed schemes has been employed in drawing up Fig. 7.3. Whichever of them is correct, on present evidence all of the deposits antedating the raised beach are Saalian.

This is as far as the stratigraphic and environmental evidence permits us to go. But (to anticipate evidence discussed elsewhere) an interesting postscript is provided by the very great similarity between the artefacts from layer H at La Cotte de St. Brelade and those from above the '18 m' beach at La Cotte à la Chèvre (Appendix C).

A transgression to this altitude is not supported for isotope stage 7 by the deep-sea core evidence; it must be at least as old as stage 9. For the subsequent occupation to have taken place obviously required a fall in sea-level, though the lapse of time is uncertain. It is clear that the possibility of a 'long chronology' for the sequence at La Cotte de St. Brelade is a line of enquiry which should be pursued when excavations are eventually reopened.

**Table 7.1**  The La Cotte Sequence in Detail

The episode number is followed by the number or letter denoting the corresponding layer (where appropriate). See also Figs. 7.1 to 7.4. The status of archaeological material in the deposits is designated by: (a) living floor in place, rich; (b) slightly disturbed, rich; (c) reworked, rich; (d) reworked, less rich.

| Episode | Layer(s) | Archaeology | Events in the ravines |
|---|---|---|---|
| *STAGE 0* | | | |
| 0 | — | — | Formation of the ravine system |
| *STAGE I (essentially cold). Isotope stage 8 (?).* | | | |
| 1 | — | — | Deposition of loess presumably (not known *in situ*). |
| INTERVAL OF UNKNOWN DURATION | | | |
| 2 | — | — | Formation of ranker soil. |
| INTERVAL OF UNKNOWN DURATION | | | |
| *STAGE II (essentially temperate). Isotope stage 7.* | | | |
| 3 | H | b | Rock falls and ponding (with diatoms) in the ravine. Ancient loess (and relict soil) washed in from the plateau. 'Temperate' pollen (*Hedera* etc.). |
| 4 | G | b | Granitic sand (fewer granite blocks) with incorporation of small amounts of reworked loess, chiefly towards the base; occasional small pebbles must have been washed in from the headland; colluviation, resulting in disturbance of occupation debris and some ponding (aquatics). Pollen suggests temperate conditions. Possibly slight ranker formation at the base, but this cannot be verified. |
| 5 | G/F | b | As before, but pollen suggests more boreal environment. Large granite blocks decreasing in abundance. |
| 6 | F | b | As before, but pollen suggests more temperate conditions as well as ponding (aquatics, especially *Callitriche*). Few large granite blocks. |
| 7 | — | — | Erosion. |
| 8 | E | a | Occupation floor unaffected by colluviation. Large amounts of burnt bone in a matrix of granitic sand. Few large granite blocks. |
| 9 | E | — | Clay illuviation resulting from soil formation. The same episode would have been responsible for phosphate migration in underlying deposits. |
| 10 | E | — | Erosion of the upper part of the soil. |
| 11 | D | b | Disturbed (colluviated) occupation floors in a granitic sand matrix with few large blocks; some layer E material incorporated. Slight deterioration of climate (pollen, gelifraction of bones). This, and the subsequent episode, are dated to 238 ± 35 kya (OX-TL 222). |
| 12 | C | b | As D, but continuing deterioration of climate. Channelling of the surface by runoff. |
| *STAGE III (essentially cold). Isotope stage 6.* | | | |
| 13 | B | ? | Loess deposition under extremely cold conditions (*Dicrostonyx*). |
| 14 | B | b/c? | Reworking, by colluviation and frost-creep, of deposits of the two previous episodes. |
| 15 | A | b | Rich occupation deposits in loessic matrix, under relatively cool conditions (bone gelifraction). Some ranker formation. Channelling of the surface by runoff. No pollen, but oak charcoal (see section 7.1.1). The fauna includes reindeer. |
| 16 | 3 (base) | b/c | Loess deposition; mudflow and surface water action incorporating material from the previous episode in the base of the deposit. Formation of the lower 'bone heap' during the final human occupation occurred either at the beginning of this episode or the end of the previous one. |
| 17 | 3 | — | Loess deposition, with frequent large blocks of granite. Extremely cold climate (*Dicrostonyx*, platy soil structure; steppe-tundra vegetation). |

**Table 7.1**   (continued)

| Episode | Layer(s) | Archaeo-ology | Events in the ravines |
|---|---|---|---|
| 18 | 4?? | — | Pollen (*Quercus, Fraxinus, Ulmus* etc. but with high herbaceous percentages) suggests a climatic amelioration, with reworking of loessic material, but it is possible that the pollen has been washed down from the topsoil during episode 19, with the first organic clay illuviation. The layer was defined during excavation on account of the darker coloration caused by this clay, and may not correspond to a discrete depositional event. |
| 19 | 5 | b | Occupation deposit in loessic matrix. Stabilisation of the surface resulted in clay illuviation in layer 4 below. Possible colluvial deposition (supported by micromorphology and by the condition of the artefacts). The base of the upper bone heap is embedded in layer 5. |
| 20 | 6.1 | b? | Loess deposition (*limon à doublet*) with frequent large blocks of granite. Extremely cold climate (*Dicrostonyx*; steppe-tundra vegetation). This loess covers the upper bone heap. |
| 21 | 6.1 | d | Solifluction and cryoturbation structures during continuing loess deposition (?). Probably some debris introduced from the headland by solifluction (e.g. artefacts). Steppe-tundra, but with juniper and sea-buckthorn pollen in the upper part of the deposit (equated in the field, probably incorrectly, with the base of 6.2) perhaps indicating the onset of a climatic amelioration. |

*STAGE IV (essentially temperate). Isotope stage 5.*

| Episode | Layer(s) | Archaeo-ology | Events in the ravines |
|---|---|---|---|
| 22 | 6.1 and below | — | Pedogenesis resulting in illuviation in underlying deposits, sometimes as deep as layer 3, and hydromorphy in the lower part of the soil (Eem *sensu stricto*). Marine transgression eroding deposits in the west ravine, with smoothing of the bedrock therein. At its maximum it had probably not yet reached the interior of the north ravine. |
| 23 | 7.1 | d | Sea-level began to fall, but marine erosion continued into the north ravine, resulting in an undercut cliff about 9 m high (cut into the earlier deposits) and deposition of a storm beach at its foot. As the lip of the cliff retreated there was probably some erosion of the ground surface further north, removing the upper part of the Eemian soil horizon. |
| 24 | 'H' | d | Continuing marine regression, leaving the cliff to begin a gradual collapse, mainly through slumping of its overhanging lower part (old layer H deposits). A lens of wood charcoal at the foot of the cliff may possibly indicate a human visit, made possible by a fall in sea-level to -20 m or thereabouts. |
| 25 | 'H', 6.2 (?) | d | Further climatic deterioration. Gelifluction at the foot of the cliff, and probably also onto the eroded ground surface at its top — with the accumulation of loessic head perhaps displaced from the top of the headland (this would have continued in the next episodes in the sequence). Though cool, the climate remained oceanic rather than continental. |
| 26 | 'H' | d | Loess from layers 3–6 was washed into the gap left under the cliff by the progressive collapse of the overhang (probably as a very short-lived episode). |
| 27 | 7.1, 6.2 | d | A new rise in sea-level resulted in the formation of a storm bar inside the cave (the second layer of beach pebbles). |
| 28 | 'H', 7.2, 6.2 (?) | d | Continuing gelifluction completed the infilling of the cavity under the cliff, leaving air spaces between the boulders in the upper part of H. Colluviation and gelifluction then continued the process of accumulation at the entrance of the north ravine, with deposits consisting primarily of loess from layers 3–6.1 with injections of ashy material from lower in the sequence. |
| 29 | 7.2, 6.2 (?) | d | Continued slumping and gelifluction. |
| 30 | — | — | Pedogenesis (*sol lessivé*) with illuviation of 6.2 and below. The exact position of this episode in the sequence south of the cliff is unclear. |
| 31 | 9, 8.1 (?) | d | Local and probably repeated erosion, removing the upper part of the preceding soil and cutting down into 6.2 and even 6.1 (especially against the west wall). This was accompanied by more or less synchronous accumulation of granitic sand and larger blocks — the former in marked contrast to surviving deposits from earlier episodes within Stage IV. This material continued to be deposited until layer 10. The granulometry of layer 8.1. as described by Burdo suggests that it is no older than the beginning of layer 9 *sensu lato* — despite its position in the numbering scheme. It includes artefacts probably derived chiefly from the headland; those in the gully along the west wall were probably incorporated in the basal deposits of layer 9 by the same process. The pollen which Leroi-Gourhan identified from 8.1 ([7A]) may possibly indicate temperate conditions, but it could well be from the top-soil of the layer 6.2 pedogenesis. |
| 32 | 9, 8.2, 8.3 | d | North of the rock arch, gradual accumulation of granitic sand. Further south, damming of the west ravine (either by rock falls or by a storm bar during a new transgression) resulted in the formation of an intermittent fresh-water pond which itself received overspill from a body of water on the plateau. Gyttja was deposited at the base; during drier periods cracks allowed diatoms and fine clay to penetrate the underlying layers. Pollen (sparse) includes some thermophilous trees. |
| 33 | 9, 9.1 | — | A ranker developed above the pond deposit; intercalated with layer 9 north of the cliff. Pollen suggests open vegetation. |
| 34 | 9, 9.2 | — | Destabilisation of the sloping surface within the north ravine allowed the layer of granitic sand with blocks to extend over a much larger area. |
| 35 | 9.3 | — | A second well-developed ranker (bipartite) formed under temperate but cooling conditions; further north, the same sandy deposit continued perhaps for some time longer. |

81

**Table 7.1** (continued)

| Episode | Layer(s) | Archaeo-ology | Events in the ravines |
|---------|----------|---------------|----------------------|
| | | | |

*STAGE V (essentially cold). Isotope stages 4–2. See Appendix E (microfiche).*

| Episode | Layer(s) | Archaeology | Events in the ravines |
|---------|----------|-------------|----------------------|
| 36 | 10 | — | Head consisting of slabs and coarse granitic sand with a minimal fine fraction; numerous air spaces. Well developed in the west ravine (dipping NW) and also in the extreme north of the site (dipping SW). Apparently missing under the rock arch — perhaps removed by later erosion. |
| 37 | — | — | Initiation of loess deposition, with rare calcareous concretions (Marett 1916, 78). |
| 38 | 11 | b | Head with loess, and hearths in the north ravine (as reported by early excavators). |
| 39 | 11 | b? | Interstadial? Pedogenesis effecting underlying deposits in the north ravine — this process continued into the beginning of the next phase of freeze-thaw. |
| 40 | — | — | Thin (30 cm) layer of aeolian (?) sand — restricted to the junction of the north and west ravines (Marett 1916, 112–3). |
| 41 | 11 | b | Loessic head; several concentrations of rodent remains (*Dicrostonyx* etc.). Human visits probably somewhat rarer than before. |
| 42 | — | — | In the west and south ravines only: pockets of rounded gravel observed by Burdo (1960, 18–19). |
| 43 | 11 | — | Loessic head (about 12.5 m thick in the west ravine — Burdo 1960, 18). Completely removed by erosion in the northern entrance? |
| 44 | — | — | In the west and south ravines: calcareous loess (Burdo 1960, 18). |
| 45(?) | 12 | — | In the northern entrance only: sandy head (related to deposition of coversands?). |
| 46(?) | 12 | — | Erosion. |
| 47(?) | 13 | — | In the northern entrance only: iron-indurated pure aeolian sand *without* granite blocks and other detritus. |
| 48 | 13 | — | Seen in the northern entrance only: podzolic soil formation on a steep slope. |

*STAGE VI (temperate). Isotope stage 1.*

| Episode | Layer(s) | Archaeology | Events in the ravines |
|---------|----------|-------------|----------------------|
| 49 | 14 | — | Modern soil. |

# SEDIMENTOLOGY

J. P. Lautridou, H. Duroy, P. Giresse, M. N. Le Coustumer and M. Levant

## 8.1 INTRODUCTION

Before discussing the results, it is appropriate to give a short account of the history of these investigations prior to the analysis undertaken by the Centre de Géomorphologie du CNRS at Caen, and of the unavoidable restrictions under which this work was carried out. Not the least of these was the fact that the samples had been taken for the most part some years previously, and from sections that had since been walled up.

Layer numbers assigned by Burdo are distinguished by the use of square brackets, as in the other chapters. The report is prefaced by a note on the numbering of the samples. A full set of tables is given in microfiche (mf 8.1–8.7).

### 8.1.1 Sample Numbering

Several different schemes had been used in the course of the excavations. In 1979, therefore, all samples (including Burdo's) were integrated into a single series and each was given a unique number, with a view to entry in a computer database. Those sent to Caen were entered also into that laboratory's numbering scheme, which has been adopted in this report. The samples are grouped into batches of 10, and are identified by the batch number followed by the individual number within the batch (giving, for example, 2661-1). The samples submitted for micromorphological examination, however, were not entered in the Caen system, but retain their computer record number (e.g. 515). A table of equivalence between the two systems is given in mf 8.7.

### 8.1.2 Data Available up to 1978 (The End of the McBurney Excavations)

Simple granulometric and chemical tests were carried out on behalf of Burdo and reported by him (1960, 15–29). At the time, the full complexity of the site was not appreciated and, moreover, some descriptions of the material are actually misleading. Burdo's 'loose grey ashes' [7F], for instance, were said to be a "*loess*, slightly coarser, with 87% of medium and coarse silt and sand" (our italics). According to Burdo's own definitions, the sample must therefore have consisted almost entirely of particles in the range 70 $\mu$m–2 mm. This misdescription, coupled with the failure to subdivide the considerable thickness of ashy deposit extending between McBurney's layers C and G, contributed to later misconceptions about the granulometric composition of these

extremely important layers.

In 1970 and 1971, when the last major review of the stratigraphy prior to the present report was being carried out, the lowest visible layer was A, which though ashy was loessic in granulometry (McBurney and Callow 1971). Together with Burdo's description, and in the absence of more recent sedimentological studies, this led to the erroneous belief that the lower layers were entirely loessic.

McBurney's 1961–1962 excavations did not attempt a re-evaluation of layers C–G, being more urgently concerned with opening the 'deep sounding' to expose the fossil beach. Subsequent backfilling, in 1966–1967, protected these layers from erosion, but their re-exposure was delayed by the exigencies of the excavation until the final season, and lasted only for a short period, when no sedimentologist was present. Though soil samples were taken, therefore, the collecting strategy did not entirely cover some of the more complex aspects of the sedimentation.

Samples had been collected in 1969 from the higher layers (3, 4, 6 and 9) then under excavation and were subjected to granulometric analysis by Dr. J. Sevink in Amsterdam. They demonstrated the contrast between layer 9 and the others which the more recent work described here confirms. (As the samples were not subjected to other tests, the results are not used in this report). The other sedimentological study undertaken while the excavations were in progress was by A. G. Cooper, of the British Museum (Natural History), who established the presence of taranakite in the phosphatic nodules typifying certain layers. This observation is also supported by the new, more extensive investigations described below.

### 8.1.3 The Circumstances of the New Investigation

J. P. Lautridou visited the cave in 1980 and discussed with Callow its potential and the problems still to be resolved. In consequence, 31 samples from the McBurney excavations (2661-1 to 2664-1) were sent to Caen for analysis between May and September 1980, together with five collected by Burdo in the 1950s (2664-2 to 2664-6). A further nine samples (2664-7 to 2665-5) were collected by Lautridou during a second visit to the site in September 1980. After a review of preliminary results in April 1981, further samples were collected where possible by Callow, with a view to micromorphological study in particular; in addition, old bulk samples were searched for pieces of deposit retaining their original

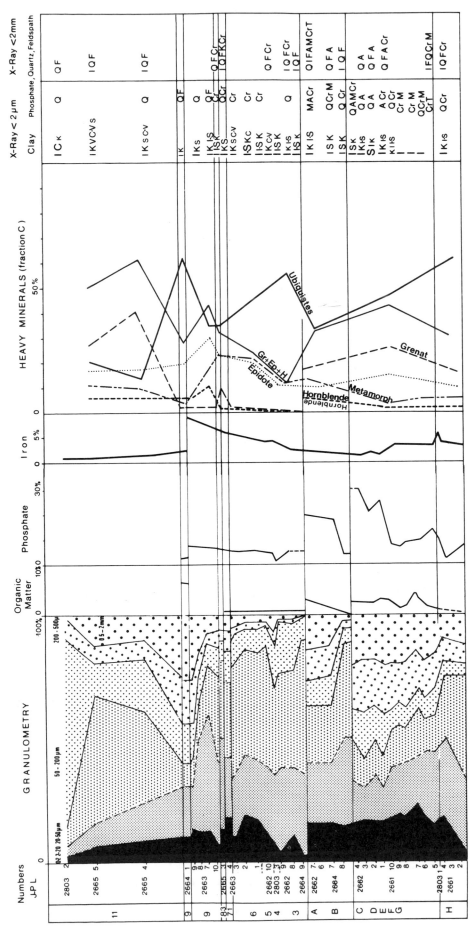

**Fig. 8.1** Sedimentology of the La Cotte de St. Brelade deposits: granulometry; percentages of organic matter, phosphates and iron; heavy minerals; X-ray diffraction (clays and powdered samples).

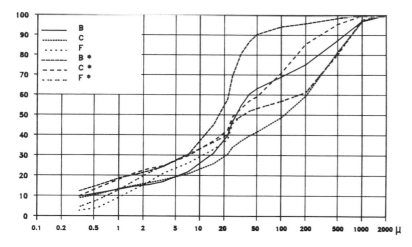

**Fig. 8.2** Cumulative curves of grain size frequencies for layers B, C and F, before and (asterisked) after treatment of samples with nitric acid (analysis by M. Levant).

structure. Six more samples (2803-1 to 2803-6) were submitted for sedimentological analysis to resolve questions arising from the work already undertaken.

The resulting sequence, though incomplete, gives fairly good coverage of the principal stratigraphic units, although the investigation was attended by a number of difficulties. In particular: (1) almost all the material was in the form of bulk samples and unsuitable for micromorphological study; (2) the importance of lateral facies variation had not been fully appreciated during excavation, and not only was close sampling of a column of sediment the exception, but also the microstratigraphic details were not always readily available where a layer had been sampled in several places; (3) the walling up of some sections made it impossible to check some results against fresh field evidence (photographs and site notebooks gave clues, but no new sampling was possible for certain layers); (4) neither the members of the team at Caen, nor those in Cambridge chiefly responsible for the research, had direct personal knowledge of some of the most important layers in the sequence (B–H). Consequently, the interpretations must be offered more tentatively than might otherwise have been the case.

## 8.2   PARTICLE SIZE (analysis by Levant)

Measurement of the relative frequency of the various grain sizes present in a sample of deposit (Fig. 8.1) is carried out for two reasons:

1.  To permit accurate description of the sample, and its comparison with others. In the field, texture may be estimated very roughly by the 'feel' of the material when rubbed between the fingers. However, precise quantitative analysis is necessary if small but perhaps significant variations are to be detected.
2.  To establish the degree of the size sorting which may have occurred, and hence to gain an insight into the depositional processes involved. For instance,

aeolian deposits are characterised by a marked predominance of a restricted range of sizes (i.e. are well sorted), whereas most slope deposits are poorly sorted (subject to the limitations imposed by the parent material). If two very different sources have supplied sediment, this may be apparent from the size distribution.

Though techniques vary in practice (as does terminology), the most convenient method of presentation for comparison is the *cumulative curve*. The percentage for each grain size interval is added to the sum of the percentages for all *smaller sizes*, to give a graph rising progressively from 0–100% in the range 0–2000 $\mu$m (being suitable for overlay, such curves are easier to use for comparative purposes than are crude percentage histograms). The shape of the curve gives a simple indication of the degree of sorting. If the majority of particles fall within a limited range, the curve is steepened locally. Thus a pronounced sigmoid graph indicates strong sorting in favour of grains of a size corresponding to the increased gradient, and a double sigmoid curve may indicate a mixture deriving from two distinct (but originally well-sorted) sources. A linear or only slightly parabolic form suggests that little sorting has occurred.

The following size classes are employed in the ensuing account: clay (less than 2 $\mu$m), *limon* or loam (2 to 50 $\mu$m) and sand (more than 50 $\mu$m). In French usage, silt implies a range slightly different from that of *limon*.

This work was carried out using A. Rivière's method (Andreasen pipettes in a thermostatically controlled water-bath) after preparation following internationally recognised practice, modified by the use of sodium tripolyphosphate as a dispersant. The fraction smaller than 2 mm was obtained from the 'raw' sample without special selection, even for the rich archaeological layers A–G. Here, the presence in quantity of bone and phosphate, as well as of other organic matter, modifies the 'natural' granulometry of the sediment (Fig. 8.2). Therefore, after presenting the general results, we shall return to these layers to compare the granulometric curves before and

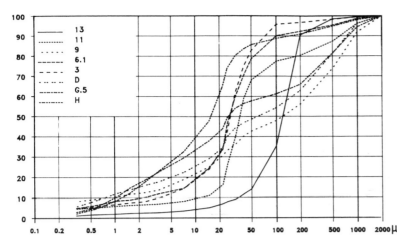

**Fig. 8.3** Cumulative curves of grain size frequencies (analysis by M. Levant). The samples used here are: 2661-2 (Layer H), 2661-6 (G), 2662-7 (D), 2662-8 (3), 2663-3 (6), 2663-9 (9), 2665-5 (11) and 2803-2 (13).

after the elimination of phosphates. For the other samples which are poor in phosphates, the granulometric curves in Fig. 8.1 are valid without dephosphatisation.

### 8.2.1 Loess and Aeolian Sands

Loessic material (*in situ* or derived) is abundant in the filling of the fissure, particularly at three levels:

| | |
|---|---|
| Layer 11 | ('upper loess') |
| Layers 3–6 | ('middle loess') |
| Layer H | ('lower loess') |

The granulometric curve is altogether typical of a loess and resembles those obtained in Jersey (see chapter 5), in the Cotentin (Coutard *et al.* 1973), in Brittany (Monnier 1980b), and extensively throughout northern France (Lautridou 1979; Jamagne *et al.* 1981). It has the following characteristics (Fig. 8.3): a very well-sorted loam with a dominant 10–50 $\mu$m fraction; a clay fraction below 20% as a rule; a weak 50–200 $\mu$m fraction (below 10%); a sigmoid curve. In the lower (H) and the upper (11) loesses, however, where the local contribution in the form of blocks and chunks from the walls is in evidence, there is a notable fraction of medium, and still more of coarse, sand. The top of the upper loess passes into a very well-sorted sand (layers 12 and 13) with a mode around 120 $\mu$m and very little material coarser than 200 $\mu$m.

The loesses are non-calcareous, as is usual in Normandy and Brittany. Nevertheless, the presence of calcareous silts locally on Jersey as well as in the Baie de St-Brieuc and in the Norman-Breton Gulf proves the initial presence of calcium carbonate (averaging 10%) which at La Cotte has been leached out either syngenetically (during deposition of the loess) or during subsequent pedogenesis.

### 8.2.2 The Occupation Layers A–G

With the exception of those for A and B, the particle size curves are linear or slightly curved in the centre, tending towards the parabolic class (Figs. 8.2 and 8.3). There is, however, a fairly clear feature around 25 $\mu$m which is characterised by an increase in gradient. In an attempt to throw light on this peak corresponding to the 20–30 $\mu$m fraction (which corresponds to a loessic origin), these samples were treated with nitric acid so as to remove the organic and anthropic contributions and permit an assessment of their initial granulometry.

This attack with acid (4N $HNO_3$, agitation for 20 min, pH approximately 2) ought in principle to eliminate the phosphates. To find out how successful this was, diffractometry was carried out on the treated sample: apatite was always destroyed, but crandallite and millisite sometimes persisted even after several treatments.

The first point to be noted is the importance of the material destroyed during treatment. The samples of A–G lost, respectively, 36, 40, 75, 47, 55, 5 and 5% of their weight; even so, some of them (notably B) had not had their phosphate entirely removed. These figures clearly exceed those obtained by chemical analysis (Fig. 8.1). One possible explanation is that the original sample included phosphatic concretions; another, that charcoal from hearths, and iron, have been incorporated in it. Moreover, in view of the high clay content which is apparent in some of the results, the possibility had to be considered that the acid altered the sediment by artificially increasing the fraction below 2 $\mu$m. The same treatment was therefore applied to a sample of B containing only a little phosphate, and to another of the upper loess (which lacks it altogether). After treatment the granulometry was unchanged, demonstrating the validity of the method.

The particle size modification is clear for A and B, which already possessed a loessic peak around 20–40 $\mu$m; locally within B there are yellow, purely loessic bands. After removal of phosphate, the curve becomes that of a loess. In contrast, for the other samples (layers C–G) there is no such clear modification. Certainly in some cases (C, D, E) there is a shift in the curve towards the fine fraction (Fig. 8.3), and the 20–40 $\mu$m fraction increases in proportion, marked in the curve by a stronger slope. But there always remains an important sandy fraction which can be studied under the binocular microscope: it consists of angular polymineral grains, sometimes with a red ferruginous cortex coming from the walls of the ravine. It is difficult to estimate the relative contributions of the walls and roof and of the loess. Taking the fraction coarser than 200 $\mu$m as an indication of the local component between A and G, this evolves as follows: 5% in A and B, 15% in C, 23% in D, 26% in E, 39% in F, 34% in G. The remainder is likely to be chiefly of loessic origin. To sum up, therefore, the principal contributing agencies are: loessic in A and B; essentially anthropic in C, D and E; and more mixed in F and G. The very high percentage of clay in the layer E sample, after treatment with acid, may well be the consequence of illuviation during soil formation (see van Vliet-Lanoë, chapter 9, this volume).

### 8.2.3 The 'Reddish' Silts 6.2 and 4

Though their slightly redder colouration (in comparison with layer 6.1, for example) led McBurney and Callow (1971) to discuss the possibility that they were soils, these layers are not distinguishable from the other loesses on the basis of particle size (but see chapters 7 and 9).

### 8.2.4 The Local Component

The purest loess, as already indicated, is the 'middle loess', from 3 to 6. Towards the top, an increase is apparent in the local coarse sand component which is related to the granular disaggregation of the wall, testifying to increasing humidity exploited by frost. The possibility that much of this material came from the headland above, as well as from the walls and roof of the cave, must be borne in mind. Thus formation of some of the weathering products could substantially antedate their deposition in the site. Equally, progressive collapse of walls and roof, after weakening of the joints by severe frost action, could be expected to continue into a subsequent period of climatic improvement.

The proportion of the local component resulting from frost action on walls and headland is weakest between A and D, as is shown by the low percentage of the sand fraction (after removal of phosphates), as well as by estimates of the incidence of large blocks based on photographic evidence (Fig. 8.4). In particular, there is a pure loess in B, under the occupation breccia. This contrasts with layers E–H, and also with the deposits above A, where the coarse component is once again very important.

### 8.2.5 The 'Fossil Beach' (Layer 7.1)

Marine reworking of the matrix was not apparent in the samples submitted for analysis, which were taken during the 1961–1962 excavations (but see chapter 9, in which the 1982 samples are discussed). The material derives, above all, from layers C, D and E (including dark grey

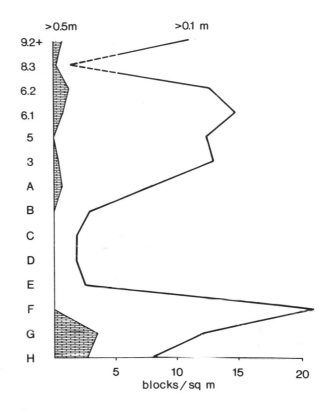

**Fig. 8.4** Incidence of granite blocks, estimated from photographs (analysis by P. J. Rose and P. Callow).

granules). This is consistent with marine erosion by progressive undercutting.

### 8.2.6 Loesses Depleted in Clay

These were observed during a visit to the site by Lautridou and Callow in 1980, and certain samples collected in the early 1970s were of this type; the phenomenon is referred to by McBurney and Callow (1971). The layers chiefly affected are 3 and 6. This impoverishment is related to leaching by rainwater in the less sheltered parts of the site, and by water circulating in contact with the walls and washing out the clay and iron over a width of several centimetres (most markedly for 1 cm or so, where the silt becomes light grey and relatively clayless).

### 8.2.7 Equation with Burdo's Layer Terminology

A number of samples taken by Burdo in the 1950s were analysed either for comparison with more recently collected samples believed to come from the same layers, or, as in the case of his [7A] and [7B], because the layers were no longer accessible.

[3A]    resembles loess of the series from layers 3–6.
[5A]    resembles 9.3 (sample 2664-1); see granulometry.
[7A]    resembles 9 (sample 2663-8).
[7B]    gives a very well-sorted loess curve of the layer 3–6 type.

The match between [3A] and the 'middle loess' was expected, as it is believed that [3A] and 6.1 are the same layer — even though the former was described by Burdo (1960, 22) as a fine aeolian *sand*. The similar results for [7B] lend support to the suggestion that this layer (renamed 7.2) is derived from the 'middle loess' and represents an early stage of cliff collapse. The granulometry of layer [7A] (renamed 8.1) suggests that the complex depositional regime with granitic sand (and blocks), characteristic of layer 9 *sensu lato*, in fact antedates the ponding recorded by layer 8.3 (see chapter 9). Burdo's layer [5A] probably amounts to no more than the base of layer 8.2 (his [6]) with less humus than usual.

However, one of Burdo's samples, purporting to be from his layer [3B] (with diatoms) proved not to provide the expected match with a fresh sample, 2665-3, from layer 8.3. Certainly the granulometric curves are identical: loessic material; rather clayey; an important coarse fraction typical of a loess-head mixture (local origin). But there is a major difference. In [3B] the sand fraction in fact consists of phosphatic concretions (such as are observed to some extent in layer 9, and more particularly 3–6) while in 2665-3 it is indeed sand. There are differences in the heavy minerals: whereas 2665-3 is identical to layer 9 (epidote, minerals of metamorphism, a little zircon), [3B] recalls more closely the earlier series, from 3–6, with a lot of zircon and less epidote. However, caution is necessary here in view of the smallness of the samples (counts of only 75 and 55 grains respectively). X-ray diffraction of the clays (see section 8.5.1) shows only illite (and phosphatic crandallite) in [3B], whereas 2665-3 is more varied: I I-S-K (and apatite and crandallite). Elsewhere in the sequence, only layer G has yielded illite on its own without other clay minerals; even so, apatite is usually recorded from C to G. Thus [3B] above all resembles layers 3–6, and particularly sample 2663-2 from layer 6 (quite clayey, with a gentler slope of the cumulative curve from 3 to 20 $\mu$m), except that 2663-2 has fewer concretions — though these are admittedly very variable in occurrence. The difference between the two samples is entirely consistent with the presence of redeposited material from different sources, with considerable lateral variation responding to local factors.

## 8.3 CHEMISTRY (analysis by Duroy)

Estimation of phosphorus was effected by colorimetry of phosphomolybdate, that of carbon by the Anne Method (reduction by the organic matter of potassium bichromate in sulphuric acid), and that of total iron by colorimetry with orthophenanthroline.

### 8.3.1 Phosphates

These are notably present in A–G, as well as in H and 6.2 (Fig. 8.1). The proportion diminishes in 3–6 and in 9. None were found higher than 9.3. These phosphates were also observed during the X-ray studies (millisite, crandallite, apatite: see chapter 10, where the matter is

discussed more fully). The low levels of phosphate in the matrix of 6 and 5, in spite of these layers' original bone content, are the result of downward migration. The destruction of carbonates in the loess has already been the subject of comment.

### 8.3.2 Organic Matter

In layers A–G (all with human occupation debris) organic matter varies between 2 and 4% (Fig. 8.1). There is very little in layers 3–6, and none higher up in the sequence (except in 9.3, and presumably in the other humic deposits such as 9.1 and 8.2, which were not analysed chemically).

### 8.3.3 Total Iron

The percentage of iron is very low in A–E, and also in the upper (11) and lower (H) loesses; it is only a little higher in F and G. On the other hand, layers 3–6 (the middle loess) and 9 are rich in iron, accompanying an increase in the contribution from the walls which provide sand with ferruginous coatings (probably the result of an earlier rubifying alteration of the granite surface). Later, in layer 11, the elements from the walls are more abundant, but fresh; perhaps the red varnish had already been completely removed, particularly during deposition of 3–9.

## 8.4 HEAVY MINERALS (analysis by Giresse)

One of the items shown in Fig. 8.1 is the evolution of certain heavy minerals (in the 50–125 $\mu$m fraction) such as epidote and garnet, and also of associations recognised in Normandy: 'ubiquitous' (zircon, tourmaline, rutile) and 'marine' (garnet, epidote, hornblende) — the latter so named because it characterises marine sediments. The respective evolution of the percentages of these two groups is very characteristic in Normandy (Lautridou 1968; Lautridou *et al.* 1976; Coutard *et al.* 1973). Dominant in the Upper Weichselian loesses, the association epidote-amphibole-garnet (less abundant) yields ground to the 'ubiquitous' minerals in the Older Loesses and in the early Weichselian formations (which are in fact derived from Saalian silts).

This mineralogical inversion is indeed found at La Cotte between the recent and earlier loesses: the epidote-amphibole-garnet stock is clearly dominant in the upper loesses and becomes weaker in the silts below layer 6. Three points should be noted, however.

1. The middle loess, 3–6, contains a notable percentage of minerals of metamorphism. This is quite an unusual feature.
2. The opposition between the two associations is not always very clear, particularly in the lower layers.
3. The lower layers behave in a distinctive manner, with an increase in garnet from H to A inclusive, while this mineral (which is usually discriminated against in the finer grades) grows in importance as

the granulometry becomes sandier: this is particularly the case for the Weichselian cover sands in the Baie de Mont-St-Michel (Lautridou *et al.* 1982). At La Cotte there is no increase in the fine sand fraction of aeolian origin, so it represents a variant which has not been found in the thick older loesses of Normandy or the Baie de St-Brieuc.

In conclusion, the classic mineralogical inversion found in Normandy occurs at La Cotte, albeit somewhat attenuated, but between the middle (3–6) and the lower (A–H) stocks there is a modification characterised by a rise in garnet of unknown origin (a shift in the prevailing wind?). The reasons for the inversion between the recent and earlier loesses are unclear. In the light of new work (Auffret and Alduc 1977; Alduc *et al.* 1979) on the subject of the Channel floor, it seems that the pre-Weichselian river alluvia (particularly those of the Seine) are very extensive; they could have supplied silt to the Older Loesses. In contrast, the Weichselian alluvia (narrower and more entrenched in the bedrock) would have contributed less to the Younger Loess, which derived chiefly from marine and estuarine sediments [comment by Lautridou].

The fact that the recent loesses contain more of the fragile heavy minerals and clays, which are sensitive to alteration, than do the earlier loesses might suggest a hypothesis of alteration of the earlier loesses, with the loss of green minerals in particular. In Normandy, however, there are thick unaltered Older Loesses which are dominated by the stock of more resistant 'ubiquitous' minerals. Such a hypothesis is therefore untenable.

## 8.5   X-RAY DIFFRACTION (analysis by Le Coustumer)

### 8.5.1   Clay Minerals

Determination of the clays was carried out by X-ray diffractometry on orientated pastes, following the classic universal procedures. The samples were decalcified and saturated with magnesium. In Fig. 8.1, the minerals are listed in order of decreasing importance, an estimate of their relative abundance having been achieved by measuring the height of peaks on the patterns so obtained.

The upper loess (11) has a mineralogy of Brittany/ Baie de Mont-St-Michel type, with vermiculite (V) and chlorite (C), as well as illite (I), smectite (S) and kaolinite (K) (illite and smectite predominant).

The middle loess (3–6) is different insofar as it has very little vermiculite and chlorite: its stock consists of illite, kaolinite and smectite.

Layers A–H differ little from the middle loess. Illite predominates except in layer D, where smectite is more important.

Thus we find once again that the young loesses possess the stock typical of the Norman-Breton Gulf, and that there is a contrast between these and the older loesses comparable to that demonstrated by the study of heavy minerals.

### 8.5.2   Silts and Sands

This work was carried out on the fraction finer than 2 mm, without orientation. As in all loesses, quartz predominates, with the presence of abundant felspar and muscovite (Fig. 8.1).

### 8.5.3   Phosphates

These were observed in both the fine and coarse fractions, and consisted of apatite, millisite, crandallite, and taranakite. Their significance is discussed in chapter 10.

# MICROMORPHOLOGY

B. van Vliet-Lanoë

## 9.1 AN INTRODUCTION TO THE MICROMORPHOLOGY OF ARCHAEOLOGICAL DEPOSITS

Until recently, only the classic sedimentological techniques have as a rule been applied to archaeological deposits. They have provided no more than a global description, limited in detail, of the complex sedimentary conditions acting in an environment where man is present. Petrographical investigations have been restricted to consolidated materials such as pottery, lithic materials and nodules.

However, over the past decade or so the micromorphological methods that have been extensively applied to hard rocks have also been of increasing value in palaeopedological, geomorphological and archaeological contexts. The development of the technique of impregnating unconsolidated material by means of plastic polymers has permitted the extension of petrographical exploration to such topics. The usual procedure entails careful field sampling into tin boxes of large lumps of unconsolidated sediments, vertically oriented and undisturbed. In routine practice, sampling locations are marked on a drawing of the profile under investigation. The samples are dried slowly to avoid stress and shrinkage; afterwards, they are impregnated with liquid plastic resin under vacuum and slowly polymerised. This leaves them rock-hard, so that thin sections of various sizes can be prepared for study under the petrographical or fluorescent microscope.

With the petrographical microscope it is possible to study the mineralogy and fabric of sediments, and to observe from a microstratigraphical viewpoint the succession of events acting before and during deposition, and diagenesis after the material has been buried. From this the succession of climatic events can be reconstructed with, under the best conditions, a precision similar to that of palynology. As a hypothetical example, we may consider the sequence of events shown in Fig. 9.1. In a sandy colluvial layer, a clay illuviation characterising a lengthy stabilisation of its surface may take place in the pores (A); subsequently a phase of frost action breaks and churns up the coatings and aggregates (B). A further phase may be characterised by an illuviation of another type, such as amorphous organic matter under a spodosol (C). Locally the spodosol is affected by a ditch, giving pronounced anthropogenic humic clay illuviation (D). Most of these features have a climatic significance. In the same way, we can distinguish a succession of different fabrics in the sediments, indicative of freeze-thaw

activity, gelifluction, biological mulching, drying, colluviation, and so on.

With a fluorescent microscope, using reflected light (blue or ultra-violet of long wave-length), it is very easy to distinguish the occurrence of phosphates, certain carbonates, fulvic acid, amorphous clays and free aluminium, and remains of cellulose, lignin and pollen. This technique is a very useful and practical complement of classical petrographical observations (van Vliet-Lanoë 1980; van Vliet et al. 1982).

## 9.2 SPECIFIC PROBLEMS IN ARCHAEOLOGY

Apart from the sedimentological and palaeoclimatological goals suggested above, micromorphology can also be used to study pottery, in order to determine manufacturing techniques and the origin of 'tempers'. Another

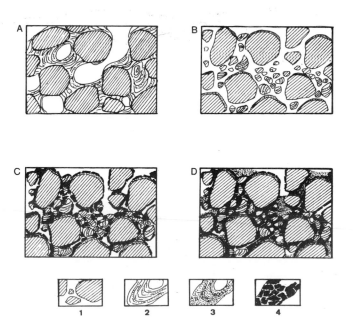

**Fig. 9.1** Hypothetical example of the application of micromorphology: evolution of a sandy sediment (width of view: 2 mm). (A) Clay illuviation on a sand, below a *sol brun lessivé*; (B) cryoturbation: destruction of the pore illuvial bodies by churning; silt cappings occur on the tops of the grains; (C) illuviation of amorphous organic matter in the Bh horizon of a podsol; (D) illuviation of humic clay, as a consequence of anthropic activity (agriculture, fire, trench etc.). 1, mineral grains; 2, clay cutans; 3, humic clay illuviation; 4, organic matter.

application is to determine the constitution of bonding media such as mortar or quicklime (Masset and van Vliet-Lanoë 1973). It is also possible to study weathering of bones and formation of phosphatic crusts in sepulchres or caves (Peneaud 1979). A recent application has been the investigation of the diagenesis of hearths (Courty 1982).

## 9.3 RESULTS FOR LA COTTE DE ST. BRELADE

Most of the observations made on excavated materials are concerned with palaeoclimatological and pedological problems. Nevertheless, the abundance of bone frag-

ments in living floors and other layers, and of secondary phosphates, permits some remarks about their occurrence.

### 9.3.1 Phosphates and Bone Fragments

Bone fragments occur in all archaeological layers at the site. In terms of conservation, they fall into two main groups:

1. *Greyish* in thin section, *microcracked*, sometimes weathered by bacterial or mycelium activity (Fig. 9.2E). The fragments are very sensitive to gelifraction and to chemical weathering, for example in layer A. This is slowly dried 'fresh' bone, probably

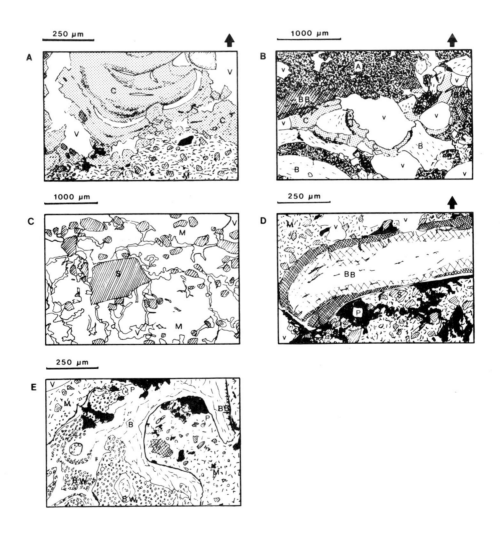

**Fig. 9.2** La Cotte de St. Brelade: thin sections (the drawings serve as a key to the colour plate). (A) Layer 6.1: clay and humic clay illuviation; (B) layer E: clay illuviation; (C) layer 9.3: ranker; micro-aggregated blackish matter (bacterial) mixed with some mineral grains; (D) layer A: burnt bone (the flecks inside the bone are small blood vessels colonised by bacteria); note the caramelised and cracked cortex; (E) layer 3: microcracked 'grey bone': dessicated bone affected by frost shattering; note neoformation of crandallite (BW) associated with bacterial precipitation; (F) (not drawn) layer 8.3: diatoms. A, ashy matrix; B, bone; BB, bone — burnt; BW, bone — weathered (crandallite pearl); C, clay coating; M, soil matrix; P, bacterial precipitation; S, skeleton; V, void.

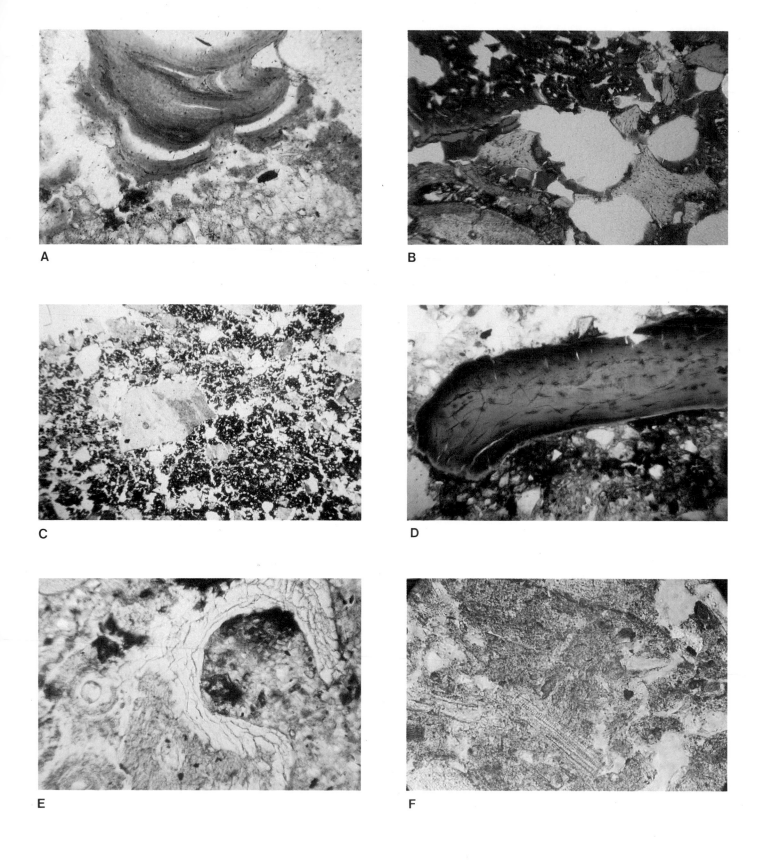

A

B

C

D

E

F

exposed on the surface for a long time and slightly weathered. It appears very similar to bones observed on arctic raised beaches, in a dry environment. Its presence in archaeological layers, when mixed with the second type, could be indicative of reworking. Teeth from the same layers appear more resistant to both gelifraction and weathering. This type of bone fluoresces well in blue light.

2. 'Caramelised' in appearance: in colour it is a hyaline honey yellow to brown or even black, *cracked*, much more resistant to weathering and gelifraction (Fig. 9.2D). Its behaviour in fluorescent light shows a more absorbent character than does the greyish type, particularly in the case of the darker fractions. It appears to relate to the thermic condensations of the organic compounds (ossein). Recent work in the field of dentistry shows that a change occurs in the crystalline structure of fluorapatite under thermic impact (using lasers). Furthermore, it seems that as a result of burning the apatite loses fluorine (Giresse 1980). This evidence, and the frequent association of wood charcoal in the same layers, suggests that many of these bones were dehydrated by burning or heat. This would explain also their cracked appearance and browning, changes in crystalline structure, and resistance to further strong weathering.

Phosphates (chapter 10) are occasionally found in the thin section preparations of sediments from La Cotte. Nodules of crandallite alone were observed in layer A, apparently *in situ* and stained with amorphous organic matter; nodules of crandallite also occur in layer H. Small pearls or thick marbling of crandallite are developed in the later Upper Temperate pedocomplex before formation of the layer 9 palaeosol.

### 9.3.2 Palaeoclimatic Interpretation

Reconstruction of the profile is incomplete, as no large fragments suitable for micromorphological study were obtained from a number of layers (the author was recruited to the La Cotte project in 1980, after the principal excavations had been completed and some of the sections walled up). Moreover, the exact provenance of some of the older samples is not always clear. The results are described here in sequence from the base to the top of the excavation. In the resulting profile, at least two important pedocomplexes were observed — one rather well preserved in layers 6 and 9, and an older one in layer E. One of the greatest potential contributions that micromorphology can make to the study of pedogenesis and sedimentation — that of recognising and distinguishing multiple episodes of soil formation and other phenomena affecting the same deposit (without being apparent at the macroscopic level) — has been very important at this site.

Different types of soil are found in this exposure, generally eroded to their lower horizons:

1. Ranker. This is a raw humus accumulation formed under an oceanic climate like that of today, associated with vegetation of heath to grassy type. Labile humus infiltration can occur below it. The pH conditions are usually very acid (around 4.5).

2. *Lessivé* soils. These are characterised by a translocation of clay minerals by leaching in neutral to weakly acid conditions (pH 7 to 5.3). The eluvial horizon is usually termed $A_2$, and the clay accumulation or illuvial horizon, Bt; the lower part of the latter, $B_3t$, exhibits the best preserved clay coatings. This type of pedogenesis can develop under a wide range of climatic conditions from boreal to tropical, if the stabilisation of the surface is long enough to allow the expression of leaching (about 1000 years).

3. Humic clay illuviation. This can be due to several factors. In archaeological sites, the most common cause is destabilisation of the clay minerals by leaching of the cations from ashes (high pH), and a local translocation of the colloids mixed with some ashes and more soluble organic compounds (Courty 1982). Other occurrences may result from destabilisation of clay by biological mulching, generally under cultivation or in prairie soils, in weakly basic to weakly acid conditions. Here, clay migrates in a complex with organic components of humic type. A third case arises in very acid conditions, such as occur in podzols or beneath rankers. Here, also, the clays can migrate as a complex with organic components of fulvic type (Guillet *et al.* 1979).

4. Periglacial disturbances. These are generally caused by the action of frost on the sediment. They usually take the form of traces of ice lenses that develop during freezing. After thawing of the segregated ice, this produces a very typical platy structure which can be deformed by frost creep or even gelifluction when the amount of available water increases during thaw. In this case, aggregates are rounded by rolling (van Vliet-Lanoë 1982). Usually silty illuviation accompanies these features, forming cappings on aggregates or coarse grains.

## 9.4 THE SEQUENCE

### 9.4.1 Lower Temperate Pedocomplex

Only one phase of *lessivé* soil formation is apparent, having been identified from clay illuviation affecting layer E. The underlying deposits lack evidence of contemporary or near-contemporary frost activity, and owe their existence in some measure to ponding and/or colluviation.

**Layer H** is a very complex unit. All the samples that have been studied in thin section as yet have proved to be of material which had previously suffered reworking, and included varying degrees of admixture of sediments from layers G–6. The original layer H was a loamy deposit derived from old loessic formation (see chapter 8, this volume) and laid down, with the addition of granitic sand, in a pool of calm water — possibly behind a beach bar. This interpretation is supported by bleaching of the mixture, the occurrence of diatoms and rare fragments of sponge spiculae. Conditions were probably rather similar to, but wetter than, those of unit 8.3; the palynological data suggest that they were temperate. All the exposures visible in the cave until the present time reflect reworking of the original H sediments. Consequently we shall use quotation marks, referring to these deposits as 'H'. They are discussed under the general heading of layer 7.1.

**Layers G–F** are disturbed living floors (rich in bone fragments, charcoal and burnt clay) associated with postdepositional bioturbation and infiltration of humic material; evidence of colluviation is provided by slight rounding of burned bone and charcoal fragments. They correspond to somewhat displaced hearth complexes with normal diagenesis (Courty 1982), suggesting stabilisation of the surface. No traces of frost activity are apparent, nor are traces of clay illuviation from layer E. These layers were not affected by the slumping of the cliff, though partly undercut during formation of 'H'.

**Layer E** is the only archaeological layer which is strictly *in situ*, being very porous and not at all displaced. It has undergone an illuviation of reddish brown clay, infiltrating the pores of the matrix and of the bones (Fig. 9.2B). These fine clay coatings have been disrupted by subsequent frost activity.

**Layers D and C** are both derived from occupation layers, but no evidence of *in situ* illuviation was found. Only a few bone fragments seem to preserve traces of the layer E illuviation, though most of them are rounded (chiefly in D). There is an erosional disconformity between E and D. Layers D and C are colluvial deposits with some hydromorphic features (iron and manganese staining in D). C seems to reflect slightly colder conditions than D; it is richer in greyish strongly weathered bones with angular breaks (due to gelifraction), indicative of a cool to mildly cold environment.

## 9.4.2  Intermediate Cold Complex

This is characterised by frost structures in material which is mostly of loessic origin, and by an absence of fine clay illuviation dating to the subsequent temperate pedocomplex.

**Layer B** consists of loessic material interstratified with bone breccia, probably displaced by sliding. Platy structures observed locally could represent freeze-thaw activity (displacement by frost creep) such as is also suggested by loamy infiltrations. A cold environment is indicated, with colluvial to creep depositional agencies.

**Layer A** is rather similar in conditions of sedimentation, but richer in bone fragments (Fig. 9.2D).

**Layer 3** is a loessic deposit affected by frost. The initial stage of sedimentation was probably accompanied by mud-flow, rill and splash action.

The above layers reveal a progressive cooling of the climate, from temperate to periglacial.

## 9.4.3  Upper Temperate Pedocomplex (Samples from the North Face)

This affects layers 4–9, and seems to be very complete in the northern part of the site. Three periods of *lessivé* soil formation have been identified: between layers 6.1 and 6.2, between 6.2 and 9, and in the lower part of 9. They are identified by their Bt horizons (the upper part of each soil has been lost through erosion). Clay illuviation extends deep into the preceding periglacial deposits (which pedologically, as opposed to depositionally, therefore belong to this pedocomplex).

**Layer 4** (base) is a foamy gravelly loam, resembling the upper layer of a gelifluction or cryoturbated soil (van Vliet-Lanoë 1983). A complex secondary clay illuviation affected the pores, and is slightly perturbated by subsequent frost action. Following deposition, silty illuviation took place ('melting water' illuviation). It is followed by coarse, humic reddish brown clay illuviation probably deriving from the leaching of the archaeological layer 5. A later yellowish brown illuviation, from layer 6, filled up the remaining pores at the surfaces of the aggregates.

**Layer 5** is a bleached loam affected by weak freeze-thaw organisation, with reworking of bone and granite fragments. The bones are 'fresh' and burnt. It is possibly colluvial.

**Layer 6.1** is a loamy solifluction layer, somewhat cryoturbated and affected by traces of ice lenses. It has been secondarily illuviated by one generation of fine, rather thick yellowish brown clay (Eem *sensu stricto*); see Fig. 9.3A. Biological activity occurred during the illuviation and weak hydromorphy developed before a later frost disturbance.

A MAJOR UNCONFORMITY OCCURS AT THIS POINT IN THE SEQUENCE.

**Layer 6.2** is a geliflucted gravelly loam, rich in granitic fragments. It is a complex soil horizon. An old generation of brownish clay illuviation has been dislocated; it is followed by another, yellowish brown, clay illuviation, *in situ* but fissured.

**Layer 7.1** is the beach complex. Above the lower unit of pebbles and marine sand (not examined in thin section) the 1982 excavations revealed a series of deposits derived largely from layer H (Fig. 9.3):

'H₁'. This was the lowest part of the exposure of H-derived deposits in the deep sounding. Resting on the lower unit of the beach, with numerous air pockets, is a mixture of H, marine sand and some burnt bone fragments (no charcoal) derived from the Eemian cliff — layers G–A, and 5. The microfabric shows that this material has been geliflucted (van Vliet 1982).

'H₂'. On top of 'H₁' a thin layer rich in large fragments of wood charcoal and pellets of baked clay, but *without* any burned bones, has been spread by frost-creep action, or even gelifluction in its lower part. It reveals the former existence of a hearth, probably in more boreal conditions.

'H₃'. During or even shortly after the dispersal of the hearth, and still under cold conditions, new lobes of gelifluction crept over the earlier deposits; they contain large quantities of ash and charcoal fragments (wood *and* bone).

'H₄'. A quieter period is represented by a layer described during excavation as 'yellow loess' — a finely stratified, silty and/or sandy colluvial deposit (with some splash effects). It is the result of gentle washing of the upper cliff (layers 3–6), with an admixture of bone fragments. Syngenetic frost activity was apparent in some thin sections.

'H₅'. A violent storm, probably implying renewed marine transgression, caused a collapse of the cliff protecting the ashy layer 'H₃' (sample 1406). The slumped material was incorporated and partly washed into the storm bar (the upper unit of the beach), as attested by rounded fragments of bone. Afterwards, the beach and the sloping surface left by the storm were somewhat affected by (chiefly) frost-creep before being buried under the gelifluction deposits of unit 7.2.

This sequence reveals a progressive cooling of the climate with only a slight lowering of the sea-level. Towards the end the sea was able to penetrate the cave during storms. Furthermore, in the light of the above, the occurrence of 'arctic' faunal remains (e.g. reindeer) mixed into the temperate 'H' is more easily understood.

**Layer 7.2.** Freeze-thaw features were observed in the deposits banked against the cliff. Diagenesis of the material in 7.2 and the preceding layers (marine and derived) is complex — amorphous clay occupies an important proportion of the pore space. In this it is rather similar to that of layer 8.3. Phosphate epigenesis first developed principally with the growth of small pearls or intense 'marbling' of crandallite and some apatite (acidular phosphate) on large crandallite nodules which have probably been reworked from layers C and B. This took place after the initial cliff collapse — and *pro parte* before units 8.3 and 6.2. Hydromorphic evolution is suggested by the occurrence of vivianite. Fine clay illuviation is observed locally after this, probably arising from the pedogenesis of layers 6.2 and 9. Later humic illuviation, associated with small diatoms, represents infiltration from layer 8.3 after dry periods (dessication cracks).

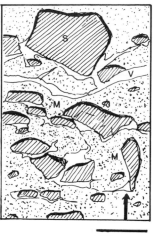

**Fig. 9.3** La Cotte de St. Brelade, layer 'H' (geliflucted): structure due to soil frost-creep. Note silt cappings on the granitic fragments and pores left by ice segregation below them. M, soil matrix; S, skeleton; V, void.

**Layer 8.1.** Not examined in thin section.

**Layer 8.2.** This is a gyttja (peaty mud) with large amounts of organic fragments and root debris.

**Layer 8.3** is a stratified clay loam, very rich in biological opal (up to 50%), laid down in a temporary pond. The whole of the layer corresponding to Burdo's (1960) "white clay" is a succession of sediments which are more or less rich in opal, probably related to discrete runoff cascading down from the plateau along a diaclase of the granite. During calm wet periods phytoliths and diatoms were abundant (Fig. 9.2F); during dry periods turbations were produced by biological activity; cracks developed, extending to the units underneath and allowing further infiltration to take place.

The predominance of (grass) phytoliths over other biological opals, and their clustering in 'life' positions or in clumps, suggests a *local* formation. The small diatoms are intact, though larger species, like the sponge spiculae, are generally broken. It is therefore possible that a proportion of the biogenic opal is reworked from another swamp resting in a depression close to the edge of the plateau, as proposed by Burdo. This temporary pond probably functioned as a 'sump' for the plateau. Shale pebbles recorded during the excavation were probably flushed down from older formations.

Some evidence as to the climatic conditions during sedimentation is provided by Burdo's record of pollen indicating mild temperate vegetation dominated by oak, *alder* and *hazel*. No traces of syngenetic frost action were observed in thin section. Stratigraphically, the deposit occurs below the ranker of layer 9.1. From a palaeopedological point of view it corresponds to the layer 9 soil formation, for which we propose a correlation with the St. Germain II Interglacial. The palynological data show, insofar as it is possible to compare La Grande Pile (Woillard 1974) with La Cotte de St. Brelade, that the conditions at the two sites were very similar; the other interglacials (St. Germain I and Eem *sensu stricto*) of the Last Interglacial complex have a quite different vegetation.

**Layer 9** as seen in the north face of the excavation is a geliflucted gravelly loam, rich in granitic fragments. A single generation of broken clay illuviation occurs at the surface of the aggregates. Accumulation of layer 9 *sensu lato* was accompanied by occasional ranker formation (e.g. 9.1 and 9.3) and erosional unconformities.

### 9.4.4 Upper Cold Complex

This comprises the heads which were excavated before World War II, and could not be studied in detail. But near the top of layer 11 in the northern entrance there occurs, locally, a *sol brun*, weakly *lessivé* with a single fine clay illuviation. In its upper horizons, developed in a loamy head, a gelifluction microstructure is already apparent; the top-soil has been displaced and partly reworked by frost-creep. Burnt bones and charcoal fragments show that a fireplace existed here *before* this soil development took place.

## 9.5 THE STRATIGRAPHIC CONTRIBUTION OF PALAEOPEDOLOGY AND MICROMORPHOLOGY

In this profile, units H–6 represent an ancient cycle ranging from temperate to periglacial climatic conditions. The long temperate period at the base of the sequence, very likely involving deposition of the original layer H, probably corresponds to an interglacial complex within the Saale glacial, with at least one long stabilisation of the soil surface after deposition of layer E.

The original layer H, which we are assuming exists at the back of the Eemian cliff, begins this mild period with a lacustrine (?) deposit, probably behind a pebble bar (beach). This would imply, as at Ecalgrain in the northern Cotentin (Clet-Pellerin 1983), a sea-level which at least at that time was very close to that of Eem. Both sites depict a very progressive cooling of the climate and probably belong to the same interglacial.

The Eemian (*sensu stricto*) soil developed at the top of unit 6.1. During Eem the sea attacked the deposit, forming a cliff. At the end of this period periglacial slumping occurred, with sporadic human occupation (the hearth represented by layer 'H2') at the time of the upper unit of the beach. Human occupation occurs in a comparable position in the northern Cotentin at St. Vaast-la-Hougue, excavated by G. Fosse, and at Port-Racine, excavated by D. Cliquet (Fosse 1982). Layer 'H' was formed with reworking of older layers in a continuing cold and humid environment (gelifluction), and was partly washed away and integrated in the upper beach material during a storm — this implies that the sea-level was rather close to that of Eem (see chapter 7).

Two other phases with pedogenesis of *lessivé* type were observed at the top of new gelifluction deposits which reworked earlier units. During the last of these temperate periods a new pond developed in the lower part of the cave, dammed by a new storm bar or a major slump. The latter is much more likely in view of the attitude of the deposits.

This sequence of events is very similar to those observed elsewhere in western Europe — La Grand Pile (Woillard 1974), Harmignies (van Vliet 1975) — and even in eastern Europe. In the Cotentin, a close analogy is provided by the succession of events observed in similar conditions at La Hague. The illuviation of layer 6.1 may correspond to the Eem soil *sensu stricto*, that of layer 6.2 to the St. Germain I and Villers St. Ghistain soils, layer 9 to St. Germain II and Malplaquet (Haesaerts and van Vliet 1981).

# 10

# PHOSPHATIC MINERALISATION

P. Giresse and B. van Vliet-Lanoë

In the La Cotte sequence, several phosphatic mineralisations (phosphorites) have been recognised and identified by X-ray spectrometry. They occur both in the form of grains or nodules in the heavy fraction of the sands (apatites, crandallites, millisite) and as small crystallites associated with the colloidal (less than $2~\mu m$) fraction (millisite, crandallite and taranakite). In addition, several powdered samples of the otherwise untreated sediment were analysed in order to determine the relative abundance of different phosphatic minerals irrespective of grain size. The results are given in Fig. 8.1 and tabulated in microfiche (mf 8.4).

**Taranakite**, $(K,NH_4)_2Al_6(PO)_6(OH)_2.18H_2O$, is a slightly evolved hydrated phosphate formed from the guano of birds or bats. The latter are less likely at La Cotte in view of the morphology of the site and the insular or peninsular character of Jersey during interglacials. This mineral occurs at the base of layer G and in layer A.

**Hydroxyapatite** derives directly from the bones, appearing in the form of pellets which are yellow, whitish or grey in colour (these last have been calcined). Where calcination is intense, the loss of $CO_2$ from the crystalline structure leads to the formation of fluorapatite. On the other hand, according to micromorphological observations, thermal caramelisation of the bones seems to inhibit their alteration and thus the liberation of phosphorus.

**Millisite**, $(Na,K)CaAl_6(PO_4)_4(OH)_9.3H_2O$, is an alkaline phosphate characteristic of gentle alteration. Its presence is consistent with leaching involving guano (P, K, Na) and the ashes of higher plants (K, Mg, Ca) from the archaeological layers.

**Crandallite** is another phosphate of aluminium and is quite frequently recorded as an alteration product of apatite in tropical zones (Flicoteaux et al. 1977). Its formula, $CaAl_3(OH)_6[PO_3O_{0,5}(OH)_{0,5}]_2$ as recently given by Blount (1974), no longer includes water but only OH ions. It is thus a more calcic phosphate than is millisite, and hence more stable (Peneaud 1978).

Phosphatogenesis seems to be connected with the evolution of the human contribution, ashes and bones, in view of the essentially aluminosilicate nature of the enclosing walls of the site. There are several possible sources for the calcium in the apatites: ashes, synsedimentary or anthropogenetic calcareous components (gastropods, shellfish, egg shells) or even loess (though now decalcified). Equally, hydrolysis of the calcic plagioclase felspar inclusions in the walls could have provided far from negligible quantities of calcium. The origin of the phosphorus is no less complex; it could derive from the bones, from guano or from organic phosphorus contained in the ashes, as noted by Courty (1982).

In this granite-walled fissure the percolating water was never sufficiently charged with calcium to permit genesis of montgomeryite, such as took place in the shelter of Caune de l'Arago at Tautavel (Courty 1982). On the other hand, it was not acidic enough to be outside the range of conditions favouring apatite and so permit the formation of wavellite, which is characteristic of the more intense hydrolyses occurring in a tropical environment (Flicoteaux et al. 1977; Giresse 1980). Co-occurrence of aluminous and calcic phosphatic mineralisations is common in the upper layers of sites containing phosphates. Here the physical and chemical characteristics of the layers allow several phases of mineralogenesis to be envisaged.

The phase with crandallite on its own is the best defined, seeming to correspond to the formation of the last interglacial pedocomplex (layers 7–9) and a lowering of pH to below 5.5 — that is to say in an environment of calcium deficit. However, the increase in pH observed today in the humus of layer 9.3 (mf 8.6) implies a state of neutrality untypical of this normally acid material, as a result of Ca resaturation. It is therefore very probable that the overlying Younger Loesses were at least weakly carbonated at the time of their deposition. This is confirmed by observations made during fluorescent microscopy, which show no significant free aluminium in the soil matrix (van Vliet-Lanoë 1980).

In the archaeological layers A–D, the simultaneous presence of calcic and aluminous phosphates was noted. The latter represents more or less advanced stages in the alteration of bones and other apatites at the interstitial level, as the phosphatisation permeated all the adsorbent surfaces of the clays.

As for the beds underlying the palaeosol E, it is very likely that the situation was similar to that existing below the Last Interglacial pedocomplex; however, it should be noted that millisite and taranakite also occur at the base of layer G, 1.5 m below layer E. This may lead one to suppose that the earlier palaeosol developed under a climate which was perhaps a little drier, and over a briefer lapse of time, than pertained during the sequence dating to the Last Interglacial.

# THE FLORA AND VEGETATION OF LA COTTE DE ST BRELADE AND ITS ENVIRONS

R. L. Jones

## 11.1 INTRODUCTION

The examination of fossil plant material traditionally plays an important role in archaeological research projects. Fossil plant remains occur mainly in sedimentary rocks, and may be either macroscopic (fruits, seeds, leaves, stems), or microscopic (pollen grains of higher plants, spores of lower plants). The aim of such investigation is to ascertain firstly the component plants (the flora), and secondly their type of cover (the vegetation) present at and near the archaeological site. In an archaeological context, studies of flora and vegetation can provide clues to the diet and economy of the people. They also allow inferences concerning other aspects of the natural environment of relevance to man (e.g. climate, soil types), and, when combined with studies of landforms and the physical and chemical characteristics of sediments, enable the reconstruction of past landscapes to be undertaken. The resource value of past landscapes, together with their specific elements of importance to early man, are of major significance to environmental archaeologists (Davidson, Jones and Renfrew 1976).

## 11.2 PALAEOBOTANICAL RESEARCH AT LA COTTE

A number of palaeobotanists have analysed the La Cotte sediments in conjunction with the archaeological excavations of Burdo, McBurney and Callow. All the samples have come from the site itself, rather than from an adjacent locality (e.g. a peat bog or lake) rich in plant fossils and of equivalent age. The majority of the investigations have been of pollen and spores (as opposed to macrofossils). Material from Burdo's excavations was submitted to C. Dubois (Burdo's layers [2A], [3B] and [6], now renumbered 9, 8.3 and 8.2) and to Arl. Leroi-Gourhan (Burdo's layers [7A] and [8], now renumbered 8.1 and H). The results of these investigations were published by Burdo (1960) and Leroi-Gourhan (1961). Samples obtained by Campbell in 1967 from layers 3, 5, 6.1 and 6.2 were analysed by Campbell and Pohl; their results were published in McBurney and Callow (1971), where they were discussed, with the earlier work, in the context of the stratigraphy as it was understood at that time. Samples from excavations by McBurney in 1972, 1973 and 1978 and by Callow in 1980 were examined by W. P. Seymour under the direction of the author, and are re-ported here for the first time.

As the account which follows is intended to synthesise the sum of palaeobotanical data from the site, a résumé of previously published results will not be given at this point. Rather, where practicable, these have been integrated with Seymour's data, both in the form of a composite pollen and spore diagram and a table of pollen and spore counts. Such integration required the recalculation of the raw data of earlier workers to fit the format used by Seymour, so that almost all the palynological evidence derived from the sedimentary sequence could be presented here and discussed in an overall context.

Plant macrofossils are scarce, but there are some early records (Nicolle and Sinel 1912; Burdo 1960), as well as later finds of wood charcoal (Cartwright, chapter 12, this volume), which add to our knowledge of the flora and vegetation.

## 11.3 METHODOLOGY

Detailed discussion of the methodology of Quaternary palaeobotany and palaeoecology can be found, for example, in Faegri and Iversen (1975), West (1977), Jones and Cundill (1978) and Birks and Birks (1980).

The sediments at La Cotte are in general highly inorganic. The various workers have employed a number of different techniques to extract pollen and spores. Campbell and Pohl used the Thoulet solution flotation/filtration method developed by Frenzel (1964) to obtain pollen from loess. Seymour carried out sodium pyrophosphate treatment to remove clay particles prior to the digestion of silicates with hydrofluoric acid (Bates, Coxon and Gibbard 1978). Save for one layer ([7A], i.e. 8.1) analysed by Leroi-Gourhan, the concentration of pollen and spores was very low. Indeed, even after Seymour's exhaustive treatment, involving multiple preparations and slide counts, some layers yielded none at all.

The number of pollen grains which should be counted per sample in order to establish statistical reliability has been a matter of some debate, as has the composition of the pollen sum (see, for example, Dimbleby 1957; Faegri and Iversen 1975; Birks and Birks 1980). The lowest figure for reliability seems to be 250, though some

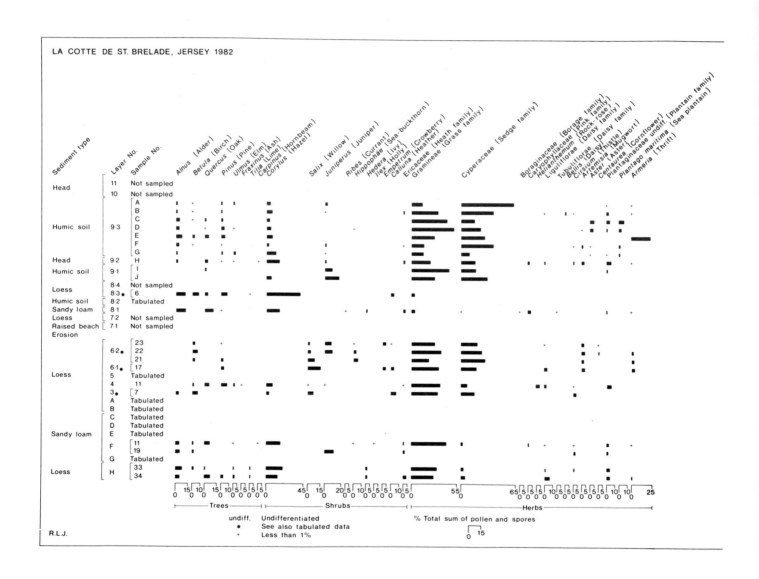

Fig. 11.1 La Côte de St. Brelade: pollen diagram.

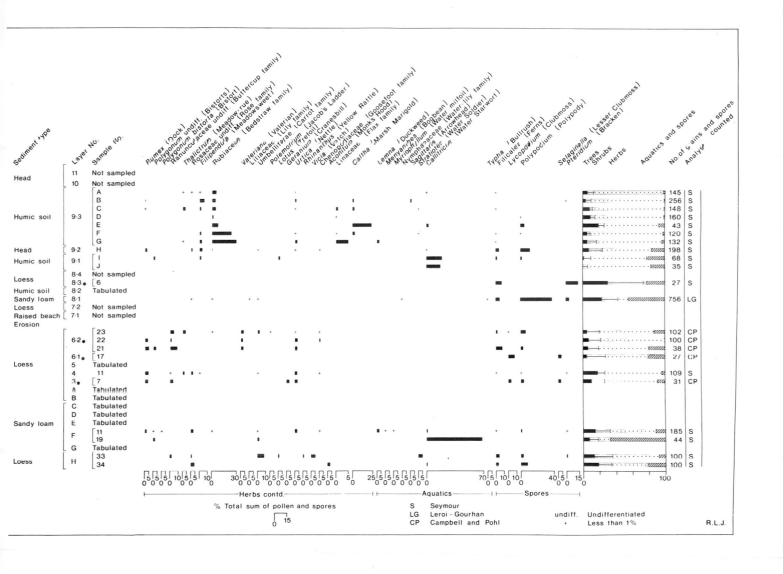

| Layer No. | Sample No. | Betula | Pinus | Quercus | Alnus | Ulmus | Carpinus | Corylus | Salix | Juniperus | Rhamnus (Alder Buckthorn) | Empetrum | Calluna | Ericaceae | Gramineae | Cyperaceae | Artemisia | Taraxacum (Dandelion) | Cirsium | Centaurea | Chenopodiaceae | Ranunculaceae undiff. | Armeria | Thalictrum | Rumex | Urtica | Gentiana (Gentian) | Plantaginaceae | Potentilla (Cinquefoil) | Rubiaceae | Valeriana | Epilobium (Rose-bay Willowherb) | Polemonium | Succisa (Devil's-bit Scabious) | Umbelliferae | Callitriche | Selaginella | Botrychium (Moonwort) | Polypodium | Pteridium | Lycopodium | Filicales | Total no. of grains and spores | Analyst |
|---|---|---|---|---|---|---|---|---|---|---|---|---|---|---|---|---|---|---|---|---|---|---|---|---|---|---|---|---|---|---|---|---|---|---|---|---|---|---|---|---|---|---|---|---|
| 8·3 | 8 | | | | | | | | | | | | | | No pollen and spores | | | | | | | | | | | | | | | | | | | | | | | | | | | | | S |
| | 7 | | | | | | | | | | | | | | No pollen and spores | | | | | | | | | | | | | | | | | | | | | | | | | | | | | S |
| | 5 | | | | | | | | | | | | | | 2 | | | | | | | | | | | | | | | | | | | | | | | | | | | | 2 | S |
| | 4 | 1 | | | | | | | | | | | | | 2 | 1 | | | | | | | | | | | | | | | | | | | | | | | | | | | 4 | S |
| | 3 | | | | | | | | | | | | | | No pollen and spores | | | | | | | | | | | | | | | | | | | | | | | | | | | | | S |
| | 2 | | | | | | | | | | | | | | No pollen and spores | | | | | | | | | | | | | | | | | | | | | | | | | | | | | S |
| 8·2 | 14 | | 1 | | | | | | | | | | | | | | | | | | | | | | | | | | | | | | | | | | | | | | | | 1 | S |
| | 13 | | 1 | | | | | | | | | | | | | | 3 | | | | | | | | | | | | | | | | | | | | | | | | | | 4 | S |
| | 12 | | | | | | | | | | | | | | | | | | | | | | | | | | | | | | | | | | | | | | | | | 1 | 1 | S |
| | 11 | | | | 1 | | | | | | | | | | | | 1 | | | | | | | | | | | | | | | | | | | | | | | | | | 2 | S |
| | 10 | | | | | | | | | | | | | | No pollen and spores | | | | | | | | | | | | | | | | | | | | | | | | | | | | | S |
| | 9 | | | | | | | | | | | | | | No pollen and spores | | | | | | | | | | | | | | | | | | | | | | | | | | | | | S |
| 6·2 | 20 | 3 | | | | | | 1 | | 1 | | | | | 6 | 5 | 1 | | | | | | | | | | | | 1 | | | | | | | | | | | 3 | | 4 | 25 | CP |
| | 19 | | 6 | | | | | | | | | | | | 6 | 3 | | | | | 2 | | | | | | | | 2 | | | | | | | | | | | | 1 | 2 | 22 | CP |
| 6·1 | 18 | 1 | | | | | | 3 | | | | | | | 2 | 1 | 1 | | | | | | | | | | | | 1 | | 2 | | | | | | | | | 1 | 1 | | 1 | 14 | CP |
| | 16 | 1 | | | | | | | | | | | | | 5 | 3 | | | 1 | | 1 | | | | | | | | 1 | | 2 | | | | | | | | | 1 | 1 | | 3 | 19 | CP |
| | 15 | 1 | | | | | | | | | | | | | 1 | 2 | | | | | | | | | | | | | | | 2 | | | | | | | | | 1 | 1 | | | 8 | CP |
| | 14 | | | | | | | | 1 | | | | | | | | 1 | | | | | | | | | | | | | | | 3 | | 1 | | | | | | | | | 6 | CP |
| | 13 | | | | | | | | | | | | | | 1 | | | | | | | | | | | | | | | | | | | | | | | | | | | | 1 | CP |
| | 12 | | | | | | | 1 | | 1 | | | | | 3 | 2 | 1 | | | | 1 | | | | | | | | | | | | | | | | | | | | | 1 | 10 | CP |
| | 11 | 1 | | | | | | | | 1 | | | | | 2 | 1 | | | | | | | | | | | | | | | | | | 1 | | | | | | | | 6 | CP |
| 5 | 10 | | | | | | | | | | | | | | | | | 1 | | | | | | | | | | | | | 1 | | | | | | | | | | | | 2 | CP |
| | 9 | | | | | | | | | | | | | | 1 | 1 | | | | | | | | | | | | | | | | | | | | | | | | | | | 2 | CP |
| 3 | 8 | 1 | 1 | | 1 | | | | | | | | | | | | 1 | | | | | | | | | | | | | | | | | | | | | | | | | | 4 | CP |
| | 6 | | | | | | | | | | | | | | 4 | 3 | 1 | | | | 2 | | | | | | | | | | 2 | | | | | | | | | | | | 12 | CP |
| | 5 | | 1 | | | | | 1 | 5 | | | | | | 1 | 1 | | | 1 | 1 | | 2 | 1 | | | | | | | | | | | | | | | | | | | | 14 | CP |
| | 4 | 1 | | | | | | 1 | 4 | | | | | | 5 | 5 | 1 | | | | 1 | 1 | 1 | | 1 | | | | | | 1 | | 1 | | | | | | | | | | 23 | CP |
| | 3 | | | 1 | | | | | | | | | | | 1 | | | | | | | | | | | | | | | | | | | | | | | | | | | | 2 | CP |
| | 2 | | | | | | | | | | | | | | | | | | | | | | | | | 1 | | | | | | | | | | | | | | | | | 1 | CP |
| | 1 | | 3 | | | | | | | 3 | | | | | 3 | 3 | 1 | 2 | | | | | | | 1 | | | | | | | | | | | | | | | 1 | | 2 | 19 | CP |
| A | 1 | | | | | | | | | | | | | | No pollen and spores | | | | | | | | | | | | | | | | | | | | | | | | | | | | | S |
| B | 4 | | | | | | | | | | | | | | No pollen and spores | | | | | | | | | | | | | | | | | | | | | | | | | | | | | S |
| C | 6 | | | 1 | | | | | | | | | | | | | | | | | | | | | | | | | | | | | | | | | | | | | | | 1 | S |
| D | 8 | 1 | 2 | | | | | | 1 | | | | | | 2 | | | | | | 1 | | | | | | | | | | | | | | | | | | | | | | 7 | S |
| E | 10 | 1 | 1 | | | | | 1 | 1 | | | | | | 5 | | 1 | | | | | | | | | | | | | | | | | | | | | | | | | | 10 | S |
| G | 18 | 1 | | | | | | | | | | | | | 1 | | | | | | | | | | | | | | | | | | | | 1 | | | | | | | | 3 | S |
| | 17 | | 1 | | | | | 1 | 1 | 2 | | | | | 7 | 2 | | | | | 1 | | | | 1 | | | | | | | | | | 1 | 1 | | | | 1 | | 2 | 22 | S |
| | 16 | | | | | | | 1 | | | | | | | 7 | 8 | | | | | | | | | | | | | 1 | | | | | | | | 5 | | | | | | 22 | S |
| | 14 | 1 | 1 | 1 | | | | | 1 | 2 | | | | | 5 | 4 | | | | | | | | | | | 1 | 1 | | | | | | | | | 1 | | | | | | 18 | S |
| | 15 | | | | | | | | | | | | | | 2 | 2 | | | | | | | | | | | | | | | | | | | | | 1 | | | | | | 5 | S |

| S | Seymour | | | | | | | | | undiff. Undifferentiated |
| CP | Campbell and Pohl | | | | | | | | | R.L.J. |

**Fig. 11.2** La Cotte de St. Brelade: samples poor in pollen and spores.

workers suggest a count of 500 or more grains of land plants per sample. If 250 is adopted, all counts but two from La Cotte are unacceptable; if 500, all but one. Nevertheless, in view of the archaeological importance of the site, and in the absence in the vicinity of any alternative, more conventional locality yielding fossil material, a pollen sum of less than 250 was employed. All samples containing over 25 grains and spores were used to calculate the percentage of each taxon in relation to the total number of pollen and spores. The results were plotted as a pollen and spore diagram. Samples with less than 25 grains and spores were tabulated numerically.

On statistical grounds alone, then, it is evident that the interpretation of almost the entire pollen and spore sequence must be extremely tentative. Furthermore, there is the problematical depositional history of the site, which may also in itself account for the paucity of grains recovered. The complexity of stratigraphy and sedimentation in caves is well known (see, for example, Sutcliffe and Zeuner 1962; Campbell 1977) and La Cotte is no exception. The reworking of older deposits and their incorporation into others will lead to a mixing of former and contemporary microfossils. Erosion and weathering will leave gaps in the sedimentary sequence, and consequently in the pollen and spore record also.

Redeposited pollen and spores sometimes exhibit signs of deterioration, a condition which hinders their identification, but can serve to distinguish them from contemporary taxa in the sediment. Some grains from La Cotte have undoubtedly been redeposited. Pollen and spores are susceptible to oxidation, which leads to their corrosion. Such corrosion is differential, and is prevalent at aerobic sites. The paucity and condition of the pollen and spores at La Cotte suggest that the sediments accumulated under predominantly aerobic conditions.

The secluded position of the cave in the promontory would have been an additional factor operating against its receiving a representative sample of the pollen and spore 'rain' from its catchment area. At La Cotte, the geomorphological situation of the fossil cliff would affect wind currents, as the sometimes near-coastal location of the site may also have done. Wind is usually the principal transporting agent of most varieties of pollen and spores, and some bias in their source of supply is therefore to be expected.

Finally, a limitation of another sort must be borne in mind. The data from La Cotte are based on analyses carried out by a number of workers, and, recent advances in technique and approach nothwithstanding, the interpretation of pollen and spore assemblages is still largely subjective. In view of all these considerations, it was thought wiser not to attempt to delimit a series of assemblage biozones (West 1977) in order to characterise changes in the pollen and spore content of the diagram. Rather, as was accomplished in McBurney and Callow (1971), description of the data is based on the pollen and spore assemblages of the various archaeological layers. Where appropriate, macrofossil data are included. An attempt is made to identify and group similar pollen and spore assemblages from the sequence of layers, and in so doing to build up a picture of floral, vegetational and environmental change.

The nature of vegetation and the mechanisms of pollen and spore dispersal are such that a pollen and spore assemblage is usually drawn from a patchwork of plant communities. The latter include those at and immediately adjacent to the site of pollen sedimentation, and others at a greater distance — perhaps some tens or hundreds of kilometres away. As a consequence, it is hard to achieve an accurate reconstruction of individual communities in such a patchwork. However, provided that there are modern analogies and that the ecological tolerances of the plant taxa in question have not changed over time, pollen and spore assemblages may be interpreted in terms of major vegetation types with comparative ease. For example, the predominance of *Quercus*, *Ulmus*, *Tilia*, *Fraxinus* and *Alnus* pollen indicates the former existence of temperate mixed deciduous forest; high values of *Pinus* and/or *Betula* pollen signify boreal forest. Abundant dwarf-shrub pollens (of Ericaceae, for example) characterise heathland, while those of taller shrubs (such as *Corylus*) often dominate scrub. Grass pollen can frequently be grouped with that of other herbs (for example, *Plantago lanceolata* and *Rumex acetosa*) to denote grassy sward. Pollen of Cyperaceae often has its provenance in terrestrial wetland, while records of aquatic taxa (*Myriophyllum* and *Typha*, for instance) enable aspects of this environment to be reconstructed.

Such reconstructions allow inferences concerning broad environmental (especially climatic) parameters of the past. If, however, a pollen and spore assemblage contains a species which requires particular ecological conditions, and, moreover, is today invariably partnered by a number of other plants, firmer and more confident assertions pertaining to past plant communities and environmental factors may be made. For example, *Hedera*, a woodland inhabitant, will not tolerate a mean temperature for the coldest month of less than -1.5°C, its fossil presence enabling former thermal conditions to be determined (Iversen 1944). Likewise, *Juniperus* and *Corylus* are shrubs which dislike shade, flowering best in the open. The presence of their pollen in considerable quantity is an indication that open areas existed in the vegetation mosaic.

## 11.4 THE POLLEN AND SPORE ASSEMBLAGES

### 11.4.1 Layers H–C

*DESCRIPTION*

**Layer H** (Fig. 11.1). The pollen and spore spectra of the two stratified samples from this loessic layer (Burdo's layer [8]) analysed by Seymour were very similar to that of the sample examined by Leroi-Gourhan, whose results are not included in Fig. 11.1 because the exact provenance of the sample within layer H was uncertain. Seymour's data show that tree pollen accounts for 18% and 13% of total pollen and spores in samples 34 and 33, the corresponding figures for shrub pollen being 16% and 22%. The combined analyses record *Quercus*, *Alnus*, *Ulmus*, *Tilia*, *Betula* and *Pinus* pollen, together with that of *Corylus*, *Hedera*, and Ericaceae. Herb pollen constitutes 54% and 57% of the total pollen and spores in Seymour's analyses. Gramineae, Cyperaceae, Liguliflorae, *Plantago maritima*, Rosaceae, Umbelliferae and Chenopodiaceae are important amongst herb pollens, while *Callitriche* and *Stratiotes* represent aquatic taxa. Fern spores present are of *Polypodium*, *Pteridium* and Filicales. The latter is a group which, following the observations of Andersen (1961), includes *Thelypteris* (Mountain, Marsh and Beech ferns), *Asplenium* (Spleenwort), *Blechnum* (Hard-fern), *Athyrium* (Lady-fern), *Cystopteris* (Brittle Bladder-fern), *Woodsia* (Alpine Woodsia), *Dryopteris* (Male Fern, Buckler-fern), *Polystichium* (Shield-fern) and *Phyllitis* (Hart's-tongue fern).

**Layer G** (Fig. 11.2). Five stratified samples were analysed from this sandy layer in which pollen and spores were very sparse. Tree pollen comprises that of *Betula*, *Pinus*, *Quercus*, *Alnus* and *Carpinus*. Shrubs are represented by *Corylus* pollen (and macrofossil charcoal), with *Salix* and *Juniperus* in sample 3 and Ericaceae in sample 6. *Callitriche* pollen is consistently recorded, and *Polypodium*, *Pteridium* and Filicales spores occur. Herb pollens are dominated by Gramineae and Cyperaceae, with Ranunculaceae, Plantaginaceae, *Rumex* and *Urtica* also present.

**Layer F** (Fig. 11.1). Two stratified samples were analysed from this layer of sandy loam. Sample 19 includes *Alnus*, *Betula*, *Juniperus* and Ericaceae, and has a peak of *Callitriche* pollen. The upper sample (11) has pollen of *Alnus*, *Betula*, *Quercus* (also macrofossil charcoal), *Ulmus*, *Carpinus*, *Corylus*, *Ilex* and *Hippophaë*. Tree pollen acounts for 14% of the total pollen and spores, shrubs for 18% and herbs for 61%, the latter dominated by Gramineae, Caryophyllaceae, Plantaginaceae, Rosaceae, *Rumex* and *Urtica*. Amongst aquatic pollens, *Callitriche*, *Lemna*, *Myriophyllum* and *Menyanthes* are recorded. *Polypodium* and Filicales spores occur.

103

**Layers E, D and C.** The sandy loams of E, D and C (E having the morphological characteristics of a palaeosol horizon — see van Vliet-Lanoë, chapter 9, this volume) were poor in pollen and spores. One sample from each layer was analysed. The overall assemblage is principally one of tree and shrub genera — *Betula, Pinus, Quercus, Ulmus, Corylus* and *Rhamnus*. Macrofossil charcoal of *Quercus* has been recorded from layer C (Cartwright, chapter 12, this volume).

## INTERPRETATION

The basal deposit, the loessic H, contains in its upper (sampled) part a temperate pollen and spore assemblage, as do the lowest three samples of layer G. The vegetation in the locality of La Cotte appears to have consisted of temperate mixed woodland with *Quercus, Alnus, Betula, Pinus, Tilia, Carpinus* (layer G), *Corylus* and *Hedera* (layer H) its major tree and shrub constituents, and maritime grass heath and marsh with Gramineae, Ericaceae, Cyperaceae, Liguliflorae, *Plantago maritima* and Chenopodiaceae important members. The ravines contained some open water with *Stratiotes* and *Callitriche*, and probably a swamp with tall herbs including Umbelliferae and Rosaceae.

A pollen assemblage such as this may belong to either an interglacial or interstadial stage of Quaternary time. If it is an interglacial that is represented here, a late stage within it is suggested by taxa such as *Carpinus* (Turner and West 1968). An interstadial is characterised by a briefer or lesser climatic amelioration than an interglacial. During the latter, climatic conditions allow temperate deciduous forest to develop in an orderly sequence (West 1977). As Woillard (1978, 15) points out, however, an interstadial, a period either too short or too cold to permit the development of temperate deciduous forest of interglacial type (West 1977), will allow a thermophilous plant assemblage if refugia for such taxa are close enough to the area in question. She further notes that ecological succession in an interstadial, as conditioned by a short-lived and moderate climatic amelioration, means that thermophiles do not immigrate from their refugia in successive well-ordered waves as they do in an interglacial. The sudden and disorderly appearance of thermophilous taxa in an interstadial is a function, according to Woillard, of the lack of competition that would exist in the landscape by comparison with that typical of an interglacial. Because of the problems outlined above, palynological arguments on their own are insufficient to establish whether the spectra should be assigned to an interglacial or to an interstadial, though, if the latter, it must be defined as such by a short duration (indeterminable here) rather than by insufficient warmth, in view of the presence of, for example, *Hedera*.

The upper samples of layer G and the lower one of the sandy loam F can be grouped, in that their pollen and spore assemblages portray a more open landscape than those of the lower units. This landscape seems to have included heath with *Betula* and Ericaceae prominent members, and may be indicative of cold conditions, perhaps of a steppic nature. The high percentage of *Callitriche* in sample 19 (layer F) probably reflects the local presence of this aquatic in a pond in the ravine.

The upper sample of layer F and those from the sandy loams E, D and C exhibit temperate pollen and spore spectra. The overall assemblage suggests thermophilous deciduous forest where *Alnus, Betula, Ulmus, Quercus, Carpinus, Corylus, Ilex* and *Rhamnus* were present. Charcoal of *Quercus* in layers F and C supports the sometime existence of temperate trees, but it must be borne in mind that in an occupation complex such as this, with its problems of reworked sediment, plant growth and firing may not have been contemporaneous. Indeed, it is possible that pollen from layer E may have been reworked into layers D and C.

As with the lower temperate episode, it is not possible to determine whether the episode just described should be referred to an interglacial or to an interstadial. However, the lack of open ground taxa in E, D and C may be indicative of a former closed forest environment, perhaps of interglacial rank. On the other hand, sample 11 (from the top of F) has a 60% herbaceous pollen component accompanying a similar tree and shrub flora.

### 11.4.2  Layers B–6.2

## DESCRIPTION

**Layers B and A.** These loessic layers were devoid of pollen and spores. However, Cartwright (chapter 12, this volume) has identified macrofossil charcoal of *Quercus* from layer A. Burdo (1960) reported comminuted charcoal from the same layer.

**Layer 3** (Fig. 11.1 and Fig. 11.2). Eight stratified samples were analysed from this loessic layer by Campbell and Pohl. In the one sample used in percentage computations (3.7) arboreal pollen accounts for 10%, shrub pollen 16% and herb pollen 65% of total pollen and spores. In general, the pollen and spore spectra of layer 3 are dominated by *Salix, Calluna*, Ericaceae, Gramineae, Cyperaceae, *Artemisia*, Compositae, Chenopodiaceae, Ranunculaceae, *Lycopodium* and *Botrychium*. There are sporadic occurrences of *Betula, Pinus, Ulmus* and *Corylus* pollen, the latter taxon especially in samples 4–8.

**Layer 4** (Fig. 11.1). One sample was analysed by Seymour from this weathered loessic layer, which was identified only locally, at the same level as (and probably equivalent to) the upper part of layer 3, analysed by Campbell and Pohl. Pollen of *Quercus, Ulmus, Fraxinus, Pinus* and *Betula* accounts for 10% of total pollen and spores, that of *Corylus, Salix* and *Juniperus* for 9% of this sum. Of the 70% herbs, Gramineae, Plantaginaceae, Rosaceae, Compositae, *Helianthemum, Urtica* and *Rumex* are most important.

**Layers 5, 6.1 and 6.2** (Fig. 11.1 and Fig. 11.2). Fifteen stratified samples from these loessic layers were analysed by Campbell and Pohl. Pollen and spores were scarce, but the flora until sample 16 layer 6.1 is dominated by *Salix, Empetrum*, Gramineae, Cyperaceae, *Artemisia, Epilobium, Botrychium* and *Selaginella*. Arboreal pollen is less than 6% of the total pollen and spores throughout this sequence. After sample 16 (layer 6.1) and through layer 6.2, *Juniperus, Hippophaë, Betula, Pinus, Salix* and *Calluna* assume importance in the pollen spectra.

## INTERPRETATION

The loessic layers B and A are devoid of pollen and spores, which may be due to one or more unfavourable depositional factors, the destruction of microfossil content after their incorporation, or the lack of vegetation in the pollen catchment area of the site. There does not seem to be evidence of a gradual cooling in the preceding layers (E, D and C), and a time of climatic severity may have been associated with a lack of vegetation in a mobile landscape dominated by solifluction and loess accumulation.

The loessic layer 3 has pollen spectra dominated by shrubs and herbs, especially *Salix*, Ericaceae, Gramineae, Cyperaceae and Compositae, indicative of a steppe-tundra flora, according to Campbell and Pohl (1971), and a landscape where periglacial activity (frost shattering, solifluction) was accompanied by loess formation. Campbell and Pohl suggest that the upper samples of layer 3 hint at a climatic amelioration towards boreal conditions.

Layer 4 is also loessic but shows signs of having incorporated in it a reworked soil (van Vliet-Lanoë, chapter 9, this volume). Its pollen content is indicative of a temperate environment with *Quercus*, *Ulmus* and *Fraxinus* accompanying *Betula*, *Pinus* and *Corylus* in woodland. There is a high herbaceous pollen count with Gramineae, Compositae, Rosaceae and Plantaginaceae most prominent, suggesting the presence of maritime grassland nearby. An interglacial or interstadial stage is possible.

For much of the time represented by the loessic layers 5, 6.1 and 6.2, the pollen and spore record, according to Campbell and Pohl, points to the existence of a steppe-tundra environment around La Cotte. Here, Gramineae, Cyperaceae, *Salix*, *Empetrum*, *Artemisia*, *Epilobium*, *Selaginella* and *Botrychium* were present. Trees were lacking and there was much open and disturbed ground. As with layer 3, the upper samples of layer 6.2 suggest a slight climatic amelioration. *Juniperus*, *Hippophaë*, *Betula* and *Pinus* now became established in a boreal environment.

### 11.4.3 Layers 7.1–11

#### DESCRIPTION

**Layer 7.1** (Fig. 11.1). This layer is composed of beach material which was unsuitable for palynological work

**Layers 7.2 and 8.1** (Fig. 11.1). Layer 7.2 is loessic — formed by collapse of the Eemian cliff — and was not originally sampled for pollen and spores, but material obtained by Callow during his excavations of 1980 is sterile. Layer 8.1 (Burdo's layer [7A]) is a sandy loam, of which the palynology of one sample was studied by Leroi-Gourhan. Arboreal pollen accounts for 23% and shrub pollen 20% of total pollen and spores. *Alnus*, *Quercus*, *Pinus* and *Corylus* are the best represented trees and shrubs, with *Ilex*, *Ribes* and Ericaceae also present. Herbs make up only 13% of total pollen and spores and consist principally of Gramineae, Caryophyllaceae and Plantaginaceae. Spore values — notably of *Polypodium* (40% of total pollen and spores), *Pteridium* and Filicales — are high. Aquatic pollen is represented by *Typha* and *Nymphaea*.

**Layer 8.2** (Fig. 11.2). Seymour analysed six stratified samples from this humic soil. All were poor in pollen and spore content, with numbers 9 and 10 being barren. The only pollen recorded was that of *Quercus*, *Alnus* and Gramineae, together with Filicales spores. Dubois also analysed one sample ([6]) from this layer and obtained pollen and spores of *Alnus*, *Quercus*, *Corylus*, Ericaceae and *Athyrium*.

**Layer 8.3** (Figs. 11.1 and 11.2). Seven stratified samples were analysed by Seymour from this loessic layer. Four were devoid of pollen and spores, two contained sparse numbers and only sample 6 was quantifiable. Here arboreal pollen comprises 30%, shrub pollen 44% and herb pollen 4% of total pollen and spores, the remaining percentage being

of spores. *Alnus*, *Quercus*, *Betula*, *Pinus*, *Tilia*, *Corylus* and *Calluna* represent trees and shrubs, with Gramineae, *Pteridium* and Filicales the only other taxa present. Dubois also looked at one sample ([3B]) from this layer, with Chenopodiaceae and *Polypodium* recorded in addition to the taxa found by Seymour.

**Layer 8.4.** No samples were available.

**Layer 9** (Fig. 11.1). Near the junction of the north and west ravines, this complex consists of two humic soils (9.1 and 9.3) separated by a head deposit (9.2). Burdo secured a sample (his layer [2A]) for pollen analysis, accomplished by Dubois. As Burdo did not recognise a three-fold sequence in this horizon, the exact provenance of the sample from [2A] is unclear. McBurney and Callow (1971) assumed it to have come from layer 9.1, but the pollen spectra obtained by Seymour from 9.1 and 9.3 suggest that it belongs to the latter, and it is discussed in this context below.

**Layer 9.1** (Fig. 11.1). Two stratified samples (I and J) were analysed from this layer. The only arboreal pollen present is *Quercus* (1% of total pollen and spores in sample I). Shrub pollen is 6% of total pollen and spores in sample J and 9% in sample I. *Corylus* occurs in I but *Juniperus* is present in greater amount in both, where Gramineae and Cyperaceae comprise the bulk of the pollen flora. Plantaginaceae, *Polygonum bistorta* and *Geranium* are other herbs recorded, while *Callitriche* and *Typha* comprise the aquatic pollen.

**Layer 9.2** (Fig. 11.1). One sample (H) was analysed, and was found to contain *Alnus*, *Quercus*, *Ulmus*, *Carpinus*, *Pinus* and *Corylus*, together with *Juniperus*, Ericaceae, Gramineae, Cyperaceae and *Artemisia*.

**Layer 9.3** (Fig. 11.1). Seven stratified samples were analysed by Seymour from this layer, within which the pollen and spore spectra show some variation. Samples G–C have arboreal pollen values of 5–9% of total pollen and spores, except for E where this value reaches 19%, although this is to be equated with a low total of grains and spores and a very restricted taxal range. *Alnus*, *Betula*, *Quercus*, *Ulmus* and *Pinus* are recorded. *Corylus* is consistently present, with *Juniperus* and Ericaceae more sporadic. Amongst herbaceous pollens, Gramineae, Cyperaceae, Compositae, Plantaginaceae, Rosaceae, Linaceae, Ranunculaceae and, most notably, Rubiaceae are important. Dubois' sample [2A] seems to be consonant with this part of 9.3, as it contains *Quercus*, *Alnus*, *Pinus* and *Corylus*. Burdo reported hollow casts of grass leaves or conifer needles from layer [2A], and Nicolle and Sinel (1912) a branch with ivy clinging to it; both macrofossil records appear to belong to this stratigraphic horizon. Samples B and A have high levels of herbaceous pollen (89% and 85% of total pollen and spores respectively). *Alnus* and *Pinus* values decline and *Quercus* is not recorded. *Juniperus* and Ericaceae reappear and Cyperaceae pollen values increase. *Thalictrum* pollen is recorded.

**Layers 10 and above.** In view of their highly inorganic nature, these layers of sandy and loessic head were not sampled for pollen and spore analysis.

#### INTERPRETATION

No pollen and spore records exist for layer 7.1 (the raised beach at c. 8 m asl). Layer 8.4 lacks palynological data, and the slumped loess 7.2 proved to be sterile. However, layers 8.1 (sandy loam), 8.2 (humic soil) and 8.3 (loessic) have temperate pollen and spore assemblages. In 8.1 43% of total pollen and spores is accounted for by trees and shrubs. The presence of *Alnus*, *Betula*, *Pinus*, *Quercus*, *Tilia*, *Corylus* and *Ilex* indicates temperate mixed woodland at this stage. The high spore count for ferns is consistent with the view that certain of them (*Polypodium*, for example) may have inhabited woodland and perhaps grew in crevices and on ledges in the cave. Elsewhere, slopes and cliff tops were probably clothed by grassland with *Pteridium*. The existence of a relatively thermophilous freshwater regime nearby (if not within the cave system itself) is indicated by the presence of *Typha* and *Nymphaea* pollen.

As noted above, the total number of grains and spores (756) counted by Leroi-Gourhan from layer 8.1 should make its spectrum the most reliable of all the analysed levels at La Cotte. However, as McBurney and Callow (1971) point out, layer 8 may be redeposited so that some or most of the pollen content of 8.1 could be reworked. If reworking took place, some pollen and spores would have come from a parent deposit. If redeposition of this was rapid, the amount of contemporary pollen and spores incorporated into the reforming sediment would have been small. The pollen spectrum of 8.1 would in that case have been similar perhaps to the upper part of layer 6.2, which antedates the raised beach and is the most likely source for redeposited material, in view of the stratigraphy at the locality of sampling for pollen and spores. If redeposition was slow, there would have been a higher chance of the incorporation of contemporary pollen into the reworked sediment. Thus the spectrum, while having a complement of reworked grains, would be more representative of the 8.1 environment than if redeposition was rapid.

Examination of the characteristics of the pollen and spore assemblages of layers 6.2 (upper) and 8.1 shows them to be dissimilar, and, as such, the second alternative (reasonable authenticity of the immediate post-beach environmental record) receives support.

Layer 8.2 (humic soil) and 8.3 (redeposited loess) also possess temperate pollen and spore assemblages, with 74% of total pollen and spores being made up of trees and shrubs in 8.3, *Alnus*, *Quercus*, *Betula*, *Pinus*, *Tilia*, *Corylus*, *Calluna* and Ericaceae accounting for this. Dubois' sample [3B] (from 8.3) has high Chenopodiaceae values, which could relate along with Gramineae, to the existence nearby of saline or brackish-water marsh. Freshwater pollen (*Typha* and *Nymphaea*), besides complementing the diatoms found in 8.3 (Fig. 9.2F), is further evidence that the pollen and spore-bearing layers of unit 8 are grossly similar and reflect a temperate forest episode which, in view of its consistency and length, may represent an interglacial.

After a hiatus caused by barren samples in upper 8.3, and an absence of pollen and spores from layer 8.4, the humic soil of layer 9.1 contains a contrasting record of flora and vegetation. The bulk of the pollen grains counted belong to *Juniperus*, Gramineae, Cyperaceae and Plantaginaceae, probably indicating the existence of tundra or steppe-like vegetation. The overall temperate pollen and spore spectrum from the head (layer 9.2) is very unlikely to have been incorporated from the vegetation existing during the period of periglacial activity which formed the deposit, and whose onset is hinted at by the vegetation of layer 9.1. While there are some representatives (e.g. Cyperaceae, Gramineae, Compositae) which may have grown in the more stable parts of a cold and mobile landscape, the *in situ* nature of thermophilous tree pollen at this stage is highly unlikely, and

may be disregarded in an ecological context.

The humic soil of layer 9.3 has a pollen and spore content which at first (samples G–C) indicates a temperate environment. *Alnus*, *Betula*, *Pinus*, *Quercus*, *Ulmus* and *Corylus* were present in woodland which, if their percentages of total pollen and spores (maximum 15–20) are considered, either did not occupy much of the landscape, or was of a very open nature. The high herbaceous pollen counts, with Compositae, Plantaginaceae, Rosaceae, Linaceae, Rubiaceae and Gramineae, suggest widespread grassland communities associated with a sparse tree cover, a situation not uncharacteristic of insular and other maritime localities at the present day. Samples B and A (of 9.3) reflect a changing environment. Arboreal pollen values fall, and shrubs increase, with *Juniperus* and Ericaceae reappearing. This, together with rising values for Cyperaceae pollen, and the appearance of that of *Thalictrum* (which has a cold climate species, *Thalictrum alpinum*) suggest that a cooler environment had developed, no doubt preparatory to the periglacial conditions that ensued and caused the head (layers 10 and 11) to be deposited at the site.

## 11.5  SUMMARY

The pollen and spore assemblages from layer H and the lower part of layer G indicate the existence of temperate conditions round La Cotte. Those of the upper part of G and the lower part of layer F reflect a cooler phase, while in the upper part of F and in layers E, D and C a return to a more temperate environment is suggested. Layers B and A are sterile, but of similar loessic lithology; 3, 4, 5, 6.1 and 6.2 have mainly cold climate pollen and spores, although a warmer phase seems to have occurred during the time of deposition of layer 4. While there is no palynological evidence from the raised beach (layer 7.1), or from the overlying layer 7.2, an alternation of temperate and cold environments seems likely on stratigraphic grounds. 8.1 and 8.2 have temperate pollen and spores; layers 8.3 and 8.4, while not analysed for pollen, are composed of loessic material, which suggests cool or cold conditions. Layer 9.1 had a pollen and spore assemblage diagnostic of cold conditions, with the head of 9.2 containing redeposited temperate pollen but reflecting a severe periglacial environment. Layer 9.3 has temperate pollen and spores and is superseded by layers 10 and 11 which, although not pollen-analysed, undoubtedly reflect a major cold stage.

The pollen and spore records from the various temperate episodes are not diagnostic of the different vegetational successions that West (1977) has demonstrated for the various interglacials of the British Quaternary sequence; and interstadial floras are more variable. The geographical position of the Channel Islands and evidence from comparative work on the Quaternary palaeoecology of the islands and the adjacent coast of France (Coope, Jones and Keen 1980; Coope *et al.* 1985; Coope *et al.* 1986) indicate that the more likely correlations are with continental rather than British Quaternary events.

# 12

# WOOD CHARCOAL IDENTIFICATION

## C. Cartwright

Over 200 samples were collected by the excavators as being possibly charcoal. After a preliminary examination in Cambridge, 76 were passed on to the author as deserving further investigation. Of these, 53 contained burnt bone only and 5 soil only. Five of the remaining 18 samples contained charcoal powder with no structure and were therefore unidentifiable. Six samples contained carbonised bark (and probable bark) fragments, one of which may be *Quercus* sp. (oak).

In the seven remaining samples (Table 12.1) the charcoal fragments were identifiable, although their overall quantity was extremely small and their condition very friable. Five samples contained *Quercus* sp. (oak), one *Corylus* sp. (hazel), and one softwood charcoal which was possibly *Juniperus* sp. (juniper).

The difficulty of interpreting these identifications is obvious. We have to consider not only how any charcoal fragments came to be present in an archaeological context, but also the possibility that the wood of some species survives better than others as charcoal in certain conditions. At La Cotte, additional uncertainties are introduced by the poor conditions for survival, and by the small number and size of usable samples. Any environmental/ecological interpretation of the charcoal results must therefore be made in conjunction with the pollen sequences from the site, as obviously one gramme (for example) of a particular species does not on its own constitute a meaningful ecological indicator.

**Table 12.1**  Wood Charcoal Identification.

| Sample reference | Coordinates (cm) | | | Layer | Weight (g) | Genus |
|---|---|---|---|---|---|---|
| | Depth | East[a] | North[a] | | | |
| 243 | 2332 | 434 | 110 | 5 | 2.0 | *Juniperus?* |
| 15 | 2489 | 100–120 | 160–180 | A | 0.5 | *Quercus* |
| 120 | 2550 | 160 | 120 | C | 2.0 | *Quercus* |
| 119 | 2555 | 80–100 | 220–240 | C?[b] | 1.0 | *Quercus* |
| 3 | 2603 | 300–320 | 20–0 S | C | 1.0 | *Quercus* |
| 1 | 2686 | 220–240 | 120–100 S | F | 0.5 | *Quercus* |
| 10 | 2750 | 20–0 W | 80–60 S | G | 0.5 | *Corylus* |

[a]The coordinates of finds from west or south of the grid origin are marked W or S.

[b]Although at the time of excavation the sample was labelled B/C, information in the field notebook for the metre square concerned makes it almost certain that it originated in the extreme top of layer C.

# 13

# THE LARGE MAMMAL FAUNA

## K. Scott

The faunal remains from La Cotte provide an insight into aspects of hominid behaviour in the late Middle Pleistocene such as is rarely preserved in the archaeological record. Two large and remarkably well-preserved bone assemblages, comprised almost exclusively of mammoth and woolly rhinoceros, shed light on the ability of hominids to exploit the topography of the headland in which the site is situated to hunt herds of such large, potentially dangerous animals. The small quantity of lithic material associated with these bone heaps contrasts sharply with the cultural horizons below, from which thousands of artefacts were recovered, but scarcely any identifiable bones. Extremely poor bone preservation accounts for this disparity, which is unfortunate in view of the great depth of the cultural horizons and the abundance of lithic material.

## 13.1 SPECIES REPRESENTATION AND ENVIRONMENTAL IMPLICATIONS OF THE FAUNA

Given below is a level by level account of the faunal remains and their significance in the interpretation of the palaeoenvironment. The species identified in each level are listed in Table 13.1, and the skeletal representation of each species is presented in Tables 13.2–13.13. A schematic version of the stratigraphy (see Fig. 6.2) and chapters 6 and 7 provide the framework within which the fauna will be discussed.

Twelve archaeological horizons were excavated at La Cotte between 1961 and 1978. These comprise Stages II and III of the stratigraphic subdivisions proposed by Callow in section 6.4.2. Stage II (layers H–C) represents

**Table 13.1**   The Minimum Number of Individuals of Each Mammalian Species Represented in the Various Horizons at La Cotte de St. Brelade (1961–1978 Excavations)[a]

| | G | F | E | D | D/C | C | B | B/A | A | A/3 | 3 | 4 | 5 | 6 |
|---|---|---|---|---|---|---|---|---|---|---|---|---|---|---|
| **Lagomorpha** | | | | | | | | | | | | | | |
| *Lepus* sp., hare | - | - | - | - | - | - | 1 | - | - | - | - | - | - | - |
| **Rodentia** (see Chaline and Brochet, chapter 14, this volume) | | | | | | | | | | | | | | |
| **Carnivora** | | | | | | | | | | | | | | |
| *Canus lupus* L., wolf | - | - | - | - | - | 1 | 1 | - | 1 | - | - | - | - | 1 |
| *Alopex lagopus* L., arctic fox | - | - | - | - | - | - | - | - | - | - | 2 | - | - | - |
| *Ursus (cf spelaeus)*, bear | - | - | - | - | - | - | - | - | 1 | - | 1 | 1 | 1 | 1 |
| **Proboscidea** | | | | | | | | | | | | | | |
| *Mammuthus primigenius* Blumenbach, mammoth | - | - | - | 1 | - | 1 | 2 | - | 1 | 1 | 7 | 1 | 1 | 11 |
| **Perissodactyla** | | | | | | | | | | | | | | |
| *Coelodonta antiquitatis* Blumenbach, woolly rhinoceros | cf | cf | cf | cf | + | 2 | 3 | - | 2 | 1 | 2 | - | 1 | 3 |
| *Equus caballus* Boddaert, horse | - | - | - | 1 | - | 1 | 2 | - | 4 | 1 | 1 | - | 1 | - |
| **Artiodactyla** | | | | | | | | | | | | | | |
| *Megaceros giganteus* Blumenbach, giant deer | - | - | - | - | - | - | - | 1 | 1 | - | - | - | - | - |
| *Cervus elaphus* L., red deer | - | - | - | - | - | 1 | 1 | - | 2 | - | 1 | - | 1 | - |
| *Rangifer tarandus* L., reindeer | - | - | - | - | - | 1 | 1 | - | 1 | - | - | - | - | 1 |
| *Bison* sp. or *Bos* sp., bison or aurochs | - | - | - | - | - | - | 1 | - | 1 | 1 | - | - | 1 | 1 |
| *Rupicapra rupicapra* L., chamois | - | - | - | - | - | 1 | 1 | - | 1 | - | - | - | - | - |

[a]cf species identification uncertain; + species present, frequency indeterminate

**Table 13.2** Hare Skeletal Elements from La Cotte de St. Brelade[a]

|  | Layer B |
| --- | --- |
| Humerus, distal | 1/1 |
| Vertebra, lumbar | 1/1 |
| Total: minimum number of individuals | 1 |

[a]Number of identifiable bones or bone fragments/minimum number of individuals represented.

**Table 13.3** Wolf Skeletal Elements from La Cotte de St. Brelade[a]

|  | Layer | | | |
| --- | --- | --- | --- | --- |
|  | C | B | A | 6 |
| Maxilla | 1/1 | – | – | – |
| Mandible | – | 1/1 | 1/1 | – |
| Canine, upper | – | 1/1 | – | – |
| lower | 1/1 | – | – | – |
| Radius, proximal | – | 1/1 | – | – |
| distal | – | – | – | 1/1 |
| Ulna, proximal | 1/1 | – | – | – |
| Metapodials | – | 1/1 | 1/1 | – |
| Phalanges, first | – | – | – | – |
| second | – | – | 1/1 | – |
| third | – | – | – | – |
| Astragalus | – | – | 1/1 | – |
| Total: minimum number of individuals | 1 | 1 | 1 | 1 |

[a]Number of identifiable bones or bone fragments/minimum number of individuals represented.

**Table 13.4** Arctic Fox Skeletal Elements from La Cotte de St. Brelade[a]

|  | Layer 3 |
| --- | --- |
| Premaxilla | 1/1 |
| Maxilla | 1/1 |
| Mandible | 2/2 |
| Canine | 1/1 |
| Total: minimum number of individuals | 2 |

[a]Number of identifiable bones or bone fragments/minimum number of individuals represented.

an interglacial complex, while Stage III (layers B–6) represents a cold stage complex. The environmental differences between these two complexes of deposits have had a significant effect on the faunal remains.

The cold stage complex has yielded a fauna in keeping with the rest of the palaeoenvironmental evidence: predominantly large ungulates that were widely distributed throughout the now extinct steppe-like habitat that extended across northern Eurasia and Beringia and into Alaska during the late Middle and Upper Pleistocene (J. V. Matthews 1982; Stuart 1982). One might envisage an open, grass-dominated landscape, including a diverse assemblage of scattered herbs, perhaps broken by gallery forests of dwarf birch, alder and willow along river banks (Olivier 1982). The loessic soils of the steppe tundra would have been alkaline and conducive to herbaceous growth of high nutritive quality, sufficient to support a varied ungulate fauna on which hominids and other carnivores (lion, hyaena, wolf) might have preyed.

A quite different fauna would be expected from the interglacial complex, but is limited as the result of poor bone preservation, which in itself is linked to environmental conditions. During cold climatic episodes loess accumulated at the site. In levels such as 6, 3 and B, where the loess is thick and has provided a protective alkaline matrix, the bones, though fragile and in need of extensive conservation, were well preserved. In levels where loess was less dense and mixed with ash and other cultural debris (such as in layers A and C), bone is poorly preserved but teeth are in reasonable condition. Where loess is absent (in the climatically temperate levels) so too is the identifiable fauna, destroyed by the acidic soils. Vestiges of hundreds of thousands of bones were visible in these levels (D–G), but generally as orange patches or burnt fragments measuring only a few millimetres. A large percentage of these were plotted on bone plans, but the majority were almost indistinguishable from the matrix and were thus irretrievable.

**Table 13.5** Bear Skeletal Elements from La Cotte de St. Brelade[a]

|  | Layer | | | | |
| --- | --- | --- | --- | --- | --- |
|  | A | 3 | 4 | 5 | 6 |
| Cranial fragment | 1/1 | – | – | – | 1/1 |
| Premaxilla | – | – | 1/1 | – | – |
| Humerus, proximal | – | – | – | – | – |
| distal | – | 1/1 | – | – | – |
| Femur, proximal | – | – | – | – | – |
| distal | – | – | – | 1/1 | – |
| Total: minimum number of individuals | 1 | 1 | 1 | 1 | 1 |

[a]Number of identifiable bones or bone fragments/minimum number of individuals represented.

**Table 13.6**  Mammoth Skeletal Elements from La Cotte de St. Brelade[a]

| | | | | | Layer | | | | | |
|---|---|---|---|---|---|---|---|---|---|---|
| | D | C | B | A | A/3 | 3 | 4 | 5 | 5/6 | 6 |
| Skulls with dentition | – | – | – | – | – | 7/7 | – | – | – | – |
| Partial skulls, no dentition | – | – | – | – | – | 4/2 | – | – | – | 3/2 |
| Skull fragments | – | – | 6 | 10 | 7 | 50 | – | 8 | 1 | 6 |
| Isolated upper molars | – | 2/1 | 6/2 | 1/1 | – | – | 1/1 | – | – | – |
| Isolated lower molars | – | 1/1 | – | 5/2 | – | 1/1 | – | – | – | 1/1 |
| Tusks | – | – | – | 6 | 1 | 7 | – | – | – | 5 |
| Tusk fragments | – | – | 1 | 14 | 5 | 14 | – | 3 | – | 8 |
| Mandibles | – | – | – | – | – | 2/2 | – | – | – | 2/2 |
| Dental fragments | 5 | 6 | 8 | 33 | 10 | 7 | – | – | – | 1 |
| Vertebrae, | | | | | | | | | | |
| atlas | – | – | – | – | – | – | – | – | – | – |
| axis | – | – | – | – | – | – | – | – | – | – |
| other cervical | – | – | – | – | – | 3 | – | – | – | – |
| thoracic | – | – | – | 1 | 1 | 2 | – | 1 | – | – |
| lumbar | – | – | – | – | – | 1 | – | – | – | – |
| Scapula, | | | | | | | | | | |
| essentially complete | – | – | – | – | – | 7/5 | – | – | – | 15/11 |
| fragment | – | – | – | – | 1 | – | – | 1 | 2 | – |
| Humerus, | | | | | | | | | | |
| proximal, no epiphysis | – | – | – | – | – | 2/1 | – | – | – | 1/1 |
| shaft, no epiphyses | – | – | – | – | – | – | – | – | – | 3/2 |
| distal with epiphysis | – | – | – | – | – | – | – | – | – | 3/2 |
| Radius, | | | | | | | | | | |
| proximal | – | – | – | – | – | – | – | – | – | – |
| distal | – | – | – | – | – | – | – | – | – | – |
| Ulna, | | | | | | | | | | |
| proximal | – | – | – | – | – | 1/1 | – | – | – | 1/1 |
| distal | – | – | – | – | – | – | – | – | – | – |
| Innominate, | | | | | | | | | | |
| complete | – | – | – | – | – | 2/1 | – | – | – | – |
| ilium | – | – | 1/1 | – | – | 2/1 | – | – | – | 14/7 |
| ischium | – | – | – | – | – | 1/1 | – | – | – | 4/3 |
| pubis | – | – | – | – | – | 2/1 | – | – | – | 1/1 |
| fragment | – | – | – | – | 1 | – | – | – | – | – |
| Femur, | | | | | | | | | | |
| proximal | – | – | – | – | – | – | – | – | – | – |
| shaft, no epiphyses | – | – | – | – | – | – | – | – | – | 9/4 |
| distal, with epiphysis | – | – | – | – | – | – | – | – | – | 4/2 |
| Tibia, | | | | | | | | | | |
| complete | – | – | – | – | – | – | – | – | – | 1/1 |
| proximal, no epiphysis | – | – | 1/1 | – | – | – | – | – | – | 2/2 |
| distal, no epiphysis | – | – | 1/1 | – | – | – | – | – | – | – |
| Fibula | – | – | – | – | – | – | – | – | – | – |
| Carpals | – | – | – | – | – | – | – | – | – | – |
| Tarsals | – | – | – | – | – | – | – | – | – | – |
| Metapodials | – | – | – | – | – | – | – | – | – | – |
| Phalanges | – | – | – | – | – | – | – | – | – | – |
| Ribs or rib fragments | – | – | + | + | – | + | – | + | + | + |
| Total: minimum number of individuals | 1 | 1 | 2 | 1 | 1 | 7 | 1 | 1 | 1 | 11 |

[a]Number of identifiable bones or bone fragments/minimum number of individuals represented.
+ present.

To a lesser extent, the preservation of bones seems to have been affected by their proximity to the west wall of the site. The deposits on the west side of the excavated area are protected to some extent by the overhanging wall, and the bones thereunder are generally in a better state of preservation than those on the eastern side, which would have been exposed to the elements through the open 'roof' (see Fig. 6.1).

## STAGE II (Layers H–C)

The lowest layer (H) is represented by a very small area of the excavations. Moreover, the fieldwork carried out in 1982 established that much of the deposit just to the north of the fossil cliff had been infiltrated by loess from higher up during a progressive collapse of the overhang cut by the Eemian sea. A few bones and dental frag-

111

**Table 13.7**  Woolly Rhinoceros Skeletal Elements from La Cotte de St. Brelade[a]

| | G | F | E | D | D/C | C | Layer B | A | A/3 | 3 | 5 | 5/6 | 6 |
|---|---|---|---|---|---|---|---|---|---|---|---|---|---|
| Skulls with dentition | - | - | - | - | - | - | - | - | - | - | - | - | 3/3 |
| Partial skulls, no dentition | - | - | - | - | - | - | - | - | - | 1/1 | - | - | - |
| Skull fragments | - | - | - | - | - | - | 20/1 | - | - | - | - | - | - |
| Maxilla with dentition | - | - | - | - | - | 1/1 | - | - | - | - | - | - | - |
| Isolated upper teeth | - | - | - | - | 1/1 | 2/2 | 7/3 | 7/2 | - | 2/1 | - | - | - |
| Mandibles with dentition | - | - | - | - | - | - | - | 1/1 | - | 1/1 | - | - | 1/1 |
| Mandibles without dentition | - | - | - | - | - | - | - | 1/1 | - | - | - | - | - |
| Isolated lower teeth | - | - | - | - | - | - | 1/1 | 3/2 | - | - | - | - | - |
| Dental fragments | 1 | 1 | 2 | 16 | 2 | 15 | 19 | 35 | - | - | 2 | 1 | - |
| Vertebrae, | | | | | | | | | | | | | |
| atlas | - | - | - | - | - | - | - | - | - | 1/1 | - | - | - |
| axis | - | - | - | - | - | - | - | - | - | - | - | - | - |
| other cervical | - | - | - | - | - | - | - | - | - | - | - | - | - |
| thoracic | - | - | - | - | - | - | - | - | - | 5/1 | - | - | - |
| lumbar | - | - | - | - | - | - | - | - | - | 1/1 | - | - | - |
| indet. fragments | - | - | - | - | - | - | - | 1/1 | - | - | - | 1/1 | - |
| Scapula | - | - | - | - | - | - | - | - | - | - | - | - | - |
| Humerus, | | | | | | | | | | | | | |
| proximal epiphysis | - | - | - | - | - | - | - | - | - | - | 2/1 | - | - |
| shaft fragment | - | - | - | - | - | - | - | 1/1 | - | - | - | - | - |
| distal epiphysis | - | - | - | - | - | - | 1/1 | 1/1 | - | 1/1 | - | - | - |
| Radius, | | | | | | | | | | | | | |
| proximal | - | - | - | - | - | - | - | - | - | 1/1 | - | - | - |
| Ulna, | | | | | | | | | | | | | |
| proximal | - | - | - | - | - | - | - | - | - | 2/2 | - | - | - |
| distal | - | - | - | - | - | - | 1/1 | - | - | - | - | - | - |
| Carpals | - | - | - | - | - | - | 2/1 | 1/1 | - | - | - | - | - |
| Innominate, | | | | | | | | | | | | | |
| complete | - | - | - | - | - | - | - | 1/1 | - | - | - | - | - |
| ilium + acetabulum | - | - | - | - | - | - | - | - | - | 1/1 | - | 1/1 | 2/2 |
| ischium + acetabulum | - | - | - | - | - | - | - | 1/1 | - | - | - | - | - |
| Femur, | | | | | | | | | | | | | |
| proximal epiphysis | - | - | - | - | - | 2/2 | - | - | - | 2/2 | - | - | - |
| shaft fragment | - | - | - | - | - | - | 1/1 | - | - | - | - | - | - |
| distal, without epiphysis | - | - | - | - | - | - | - | 1 | - | - | - | - | 1/1 |
| Tibia, | | | | | | | | | | | | | |
| proximal | - | - | - | - | - | - | - | 1/1 | - | - | - | - | - |
| Astragalus | - | - | - | - | - | - | - | - | - | - | 1/1 | - | - |
| Metapodia | - | - | - | - | - | - | 1/1 | 1/1 | 2/1 | - | - | - | - |
| Phalanges, | | | | | | | | | | | | | |
| first, lateral | - | - | - | - | - | - | 1/1 | - | - | - | - | - | - |
| second | - | - | - | - | - | - | 1/1 | - | - | - | - | - | - |
| Total: minimum number of individuals | 1 | 1 | 1 | 1 | 1 | 2 | 3 | 2 | 1 | 2 | 1 | 1 | 3 |

[a]Number of identifiable bones or bone fragments/minimum number of individuals represented.

ments of mammoth, woolly rhinoceros and horse, and most of the bones of the lower limb of a reindeer, were recorded as being found in H during the 1961–1962 excavations. Their state of preservation is exceptionally good, particularly as regards hardness. The new evidence about the nature of 'H' (see especially section 9.4) indicates that it is likely that these remains became incorporated in the deposit during the collapses, and that they originated in layers D–6. Consequently, the fauna originally attributed to layer H has been excluded from the tables and discussion in this chapter.

Layers E, F and G are characterised by vast quantities of bone in a sandy ash-rich matrix. Artefacts are abundant but, as mentioned, bone preservation is poor. The only animal that could be identified from a few

dental fragments is rhinoceros, and even these could not be identified to species.

In layers C and D the fauna is slightly better preserved, although species and body-part representation is still poor. In layer D, the occurrence of mammoth and woolly rhinoceros probably indicates open grassland in the vicinity of the site, although mammoth is also known from deposits associated with boreal forest in the Chelford Interstadial of the Last Cold Stage in Britain (Stuart 1982). Similarly, horse and red deer (both highly adaptable species today) occurred in the European Middle and Upper Palaeolithic, both in interglacial faunas associated with temperate and coniferous forests and in cold-stage faunas that accumulated under treeless, herb-dominated vegetational conditions (Stuart 1982). Mam-

**Table 13.8**  Horse Skeletal Elements from La Cotte de St. Brelade[a]

| | D | C | B | Layer A | A/3 | 3 | 5 |
|---|---|---|---|---|---|---|---|
| Maxillary dentition | 1/1 | – | 5/2 | 12/3 | – | 1/1 | 1/1 |
| Mandibular dentition | 1/1 | 2/1 | 1/1 | 20/4 | – | 1/1 | – |
| Dental fragments | 7 | 2 | 2 | 20 | – | 2 | – |
| Humerus, distal | – | – | 1/1 | – | 1/1 | – | – |
| Femur, proximal | – | – | – | – | 1/1 | – | – |
| distal | – | – | – | 1/1 | – | – | – |
| Tibia, distal | – | – | – | 1/1 | – | – | – |
| Metacarpal, proximal | – | – | – | 1/1 | – | – | – |
| Metapodial, distal | – | – | – | – | – | – | – |
| Astragalus | – | – | – | – | – | – | 1/1 |
| Total: minimum number of individuals | 1 | 1 | 2 | 4 | 1 | 1 | 1 |

[a]Number of identifiable bones or bone fragments/minimum number of individuals represented.

**Table 13.9**  Giant Deer Skeletal Elements from La Cotte de St. Brelade[a]

| | Layer B/A | A |
|---|---|---|
| Phalanges, first | 1/1 | – |
| second | – | 1/1 |
| Total: minimum number of individuals | 1 | 1 |

[a]Number of identifiable bones or bone fragments/minimum number of individuals represented.

**Table 13.10**  Red Deer Skeletal Elements from La Cotte de St. Brelade[a]

| | D | C | B | Layer A | 3 | 5 |
|---|---|---|---|---|---|---|
| Cranial fragment | – | – | – | 1/1 | – | – |
| Antler | 1/1 | – | – | 3/2 | – | 1/1 |
| Maxillary dentition | – | – | 1/1 | – | – | – |
| Vertebra, fragment | – | – | – | 1/1 | – | 1/1 |
| Scapula | – | – | – | 1/1 | – | – |
| Humerus, proximal | – | 1/1 | – | – | – | 1/1 |
| distal | – | – | – | – | 1/1 | 1/1 |
| Radius, distal | – | – | 1/1 | – | – | 1/1 |
| Metatarsal | – | – | – | 1/1 | – | – |
| Phalanges, first | – | – | – | 1/1 | – | – |
| Femur, distal | – | – | 1/1 | 1/1 | – | – |
| Tibia, proximal | – | – | 1/1 | – | – | – |
| distal | – | – | – | 1/1 | – | – |
| Patella | – | 1/1 | – | – | – | 1/1 |
| Astragalus | – | 3/1 | – | 1/1 | – | – |
| Total: minimum number of individuals | 1 | 1 | 1 | 2 | 1 | 1 |

[a]Number of identifiable bones or bone fragments/minimum number of individuals represented.

113

**Table 13.11**  Reindeer Skeletal Elements from La Cotte de St. Brelade[a]

| | Layer | | | |
| | C | B | A | 6 |
|---|---|---|---|---|
| Maxillary dentition | - | - | 1/1 | - |
| Mandibular dentition | - | 1/1 | - | - |
| Sacral vertebra | - | - | 1/1 | - |
| Humerus, distal | - | 1/1 | - | - |
| Femur, fragment | - | 1/1 | - | - |
| Metacarpal, distal | - | - | - | - |
| Metapodial, indet. frag. | - | - | - | - |
| Accessory metapodial | - | - | - | - |
| Phalanges, first | - | - | - | - |
| second | - | - | - | - |
| Accessory phalanx | - | - | - | - |
| Sesamoid | - | - | - | - |
| Calcaneum | - | - | 1/1 | 1/1 |
| Astragalus | 1/1 | - | - | - |
| Total: minimum number of individuals | 1 | 1 | 1 | 1 |

[a]Number of identifiable bones or bone fragments/minimum number of individuals represented.

**Table 13.12**  Bos/Bison Skeletal Elements from La Cotte de St. Brelade[a]

| | Layer | | | | |
| | B | A | A/3 | 5 | 6 |
|---|---|---|---|---|---|
| Maxillary dentition | - | 1/1 | - | - | 1/1 |
| Dental fragment | 1/1 | - | - | - | - |
| Humerus, diaphysis | 1 | 1/1 | - | 1/1 | - |
| distal | - | - | - | - | - |
| Ulna, proximal | - | 1/1 | - | - | - |
| Femur, diaphysis | - | - | 1/1 | - | - |
| Metatarsal | - | 1/1 | - | - | - |
| Phalanges, first | 1/1 | - | - | - | - |
| Astragalus | - | - | 1/1 | - | - |
| Total: minimum number of individuals | 1 | 1 | 1 | 1 | 1 |

[a]Number of identifiable bones or bone fragments/minimum number of individuals represented.

moth, woolly rhino, red deer and horse are also present in layer C, but in addition there are a few bones of wolf, reindeer and chamois, and at the C/B interface, one phalanx of a giant deer.

The presence of reindeer in layer C is suggestive of a cold environment at the time of its deposition; it is known only from cold stages in Pleistocene Europe, associated with evidence for steppe-tundra and boreal forest. The present day distribution of the species is in the circumpolar regions of Eurasia and North America, where it ranges between tundra, taiga and boreal forest environments.

The chamois, represented by an upper molar (Fig. 13.1) and part of a mandible with dentition in layer C, is especially interesting. It is well known from 'Saalian/Rissian' deposits in France (de Lumley 1975) and Spain (Kahlke 1975), but to date there is no Pleistocene record of chamois as far to the northwest as the Channel Islands. In western Europe it is generally associated with mountainous areas, but the cliffs of Jersey possibly provided a suitable enough habitat. It is not a useful climatic indicator in that it occurs as a member of interglacial as well as of cold-stage faunas.

## STAGE III (layers B–6)

From the point of view both of archaeology and of palaeontology, this complex of deposits contains the most interesting faunal material from La Cotte. The onset of extremely cold, dry conditions after the occupation of layer C is indicated by the loessic character of layer B. There are artefacts and well-preserved bone fragments in this layer as well as a number of ash lenses, but as Callow points out (section 7.1.3, this volume), there is evidence to suggest that layer B has been considerably disturbed, and it is probable that some of the occupational debris was incorporated from the underlying layer C during reworking of the loess. There are mammoth and woolly rhino in B, as well as wolf, hare, reindeer, chamois and a large bovid, but the total number of bones is small.

**Table 13.13**  Chamois Skeletal Elements from La Cotte de St. Brelade[a]

| | Layer | | |
| | C | B | A |
|---|---|---|---|
| Maxillary dentition | 1/1 | - | - |
| Mandibular dentition | 1/1 | 1/1 | - |
| Femur, proximal | - | 1/1 | 1/1 |
| distal | - | - | 1/1 |
| Tibia, proximal | - | - | 1/1 |
| Astragalus | - | 1/1 | 2/1 |
| Total: minimum number of individuals | 1 | 1 | 1 |

[a]Number of identifiable bones or bone fragments/minimum number of individuals represented.

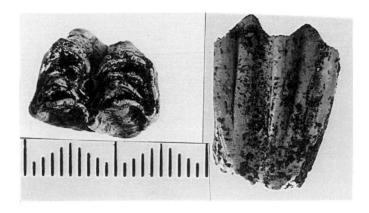

**Fig. 13.1**  Chamois from layer C (78/811). Right upper molar (M2). Scale in mm.

Layer A is extremely rich in artefacts, ash and bone, although the state of preservation of the bones is generally poor. The fact that bone is preserved at all is undoubtedly due to the high loess content of this level, but in some areas the deposit was so cemented that the bones could not be extracted. When blocks of this breccia were broken open, casts of bones and teeth were frequently observed amid masses of yellow-orange bone fragments, and much of the bone in this level had been burnt. The interpretation of the climatic conditions prevailing at the time of deposition of layer A is complex (see chapter 7). Although a high loess content in some samples would suggest a cold environment, there is also evidence of a period of climatic amelioration — possibly an interstadial. The species represented are much the same as in layer C (with the addition of giant deer), but the number of identifiable bones is considerably greater. Horse is more abundant in A than in any level at the site, and mammoth and rhino are more numerous than in layer C, although overall numbers are still low and identification is based primarily on teeth and dental fragments. Identifiable post-cranial material is rare. A small fragment of a bear frontal proved interesting in that it is only part of this animal preserved at La Cotte that allows specific identification to *Ursus spelaeus*, the cave bear.

After the uppermost occupation of layer A, a considerable quantity of loess (layer 3) was deposited, and within this loess a most remarkable fauna was discovered. It is comprised almost exclusively of the remains of a number of mammoth and woolly rhinoceros. Palaeontologically, this is a significant occurrence since, although both species are known from a variety of Pleistocene deposits in Europe, it is rare to find a great number of these animals in a well-stratified context. Archaeologically, this is a most interesting and important assemblage because the presence of man is indicated by the stacking of, and damage to, some of the bones. Prehistoric kill/butchery sites of elephants and other large mammals are known from various regions of the world, but this occurrence is exceptional in that there are so many animals — and in particular, so many complete skulls and tusks — in a cave rather than an open situation. The archaeological implications of this heap of mammoths and woolly rhinos are discussed further in chapters 18 and 31, but one point that ought to be clarified at this juncture is the extent to which this assemblage might represent a single kill/butchery 'event' rather than be part of layer A, and thus part of a larger and more diverse fauna.

There are two reasons to suggest that the faunal remains in layer 3 are distinct from those in layer A. Firstly, layer A in the area underlying the bone heap was, for the most part, a cemented breccia. Although many of the mammoth and rhino bones lay directly on the surface of A, they did not penetrate this extremely hard matrix. Secondly, horse is abundant in layer A, whereas only two teeth of horse were found in the loess of 3, and these at the 2/3 interface; horse was not distributed among the mammoth and rhino. The question of how much time elapsed between the faunal and artefactual accumulations of layers A and 3 may never be resolved. The absence of a sterile deposit between the two might be indicative of a relatively short interval (archaeologically speaking), but one long enough to have given the impression at the time of excavation that the bone heap represents a discrete event.

After the bone heap episode, the site was abandoned and a considerable deposit of aeolian loess covered the bones. The climate deteriorated in the region of La Cotte, and frost action evidently caused exfoliation of blocks of granite from the walls of the cave during the loess deposition. Directly above the bone heap, a shattered slab some 2 x 3 m became detached from the west wall (Fig. 2.7b). This slab appears to have done little damage to the bones and, in fact, probably contributed to their preservation by protecting them from water percolating through the overlying deposits.

Once again, the time interval between layer 3 and the next faunal accumulation at La Cotte cannot be established. A thin layer (5) provides evidence of hominid reappearance at the site during a milder climatic interval (chapter 7). Layer 5 is not a very rich occupation horizon, and most of the bones, though well preserved, are unidentifiable post-cranial fragments; however, mammoth, woolly rhino, red deer, reindeer and bison (or aurochs) have been identified.

A particularly remarkable feature of this site is that, directly on top of layer 5, there is another great pile of bones of woolly rhinoceros and mammoth, the latter being represented by at least 11 animals. As in the case of the layer 3 assemblage, this heap of bones is also protected and preserved by a thick deposit of loess (layer 6.1). The distinction between this assemblage and the bones of layer 5 is not always as clear as it is between layers A and 3, and many of the mammoth and rhino remains were partly situated in layer 5. Again, hominid activity is evident in the form of artefacts, although these are quite different from those associated with the lower bone heap (Callow, chapter 24, this volume), and some of the bones seem to have been stacked near the west wall of the site (see chapter 18). The influx of loess that buried this upper assemblage denotes further climatic deterioration and the site was then abandoned by hominids for perhaps tens of thousands of years.

The two heaps of mammoth and woolly rhinoceros from layers 3 and 6 are discussed further in chapter 18, where it is proposed that such a number of animals (at least 18 mammoths and five rhinos) were not brought to the site but were driven into it. It is possible that the mammoths and rhinos in other levels were also acquired in this manner, but poor bone preservation allows no more than speculation on this point. The only exception is in the case of layer B, where preservation is good and mammoth and rhino occur. Their remains are few but include a complete mammoth innominate and a rhino skull (most of which is still in the unexcavated deposit). It is therefore not inconceivable that a larger bone heap in layer B remains to be excavated.

From the above discussion and from the tables it will be evident that only three species are sufficiently well represented to allow detailed analysis: horse, woolly rhinoceros and mammoth. Each of these species is discussed in more detail below, and data presented which include measurements, and body-part and age representation.

## 13.2  THE HORSE REMAINS

Horse is best represented in layer A, and in all levels where it occurs teeth and dental fragments are more numerous than post-cranial remains (Table 13.8). The measurements taken on more or less complete teeth are given in microfiche 13.2 and 13.3 (see also mf 13.1). It was not thought possible to be consistent in distinguishing P3 from P4 and M1 from M2, hence the designation of these teeth as P3/4 and M1/2.

J. Eggington (Dept. of Zoology, University of Cambridge) has compared these specimens from La Cotte with those from other west European Pleistocene deposits and considers them (pers. comm.) to be well within the size range of Hoxnian-Wolstonian-Ipswichian *Equus caballus*; that is to say, not large enough to have occurred earlier than the Hoxnian, nor small enough to be of Devensian age.

## 13.3  THE RHINOCEROS REMAINS

The rhinoceros remains from La Cotte are assigned to the genus and species *Coelodonta antiquitatis*, the woolly rhinoceros. It is not possible, however, to discuss the extent to which these specimens are typical of *C. antiquitatis* of comparable stratigraphic age, as relatively few of the numerous European specimens identified as woolly rhinoceros come from reliably dated deposits. *C. antiquitatis* is known in western Europe from the late Middle Pleistocene until the latter part of the Last

Glacial (Belyaeva *et al.* 1968; Kurtén 1972; Guérin 1976 and 1978). Guérin (1978) makes a number of distinctions between the various forms of woolly rhinoceros of 'Rissian', 'Eemian' and 'Wurmian' age (his terminology). These differences are in the relative size of bones and teeth rather than in their morphology. According to Guérin, the 'Rissian' form of *C. antiquitatis* is taller and less massive than later forms, while the later ('Wurmian') specimens have relatively narrower, more hypsodont teeth.

There is too little complete post-cranial material from La Cotte to enable comment to be made on the relative proportions of the La Cotte rhinos, and in any event, much of the material belongs to subadult animals. As regards the question of relative dental proportions, no significant differences were evident when the measurements taken on the La Cotte rhino teeth were compared with those taken on dozens of woolly rhinoceros teeth from Middle Devensian (Last Cold Stage) British sites such as Coygan Cave and Picken's Hole (Scott, in prep.). It is probable that regional variation accounts for differences among populations of *C. antiquitatis* as much as temporal differences might, and indeed this has been demonstrated to be true of other mammalian species. Thus, in view of the fact that the effects of regional and temporal variation are not well known for the woolly rhinoceros, and as so few specimens can be reliably assigned to the various stages of the late Middle and Upper Pleistocene, no attempt is made here to place the La Cotte specimens within the as yet poorly established evolutionary framework. Instead, the La Cotte material is described and illustrated, and measurements provided, which it is hoped will be useful to other researchers for comparative purposes.

### 13.3.1  Cranial and Dental Characteristics of *Coelodonta antiquitatis*

The dental and cranial characteristics of the European Dicerorhininae are described by various authors (viz. Bonifay 1961; Borsuk-Bialynicka 1973; Loose 1975; Guérin 1978). It is possible to distinguish the woolly rhinoceros *Coelodonta antiquitatis* from other extinct members of the European Pleistocene Dicerorhininae (*D. kirchbergensis* and *D. hemitoechus*) on the basis of dental characteristics, relative proportions of the skull, and various cranial angles which describe the skull morphology.

On the basis of cranial and dental characteristics outlined below, the remains of rhinoceros from La Cotte are assigned to the genus and species *C. antiquitatis*. In *C. antiquitatis* the occiput is square and the occipital crest heavy. In *D. hemitoechus* the occiput is trapezium shaped; in *D. kirchbergensis* it is slightly trapezoidal without a pronounced occipital crest (Figs. 13.2 and 13.3).

The processus paroccipitalis and the posttympanicus are one entity in the woolly rhino. This is also true of *D. hemitoechus*, although in *D. kirchbergensis* the processus posttympanicus is a separate element (Figs. 13.4 and

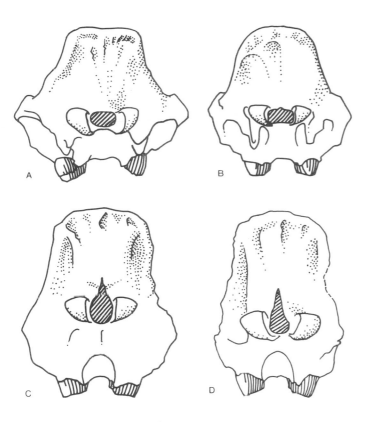

**Fig. 13.2** Occipital regions of Pleistocene rhinoceros skulls. (A) *D. kirchbergensis* (Jäger); (B) *D. hemitoechus* (Falc.); (C) *C. antiquitatis* (Blum.); (D) *C. antiquitatis* (Blum.), La Cotte de St. Brelade 69/22. (A–C after Loose 1975).

of all relatively complete teeth are presented in Table 13.14 and 13.15 (see Fig. 13.7 for key to measurements). The use of these measurements in the construction of age profiles of the woolly rhinoceros from La Cotte is discussed later in this chapter.

### 13.3.2 Woolly Rhinoceros Post-Cranial Remains

The more complete elements have been measured to allow comparison to be made of the relative sizes of individuals. These measurements are presented in mf 13.5. For key to measurements see mf 13.4. Measurements of vertebrae, ribs, carpals, tarsals, metapodials and phalanges are not included.

13.5). The zygomatic arch curves gradually forward and down, its anterior part thickening with a slight curve of the upper edge. In *D. kirchbergensis* the zygomatic arch plunges forward and down, and even more steeply so in *D. hemitoechus* (Figs. 13.4 and 13.5). *C. antiquitatis* differs most noticeably from the other species of rhinoceros in that the nasal septum (Fig. 13.4d) is completely ossified in adult specimens. This is not evident from skull 69/22 (illustrated in Fig. 13.5), as it is a relatively young individual. However, ossification of the nasal septum has begun in another less well-preserved specimen (68/3).

The teeth of all species are high-crowned, but the woolly rhino has distinctive corrugated enamel. The upper teeth are characterised by a very slender fold of the paracone, separated from the parastyle by a marked depression. The fourth deciduous premolar (dp4) is particularly distinctive in that the crista and crochet merge, enclosing the medifossa (Fig. 13.6). This specimen is of particular interest in that the first deciduous premolar (dp1) is present. This tooth is lost at a young age in living rhinos and is rarely preserved in the fossil record.

There are three skulls from La Cotte which serve to illustrate the above-mentioned cranial and dental characteristics, the best preserved of which (69/22) is illustrated in Figs. 13.3 and 13.4. In addition to these skulls (and also one without dentition) there are many fragments of dentition and a number of isolated teeth. Measurements

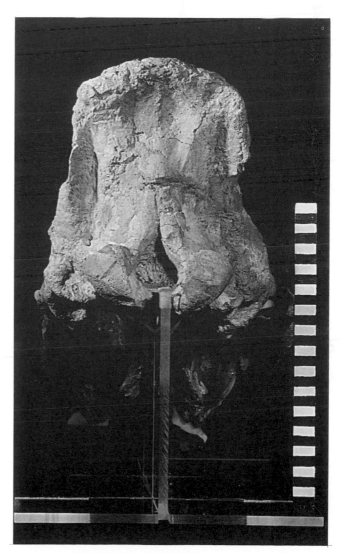

**Fig. 13.3** Skull of *Coelodonta antiquitatis*, occipital view (69/22). Scale in cm.

117

**Table 13.14**  Measurements of Upper Dentition of Woolly Rhinoceros from La Cotte de St. Brelade[a]

| Layer | Ref. no. | Tooth | Wear | L/R | 1 Maximum length | 2 Wear length | 3 Height of anterior lobe | 4 Height of posterior lobe | 5 Minimum ectoloph width | 6 Maximum tooth width | 7 Maximum crown width |
|---|---|---|---|---|---|---|---|---|---|---|---|
| C | 78/372 | dp1 | + | L | 19.0 | 17.0 | 12.5 | 10.5 | 2.0 | 18.5 | 14.5 |
| C | 78/896 | dp2 | vew | L | 32.5 | c25.0 | 35.0 | 35.0 | 2.5 | 29.5 | 24.0 |
| u/s | 77/753 | dp2 | x | R | – | – | 31.5 | 29.0 | 5.5 | >23.0 | – |
| B | 78/668 | dp2 | x | L | – | – | c28.0 | – | – | 30.0 | – |
| C | 78/937 | dp2 | + | R | 31.0 | 31.0 | 23.5 | 22.0 | – | – | – |
| C | 78/372 | dp2 | ++ | L | 31.5 | 30.5 | 18.0 | 17.5 | 6.5 | 33.0 | 26.5 |
| B | 78/962 | dp2 | ++ | L | – | – | – | – | – | – | – |
| B | 78/602 | dp3 | uw | L | – | uw | 34.5 | – | 2.0 | – | – |
| A/3 | 77/672 | dp3 | + | L | >30.0 | – | c16.0 | c17.0 | 3.0 | – | – |
| B | 78/785 | dp3 | + | L | – | – | – | 26.5 | 8.0 | – | – |
| C | 78/372 | dp3 | ++ | L | 41.5 | c38.5 | 15.5 | 18.0 | 8.5 | 38.5 | 28.5 |
| B | 78/661 | dp4 | uw | R | – | – | – | 38.0 | – | – | – |
| C | 78/372 | dp4 | ++ | L | 47.0 | c47.0 | 26.5 | 30.0 | 3.5 | 43.5 | 30.5 |
| 6 | 69/22* | dp4 | +++ | R | 47.0 | 47.0 | 20.5 | 25.5 | 13.5 | 49.0 | 36.5 |
| A/3 | 77/314 | dp4 | ++ | – | – | – | – | – | – | – | – |
| 6 | 69/22* | P2 | ew | R | 38.0 | 32.5 | / | 28.5 | 3.5 | 34.5 | 25.0 |
| B | 78/1187 | P3 | uw | L | 41.0 | 39.0 | / | – | 2.0 | – | – |
| 6 | 69/22* | P3 | ew/x | L | 46.5 | 40.5 | / | 36.5 | 2.5 | 45.0 | 27.0 |
|  |  |  | ew/x | R | 46.0 | 37.5 | / | 29.0 | 4.0 | 42.5 | 28.0 |
| D/C | 61/45 | P3 | ew/x | R | 40.0 | 37.0 | / | 51.0 | 4.5 | 40.0 | 30.0 |
| 6 | 68/12* | P3 | x | L | 39.0 | 40.5 | / | 48.5 | 8.5 | 43.0 | 31.0 |
| 6 | 68/12* | P4 | uw | L | 45.5 | 29.5 | / | ic | 3.0 | ic | 24.5 |
|  |  |  | ew | R | – | – | / | – | – | – | – |
| A/3 | 77/570 | P4 | vew | L | 49.5 | – | / | 61.0 | 5.5 | 51.5 | 29.0 |
| A | 78/303 | P4 | x | R | 50.5 | 50.5 | / | 36.0 | 14.5 | – | – |
| A | 78/544 | M1 | uw | L | 53.5 | c29.0 | c56.0 | 45.0 | 2.0 | c41.0 | 19.5 |
| 6 | 69/22* | M1 | ew | L | 65.0 | 50.0 | 63.5 | 69.5 | 5.5 | 58.5 | 31.5 |
|  |  |  | ew/x | R | 63.5 | 48.5 | 66.0 | 66.5 | 3.5 | 60.0 | – |
| 6 | 68/12* | M1 | x | R | 56.0 | 53.5 | 44.5 | 49.0 | 10.0 | 55.0 | 40.5 |
| u/s | 61/32 | M1 | + | L | – | – | 29.0 | – | – | – | – |
| A | 77/289 | M1 | x | L | – | – | – | – | – | – | – |
| 6 | 69/22* | M2 | uw | L | 74.5 | uw | ic | 69.5 | 3.5 | ic | 21.5 |
|  |  |  | uw | R | 68.5 | uw | ic | 71.5 | 3.5 | ic | 21.0 |
| 6 | 68/12* | M2 | ew | R | 62.5 | c41.0 | 60.0 | >52.0 | 7.5 | 57.0 | 32.5 |
| 6 | 68/3* | M2 | x | L | 55.0 | 52.5 | 38.0 | 32.5 | bro | 58.0 | >40.0 |
|  |  |  | x | R | 55.5 | 53.5 | 44.0 | 34.0 | 12.0 | 60.0 | >40.0 |
| B | 78/658 | M2 | ++ | L | c45.0 | c44.0 | c35.0 | c32.0 | 10.0 | >40.0 | >40.0 |
| A | 78/460 | M2 | + | L | – | – | – | – | – | – | – |
| 3 | 77/511 | M3 | uw | R | – | – | – | 55.5 | 3.0 | 47.0 | 12.0 |
| 6 | 68/12* | M3 | uw | L | e | e | – | – | – | – | – |
|  |  |  | uw | R | 55.5 | 60.0 | ic | ic | 4.0 | 52.0 | 19.5 |
| 6 | 68/3* | M3 | ew | L | 54.0 | 58.0 | – | – | – | 52.5 | – |
|  |  |  | ew | R | – | 65.0 | 45.0 | 40.5 | – | 46.0 | – |
| A | 78/699 | M3 | x | L | 55.0 | 59.0 | 36.5 | c39.5 | 8.0 | c46.5 | 37.0 |

| | | | |
|---|---|---|---|
| ic | in crypt | vew | very early wear |
| uw | unworn | ew | early wear |
| e | erupting | x | medium wear |
| u/s | unstratified | — | broken |
| * | tooth in mandible | | |

+ worn
+ + very worn
c measurement approximate
/ measurement not applicable

[a]In millimetres. For illustration of measurements see Fig. 13.7.

**Table 13.15**  Measurements of Lower Dentition of Woolly Rhinoceros from La Cotte de St. Brelade[a]

| Layer | Ref. no. | Tooth | Wear | L/R | 1 Maximum length | 2 Height of anterior lobe | 3 Height of posterior lobe | 4 Minimum ectoloph width | 5 Maximum basal width | 6 Occlusal width |
|---|---|---|---|---|---|---|---|---|---|---|
| A | 77/1115 | dp3 | x | L | – | 24.0 | 24.0 | 7.5 | 18.5 | 13.5 |
| u/s | 77/1092 | dp3 | + | R | – | – | 20.5 | c5.0 | 18.5 | – |
| A | 77/617 | dp4 | + | R | 44.0 | 20.0 | 18.5 | 6.5 | 22.5 | 19.0 |
| 6 | 68/13* | P2 | x | R | – | – | – | – | – | – |
| A | 78/724 | P3 | x | L | 31.0 | 32.0 | 28.5 | 14.5 | 24.5 | 21.0 |
| 6 | 68/13* | P3 | x | R | – | – | – | – | – | – |
| 6 | 68/13* | P4 | ew | | 43.5 | 55.0 | ic | 9.0 | c26.5 | 22.5 |
| 5 | 73/82 | P4 | ew/x | L | 36.0 | 35.5 | 34.0 | 8.0 | 22.0 | 18.0 |
| B | 78/908 | M1 | uw | L | 50.5 | 52.0 | c47.5 | 4.0 | 23.0 | 16.0 |
| A | 77/617 | M1 | ew | R | 56.3 | 50.0 | 51.5 | 8.5 | 28.0 | 19.5 |
| 6 | 68/13* | M1 | x | R | 49.0 | 42.0 | 43.0 | c8.5 | 30.5 | 23.5 |
| A | 77/665 | M1 | x | R | 40.5 | 45.5 | 39.5 | 9.5 | 22.5 | 20.5 |
| u/s | 61/13 | M1 | uw | L | 51.5 | 52.5 | – | 4.5 | 26.0 | 18.0 |
| 6 | 68/13* | M2 | vew | R | 52.0 | 56.0 | – | 6.5 | 33.0 | – |
| 3 | 76/87 | M2 | x | L | 58.6 | 37.0 | 32.5 | 12.5 | 35.5 | 25.0 |
| 6 | 68/13* | M3 | ic | R | / | / | / | / | / | / |

| | | | |
|---|---|---|---|
| ic | in crypt | vew | very early wear |
| uw | unworn | ew | early wear |
| e | erupting | x | medium wear |
| u/s | unstratified | — | broken |
| * | tooth in mandible | | |

| | |
|---|---|
| + | worn |
| + + | very worn |
| c | measurement approximate |
| / | measurement not applicable |

[a]In millimetres. For illustration of measurements see Fig. 13.7.

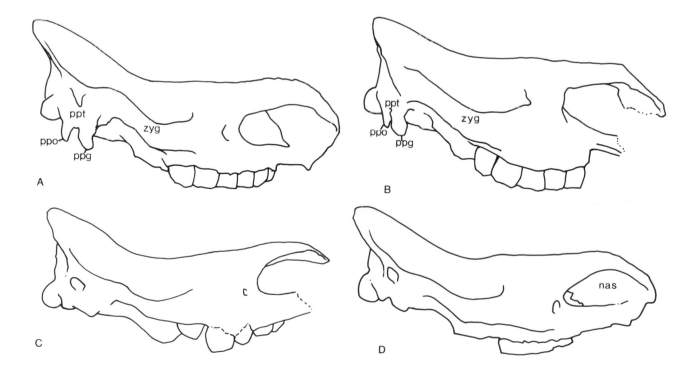

**Fig. 13.4**  Lateral views of skulls of (A) *D. hemitoechus* (Falc.); (B) *D. kirchbergensis* (Jäger); (C) *C. antiquitatis* (Blum.), La Cotte de St. Brelade 69/22; (D) *C. antiquitatis* (Blum.). zyg, zygomatic arch; ppg, processus postglenoideus; ppt, processus posttympanicus; ppo, processus paroccipitalis (A, B and D are after Loose 1975).

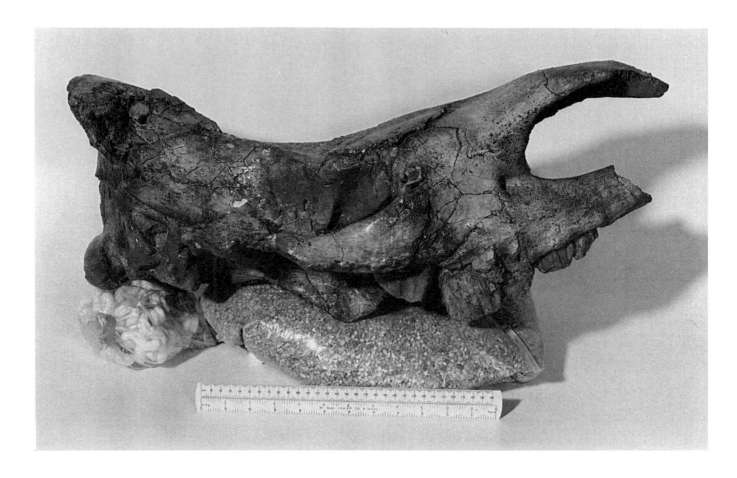

**Fig. 13.5** Skull of *Coelodonta antiquitatis*, lateral view (69/22).

**Fig. 13.6** Deciduous dentition of *Coelodonta antiquitatis*, left maxilla, occlusal view (78/372).

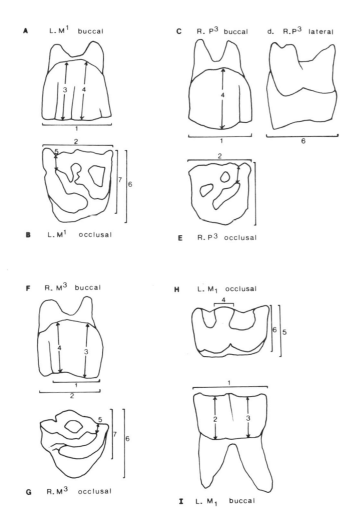

**Fig. 13.7** Measurements taken on dentition of woolly rhinoceros; (A) and (B) are upper deciduous teeth and first and second upper molars (M1 and M2), and (C), (D) and (E) are upper premolars (P2, P3 and P4): 1 maximum length, buccal; 2 maximum length of wear, occlusal surface, buccal; 3 crown height, anterior lobe (not applicable to premolars); 4 crown height, posterior lobe (maximum crown height in the case of premolars); 5 minimum width of ectoloph; 6 maximum width of tooth; 7 width of occlusal surface. (F) and (G) are the third upper molar (M3): 1 length of buccal edge; 2 total length including posterior fold; 3 crown height, anterior lobe; 4 crown height, posterior lobe; 5 minimum width of ectoloph; 6 maximum width of tooth; 7 width of occlusal surface. (H) and (I) show lower dentition: 1 maximum length; 2 height of anterior lobe; 3 height of posterior lobe; 4 minimum width of endoloph; 5 maximum basal width; 6 width of occlusal surface.

## 13.4  THE MAMMOTH REMAINS

The remains of mammoth are assigned to the genus and species *Mammuthus primigenius*, the woolly mammoth. In Europe, *M. primigenius* is represented by thousands of teeth and many skulls and post-cranial bones from Upper Pleistocene localities. Descriptions and illustrations of a great number of these specimens are provided by various authors (Toepfer 1957; Osborn 1936 and 1942; Hooijer 1955; Maglio 1973; Kubiak 1980 and 1982; Vereschagin and Baryshnikov 1982). It is on the basis of distinctive cranial and dental criteria outlined by these authors that the La Cotte material is identified as *M. primigenius*. These criteria are outlined below, and illustrations and measurements of the La Cotte mammoths are provided to confirm their taxonomic identification and for use by other workers.

As the majority of European mammoth remains come from unreliably dated localities, the degree of variation within the species is poorly known. Thus it is not possible to say whether the mammoths from La Cotte are typical of mammoths of comparable stratigraphic age. It is generally considered that Last Glaciation representatives of the species were shorter and more robust than their predecessors. However, it is not possible to discuss the size of the La Cotte mammoths in relation to other Pleistocene mammoths as almost all the bones, skulls and teeth belong to immature individuals.

### 13.4.1  Cranial and Dental Characteristics of *Mammuthus primigenius*

A number of criteria serve to distinguish the woolly mammoth from other elephants, and in particular from *Mammuthus trogontherii* (also referred to as *Mammuthus armenaicus*), which shares several characteristics with *M. primigenius*. Although the molar teeth are somewhat similar in structure, and the skulls of the two species show fairly similar adaptations to accomodate the high-crowned teeth, *M. primigenius* carries these adaptations to the extreme (Maglio 1973).

In the woolly mammoth the skull is extremely compressed anteroposteriorly, the parietals are greatly elevated, and the occipital plane is flat, meeting the parietals at a sharp acute angle. The forehead is concave fore and aft and convex laterally, the premaxillae are strongly directed downwards, and the tusk sheaths (alveoli) are closely spaced (Fig. 13.8). Some of these characteristics may be observed in the specimens from La Cotte (Fig. 13.9), although, unfortunately, the parietal region is damaged or missing in all cases. However, the molar teeth of *M. primigenius* are so distinctive that they are rarely mistaken for those of any other elephant, and there are many of these from La Cotte. Three characteristics in particular serve to distinguish them from those of other species:

1.  The plates (lamellae) of which the tooth is comprised are thinner and more closely spaced than in any other elephant. In elephants this characteristic is

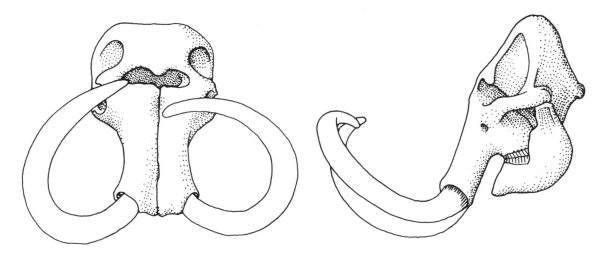

**Fig. 13.8** Frontal and lateral views of skull of *M. primigenius* (after Toepfer 1957).

expressed as the lamellar frequency: a calculation of the number of plates in a distance of 10 cm along the anteroposterior axis of the tooth. In *M. trogontherii* the plates are thin and closely spaced with a lamellar frequency of 5–8 for the last three molars; in *M. primigenius* the lamellar frequency for these teeth is 7–12 (Maglio 1973). In the case of La Cotte mammoths, the lamellar frequency of these teeth ranges between 8 and 13. This high degree of lamellar compression may be observed, for example, in Fig. 13.10.

2. The enamel is usually extremely thin and finely ribbed. In *M. trogontherii* enamel thickness ranges between 1.0 and 3.0 mm in the last three molars, in *M. primigenius* between 1.0 and 2.0 mm (Maglio 1973). The enamel thickness of the last three molars in the La Cotte specimens ranges between 1.1 and 2.1 mm.

3. Crown height relative to crown width is greater than in any other elephants.

These dental characteristics are best described by measurement as outlined below.

### 13.4.2 Measurement of Mammoth Molars

In the mammoth there are six molariform teeth in each jaw. For the purposes of this study they are referred to as M1–M6 (see section 13.5.5). A number of measurements (given in Tables 13.16 and 13.17) were taken on the more complete mammoth molars from La Cotte following Maglio (1973) and Archipov (1976): overall length, number of plates (lamellae), height of crown and width of tooth. The method of taking these measurements may be summarised as follows.

**Length (L).** This is a difficult measurement to standardise because of the curvature of the teeth sagittally and horizontally. Measurement taken along the occlusal surface does not necessarily give an accurate account of length

as plates come into wear successively, and the tooth is therefore frequently longer than the wear surface. The most accurate measurement of length is taken along the axis of growth and perpendicular to the average lamellar plane (Fig. 13.11).

**Plate number (P).** As each molar becomes worn it is forced forward and out of the alveolus by the subsequent tooth. In cases where anterior plates have worn away, a plus sign (+) precedes the plate number to indicate that the original number was greater. In the case of broken teeth, plates are counted from the front as P1, P2 and so on, and as PI, PII when counted from behind. The plate formula for the European woolly mammoth (Osborn 1942) is

$$M1\tfrac{4}{4} \quad M2\tfrac{8}{8} \quad M3\tfrac{12}{12} \quad M4\tfrac{12}{12} \quad M5\tfrac{16}{16} \quad M6\tfrac{24}{24}$$

**Width (W).** This is simply the widest part of the tooth. In order to make this measurement more useful for comparative purposes it is important to know on which plate the measurement was taken. The plate number is given in parentheses after the measurement (P1, P2 etc. if counted from the front, and PI, PII etc. if counted from the back).

**Crown height (H).** Maximum crown height is measured vertically along a plate from the base of its enamel sheath to the highest point of the tooth. This measurement should be taken parallel to the vertical axis of the plate (Fig. 13.11). When the plate is in wear the measurement is followed by a plus sign (+). Maximum height is accompanied by the plate number on which it was taken.

### 13.4.3 Measurement of Tusks

At birth the woolly mammoth had a pair of deciduous tusks (tushes) which reached a length of 30–60 mm. None of these is represented at La Cotte. The tushes

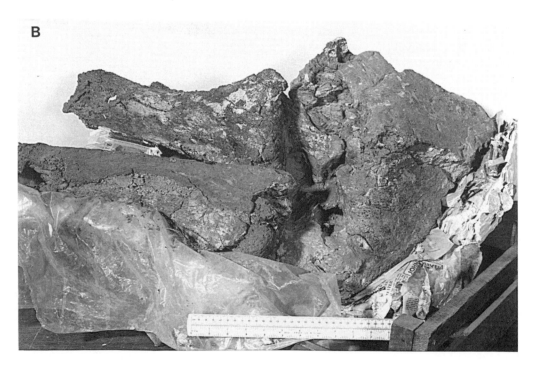

**Fig. 13.9** Mammoth skulls from layer 3 undergoing conservation in the laboratory. (A) 76/175 (lateral view, upside down); faulting has caused M5 to be pushed out of alignment with M6. (B) 76/429 (frontal view); note that, as this is an immature individual, the premaxillae had not fused and have separated.

B

A

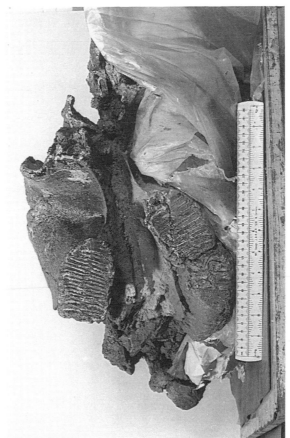

C

**Fig. 13.10** *M. primigenius.* (A) Left mandible from layer 6; occlusal view, with M4 in wear and part of unerupted M5 visible in the jaw (72/47). (B) Skull from layer 3; occlusal view showing M5 and M6 (76/98). (C) Skull from layer 3; occlusal view showing M4 and M5 (74/434).

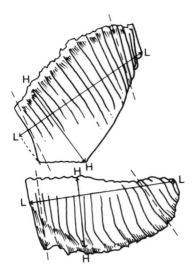

**Fig. 13.11** Upper and lower molars of *M. primigenius* in sagittal section. Plane of measurement (L) shown perpendicular to average plate direction.

**Table 13.16** Measurements of Dentition of Mammoth from La Cotte de St. Brelade: Layers 6, 4 and 3[a]

| Layer | Ref.no. | Tooth | L/R | No. of lamellae | Length | | Width | | Height | |
|-------|---------|-------|-----|-----------------|--------|---|-------|-----------|--------|-----------|
| 6 | 70/59 | u M4 | R | +3 | – | | 64.0 | (PIII) | 110.0 | (PIII) |
| 5/6 | 73/264 | u M4 | – | +5 | – | | 59.5 | (PV) | 102.5 | (PV) |
| 6 | 68/40 | l M3 | – | +6 | 59.0 | | 39.0 | (PVI) | 46.0 | e |
| 6 | 70/47 | l M4 | L | +11 | 113.5 | | 55.0 | (P5) | – | |
| 6 | 70/49 | l M4 | L | 12–14 | 118.5 | | 61.5 | (PV) | – | |
| 4 | 73/246 | u M6 | L | +13 | 240.0 | e | 85.0 | e | 228.0 | (PXIII) |
| 3 | 76/222 | u M3 | L | +7 | 57.5 | | 34.0 | (P4) | 32.5 | (PI) |
| 3 | 76/85 | u M3 | – | +8 | – | | 30.0 | e(P2) | 100.0 | (P3) |
| 3 | 78/9 | u M4 | – | +10 | >100.0 | | 71.0 | (PVII) | 117.0 | (PX) |
| 3 | 77/33 | u M4 | R | +7 | >67.0 | | 51.5 | (PVII) | 52.0 | (PI) |
| 3 | 74/434 | u M4 | L | 10–12 | 103.0 | | 67.5 | (PVIII) | 44.5 | (PI) |
| 3 | 74/434 | u M4 | R | 10–12 | 93.0 | | 69.0 | (PVII) | 42.5 | e |
| 3 | 74/429 | u M4 | L | 12 | 130.5 | | 60.0 | (P4) | 60.0 | e |
| 3 | 74/429 | u M4 | R | 12 | 134.0 | | 60.5 | (P4) | 60.0 | e |
| 3 | 76/280 | u M4 | R | +8 | >77.0 | | 55.5 | (PIII) | 32.0 | e |
| 3 | 76/501 | u M5 | – | +5 | – | | 70.0 | | 155.0 | e |
| 3 | 74/434 | u M5 | L | 19–20 | 173.0 | e | 63.0 | (P3) | 140.0 | e |
| 3 | 76/98 | u M5 | L | +10 | 140.0 | | 77.0 | (PVIII) | 28.5 | (PV) |
| 3 | 76/98* | u M5 | R | +10 | 140.5 | | 78.0 | (PVIII) | 29.0 | (PV) |
| 3 | 76/175 | u M5 | L | +9 | 81.0 | | 76.0 | (PIV) | 30.5 | e |
| 3 | 76/175 | u M5 | R | +10 | 90.0 | | 77.5 | (PV) | 33.5 | e |
| 3 | 76/121 | u M5 | L | +16 | 179.0 | | 64.5 | (P4) | 147.0 | e |
| 3 | 76/98 | u M6 | L | – | – | | 76.0 | (P3) | – | |
| 3 | 76/98 | u M6 | R | – | – | | 76.0 | (P3) | – | |
| 3 | 76/175 | u M6 | L | 17 e | 180.0 | e | 74.0 | (P4) | 195.0 | e(P10) |
| 3 | 76/175 | u M6 | R | 17 e | 180.0 | e | 70.5 | (P4) | 195.0 | e(P10) |
| 3 | 76/88 | l M4 | L | +7 | 110.0 | e | 54.5 | (P6) | 97.0 | (PI) |
| 3 | 76/189 | l M4 | R | +7 | 110.0 | e | 57.0 | (P6) | 97.0 | (PI) |

| | | | |
|---|---|---|---|
| * | tooth in mandible or maxilla | u | upper dentition |
| e | estimate | l | lower dentition |
| – | broken | | |

[a]Measurements in millimetres. For illustration of measurements see Fig. 13.11.

**Table 13.17**  Measurements of Dentition of Mammoth from La Cotte de St. Brelade: Layers A, B, and C[a]

| Layer | Ref.no. | Tooth | L/R | No. of lamellae | Length | Width | | Height | |
|-------|---------|-------|-----|-----------------|--------|-------|---|--------|---|
| A | 78/701 | 1 M1 | – | 4 | 29.0 | 24.0 | (P2) | 26.8 | (P2) |
| A | 78/683 | 1 M2 | – | – | 50.0 | 33.0 | | – | |
| A | 78/491 | 1 M3 | – | 7 | – | 40.0 | e | 27.4 | e |
| A | 77/786 | 1 M4 | L | +8 | – | 51.9 | (P4) | 50.0 | (P8) |
| A | 78/877 | 1 M4 | – | +3 | – | 40.0 | e | – | |
| A | 77/1033 | u M6 | L | +17–18 | 210.0 | 71.0 | (P3) | 190.0 | (P4) |
| A | 76/249 | fragment M1/2/3 | – | – | 40.0 | 30.0 | | – | |
| A | 78/745 | fragment M1/2/3 | – | – | 50.0 | 40.0 | | – | |
| A | 76/430 | fragment M1/2/3 | – | – | 80.0 | 60.0 | | – | |
| A | 78/478 | fragment M1/2/3 | – | – | 105.0 | 40.0 | | – | |
| A | 76/230 | fragment | – | – | 140.0 | 105.0 | | – | |
| B | 78/917 | fragment u M2++ | – | +4 | – | 33.5 | | 12.2 | |
| B | 78/533 | u M2 | R | 6–7 | 44.5 | 28.5 | (P5) | 22.5 | (P5) |
| B | 78/877 | u M2//M3 | | | 25.0 | 40.0 | e | | |
| B | 78/682 | posterior part u M4uw | – | – | 45.0 | 47.0 | | – | |
| B | 78/887 | u M4 | R | +10 | – | 46.6 | (P4) | 125.6 | (P6) |
| B | 78/871 | posterior part u M4/M5uw | – | +7 | – | 61.0 | (PVI) | 109.6 | (PVI) |
| B | 78/1066 | fragment M4/M5 | – | – | – | 65.0 | | 100.0 | |
| C/B | 78/834 | u M5 | L | +14 | >142.5 | 77.5 | (P3) | 173.0 | (P10) |
| C | 78/685 | u fragment M4/M5 | – | – | – | 82.0 | | 150.0 | e |
| C | 78/519 | 1 M3 | – | +9 | – | 31.1 | (P2) | 59.0 | (P8) |

| | | | |
|---|---|---|---|
| u | upper dentition | l | lower dentition |
| — | broken | e | estimate |
| + + | tooth very worn | uw | unworn |

[a]Measurements in millimetres. For illustration of measurements see Fig. 13.11.

were replaced after one year by permanent tusks which grew throughout the life of the animal, becoming more strongly curved and spirally twisted with age, turning forward and inward to form a huge circle. As in the case of extant elephants, the tusks of males attained an average thickness and length greater than those of females. Tusks of old males weighed up to 100–120 kg and were 3.5–4.0 m in length (Vereschagin and Baryshnikov 1982).

Measurements taken on the mammoth tusks from La Cotte are presented in Table 13.18 (see also Fig. 13.12). Few of these measurements indicate more than the dimension of the portion of the tusk that survives, and as many were crushed, it was not always possible to distinguish left from right. The most useful measurement on this material is the maximum depth of the proximal end of the tusk, from which it is apparent that the majority of tusks belonged to relatively small (i.e. young) individuals.

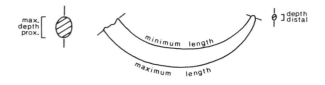

**Fig. 13.12**  Measurements taken on mammoth tusks.

None of the La Cotte tusks was recovered in the alveolus of a skull. In order to ascertain the likelihood that some of the tusks might belong to some of the skulls, a comparison is made between the measurements of the tusk sheaths (alveoli) and the proximal ends of the isolated tusks (Fig. 13.13). It is interesting to note that four of the tusks in layer 3 belong to animals that are larger than are represented by skulls. This is important as it raises the estimated minimum number of mammoths represented by heads in this level from 9 to 11 or 12. Similarly, in layer 6 there are only two skulls but four tusks of different sizes indicating that a minimum of four mammoths are represented by cranial material.

### 13.4.4  Mammoth Post-cranial Remains

Those bones identifiable to skeletal element have been measured to allow comparison to be made of the relative sizes of the individuals represented (for illustration of these measurements, see mf 13.6).

**Scapulae** (mf 13.7). The length of the glenoid (1) and the width of the proximal end of the scapula (2) give the clearest idea of the size of the animals represented at La Cotte. As many specimens are obviously subadult, the distal margin is unfused, relatively thin and therefore easily damaged. Thus length (4) and distal width (5) as shown in mf 13.7 are generally minimum dimensions.

**Table 13.18**  Measurements of Mammoth Tusks from La Cotte de St. Brelade[a]

| Layer | Ref. no. | L/R | Maximum proximal end | | Maximum (outer) length | | Minimum (inner) length | | Distal depth | |
|---|---|---|---|---|---|---|---|---|---|---|
| 6 | 68/18 | – | 81.5 | e | 940.0 | | 865.0 | | 23.0 | e |
| 6 | 66/4 | L | 58.0 | | 960.0 | | 860.0 | | 28.5 | |
| 6 | 70/66 | L | 87.5 | | 620.0 | | 550.0 | | 27.0 | |
| 6 | 68/8 | – | bro | | 360.0 | e | 360.0 | e | 67.0 | |
| 6 | 68/4 | R | 101.0 | | 620.0 | | 550.0 | | 71.0 | |
| 3 | 73/292 | R | 110.0 | | 1130.0 | | 890.0 | | 75.0 | e |
| 3 | 74/426 | R | 110.0 | | 1600.0 | | 1400.0 | | 66.5 | |
| 3 | 77/256 | R | 36.5 | | 400.0 | | 360.0 | | bro | |
| 3 | 77/539 | – | 28.5 | | 200.0 | | 190.0 | | bro | |
| 3 | 76/274 | – | 72.0 | | 870.0 | | 760.0 | | bro | |
| 3 | 76/418 | – | 85.5 | | 450.0 | e | 380.0 | e | 82.0 | |
| 3 | 76/434 | L | 25.5 | | 230.0 | | 220.0 | | 19.5 | |
| A | 77/450 | R | 76.5 | | 800.0 | | 670.0 | | 40.0 | |
| A | 77/1123 | – | 188.5 | | 670.0 | e | 610.0 | e | bro | |
| A | 77/834 | – | 51.5 | | 340.0 | | 340.0 | | 18.5 | |
| A | 77/486 | – | 51.5 | | 410.0 | | 360.0 | | bro | |
| A | 78/685 | – | 65.0 | | 180.0 | | 180.0 | | 45.9 | e |
| A | 76/302 | – | 97.5 | | 1140.0 | | 940.0 | | 69.0 | e |

bro  broken               e  estimate
—  L/R indeterminate

[a]In millimetres. For illustration of measurements see Fig. 13.12.

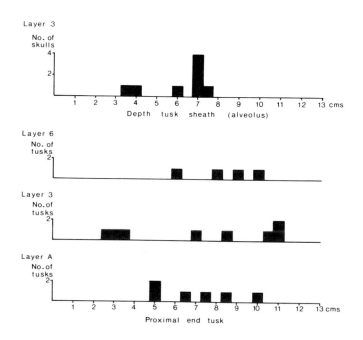

**Fig. 13.13**  Comparison of measurements of tusks and tusk sheaths (alveoli) of *M. primigenius* from La Cotte de St. Brelade.

**Humeri** (mf 13.8). None of the humeri is complete, and in no case does the proximal epiphysis survive. Three of the specimens have fused distal epiphyses, and the others are immature diaphyses without epiphyses.

**Ulnae** (mf 13.9). Only one ulna could be measured; the other three are diaphyseal fragments.

**Innominates** (mf 13.10). Three innominates are complete. The others are best represented by the ilium and part of the acetabulum. Where preservation is good, it can be seen that the ilium has not been broken at this point, but rather that fusion with the ischium and pubis has not taken place. These latter are present in the assemblages, but not as well represented as the ilia. As in the case of the vertebral border of the scapulae, the outer margin of the iliac crest is one of the last bones to fuse; this makes it difficult to obtain an accurate measurement of the ilium (4), as it is almost invariably broken. The most useful measurements appear to be width (1) and length (2) of the acetabulum, and the base of the ilium (3). This minimum distance can be taken on the lateral and medial side of the pelvis, but it is important to be consistent with this measurement, as measurements of the lateral and medial sides give very different readings.

**Femora** (mf 13.11). The femur is represented by several diaphyses of varying size. There are no proximal epiphyses, but there are a few distal epiphyseal condyles.

**Tibiae** (mf 13.12). There is only one complete tibia with fused epiphyses. Two other tibiae are represented by diaphyses of immature animals.

It is evident from the measurements given in mf 13.7–13.12 and from the state of fusion of epiphyses that a range of young and adult mammoths is represented at La Cotte. However, it should be remembered that although differences in size may indicate differences in age, sexual dimorphism also accounts for size differences among mammoth (Hooijer 1955; Skarlato 1977), and among elephants after the age of about 10 years (Laws 1966; Hanks 1979). It is also important to realise that the incomplete fusion of the epiphyses of long bones is not necessarily indicative of a young animal. The living elephants continue to grow throughout their lives, and the epiphyses of limb bones and vertebrae only begin to fuse after the appearance of the last molar (M6) when the animal is about 30 years of age (Deraniyagala 1955; Laws 1966).

## 13.5 AGE AT DEATH OF THE LA COTTE MAMMOTHS AND RHINOCEROSES

It is clear from the preceding pages that although mammoth and woolly rhino are represented in several layers at La Cotte, they are extraordinarily represented in Layers 3 and 6. The significance of these two principal accumulations is discussed in chapter 31, where the hypothesis is put forward that on two occasions herds of these animals were driven over the headland (La Cotte Point) into the gullies below, one of which is part of the excavated site. A substantial part of the evidence for two such 'events' is suggested by the condition of the bones and the nature of the matrix in which they were recovered. However, the mortality profiles constructed for both species also seem to lend weight to such an interpretation, rather than one favouring the sporadic hunting or scavenging of individual animals over a long period.

The methods used in the construction of the age profiles of the La Cotte mammoths and woolly rhinos are detailed below. Both profiles are based on analyses of dentition. However, as dental eruption and attrition differs in these two species, the mortality profiles have been constructed somewhat differently, as will be discussed below. In both cases the estimated age of death of all specimens is inferred from data on living counterparts: the black rhinoceros *Diceros bicornis* and the African elephant *Loxodonta africana*.

### 13.5.1 The Construction of Age (Mortality) Profiles Based on Woolly Rhinoceros Dentition

Two methods are used in conjunction in order to estimate the age at death of the La Cotte rhinos: (1) the measurement of the dentition, and (2) the subjective evaluation of dental eruption and wear.

The most practical and accurate of these methods is the measurement of dental crown height. The construction of age profiles from crown height measurements is based on the fact that the crown height of ungulate teeth steadily decreases with age. In order to construct a profile that will include individuals of all ages, crown height must be measured on a category of deciduous teeth as well as on a category of permanent teeth (Klein 1981). Unfortunately, there are too few isolated teeth per layer in the La Cotte assemblages to allow such systematic selection. If, for example, the deciduous premolar (dp4) and the permanent premolar (P4) are selected for the construction of mortality profiles, the total number of teeth for all layers is eight.

Instead, age distribution of the La Cotte rhinos is arrived at by the synthesis of measurements made on all the dental material and the visual evaluation of dental eruption and wear. As Klein (1981) points out, visual evaluation is generally a less satisfactory method of analysis than is systematic measurement, particularly when teeth are recovered in quantity. In addition, it is a method most accurately employed on that material least likely to survive on archaeological sites: erupting and newly erupted teeth, which are delicate and easily destroyed, and more or less complete mandibles and maxillae, which are seldom recovered from archaeological deposits. However, it is a method which is applicable to the analysis of the La Cotte rhinos because there are several virtually complete skulls and mandibles. These have been used to construct an age framework within which isolated teeth may be evaluated. It is particularly fortuitous that all these specimens are immature, with dentition in various stages of eruption and wear.

### 13.5.2 Dental Eruption and Replacement in the Woolly Rhinoceros

The development, eruption and subsequent attrition of teeth follows the same pattern in both mandibles and maxillae, although the maxillary dentition usually develops slightly ahead of the mandibular dentition. A number of studies of extinct and extant members of the Dicerorhininae serve to illustrate that there is a fairly consistent pattern to the eruption and replacement of deciduous and permanent teeth. These may be summarised as follows. The animal is born with four deciduous teeth (dp1–dp4) which erupt during the first few months of life. The first molar (M1) is the next tooth to appear. This tooth is in marked wear and the second molar (M2) is erupting but unworn when the deciduous premolars are replaced by the permanent premolars. This stage of dental development is well illustrated by skull 69/22 (see Fig. 13.14). Dp2, dp3 and dp4 are generally replaced in

**Fig. 13.14** Skull of *Coelodonta antiquitatis*, occlusal view (69/22). The left maxilla has a broken anterior premolar (P$^2$), and P$^3$ in early wear. P$^4$ is in the process of erupting and therefore is unworn. The first molar (M$^1$) is in wear, and M$^2$ is erupting but unworn. The right maxilla is in a similar state of eruption and wear with the exception of the posterior deciduous tooth; dp$^4$ is still *in situ*, but very worn, and its permanent counterpart is visible in the maxilla.

**Table 13.19**  Allocation of Rhinoceros Cranial Material from La Cotte to Ontogenetic Groups[a]

| Age category | State of dentition in complete skulls | State of dentition of La Cotte rhinos | Ref. no. |
|---|---|---|---|
| 'Young' | Deciduous dentition or replacing deciduous dentition up until the loss of dp4 | L. maxilla with dp1-4 in wear | 78/372 |
|  |  | L. P2 broken; R. P2 early wear |  |
|  |  | L. and R. P3 early wear |  |
|  |  | L. dp4 broken; R. dp4 very worn with P4 visible below | 69/22 |
|  |  | L. and R. M1 early to medium wear |  |
|  |  | L. and R. M2 erupting, unworn |  |
| 'Young adult' | P2 and P3 in wear M1 and M2 in wear P4 and M3 erupting, but unworn | anterior of skull missing; L. P3 medium wear; R. P3 broken L. and R. P4 erupted, unworn L. M1 missing; R. M1 medium wear L. M2 missing; R. M2 early wear L. M3 erupting, unworn R. M3 erupted, unworn | 68/12 |
| 'Adult' | P4 and M3 in early wear | anterior of skull missing; L. and R. M2 in medium wear L. and R. M3 in early wear | 68/3 |
| 'Old adult' | M3 worn to two-thirds of initial crown height |  |  |
| 'Old' | M3 ½ or more than ½ worn |  |  |

[a]Modified from data presented by Borsuk-Bialynicka 1973.

**Table 13.20**   Allocation of Rhinoceros Mandibles from La Cotte to Age Groups[a]

| Age category | State of dentition in complete mandibles | State of dentition of La Cotte rhinos | Ref. no. |
|---|---|---|---|
| 'Young' | Deciduous dentition or replacing deciduous dentition up until loss of dp4 | R. mandible, dp4 worn; M1 early wear | 77/617 |
| 'Young adult' | P2 and P3 in wear; M1 and M2 in wear; P4 and M3 erupting but unworn | R. mandible, P2 and P3 broken; P4 early wear; M1 medium wear; M2 very early wear; M3 erupting, unworn | 68/13 |
| 'Adult' | P4 and M3 in early wear | | |
| 'Old adult' | M3 worn to two-thirds of initial crown height | | |
| 'Old' | M3 ½ or more than ½ worn | | |

[a]Partly based on data presented by by Borsuk-Bialynicka 1973.

that order by P2, P3 and P4. The first deciduous premolar (dp1) is a vestigial tooth and not replaced by a permanent premolar. M2 may erupt slightly before or slightly after the loss of dp2 and dp3. P4 then appears between P3 and M1. The eruption of P4 is almost invariably simultaneous with that of the third molar (M3).

### 13.5.3   Allocation of the La Cotte Rhinos to Age Classes Based on Cranial and Dental Development

Progressive changes in the dentition of the woolly rhinoceros are well illustrated by the three skulls from layer 6 at La Cotte, each of which is at a different stage of development. The state of eruption of the dentition of these specimens is outlined in Table 13.19. Here they are allocated to age categories devised by Borsuk-Bialynicka (1973). On the basis of a combination of dental and cranial criteria she allocates her study material to four categories: 'young', 'young adult', 'old adult' and 'old'. By the same criteria the skulls from La Cotte are placed within these broad age groups, but in order to accommodate a specimen at a stage of development between 'young adult' and 'old adult', an additional category 'adult' is included here. This is defined solely by the state of eruption and wear of the teeth and not by cranial characters.

In addition to the skulls there are three mandibles, two of which have dentition. These are allocated to age groups on the basis of the state of eruption of the teeth, as described for the skulls (Table 13.20). Specimen 68/13 (a right mandible) was found with, and almost certainly belongs to, a skull with dentition in the same state of wear (68/12).

Once the virtually complete skulls and mandibles had been assigned to relative age categories, the teeth were measured. These specimens then constituted the basis of an age framework within which the isolated teeth (also measured) could be assessed.

The measurements made on the rhinoceros dentition from La Cotte (all levels) are presented in Tables 13.14 and 13.15 (for description see Fig. 13.7). Two pertaining to the upper dentition are considered to be most useful for describing progressive wear (age) of the tooth: maximum crown height (MCH) and minimum ectoloph width (MEW). As the height of the crown decreases with wear, so the width of the ectoloph increases.

Either MCH or MEW can be used as the basis for relative ageing. MEW is selected here because crown height cannot always be measured when teeth are in jaws, as they are in the best specimens from La Cotte. In Table 13.21 all the teeth from La Cotte are arranged in order of eruption and of increasing wear on the basis of the measurement MEW. An indication of the relative degree of wear by subjective evaluation accompanies this measurement. The specimens delineated are those contained in a mandible or maxilla.

It is clear from the tables and illustrations that most individuals represented at La Cotte (whether by skulls, mandibles or isolated teeth) are comparatively young. With the exception of one M3 (78/699), the oldest animal is represented by a skull (68/3) which is assigned to the category 'adult'. Animals in the category 'old' are not represented.

In order to refine these broad age categories and to define more clearly the age distribution of individuals within each level, the age data from La Cotte is compared with that available for the East African black rhino *Diceros bicornis*.

**Table 13.21** Dentition of *C. antiquitatis* from La Cotte, Arranged in Order of Progressive Age as Defined by Minimum Ectoloph/Endoloph Width (MEW) and the Subjective Evaluation of Wear[a]

| Ref. no. | Layer | L/R | dp1 | dp2/P2 | dp3/P3 | dp4/P4 | M1 | M2 | M3 |
|---|---|---|---|---|---|---|---|---|---|
| **UPPER DENTITION** | | | | | | | | | |
| *Young* | | | | | | | | | |
| 78/896 | C | L | dp02.5 vew | | | | | | |
| 77/753 | C | R | dp05.5 x | | | | | | |
| 78/668 | B | L | dp05.5 x | | | | | | |
| 78/937 | C | R | dp06.0 + | | | | | | |
| 78/962 | B | L | dp06.5 ++ | | | | | | |
| 78/602 | B | L | | dp02.0 uw | | | | | |
| 77/672 | A/3 | L | | dp03.0 x/+ | | | | | |
| 78/785 | B | L | | dp08.0 x/+ | | | | | |
| 78/661 | C/B | R | | | dp02.0 uw | | | | |
| 77/314 | A/3 | – | | | dp03.5 ++ | | | | |
| 78/372 | C | L | dp02.0 + | dp06.5 ++ | dp08.5 ++ | dp03.5 ++ | | | |
| 78/1187 | B | L | | | P02.0 uw | | | | |
| 69/22 | 6 | L | | P03.5 ew | P02.5 ew/x | $\frac{dp13.5}{p/ic}$ +++ | 05.5 ew/x | 03.5 uw | |
| | | R | | P/bro | P04.0 ew/x | P/bro | 03.5 ew/x | 03.5 uw | |
| *Young adult* | | | | | | | | | |
| 61/45 | D/C | R | | | P04.5 ew/x | | | | |
| 68/12 | 6 | L | | | P08.5 x | P03.0 uw | bro x | bro ew | er uw |
| | | R | | | P/bro | P/bro | 10.0 x | 07.5 ew | 04.0 uw |
| *Adult* | | | | | | | | | |
| 77/570 | A/3 | L | | | P05.5 vew | | | | |
| 78/303 | A | R | | | P14.5 x | | | | |
| 78/544 | A | L | | | | | 02.0 uw | | |
| 77/289 | A | L | | | | | 10.0 x | | |
| 78/658 | B | L | | | | | | c10.0 x/+ | |
| 78/460 | A | L | | | | | | 10.0 x/+ | |
| 77/511 | A/3 | R | | | | | | | 03.0 uw |
| 68/3 | 6 | L | | | | | | bro x | bro ew |
| | | R | | | | | | 12.0 x | bro ew |
| *Old adult* | | | | | | | | | |
| 78/699 | A | L | | | | | | 08.0 x | |
| **LOWER DENTITION** | | | | | | | | | |
| *Young* | | | | | | | | | |
| 77/1115 | A | L | | | dp07.5 x | | | | |
| 77/617 | A | R | | | | dp06.5 + | 08.5 ew | | |
| 78/724 | A | L | | | dp06.0 + | | | | |
| *Young adult* | | | | | | | | | |
| 73/82 | 5 | L | | | | P08.0 ew/x | | | |
| 68/13 | 6 | R | | | | P09.0 ew | 08.5 x | 06.5 vew | ic uw |
| 78/908 | B | L | | | | | 04.0 uw | | |
| 61/13 | G/H | L | | | | | 04.5 uw | | |
| 77/665 | A | R | | | | | 09.5 x | | |
| *Old adult* | | | | | | | | | |
| 76/87 | A | L | | | | | | 12.5 x | |

| | | | | | | |
|---|---|---|---|---|---|---|
| ic | in crypt | ew | early wear | uw | unworn |
| x | moderate wear | er | erupting | vew | very early wear |
| + + + + | worn/very worn | bro | broken | | |

[a]Measurements in millimetres; a leading zero has been introduced where the value is below 10, in order to simplify the appearance of the table. Delineation indicates a mandible or maxilla.

### 13.5.4 Estimated Age at Death of the La Cotte Rhinos Inferred from Data on the Black Rhino *Diceros bicornis*

This is not an attempt to establish absolute ages in years for the La Cotte rhinos, but rather to indicate more clearly the proportion of their potential life-span expended by the time death occurred. In herbivores potential longevity is effectively the time it takes from birth for the deciduous teeth to be replaced and the permanent teeth to be worn down to the gums so that the animal can no longer sustain itself. Detailed information on this process is available for the black rhino (Schenkel and Schenkel-Hulliger 1969; Goddard 1970).

It may be argued that a comparison is inadmissable between two species subject to entirely different environmental influences. The potential life-span of the woolly rhino may well have been different from that of the extant rhino, but this is unimportant in this context. What is important is that the *sequence* of dental eruption and replacement is observed to be the same in both species, and it is assumed that tooth replacement occurs at a similar point in the life cycle of these related species, regardless of actual age.

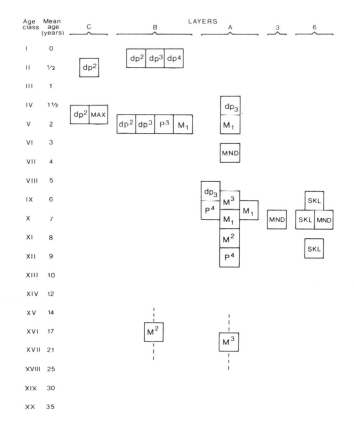

**Fig. 13.15** La Cotte woolly rhinoceros dentition assigned to age classes on the basis of age data for the black rhinoceros. MAX, dentition in maxilla; SKL, dentition in virtually complete skull; MND, dentition in mandible. Vertical broken lines signify an approximate age category.

The rhinoceros dental material from La Cotte is assigned to age classes following criteria established for the black rhino by Goddard (1970). Goddard collected the crania and mandibles of over 500 black rhinos from the Tsavo National Park in Kenya. This material was divided into 20 relative age classes based on the eruption and wear of the complete series of teeth. The characteristics for each class are outlined in Table 13.22. Goddard then assigned crude ages in years based on an estimate of life expectancy and the examination of the dental records of captive animals of known age. As the inherent effects of captivity and nutrition on the rate of wear of the teeth of captive animals are well known, these crude ages were then refined by the examination of the dental records of wild rhinoceros of known age, and an age scale established. As the sample of young wild animals of precisely known age was fairly large, Goddard considered that up to Class XI the ages assigned are very accurate.

Using the criteria outlined in Table 13.22, the specimens from La Cotte were assigned to age classes as illustrated in Fig. 13.15. It is evident that most teeth belong to fairly young animals. This predominance of young animals is better illustrated in Fig. 13.18, where the age classes are given their correct relative intervals. As there are relatively few teeth, it was possible to calculate a minimum number of individuals (as shown in the same illustration) by matching teeth in a comparable state of wear and by gauging the possibility that several isolated teeth might belong in one tooth row, or that maxillary teeth might have counterparts among mandibular teeth.

### 13.5.5 Estimated Age at Death of the La Cotte Mammoths Inferred from Data on the African Elephant *Loxodonta africana*

The ageing of the La Cotte mammoths is based on the fact that growth and replacement of teeth of *M. primigenius* take place in the same way as in living elephants. There are 24 molariform teeth (6 in each jaw), a pair of tusks (the upper incisors) which are replaced once, and no canines. There is some discussion as to whether the first three of the six molariform teeth are true milk molars or premolars, and they are referred to by various authors as milk molars (mm2–mm4), as deciduous molars (dm2–dm4), and as molars (M1–M3). As the La Cotte mammoths are assigned to age groups on the basis of work on the African elephant carried out by Laws (1966), his nomenclature has been adopted: M1–M6.

The mammoth tooth is composed of a variable number of parallel plates (lamellae) of dentine, covered with enamel and joined by cement. The six molars develop in sequence, the first forming towards the anterior of each jaw and others developing progressively behind. Of these six teeth, three are already formed at birth, the first two of which erupt and come into wear within the first few weeks of life (Fig. 13.16). Only one or two molars in each jaw are in use at any one time. As the anterior tooth moves forward, the occlusal surface is worn down

**Table 13.22**  Tooth Wear Related to Age in the Black Rhinoceros[a]

| Age class | Mean age (years) | Probable range % | State of dentition |
|---|---|---|---|
| I | 0 | – | dp1 and dp2 just erupting; dp3 and dp4 not usually erupted |
| II | 0.5 | – | dp2 and dp3 fully erupted, slight wear; dp4 has pierced gum |
| III | 1 | – | dp2 and dp3 in wear dp4 fully erupted but little or no wear; crests of M1 visible just below bone level |
| IV | 1.5 | – | dp2, dp3 and dp4 show marked wear; anterior crest of M1 erupted |
| V | 2 | 0.5 | M1 erupted, sometimes fully, but not in wear |
| VI | 3 | 0.5 | M1 slight to marked wear; M2 about to erupt |
| VII | 4 | 0.5 | deciduous premolars very worn and permanent premolars erupting above bone level; M1 in marked wear; M2 usually erupted |
| VIII | 5 | 0.5 | P2 and P3 sometimes replaced dp2 and dp3; M2 fully erupted, not usually in wear |
| IX | 6 | 0.5 | P2 and P3 invariably replaced dp2 and dp3; P4 usually replaced dp4; M1 in marked wear; M2 in slight to medium wear; M3 just erupted above bone |
| X | 7 | 1 | P2-4 fully erupted and in wear; M3 approx. ½- erupted, not in wear |
| XI | 8 | 1 | P1 usually lost; M3 fully erupted, sometimes in early wear |
| XII | 9 | 1 | P1 missing in 90% of specimens; M1 very worn; M3 early wear |
| XIII | 10 | 1 | |
| XIV | 12 | 1 | permanent dentition continuing to wear |
| XV | 14 | 2 | |
| XVI | 17 | 2 | P2 sometimes missing; M1 occlusal surface very worn, tooth often missing |
| XVII | 21 | 3 | P2 and P3 sometimes missing; M1 sometimes |
| XVIII | 25 | 3 | worn to gumline and occlusal surface |
| XIX | 30 | 4 | sloping markedly towards lingual edge; M1 sometimes missing and M2 worn to gumline |
| XX | 35 | 4 | Several teeth often missing or reduced to stubs; M1 invariably missing |

[a]From Goddard 1970.

**Fig. 13.16** Unworn lower M₁ of mammoth (78/701) from layer A. Scale in mm.

**Table 13.23** Tooth Wear Related to Age in the African Elephant[a]

| Age class | Mean age | State of eruption and wear of teeth (years) |
|---|---|---|
| I | 0 | No teeth worn, M1 protruding |
| II | 0.5 | Slight wear of M1 and M2 |
| III | 1 | M1 well worn; M2 moderate wear; M3 slight wear |
| IV | 2 | M1 lost; M2 well worn; M3 early wear |
| V | 3 | M2 well worn; M3 worn; M4 in crypt |
| VI | 4 | M2 almost disappeared; M3 well worn; M4 unworn |
| VII | 6 | M2 lost; M4 erupted but unworn |
| VIII | 8 | M3 anterior edge eroding |
| IX | 10 | M3 almost completely eroded: M4 in wear |
| X | 13 | M3 almost disappeared; almost all of M4 in wear |
| XI | 15 | M3 usually lost; M4 complete with all lamellae in wear |
| XII | 18 | M4 erosion of anterior edge; M5 visible or just in wear |
| XIII | 20 | M4 progressive erosion: M5 in wear |
| XIV | 22 | M4 erosion from both ends |
| XV | 24 | M4 as in XIV; M5 enamel loops complete |
| XVI | 26 | M4 almost disappeared: M5 all lamellae in wear |
| XVII | 28 | M4 socket; M5 in wear; M6 visible |
| XVIII | 30 | M5 slight erosion of anterior edge; M6 just in wear |
| XIX | 32 | M5 eroding; M6 in wear (2-3 lamellae) |
| XX | 34 | M5 eroding; M6 further worn |
| XXI | 36 | M5 almost gone; M6 half worn |
| XXII | 39 | M5 socket only |
| XXIII | 43 | M6 erosion of anterior edge begun |
| XXIV | 45 | M5 vestige of socket; all except last few lamellae of M6 in wear |
| XXV | 47 | M6 slight erosion anterior edge |
| XXVI | 49 | M6 eroding anterior edge |
| XXVII | 53 | M6 anterior one-third eroded away |
| XXVIII | 55 | M6 few enamel loops remain |
| XXIX | 57 | M6 all but 1 or 2 enamel loops confluent |
| XXX | 60 | M6 less than 15 cm remains rooted |

[a]After Laws 1966.

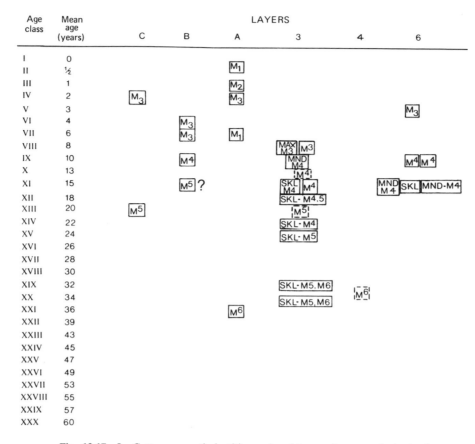

**Fig. 13.17** La Cotte mammoth dentition assigned to age classes on the basis of age data for the African elephant. MAX, dentition in maxilla; SKL, dentition in virtually complete skull; MND, dentition in mandible; ?, specimen from B/C interface. A broken box indicates an approximate placement.

and resorbed at the roots to be replaced by the successive tooth (Osborn 1936 and 1942; Kubiak 1982). Each successive tooth is larger than its predecessor and has a greater number of lamellae (see section 13.4.2 for measurement of mammoth dentition).

The age categories to which the La Cotte mammoths have been allocated are those devised by Laws (1966). He divided a sample of nearly 400 lower jaws of the African elephant into 30 age classes, based on the state of eruption and wear of all the teeth (M1–M6). On the basis of the data collected on Indian and African elephants, both wild and in captivity, he assigned chronological ages to the 30 groups, estimating a potential life span of 60–70 years (see Table 13.23). While it is not possible to suggest life expectancy for the mammoth, it is possible to obtain an impression of the range of ages of the La Cotte specimens by allocating them to Laws' categories, as illustrated in Fig. 13.17. The dentition from layers 3 and 6 was relatively easy to categorise in this way, as the majority of teeth are in mandibles or maxillae. Isolated teeth (such as recovered in the other levels at La Cotte) tend to have broken into their component lamellae, or to have broken into an incomplete series of lamellae. As each tooth wears successively from anterior to posterior, it is generally difficult to give an

age to it if anterior lamellae are missing and posterior lamellae are unworn. Thus teeth from levels other than 3 and 6 are assigned to age categories by using comparative measurements in addition to assessment of state of wear. A number of fragmentary teeth could not be categorised in this manner. If they could have been included, the number of individuals represented might have been slightly higher; however, their general dimensions and state of wear indicate that they fall within the age range of the more complete material.

Fig. 13.18 shows the minimum number of mammoths and woolly rhinoceros in all levels at La Cotte, as calculated on the basis of dentition. These individuals have been assigned to age categories as defined above. It will be observed that, in all levels, the woolly rhinos represented are quite young, but particularly so in the lower layers: A, B and C. This tendency for young animals to be represented in the lower levels is also true of the mammoths, whereas in Layers 3 and 6 a wider range of ages is represented. This is even more evident if one takes into account also the post-cranial bones, which are particularly numerous in Layer 6. The possible significance of the age distribution of these two species, and its implication for hominid behaviour at or near the site, is discussed in chapter 18.

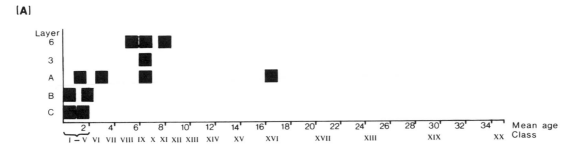

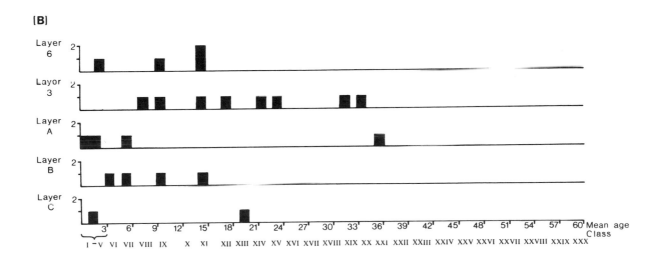

**Fig. 13.18** La Cotte woolly rhinoceros and mammoth dentition assigned to age classes, based on studies of (A) the black rhinoceros (Goddard 1970) and (B) the African elephant (Laws 1966). Each square represents one individual.

# 14

# THE RODENT FAUNA

J. Chaline and G. Brochet

## 14.1 INTRODUCTION

Rodents, especially voles, are among the most common mammals in the wild, and are an important prey for nocturnal hunting birds, especially owls. The indigestible fur and skeleton result in rejection pellets, which are regurgitated by the birds to form accumulations at the foot of nesting areas. The bones, and more particularly the teeth, are frequently fossilised in huge numbers to form 'rodent beds' in prehistoric deposits. These can provide valuable information of two kinds.

An important characteristic of the voles is their very rapid evolution in the course of the Quaternary era. Consequently, the occurrence of different genera and species, and also the detailed morphology of their teeth, can often be used to assign geological layers to a position in the evolutionary sequence and hence (insofar as this can be calibrated against the Quaternary time-scale) to estimate their age. In extreme cases, a single rodent tooth may give an indication of the age of a human skull or a prehistoric settlement and throw light on the environment and climate of the time.

A second line of enquiry is provided by the close adaptation of rodents to their biotopes; the fossil associations may thus be used as a basis for reconstructing prehistoric landscapes. As rodents live under well-defined climatic conditions, the climatic fluctuations during the Quaternary era led to important changes in their area of distribution (migration, or 'faunal transgressions'). The demonstration of shifts in rodent distribution now appears to be one of the most reliable sources of evidence for reconstructing climatic changes in the Pleistocene period. For reviews of the British and French rodent faunas see Sutcliffe and Kowalski (1976) and Chaline (1972).

## 14.2 RODENT REMAINS FROM LA COTTE DE ST. BRELADE

In 1980 the rodent remains were sent to the authors for study, after extraction at the University of Cambridge. These had been recovered in the course of the 1961–1978 excavations from deposits believed to be anterior to the last interglacial. The majority came from the upper part of this sequence (chiefly the loessic layers 6.1, 3 and B), though a few specimens had been found in 1961 in the deep sounding (in deposits partly derived from these same layers). The material had been divided into 341 small packets (some containing only a single tooth,

**Fig. 14.1** Present day distribution of *Dicrostonyx torquatus* (after Brunnacker 1978).

**Fig. 14.2** Present day distribution of *Microtus gregalis* (after Brunnacker 1978).

others several jaws) for ease of reference, and to allow correction where required by later minor changes in the interpretation of stratigraphic boundaries.

The deposits of Weichselian age had provided an important rodent fauna (apparently from several different beds, though the stratigraphic relationships are not always clear), which had been studied by M. A. C. Hinton (in Marett 1918; 1919). A small sample of rodent remains from the early excavations, belonging to the Société Jersiaise, was forwarded to the authors in 1982. Its exact provenance is unknown, however, and, as it had clearly been selected, it must be treated with caution and has not been included in the tabulations below.

The rodents from the fossiliferous Saalian layers excavated by McBurney exhibit great homogeneity. The assemblages are all of a typical cold fauna and include only five species. The collared lemming (*Dicrostonyx torquatus*), which today lives in the area eastwards from the Kola peninsula in the extreme northwest of the USSR (Fig. 14.1), is by far the most abundant species; the hill vole (*Microtus gregalis*), an inhabitant of the Siberian steppes (Fig. 14.2), is also quite numerous. Numerically far less important are the steppe birch mouse (*Sicista* sp.), the root vole (*Microtus malei*) and the common vole (*Microtus arvalis*). It appears that this is the first record of *Sicista* from a British site (Fig. 14.3).

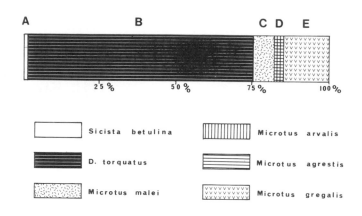

Fig. 14.4 Environmental preferences of the La Cotte microfaunal assemblage: (A) boreal forest, (B) arctic cold steppe, (C) boreal open marshland, (D) continental steppe or woodland, (E) continental steppe.

Fig. 14.3 *Sicista* sp. left M₁ (layer 6.1, no. 381).

## 14.3 LANDSCAPE RECONSTRUCTION

The five species recorded at La Cotte de St. Brelade provide us with an accurate reconstruction of landscape. The area was covered by a cold windswept steppe frequented by collared lemming, hill vole, common vole and steppe birch mouse. Some marshy areas with ponds existed in the neighbourhood of the cave and were inhabited by root vole and collared lemming. It was a typical environment of the periglacial zone as shown by the climatogram (Fig. 14.4).

## 14.4 QUANTITATIVE PALAEOCLIMATOLOGY

Hokr (1951) has proposed a quantitive approach to palaeoclimatology which has since been amended by Brochet (1981). Taking the area of distribution of each present day species, the following statistics are employed:

1. mean annual rainfall in mm,
2. mean temperature for January,
3. mean temperature for July,
4. number of days with temperature over 5°C, and
5. number of days with temperature over 10°C.

With this information for the five species of microfauna from La Cotte de St. Brelade, and taking relative abundance into account, we have constructed a curve which synthesizes the climatic conditions contemporary with the deposition of at least layers B, 3 and 6.1 (Fig. 14.5).

The January temperature varied between -30 and -45°C. The July temperature varied between 0 and 15°C. There were about 100 days with a temperature over 5°C and 75 days over 10°C. Monthly rainfall from November to February varied between 15 and 35 mm, from March to June between 15 and 35 mm, and from July to October between 25 and 50 mm.

## 14.5 AGE OF THE CAVE FILLING

The European record of *Dicrostonyx* extends back to the Lower Pleistocene. Its occurrence with *M. gregalis* is recorded in the Middle Pleistocene at Cagny, late in the Mindel glaciation (Bourdier *et al.* 1974b). The extinct species *Arvicola cantiana* is present in the same assemblage, as it is in the (probably rather earlier) sequence of Westbury-sub-Mendip (Stuart 1982).

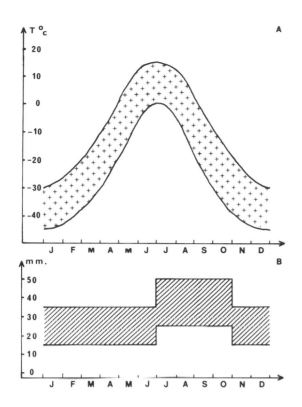

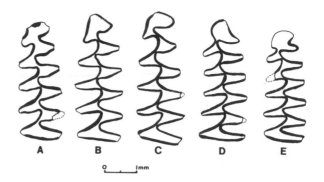

**Fig. 14.6** *Dicrostonyx torquatus.* Left M¹: (A) Morphotype I? (deep sounding, reworked layers B–6 (?), no. 147); (B) Morphotype II (layer 3, no. 200); (C) Morphotype II (layer 6.1, no. 395); (D) Morphotype III (layer 6.1, no. 388); (E) Morphotype IV (layer 3, no. 326).

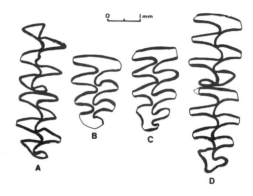

**Fig. 14.5** Climatic curves based on the amended Hokr method. (A) Mean annual temperature in degrees centigrade; (B) mean annual rainfall in mm.

**Fig. 14.7** *Dicrostonyx torquatus* dentition. (A) Left M¹ morphotype I and M² morphotype II (layer 6.1, no. 356); (B) Right M³ morphotype I (layer B, no. 117); (C) Right M³ morphotype IV (layer B, no. 117B); (D) Left M¹ morphotype I and M² morphotype II (layer 6.1, no. 356B).

The pairing of *D. torquatus* and *M. gregalis* is documented for southern France by Chaline (1976). They occur again in the later Middle Pleistocene Aldènian microfaunal complex at such sites as Caune de l'Arago, Orgnac and Aldène; this is the most southerly extension of the distribution of collared lemming. They are found together once more during the last glaciation (the Régourdian microfaunal complex). In the absence of any extinct species, therefore, the entirety of the La Cotte rodent fauna could be attributable to the later Middle or the Upper Pleistocene.

Following the work of Agadjanian and von Koenigswald (1977) on the morphological variation of *D. torquatus* during the Pleistocene, the dental morphotypes they defined have been counted for the La Cotte specimens. The teeth used are M₁, M¹, M² and M³. The four morphotypes are defined by the development of the anterior complex on the M₁, and the growth of a posterior loop on the upper teeth (Figs. 14.6 and 14.7). A histogram is then plotted for the percentage frequency of the various morphotypes and can be used for comparative purposes. As no significant differences were apparent between the assemblages from the McBurney excavations at La Cotte, the material has been treated as a single sample, in compiling Fig. 14.8 and Table 14.1, for comparison with the published sequences from southern Germany and the USSR (Figs. 14.9 and 14.10). To obtain

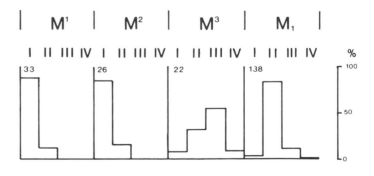

**Fig. 14.8** Frequency of the various morphotypes of *Dicrostonyx torquatus* at La Cotte de St. Brelade.

**Table 14.1** Fossil Record of the Rodents of La Cotte de St. Brelade[a]

| Species | 6.1 % | N | 5 % | N | 3 % | N | A % | N | B % | N | C % | N |
|---------|-------|---|-----|---|-----|---|-----|---|-----|---|-----|---|
| *Sicista* sp. | 2.1 | 1 | 0.0 | 0 | 0.0 | 0 | 1.0 | 1 | 0.0 | 0 | 0.0 | 0 |
| *Dicrostonyx torquatus* | 68.8 | 33 | 60.0 | 3 | 73.3 | 44 | 50.0 | 1 | 76.5 | 75 | 83.3 | 5 |
| *Microtus malei* | 0.0 | 0 | 0.0 | 0 | 1.7 | 1 | 0.0 | 0 | 11.2 | 11 | 0.0 | 0 |
| *Microtus arvalis* | 0.0 | 0 | 0.0 | 0 | 0.0 | 0 | 0.0 | 0 | 4.1 | 4 | 0.0 | 0 |
| *Microtus gregalis* | 29.2 | 14 | 40.0 | 2 | 25.0 | 15 | 50.0 | 1 | 7.1 | 7 | 16.7 | 1 |

[a]$N$ is the minimum number of individuals represented.

the frequencies, the count for left or right teeth, whichever is the higher, has been used (italicised in Table 14.2).

$M_1$. All four morphotypes are present, with type II being the most frequent. The situation parallels that at the Middle Pleistocene site at Kipiewo II.

$M^1$. Only morphotypes I and II are present, the former predominating in the USSR during the Middle Pleistocene and in the middle Würm in Germany.

$M^2$. This is comparable to $M^1$.

$M^3$. Morphotypes II and III are the most frequent. This situation exists from the Middle Pleistocene onwards in the USSR and in the Late Glacial in Germany. La Cotte also possesses morphotype IV, which is absent from the German sites and according to Agadjanian occurs only at the present day.

The study of morphotypes does not provide definitive evidence of the exact age of La Cotte, though there is a strong resemblance to the Russian Middle Pleistocene assemblages.

Two examples of 'disharmony' were observed in the distribution of morphotypes: in one instance (Fig. 14.7D) a $M^1$ I was associated on the same jaw with a $M^2$ II, and in another a $M^2$ II occurred with a $M^3$ III. The modification of the teeth was thus not strictly synchronous.

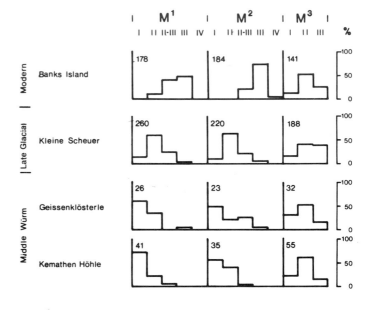

**Fig. 14.9** *Dicrostonyx* morphotype frequencies at three German sites (after Agadjanian and von Koenigswald 1977).

## 14.6 CONCLUSION

The rodent fauna of La Cotte de St. Brelade is related to a faunal shift originating in the area bordering the Arctic Ocean and Siberia during a glacial period. Parallels with sites in Germany and the USSR, based on the morphological evolution of *Dicrostonyx torquatus*, suggest a dating to the Middle Pleistocene, or at the latest the middle of the Weichselian glaciation. The dominance of *D. torquatus* and *M. gregalis* points to the same conclusion.

Morphometric studies may one day provide better resolution, but the existing evidence indicates that the microfauna from the McBurney excavations are at least not inconsistent with a Saalian age, as proposed elsewhere in this volume. It should be noted that the rodents from the Marett excavations belong to the same species, for which a Weichselian date is also permissible; regrettably, the specimens available for study were not sufficiently numerous for comparison with the older series.

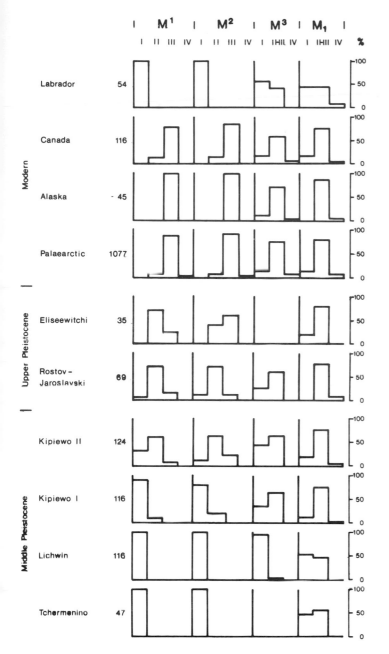

**Fig. 14.10** *Dicrostonyx* morphotype frequencies at six Russian sites (after Agadjanian 1976).

**Table 14.2** *Dicrostonyx* Dental Morphology[a]

|  | Left | Right | % |
|---|---|---|---|
| **M₁ (N = 138)** | | | |
| I | 5 | 1 | 4 |
| II | 113 | *116* | 84 |
| III | 5 | *15* | 11 |
| **M₁ (N = 33)** | | | |
| I | 23 | 29 | 88 |
| II | 3 | 4 | 12 |
| III | – | – | – |
| IV | – | – | – |
| **M₂ (N = 26)** | | | |
| I | 20 | 22 | 85 |
| II | 4 | 4 | 15 |
| III | – | – | – |
| IV | – | – | – |
| **M₃ (N = 22)** | | | |
| I | – | 2 | 9 |
| II | 7 | 4 | 32 |
| III | 6 | *12* | 55 |
| IV | – | *1* | 5 |

[a]The percentages for each morphotype are calculated from the italicised figures.

# THE THERMOLUMINESCENCE DATES

J. Huxtable

## 15.1  INTRODUCTION

When buried in a soil, a mineral specimen (e.g. a flint tool, or an inclusion in pottery) is subject to bombardment by ionising radiation; some of this is the product of radioactive elements in the body of the material, while the remainder originates in the object's surroundings. A small proportion of this radiation has the effect of knocking electrons from their original sites in the crystal lattice of the mineral(s). A few of the electrons are fully displaced, leaving electron-deficient sites, and are subsequently trapped elsewhere in imperfections in the lattice (e.g. where impurities occur). The trapped electrons can be freed by heating to at least 400°C, in which case some of them will recombine at the electron-deficient sites, with the emission of light.

Following such a heating episode, accumulation of trapped electrons begins again. After a period of time has elapsed, further heating will release an amount of thermoluminescence (TL) energy which is directly proportional to the time during which irradiation has occurred; the other elements in the equation are, principally, the radioactive dose rate and the amount of TL energy which the specimen is capable of emitting after unit received dose (the latter quantity may be determined by deliberate irradiation under laboratory conditions).

Thus, where the TL record has been erased by heating, the age of that event (i.e. TL age) may be computed using the formula

$$\frac{\text{archaeological dose (AD)}}{\text{annual dose rate}}$$

where AD is

$$\frac{\text{naturally produced TL energy}}{\text{TL energy per rad of artifical irradiation}}$$

Pottery is an ideal subject for dating by this technique, as the firing resets the TL signal to zero. At pre-pottery sites such as La Cotte, flint may be employed provided it has been sufficiently heated — typically, by being accidentally burnt in a hearth.

In the case of flint, the internal radioactivity is usually low, and the environmental contribution is often 80% or more of the total dose. In the ideal situation it is therefore essential to collect and analyse the sediment which is immediately surrounding the sample to be dated. At La Cotte this was not possible since the burned flints had already been excavated and the soil samples collected from different locations at the site.

Also, in the ideal situation the surroundings should be uniform to a radius of 30 cm so that the soil samples analysed in the laboratory are representative of the sphere of radioactivity surrounding the sample. If this is not the case, then on-site measurement of the radioactivity should be carried out. Once again this was not possible at La Cotte, since excavation had been completed (and the critical sections walled up) before the difficulties of the TL dating at this site had been appreciated. These problems are compounded by the very great geological and physical complexity of the environment at La Cotte. The granite bedrock is highly radioactive, as are fragments derived from it. Where the matrix is loessic, a lower level of radiation may be expected, but this is li able to be offset by the presence of large granite blocks. Moreover, leaching (which was of varying importance locally) and reworking of the deposits, in the course of their history, contribute to the uncertainties (see below). Inspection of the photographs and drawings of the sections is sufficient to emphasise forcefully the problem of finding flints with homogeneous surroundings.

Initial analysis of the 37 soil samples in the preliminary work on the site used thick source alpha counting (a 42 mm diameter ring; Tite and Waine 1962) and flame photometry to establish the uranium, thorium and potassium oxide content of the soils. This work showed striking differences in radioactive content for the soils from the site. The least active soil (sample 29) would have contributed 80 mrad per year environmental dose to objects buried in it, the most active (sample 34) 1 rad per year of environmental dose, which is most likely to be due to the higher radioactive content of the granite.

In the ideal TL dating study, the soil dose rate measured today is the same as it has been in the past. However, in some environments it may have changed; in particular, various nuclei in the uranium decay chain may be leached out of one deposit by ground water and deposited in another. This causes disequilibrium of the decay chain. For example, disturbance of ground water flow due to plant growth can be fast and localised and can deplete the soil of many trace elements.

In order to establish whether there has been disturbance of the radiation environment at La Cotte, five soil samples were chosen for high resolution gamma spectrometry using a germanium detector (Murray 1981). This technique measures the state of equilibrium, or otherwise, in the soils. From Table 15.1 it can be seen that the soil samples form two groups, one of high and one of low $^{238}U$ activity. There was a considerable de-

**Table 15.1**  Group Activities (in pCi g$^{-1}$) before and after Exchange, for Samples from La Cotte de St. Brelade[a]

| Group | Long-lived isotope | 222/2 Before | 222/2 After 75 days | 222/10 Before | 222/10 After 80 days | 222/23 Before | 222/23 After 81 days | 222/28 Before | 222/28 After 84 days | 222/29 |
|---|---|---|---|---|---|---|---|---|---|---|
| 1 | $^{238}$U | $1.41_7$ | $0.60_6$ | $11.2_4$ | $5.9_2$ | $13.1_4$ | $9.4_3$ | $0.69_{11}$ | $0.39_{11}$ | $0.74_7$ |
| 4 | $^{226}$Ra | $1.00_{14}$ | $1.23_{12}$ | $6.0_3$ | $6.4_2$ | $12.7_4$ | $10.4_3$ | $0.82_{15}$ | $0.79_{12}$ | $0.81_{13}$ |
| 5 | $^{222}$Rn | $0.77_2$ | $0.68_5$ | $3.30_7$ | $3.61_6$ | $3.71_8$ | $5.6_1$ | $0.42_3$ | $0.52_2$ | $0.44_3$ |
| 6 | $^{210}$Pb | $1.16_{10}$ | $1.11_{10}$ | $3.8_2$ | $2.07_{11}$ | $6.4_2$ | $3.3_2$ | $0.94_{11}$ | $1.0_1$ | $0.91_9$ |
| 7 | $^{232}$Th | $1.24_7$ | $0.65_5$ | $2.43_{10}$ | $1.43_6$ | $1.24_7$ | $0.97_7$ | $0.47_4$ | $0.46_8$ | $0.65_5$ |
| 8 | $^{220}$Rn | $1.11_3$ | $1.63_5$ | $2.16_6$ | $2.55_6$ | $1.04_4$ | $1.92_5$ | $0.44_2$ | $1.79_5$ | $0.53_2$ |
|  | $^{40}$K | $19.1_7$ | $7.28_{10}$ | $22.4_8$ | $19.9_9$ | $34_1$ | $34.5_{11}$ | $4.41_{10}$ | $6.2_9$ | $7.6_9$ |

Ratios derived from the above

| Ratio | 222/2 | 222/10 | 222/23 | 222/28 | 222/29 |
|---|---|---|---|---|---|
| $^{232}$Th/$^{238}$U | $0.88_7$ | $0.22_1$ | $0.095_6$ | $0.68_{12}$ | $0.88_{11}$ |
| $^{226}$Ra/$^{238}$U | $0.71_{11}$ | $0.54_3$ | $0.97_4$ | $1.19_{29}$ | $1.09_{20}$ |
| $^{222}$Rn/$^{226}$Ra | $0.77_{11}$ | $0.55_3$ | $0.29_1$ | $0.51_{11}$ | $0.54_9$ |
| $^{210}$Pb/$^{226}$Ra | $1.16_{16}$ | $0.63_5$ | $0.50_2$ | $1.15_{25}$ | $1.12_{21}$ |
| $^{220}$Rn/$^{232}$Th | $0.90_6$ | $0.89_5$ | $0.84_6$ | $0.94_9$ | $0.82_7$ |

[a]Laboratory reference 222 (from Murray 1981). The unit used for errors (subscripted) is that of the least significant figure.

gree of $^{222}$Rn and $^{220}$Rn escape from all the samples. Four of the soils were further investigated to see whether a significant proportion of the activity was potentially mobile, or whether the activity was firmly bound to the mineral matrix. A portion of each soil sample had all the activity which was retained on ion exchange sites replaced with an inert ion (Ba$^{+2}$) as described in Hedges and McLellan (1976). All four samples lost between 30 and 50% of the initial $^{238}$U concentration in the exchange process.

Murray concluded that uncertainties in the $\gamma$ dose rates derived from the high activity samples (10 and 23) were about $\pm 25\%$, presumably due to ion mobility effects, and were therefore much less reliable than those from the other three soils, with associated uncertainties of about $\pm 10\%$. The dating effort was therefore directed towards flints associated with the latter soil samples of lower activity. It would be very unwise to attempt to date any flints associated with the high activity soil samples until the geochemistry of the soils was fully understood.

The soil samples were then sorted into groups of high and low $^{238}$U activity on the basis of the alpha counting and flame photometry results. Callow then analysed the distribution of the samples falling within these groups and located the region which was likely to contain flints associated with the low activity soils. He suggested the area bounded by co-ordinates between approximately 0 and 1.5 m east. The TL effort was then concentrated on well-burned samples from three consecutive layers, C, D

and E, within this area of the site. No attempt was made to extend the dating to the admittedly extremely important layers outside this range, as either no soil samples were available for the more promising area of the site or the deposits were characterised by the presence of numerous granite boulders.

## 15.2  THE THERMOLUMINESCENCE SAMPLES AND MEASUREMENTS

The method of sample preparation and measurement has been described in detail before (Huxtable and Jacobi 1982). A total of 13 flints were measured from the site, 5 from layer C and 4 each from layers D and E. The light output from these flints was usually of the same order of intensity for fine (1–8 $\mu$m) grains and coarse (90–150 $\mu$m) grains from the same flint. The sensitivities ranged from about 50 kHz to about 250 kHz for the 375°C peak ordinate for the natural glows from the flints (violet and blue filters, Corning 7/59 and 5/60, heating rate 10°C s$^{-1}$). The glow shapes were all similar with one high temperature peak at 375°C, very like quartz glow curves. The plateau region of the glow curve, that is, where the ratio of natural TL to combined natural and artificial TL becomes constant (within 5%) was around the 375°C peak; usually from 350°C to 425°C, occasionally reaching as high as 450°C. The archaeological dose (AD) was evaluated in the plateau region of the glow curve, using the additive dose method to produce the growth curve in the usual way.

**Table 15.2** Archaeological Doses Measured for Burnt Flints from La Cotte de St. Brelade

| Sample number | Fine grain AD (krad) | Coarse grain AD (krad) | a value* |
|---|---|---|---|
| **Layer C** | | | |
| 221 | 21.0 ± 1.0 | 19.9 ± 1.5 | 0.130 |
| 229 | 22.0 ± 1.5 | 24.3 ± 2.5 | 0.077 |
| 278 | 21.4 ± 0.5 | 23.6 ± 1.7 | 0.079 |
| 283 | 27.5 ± 1.5 | 25.0 ± 1.5 | 0.099 |
| 284 | 28.0 ± 1.0 | 27.8 ± 3.0 | 0.062 |
| **Layer D** | | | |
| 299 | 37.5 ± 1.0 | 38.2 ± 1.5 | 0.094 |
| 301 | 37.2 ± 1.5 | 38.2 ± 4.0 | 0.093 |
| 302 | 35.7 ± 2.0 | 34.7 ± 3.0 | 0.068 |
| 318 | 35.7 ± 1.5 | 36.1 ± 4.0 | 0.130 |
| **Layer E** | | | |
| 247 | 49.0 ± 2.0 | 48.6 ± 2.0 | 0.108 |
| 263 | 49.0 ± 1.0 | 48.6 ± 1.0 | 0.137 |
| 266 | 47.9 ± 2.5 | 45.9 ± 3.5 | 0.091 |
| 297 | 50.0 ± 2.0 | 48.6 ± 1.5 | 0.105 |

*As described by Aitken and Bowman (1975).

The TL parameters are shown in Table 15.2. For three of the flints there was no sensitivity change from first to second heating ($\pm 2\%$). Nine of the others showed a decrease in sensitivity varying from 5% to 40%, one flint showed an increase in sensitivity of 30%. In all cases the supralinearity correction (I) obtained from the second glow growth curve was zero ($\pm 2\%$ of the AD). The good agreement between the fine and coarse grain ADs is evidence that there has been no contamination of the fine grain TL by spurious luminescence (one would expect the former to show its presence more readily). It also shows there has been no preferential radioactive separation between grain sizes.

In order to test for any adverse effects due to crushing, a piece of unburned flint from the site was drained of TL signal at 500°C and a $\gamma$ dose (from a $^{60}$Co source) of 15 krad was given to the drained piece. It was then crushed and the dose evaluated by additive $\beta$ dose was found to be (15.16 $\pm$ 0.2) krad in the fine grains and (15.02 $\pm$ 0.2) krad in the 100 $\mu$m grains from the flint. This was considered satisfactory and showed there was no adverse effect induced by crushing the material. The experiment was repeated on pieces of two of the dated flints — sample 221 with a $\gamma$ dose of 7.5 krad, and sample 229 with a $\gamma$ dose of 15 krad — and the same satisfactory agreement was obtained.

Tests on stored (natural + $\beta$ dose) material with storage times in excess of a year were used to check for the presence of anomalous fading (Wintle 1973) and none was detected (less than 3%). This storage time is short in comparison with the age of the material and cannot exclude the possibility of a slow-fading component. It is thought unlikely, though, that a slow-fading component would exist without the presence of a fast component (Wintle 1977), and the test applied in this work is adequate to show there is none.

## 15.3 RADIOACTIVITY MEASUREMENTS

The combined internal dose rate due to uranium and thorium content of the flints was measured by thick source alpha counting, using the conversion factors of Bell (1976), and their potassium contents by flame photometry. The alpha counts for the flints varied from 0.4 to 1.7 counts per ksec, for a 42 mm diameter alpha counting ring, and their potassium oxide contents from 0.05 to 0.3%. There was no detectable emanation of radon on sealing the alpha counting ring.

The environmental dose received by the flints has been calculated from an analysis of the gamma spectrometer data given in Table 15.3. The soils were counted both unsealed and cast in resin. The latter were left for three weeks before being analysed so that radon could build up. Dose rates from these sealed samples have been used to calculate the ages.

The predicted dose rates were checked as far as possible by other methods, for example, predicted $\beta$ dose by $\beta$ TLD measurements (see Bailiff and Aitken 1980) on all the soils used and $^{210}$Pb predictions 'gas cell' measurements and $^{210}$Po determinations (see Aitken 1978) using alpha spectrometry. The results are given in Table 15.4. If the predicted and observed doses were in reasonable agreement (within 2 standard deviations) the soil was classed as satisfactory and its dose rate was used in the age calculations. One soil was excluded from this final analysis (soil 17, associated with sample 284 layer C) because repeat analyses were inconsistent. This could have been due to the lack of homogeneity of the bulk soil sample from which 8 g aliquots were taken for the gamma spectrometer analyses. A more recent detailed study of Layer E by Callow revealed that the burnt flints used were in close proximity to large granite blocks in the underlying Layer F. As a result, their environmental dose rate is subject to very great uncertainty.

It has thus been unacceptable to calculate an age for layer E. All that can be done is to calculate a maximum age for the level, assuming that it contains all soil and no granite blocks, and a minimum age assuming the reverse situation applies; the truth lies somewhere between these two limits. An estimate of the amount of granite rubble close to the samples has been made by Callow,

**Table 15.3**  Gamma Spectrometer Activity Measurements in pCi g$^{-1}$ for Soils Used in the Dating Programme[a]

| Long-lived isotope | Layer C | | Layer D | | Layer E | | |
|---|---|---|---|---|---|---|---|
| | Soil 15 | Soil 29 | Soil 2 | Soil 3 | Soil 4 | Soil 5 | Soil 37 |
| $^{238}$U | $0.83_{12}$ | $0.39_{22}$ | $1.57_{15}$ | $1.44_{25}$ | $0.96_{24}$ | $0.94_{23}$ | $1.64_{17}$ |
| $^{222}$Rn | $1.24_{4}$ | $0.66_{5}$ | $1.51_{4}$ | $1.31_{6}$ | $1.44_{7}$ | $1.37_{6}$ | $1.88_{6}$ |
| $^{210}$Pb | $0.94_{14}$ | $0.89_{30}$ | $1.26_{21}$ | $0.99_{25}$ | $1.09_{29}$ | $1.15_{27}$ | $1.22_{22}$ |
| $^{232}$Th | $0.94_{8}$ | $0.91_{16}$ | $1.48_{10}$ | $1.48_{17}$ | $0.66_{17}$ | $0.94_{13}$ | $3.24_{11}$ |
| $^{220}$Rn | $1.05_{8}$ | $0.53_{13}$ | $1.39_{10}$ | $1.55_{8}$ | $1.04_{14}$ | $0.94_{9}$ | $3.38_{15}$ |
| $^{40}$K | $8.1_{16}$ | $3.6_{8}$ | $12.8_{22}$ | $16.5_{19}$ | $12.5_{28}$ | $11.3_{21}$ | $29.8_{14}$ |

[a]The unit used for errors — for counting statistics only — is that of the least significant figure.

**Table 15.4**  Other Radioactivity Measurements on Soils[a]

| Method | Layer C | | Layer D | | Layer E | | |
|---|---|---|---|---|---|---|---|
| | Soil 15 | Soil 29 | Soil 2 | Soil 3 | Soil 4 | Soil 5 | Soil 37 |
| Flame photometry K (pCi g$^{-1}$) | $8.72_{44}$ | $5.89_{29}$ | $6.67_{33}$ | $7.87_{39}$ | $7.66_{38}$ | $5.67_{28}$ | $18.5_{9}$ |
| $\alpha$ spectrometry $^{210}$Po (pCi g$^{-1}$) | $1.22_{6}$ | $0.80_{4}$ | $0.85_{4}$ | $0.95_{5}$ | $1.49_{8}$ | $1.37_{7}$ | $3.12_{16}$ |
| $\beta$ dose predicted (mrad yr$^{-1}$) | 166 | 76 | 212 | 255 | 225 | 204 | 452 |
| $\beta$ TLD measured (mrad yr$^{-1}$) | $164_{2}$ | $74_{5}$ | $210_{6}$ | $273_{15}$ | $222_{8}$ | $196_{3}$ | $434_{20}$ |
| $\alpha$ count predicted (cts ks$^{-1}$) for 42 mm diameter ring) | 10.0 | 7.1 | 13.4 | 13.0 | 10.7 | 11.2 | 23.0 |
| $\alpha$ count measured (cts ks$^{-1}$) | $11.6_{2}$ | $7.5_{2}$ | $13.3_{4}$ | $13.2_{4}$ | $12.4_{2}$ | $12.6_{2}$ | $30.5_{6}$ |
| Gas cell measurements: "lost counts" due to | | | | | | | |
| radon gr | 0.7 | 0.6 | 1.4 | 1.6 | 0.9 | 1.3 | 1.5 |
| thoron gh | 1.1 | 0.2 | 0.6 | 0.2 | 0.3 | 0.5 | 0.6 |

[a]The unit used for errors is that of the least significant figure.

and is included here with the other data for completeness (see Appendix H, in microfiche).

None of the methods of dose rate determination mentioned above allow for the absorption of energy by water in the soil. The water contents (calculated as a percentage of dry weight) were measured for the soils as they were taken from the site. The saturation water content was measured in the laboratory, and comparison of the water contents indicated that the soils were 70% saturated. After discussion with Callow and Lautridou a water content of 0.7 ± 0.3 of the saturation value was applied. Soil saturation values are given in Table 15.5; there was no detectable water uptake for the granite fragments or the flints.

The radioactivity data is summarised in Table 15.4. The poor correlation of the potassium activities measured by flame photometry and gamma spectrometry is thought to be due to the small sample size (100 mg) used in flame photometry. Gamma spectrometry gives the more representative estimate. With the exception of soil 37, the predictions from the $^{210}$Pb activity measurements agree (within 2 s.d.) with the results obtained by $^{210}$Po measurements and alpha spectrometry. Discrepancies are probably explained by lack of sample homogeneity and the gamma spectrometry measurements are used as they are more representative (alpha spectrometry uses only 0.25 g).

There is an unexplained excess $^{226}$Ra measurement for soils 15 and 5. Soil sample 15 associates with flint sample 283 (layer C), which was rejected on other grounds; soil 5 is from Layer E, which does not appear in the final calculation of the date.

**Table 15.5** The Thermoluminescence Ages

| Sample number | Age (kyr) | Dose rate (mrad yr$^{-1}$) | Internal dose rate contribution % α | Internal dose rate contribution % β | Soil saturation wetness (% of dry weight) |
|---|---|---|---|---|---|
| Layer C | | | | | |
| 221 | 210 | 100 | 15 | 12 | 26 |
| 229 | 234 | 94 | 11 | 11 | 26 |
| 278 | 243 | 88 | 5 | 12 | 26 |
| 283 | 211 | 130 | 16 | 14 | 40 |
| Layer D | | | | | |
| 299 | 250 | 150 | 8 | 7 | 34 |
| 301 | 230 | 162 | 12 | 9 | 34 |
| 302 | 248 | 144 | 5 | 6 | 34 |
| 318 | 225 | 159 | 7 | 7 | 25 |
| Layer E (maximum) | | | | | |
| 247 | 388 | 126 | 10 | 17 | 34 |
| 297 | 385 | 130 | 9 | 20 | 34 |
| 263 | 348 | 141 | 15 | 8 | 28 |
| 266 | 368 | 130 | 7 | 10 | 28 |
| Layer E (minimum) | | | | | |
| 247 | 140 | 350 | 4 | 6 | 0 |
| 297 | 141 | 354 | 3 | 8 | 0 |
| 263 | 140 | 350 | 6 | 3 | 0 |
| 266 | 142 | 338 | 3 | 3 | 0 |

## 15.4 OPTICAL BLEACHING

All the samples of flint in the dating programme were tested to see whether there had been any optical bleaching effects, that is, lowering of TL signal by exposure to light. It was feared that as some of the dated material had not been collected primarily for TL dating it might have been exposed to unknown amounts of light before arriving at the laboratory (for classification or drawing, for example). The experiment involved comparing the amount of bleaching of natural and artificial TL signals induced by a solar simulator lamp and has been fully reported elsewhere (Huxtable 1981). Three of the 13 flints involved in the programme appeared to have been bleached before they arrived in the laboratory. They were all from Layer C: sample 221 had been bleached by 13%, sample 283 by 12%, and sample 284 by 6%. The last sample was associated with soil 17, and so it has not been possible to calculate an age for it, but it is interesting to see that the ages calculated for samples 221 and 283 are lower than the other two samples from layer C by about 12%. As the experimental data collected on optically bleached flints is very sparse, it is not proposed to apply a correction to these two flints. However, since the evidence suggests that bleaching has occurred they have been excluded from the final age average.

## 15.5 AGE DETERMINATION

The ages calculated for the individual flints are given in Table 15.5. The maximum and minimum ages for four flints from Layer E are included for completeness. Once the two light-bleached flints are excluded, there appears to be no difference in TL age between the two remaining samples in Layer C and the four samples in Layer D, so an average TL age for these six samples has been calculated.

The average TL age for La Cotte Layers C and D is 238 ± 35 x 10³ years B.P. (OxTL222).

The dominant source of error is the uncertainty in the environmental dose rate; for levels C and D the uncertainty in the γ dose rate from the soils is ±10%, and this is included in the total error (at the 65% level of confidence) of ±35 x 10³ yrs.

## 15.6 ACKNOWLEDGMENTS

Various aspects of this work were supported by a grant from the Science Based Archaeology committee of the Science Research Council. My thanks are due to M. J. Aitken for helpful discussion at all stages of the project, to C. J. Shaw for initiation into the techniques of gamma spectrometry and to A. S. Murray for the initial soil analyses reproduced here. I also wish to thank N. C. Debenham for doing the ⁶⁰Co irradiation of the flint samples, N. Garton for preparation of the discs for alpha spectrometry measurements and P. Callow for suggesting the area of the site most profitable for study. Finally I wish to thank A. G. Wintle, who collected the first samples for TL dating of this site and without whom none of this would have happened.

149

# 16

# URANIUM-SERIES DATES ON BONE

B. J. Szabo

Uranium-series dating of fossil vertebrates from other archaeological sites has been attempted previously (Szabo et al. 1969; Howell et al. 1972; Sakanoue and Yoshioka 1974; Szabo and Collins 1975). The dates are obtained by the measurement of the amount of $^{230}$Th nuclide that is produced by its radioactive parent and grandparent, $^{234}$U and $^{238}$U, respectively. Successful application of the method requires that the fossils initially take up uranium but no thorium from the ground water and that the bones subsequently neither lose nor gain isotopes of uranium and thorium. The assimilation of uranium by fossil bones is usually rapid as shown by concordant uranium-series and radiocarbon dates (Bischoff and Rosenbauer 1981). However, occasionally the time-lag due to slow uranium assimilation by fossil bones may be several thousands of years, depending on their degree of preservation (Szabo 1980).

The bones for this study were crushed to a fine powder and heated at 80°C for about 6 hours. Uranium and thorium isotopes were isolated by anion exchange procedure, and the uranium concentrations and the thorium activities were determined by alpha spectrometry, using standard spike solutions of $^{236}$U, $^{229}$Th and $^{228}$Th. The results of the analyses and calculated $^{230}$Th ages for each sample are shown in Table 16.1.

A single bone fragment from layer G (sample 1A) yielded a $^{230}$Th date of 140 ± 20 kya, and the combined sample of three other bone fragments from the same layer (sample 1B) yielded a $^{230}$Th date of 120 ± 8 kya. Although these dates are concordant within limits of experimental error, they are in conflict with geomorphic and thermoluminescence evidence from which the age of layer G is estimated to be greater than 240 kya. Because these fragments of sample 1 were collected from a small area, the large variation of the $^{234}$U/$^{238}$U activity ratios of samples 1A and 1B (1.00 and 1.59, respectively) is rather unexpected. Perhaps these strongly weathered bone fragments underwent complex nuclide migration after burial because of the action of percolating water, resulting in minimum ages. Sedimentary studies in the cave indicate a great deal of post-depositional change, including phosphate dissolution and reprecipitation.

The weathered bone of sample 3 from layer B yielded a $^{230}$Th date of 87 kya that appears to be too young when compared with the thermoluminescence age estimation for layers C and D of about 240 kya. As layer B is

**Table 16.1**  Analytical Data and Apparent Uranium-Series Ages of Fossil Bones

| Sample number | Layer | Uranium (ppm) | | | $^{234}$U/$^{238}$U | | Activity ratios $^{230}$Th/$^{232}$Th | | $^{230}$Th/$^{234}$U | | Apparent age[a] (ka) | |
|---|---|---|---|---|---|---|---|---|---|---|---|---|
| 1A[b] | G | 1.62 | ± | 0.05 | 1.00 ± 0.03 | | 7.8 ± 1.6 | | 0.726 ± 0.044 | | 140 ± | 20 |
| 1B[c] | G | 0.579 | ± | 0.017 | 1.59 ± 0.03 | | 6.0 ± 1.2 | | 0.707 ± 0.028 | | 120 ± | 8 |
| 3[d] | B | 5.11 | ± | 0.10 | 1.31 ± 0.02 | | 2.6 ± 0.2 | | 0.567 ± 0.028 | | 87 ± | 7 |
| 5[e] | 11 | 24.6 | ± | 0.4 | 1.37 ± 0.02 | | 101 ± 50 | | 0.058 ± 0.006 | | 6.5 ± | 1 |

[a]The $^{230}$Th date is calculated using half-lives for $^{230}$Th and $^{234}$U of 75,200 and 244,000 years respectively.

[b]Unidentified fragments (burnt bone sample no. 57). One fragment, weighing about 1 g, was selected for the first analysis.

[c]Unidentified fragments (burnt bone sample no. 57). Three small fragments, weighing about 2 g, were combined for the second analysis.

[d]Mammoth limb fragments from bone 78/837.

[e]Horse limb bone; Marett collection B1360.

considered older than the last interglacial (and younger than C), its age ought to fall in the range 130–230 kya, probably tending towards the older age. Therefore, the apparent $^{230}$Th age of sample 3 is at least 50 (and possibly 150) ky too young, indicating that sample 3 also has been affected by post-depositional changes.

The well-preserved bone sample 5 from layer 11 that yielded a $^{230}$Th date of 6.5 kya is considered too young on sedimentological and archaeological grounds. Deposition of the loessic layer 11 is estimated somewhere within the last glaciation. The sediment contains a Mousterian industry, and therefore any associated bones should be older than 30 kya. The deposit was excavated at the beginning of the century, and unfortunately the exact provenance of sample 5 is unknown. It is conceivable that it came from the recent soil at the site and had been wrongly recorded. Another possible explanation for such a young age is that the sample underwent extreme post-depositional alteration.

### Acknowledgement

I thank P. Callow for supplying the fossil bone samples and for his assistance in evaluating the results.

# PART III

## NON-INDUSTRIAL ASPECTS OF THE HUMAN OCCUPATION

# HOMINID SPECIMENS FROM LA COTTE DE ST. BRELADE

C. B. Stringer and A. P. Currant

## 17.1  THE 1910–1911 HOMINID FINDS

### 17.1.1  Earlier Accounts

Twelve human teeth and the root of an incisor were re-covered in association with a 'hearth midden' about 2.4 m from the western entrance at a height of 22.15 m O.D. during excavation in 1910–1911. The nine teeth found in 1910 were described in detail by Keith and Knowles (1912a), with a further short note added on the 1911 finds (1912b). The teeth were identified mainly by comparison with the teeth of the Mauer mandible and the Gibraltar 1 palate, since few other Neanderthal dentitions were available for study at that time. In the 1912 studies much attention was paid to the unusual root morphology of the teeth, where it was noted that they resembled the Krapina dentitions in the presence of very thick prismatic roots and large pulp cavities. In 1913 Keith further discussed the La Cotte teeth and first applied the term 'taurodontism' to this type of root structure, which he considered resembled that of ungulates, such as the ox. The biological significance of this condition has been the subject of much discussion (Blumberg *et al.* 1971; Jaspers and Witkop 1980), and it is evident that, although taurodontism is relatively common in Neanderthal teeth, it is not exclusive to them, nor even typical of them in general.

### 17.1.2  The Present Study

When the present authors visited Jersey in 1976, Stringer took the opportunity to examine the original specimens at the Jersey Museum (Figs. 17.1 and 17.2). It was apparent, however, that the set of 13 teeth curated as the original specimens actually consisted of only 10 genuine fossils and three casts. On further study it was ascertained that one of the casts certainly represents the missing right $M^3$, one may represent the missing lower incisor (but compares badly with another cast made in 1911) and one appears to represent a very poor cast of the existing left $M_2$. Thus the right $M^3$, a lower incisor and a claimed upper incisor root are missing. Searches for the missing fossils at the Jersey Museum have been made without success. It is conceivable that the missing teeth were mistakenly sent out as part of a set of casts, or were borrowed by a research worker, so it would be useful to know when the loss occurred. Perhaps other workers who have studied the specimens or have photographs of them could help in establishing this.

From the preserved teeth it is possible to agree with the original identifications by Keith and Knowles in every case. The missing right $M^3$ must also have been correctly identified since the cast of it provides a good counterpart to the surviving antimere. Both show similar unusually heavy distal wear. The surviving casts of the missing lower incisor are too poor and too different

**Fig. 17.1**  Occlusal views of the La Cotte de St. Brelade Neanderthal teeth. Upper dentition (A) and lower dentition (B). The teeth have been located as originally placed by Keith and Knowles (1912b), but the right $M_3$ has been reorientated. The lower incisor and right $M^3$ can only be represented by casts.

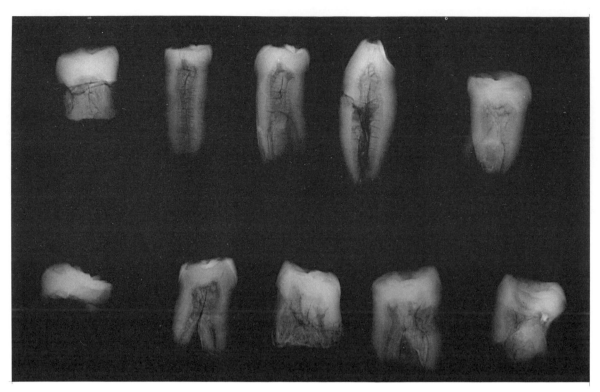

**Fig. 17.2** Radiographs of the Neanderthal teeth from La Cotte de St. Brelade. *Upper row* (left to right): l. $M_2$, l. $P_4$, l. $P_3$, l. C, r. $M_2$. *Lower row* (left to right): r. $M_3$, l. $P^4$, r. $M^1$, l. $M^2$, l. $M^3$.

from each other to allow confirmation of its identification as a right $I_2$. It seems likely that Keith and Knowles' orientation of the right $M_3$ is incorrect. There is a contact facet on the right $M_2$ in an unusual distolingual position, and the only part of the right $M_3$ which seems to preserve a corresponding facet is the damaged margin, which they placed lingually. Establishing contact between the facets reorientates the main area of wear on the right $M_3$ buccally, rather than mesially, and suggests that the long axis of the right $M_2$ was somewhat rotated. The left $M_2$ has mesial and distal contact facets in the expected positions, but because of the lack of the left $M_3$ it is impossible to explain the remarkable dental wear on the opposing $M^3$.

The surviving 10 teeth were measured, and Table 17.1 gives data for them, together with comparative data from other Neanderthal samples. The La Cotte de St. Brelade teeth fall within the expected ranges of variation for European Upper Pleistocene Neanderthal samples. They are relatively smaller than the mean values of these samples in the dimensions of $P_4$, $M^1$ and (if the original data are accepted) $I_2$. They are relatively larger than average in C (lower), $M_2$ and $M_3$ dimensions compared with the 'Würm' sample, but smaller in most respects than the Krapina average, although the broad $P_3$, $M_2$, $M_3$ and $M^3$ dimensions are exceptions. The breadth of $M_3$ actually falls outside the reported Krapina range (Wolpoff 1979).

## 17.2 THE 1915 FINDS

In 1915 what were claimed to be three further hominid fragments were recovered 1.8 m outside the cave entrance, in or under deposits described as yellow loessic clay and rock fragments ('head'), at a height of about 24 m O.D. (Marett 1916). Keith (in Marett 1916, 83) considered the three fragments to represent an occipital fragment and a "somewhat anomalous" malar and mandibular fragment of a single child's skull. Angel and Coon (1954) restudied the specimens and concluded that only the occipital fragment was from a human skull. We concur with this judgement and are unable to identify the other two fragments as representing any part of a human skeleton. The occipital fragment derives from the left occipital squama, including a 49 mm portion of the lambdoid suture running from the asterion. Internally, there is a well-marked and wide groove for the lateral venous dural sinus. The bone is thin with gracile surface markings and may derive from a child of about 7 years, as implied by Angel and Coon (1954). In their view, the surviving occipital fragment suggested the presence of a protruding occiput, and the supreme nuchal lines were placed very superiorly in relation to the superior nuchal lines and the upper border of the internal transverse sinus groove. Both these characteristics allied the St. Brelade 2 individual with Neanderthal rather than mod-

**Table 17.1** Measurements on the La Cotte de St. Brelade Teeth[a]

| Measurement | r. $I_2$? | l. $\overline{C}$ | l. $P_3$ | l. $P_4$ | r. $M_2$ | l. $M_2$ | r. $M_3$ | l. $P^4$ | r. $M^1$ | l. $M^2$ | r. $M^3$ | l. $M^3$ |
|---|---|---|---|---|---|---|---|---|---|---|---|---|
| *St. Brelade 1* | | | | | | | | | | | | |
| Mesio-distal length | – | 8.2 | 7.6 | 6.9 | 11.9 12.1 | 12.1 12.2 | (12.0) | (6.8) | 10.6 10.8 | 11.7 | – | 10.0 10.1 |
| Bucco/labio-lingual breadth | – | 9.6 | 10.3 | 9.0 | 11.5 | 11.9 | 12.3 | 10.7 | 11.9 | 12.7 | – | 13.4 |
| length (Keith & Knowles 1912b) | 5.75 | 8.0 | 7.75 | 7.0 | 12.0 | 12.0 | 12.0? | 6.5? | 11.0 | 10.5 | 9.3 | 9.2 |
| breadth (Keith & Knowles 1912b) | 7.0 | 8.0 | 10.0 | 9.0 | 11.5 | 11.75 | 9.2 | 10.5 | 12.0 | 13.0 | 13.0 | 12.0 |
| Crude crown area (new data where possible) | – | 78.7 | 78.3 | 62.1 | 139.2 | 145.2 | (148.0) | (82.8) | 128.5 | 148.6 | (120.9) | 135.3 |
| *Last glaciation Neanderthal* | | | | | | | | | | | | |
| length – mean | 6.8 | 8.1 | 7.7 | 7.6 | 11.8 | 11.8 | 11.6 | 7.3 | 11.6 | 10.6 | 9.9 | 9.9 |
| s.d. | 0.7 | 0.5 | 0.9 | 0.8 | 0.7 | 0.7 | 0.6 | 0.6 | 1.0 | 0.8 | 1.0 | 1.0 |
| sample | 11 | 12 | 12 | 15 | 17 | 17 | 14 | 7 | 17 | 10 | 8 | 8 |
| breadth – mean | 7.9 | 9.1 | 9.2 | 9.0 | 11.3 | 11.3 | 11.3 | 10.6 | 12.0 | 12.7 | 12.1 | 12.1 |
| s.d. | 0.4 | 0.9 | 1.0 | 0.9 | 0.6 | 0.6 | 0.8 | 0.7 | 0.7 | 1.2 | 1.1 | 1.1 |
| sample | 12 | 10 | 12 | 15 | 17 | 17 | 14 | 8 | 18 | 10 | 8 | 8 |
| area – mean | – | 73.3 | 73.4 | 69.0 | 133.0 | 133.0 | 131.7 | 77.6 | 140.1 | 134.8 | 119.2 | 119.2 |
| s.d. | – | 9.8 | 14.3 | 10.6 | 12.9 | 12.9 | 14.4 | 8.1 | 14.4 | 17.5 | 11.1 | 11.1 |
| sample | – | 10 | 12 | 15 | 17 | 17 | 14 | 7 | 17 | 10 | 8 | 8 |
| *Krapina* | | | | | | | | | | | | |
| length – mean | 6.8 | 8.2 | 8.3 | 8.1 | 12.7 | 12.7 | 12.2 | 8.1 | 12.4 | 11.3 | 10.4 | 10.4 |
| s.d. | 0.2 | 0.1 | 0.1 | 0.2 | 0.2 | 0.2 | 0.2 | 0.2 | 0.3 | 0.4 | 0.2 | 0.2 |
| sample | 6 | 11 | 11 | 13 | 12 | 12 | 11 | 12 | 9 | 10 | 10 | 10 |
| breadth – mean | 8.0 | 9.4 | 9.4 | 9.6 | 11.5 | 11.5 | 10.8 | 10.9 | 12.6 | 12.8 | 12.5 | 12.5 |
| s.d. | 0.2 | 0.2 | 0.2 | 0.1 | 0.2 | 0.2 | 0.2 | 0.1 | 0.3 | 0.2 | 0.2 | 0.2 |
| sample | 7 | 11 | 10 | 13 | 12 | 12 | 10 | 11 | 9 | 10 | 9 | 9 |
| area – mean | – | 78.6 | 78.1 | 77.5 | 146.8 | 146.8 | 132.2 | 88.5 | 156.5 | 144.3 | 130.3 | 130.3 |
| s.d. | – | 2.4 | 2.6 | 1.9 | 4.9 | 4.9 | 3.6 | 3.1 | 7.3 | 7.2 | 4.2 | 4.2 |
| sample | – | 11 | 10 | 13 | 12 | 12 | 10 | 11 | 9 | 10 | 9 | 9 |

[a]Where two figures are given, the second allows for interstitial wear. Comparative data for means of a last glaciation Neanderthal sample (recomputed from data in Frayer 1978) and a Krapina sample (Wolpoff 1979) are given, left and right sides pooled. Data on the St. Brelade $M_3$ apply to the re-orientated specimen.

[b]Keith and Knowles (1912b).

ern hominids according to them. However, the specimen is so fragmentary, comparative fossil material so sparse, and ontogenetic variation in such characters so great amongst anatomically modern humans that it seems preferable to maintain caution about the firm attribution of the occipital fragment to a Neanderthal child, while admitting its possibility.

## 17.3   THE BURDO FINDS (1954–1958)

Ten bone fragments from the Black Ashes at about 17 m O.D. (Burdo's layer [7E], probably equivalent to McBurney's C–E) were described as hominid by Burdo (1960, 73–74 and pl. VIII); they are preserved at the British Museum (Natural History), where they are catalogued as E.M. 511–520. They have been carefully re-examined and, as previously stated in Oakley, Campbell and Molleson (1971), no support can be given to the proposal that they are human; indeed they are so fragmentary as to preclude any reliable specific determination. From the general account of the fauna the amount of identifiable material from this layer appears to have been small, yet Burdo notes that the supposed human remains "were mixed up promiscuously with a mass of animal fragments", suggesting that many pieces of strictly indeterminate bone were recovered. In the event it is perhaps not surprising that the uncritical eye could attribute some of them to parts of the human skeleton, particularly in the light of previous discoveries.

### Acknowledgements

We are grateful to the Société Jersiaise and to G. P. Drew, Acting Curator, for the opportunity to study and figure the original specimens curated at The Jersey Museum in St. Helier. We also thank L. Martin for useful discussion about the dental material, and to R. Kruszynski and R. Eames for research assistance. Photographs were taken by Stringer and radiographs by T. I. Molleson, and they were printed by the photographic studio of the British Museum (Natural History).

# THE BONE ASSEMBLAGES OF LAYERS 3 AND 6

K. Scott

## 18.1  INTRODUCTION

The faunal remains from all levels at La Cotte are described in chapter 13, with special reference to the most abundantly represented species: mammoth and woolly rhinoceros. The remains of these animals occur predominantly in two levels (layers 3 and 6), in which a number of artefacts were also discovered. This is a remarkable situation both in terms of the quantity and state of preservation of the mammoth and woolly rhino remains and in terms of the context in which they were found.

Throughout Europe, caves have been found to contain palaeolithic material associated with the remains of animals which were either hunted or scavenged for meat, fur or other raw materials. Mammoth and rhino are rarely represented in these archaeological cave deposits, and then only by the occasional bone, tooth or tusk fragment. In fact, such poor representation has led to the widespread opinion that these species were of little economic importance to prehistoric man (Bay-Petersen 1975), or that they were not hunted by man at all (Straus 1977).

The fact that at least 20 mammoths and five rhinos comprise virtually the entire fauna in these two levels at La Cotte is therefore particularly striking and raises several questions concerning how and why they came to be there. In terms of prehistoric behaviour, the most pertinent of these questions is clearly the extent to which man may have been responsible for the accumulations.

This chapter reviews the evidence that led the author to the following conclusions. Firstly, that the two assemblages of mammoth and woolly rhinoceros accumulated on specific occasions rather than as the result either of long-term predation or of gradual sedimentary infilling of the site. Secondly, that hominids caused the death of these animals at the site. Thirdly, that the disparate representation of parts of the skeletons of both species are the result, in part at least, of butchery activities.

## 18.2  GENERAL DESCRIPTION OF LAYERS 3 AND 6

Fig. 18.1 provides a summary of the skeletal elements by which mammoth and rhino are represented in these two levels (this information is tabulated, and measurements of bones given, in chapter 13). Both levels are remarkably similar in that they contain large numbers of mammoth and several woolly rhinos to the almost total exclusion of any other species; there are a few bones of other species, but these occur only at the base of each level and may well have been incorporated from the underlying cultural levels during the deposition of the loesses of layers 3 and 6. However, it will be apparent from Fig. 18.1 that the two assemblages differ markedly in the parts of the skeleton represented, particularly in the case of the mammoths. These differences can also be observed by reference to the plans (Figs. 18.2–18.7).

Figs. 18.2 and 18.3 are composite illustrations of the two piles of bones as each would have appeared if uncovered in one excavation. In reality, the heaps were dismantled during the course of many field seasons (see, for example, Fig. 18.8) and the reconstructed plans were compiled from field drawings and photographs. For easier discernment of the contents of the two assemblages, Figs. 18.4–18.7 represent a breakdown of the heaps into their component skeletal elements. In these illustrations the bones are numbered to allow comparison with the text and also with the tables in chapter 9. As will be observed, woolly rhinoceros is represented in both levels by parts of the body of a few individuals, the most striking occurrence being that of three virtually complete skulls in layer 6, two of which are on top of the pile of bones. The representation of mammoth, however, is more curious. In the lower of the two heaps (layer 3), there are at least nine skulls, two mandibles, several tusks, scapulae and innominates, but few limb bones (Fig. 18.1). By contrast, the dental and cranial material in 6 represents only two mammoths, while a minimum of 11 individuals are represented by large post-cranial bones — scapulae, femora, humeri and innominates. There are numerous ribs in layer 3 and few in 6, and several vertebrae in 3 but none in 6. It is also odd that there are so few bones of the lower limbs in either assemblage, particularly considering that a minimum of 20 mammoths are represented in total by other parts of the skeleton.

To what extent this disparity might have been due to selectivity on the part of predators or to post-depositional disturbance and destruction will be discussed below, but at the outset it must be stressed that both assemblages are obviously samples of larger accumulations. A relatively small area of the total site was excavated during the 1961–1978 excavations, and it was evident that both piles of bones extended into the unexcavated north section of the deposit. It is probable too that there were more remains further south, where they

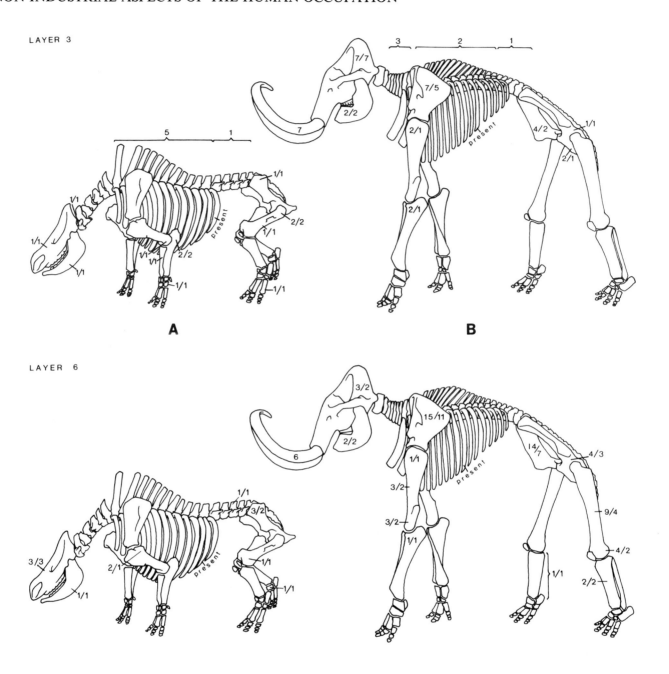

**Fig. 18.1** Schematic representation of skeletons of woolly rhinoceros (A) and mammoth (B) to illustrate the numbers of bones of these species from layers 3 and 6 at La Cotte. Where it was possible to calculate the minimum number of individuals represented by any skeletal element, this number follows the number of bones (i.e. 15/11 — fifteen scapulae representing at least eleven individuals). The rhinoceros and mammoth skeletons are modified from illustrations in Borsuk-Bialynicka (1973) and Toepfer (1957).

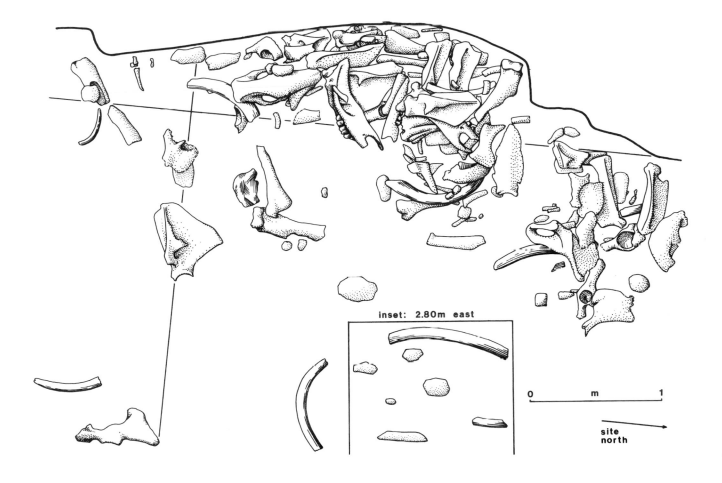

**Fig. 18.2** Composite plan of bones in layer 6. The axes in all bone plans in this chapter intersect at (0.5 N, 0.0 E) rather than at the grid origin.

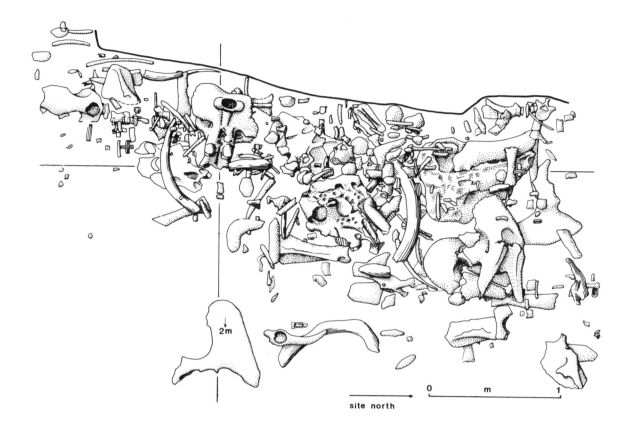

**Fig. 18.3** Composite plan of bones in layer 3.

**Fig. 18.4** Skulls and dentition (A) from layer 6 and (B) from layer 3; specimens whose original orientation is unknown are marked X in these and subsequent plans.

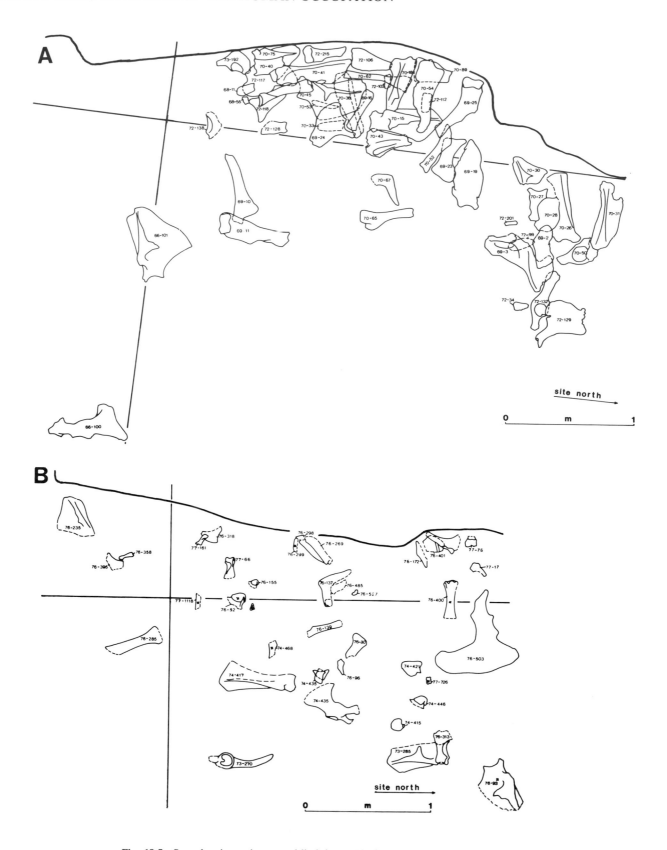

**Fig. 18.5** Scapulae, innominates and limb bones (A) from layer 6 and (B) from layer 3.

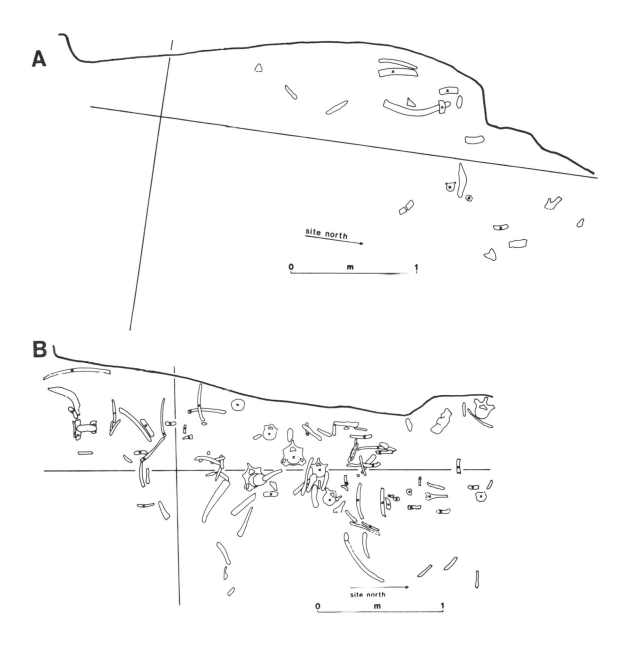

**Fig. 18.6** Ribs and vertebrae (A) from layer 6 and (B) from layer 3.

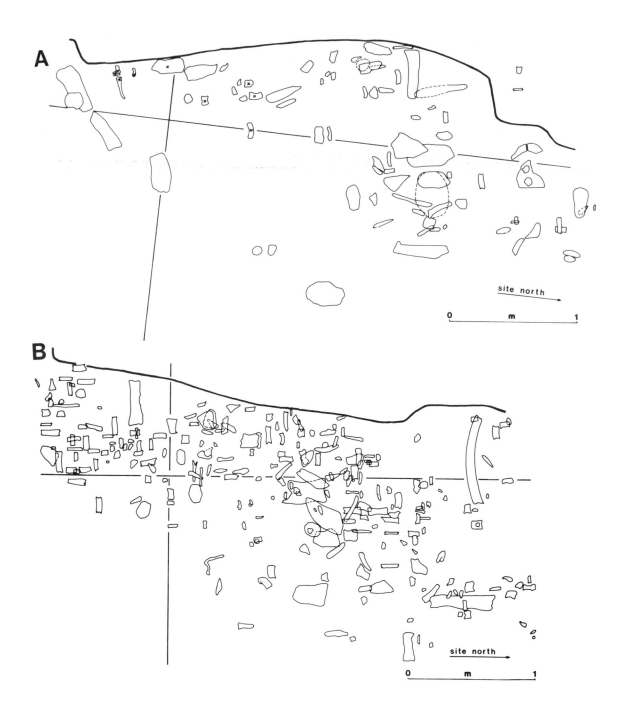

**Fig. 18.7** Unidentified bones and fragments (A) from layer 6 and (B) from layer 3.

**Fig. 18.8**  Excavation of the lower bone heap in progress (the upper bone heap is shown in Fig. 2.7A).

would have been destroyed by interglacial marine action, and Burdo's diaries refer to numerous, irretrievable large bones in the areas he excavated in the 1950s. Another consideration is the effect of the overhang of the west wall on the preservation of the bone heaps; their eastern limits are roughly coincident. It is possible that the bones originally extended beyond the drip line, but were subsequently destroyed by water filtering through the overlying deposits. Certainly those bones that were recovered from the margins of the area protected by the overhang were water-logged and often unidentifiable.

The predominance of mammoth skulls in one level and post-cranial bones in the other may therefore signify nothing more than the fact that two samples of much larger assemblages have been recovered. In fact, in this exposed, only partly roofed granite shelter, it is quite remarkable that any bones survived at all. Undoubtedly, they would have suffered the same fate as the bones of the other levels had it not been for the fortuitous deposition of calcareous loess (the matrix of layers 3 and 6) in the fissures of the headland.

Before considering how these assemblages might have accumulated, it is necessary to review the evidence suggesting that these were rapid rather than long-term accumulations. There are several indications that the bones of each level accumulated on one particular occasion and were covered by loess relatively soon after depo-

sition. The bones of each level occurred on one plane and were not distributed throughout the loess. Where bones occurred one above the other, they were in direct contact with each other rather than separated by loess and debris. A number of bones such as scapulae and innominates were found lying 'on end' rather than flat, and some of the ribs were found resting vertically against other bones (as in Fig. 18.9). It seems unlikely that they would have remained in these positions if they had been exposed and accumulating over a long period.

It would be impossible to estimate how much time elapsed between the death of the animals represented at La Cotte and the subsequent deposition of loess. From a number of field studies that have been conducted in Africa on the rate of disintegration of animal carcasses (e.g. Behrensmeyer 1978; Hill 1979; Crader 1983,) it is evident that decay varies according to a number of factors: age and size of the animal, degree of destruction caused by carnivores, and local environmental conditions all play a part. Crader reports considerable dispersal and destruction of one elephant carcass after only 6 weeks, whereas others, ranging up to 10 years since death, are less scattered. The effects of the Saalian climate on accelerating or delaying the decay of animal carcasses can only be surmised. On the one hand, periods of intense cold might have caused the freezing and preservation of flesh; on the other, an arid, windy environment might

167

**Fig. 18.9** (A) Mammoth skull (76/175) from layer 3; ribs were found resting vertically against either maxilla; one of these may be seen on the left of the photograph, which was taken *in situ*. (B) Mammoth skull (76/280) also from layer 3, photographed during preparation; note the rib against the right maxilla (on the right in the photograph).

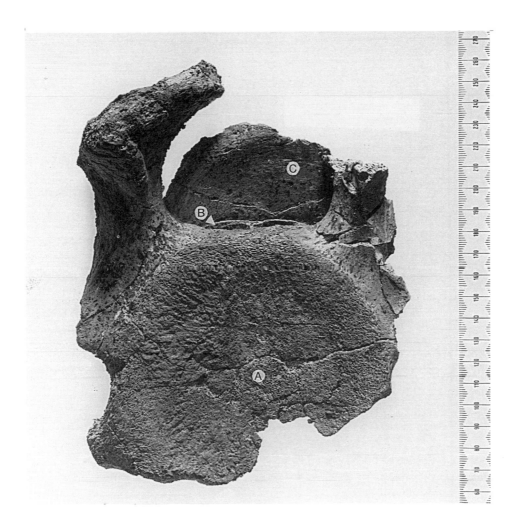

**Fig. 18.10** Mammoth cervical vertebrae (74/432) showing centrum (A) from which the unfused epiphysis (found alongside) has detached. The posterior epiphysis (B) has slipped upwards slightly, as has the anterior epiphysis (C) of the next vertebra.

have caused rapid surface weathering and disintegration of the bones. It can only be said of the La Cotte skeletons that they appear to have been exposed long enough to have become disarticulated and substantially dispersed, yet to have been blanketed with loess while the bone surfaces were still in reasonable condition and while some elements remained in articulation. There are several instances of bones that must have been buried while there was still muscle or ligament holding them together. For example:

1. Two immature mammoth cervical vertebrae (illustrated in Fig. 18.10) have separated by a few centimetres but are still in contact, and the unfused epiphyses are still attached to the centra. If these bones had been exposed for long, or had been disturbed by carnivores, the epiphyses would have become detached and the two vertebrae separated.

2. Three of the mammoth skulls are of immature animals as shown by their unfused nasomaxillary sutures (specimens 74/434, 74/429, 76/121). Once the flesh has been removed or has decayed from an im-

mature skull, it soon falls apart at the sutures. In the case of two of these skulls, each found resting on its left maxilla, the right (upper) half of the skull had slipped slightly out of alignment (Fig. 18.11). The third skull was discovered resting on the occlusal surfaces of the maxillary teeth; the two halves of the skull had separated slightly at the nasomaxillary suture, but the skull had not collapsed.

3. The atlas vertebra of a rhino was found in articulation with the occipital condyles of a skull (specimen 77/25). This is a shallow articulation, and in exposed conditions the atlas detaches from the skull after the flesh is removed. This suggests that the skull was covered by loess while flesh still held it in contact with the atlas.

If the bones from La Cotte had been exposed to the elements for any length of time one would expect to see the effects of weathering on the surfaces. Field studies available on the Indian elephant (Deraniyagala 1955) and the African elephant (Hanks 1979; Coe 1980) indicate that their skeletons disintegrate rapidly once exposed,

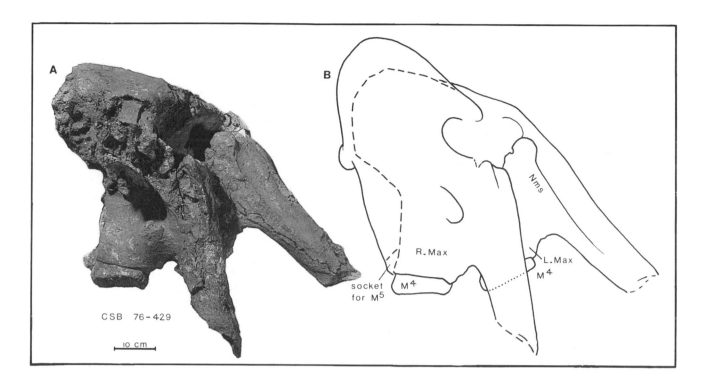

**Fig. 18.11** (A) Skull of mammoth (CSB 76/429). (B) Outline of a mammoth skull in profile illustrating the part of the skull represented by CSB 76/429 and showing the extent to which the left half of this skull is out of alignment (Max, maxilla; Nms, nasomaxillary suture).

particularly those of subadult animals, such as are well represented at La Cotte. The long bones of elephants do not have marrow cavities but are composed of porous, spongy bone tissue which is surprisingly fragile and soon cracks and flakes when exposed. The teeth, which are composed of a series of lamellae held together with dentine, break up very quickly on exposure; the younger the animal, the more rapidly the lamellae separate and disintegrate. At La Cotte it is remarkable that there are many isolated teeth of young animals in almost perfect condition, showing no indications of weathering (Fig. 18.12).

Young rhinoceros skulls are fairly fragile, and are soon destroyed in the open (Goddard 1970). The rhinos from layers 3 and 6 are immature though not very young, and might have been able to survive a moderate degree of exposure; however, like the mammoth teeth, the rhino dentition is in excellent condition.

It might appear on examination that some of the La Cotte bones did in fact suffer severe exposure before burial. However, the poor condition of a great many of the bone surfaces is the inevitable result of the enormous technical difficulties encountered in the retrieval of the material during excavation. The loess (densely packed in and around the bones) was moist and easily removed, revealing smooth, unweathered surfaces. Leaching through this porous matrix in antiquity resulted in the total decalcification of the bones in some areas of the site, making it almost impossible to move them without breakage, particularly as many of them were very large. This necessitated leaving the bones *in situ* while chemical hardeners were applied (conservation materials are listed in Appendix G, in microfiche). In ideal conditions, the hardeners should be allowed to take gradual effect in controlled atmospheric conditions; rain hinders the hardening process, and wind results in the rapid drying of surface chemicals to form a skin which contracts and cracks, lifting the surface of the bone. Conditions in the field were far from ideal, and although care was taken to protect the bones during treatment, they were often unavoidably exposed to wind and rain.

Another field process involved jacketing the larger specimens with glass fibre to enable them to be lifted in a rigid case. Ideally, the specimen should be dry before the application of a chemical separator and fibreglass, but it was seldom possible to meet this requirement. As a result, the separator was sometimes inefficient, and the surface of the bone unavoidably damaged in the subsequent removal of the fibreglass in the laboratory.

The lengthy preparatory processes often necessitated leaving larger specimens in the matrix through several seasons, and although every effort was made to protect the bones, many began to show signs of weathering not observed when they were discovered. However, despite the above-mentioned difficulties, the surface condition of many of the bones (particularly those in the protected western area and those underlying subsequent rock-falls) remains excellent.

**Fig. 18.12** Teeth of young mammoths from layer 3 (scale in mm). (A) Anterior part of an uncrupted lower right molar (M4) (76/189). (B) Upper left molar (M3) in maxilla (76/222).

## 18.3 ORIGINS OF THE BONE ASSEMBLAGES IN LAYERS 3 AND 6

There are various ways in which animal bones and teeth may be incorporated in cave deposits. Some animals hibernate and die in caves, some are brought in by predators; others may be trapped in holes and fissures in a cave system, or their bones may be included amongst other debris that becomes part of a cave deposit through sedimentary processes.

The identification of the agent of accumulation of bone assemblages has been the subject of many papers, particularly in the past decade or so; see, for example, studies by Klein (1975 and 1980a), Binford and Bertram (1977), Vrba (1980) and Brain (1981 with references). As the result of the work of these and other authors, a number of characteristics are now recognised by which various types of bone assemblage may be differentiated. These characteristics refer to the range of species in the accumulation, to the parts of the skeleton and age range of the animals represented, to the condition of the bones (signs of gnawing or acid action from gastric juices, cut marks, degree of breakage). Other clues may be provided by analysis of the spatial distribution of the bones and enclosing matrix, and by associated debris such as artefacts, hearth material and coprolites.

One of two possibilities could account for the mammoth/rhino assemblages at La Cotte: either a predator was involved in some way, or the bones were deposited by sedimentary processes. Mention has already been made in Chapter 6 of the fact that the site is a partially roofed chasm and was almost certainly so during the later Saalian; this suggests either that animals fell in at various times, or that bones of animals that died on the headland became incorporated in the sediments that gradually filled the chasm. However, there are two reasons for refuting both suggestions in the case of these two bone heaps. Firstly, it would be reasonable to expect species in addition to mammoth and rhino to have fallen in if this were a gradual fissure infill and, secondly, one would expect the bones to be distributed throughout the deposit, which was not the case at La Cotte.

If these assemblages did not accumulate by gradual sedimentary process, then an alternative explanation is that they are the result of predator activity. The spotted hyaena *Crocuta crocuta* was evidently responsible for

171

**Fig. 18.13** View into the west ravine from the top of the headland, showing a drop of approximately 50 m. An entrance to the site (A) can be seen at the bottom of the photograph. This entrance is about halfway up the cliff. Concealed from view by the rock arch is a large opening, indicated by arrow (B), giving access to the occavated deposits.

many vast accumulations of bones in European caves; young mammoth and, in particular, young woolly rhinoceros are well represented in these deposits. It is also well known that the extant spotted hyaena is capable of carrying heavy parts of carcasses over a considerable distance (Kruuk 1972). Nevertheless, it is clear that the majority of bones at La Cotte would have been beyond the carrying ability of any carnivore other than man. Even for man, the transportation of so many carcasses would have presented enormous difficulties. In fact, the suggestion that these bone heaps were the result of hominid activity only becomes tenable if the animals were killed at, or in the immediate vicinity of, the site. The idea of a mammoth/rhino kill site in the confined and fairly inaccessible situation at La Cotte may seem extraordinary, but it is perhaps less surprising if one considers that, by virtue of its particular structure and setting, La Cotte is (and presumably would have been) a ready-made trap (see Fig. 18.13).

## 18.4  LA COTTE AS A KILL SITE

Fig. 18.13 shows that the site is at the end of a headland and would have provided an ideal situation for a game drive. Animals grazing on the upland plateau could have been driven towards La Cotte Point and off the steep-sided cliff. Some may have fallen through the open 'roof' of the site, or into adjacent chasms from which parts of carcasses may have been dragged into the shelter. The feasibility of this method of hunting would seem to be corroborated by both archaeological and ethnographic literature.

There is substantial archaeological evidence that prehistoric hunters took advantage of the gregarious nature of some species by driving herds into traps. For example, ibex seem to have been regularly killed by driving them up blind passages in the Dordogne region of France (Bay-Petersen 1975). Also in France is the famous 'horse magma' horizon at Le Solutré, where it is suggested that great numbers of horses were driven to the base of an inaccessible cliff and then killed (Bay-Petersen 1975; Levine 1979). Bison in America were evidently hunted on a regular and massive scale by stampeding herds into chasms, as described, for example, by Wheat (1972), Kehoe (1973), Frison (1974) and Reher (1977).

A number of travellers in Africa in the 19th century give vivid accounts of the organisation of such game drives by indigenous people and their methods of constructing of traps for large wild animals. For example:

Almost all the tribes of southern Africa avail themselves of pitfalls (often on the most gigantic scale) for the capture of game. These traps, or rather, these lines of pitfalls, are either constructed in the shape of very obtuse triangles, open at the base and gradually tapering to a point, where a single, double, or treble row of pits are dug, into which the game is driven by shouts and yells; or they are formed in the shape of a crescent — often miles in extent — usually shutting out a valley or defile, with pits at every fifty or a hundred paces apart, artfully concealed with grass, sand, etc., the intervening spaces being planted and filled up with stout palisades, closely interwoven with boughs and branches of thorn trees. (Andersson 1856, 376–377.)

In 1881 Oates wrote: "During a ride I saw a big game-drive made by the Makalakas, consisting of a long broad alley, the sides composed of large tree branches, forming a strong hedge. At the end were three pits side by side, walled round with stakes. On the top were placed light stakes, and long grass was laid over all" (Oates 1881, 233).

W. C. Harris is one of many others to give similar accounts of game drives in southern Africa. In one passage of his journal he wrote:

And I for the first time observed several pitfalls constructed purposely for taking the rhinoceros. They differed from the others in being dug singly instead of in groups — very deep and very large — at the extremity of a narrow path cut through the bushes, and fenced outside with thorns — a sharp turn leading directly upon the trap, so that an unwieldy animal being driven furiously down the avenue could have no chance of avoiding the snare. Many skulls and bones of these huge beasts were lying at the bottom of these sepulchres that had swallowed them up alive. (Harris 1838, 149.)

On the same page he mentions that the pitfalls were "generally provided with a sharp stake at the bottom, on which he [the rhino] is impaled".

Reference to the above ethnographic accounts is not intended to suggest that the mammoths and rhinos at La Cotte were necessarily the result of such organised and elaborate hunting methods, but rather to make the point that, despite the unlimited availability of smaller, more easily obtained game, southern African hunters with essentially prehistoric technology made regular drives of elephant and rhinoceros. Although the above authors do not mention whether a game drive included both elephants and rhinos at one time, they do record these species within close proximity of one another. Thus the fact that mammoth and woolly rhino should occur in the same assemblages at La Cotte may perhaps be explained with reference to the behaviour and social organisation of living elephants and rhinos. Elephants are gregarious animals moving in herds which vary in size from a few individuals to a few hundred. Rhinos, on the other hand, are principally solitary, occasionally sharing territory with one or two others, or forming units of mothers with offspring. They are frequently observed on the fringes of groups of elephant, and in the case of the black rhino and the African elephant, the two are ecologically interdependent: they use the same feeding tracks, and there is a considerable overlap of food preferences (Schenkel and Schenkel-Hulliger 1969). This association might have been true of the extinct species: mammoth and woolly rhino are frequently associated in fossil deposits of Europe and Russia (e.g. Pfizenmayer 1939; Heintz and Garutt 1965; Kurtén 1968; Stuart 1982). Hence it might not be surprising to find the remains of a few rhino amongst a larger quantity of mammoth bones if such associated species had been driven *en masse* off the headland.

## 18.5 LARGE MAMMAL BUTCHERY SITES: ARCHAEOLOGICAL AND ETHNOGRAPHIC PARALLELS

If one compares these two assemblages from La Cotte with other archaeological faunas from cave sites, and considers them also in the light of ethnographic observation on hunting and butchery, the possibility that they represent two kill/butchery incidents seems quite plausible.

In the case of this particular site, there are several reasons why this comparative data is perhaps best drawn from the archaeological and ethnographic literature of southern Africa. Firstly, at the time when the southern African Middle and Later Stone Age sites were occupied, elephants and rhinos were among a wide range of large mammals potentially available to prehistoric man. Secondly, early European travellers in Africa observed and recorded the hunting and butchery of these species by the indigenous stone tool-using peoples. Thirdly, the detailed faunal analyses from more than 40 palaeolithic sites where man was evidently the main contributor to the faunal assemblage provide more detail on prehistoric hunting and butchery than is perhaps available from anywhere in the world. This body of data, published principally by R. G. Klein during the past decade or so, is mainly concerned with hominid activity in caves — whether temporary or long-term occupation sites (e.g. Klein 1976; 1980a; 1980b). Size is one of the factors that appears to determine species and body-part representation at these cave sites. The faunal remains are characterised by a range of small and medium-sized herbivores. Analysis of body parts of the prey species represented shows that the smaller the animal, the more likely it is to be represented by all parts of the skeleton, indicating that the carcasses were brought whole from the kill site to the cave for butchery. The larger the prey species, the smaller the range of body-parts from each carcass: heads, scapulae, vertebrae, ribs and the pelvic girdle tend to be poorly represented, whereas the more easily transported meat-bearing limbs are better represented. The very large herbivores are generally best represented by their smallest bones: those of the lower limbs and feet, which suggests that the bulk of meat was cut from the carcass at the scene of the kill, and the skeleton left behind.

It is interesting to compare the southern African archaeological evidence from caves with ethnographic observations from various regions of sub-Saharan Africa. The archaeological faunas are virtually devoid of remains of elephant and rhino, yet the first European travellers in southern Africa record that both species were hunted by the indigenous hunter-gatherers. The fact that elephant and rhino so seldom became incorporated in cave deposits would seem to be a function of size, as is well substantiated by both the early and more recent observations of the butchery of these and smaller species. The general tendency is for smaller animals not eaten at the kill site to be transported whole to the home base or camp. Larger animals are dismembered at the kill site, and the more easily transported limbs taken away. The head, vertebrae, scapulae and unwieldy pelvic girdle are generally left at the kill site. Some accounts refer specifically to the problems of dealing with the carcasses of such large animals as rhino and elephant. For example, Cumming (1850) watched the Bechuana in South Africa butcher an elephant, first removing the flesh in enormous sheets from the ribs and then climbing into the body to get at the bowels. The trunk and feet, considered to be delicacies, were baked at the kill site, and then the rest of the flesh was carried away to temporary camp sites.

Marks (1976) noted that the Bisa people in East Zambia disarticulate and dismember large animals, cook the liver and heart at the kill site, remove the meat from the heavy bones, cut it into strips and carry it back to the village on a pole. Crader's observations also concern the Bisa people. While smaller animals such as impala or warthog are taken back to the village as complete carcasses, the bones of very large animals such as elephant and, rarely, buffalo are left at the kill site. With regard to the elephant, the only exception is that the feet are cut off and carried back to the village "for the much-prized wedge of fat they contain" (Crader 1983, 126).

Marshall observed the way in which the !Kung bushmen deal with large game:

> When the kill is made, the hunters eat the liver on the spot, and more if hungry ... they may eat the parts that are especially perishable or awkward to carry, for example, the head .... if the animal is very big, they leave most of the bones and cut the meat into strips. The strips dry to biltong quickly and are thus preserved before they decay, and they can be hung on carrying sticks and transported more easily than big chunks. (Marshall 1976, 358.)

The remains of such butchery activities as described above are obviously less likely to become part of the archaeological record than are the carcasses or parts of carcasses carried to a cave site, simply because bones have little chance of long-term survival in an open context unless buried in favourable conditions relatively soon after the death of the animal. Then they are less likely to be discovered. Nonetheless, such sites are known from various localities in the world, and the representation of parts of the skeleton is generally as might be expected in view of ethnographic observation: skulls, vertebrae, ribs, pelvic bones, scapulae, and the larger limbs tend to be better represented than the smal-

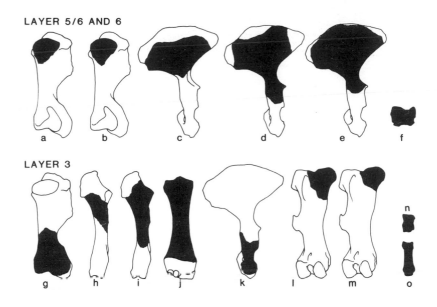

**Fig. 18.14** Schematic representation of surviving parts (shaded) of post-cranial bones of woolly rhinoceros in layers 3 and 6. For key to specimen numbers in this and other illustrations see mf 18.1.

ler (more portable) bones. See, for example, discussion on the kill/butchery sites of bison (Frison 1974), buffalo and eland (Klein 1978 and 1983), elephant (MacCalman 1967; Clark and Haynes 1969), and mammoth (Chmielewski and Kubiak 1962; Frison 1974).

It would appear then that the high numbers of heavy, cumbersome elements found at La Cotte — scapulae, pelvic bones, skulls and tusks — resemble body-part representation observed at large mammal butchery sites. However, it is also evident (see Figs. 18.1–18.7) that body-part representation in layer 3 is noticeably different from that in layer 6: there are nine skulls and seven tusks in layer 3, and five individuals represented by scapulae, but there are scarcely any limb bones. In layer 6, although there are only two partial skulls, there are five tusks, and at least 11 mammoths represented by post-cranial bones. In both levels, bones of the lower limbs and feet are virtually absent. Possible reasons for these differences are discussed below, although, as the remains of rhino are few (see Fig. 18.14), the discussion refers principally to the mammoth remains.

## 18.6 SKELETAL REPRESENTATION AND HOMINID ACTIVITY AT LA COTTE

There is no obvious single explanation for the somewhat peculiar skeletal representation of the two assemblages. Each lacks what is abundant in the other, and yet they cannot possibly be components of a larger, single kill/butchery assemblage, as they were separated by a considerable quantity of loess and a substantial rock-fall. In all probability, a combination of post-depositional activities and natural processes affected the two heaps

from the time of death of the animals through to the present day, but before some of these are considered, two points made earlier should be be stressed. Firstly, the excavated area under consideration is only a sample of the whole site and, secondly, bones that might have been on the eastern side of this excavated area would have been exposed constantly to water percolating down through the deposits from the open 'roof', and thus destroyed. If butchery had in any way been organised to utilise particular areas of the site or to discard unwanted parts of the carcasses in heaps, then the differences in assemblage content might be the result of such spatial differentiation. Certainly it is evident that hominids were involved in post-kill activities at La Cotte. In layer 6, for example, several mammoth scapulae have been stacked at one side of the cave, and a rhino skull left on top of the pile (Fig. 18.2). In layer 3, the concentration of skulls in one area, some on top of others, and the absence of corresponding body-parts, strongly suggests hominid activity (Fig. 18.3). One of these skulls is particularly interesting in that it has a rib apparently driven right through it from the parietal to the squamosal region (Fig. 18.15). Ribs also appear to have been pushed into the ground beside two other mammoth skulls (Fig. 18.8). Such indications that the bones were moved about might be one explanation for the missing lower limbs in both levels, and in the case of layer 6, might also account for the absence of vertebrae and ribs: perhaps different parts of the carcasses were being butchered in different areas of the site.

P. R. Jones (pers. comm.) makes an interesting observation with regard to this possibility. He noted incidences of elephant butchery in Tanzania where the hind limbs had meat removed from the top, were disar-

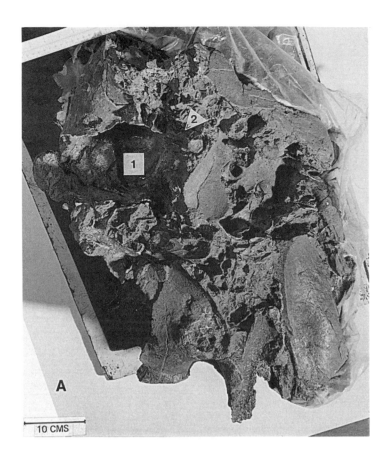

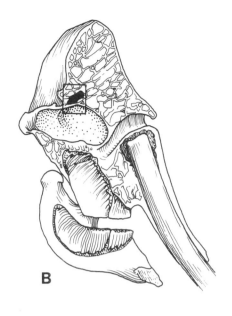

**Fig. 18.15** (A) Mammoth skull (74/434) during laboratory preparation, showing the brain case (1) and the head of a mammoth rib (2) evidently driven into the skull. (B) Diagrammatic section of a mammoth skull and mandible, showing the situation of the rib in the La Cotte specimen. (C) Detail of 74/434 showing the emergence of the rib on the other side of the skull in the squamous region; when discovered the rib was virtually complete, though broken in several places.

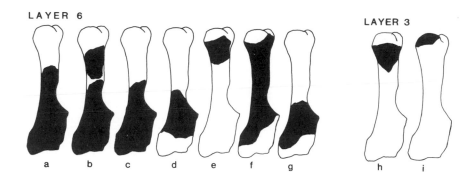

**Fig. 18.16**  Schematic representation of surviving parts (shaded) of mammoth
humeri.

ticulated at the femur/pelvis (joint), and dragged away to be butchered. The forelimb had meat removed from the top, was disarticulated so as to leave the scapula in place, and dragged off for others to work on. The scapula was then lifted off (without much meat on it). This was done to allow more people to work at the same time on one carcass.

Crader (1983) also observed (among the Bisa in Zambia) that dismemberment of an elephant is effected so that several individuals can work on a carcass simultaneously. As Jones suggested (pers. comm.), this might be even more necessary in the confines of a cave than in the open. It would certainly explain the stacking of bones at La Cotte, and perhaps also the concentrations of different parts of the mammoths in the two levels.

Apart from dismemberment and damage that might have been caused by hominid activity, it is highly likely that substantial damage was caused by large carnivores such as hyaena, lion and wolf. In this respect, a most interesting study by Crader (1983) is relevant. It concerns observations on the damage to, and dispersal of, seven elephant carcasses killed and butchered by humans over a period of about 10 years. A consistent feature of these sites is the survival of the skull and pelvis, and although a certain amount of damage was caused to some bones during the butchery process, the most extensive damage to individual bones was caused by scavenging carnivores — probably lion and hyaena. They were also responsible for totally removing some body parts, and for scattering others over a considerable distance. In one instance, bones of a single elephant were found as far apart as 150 m. Gnawing of bones by carnivores certainly cannot be ruled out at La Cotte, but is not obvious. Some bones have battered surfaces that might have resulted from chewing, but the chemical treatment necessary for the preservation of this material has made it difficult to identify the cause of damage to a bone with any degree of certainty.

Even if tooth marks are difficult to detect on the La Cotte specimens, the low frequencies of ribs, vertebrae and lower limbs, the dispersal of the skeletons and the fairly consistent pattern of damage to surviving elements are strongly suggestive of carnivore in addition to hominid activity at the site. The degree of damage that

carnivores or stone tool-using hominids could inflict on a mammoth skeleton would be determined largely by the differential durability of bones, depending on the structure of the particular bone and the age of the animal. Experiment has shown that there is a fairly predictable pattern of survival in limb bones: those bones or parts of bones with the highest specific gravity and those characterised by early epiphyseal fusion tend to survive butchery, chewing and post-depositional destruction in general, better than do bones without those properties (Brain 1967; 1969; 1981). This is an important point with respect to body-part representation at La Cotte, particularly of the mammoth. Most of the surviving mammoth bones have not fully matured: that is to say that epiphyses and cranial sutures have not yet fused in the majority of specimens. Fusion of most bones begins to take place relatively late in the life of living elephants: after the last molar (M6) comes into wear, when the animal is approximately 30 years of age (Deraniyagala 1955). The growth of females slows down after about this age; males, however, continue to grow and some long bone epiphyses remain unfused throughout most of their lives (Laws 1966). The distal humerus, distal femur and bones of the pelvis are among the first to fuse. A comparable state of fusion in relation to dental development was also observed by the author, of a reconstructed mammoth skeleton in the Musée d'Histoire Naturelle, Paris: M6 was in early wear, the proximal and distal epiphyses of the humerus had fused, as had the distal femur, the component parts of the pelvis, and the proximal ulna. The other bones of the lower limbs had not fused, neither had the margins of the iliac crest, nor the vertebral epiphyses.

In view of the above, the evident correlation between the representation of body parts at La Cotte and the likelihood that the bone or part of the bone had completed growth and fusion might be expected. Whether the missing skeletal elements (being more cancellous and fragile) were removed by carnivores or man, or whether they were less able to withstand damage from water percolating through the loess, cannot be ascertained, but as may be seen from Figs. 18.16–18.22, the pattern of damage and loss to those bones that have survived is fairly consistent.

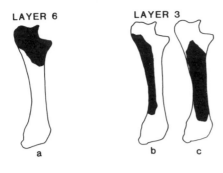

**Fig. 18.17** *Schematic representation of surviving parts (shaded) of mammoth ulnae.*

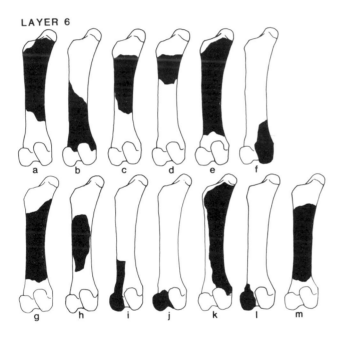

**Fig. 18.18** *Schematic representation of surviving parts (shaded) of mammoth femora.*

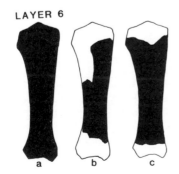

**Fig. 18.19** *Schematic representation of surviving parts (shaded) of mammoth tibiae.*

Femora and humeri are well represented by diaphyses, some with fused distal epiphyses, but in no case have the proximal ends survived. Crader (1983) observed that carnivores had chewed these bones of elephants extensively, and scooped out the cancellous tissue. The innominate is a particularly well represented bone but, apart from three virtually undamaged specimens, is best represented by the acetabulum and ilium. As mentioned above, the iliac crest fuses relatively late so that damage to this margin (as in the La Cotte specimens) might have occurred fairly easily. Many of these ilia had evidently not fused with the ischium and pubis, although several fragments of these parts of the pelvis were recovered. Scapulae are as well represented as the innominates and, as is the case with the latter, damage appears to have occurred to the most fragile regions of the bone: the distal vertebral border and the spine. Here again, Crader (1983) reports a comparable situation at the elephant kill sites: carnivores had chewed away the edges of flat bones such as the scapulae, sometimes leaving a scalloped edge. Vertebrae are few, and relatively undamaged. Ribs, however, are numerous in layer 3; some of these are virtually complete and many others very fragmented, although whether these were broken in the process of butchery, or subsequently, is debatable. Ulnae are represented only by a couple of immature diaphyses and one proximal end with fused epiphysis. There is one complete tibia with fused epiphyses and two immature diaphyses. The radius is not represented. It is improbable that the virtual absence of bones of the lower limbs is due to their being smaller and therefore more easily destroyed by chemical leaching through the matrix than the bones of the upper limbs, pelvis and scapulae. In fact, tarsals and carpals are very compact and dense, and mid-shaft density of cancellous trabeculae is greatest in the radius, ulna and tibia of the elephant (Morlan 1983). Either they were left in some other part of the site, or they were removed by hominids and/or scavengers. Whatever the cause, it is nonetheless rather remarkable that virtually every one of as many as 80 lower limbs is missing.

As regards the thickness and durability of elephant bones, the only exception is in the structure of the cranium, which is thin-walled and almost entirely filled with sinuses formed by a network of extremely thin bone (c. 1 mm thick). Without exception, the La Cotte crania are extensively damaged, although the maxillary and premaxillary regions are relatively complete (Fig. 18.23). Dozens of skull fragments were found in the areas around the skulls during excavation. Some of this damage could have been caused by the infilling and weight of the surrounding matrix, or perhaps the skulls were deliberately broken open to get at the brain — a much sought after delicacy among many contemporary hunting communities.

Evidence of butchery in the form of cut marks on post-cranial bones is scarce (see Jones and Vincent, chapter 19, this volume), but this is perhaps not surprising. The soft state of many of the bones and their subsequent chemical treatment may have obscured prehistoric cut marks, but it is as likely that there never were

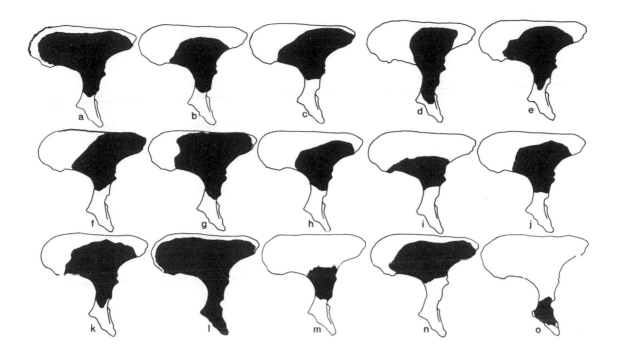

**Fig. 18.20** Schematic representation of surviving parts (shaded) of mammoth innominates, layer 6.

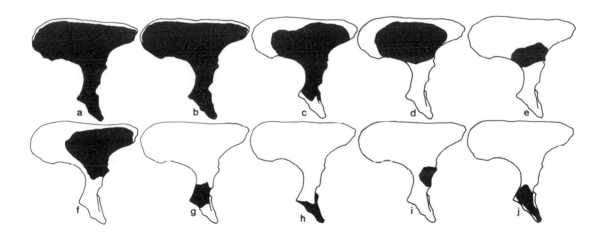

**Fig. 18.21** Schematic representation of surviving parts (shaded) of mammoth innominates, layer 3.

many cuts on these bones. Although J. D. Clark and Haynes (1969) cite a number of reports of prehistoric elephant butchery sites in various parts of the world at which artefacts are associated with elephant or mammoth skeletons apparently smashed by man as part of the butchery process, it might also be argued that the fragmentation observed at these sites is more probably the result of later disturbance, weathering, or compaction of overlying deposits than of butchery.

Ethnographic observation among living hunter-gatherers shows that the skeleton of an elephant is seldom reached with the cutting implements because there

is so much meat on the carcass (Yellen 1977; Crader 1983; P. R. Jones, pers. comm.). It has been demonstrated experimentally that it is virtually impossible to break a 'green' elephant bone (unless it is very young) with only lithic technology at one's disposal (Morlan 1983). The shaft walls are extremely thick, and there is no weak point in the shaft in the form of a marrow cavity as there is in most vertebrates: in elephants the marrow cavity is replaced by cancellous bone through which the marrow is disseminated. In this manner, the strength of the bone is increased without augmenting its diameter (W. B. Scott 1937). P. R. Jones, however, has demon-

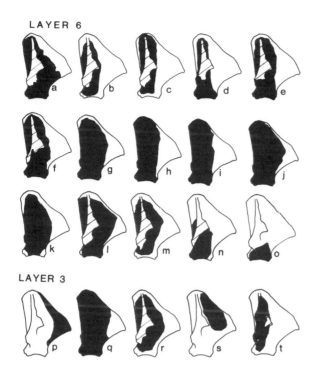

Fig. 18.22 Schematic representation of surviving parts (shaded) of mammoth scapulae.

Fig. 18.24 Elephant long bone fractured experimentally by vigorous hammering against a rock (photo P. R. Jones).

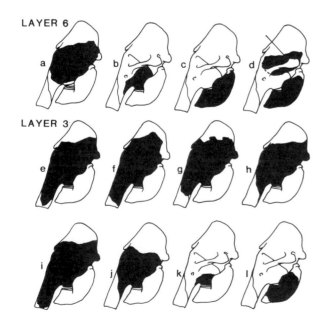

Fig. 18.23 Schematic representation of surviving parts (shaded) of mammoth skulls.

strated (pers. comm.) that it is possible to smash the limb bones of an elephant that has been dead for a few months (see Fig. 18.24). Various researchers (e.g. Myers *et al.* 1980; Binford 1981; Haynes 1983; Morlan 1983), attempting to identify the origin of bone assemblages and the causes of bone breakage, describe the attributes of bones broken when fresh. These include

1. relatively smooth fracture surfaces,
2. presence of acute or obtuse angles formed by the intersection of fracture surfaces with outer surfaces of the shaft, and
3. no difference in the colour of the fracture surfaces and the surface of the outer compacta.

These attributes characterise a number of the La Cotte mammoth limb bones, one of which is illustrated in Fig. 18.25. Such a fracture is considered by some to be indicative of human activity. However, from the literature cited above, it is evident that fractures may also be caused by carnivores, sediment loading, fluvial transport, trampling, and accident during the life of the animal. In the case of the La Cotte mammoth bones, some of these causes are unlikely to be relevant; there is no indication that the bones were washed into the site, and it is highly improbable that they were broken mid-shaft by carnivores or by trampling, given their size and the situation of the deposits. Sediment loading has indeed resulted in crushing and faulting of many of the bones from this site, but cannot account for fractures such as those illustrated, because in most cases, the rest of the bone was not found in the deposit. This suggests that breakage occurred before burial, and, if the unweathered surface condition of many bones is any guidance, burial must have taken place fairly soon after the death of the animals. While Morlan (1983) considers that it is extremely difficult to fracture 'green' bones of elephant, he points out that it is possible for spiral fractures to be caused by torsional loading. He refers, for example, to a situation in which a horse stumbles as it plants its forefoot, and the weight of its trunk and hindquarters supplies the torsional loading during the fall which fractures the limb. It is also evident that end-on impact of the femur and tibia can induce spiral fractures. These "occur most often when people jump from high places or when they are involved in head-on automobile collisions in which the feet are planted against the floorboards just before impact". As Morlan suggests, "we might expect to find counterparts to such fractures on ungulate limb bones found in natural traps and jump sites".

The possibility that the bones were brought to the site for construction purposes seems unlikely. The famous mammoth houses of the USSR and as far west as Poland (Kozlowski and Kubiak 1972; Formazov 1975; Kubiak 1976) testify to cooperation on the part of man on a massive scale. Hundreds of mammoth bones were accumulated and stacked in order to construct these houses, and occupational debris is evident within them. At La Cotte no such pattern is evident in the positioning of the bones and neither layer shows signs of occupation in the accepted sense in the form of hearth debris.

The question of association of artefacts with these two bone heaps raises some interesting issues as well as problems of interpretation. Isaac (1971) suggests that a low ratio of artefacts to bones might serve as a guide to differentiating kill/butchery sites from transitory camps, permanent occupation sites and quarry/workshop sites. At La Cotte, in the case of layer 6.1, there is some uncertainty as to how many artefacts were actually in the loess, and how many are attributable to layer 5, but in layer 3 there are over 1100 artefacts (see also chapter 24). This is a much larger assemblage than is generally associated with a kill/butchery site. Furthermore, various authors have commented on the fact that a high percentage of waste flakes and a low percentage of retouched or heavy-duty tools usually typify large mammal butchery sites. East African occurrences include the elephant butchery site at Olduvai Gorge, Bed 1 (FLK N1, level 6), where only 5 out of more than 100 artefacts had been

**Fig. 18.25** Mammoth limb bone with ancient fracture (A). Other breaks (B) and (C) occurred during excavation and preparation.

retouched, the majority being waste (Leakey 1971). At Olorgesailie, in the Gregory Rift Valley, a disarticulated hippo carcass was found with a few tools and almost 100 flakes (Isaac 1977). (One might also refer to the discussion on these and other large mammal butchery sites provided by Jones [1980] who carried out experimental butchery using stone tools). In Hungary, Vértes (1966) reports a similarly high percentage of waste associated with the remains of a mammoth, as does Movius (1950) with reference to an elephant butchery site at Lehringen, Germany.

The assemblage from layer 3 at La Cotte differs markedly from those referred to above above in that some 23.0% of the artefacts have been retouched. One of two possibilities might account for this: either meat-processing activities were intensive during the formation of the bone heap (which would account for the large assemblage size as well as the high proportion of retouched tools) or some of the artefacts were originally from layer A and have become incorporated into the base of the loess. Typologically and technologically the artefacts from 3 and A are strikingly similar (chapters 24

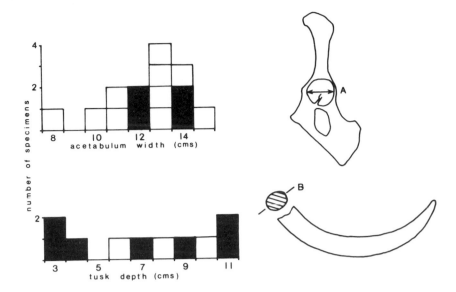

**Fig. 18.26** Comparison of measurements of (A) mammoth acetabula and (B) mammoth tusks. Specimens from layer 6 are shown in white, and those from 3 in black.

and 26); there is therefore no means of distinguishing activities carried out during the two bone heap events from those associated with the underlying occupation horizons, and uncertainty remains as to the extent of artefact manufacture associated with the former.

Much of the evidence presented above in support of the hypothesis that these two assemblages from La Cotte are the result of driving animals off the headland relates to the condition, representation and situation of the bones. Finally, it is interesting to consider the age distributions of the mammoths and rhinos, as they appear to corroborate the other evidence for the proposed kill/butchery 'events'.

## 18.7 IMPLICATIONS OF THE AGE DISTRIBUTION OF MAMMOTH AND WOOLLY RHINOCEROS FROM LA COTTE

Age distribution based on the dentition of both species from all levels is detailed in chapter 13, a summary of which data is presented in Fig. 13.18. In both levels it will be observed that the rhinos are subadult. In layer 3, the mammoths are best represented by skulls with dentition, from which it is evident that very young and old individuals are absent. The post-cranial material suggests a similar age distribution. Unfortunately, mammoths in layer 6 are poorly represented by teeth, although those which are represented are fairly young animals. However, a minimum of 11 mammoths are represented by post-cranial bones in layer 6, and a comparison of the measurements of limb bones from both levels indicates a generally comparable age distribution of relatively young animals and prime age adults, with a lack of very young and full-grown individuals. To illustrate that the size range of the mammoths is similar in both levels, meas-

urements taken on the pelvic bones and tusks in the two levels are compared in Fig. 18.26. The latter appear also to represent a range of fairly young animals, with few very small individuals and no very large ones. Tusks of old males (from other Upper Pleistocene deposits) are strongly spiralled attaining a length of between 3.5 and 4.0 m and a weight of up to 120 kg. The largest of those from La Cotte is only 1.5 m and, even allowing for the fact that it is damaged, would not have exceeded 2 m in length. Tusks of females and younger animals are thinner, shorter and straighter, as are most of those from La Cotte.

As mammoth and rhino occur not only in these bone heaps but also in the lower levels, it is interesting to compare the age distribution of animals in layers 3 and 6, with those in the other levels (Fig. 18.26). The mammoths in the lower levels tend to be younger than those in 3 and 6, and some are undoubtedly new-born, but the difference in age distribution is particularly marked in the case of the rhinos. It may suggest that the hunting of large, potentially dangerous species depended primarily upon finding isolated, unprotected young animals, or upon scavenging young animals among which group mortality is very high. Of living elephants and rhinos, two out of every three may be expected to die in the first few years of life from predation, disease or accident (Bere 1966; Goddard 1970). The animals represented in the two bone heaps at La Cotte, on the other hand, are the relatively young and the prime adults — those least likely to suffer mortality and predation. They would also have posed the greatest threat to hunters attempting to select for these age groups, particularly in view of the fiercely protective behaviour displayed by elephants for other members of the herd. An explanation of the age representation of the mammoth and woolly rhino in layer 3 and 6 might lie in the behaviour and social organisation of their living counterparts. Rhinos of the age represented at La Cotte, having left their mothers at the age of 2–3 years (shortly before the birth of the next offspring) are solitary, or found sometimes in the company of a few other individuals (Schenkel and Schenkel-Hulliger 1969). As mentioned earlier in this chapter, they are frequently found on the fringes of herds of elephant. Young elephants, however, stay with their mothers long after the birth of successive offspring, forming various sized groups of females with young. At puberty, the males are generally driven out of the herd by older females, and they then join bachelor herds (Douglas-Hamilton and Douglas-Hamilton 1975; Hanks 1979). Deraniyagala (1955) records similar social organisation among Indian elephants, mentioning also that individuals of similar age, and at times of the same sex, appear to keep together, and occasionally such parties leave the herd in small groups to go out foraging.

There is no reason to suppose that the behaviour and social organisation of the mammoth and woolly rhinoceros was markedly different from that outlined above. In that case, what we might have in layers 3 and 6 at La Cotte are the remains of two herds of mammoth (accompanied by a few rhinos). These might have been bachelor herds, or perhaps a number of females with a range of offspring. In any event, they were animals predominantly in their prime — herds that would have been dangerous to man and probably impossible to kill *en masse* without the use of some kind of trap or pit-fall. The chasms at La Cotte would have provided just such a kill site.

Whether this was organised hunting, or whether hominids in the vicinity of La Cotte Point seized the opportunity to drive animals off the headland, can never be known. Neither can we do more than speculate on whether such a hunt involved a large number of people or a few. If the site could be used twice for the same purpose, then it is possible that the mammoths and woolly rhinos in levels other than 3 and 6 were also acquired by driving them over the cliffs. Unfortunately, bone preservation is so poor and the site utilisation apparently so intensive in the lower levels that such a possibility must remain conjectural. What is most interesting, however, is that the two bone heaps cannot be attributed to the same group of hominids. The time lapse between levels 3 and 6 is not known, but evidence for intervening interstadial conditions (chapter 7), indicates that it must have been considerable. It is thus remarkable that the same two species were hunted on the two occasions, and even more remarkable that the climatic conditions responsible for the influx of loess into the site after the deposition of the older bone heap should have recurred after the deposition of the later heap.

Undoubtedly, there is much that is not known about the depositional and post-depositional history of this complex site, and there is much that remains speculative concerning the activities of the hominids who visited it. It certainly seems that they were present soon after the death of the mammoths and rhinos. If cause of death may be attributed to them too, as seems likely, then La Cotte perhaps provides the earliest evidence in Europe for human cooperation in effecting a mass kill.

# A STUDY OF BONE SURFACES FROM LA COTTE DE ST. BRELADE

P. R. Jones and A. S. Vincent

## 19.1 INTRODUCTION

At the invitation of the members of the La Cotte project, the authors undertook a study of the different kinds of marks preserved on the fossil bone surfaces from this site. Particular attention was paid to the remains from the bone heaps in layers 3 and 6 (see Scott, chapter 18, this volume), although many other bone surfaces, some of which showed marks which had previously been noted by Scott, were examined as well. These marks constitute one line of evidence that could indicate the agencies involved in the accumulation of the bones.

Subsequent to our study of the material from La Cotte, Jones had the opportunity to examine a large sample of tusks ($n = 1342$) that had been recently removed from elephant carcasses in the Sudan. The location and morphology of cut marks found on these tusks were observed to be identical to cut marks found on some of the La Cotte specimens.

## 19.2 EXAMINATION OF LA COTTE BONE SURFACES

### 19.2.1 Methods and Problems

Our aim was to examine the well-preserved bone surfaces and to make moulds of each type of mark and feature found on them. The agencies responsible for these might then be identified on the basis of microscopic examination, experimental work, and comparison with modern bone surfaces. The relatively large sample of particular body parts, especially scapulae, tusks and pelvic fragments, led us to concentrate on an examination of these, in the hope of documenting a consistent pattern in the location of marks.

The major problem encountered was the poor condition of the bone. As it was very soft it had been heavily impregnated with preservative, and this filled up surface marks and obscured the detail. Many surfaces were therefore eliminated from our consideration immediately. As time did not allow the cleaning and examination of all bone surfaces we decided to select for study only those surfaces on which marks could already be detected. A distinction was also made between bone surfaces that had been treated in the field and those that had been cleaned in the laboratory. Marks found on the latter would be of greater significance for our purposes, as more care could be taken to remove the adhering sediment.

A high quality, catalyst-setting silicone rubber was used to make moulds of all the types of marks that we observed and cleaned. Casts were made from these moulds using an epoxy resin and powdered colouring. The marks were described and their location on the bones was recorded. The moulds and casts thus provide a permanent record of the types of marks that were found.

### 19.2.2 Results

In all, over 70 bones (these include tusks) were examined. Moulds were made of 42 areas on 25 of them. The marks noted can be provisionally assigned to one of the following types:

1. Short 'gouge' marks (10–20 mm long and a few mm wide). These have a shallow entry point, get deeper (up to 2 mm) and end abruptly. They were commonly noted on scapulae and pelves.
2. Long linear 'scratches' occurring singly or in pairs. These were rarely encountered.
3. Short 'scratches' (10 mm) in twos or threes or in patches, often oriented in different directions on a surface.
4. 'Pitting' marks covering quite extensive areas of bone surface.
5. Short 'cuts' with well-defined edges, generally found on well-preserved hard bones and tusks.

A variety of agencies could have produced these marks: natural nutrient vessel channels bisecting the bone surface; carnivore teeth and claws; cutting and scraping by hominids; surface erosion; abrasion within the deposit; damage during excavation or during preparation and storage in the laboratory. At present very few casts have been examined under a microscope. It is nevertheless possible to make some general observations on the nature and location of the marks and to offer some suggestions as to their cause.

Typical gouge marks of type 1 were often found on the softer surfaces of scapulae and pelvic fragments. Quite often these occurred where there was some change in the surface contour, and are consistent with their having been made accidentally in the field before the excavator was aware that a bone surface was coming to light. Furthermore, the naturally occurring nutrient channels that can be observed on the pelves and scapulae of modern African elephants give rise to a very similar feature. Marks of type 1, as in Fig. 19.1, could be attributed to either of these two agencies. Marks of

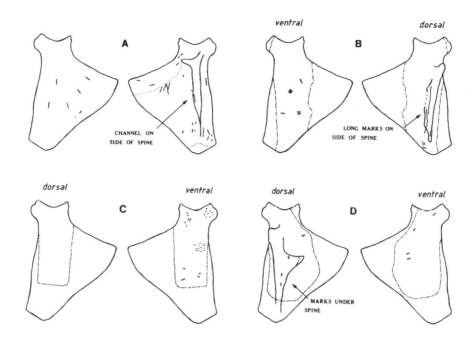

**Fig. 19.1** Cut marks on a modern elephant scapula and on mammoth scapulae from La Cotte de St. Brelade: (A) modern, (B) 70/15, (C) 70/18, (D) 70/41.

type 2 were rare; while many of them could have been made by stone tools, this was impossible to prove, and no significant pattern was seen in their occurrence.

Marks of type 3 have been noted by the authors on modern bones found in a seasonal river bed and on rocky slopes, and could well have been produced by abrasion in wet or dry sediments. These were noted on two bones from layer 6, and contradict the excavators' initial impression that they had not been reworked to any great extent.

The marks of type 4 have not yet been fully explained. Hyena canine puncture marks have a similar basic morphology, but the locations of the fossil marks are inconsistent with that possibility.

Marks of type 5 are of outstanding interest, as their characteristic features identify them as being the products of using stone tools. Seven marks of this type have been identified, four on long shaft fragments and three on tusks. An example on a limb bone is shown in Fig. 19.2A (bone 78/999). The mark on a modern sheep radius shown in Fig. 19.2B was made in the course of removing meat with an unretouched flint flake. It is remarkably similar to the mark on the illustrated specimen from La Cotte, The characteristic features of this type of oblique slice mark are: (1) one side of the cut has a smooth clean-cut edge indicating the angle of slicing; (2) the flint edge has then dug into the surface lifting a flap of bone from the other side which subsequently breaks off giving the irregular edge to the other side of the cut; (3) striations made by irregularities on the tool's edge are preserved at the base of the cut.

Other cut marks of particular interest were found on tusks 76/418 (Fig. 19.3), 77/256 (Fig. 19.4) and possibly on 74/426. These marks are oriented at right-angles to the length of the tusk and are located between its proximal extremity and the end of the pulp cavity, approximately at the point where the tusk leaves the bone sheath of the skull (see Fig. 19.5). The significance of this orientation and location is discussed in the following section.

## 19.3 CUT MARKS ON MODERN TUSKS

A large sample of tusks of the modern savannah elephant (*Loxodonta africana africana*) was examined at the Sudan National Parks headquarters in Khartoum. It was all poached ivory that had been confiscated by the Sudanese authorities and consisted on the whole of tusks that had been removed from newly killed elephant carcasses, rather than of weathered ivory collected from skeletons. Many of the tusks had been damaged and marked by a variety of natural and human agencies. A random subsample of 663 of them (49.4% of the total sample) was examined to collect information for comparison with the cut marks that had been found on the three mammoth tusks from La Cotte.

Two of the kinds of damage seen were obviously made by man:

1. shallow scalloped areas cut into the ivory between the proximal extremity of the tusk and the area

186

Fig. 19.2  (A) SEM photograph of an epoxy resin cast of a shaft fragment from a long bone of a deer-sized animal, 78/999, from La Cotte de St. Brelade layer B. (B) SEM photograph of an epoxy resin cast of an experimental cut mark made using a flint flake on a sheep radius.

**Fig. 19.4** Two views of a mammoth tusk, 77/256, from La Cotte de St. Brelade layer 3. Note the curvature, and the orientation of the cut marks.

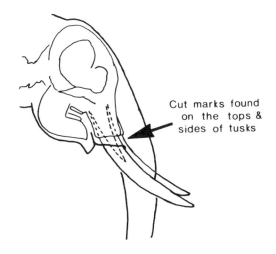

Cut marks found on the tops & sides of tusks

**Fig. 19.3** Mammoth tusk, 76/418, from La Cotte de St. Brelade layer 3. (A) The distal end, showing the pulp cavity. (B) Detail of cut marks. (C) SEM photograph of the area indicated.

**Fig. 19.5** Skull and head outline of African elephant, showing the position of the cut marks found on *Mammuthus primigenius* tusks, transverse to the length of the tusk and close to its exit from the skull.

**Fig. 19.6**   Modern elephant tusks, showing the marks made by ivory poachers.

where the tusk emerges from the skull (the 'exit line'); and

2. short deep cuts, occurring singly and in groups, found mainly within a 10 cm band around the tusk, on and outside (distal side) the exit line.

Cut marks of the first type were found on all portions of the tusk that are normally inside the skull. They were usually no more than 2 mm deep but often as much as 20 mm wide and 30 mm long (Fig. 19.6A). Anyone who has removed or observed the removal of elephant tusks

from a fresh carcass will recognize these as having been made by an axe or large bush knife at the stage of the operation when the final pieces of skull adhering to the tusk are removed from it. While marks of this type were seen on 75% of the modern tusks, none were noted on those from La Cotte.

The second type of humanly inflicted damage was seen on 21% of the modern tusks. In all cases it was found close to the exit line, on the distal side, and consisted of cut marks that had undoubtedly been made by a small hand-knife. Figs. 19.6A and B show these in rel-

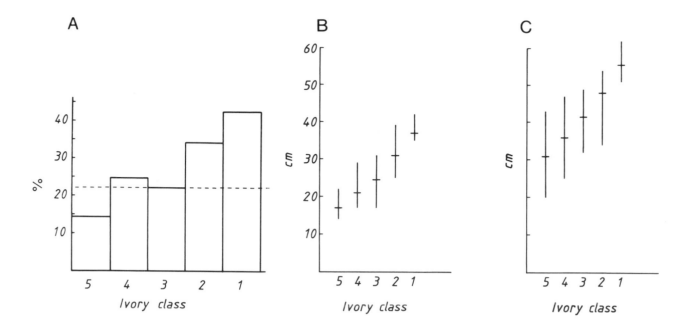

**Fig. 19.7** Characteristics of the sample of modern elephant tusks. (A) The percentage of tusks within each ivory class which preserve cut marks made by steel knives; the larger tusks would have come from skulls on which there was much more meat, as well as larger areas of tusk, which accounts for the high percentage of examples showing cuts (the dotted line indicates the average for the entire sample). (B) The circumference of the tusks at the 'exit line' from the skull. (C) The distance of the 'exit line' from the proximal extremity of the tusks (the pulp cavities of modern elephants extend beyond the 'exit line'). The cross bars indicate the mean value, the vertical bars the range of the observations. See also Table 19.1.

ation to the exit line, which in Fig. 19.6B appears as a lighter band separating the rough textured ivory normally inside the skull from the smoother, exposed, ivory. These cut marks occurred singly or in patches. They were commonly between 10 and 30 mm long, and all were oriented between 45 and 90° to the length of the tusk. Figs. 19.7–19.8 illustrate various features of the modern tusks and the position of their cut marks. These cut marks were made by small hand knives as meat was being removed from the elephant's head.

Several types of natural (non-human) damage were noted on the tusks, the features of one of which should be described here, since they could be confused with marks made by a knife. These marks were transversely

oriented and ranged from about the middle of the exposed length of the tusk to near the exit line. They were also generally shallow (barely discernible with the finger tips) and differed significantly from the knife marks in being long and extending usually *around* the curvature of the tusk, sometimes for as much as a third of its circumference. Examples of these are visible in Fig. 19.9. The most likely explanation of them is that they are made by contact with some flexible material, perhaps bark, as the elephant strips it from a tree.

The modern, man-made cut marks of the second type described above are identical in size and orientation to those found on the mammoth tusks 76/418 and 77/256 from La Cotte. The location of the palaeolithic cut marks (estimated on the basis of the curvature of the tusk section and the diameter of the visible pulp cavity) is also identical to the location of the modern cut marks (see Fig. 19.8). The modern sample of tusks came from an area stretching from southern Sudan to western Sudan and Chad; the elephants must have been killed and their tusks removed by people of many different ethnic groups. The patterning and location of the cut marks on these tusks cannot therefore have been determined by local cultural practice only, but must also represent techniques of butchery which relate to elephant physiology and therefore cut across cultural boundaries. The evidence therefore suggests that the marks on at least two of the mammoth tusks from La Cotte — both in layer 3 — were made in the course of butchering by palaeolithic peoples.

**Table 19.1** Cut Marks on Modern Elephant Tusks.

| Ivory class | Weight range (kg) | Sample size | Tusks with cut marks | % of sample with cut marks |
|---|---|---|---|---|
| 1 | 12.5-28.0 | 57 | 24 | 42.1 |
| 2 | 6.5-12.5 | 80 | 27 | 33.7 |
| 3 | 4.5-6.5 | 86 | 19 | 22.0 |
| 4 | 2.5-4.5 | 138 | 34 | 24.6 |
| 5 | less than 2.5 | 302 | 43 | 14.2 |
| Totals | | 662 | 147 | 22.1 |

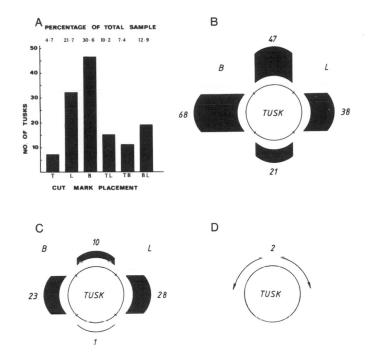

**Fig. 19.8** Location of the cut marks on modern (A–C) and ancient (D) tusks. T, top of the tusk (as it sits in the skull; L, lingual; B, buccal. (A) Frequencies and percentages of the most common occurrences of cut marks in the various locations. (B) The numbers and position of cut marks found within 50 mm of the tusk exit line. (C) The numbers and position of cut marks found between 50 and 110 mm from the exit line. (D) The position of cut marks found on tusks 76/418 and 77/256 from La Cotte de St. Brelade layer 3, based on the observed curvature of the tusks; this compares well with the majority of the modern cut marks.

## 19.4  DISCUSSION

A major problem in identifying cut marks on the La Cotte specimens from layers 3 and 6 was that a chemical preservative had been extensively used. This was very hard to remove from within many of the marks without also destroying the surrounding bone surfaces. As a result, the value of microscopic analysis may be reduced and the use of the scanning electron microscope restricted to low magnifications. Scanning electron microscopy has been used mainly as an instrument for taking high detail macrophotographs. Further research by ourselves and others may lead to more information on the agencies responsible for many of the observed marks. But while it may eventually be possible to be fairly sure that some of the marks were not made by tools during excavation, or are not nutrient vessel channels, it will probably not be possible to prove that others can *only* have been made by stone artefacts in antiquity.

Hard bone surfaces were more useful for this analysis, since they required less preservative during excavation and the marks on them retained more detail. The hard bone surfaces from layers 5, A, B and C, as well as the hard tusk surface from layer 3, preserve man-made cut marks that are identifiable by the naked eye. That cut marks were found from layers C, B, A and 5 is not surprising, since these layers are rich in artefacts, contain charcoal, and were always presumed to have been living horizons.

The cut marks from layer 3 were found on the short section of mammoth tusk adjacent to the point where it leaves the skull. A modern sample of over 100 elephant

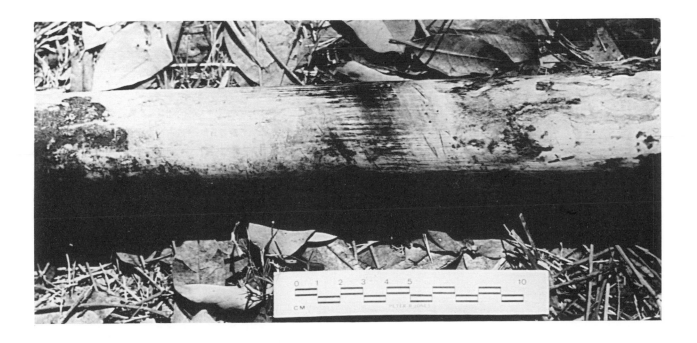

**Fig. 19.9**  Knife marks and (larger) natural scratches on a modern tusk.

tusks from the Sudan preserve steel cut marks which are identical in morphology, location and orientation to the cut marks from layer 3. Butchery marks were not found on any of the post-cranial bones from this level, owing to their poor state of preservation. Tusks are an unexpected location for the occurrence of evidence of palaeolithic butchery, and similar marks from other sites may have escaped notice. The cut marks from layer 3 are of particular interest, since this layer contains a high concentration of mammoth and woolly rhino bones but relatively few artefacts. There has been much discussion on the processes by which this material accumulated (K. Scott 1980; chapters 18 and 31, this volume). The finding of cut marks from layer 3 supports the conclusion reached on other grounds that palaeolithic people were hunting and butchering the mammoths at La Cotte.

# THE USE OF FIRE AT LA COTTE DE ST. BRELADE

P. Callow, D. Walton and C. A. Shell

## 20.1 MAN AND FIRE IN THE LOWER PALAEOLITHIC

This subject has been comprehensively reviewed in studies by Oakley (1955, 1958 and 1961), and more recently Perlès (1976, 1977 and 1981). At the time when these were written, the earliest reported occurrences were assigned to the Elster glaciation or its equivalent elsewhere (see below). However, it has been claimed by Gowlett et al. (1981) that in Africa at Chesowanja, Kenya, a volcanic tuff dated to >1.42 mya overlies a site containing fragments of burnt clay thought to be the result of controlled combustion. If this is so, it enormously increases the age of man's use of fire, but the scarcity of other early records suggests that this must have been infrequent; evidence from other ancient East African sites is inconclusive.

In Europe, the oldest sites claimed to provide unequivocal evidence of burning include: Escale, Tautavel and Terra Amata, all in southern France (de Lumley 1976d); and Torralba and Ambrona in Spain (Freeman 1976, 674–5), where it has been suggested that the charcoal is the result of grass fires used in hunting elephants. All these sites are dated by their excavators to the Mindel (Elster) glaciation, though this has been challenged by other commentators. While equivalence is difficult to establish over so great a distance, an age in the same general range is just possible for locality 13 at Zhoukoudian (formerly Choukoutien) in north China (Wu and Lin 1983).

Whatever the exact date of the above, it is only with the onset of the Saale glaciation *sensu lato* that we find extensive traces of the use of fire at a high proportion of sites. A contributory factor here may be an apparent increase in the use of caves, which tend to concentrate the burnt remains and reduce the likelihood of their dispersal, though this is scarcely sufficient to explain the increase completely. An important innovation is the development of a range of more elaborate types of hearth (Perlès 1976, 680). Apart from the simple (open) type these include: dished hearths (scooped out of the ground), hearths surrounded by stones, hearths resting on a stone pavement, and hearths 'walled' with stones set on edge, or with earth.

In the course of excavation, such structures usually announce themselves with a concentration of burnt material, of which charcoal fragments are the most conspicuous component. Confirmation that they are truly hearths in place then depends upon their appearance as work continues. The following questions, in particular, have to be satisfactorily answered: (1) Is the concentration clearly defined in extent? (2) Does the substrate show signs of the effects of heat?

An affirmative to question (1) may indicate a fireplace bounded by stones, or a 'dished' hearth if the concentration proves to be lenticular in cross-section (see Laloy 1980, 107–8). In the case of simple hearths — characterised by a thin, rather extensive layer of combustion products — the appearance of the underlying soil assumes critical importance; in the absence of rubefaction, it is unsafe to assume that the concentration represents a true hearth, as there is no guarantee that the burnt material originated at this precise locality within the site. Indeed, as a general rule it is essential that natural agencies with an ability to concentrate the lighter combustion products be shown not to be responsible for the feature. Even when the presence of a hearth seems certain according to stratigraphic criteria, and a layer of burnt soil is found underneath, it is still desirable to show that the ash and burnt soil are related.

## 20.2 LA COTTE DE ST BRELADE

### 20.2.1 Evidence for the Use of Fire at La Cotte

There are indications of burning in every occupation layer at La Cotte. That this was the result of natural combustion (forest or grass fires) can be ruled out by the extent and density of the burnt material, and by the predominance of burnt bone over wood charcoal. Moreover, even if the ravines (as opposed to the plateau or the exposed coastal plain) were liable to such catastrophes, the absence of burnt material in those last interglacial deposits which are archaeologically sterile contrasts with its abundance in similarly temperate layers (H–E) when man is known to have been present.

The action of fire at the site is attested by the following:

1. Macroscopic remains of wood charcoal (Cartwright, chapter 12, this volume). Identifiable pieces are rare, no doubt in part as a result of mechanical damage caused by reworking of the deposits (as discussed in chapter 6).
2. Innumerable fragments of bone showing varying degrees of burning. Many are so well carbonised as to have been mistaken for wood charcoal during excavation.
3. Abundant comminuted carbonised material from the combustion of both bone and wood, consisting of

the unleached finer residues (some owing its fragmented condition to trampling). It is partly responsible for the dark coloration of the richest archaeological layers (see Fig. 6.16 and other site photographs).

4. Stone artefacts exhibiting heat-crazing (implying rapid heating or cooling) or the effects of more prolonged 'heat treatment'. In the latter case, the interior of the flint acquires a lustrous appearance which is revealed if the piece is subsequently reflaked. At La Cotte there is no reason to suppose that this was a deliberate practice; accidental heating of pieces just below the surface of the ground, and their subsequent disturbance and modification, is sufficient to explain the few cases noted. It is the presence of the more thoroughly heated fragments that has made possible the thermoluminescence dating, described in chapter 15.

5. Fire-reddened earth.

6. Burnt pieces of roof granite.

### 20.2.2 Hearths

The quantity of fire-altered material at La Cotte is so great that one might hope for evidence of formal hearths, in the sense of distinct localisations of ashes, burnt soil etc. with, in the immediate vicinity, the artefacts and food remains relating to the same occupation event. Accordingly, in the course of the excavation a considerable amount of effort went into attempting to isolate such structures, taking as a starting point any concentrations of burnt debris.

Possible hearths were initially noted in all layers from G to A. But detailed analysis indicates that not one of these can be confirmed as a true hearth *in situ*. All may be explained as geomorphologically or geochemically produced features, or products of interaction between the excavation process and site topography. Nevertheless, consideration of these occurrences is instructive.

In the areas of the site most affected by runoff the dissection of particularly ashy stratigraphic subunits, with subsequent deposition of less conspicuously burnt material in the intervening areas, resulted in local remnants of the former which were easily mistaken for man-made concentrations. They readily assume a lenticular form (even markedly biconvex where the original surface was irregular). The natural origins of these features are usually indicated by the contact with the overlying layer, particularly where this continues a downward trend beyond the margin of the 'hearth', and cuts into the underlying deposits. Features of this kind are particularly characteristic of layers C to A; in fact, during the excavation there was speculation that ashy pinnacles of the type shown in Fig. 6.11 represented the interiors of 'ovens' cut into a bank of loess against the eastern rock wall.

A less common phenomenon (though presenting very real problems at La Cotte) is local variation in the degree of post-depositional leaching by subsurface water.

Textural changes and the removal of humus and mineral salts from the affected area may have given a 'patchy' aspect to the excavated surface. As the same areas are likely to have been affected previously by surface water to the extent (at least) of mobilising the finer charcoal fragments, these leached zones can give the impression of being the 'non-hearth' parts of the site, whereas in reality they may be just as rich in coarser and less strongly coloured burnt material.

Where the excavation was conducted horizontally rather than so as to follow the dip of the strata (as was the case within individual metre squares when the stratigraphy was unclear), a 'pseudo-hearth' could be revealed if a lower, rather ashy unit exhibited a convex surface or was banked up against the cave wall, thereby protruding into the upper unit. A circular (or semi-circular) patch of burnt material would be uncovered, looking superficially very like a true hearth (Fig. 20.1). Only during further excavation did it become apparent that it joined up with other 'hearths', and extended right across the site! Not the least problem that this caused was the temporary mis-identification of the layer itself.

### 20.2.3 Burnt Sediment

On the west side of the north ravine, at the interface between layers D and C, a number of small patches of fire-reddened earth (typically 10 cm long and 3 cm thick, 7.5YR to 2.5R in colour) were collected; smaller but similar fragments had been observed elsewhere and were initially investigated as possible red ochre. As the C/D examples appeared to come from a single level and were overlain by a deposit rich in burnt bone etc. — some were even capped with a thin (2 mm) layer of dense black material — there appeared to be a good chance that they represented the base of a hearth. However, the discontinuous distribution of the burnt soil and the ab-

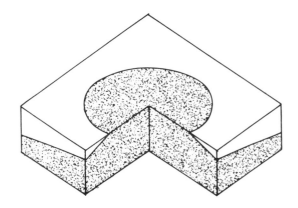

**Fig. 20.1** A situation which can lead to the erroneous identification of a natural feature as a 'hearth'. The protruding 'boss' of lower (ashy) deposit, when first encountered during excavation, has the appearance of a fireplace. Continued excavation should reveal its true nature, but a change of excavation personnel (especially between seasons) and a heavy work load could easily result in non-correction of the initial field notes about it.

rupt edges of the fragments gave cause for concern. Some independent test was therefore required in order to establish whether the material had been fired in place.

Depending on its composition, a sample of sediment exhibiting such a degree of fire-reddening might be expected when it cooled to have acquired a magnetic moment from the earth's field — a phenomenon exploited in archaeomagnetic and palaeomagnetic dating (Fleming 1976). If the sediment were undisturbed, the moment should be detectable; moreover, by slowly heating the sample under laboratory conditions and measuring the temperature at which it loses its magnetic signal, an estimate may be made of the original history of firing.

Three samples from the C/D boundary at La Cotte were submitted to Walton, who measured their magnetic moment using a Squid magnetometer. They proved to have no moment at all; nevertheless, X-ray diffraction analysis arranged by A. Putnis (Department of Earth Sciences, University of Cambridge) showed that an appreciable amount of haematite is present. The lack of a magnetic signal is probably due to small grain size (even after gentle heating and cooling in the laboratory no moment was found). It may be, therefore, that the heating episode responsible for oxidation occurred when the samples were already well buried, where the temperature rise would have been insufficient to cause an increase in the grain size of the haematite. This area of the site is now encased in concrete, so a more comprehensive study of the context of the reddened earth could not be performed.

### 20.2.4 Bone as a Fuel

La Cotte is rather unusual among prehistoric sites in having a very high incidence of burnt bone fragments compared to that of charcoal. Was this the result of cooking, or was it caused by deliberately burning bones as a fuel? The quantities involved preclude accidental burning as the primary cause.

Experiments suggest that cooking usually leaves no visible traces on bone, as at temperatures lower than 300°C colour changes do not occur (Laloy 1980, 82–3). Above this temperature, browning occurs, while at temperatures of 700°C or more (implying direct contact with the flames, with no meat to protect its surface) the bone becomes white or ivory in colour. Furthermore, Laloy's work showed that bone is an excellent fuel, but needs a relatively high thermal flux before ignition can take place, and requires therefore some sort of kindling. La Cotte poses a serious problem inasmuch as fresh mammoth bone is not available for experiment. However, mammoth long bones differ from those of most other animals in having a spongey structure and, presumably, a very high fat content when fresh. Their large surface area when broken and their calorific value would therefore make them especially suitable as a fuel, and probably more easily set alight than experiments on the bones of other species might suggest.

No statistics are available for La Cotte (small unidentifiable burnt bone fragments were not systematically collected during the excavation), but enough of the retrieved pieces exhibit marked thermal alteration for it to be clear that *burning* of the bone, as opposed to cooking, was a regular practice.

### 20.3 CONCLUSIONS

The evidence from La Cotte de St. Brelade shows that (1) fire was extensively employed; (2) conditions at the site have not favoured the preservation of hearths; and (3) besides wood, bone was burnt as a fuel.

The disappointing results of the attempts to identify hearths in the ashy layers, despite initial optimism, serve as a reminder of the extreme fragility of such features. During the past decade or so, discussion of site formation processes has highlighted the dangers of over-hasty identification of individual occupation floors and related structures in complex sedimentary environments, and the need to treat with caution many such claims in older and no longer verifiable reports.

# PART IV

## THE LITHIC INDUSTRIES

# THE ARTEFACTS: INTRODUCTORY REMARKS

P. Callow

## 21.1 DEFINITION OF ASSEMBLAGES

Excavation techniques, and ideas about the stratigraphy, evolved very greatly in the course of the excavations from 1961 to 1978. Layer allocations in 1961–1962 are now thought to have been too uncertain for the pieces recovered at that time to be amalgamated with those from later years. The problem stems from the employment of a system of approximately horizontal excavation units (spits), whereas the strata later proved to dip markedly to the south (previously, Burdo had not attempted to draw a clear boundary between his layers [7E] and [7F], redefined as layers C–G, and the complexity of the deposits was not then appreciated). Layers were assigned according to the position of the 'spit' boundaries as located on the section. The resulting assemblages formed the basis for the 1971 report by McBurney and Callow, at least for layers A–H, but it now appears likely that a proportion of the material assigned to a given layer should rather have been assigned to the one immediately above. Thus the local peak in the percentage of quartz shown in their Table III as occurring in layer E is now seen in layer D (Tables 22.1 and 22.2, and Fig. 22.2, this volume). Again, the number of finds from layer B reported in their Table VI is surprisingly large, given Burdo's comments (in his diary) about the relative poverty of this unit. Burdo was digging in the southern part of the bank of Saalian deposits where, as in 1961–62, B consisted of much 'cleaner' loess than that in the area excavated from 1976 onwards. This is probably explicable by assignment to it, during excavation, of material from the very rich layer A. (The above points should be borne in mind by any researcher considering making use of the collections in London and Cambridge.)

The period 1966–1969 was one of experiment in methods of on-site recording because of rapid depth changes during removal of relatively sterile deposits, and dissatisfaction with the 'spit' system. As a result it has not been possible to establish with confidence a distinction between artefacts from layer 5 and those — certainly far fewer — from the extreme base of layer 6.1 (equivalent to the upper 'bone heap') except in a very few cases. All these have been grouped as if entirely from layer 5, reserving the designation 'layer 6' for the material caught up in the latter's soliflucted upper part.

Despite its thickness layer A has been left undivided, as in the field it proved impossible to follow the finer details of the stratigraphy over large areas. Material from a block of deposit in the southeast corner exc-avated in 1977 has been excluded altogether, as the boundaries between A, B and C could not be established; in this area the effects of solifluction and colluviation, as well as colour changes due to percolating groundwater, were particularly confusing. Similarly, certain deposits which were assigned to G and H in 1978 may very well have been 'trample' or 'backfill' from previous excavations, and, moreover, some doubt remains as to the exact position of the interface between these layers in some areas. In these instances, also, the artefacts have been discounted. The extent of these problems became apparent only during post-excavation analysis, so the field records must be treated with caution.

More generally, human error (in the form of incorrectly recorded coordinates or misconceptions about the identity of the layer being excavated) proved to be a considerable source of difficulty in defining the relationship between artefacts and geological layers. It should be said that comparable discrepancies are likely to be common in the records of other excavations, but generally do not come to light because, without the aid of a computer, the labour of cross-checking is too great to be contemplated, particularly when tens of thousands of findspots are involved.

In all, about 15% of provenance data proved to contain errors or to be deficient in some respect (not necessarily seriously). But after an exhaustive investigation the proportion of the lithic material that could not be assigned to one or other of the layers with reasonable confidence, and had therefore to be omitted, was very small because the detail with which provenances and stratigraphy had been recorded provided enough 'redundant' information for errors to be remediable (the techniques used are described in Appendix A).

Although the verification operation was time-consuming, and greatly complicated the study of the collections, there can be no doubt that it has enormously increased the quality of the results. It was carried out entirely on the basis of provenance data, without feedback from typological analysis, and it became apparent in the sorting of the collection according to the redefined layers that the resulting assemblages were internally much more homogeneous than before, and better differentiated. This is of great significance when questions of interpretation are considered, as any mechanism which results in a mixing of assemblages (be it in antiquity or as part of the excavation process) can generate 'trends' in the archaeological record. A classic example of this phenomenon is the 'Emiran' industry, which was thought to represent transition in the Levant between the Middle and Upper

Palaeolithic (Garrod 1951). Consisting of side-scrapers and points made by Levallois technique, and end-scrapers and burins made on blades of Upper Palaeolithic type, this is now generally regarded as a geologically produced mixture of Mousterian and Aurignacian artefacts (Bar Yosef and Vandermeersch 1969). At La Cotte, also, some admixture as a result of natural agencies must have occurred (see chapter 6); its probable extent, and effects, are discussed during the description and interpretation of the industrial sequence (chapters 24 and 26).

## 21.2    DIVISION OF LABOUR DURING LITHIC ANALYSIS

The study of so large a collection as that from La Cotte would have been impossible in the time available had it not been undertaken on a co-operative basis, though the difficulties inherent in an investigation by several workers are numerous. As far as possible, the pieces were re-examined as the work proceeded and new questions arose, though some omissions were inevitable. This process of reclassification, coupled with the fact that several studies which would have been better conducted in sequence had (in the interests of speed) to be conducted simultaneously, enormously complicated the logistics of the operation. In particular, it was not always possible to incorporate some of the latest — and minor — typological revisions in some of the tables. The reader is asked to view any small discrepancies with tolerance, as to have repeated all the calculations for the sake of these would have caused serious delays.

When the finds were brought back from the field they were washed and marked with a unique reference number, beginning with the year of excavation. Length, breadth and weight were recorded, together with the raw material — at this stage, flint, quartz and 'stone' (implying other materials, as discussed in chapter 22). Pieces less than 20 mm in length were treated differently: for each provenance they were sorted according to raw material, counted, weighed collectively and bagged, and then put on one side until such time as it might be possible to examine them in detail. Though it could not be included in the present report, this component of the collection is one the desirability of whose study (particularly for evidence of the nature of on-site knapping activity) has become increasingly apparent in the course of the investigation. Meanwhile it must be emphasised that the discussion and tabulation of artefacts is concerned only with pieces with a length of at least 20 mm.

Initial technological and typological classification was based upon the McBurney system (section 21.4), and was performed by McBurney himself, and, later, by J. M. Cornford, J. Hutcheson and to a lesser extent by P. Callow; it was never applied to the quartz artefacts, however. This information, with provenance data, formed the main content of a computer-based master catalogue consisting of almost 100,000 records.

Further work on the butts of the retouched pieces was done by M. Hemingway and F. Hivernel, while Callow and Hutcheson re-examined much of the waste in the course of updating the master catalogue. Callow also searched the waste for unretouched 'Bordes types' (e.g. Levallois flakes), which were not isolated in the McBurney scheme. Most of the raw material identification was carried out by Cornford, while R. Burnell was instrumental, during the preliminary laboratory processing and later, in identifying the more unusual kinds of flint and in refitting. The Bordes classification of flint and 'stone' tools, and associated metrical work, was the responsibility of Callow. Cornford undertook the detailed study of the resharpening flakes and burin spalls, Hutcheson undertook the measurement and classification of the flint waste and cores, and to Hivernel fell the difficult task of studying the quartz tools and cores. H. Frame kindly performed the laborious examination of selected pieces for microscopic wear traces.

## 21.3    SAMPLING PROCEDURES

Many categories of artefact were too numerous (especially in layer A) for it to be possible or necessary to study all examples in the course of detailed analysis, but a policy of sampling at a fixed percentage for the entirety of an assemblage would have reduced the representation of the rarer types below a useful figure. It was therefore decided to reduce only the commoner classes, by varying degrees according to their representation in each layer.

Computerisation of the artefact inventories made it possible to generate a pseudo-random number (in practice, the range 0–9999 proved sufficient) which was permanently associated with each artefact record, both in the master catalogue and in derivative files; this offered the advantage of allowing sample lists to be regenerated easily. To recover a 20% sample, for example, it was sufficient to use artefacts whose records lay in the first 20% when sorted according to this variable. As it was not usually important to meet the target percentage exactly, a more convenient procedure was to take pieces associated with a random number below 2000. If this yielded too few pieces, the acceptance limit was raised to enlarge the sample on a random basis. Where it was necessary, during statistical computation, to group classes which had been sampled differently, cases were weighted with the reciprocal of the sampling proportion, in order to estimate means and other statistics. In the instances where sampling has been employed, this is indicated in the discussion.

## 21.4    THE MCBURNEY CLASSIFICATION SCHEME

With the exception of artefacts made of quartz (whose study was postponed), all lithic finds from La Cotte were classified according to a scheme devised by McBurney. Although it has not been practicable to use his type list for the description given in this publication (the reasons are set out below), his method is important for two reasons.

1. In stressing the desirability of employing a classification and system of attributes chosen in such a way that the features of the assemblages are best delineated, he pointed out a particular disadvantage of "universal systems of classification which, although they may give an apparently impartial description of two assemblages, have the danger of forcing one or both into a preconceived framework which may end by minimising rather than bringing out their true character and their differences" (McBurney 1967a, 13). Designing and applying a 'purpose-built' typology and attribute list presents enormous problems. Indeed, true objectivity can never be achieved, but the attractions of such a philosophy are considerable even if compromises must be made in practice. At La Cotte, the peculiarities of the industries are such that some special treatment is essential.

2. The McBurney scheme provided the principle upon which a storage system was developed for the collections at an early stage of study. Anyone contemplating research on this material will therefore have to become acquainted with it. Full documentation is to be lodged with the collections in Jersey.

Fundamental concepts underlying the McBurney scheme are that simplicity should be retained as far as possible, that no more subdivisions should be created than are likely to be statistically useful, and that type definitions should be entirely explicit. As applied to essentially flake-tool industries of Middle Palaeolithic type, as at La Cotte, these have the effect of ensuring that the only group of tools which was extensively partitioned was that possessing scraper-type edges; this effectively corresponds to Bordes's 'Mousterian' Group II, consisting of points and side-scrapers (Bordes 1972). Other categories were end-scrapers, burins, notches, denticulates, awls and 'bifacially retouched' tools, which included handaxes and (more or less) fully bifacial scrapers. Where an artefact did not fit easily into one of these categories it was treated as 'miscellaneous'; the intention was to refine the scheme later in order to clarify the character of such pieces, many of which could be regarded as composite tools, or as 'normal' tools which had been heavily modified during re-use. A digit was included in the identifying numerical code to indicate the presence of retouch on the bulbar surface, or of 'plano-convex' retouch (McBurney 1950, 176). In this way a three-digit code was devised to describe the main characteristics of any tool at this stage in the development of the typology.

In essence McBurney's provisional classification may be thought of as combining the morphological bias of the Bordes method, as regards the scraper-like tools (other categories being much reduced in number), with the rigorous approach to denomination attempted by Laplace (1964). Nevertheless, it is clear from his teaching and writings that, while deploring a wholly subjective approach to description, McBurney was also strongly opposed to the creation of a lexicon of types without reference to a defined body of archaeological material.

The most significant feature of his approach, clearly apparent (in an early form) in his work on the Haua Fteah, is the importance attached to quantitative and qualitative observations as an adjunct to the typological method: "the true *modal* values of the measurable characters of a type require to be specifically established even after the type itself has been independently shown to have separate existence" (McBurney 1967a, 13). At the time when he was writing, such concepts were far from being accepted by those who relied more heavily on typology, and the publication of the Haua Fteah report played an important part in developing modern practice — though the beginnings may be found at least as far back as in the metrical work of Solomon (1933) for example. Had McBurney lived long enough, the La Cotte material would have given him an excellent opportunity to develop his ideas.

At the time of his death, McBurney's classification had not been fully worked out. In particular, a large proportion of the lithic material remained in the miscellaneous category, and the attributions certainly included many errors, as there had been no time for checking. Even if guidelines had already been laid down for the next stage of the operation, this would have been a lengthy business. Given that the application of the Bordes typology (whatever its limitations) was seen as essential to the usefulness of this report, it was reluctantly decided that to work with two separate classificatory schemes was impracticable. That of Bordes has therefore been employed to allow comparison with published data from other sites, but the investigation has been carried much further to take account of the particular aspects of the La Cotte industries. In this way, it is hoped that some elements of McBurney's philosophy towards stone assemblages have been realised.

# RAW MATERIALS AND SOURCES

P. Callow

## 22.1  VARIATION IN THE RAW MATERIALS USED

A feature of the La Cotte lithic industries which was particularly stressed in the preliminary report (McBurney and Callow 1971) is the strong overall diminishing trend in the use of flint in the course of the Saalian layers (Fig. 22.1, and Tables 22.1 and 22.2). After an initial period of very high percentage in layers H–F, it is supplemented by quartz and by other hard rocks which are referred to collectively as 'stone' (see next section). In the assemblage from layer 5, the frequency of flint artefacts drops to less than 40%, with quartz at about the same level. The changes in the proportion by weight are even more dramatic; by layer 5 that of flint has fallen to less than a quarter of the total. There is then a remarkable reversal of the trend, with almost exclusive use of flint in the Weichselian layers.

As a general rule, quartz seems to have been used in greater quantities than stone. However, the ratios between these materials are not constant; after a very high value (almost 3) in layer D, that based on frequency drops to only 0.4 in C. No satisfactory explanation has been found for this; the significance of larger-scale variation in raw material is considered in section 22.5.

## 22.2  CLASSIFICATION OF HARD ROCKS OTHER THAN QUARTZ

The materials grouped under the heading 'stone' are in fact of several types. Considerable importance has been attached to identifying these, with the particular purpose of discovering whether some of them were exotic and hence perhaps indicative of the movements of prehistoric man. Of special concern were certain quartzites thought to resemble the *grès lustré* (see below) which is the predominant rock used at the very rich site of Bois-du-Rocher in Brittany. Preliminary discussions with J. T. Renouf were followed up by enlisting the aid of A. C. Bishop, of the British Museum (Natural History) to undertake a systematic hand classification of some 400 pieces recovered during the excavations. In addition, seventy archaeological specimens and ten beach pebbles were submitted for slicing and examination under the petrological microscope by W. E. Cameron, with funding from the Science Research Council. These investigations gave an indication of the range of rock types present in the collection and the frequency with which the principal categories occurred. They established that many of the materials were of local origin, the most conspicuous exceptions being flint and chert, and some of

**Table 22.1**  Total Weight of Principal Raw Materials, excluding Hammerstones[a]

| Layer | Flint kg | % | Quartz kg | % | Stone kg | % | Total kg |
|-------|------|------|------|------|------|------|------|
| 9 | 1.2 | 16.7 | 4.2 | 59.3 | 1.8 | 25.5 | 7.1 |
| 6 | 0.2 | 8.5 | 1.9 | 78.6 | 0.3 | 12.9 | 2.4 |
| 5 | 15.4 | 23.2 | 34.8 | 52.5 | 16.1 | 24.3 | 66.3 |
| 3 | 5.9 | 48.9 | 2.2 | 18.4 | 3.9 | 32.6 | 12.0 |
| A | 146.3 | 53.4 | 64.8 | 23.7 | 62.9 | 23.0 | 274.0 |
| B | 27.4 | 70.6 | 5.2 | 13.5 | 6.2 | 16.0 | 38.8 |
| C | 46.8 | 72.4 | 3.8 | 5.9 | 14.0 | 21.7 | 64.5 |
| D | 39.0 | 65.9 | 12.5 | 21.2 | 7.7 | 13.0 | 59.2 |
| E | 36.9 | 80.6 | 5.0 | 10.8 | 3.9 | 8.6 | 45.8 |
| F | 48.5 | 92.3 | 1.7 | 3.3 | 2.3 | 4.4 | 52.5 |
| G | 39.9 | 88.9 | 1.0 | 2.2 | 4.0 | 9.0 | 44.9 |
| H | 17.6 | 92.8 | 0.7 | 3.7 | 0.7 | 3.5 | 19.0 |

**Table 22.2**  Frequency of Principal Raw Materials, excluding Hammerstones

| Layer | Flint n | % | Quartz n | % | Stone n | % | Total |
|-------|------|------|------|------|------|------|------|
| 11 | 3526 | 98.5 | 11 | 0.3 | 42 | 1.2 | 3579 |
| 9 | 87 | 45.5 | 81 | 42.4 | 23 | 12.0 | 191 |
| 6 | 36 | 37.9 | 43 | 45.3 | 16 | 16.8 | 95 |
| 5 | 1349 | 39.9 | 1374 | 40.7 | 655 | 19.4 | 3378 |
| 3 | 843 | 71.1 | 212 | 17.9 | 130 | 11.0 | 1185 |
| A | 27490 | 69.9 | 7325 | 18.6 | 4497 | 11.4 | 39312 |
| B | 4691 | 80.6 | 611 | 10.5 | 519 | 8.9 | 5821 |
| C | 7981 | 82.9 | 476 | 4.9 | 1166 | 12.1 | 9623 |
| D | 6004 | 78.9 | 1200 | 15.8 | 406 | 5.3 | 7610 |
| E | 5590 | 87.3 | 564 | 8.8 | 247 | 3.9 | 640 |
| F | 6030 | 95.3 | 195 | 3.1 | 103 | 1.6 | 6328 |
| G | 4603 | 95.4 | 114 | 2.4 | 108 | 2.2 | 4825 |
| H | 2024 | 94.4 | 74 | 3.5 | 46 | 2.1 | 2144 |

[a]This information is not available for the Weichselian material (layer 11).

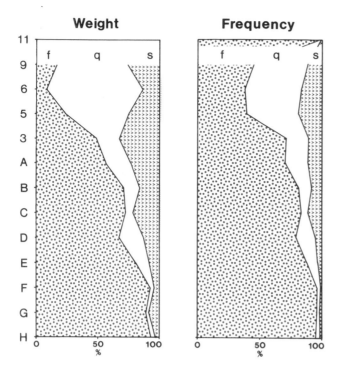

**Weight**

**Frequency**

Fig. 22.1 Frequency of occurrence of the principal raw materials.

importance to the knapper, such as the grain size of quartzites. Also, in the case of siltstones it was thought desirable to distinguish the more weathered pieces, as likely to be 'unreadable' for retouch.

The categories that were eventually selected, and employed during initial processing of the collection by J. M. Cornford, reflect flaking quality more exactly than geological origin; petrologically, the scheme involves a considerable amount of 'lumping':

1. **Granite (1.5%).** Some fragments of the coarse grained granite (of which much of the cave is formed) have clearly been flaked, but most of the worked pieces are of microgranite, which is also local.
2. **Basic igneous rocks (28.6%).** Dolerite, basalt etc. Often in poor condition due to chemical weathering, so that precise determination of the rock type is rarely possible without fresh breaks or sectioning. Flaking characteristics: medium to coarse grained, sometimes resulting in irregular conchoidal fractures; working edges of only moderate strength and durability.
3. **Siltstones (and greywackes) (13.0%).** Fine to very fine grained. Flaking characteristics: inclined to fracture along cleavage planes rather than conchoidally; usually rather soft for tool-making (though metamorphism has sometimes produced a harder material).

   (3₁) Unweathered siltstone (9.0%);
   (3₂) Weathered siltstones (4.0%). As above, but these are always unmetamorphosed and therefore more susceptible to attack.

4. **Sandstones (17.1%).** Medium to coarse grain. Flaking characteristics: a fairly tough material with good

the quartzites.

In the design of a classification scheme to be applied to the stone artefact finds, two considerations played an important part. Firstly that over 7,000 pieces were to be examined, and the scheme had to be sufficiently robust for this to be done without resort to specialist petrographic study on a large scale. Secondly that conventional petrological groupings might obscure attributes of

Table 22.3  Frequency of Raw Materials, other than Flint and Quartz

| Layer | Granite (1) % | Basic igneous (2) % | Siltstone fresh (3₁) % | Siltstone altered (3₂) % | Sandstone (4) % | Quartzite dark (5₁) % | Quartzite medium (5₂) % | Grès lustré (5₃) % | Other (6) % | Unident-ified (7) % | Sample |
|---|---|---|---|---|---|---|---|---|---|---|---|
| 9 | 0.0 | 42.9 | 28.6 | 4.8 | 4.8 | 0.0 | 0.0 | 19.0 | 0.0 | 0.0 | 21 |
| 6 | 0.0 | 12.5 | 31.3 | 12.5 | 25.0 | 0.0 | 12.5 | 6.3 | 0.0 | 0.0 | 16 |
| 5 | 0.0 | 31.1 | 18.0 | 8.1 | 22.1 | 3.3 | 15.8 | 1.4 | 0.2 | 0.0 | 628 |
| 3 | 0.0 | 22.1 | 6.6 | 9.8 | 19.7 | 15.6 | 11.5 | 13.9 | 0.0 | 0.8 | 122 |
| A | 0.7 | 25.2 | 7.3 | 4.2 | 18.5 | 14.2 | 14.2 | 14.7 | 0.9 | 0.1 | 4258 |
| B | 3.5 | 28.6 | 13.5 | 1.0 | 11.8 | 8.8 | 6.7 | 24.3 | 1.8 | 0.0 | 490 |
| C | 3.1 | 34.5 | 10.1 | 1.7 | 13.2 | 3.9 | 13.2 | 19.2 | 1.0 | 0.0 | 1148 |
| D | 2.5 | 41.6 | 6.2 | 2.2 | 13.0 | 2.0 | 23.2 | 9.0 | 0.2 | 0.0 | 401 |
| E | 0.8 | 32.5 | 5.3 | 3.7 | 9.9 | 1.6 | 27.6 | 16.5 | 2.1 | 0.0 | 243 |
| F | 3.2 | 34.4 | 9.7 | 4.3 | 23.7 | 4.3 | 15.1 | 4.3 | 1.1 | 0.0 | 93 |
| G | 17.5 | 18.8 | 5.0 | 13.8 | 26.3 | 0.0 | 10.0 | 7.5 | 1.3 | 0.0 | 80 |
| H | 5.4 | 45.9 | 2.7 | 2.7 | 8.1 | 2.7 | 16.2 | 5.4 | 10.8 | 0.0 | 37 |
| Combined | 1.5 | 28.6 | 9.0 | 4.0 | 17.1 | 9.9 | 14.5 | 14.4 | 0.9 | 0.1 | |
| Sample | 114 | 2153 | 679 | 305 | 1287 | 749 | 1091 | 1083 | 70 | 5 | 7537 |

conchoidal fracture, though not as good as the quartzite of similar grain size (from which it is not always easily distinguished).

5. **Quartzites** (38.8%). Very variable grain size. Flaking characteristics: tough, with a good working edge and clean conchoidal fracture as a result of the interlocking quartz grains. Certain unmetamorphosed sandstones and orthoquartzites have been included because their flaking characteristics are similar to those in this group. The quartzites were subdivided into three categories:

(5₁) Medium grained, dark (9.9%). Chiefly quartzitic sandstones;

(5₂) Medium grained, well cemented (14.5%). Hard quartzites with a characteristic fracture which cuts across the original sand grains. A few pieces are clearly *Grès Armoricain* (see below);

(5₃) Fine grained and very fine grained (14.4%). These are the *grès lustrés* referred to above. They flake like flint, and give a similarly keen, durable but brittle edge.

6. **Other rocks** (0.9%). A few rare types were quite distinctive, but did not fit well into the major categories (for example, certain schists of presumed Brioverian age).

7. **Unidentifiable** (0.1%). In a few instances surface alteration precluded confident attribution to the above categories; there were also a few 'borderline' cases.

Tables 22.3 and 22.4 show the frequency of occurrence of the above rock types in respect of the totality of the material and of the retouched tools.

## 22.3 POSSIBLE SOURCES

### 22.3.1 Flint

One of the most remarkable features of La Cotte is the high percentage of flint used (layer 5 excepted) even though there is is no native flint on the island. The nearest outcrop of Chalk is about 10 km north of Jersey, and 20 km from La Cotte de St. Brelade (Fig. 5.2); for it to be exposed a 25 m fall in sea-level would be required (when it would become an offshore island). Graindor and Roblot (1976) mention a very small outcrop close to the Minquiers lighthouse, about 25 km south of La Cotte. Apart from these exposures, the Cretaceous deposits are 40 km and more away from Jersey, between Guernsey and Alderney. To become accessible, these very much more extensive exposures nearer the centre of the English Channel would require a considerable sea-level fall, of at least 50 m, which would imply fully glacial conditions; they are unlikely to have been available during most of the La Cotte occupations.

Examination of residual cortex on the flint artefacts suggests that these were made almost exclusively from water-worn pebbles, rather than from tabular or nodular flint from the Chalk or from slope deposits. Only convex surfaces can yield reliable evidence, as concave cortical surfaces do not always exhibit clear signs of battering even on quite well rolled pebbles. Thus of 146 cortex-backed knives, 87% exhibit pebble cortex, 9% have a 'back' formed by an old patinated/stained surface, and the remainder are indeterminable. This provides confirmation that prehistoric man was not exploiting the Cretaceous deposits directly. Dredging shows that flint pebbles are distributed in considerable quantities on the sea bed, beyond the immediate area of the Chalk; Giresse *et al.* (1972) record a thickness of 6–12 m of sili-

**Table 22.4** Frequency of Raw Materials, other than Flint and Quartz (Well-retouched Tools only)[a]

| Layer | Basic igneous (2) % | Siltstone fresh (3₁) % | altered (3₂) % | Sandstone (4) % | Quartzite dark (5₁) % | medium (5₂) % | Grès lustré (5₃) % | Unidentified (7) % | Sample |
|---|---|---|---|---|---|---|---|---|---|
| 9 | 50.0 | 0.0 | 0.0 | 0.0 | 0.0 | 0.0 | 50.0 | 0.0 | 2 |
| 6 | 0.0 | 0.0 | 0.0 | 0.0 | 0.0 | 100.0 | 0.0 | 0.0 | 1 |
| 5 | 17.2 | 34.5 | 3.4 | 20.7 | 3.4 | 17.2 | 3.4 | 0.0 | 29 |
| 3 | 12.5 | 12.5 | 12.5 | 25.0 | 25.0 | 12.5 | 0.0 | 0.0 | 8 |
| A | 23.5 | 5.1 | 0.0 | 14.0 | 8.8 | 8.1 | 39.7 | 0.7 | 137 |
| B | 21.4 | 14.3 | 0.0 | 14.3 | 0.0 | 0.0 | 50.0 | 0.0 | 14 |
| C | 22.7 | 27.3 | 0.0 | 9.1 | 0.0 | 0.0 | 40.9 | 0.0 | 22 |
| D | 30.0 | 10.0 | 0.0 | 0.0 | 0.0 | 40.0 | 20.0 | 0.0 | 10 |
| E | 16.7 | 16.7 | 0.0 | 16.7 | 0.0 | 16.7 | 33.3 | 0.0 | 6 |
| F | 0.0 | 0.0 | 0.0 | 0.0 | 0.0 | 100.0 | 0.0 | 0.0 | 1 |
| G | 100.0 | 0.0 | 0.0 | 0.0 | 0.0 | 0.0 | 0.0 | 0.0 | 1 |
| H | 0.0 | 50.0 | 0.0 | 0.0 | 0.0 | 50.0 | 0.0 | 0.0 | 2 |
| Combined | 22.1 | 12.1 | 0.8 | 13.8 | 7.5 | 10.8 | 32.5 | 0.4 | |
| Sample | 53 | 29 | 2 | 33 | 18 | 26 | 78 | 1 | 240 |

[a]Based on the 'reduced' count.

ceous gravel less than 10 km south of La Cotte. The modern and fossil beaches of Jersey contain very little flint, however; in the former, some of the pebbles are likely to be from gravel used as ships' ballast. The percentage is only 1.4 in the 8 m beach exposures on the south coast (Keen 1978). Though it might be supposed that the desirability of these pebbles as raw material might outweigh their scarcity, they are in fact mostly too small to account for the industry at La Cotte (it is rare to find one as much as 100 mm long). Further consideration is given to this problem in section 22.4.5.

More than one original source is required to account for the considerable variation in colour, opacity and texture that is apparent among the flints from La Cotte. The location of these sources has not been determined, but it is likely that most of the darker grey to black flint is originally from the Chalk, and that the brown and lighter shades of grey (and also the cherts) are Jurassic varieties from the limestones in the Cotentin.

It had been suspected during other studies of the artefacts that the character of the material varied from layer to layer. To verify this, an experimental hand classification was carried out by R. Burnell (one of the staff of the research project). The samples used were those of waste flakes with plain butts which had been extracted for the work described in chapter 25; for layer 5, because of the small number of pieces, broken examples with plain butts were also included. The percentages obtained, listed in microfiche 22.1, were then analysed using multivariate techniques.

The results confirmed the earlier impression that the types of flint used changed through time. Figure 22.2, summarising the most important features, shows that layers H–F form a group dominated by grey opaque flint, whereas in the later layers this declines in importance, and off-white (or pale grey) flint becomes one of the most common. A third major category, medium-grey (semi-translucent), peaks strongly in F. Layer 5 is distinguished from the rest of the upper part of the sequence by the virtual disappearance of medium-grey, and by a high incidence of brown and grey-brown flint.

It would be unwise to place too much reliance upon these results in view of the provisional nature of the classification, the occurrence of pieces of variable composition (e.g. lighter grey cherty inclusions in dark grey translucent flint), and the limitation of the study to a single class of object. Nevertheless, the conclusion seems unavoidable that in the lower layers (when the sea-level is presumed to have been relatively high) the flint used for tool-making was different from that employed later. The implications are considered in section 22.5 below. A more exhaustive investigation in the future would be rewarding.

### 22.3.2 Quartz

The local granites contain quartz veins and pegmatitic quartz, which could have been exploited directly by prehistoric man, though it would have been obtained more easily from indissoluble residues remaining after weathering of the granite. At the present time the veins known from the vicinity of La Cotte are not very large, but it is possible that wider ones are now submerged. Beach pebbles do not appear to have been extensively used as a source of this raw material (they would in any case have been more difficult to break up than would unrolled fragments).

### 22.3.3 Granite

The granites of southwest Jersey have been discussed at length in chapter 4. Because this is the rock of which the site itself is formed, caution must be exercised in accepting granite splinters as man-made, particularly where the material is of the porphyritic type. Only the most convincing pieces have been included, therefore, and this may have introduced a bias towards the fine-grained Beauport variety with its cleaner conchoidal fracture and ability to take retouch. Even so, none showed any signs of secondary flaking.

### 22.3.4 Dolerite and Other Dyke Rocks

These are available locally; thus Fig. 4.2 shows a dolerite dyke intruded into the Beauport granite about 30 m north of La Cotte Point.

### 22.3.5 Siltstones and Greywackes

The vast majority of the fine grained sedimentary rocks used at La Cotte are clearly attributable to the Jersey Shale Formation of Brioverian (late Pre-Cambrian) age

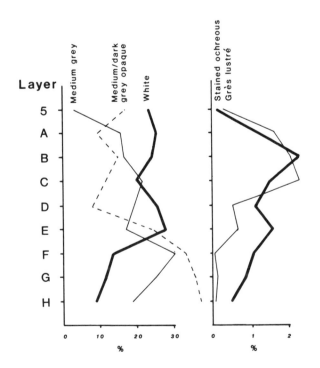

**Fig. 22.2** Frequency of occurrence of principal flint colour categories (based on waste flakes with plain butts), of re-used flint with an ancient ochreous patina, and of *grès lustré*; the last two are calculated as a percentage of all materials, the others of flint only.

which extends over much of the western half of the island, flanked by the southwestern and northwestern granites. However, many of the artefacts exhibit traces of the rounded and very smoothed surface which such materials acquire in marine and fluviatile deposits, and it is likely that in spite of their unpromising quality they were employed because they were readily to hand in the form of beach pebbles. In certain layers (notably G and 8.3) the presence of Brioverian pebbles which seem too small to have been brought in by human agencies may imply the former existence of suitable gravels on the plateau above, in addition to the marine beaches at the foot of the headland.

### 22.3.6 Sandstones

These are not easy to attribute to specific sources, though many could be Brioverian in origin, and hence local. Some 'clean' sandstones are poorly, if at all, represented in Jersey (depending on their precise character).

### 22.3.7 Quartzites

As already mentioned, the term 'quartzite' has been used in a rather catholic sense in the applied classification, on technological grounds. The well-sorted dark quartzitic sandstones (class 5₁) could be derived equally from the Brioverian or elsewhere. The massive quartzites of class 5₂ are not known to be native to the island, though similar material occurs in the beaches. One grey Ordovician variety, known as *Grès Armoricain*, crops out from Lower Normandy to Finistère (the nearest occurrence is in the Anse de Vauville, north of Flamanville). In Brittany it was exploited at the factory sites of Forest-Landernau (Finistère) and Saint-Congard (Morbihan), though in most other instances it was used in the form of beach pebbles (Monnier 1980b, 122).

The fine grained *grès lustrés* (class 5₃), of glassy appearance, are particularly interesting. Again, they do not crop out in Jersey, but comparable rocks are well known on adjacent areas of the French mainland. They have been studied in detail by Monnier (1980b, 116–122) in his examination of the Palaeolithic of Brittany, where they are the principal materials used at three sites: Bois-du-Rocher and Clos-Rouge in Côtes-du-Nord, and Kervouster in Finistère. The Breton materials were divided into seven groups after thin sectioning; all are formed by secondary silicification, though of slightly different parent rocks (clayey sands or sandstones, or sandy clays), probably during the Tertiary Era. Ten thin sections and the artefacts from which they were cut were taken to Rennes for comparison with the Breton samples. Monnier's group VI provided a fairly close parallel:

> *La roche a un aspect de silex d'autant plus accentué qu'une épaisse patine blanche et une pseudo-cortex apparaissent très souvent. En fait ce groupe recouvre de nombreuses variétés, particulièrement en ce qui concerne les couleurs (blanc, jaune, brun, brun-rouge, rose violacé, translucide). C'est une bonne matière première pour la taille, rencontrée aussi bien à Ker-*

> *vouster qu'au Bois-du-Rocher. Au microscope, on observe des grains détritiques (quartz) très dispersés dans une matrice finement silicifiée (texture mudstone ou wackstone). Il y a en outre de nombreuses géodes silicifiées en calcédoine et en quartz (Monnier 1980, 119).*

Nevertheless, though the above description fits them in a general sense, no exact match was found for the La Cotte specimens. The same specimens were examined by Professor F. Doré (Geology Department, University of Caen), but once again no parallel was found among the *grès lustrés* of Normandy. Indeed, Professor David Evans informed Burdo (1960, 60) that this material "has not got sufficient felspar in it to suggest that it might have been derived from Brittany or Normandy. In England, the nearest source of such material would be in the neighbourhood of Torquay, or possibly from boulders near Budleigh Salterton". The most likely explanation is that the examples from La Cotte were from pebbles originating at outcrops which are now submerged.

In using the *grès lustré*, prehistoric man seems to have treated it as though it were flint. It provides a far higher percentage of finished tools than its gross frequency might lead one to expect, and was clearly an entirely acceptable substitute. In spite of this, its contribution to the total raw material supply is extremely variable (Fig. 22.2), being negligible in layers H–F and 5, slight in D and E, and at a maximum in C–A, suggesting that its availability was very different from that of flint.

## 22.4 CONSEQUENCES OF CHANGING SEA-LEVEL AND ENVIRONMENT

### 22.4.1 Modes of Occurrence of Raw Materials

The raw material resources for La Cotte may be divided into four major categories: artefacts picked up at the site or elsewhere, and reflaked; outcrops and dykes fairly close at hand (with fragments perhaps already detached by frost action, and lying close to the source or else incorporated in slope deposits); more remote occurrences of rocks not represented in Jersey, accessible only with a substantial fall in sea-level; pebbles (marine or fluviatile, or derived in slope deposits).

### 22.4.2 Re-used Artefacts

This source may be regarded as opportunistic; nevertheless, at times when flint was scarce, the ready availability of a supply at the site itself may have been seen as an asset. The incidence of double-patinated pieces (flint and *grès lustré*) is small but significant. However, in the course of the wear-trace analysis carried out by H. Frame (chapter 30, this volume), it became apparent that hand sorting was inadequate for the identification of all such instances of re-use, as the differential patination is often determinable only with the aid of a microscope. It would therefore be misleading to attempt to document this phenomenon statistically without substantial further

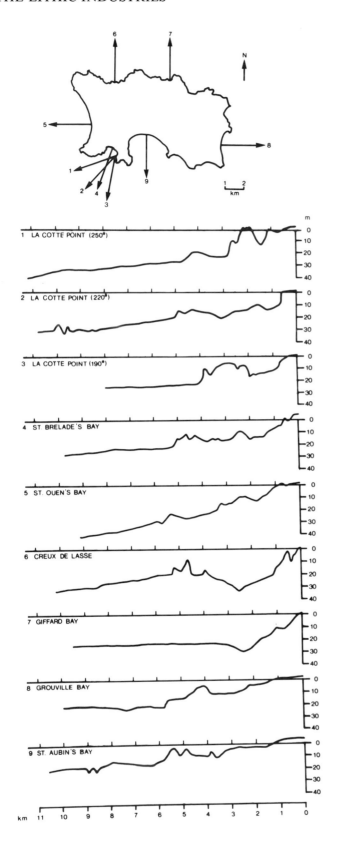

**Fig. 22.3** Sea-bed profiles around Jersey, compiled from information in British Admiralty charts 1136, 1137 and 1138 (Crown Copyright), by permission of the Controller of Her Majesty's Stationery Office and of the Hydrographer of the Navy.

investigation, a high priority for continuing research on the collections.

Ochreous staining of the flint artefacts does not appear to have occurred *in situ* at La Cotte; nevertheless, 1.6% of the flints do exhibit this on flake surfaces predating the primary removal or the retouch. The best explanation may be that they were picked up at other sites or in fluviatile deposits (in the latter case, relatively low-energy transport agencies must have been involved in view of the lack of rolling marks). There was an upwards trend in the extent to which such material was employed, until its virtual disappearance in layer 5 (Fig. 22.2).

### 22.4.3 Local Outcrops and Dykes, and Derivative Deposits

The materials in this category are notably quartz, basic igneous rocks, and granite. All these were of course available as pebbles, and were exploited in that form also. Being tough materials, they survive the mechanical shocks of marine transport very well, and retain their usefulness. The pebbles are not easily quartered in order to initiate the flaking process, though, and angular fragments are generally preferable for knapping if they are available. The enormous increase in the percentage of quartz in the upper Saalian layers clearly implies a greater reliance on local rocks during this period, just as the rise in the other poorer quality materials indicates a greater acceptance of whatever came to hand.

### 22.4.4 More Remote Terrestrial Sources

Some of the quartzites and sandstones were evidently collected as pebbles. Whatever their source, the sandstones seem to have been available throughout the Saalian occupation. It may, however, be significant that these 'peak' twice, once in layers G and F (where a slight deterioration of climate has been very tentatively postulated in chapter 11 on the basis of admittedly scant palynological data), and again in layer A and above, when a far better documented climatic deterioration occurred. This suggests the exposure, as a result of somewhat lowered sea-level, of outcrops now unknown, or of beach deposits rather different in composition from those available at other times.

The most striking fluctuation of materials in this category is of dark quartzitic sandstone (quartzite $5_1$ of Table 22.3) which is important only in layers B and (more particularly) A and 3 — that is, from the outset of late Saalian loess formation (it is thought that B contains a mixture of material from A and C). Again, exposure of formerly submerged outcrops or concentrations may be the explanation, though in that case its scarcity in layer 5 seems surprising; in view of the shift in flint procurement strategies postulated in section 22.5, the need for non-flint raw material from possibly quite remote sources may no longer have existed.

### 22.4.5 Pebbles

At one level of investigation, identification and quantification of the use of pebbles as a raw material seems straightforward; for most rock types, the cortex is quite distinctive. Nevertheless, unrolled cortex may be preserved in concavities on the pebble (see 22.3.1 above); also, different techniques of debitage result in varying percentages of decortication flakes. A less obvious source of difficulty is the possibility that large boulders will have been worked on the spot, and only the usable flakes, finished tools and partly reduced cores brought back to the living site. This consideration is very likely to apply when the size of the required artefacts forces the selection of a large piece of material.

In spite of these difficulties, it is clear that throughout the archaeological sequence the greater part of the artefacts were made from beach pebbles (most of the exceptions being of quartz). In view of the rarity of large flint pebbles in particular, in both the modern and the raised beaches of Jersey, the occurrence of flint as an important rock type in most of the industries of La Cotte de St. Brelade (and of La Cotte à la Chèvre) indicates that it must have been very much more readily available then than now. This in turn would suggest that there is at work today a mechanism preventing flint from reaching the active beaches which was not effective during the prehistoric occupation. Such a mechanism is provided by the form of the sea bed around the island (Fig. 22.3).

Where the sea floor consists of Tertiary sediments, it exhibits a very gentle overall gradient — typically 0.5% or less. The contact with the older rocks of which the island is formed often features a dramatic increase in slope, and even in low cliffs. This change in relief is usually to be found 15 to 25 m below the present low tide level; its significance is that much less energy would be required for the sea to transport exotic pebbles at and below this range of altitudes than to raise them over the more rugged topography to the present level. Moreover, a lowered sea-level would result in a much reduced content in the beaches of the durable native rocks of Jersey itself, and thus an impoverishment of the pebble component of the strandline as a whole; the flint pebbles would thereby become a far more visible resource. The implications of this for any flint-using occupant of Jersey are important: so long as Jersey is connected to the mainland by even the narrowest of land-bridges, good flint ought to be reasonably available in the active marine beaches.

With a rise in the sea-level to above about -20 m, flint would become progressively scarcer as the percentage of the coastline receptive to exotic pebbles diminished. Thus, during the Mesolithic, and even the Neolithic, it should have been easier to obtain flint than is the case today. With a very substantial fall in sea-level, however, the active beaches would retreat very rapidly from the island on account of the low gradient, while fossil deposits may have been blanketed by loess and solifluction and exposed only by fluviatile activity, freeze-thaw effects and other causes (Figs. 22.4 and 22.5).

## 22.5 AN INTERPRETATION OF RAW MATERIAL VARIATION AT LA COTTE

In attempting to explain the changes in raw materials, the following premises have been employed:

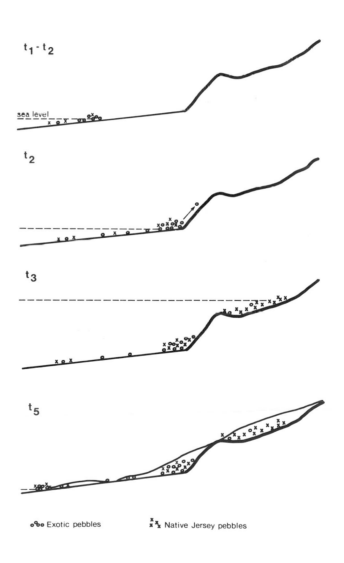

**Fig. 22.4** Mechanisms controlling the availability of pebbles of exotic raw materials during transgression and regression, based on the sea bed profiles of Fig. 22.3. Times $t_{1-5}$ represent the following conditions: $t_1$ sea-level low, rising; $t_2$ sea-level critical, rising, i.e. with the shore-line at the margin of the local (hard) rocks; $t_3$ sea-level maximum; $t_4$ sea-level critical, falling; $t_5$ sea-level low, falling.

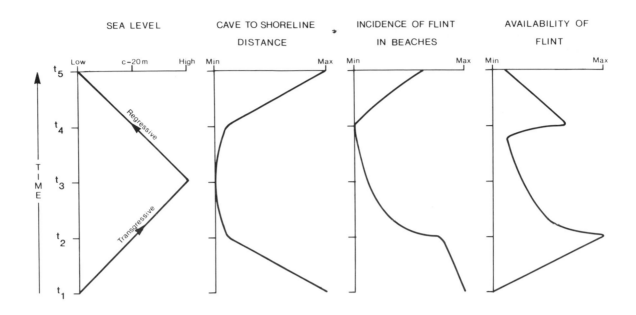

**Fig. 22.5** Flint availability as a function of sea-level (compare Figs. 22.3 and 22.4).

1. That the most suitable raw material would be selected and that normally flint would be preferred to other rocks, except where the tool was to be used for exceptionally heavy work, or the size of the desired tool was greater than that which could be produced from the available flint.
2. That less desirable raw materials might sometimes be used, if they were ready to hand.
3. That departure from the behaviour suggested by (1) is primarily due to restriction in the availability of the preferred material — apart from the 'noise' introduced by (2).

Of the other raw materials employed, some in the 'stone' category were favoured (during the later Saalian occupations) for the manufacture of heavier duty tools such as handaxes and flake cleavers, as well as of large Levallois flakes. However, apart from *grès lustré*, which evidently served as an entirely satisfactory substitute for flint (see chapter 28), it is difficult to believe that the other hard rocks were preferred for the making of flake tools, except where size was at a premium. Still less would quartz have offered any advantage over flint for light-duty tasks.

If this reasoning is correct, it follows that any reduction in the amount of flint from the percentages recorded in layers H, G and F is likely to have been imposed rather than voluntary. The assemblages from these layers, and that from the Marett excavations (Appendix B, this volume) provide an estimate, of about 5% of non-flint rocks, for the residual effect suggested by premise (2). The Weichselian material is also important in that it shows that an industry dominated by well-made flint side-scrapers, comparable to those of the later Saalian occupations and probably made under not dissimilar environmental conditions, had no need of quartz

— thus tending to confirm the view that the earlier abundance of quartz does not imply some unidentified activity for which flint was inappropriate.

The model introduced in section 22.4.5 to relate flint availability to sea-level is the antithesis of that proposed in the preliminary report (McBurney and Callow 1971), which supposed that the variation in the flint percentage during the Saalian layers was due to flooding of the source of flint (then assumed to be the Chalk). Such a view is no longer tenable, given the expanded chronology, the evidence of continuing extremely cold conditions in layer 6.1 and the employment of flint pebbles. Instead, supplies of exotic pebbles may be seen as limited to the extremes of excessively high or excessively low sea-level (with the additional restrictions imposed by a loess mantle in the latter case), but only the second of these is likely to have occurred during occupation at La Cotte. The explanatory model is somewhat complicated by the 'threshold effect' which is postulated in the relationship between sea-level and beach composition in Figs. 22.4 and 22.5. On the other hand, the intensive tool resharpening during the later Saalian and the change-over to the importing of finished tools in layer 5 (chapters 24 and 29, this volume) are strong corroborative evidence that flint sources were less accessible.

On a simple sea-level/flint availability model, the slight increase in exploitation of other materials in layer E might suggest that it was at this point in the sequence that one might expect the beginnings of a marine regression — particularly since the waste from F indicates that decortication of pebbles had been taking place (chapter 25), implying optimum conditions. However, the pedological evidence (chapters 7 and 9) suggests otherwise, as a *lessivé* soil developed after deposition of E — the presence lower down of taranakite derived from guano may indicate abandonment of the site during the sea-level maximum. A possible explanation for the reduction in flint content in E is that the situation then corresponded to just after time $t_2$ in Fig. 22.4, when the percentage of flint in the nearby beaches was beginning to diminish. The same model could account for the continuing fall in D and C, when the retreating beaches would have contained more local rocks than in the transgressive phase.

By the time of deposition of layer 5, good raw material sources must have been so far away that the practice adopted in layers C–A (intensive resharpening) no longer sufficed, and most tools were made elsewhere: hence the very low incidence of manufacturing debris (chapter 24, this volume). Heavy reliance on new and probably distant sources of pebbles would explain the changed character of the non-local material used (virtual absence of *grès lustré* and of medium-grey flint) and the extreme rarity of re-used ochreous-stained artefacts.

The high percentage of flint in the Weichselian layers might appear anomalous, since the sea may well by then have been some distance away. However, the possibility cannot be ruled out that during isotope stage 5 the sea was more effective in supplying flint in an accessible form (e.g. by laying down substantial flint-rich pebble bars which were later exposed by fluviatile action or other processes) than it was during isotope stage 7. In any case, there must always have been a period of time after a transgression when nearby exposures had not yet been exhausted by the flint-knappers nor yet completely buried. Whatever the cause, though, there is some evidence (e.g. the working down of Levallois cores into discs, and the scarcity of unmodified Levallois flakes) that the flint supply was only just adequate.

# METHODS OF STUDY EMPLOYED FOR THE ARTEFACTS

P. Callow

## 23.1  CLASSIFICATION OF TOOLS

The typology devised by the late Professor F. Bordes (1961b) has been used in the interests of comparability with data from other sites. Its application was not entirely straightforward; there are important aspects of the La Cotte industries which cut across the types and force one to question the standard order of attribute priorities when dealing with some of the more elaborate pieces. At first the more 'difficult' artefacts were treated as miscellaneous (type 62) but this category became too numerous, especially in layer A. A second attempt at classification of the members of this group, according to their dominant features, reduced their numbers to the level given in the tables in chapter 26; the adoption of a 'multiple-response' description (observation 24 below), i.e. one allowing a piece to exhibit several states of a single attribute, has permitted the remaining 'type 62s' to be included in the later stages of analysis.

Frequency tables based on the full type-list are too cumbersome to allow easy comparison between industries, and, except where the assemblages are very large indeed, they include many types whose scarcity renders them subject to considerable sampling error. Bordes therefore introduced the practice of supplying a number of summary indices, which are the percentages obtained by grouping types which are closely related taxonomically ('typological indices') or otherwise seem often to be associated statistically ('typological groups'). Moreover, these calculations are usually made both on the full set of types and on a subset or subsets, whose exact composition varies according to the intentions of the investigator.

The first two (real and essential) were devised by Bordes and now feature regularly in descriptions of Lower and Middle Palaeolithic assemblages. The purpose of the essential counts and indices is to allow comparison of series with varying amounts of Levallois debitage and irregularly retouched tools; the latter seem in many sites to be at least partly the result of soil movement rather than deliberate shaping (Bordes 1972, 52). The third count (reduced) was introduced by Girard (1978) on the grounds that types 5 and 38 (pseudo-Levallois points and naturally backed knives) are unretouched — unlike all the other types included in the essential count — and are only doubtfully tools. The reduced count is thus based on well-retouched pieces only, and is probably the most reliable of the three for comparing assemblages on the basis of shaped tools; it has been preferred in the account of the La Cotte finds, though the real and essential values are also tabulated for use in conjunction with published data from other sites.

The indices used, and the tool types included in the denominator during their computation, are:

| Real | 1–63 |
| Essential | 4–44, 51–63 |
| Reduced | 4, 6–37, 39–44, 51–63 |

The basic typological indices and groups devised by Bordes are:

Indices

| ILty | Levallois (types 1–3) |
| IR | Side-scrapers (types 9–29) |
| IC | Charentian (types 8, 10, 22–24, 28) |
| IAu | Acheulian unifacial, i.e. knives (types 36–37) |
| IAt | Acheulian total, i.e. knives and bifaces |
| IB | Bifaces |

Groups

| I | Levallois (types 1–4) |
| II | Mousterian (types 6–29) |
| III | Upper Palaeolithic (types 30–37, 40) |
| IV | Denticulates (type 43) |

For the calculation of IAt and IB the total number of handaxes is added to that of types 1–63 to obtain the denominator. Depending upon the characteristics of the assemblages under study, further indices are often employed. For La Cotte de St. Brelade the following have been used:

| 5 | Unretouched pseudo–Levallois points (type 5) |
| 38 | Naturally backed knives (type 38) |
| I rc | Retouched pieces with convergent edges (6, 7, 18–21). After de Lumley (1969, 147). |
| 6–7 | Mousterian points (6–7) |
| Rsingle | Single side-scrapers (9–11) |
| Rdouble | Double side-scrapers (12–17) |
| Rtransv | Transverse side-scrapers (22–24) |
| Rinv | Inverse and alternate side-scrapers (25, 29) |
| 42 | Notched pieces (42) |
| 42–43 | Notched and denticulates (42, 43) |
| Ident | Extended denticulate group (42, 43, 51, 54) |
| Iirr | Irregularly retouched pieces (54–50) |

Some indices are computed with a limited range of tool types in the denominator; these do not vary as do the principal indices, and so are given once only.

For side–scrapers (9-29)
| IQ | Scrapers with Quina retouch |
| IQ + ½Q | Scrapers with Quina or 'half-Quina' retouch |

For notched pieces (42)
| ICl | Pieces with notches of 'Clactonian' (single-blow) type. |

One feature of some of the La Cotte industries which gives pause for thought is the extent of resharpening and re-use which occurred, the most striking evidence of which is the quantity of specialised waste products (LSFs) described in chapter 29. Such behaviour calls into question the whole basis of conventional typology, and in particular the true significance of the 'type'. One would like to have confidence that a tool form was the product of a single design process; if this were so, the interpretation of type-frequencies would be very much simpler. However, modification of an artefact for the purpose of re-use may transform it from one category to another, possibly leaving no traces of its previous character. This phenomenon is documented in the ethno-archaeological literature, the best-known case being that of the hafted Australian adze flakes (Gould 1980, 127 ff.) which develop through end- or side-scrapers into slug-shaped objects in the course of perhaps twenty rejuvenations.

In purely archaeological contexts the reconstruction of such a series of events is not always possible; however, a combination of refitting and microwear studies on artefacts from the late Upper Palaeolithic site of Meer, in Belgium, enabled Cahen, Keeley and van Noten (1979) to establish the complex histories of a number of pieces. Perhaps their most significant result, in this respect, was the discovery that one artefact began as a *bec* and was converted into a burin (with two episodes of resharpening), being used throughout its life for boring antler. Thus it is clear that the classification process may separate tools which are functionally equivalent or which may be no more than the different evolutionary stages of a single functional category. The converse, that a single tool-type may have possessed several functions, has also

been established at Meer, and indeed some of the La Cotte artefacts proved to have been used for working more than one material. This should be borne in mind throughout the discussion based on the Bordes typology — a typology which is nevertheless of great value as a rapid and fairly repeatable means of describing large amounts of material in a widely understood manner.

Reference has already been made to resharpening by the LSF technique. In the classification of the tools from La Cotte, pieces exhibiting the distinctive removal scars have usually been classed as scrapers of the appropriate type provided that some trace of the retouched edge remains. In a number of instances, however, the removal was effective enough to leave no trace of the original working edge (which may not have been previously retouched); the absence of later retouch could be due either to abandonment of the piece or, say, to its intended use as a knife (see Fig. 29.15). Such pieces have been placed in the miscellaneous category (in all cases, however, information about the rejuvenation is recorded in the database).

## 23.2 OTHER OBSERVATIONS ON TOOLS

The following were applied to pieces made of flint or stone; a more restricted set was used for the quartz (see chapter 27).

**Table 23.1** Application of Qualitative Observations to Tools Made of Flint and Stone[a]

| | Observation | Bordes type | | | | | | | | | | | | |
|---|---|---|---|---|---|---|---|---|---|---|---|---|---|---|
| | | 6 to 7 | 8 | 9 to 11 | 12 to 17 | 18 to 20 | 21 | 22 to 24 | 25 to 27 | 28 to 29 | 30 to 31 | 32 to 33 | 42 | 43 | 62 |
| 1 | Bulb removal etc. | 1 | 1 | 1 | 1 | 1 | 1 | 1 | 1 | 1 | 1 | – | – | 1 | – |
| 2 | Retouch – type | 2 | 2 | 1 | 2 | 2 | 2 | 1 | 1 | 1 | 1 | – | – | – | – |
| 3 | – edge shape | 2 | 2 | 1 | 2 | 2 | 2 | 1 | 1 | 1 | – | – | – | – | – |
| 4 | – position | 2 | 2 | 1 | 2 | 2 | 2 | 1 | 1 | 1 | 1 | – | 1 | 1 | – |
| 5 | – Quina | 2 | 2 | 1 | 1 | 1 | 1 | 1 | 1 | 1 | 1 | – | – | – | 1 |
| 6 | *Déjeté* scraper subtype | – | – | – | – | – | 1 | – | – | – | – | – | – | – | – |
| 7 | Bifacial/alternate scraper subtype | – | – | – | – | – | – | – | – | 1 | – | – | – | – | – |
| 8 | Equivalent scraper type | – | – | – | – | – | – | – | 1 | 1 | – | – | – | – | – |
| 9 | End-scraper – subtype | – | – | – | – | – | – | – | – | – | 1 | – | – | – | – |
| 10 | – shape of blank | – | – | – | – | – | – | – | – | – | 1 | – | – | – | – |
| 11 | – symmetry | – | – | – | – | – | – | – | – | – | 1 | – | – | – | – |
| 12 | Other retouched edges | – | – | – | – | – | – | – | – | – | 1 | 1 | 1 | – | – |
| 13 | Burin – subtype | – | – | – | – | – | – | – | – | – | – | 1 | – | – | – |
| 14 | – direction of blow(s) | – | – | – | – | – | – | – | – | – | – | 1 | – | – | – |
| 15 | single/multiple blow(s) | – | – | – | – | – | – | – | – | – | – | 1 | – | – | – |
| 16 | – termination | – | – | – | – | – | – | – | – | – | – | 1 | – | – | – |
| 17 | – position | – | – | – | – | – | – | – | – | – | – | 1 | – | – | – |
| 18 | – truncation shape (if applicable) | – | – | – | – | – | – | – | – | – | – | 1 | – | – | – |
| 19 | Notch – number of notches | – | – | – | – | – | – | – | – | – | – | – | 1 | – | – |
| 20 | – type | – | – | – | – | – | – | – | – | – | – | – | 1 | – | – |
| 21 | Denticulate – coarseness | – | – | – | – | – | – | – | – | – | – | – | – | 1 | – |
| 22 | – subtype | – | – | – | – | – | – | – | – | – | – | – | – | 1 | – |
| 23 | – equivalent scraper type | – | – | – | – | – | – | – | – | – | – | – | – | 1 | – |
| 24 | Miscellaneous – classification | – | – | – | – | – | – | – | – | – | – | – | – | – | 1 |

[a]The figures in this table and the next indicate the number of times the observation is recorded for each type (once for single-edged and twice for multiple-edged pieces).

214

### 23.2.1  For All Tools

**Length, breadth, weight**. These had been recorded, for all artefacts, in the master catalogue during initial post-excavation processing of the finds. The length and breadth measurements were not based on a systematic technological or typological orientation (e.g. tool axis or percussion axis); instead, the pieces were laid on a measuring-board with the bulbar surface downwards and rotated to minimise the breadth. These measurements should be viewed as providing only a gross indication of size and relative elongation.

**Type of blank**. Because, in many instances, extensive re-touch prevents classification according to the shape of the original flake or blade, the categories employed are technological rather than morphological: Levallois (typical/atypical), pseudo-Levallois point, sharpening flake (LSF/TSF — see chapter 29), handaxe thinning flake, tranchet flake, naturally-backed knife, core, pebble. In the case of end-scrapers, however, a shape classification was also attempted as described below.

**Butt** (where applicable). The categories listed in section 25.2.2 were used, with the addition of: removed (by retouch), missing (broken), indeterminable.

**Special features**. The occurrence of LSF removals and technologically (not necessarily functionally) related treatment was recorded:

Nature of the feature(s): LSF or burin spall removal, dorsal scaling from an inverse truncation, or some combination of these.

Location of the feature(s): distal, proximal, lateral etc.

The platform used in their production: hinge, break, normal or inverse truncation, retouched surface, other.

### 23.2.2  Specific to Certain Types of Tool

Some of the rarer tool-types were considered to be too scarce to merit detailed statistical consideration. The following broad categories were examined more closely, however: handaxes (bifaces and flake cleavers), the 'Mousterian' group (i.e. points, *limaces*, and side-scrapers), end-scrapers, burins, notches, denticulates, and 'miscellaneous' tools.

The list of attributes upon which the study has been based derives in part from one devised by the author when working on French Lower and Middle Palaeolithic flake-tools during the period 1971–2, and owes much to discussions in Bordeaux with Bordes and with P. Timms. The original scheme has been slightly modified in the light of subsequent experience, the work of others (e.g. Girard 1978), and the limited time available.

Although the observations which were employed varied according to the type under consideration, types 6–29 form a natural group in view of their possession of 'scraper' edges and, in some cases, pointed tips formed by the convergence of two such edges. In order to standardise the treatment given to the pieces which fall into types 25–29 (in which technology is allowed to outweigh form), the appropriate morphological type (9–24) has been noted. For the denticulates a morphologically equivalent scraper type has also been recorded. The measurements used for handaxes are essentially those proposed by Bordes (1961b) and by Roe (1968), and need not be described here.

The tool-types for which each observation has been recorded are summarised in Tables 23.1 and 23.2.

#### Discrete variables

*Various tool types*

1. 'Non-standard' treatment: thinned bulb, thinned back, additional retouch at the distal end, presence of notches (Clactonian or retouched) on a side- or end-scraper. In the very few instances where more than one of these was encountered, the first encountered category was recorded.
2. Retouch type: scalar, subparallel, parallel, step, denticulate, abrupt, *raclette*, *surélevée* (defined according to standard usage). Semi-abrupt marginal retouch was included with the scalar category, as no clear distinction was apparent.
3. Edge shape: straight, convex, concave.
4. Position of retouched edge: left, right, distal, multiple (this last is used only where the measurements are taken on one of several retouched edges, rather than on each of them).
5. Quina retouch: absent, demi-Quina, Quina (used in computing IQ).

*Side-scrapers*

6. *Déjeté* side-scraper subtype: single or double.
7. Bifacial/alternate side-scraper subtype: indicates completely bifacial scrapers (as opposed to those with bifacial retouch) or alternating retouch on alternate scrapers.
8. Equivalent type, normally in the range 9–24 (used only for side-scrapers of types 25–29, ignoring the features which led to their original classification). One piece had to be classed as a *limace* (type 8).

*End-scrapers*

9. Subtype: simple, ogival, carinated, denticulated, fan-shaped, single-shouldered, double-shouldered, on the butt, double, double alternate, thumbnail, circular.
10. Blank shape: flake, blade (length/breadth ratio exceeding 2), transverse flake (length/breadth ratio less than 1).
11. Symmetry: symmetrical, left-canted, right-canted.

*End-scrapers, burins, notched pieces*

12. Presence of retouched (scraper-type) edges.

*Burins*

13. Subtype: single, dihedral, busked, mixed, *plan*, double (opposed), double (at the same end), other.
14. Direction of blows: symmetrical about the axis, *déjeté*, transverse,

longitudinal, transverse + longitudinal.
15. Single or multiple blows from a single platform.
16. Termination of burin scar: feathered, hinged, overstruck.
17. Burin position: axial, left, right, left + right (orientated with the bulbar surface away from the viewer, and the burin point at the top).
18. Shape of truncation from which the burin was struck (if applicable): straight, convex, concave.

*Notched pieces*

19. Number of notches.
20. Type of notch: retouched or Clactonian (single-blow).
21. Coarseness: multiple adjacent Clactonian notches, or retouched (coarse, fine, microdenticulate).
22. Direction: normal, inverse, alternate, bifacial.
23. Scraper equivalent type, in the range 9–30 — the type of scraper to which the piece would have been assigned had the retouch been other than denticulated.

*Miscellaneous*

24. 'Multiple response' classification, allowing for more than one attribute state for a single artefact. Any number of the following may be recorded: scraper, notch, denticulate, overstruck removal on the dorsal surface (see section 26.5.6), scaled piece, burin.

**Continuous variables** (see mf 23.1)

*Retouched edges*: points, end- and side-scrapers, denticulates.

a. Chord length: distance between the *ends of the retouch*.
b. Maximum displacement of the retouched edge from the chord — positive for convex, negative for concave pieces (used for computing curvature).
c. Retouch angle (not recorded for denticulates); measured at about 1 mm from the edge, to avoid rounding due to use-wear.
d. Invasiveness of retouch (estimated mean length of scars, measured in the direction of the blow).

*Points*

e. Width 10 mm from tip.
f. Width 20 mm from tip.
g. Thickness 10 mm from tip.

*Burins*

h. Length of principal burin scar.
i. Width of burin chisel-point.
j. Angle between the principal burin scar and the ventral surface (measured at the point of impact, to avoid the effect of twisted scars).
k. Angle of the chisel-point (i.e. between burin scar and break/truncation/other burin scar).

*Notches*

l. Chord length ('notch width') — as (a) above.
m. Maximum displacement from the chord ('notch depth') — compare (b) above. This is, by definition, negative.
n. Invasiveness, i.e. length of scar(s).

**Derived variables (continuous)**

Several compound attributes were computed from the measurements described above. They include:

I. Length/breadth ratio (primarily used for estimating the incidence of blades in the unretouched Levallois material, and for the Mousterian points).
II. Estimated thickness. Thickness had not been recorded with length and breadth during compilation of the master catalogue for the site. An index was therefore computed by means of the formula 100 x weight/(length x breadth); it roughly approximates the thickness obtainable using callipers at a measuring board. In passing, it may be noted that the same calculation, when performed for flint handaxes (admittedly more regular in

**Table 23.2** Application of Quantitative Observations to Tools Made of Flint and Stone

| | Observation | Bordes type 6 to 7 | 8 | 9 to 11 | 12 to 17 | 18 to 20 | 21 | 22 to 24 | 25 to 27 | 28 to 29 | 30 to 31 | 32 to 33 | 42 | 43 |
|---|---|---|---|---|---|---|---|---|---|---|---|---|---|---|
| a | Retouched edge - chord length | 2 | 2 | 1 | 2 | 2 | 2 | 1 | 1 | 1 | 1 | - | - | 1 |
| b | - perpendicular | 2 | 2 | 1 | 2 | 2 | 2 | 1 | 1 | 1 | 1 | - | - | 1 |
| c | - edge angle | 2 | 2 | 1 | 2 | 2 | 2 | 1 | 1 | 1 | 1 | - | - | - |
| d | - invasiveness | 2 | 2 | 1 | 2 | 2 | 2 | 1 | 1 | 1 | 1 | - | - | 1 |
| e | Point - width at 10 mm | 1 | 2 | - | 1 | 1 | 1 | - | - | - | - | - | - | - |
| f | - width at 20 mm | 1 | 2 | - | 1 | 1 | 1 | - | - | - | - | - | - | - |
| g | - thickness at 10 mm | 1 | 2 | - | 1 | 1 | 1 | - | - | - | - | - | - | - |
| h | Burin - removal length | - | - | - | - | - | - | - | - | - | - | 1 | - | - |
| i | - removal width | - | - | - | - | - | - | - | - | - | - | 1 | - | - |
| j | - angle with ventral face | - | - | - | - | - | - | - | - | - | - | 1 | - | - |
| k | - angle of tip | - | - | - | - | - | - | - | - | - | - | 1 | - | - |
| l | Notch - chord length ('width') | - | - | - | - | - | - | - | - | - | - | - | 1 | - |
| m | - perpendicular ('depth') | - | - | - | - | - | - | - | - | - | - | - | 1 | - |
| n | - invasiveness of scar(s) | - | - | - | - | - | - | - | - | - | - | - | 1 | - |

shape than the La Cotte flake tools) from other sites, gave a result which was usually within 5% of the measured value — apparently because the specific gravity of flint is very close to the ratio between a handaxe's containing cuboid and its true volume (Callow 1976). An incidental advantage of using such a computed figure is that it compensates for minor irregularities in the profile of the piece.

III. Relative thickness. This is obtained by dividing the estimated thickness by length x breadth. It is therefore more generalised than a simple T/B ratio.

IV. Curvature of retouched edges. This is not calculated according to the strict mathematical definition (i.e. the reciprocal of the radius of curvature) but is obtained by dividing the maximum displacement of the retouched edge from the chord by the chord length, i.e. variables (b) and (a) above. Concavity is distinguished from convexity by a negative curvature.

V. Curvature of notches: calculated in a similar manner to that for retouched edges, using variables (m) and (l). As there is no need to distinguish concavity from convexity in this case, the curvature is always quoted as positive, despite the inconsistency with (IV) above.

VI. Relative thickness of points: the ratio between thickness and breadth measured at 10 mm from the tip, i.e. variables (g) and (e).

## 23.3  TECHNOLOGY

The standard Bordes technological indices were employed, with the addition of an extra facetting index, IFss. It was not possible to compute the blade index, Ilam, for the entire assemblages because of the manner in which artefact dimensions had been measured (see above), but estimates based on sampled unretouched material are given in chapter 25. Indices are expressed as percentages of the number of flakes and flake-tools for which the relevant feature is determinable (for definitions of the butt categories see section 25.2.2).

IL  Levallois flakes, blades and points (whether retouched or unretouched)
IF  Facetted, polyhedral and dihedral butts
IFs  Facetted and polyhedral butts
IFss  Facetted butts only
Ilam  Blades (flakes with a length/breadth ratio of at least 2)

All cores were classified according to Bordes's scheme. For flint waste and cores, much more detailed information was collected and is discussed in chapter 25.

## 23.4  DATA PROCESSING

The computer employed was Cambridge University's IBM 370/165, replaced in 1983 by an IBM 3081D. Data management and graphics software was written by this author (in the latter case making use of the University's CAMPLOT subroutine libarary). The SPSS (Nie et al. 1975) and CLUSTAN (Wishart 1978) program packages were used for the statistical analyses.

# AN OVERVIEW OF THE INDUSTRIAL SUCCESSION

P. Callow

## 24.1 INTRODUCTION

Much of what follows is a précis of other chapters in this part of the report; it is primarily intended to be read as an introduction to the more lengthy description and discussion provided elsewhere. In addition the opportunity has been taken to link elements which are otherwise dispersed through the book by the exigencies of a structured presentation and by multiple authorship. This is also the most suitable point at which to consider the results of multivariate analysis.

In addition to providing a long succession of rich and comparatively well-dated lithic assemblages, La Cotte documents early man's responses to variation in raw material supply much more clearly than is usually discernible at individual sites. It has also yielded the most conclusive evidence so far available that the present tendency in man to favour the right hand over the left, in a ratio of about 4:1, already existed in the Middle Pleistocene, with all that this implies for the development of the human brain.

Consideration of the lithic industries of La Cotte de St. Brelade is complicated by the range of materials used. Any review of the literature will show that it is quite common practice to ignore this issue when giving statistical information about an assemblage, or else to include additional data about features related to raw material almost as an afterthought. It became clear at an early stage of the investigation that at La Cotte variation from this cause is too extreme for a 'lumping' approach to be meaningful. In particular the technology of the

quartz debitage, though not formally investigated as yet, is seen to give rise to waste products very different from those produced from other rocks: good flakes, as opposed to 'chips and chunks', are comparatively scarce. The morphology of the quartz tools is likewise idiosyncratic, with a strong component of pieces classed as end-scrapers (entirely at variance with the character of the assemblages in respect of other materials). For this reason the artefacts made of the three principal types of material (flint, quartz and 'stone') have as a rule been dealt with separately; inventories for the first and last of these are given in Tables 24.1–24.2, while that for quartz is to be found in chapter 27.

Apart from stratigraphic problems mentioned in chapter 20 as being associated with the 1961–1962 finds, there are very considerable differences between the typological data published in this account and those given in the interim report of 1971; these arise from the classificatory systems employed. In the early 1970s McBurney was grouping pieces with notches (whether single-blow or made by retouch) with scrapers, on the grounds that it had not been demonstrated that they were functionally distinct. All pieces with any bifacial secondary work, irrespective of their shape or technological details, were assigned to a single category. Moreover the special resharpening flakes described in chapter 29 had not been identified, and the pieces from which they had been removed were therefore described as burins (or *burins plans*) — hence the extraordinarily high frequency formerly recorded for this class in layer A.

Table 24.1 La Cotte de St. Brelade 1966–1978. Inventory of Flint Finds [a]

| Type | Layer | | | | | | | | | | | | Total |
| | H | G | F | E | D | C | B | A | 3 | 5 | 6 | 9 | |
|---|---|---|---|---|---|---|---|---|---|---|---|---|---|
| Tools (whole) | 296 | 685 | 917 | 650 | 757 | 948 | 639 | 3561 | 130 | 338 | 8 | 11 | 8940 |
| Handaxes (whole) | 0 | 0 | 0 | 0 | 4 | 6 | 3 | 66 | 1 | 7 | 0 | 0 | 8 |
| Handaxes (broken) | 0 | 0 | 0 | 0 | 0 | 1 | 0 | 4 | 0 | 0 | 0 | 0 | 5 |
| Retouched fragments | 141 | 333 | 392 | 403 | 632 | 879 | 541 | 3201 | 113 | 257 | 3 | 9 | 6904 |
| Waste flakes (whole) | 838 | 1910 | 2546 | 2338 | 2420 | 3113 | 1813 | 10431 | 321 | 349 | 8 | 20 | 26119 |
| Waste flakes (broken) | 547 | 1358 | 1722 | 1698 | 1686 | 2460 | 1405 | 8680 | 236 | 317 | 14 | 35 | 20158 |
| Chips and chunks | 136 | 166 | 291 | 356 | 370 | 410 | 210 | 920 | 20 | 51 | 3 | 4 | 2937 |
| Cores (whole) | 56 | 123 | 132 | 112 | 91 | 97 | 51 | 408 | 19 | 26 | 0 | 5 | 1120 |
| Cores (broken) | 8 | 25 | 26 | 30 | 42 | 46 | 25 | 166 | 3 | 3 | 0 | 1 | 375 |
| Hammerstones (broken) | 0 | 0 | 0 | 0 | 0 | 1 | 0 | 0 | 0 | 0 | 0 | 0 | 1 |
| Manuports | 0 | 0 | 0 | 0 | 0 | 2 | 0 | 0 | 0 | 0 | 0 | 0 | 2 |
| Total | 2022 | 4600 | 6026 | 5587 | 6002 | 7969 | 4687 | 27437 | 843 | 1348 | 36 | 85 | 66642 |

[a]Pieces at least 20 mm long only.

**Table 24.2** La Cotte de St. Brelade 1966–1978. Inventory of Stone Finds [a]

| Type | H | G | F | E | D | C | B | A | 3 | 5 | 6 | 9 | Total |
|---|---|---|---|---|---|---|---|---|---|---|---|---|---|
| | | | | | | Layer | | | | | | | Total |
| Tools (whole) | 5 | 4 | 3 | 14 | 22 | 45 | 19 | 268 | 10 | 61 | 2 | 5 | 458 |
| Handaxes (whole) | 0 | 0 | 0 | 0 | 0 | 3 | 2 | 20 | 1 | 4 | 0 | 0 | 30 |
| Handaxes (broken) | 0 | 0 | 0 | 0 | 0 | 1 | 0 | 0 | 0 | 0 | 0 | 0 | 1 |
| Retouched fragments | 3 | 3 | 3 | 18 | 18 | 84 | 41 | 376 | 16 | 108 | 2 | 5 | 677 |
| Waste flakes (whole) | 9 | 20 | 44 | 85 | 176 | 427 | 163 | 1561 | 29 | 149 | 1 | 3 | 2667 |
| Waste flakes (broken) | 12 | 31 | 37 | 106 | 148 | 474 | 220 | 1750 | 55 | 255 | 8 | 4 | 3100 |
| Chips and chunks | 8 | 20 | 7 | 17 | 34 | 112 | 42 | 270 | 6 | 44 | 3 | 2 | 565 |
| Cores (whole) | 1 | 1 | 0 | 1 | 2 | 0 | 4 | 28 | 3 | 6 | 0 | 1 | 47 |
| Cores (broken) | 0 | 1 | 0 | 1 | 1 | 4 | 0 | 10 | 2 | 2 | 0 | 1 | 22 |
| Anvil stone | 0 | 0 | 0 | 0 | 0 | 0 | 1 | 0 | 0 | 0 | 0 | 0 | 1 |
| Hammerstones (whole) | 0 | 5 | 3 | 1 | 0 | 2 | 2 | 15 | 2 | 3 | 0 | 1 | 34 |
| Hammerstones (broken) | 0 | 2 | 1 | 0 | 1 | 3 | 5 | 53 | 3 | 5 | 0 | 0 | 73 |
| Manuports | 1 | 3 | 1 | 1 | 0 | 2 | 6 | 59 | 0 | 7 | 0 | 0 | 80 |
| Total | 39 | 90 | 99 | 244 | 402 | 1157 | 505 | 4410 | 127 | 644 | 16 | 22 | 7755 |

[a]Pieces at least 20 mm long only.

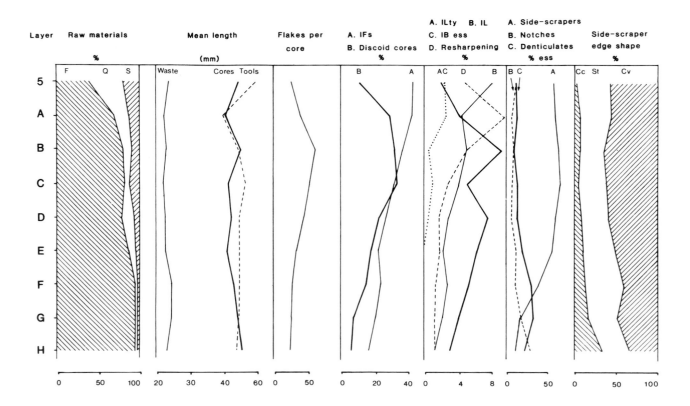

**Fig. 24.1** Variation in some of the artefact assemblage attributes. Except for the first column, the diagram is based on flint artefacts only; hammerstones have been excluded from the raw material counts (F, flint; Q, quartz; S, stone). In computing mean length the reduced type-list has been used. The 'resharpening' curve gives the frequency of all types of tool rejuvenation waste (LSF, handaxe tranchet flake etc.) as a percentage of flakes and flake tools (unbroken pieces only).

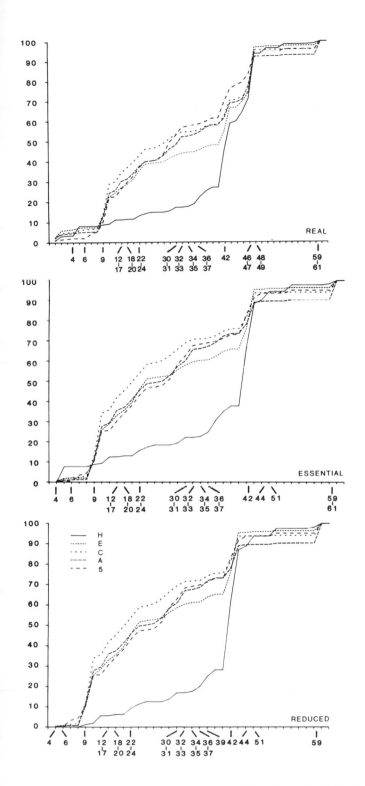

**Fig. 24.2** Cumulative curves for the flint tools from layers H, E, C, A and 5.

## 24.2 THE RAW MATERIALS

The classification scheme employed is described in chapter 22, in which the sources are also discussed. With the passage of time there is a dramatic fall in the amount of flint in the assemblages, from 92.8% by weight in layer H, to 23.2% in layer 5, by which time over half of the raw material being brought into the site was vein quartz of local origin (see Fig. 22.1). This change was formerly believed to be due probably to the submergence of flint sources as a result of rising sea-level (McBurney and Callow 1971). As explained in the chapter on raw materials, this view can no longer be sustained in the light of the environmental data now available; rather, the principal cause is likely to be marine *regression* beginning around the time of deposition of layer D, with a consequent distancing from the site of active beaches and the supply of flint pebbles which they provided, as well as burial of the fossil beaches by periglacial aeolian and slope deposits.

No conclusive examples of artefacts made from nodular flint straight from the Chalk have yet been identified. It seems that the vast majority were made from pebbles, and a few more from older artefacts picked up in or around the site (see also section 22.4.2). The quartz, which originated as veins in the local granites, is hardly ever waterworn, presumably because quartz pebbles are particularly difficult to work. The other rocks (various quartzites, sandstones, siltstones, and local dolerites and microgranite) are from various sources and in some cases were obviously collected as pebbles; some of the larger pieces may equally have been obtained directly from outcrops or (perhaps more likely) after spalling through frost action.

An alternative hypothesis, that the variation in abundance of different raw materials was the result of the knappers' changing preferences rather than of environmental constraints, may be dismissed because of the evidence of more economical use of flint when this rock was particularly poorly represented in the assemblages. This is considered in more detail in section 24.5.

## 24.3 TECHNOLOGY AND TYPOLOGY

There are very great differences between the earliest and latest of the Saalian industries (see Figs. 24.1 and 24.2). The sequence begins, in layer H, with an assemblage dominated by coarse but confidently made notches and denticulates on blanks produced by simple techniques. True handaxes appear to be lacking. By the time of the deposition of layer 5 there had been a succession of assemblages dominated by carefully retouched sidescrapers, culminating in one with fairly frequent step (Quina) retouch and a number of elegant, thick elongated points. The handaxes which occur in these later layers include examples of classic Acheulian form as well as the morphologically less well characterised types which are a not uncommon feature in other cave deposits of about the same age. The technology employed in the manufacture of the tools is generally more advanced than that in the earliest assemblages, including Levallois and disc-core techniques, but the extent to which this is apparent from the debitage and cores (as opposed to the tools themselves) is governed by the introduction of alternative approaches to tool manufacture and re-use.

If there is a strong time-related aspect to the variation, as discussed in section 24.5, it is also clear that many of the assemblages possess idiosyncratic features that cannot be adequately explained in terms of simple linear development within a single industrial tradition. In the interpretation of the variation from assemblage to assemblage, an explanation based on style has been considered acceptable only for those aspects which cannot be accounted for either by sampling error or by the constraints imposed by supposed fluctuation in raw material availability. Into this 'residual' category may be placed, for example, the high percentage in F of *déjeté* side-scrapers, and the extensive employment of inverse retouch on the scrapers from the two assemblages immediately following. On the other hand, the extent to which resharpening has taken place, with the aid of a highly specialised technique, is strongly linked to a number of other parameters explained as related to flint shortage.

### 24.3.1 Technology

This topic is discussed more fully in chapters 25 (flint) and 28 (stone). In the case of quartz (chapter 27) no more has been attempted for the present than classification and simple measurement of cores, and examination of the butts of the flake tools. In all but layer 5, the largest group of artefacts, and the one for which the technology is the most easily discerned, because uncomplicated by cleavage planes, is that made on flint. The following remarks, and also those on typology, are therefore concerned with this material alone.

The flakes and flake tools exhibit a strong upwards trend in the use of facetted butts (*sensu lato*) from 33.4 to 53.9%; true facetted butts — those with preparation immediately prior to detachment of the flake — show an even stronger increase, in frequency, from 11.4% in layer H to 33.3% in layer 5 (Fig. 24.1). The frequency of Levallois flakes (retouched and unretouched) also rises, from 1.6 to 8.2%. As these figures indicate, Levallois technique is never a very conspicuous element in the knapping technology practised on site, and indeed Levallois cores are always scarce. But from layer C onwards the percentage of retouched pieces made on Levallois blanks approaches 20. Since *unmodified* Levallois pieces are much rarer, such blanks of this type as were produced were evidently favoured for tool-making. The most extreme case is layer 5, in which unmodified Levallois flakes, blades and points are virtually lacking, as are the more sophisticated types of core.

Other strongly varying attributes of the assemblages include:

The ratio of flakes to cores (based on complete examples, and including tools on both kinds of blank); this is less than 22 in H, peaks at 51 in B, and returns to nearly its original level in 5. The lowered value in A appears to be due to the use of flakes as cores (with only a few removals), and also perhaps to a decline in the size of the pebbles used as raw material.

Disc cores (in the sense used by Bordes) reach their maximum frequency in C, at 34% of the cores, and then decline slightly before a more abrupt fall in layer 5. The lowered incidence in A and 3 is less a reflection of a decline in this technique of primary flaking than of the re-use of flakes as cores to supplement it.

The percentage of tools (whole and broken); initially fairly stable at around 20%, it rises slightly in D before a remarkable increase in layer 5 to 44%.

Size; weight (particularly) and length exhibit an erratic but generally downward trend until and including layer A, and increase sharply in 5. Core length behaves in a more complicated manner than core weight because of morphological variation.

The ratio of whole to broken tools is interesting (the ratio for waste flakes is more uniform): beginning at 2.6, it remains high until E. Thereafter it remains at about 1, rising somewhat in 5. This is clearly due to the relatively thin and more easily broken blanks employed in the intermediate layers, though the value for H may have been slightly augmented by the difficulties of identifying fragments of notched pieces.

A technique which was of particular significance in the later Saalian assemblages at La Cotte was tool rejuvenation by the LSF (long sharpening flake) technique discussed in chapter 29. This involved fabrication of a *burin plan* along the working edge of a side-scraper, or more rarely another type of tool, in order to remove the retouch. If successful it resulted in a fresh sharp margin which could then be treated as if primary, and used directly or after application of more retouch. In layer A 6.8% of the tools (in real count) preserve evidence of the technique, but because of subsequent retouch this must greatly under-represent the extent to which it was practised. The characteristic waste products — the LSFs — one of whose edges consists of the former tool margin, exceed 1000 in number even after exclusion of fragmentary examples. In some cases they were themselves retouched as scrapers. Because the technique involves very precise control and the LSFs are asymmetrical, they have the important property of reflecting very accurately the preferred handedness of the knapper; consequently it is possible to state that at this time the ratio of right- to left-handed people was much the same as today's.

## 24.3.2 Typology

Handaxes are entirely lacking in the lower layers, apart from a single (questionable) partial biface of quartz in H. Layers H and G are dominated by notched pieces and denticulates (the latter coarsely toothed). In H the notched pieces are in the majority; most of the notches are Clactonian. There is one chopping tool. This is also the only series in which Tayac points play a significant part, as also do thick pseudo-Levallois points, both unretouched and as blanks for retouched tools. On the other hand, side-scrapers are quite rare. Denticulates outnumber notched pieces in G; although there are more side-scrapers than before, their relative frequency is still low, and many of the single lateral examples have only marginal retouch (though the other types are better characterised). There are several excellent knives, end-scrapers and burins.

The assemblage from layer F is the first in which side-scrapers are generally well made and numerically important, though they are still less common than the combination of notched and end-notched pieces, denticulates and Tayac points. No fewer than 16.3% of the side-scrapers are *déjeté*, and it is notable that the predominant edge shape is straight. Denticulates remain numerous — it is the notched pieces and Tayac points whose frequencies decline most markedly.

From layer E onwards, the side-scrapers (more usually with convex edges and of improved quality) exceed 50% of the reduced total, with a maximum of 64.7% in layer C. The much scarcer denticulates are more finely toothed than before. In E itself there is increased employment of inverse and alternate retouch on the side-scrapers; this feature is even more apparent in the succeeding layer D, in which 23.3% of the scrapers are so treated.

Typologically the assemblages from C to A form a well-defined group, though one in which changes take place through time. Side-scrapers, (particularly those with convex edges) are always the principal element, and bifaces of classic Acheulian type appear for the first time (the four examples in layer D are nucleiforms). Levallois blanks have frequently been used during tool manufacture but unretouched Levallois flakes are rare, as mentioned above. The inverse and alternate scrapers referred to above are much rarer in these layers. Inverse truncations recalling those on Kostienki knives become increasingly common (by layer A they are present on 10.7% of the retouched tools) and are often associated with dorsal scaling or with resharpening by means of the LSF technique (see above); it may be that these features are causally linked to the developing shortage of flint, being adopted to facilitate both the hafting of ever smaller scrapers and the re-use of rejuvenated tools. Burins, chiefly angle burins on breaks, are particularly common in A (5.9%), and as use-wear studies suggest that they were not employed for engraving or cutting this technique, too, may have been an adaptation for gripping or hafting.

The flint assemblage from layer 5 possesses (though to a lesser degree) such characteristics of A as LSF technique and the use of inverse truncations. Side-scrapers continue to be the most numerous class of tool but are often much larger than before, and in some cases are made with Quina retouch; denticulates and notches become rather more common. The handaxes include a small lanceolate or *ficron* and, from the Burdo excavations, a fine twisted cordiform. But one of the most striking tool types, unparalleled in earlier layers, is a group of thick elongated Mousterian points, several of which bear indubitable impact damage at the tip, and which were no doubt discarded when spears were repaired.

### 24.3.3 Non-flint Artefacts

Only layers D, A and 5 have yielded enough quartz tools to permit statistical treatment    only A and 5 in the case of stone. The cumulative curves for the tools of a particular material (Fig. 27.1 for quartz and Fig. 28.1 for stone) show that the differences between the assemblages are not very great. But when the curves for a single assemblage are plotted in one diagram it is clear that typology is strongly linked to raw material (Fig. 24.3).

Among the quartz tools denticulates and notches are much more common than for flint, and side-scrapers are much rarer. End-scrapers are also extremely numerous, but their purpose is uncertain. Were they used conventionally, or was the retouched end intended to serve as a finger-rest while the unretouched lateral margin served as the working edge? Bifaces are both rare and atypical, and Levallois technique is virtually absent.

Stone falls midway between quartz and flint in respect of the relative importance of side-scrapers and denticulates/notches (end-scrapers are comparatively scarce). Levallois technique is used not infrequently and unmodified Levallois flakes are in fact relatively more common in stone than in flint. Also the very high ratio of flakes to cores in layer A (56) suggests, in view of the nature of the pieces themselves, that some of the flakes were manufactured elsewhere. There are many fine bifaces; some fine flake-cleavers are of interest as the type does not exist in flint at the site. One of the exotic stone varieties, *grès lustré* (a fine grained quartzite), was evidently regarded as a good substitute for flint and was particularly favoured for elaborate though small side-scrapers. In layer 5 *grès lustré* was not employed, but advantage was taken of the cleavage of a locally obtained siltstone to manufacture side-scrapers with a rather flat step retouch which echoes the comparative abundance of Quina flaking on the flint tools from the same series.

Both the general character of the artefacts and the preference shown for certain materials in the manufacture of different types of tool show that the La Cotte knappers had a clear appreciation of the flaking properties of the various rocks. Not the least important of the lessons to be learnt from studying these series is that the typology and technology of an assemblage may be greatly influenced by the available rocks, and that assessment of the significance of variation between assem-

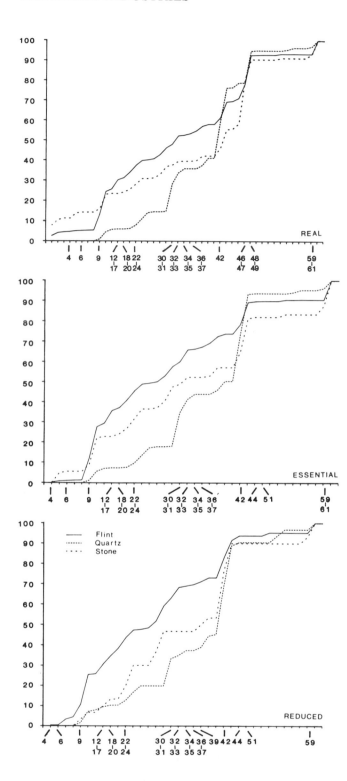

**Fig. 24.3** Cumulative curves for flint, stone and quartz tools from layer A.

blages of different composition is therefore unlikely to be straightforward. In particular, it is worth noting that most of the types of tool present at La Cotte could in theory have been made using any of the raw materials present; that marked preferences are apparent must be a matter of the ease with which a desired result could be obtained and the durability of the resulting working edge.

## 24.4 MULTIVARIATE ANALYSIS OF THE TYPOLOGICAL AND TECHNOLOGICAL DATA (FLINT ARTEFACTS ONLY)

### 24.4.1 Analytical Methods

It has not been difficult to identify a number of obviously important changes in the La Cotte assemblages with respect to time, as well as some clear correlations between different aspects of the industries. But the very large number of strongly varying attributes means that systematic evaluation of all possible relationships, and their resolution into a simpler structure, cannot be performed manually. The very extensive information available for the site can only be satisfactorily dealt with by multivariate data reduction techniques. The procedures used here are very well known (see for instance Doran and Hodson 1975): principal components analysis (PCA) followed by varimax rotation of the significant components in order to extract factors.

The principal component and factor loadings and the corresponding assemblage scores are given in mf 24.1–12.

### 24.4.2 The Data

The variables employed in the multivariate investigation of the La Cotte flint series were drawn from the following.

1. The expanded set of Bordes's typological groups and indices: ILty, IR, IC, IAu, IAt, IB, I, II, III, IV, 5, 38, I rc, 6-7, Rsingle, Rdouble, Rtransv, Rinv, 42, 42-43, Ident, Iirr. 'Reduced' values have been used except in the case of ILty, 5, 38 and Iirr (for the meaning of the abbreviations see chapter 23 and the typological tables in chapter 26).
2. The technological indices: IL, Ilam, IF, IFs and IQ.
3. Mean weights:
   WTTOOL      Reduced tools
   WTCORE     Cores
   WTWASTE   Waste.
4. Technological group percentages:
   PERTOOL    'Real' tools/whole assemblage
   PERCORE    Cores/whole assemblage.
5. Details of specific tool classes:
   CONVSCR    Side-scrapers with convex edges
   CLDENT      Denticulates made with multiple Clactonian notches
   CLNOTCH    Notched pieces with Clactonian notches.
6. Percentage of tools exhibiting special treatment (section 26.5):
   INVTRUNC Inverse truncations
   LSFREM      LSF removals
   DORSREM    Dorsal scaling.
7. Other technical data:
   FLCORRAT Ratio flakes/cores (irrespective of retouch)
   PSLEV         Pseudo-Levallois points (retouched or

LEV          Levallois pieces (retouched or unretouched)
LSF          Sharpening flakes (subsequently retouched or not).
8. Technical data based on waste and cores only (see chapter 25):
     FLLENG     Mean length of waste flakes
     FLTBRAT    Mean thickness/breadth ratio of flakes
     FLANG      Mean flaking angle
     CTFLAKE    % of flakes with not less than 50% cortex
     CTCORE     % of cores with cortex
     DISC        % of disc cores/total cores
     CORVOL     Mean (containing) volume of cores
     CORSECT    Mean cross-section index for cores.

The results of three analyses are described here. They are based on the following combinations of variables from the above list:

Dataset A: groups (1) and (2), i.e. Bordes indices only. Because of the way that the list is constructed the weighting is in favour of typology over technology.

Dataset B: groups (2)–(4), (6)–(8) and ILty, but excluding IQ; i.e. technology only.

Dataset C: all of the variables listed.

### 24.4.3 Extraction of Factors

**Dataset A** (Bordes indices). Though five principal components have eigenvalues exceeding 1, only three exhibit strong loadings on the input variables, between them accounting for 82.4% of the total variance. Varimax rotation gives as factors:

1. [Notches, denticulates etc.; knives with retouched or cortical backs; irregularly retouched pieces; pseudo-Levallois points] against [side-scrapers, especially single, double, or with inverse or alternate retouch];
2. [Bifaces; 'Upper Palaeolithic' tools, especially burins; blades; facetted butts];
3. [Mousterian points; Levallois technique: Quina retouch].

The first factor corresponds in fact to the classic Middle Palaeolithic opposition of denticulates and notches to 'Mousterian' tools. The second to some degree records 'Acheulian-ness' as well as technological features, whilst the third represents a more unusual combination of characters. We may note, in passing, that factoring *on typological indices only* yields very similar results.

**Dataset B** (technology only). Again, only the first three of the five 'significant' principal components are of particular interest; together they are equivalent to 86.6% of the original information in the data. The first three of the rotated factors are

1. [(for cores and flakes) size, relative thickness and presence of cortex] against [Levallois blanks; disc cores; ratio of flakes to cores];

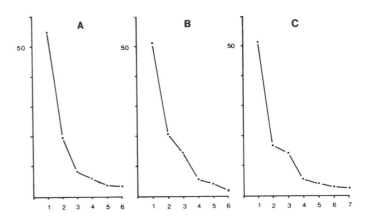

**Fig. 24.4** The changes in percentage variance explained during extraction of principal components for datasets A, B and C.

2. [resharpening and elaboration of tools; blades; facetted butts; use of flakes as cores];
3. [Levallois technique; Quina retouch; percentage of tools in the assemblage].

Thus the first factor is a measure of skill and efficiency in primary flaking, and the second largely of optimisation of the use made of flakes and tools. The third is less simply interpreted and we shall return to it in due course.

**Dataset C** (all variables). Seven principal components have eigenvalues exceeding 1; however, it is clear from Fig. 24.4 that only the first three are of any importance. They explain 82.0% of the total variance in the input matrix. This percentage may seem surprisingly high, given the large number of variables, but it should be remembered that this data matrix, like the others, is singular (that is, it possesses fewer rows than columns); the maximum number of components that can be extracted is therefore limited by the number of assemblages. It will be seen from the labelling of the axes in Figs. 24.5–24.7 that the factors are in effect constructed by combining those from the two analyses already described: the variables which are most highly correlated with the first factor are those making up the first factors of datasets A and B, and so on.

### 24.4.4 Distribution of Assemblages in Factor Space

The scattergrams of the first two principal components for the three datasets show considerable structural uniformity; the assemblages are ordered in perfect or near-perfect chronological sequence in the form of a letter U. In each case it is the first component that is the more obviously time-related, while the scores for the second fall to a minimum in layers F–D, and then rise again.

A somewhat similar (though less consistent) pattern is produced by the factor scores for dataset A; in fact what little rotation has occurred serves chiefly to emphasise

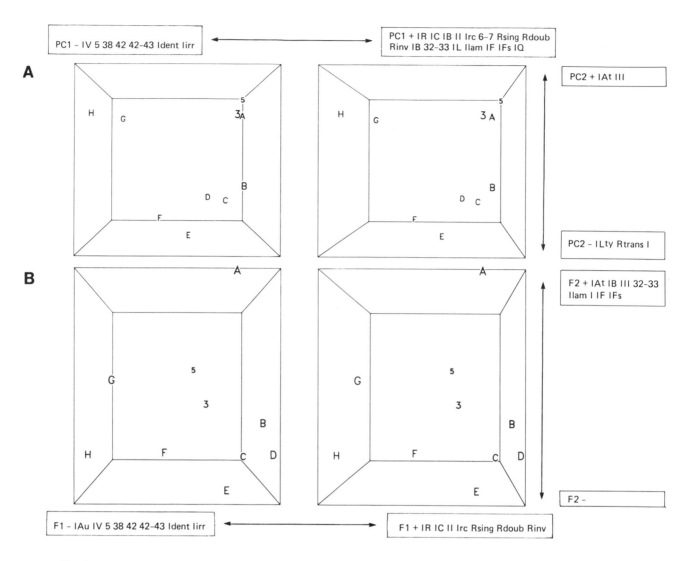

**Fig. 24.5** Dataset A (Bordes indices): stereoscopic pairs of scattergrams of (A) principal component and (B) factor scores.

the unique position of layer 5 on the third factor. The factor scores for dataset B (technology) show that sophistication of primary flaking technique is at a maximum in layer C and that layer 5 is more like the earliest series for size and for low incidence of resharpening (and associated features). But 5 is once again isolated on the third factor because of its combination of high IL, IQ and overall percentage of tools. A very similar pattern emerges from the factors for dataset C, based on the totality of the data. If 5 is excluded, the variation in the scores on the third factor is so restricted that we may visualise all of the other Saalian assemblages in terms of the plane defined by factors 1 and 2.

The distributions are such that formal clustering tech-

niques are not very productive because of chaining. The scattergrams of scores (Figs. 24.5–24.7) show that the process of 'Mousterianisation' advances from H to C. Nevertheless the position of H and G on the PCA scattergrams for dataset A, some uncertainty over the definition and significance of the G/F boundary (chapter 26), and the Rinv peak in layer D imply that simple continuity cannot be safely assumed. Roughly coincident with C there is a clear change of direction, in that the earlier tendency shows signs of reversal thereafter, while the second factor becomes the axis along which development takes place. And with 5 we see yet another directional alteration. The significance of these events is considered in the next section.

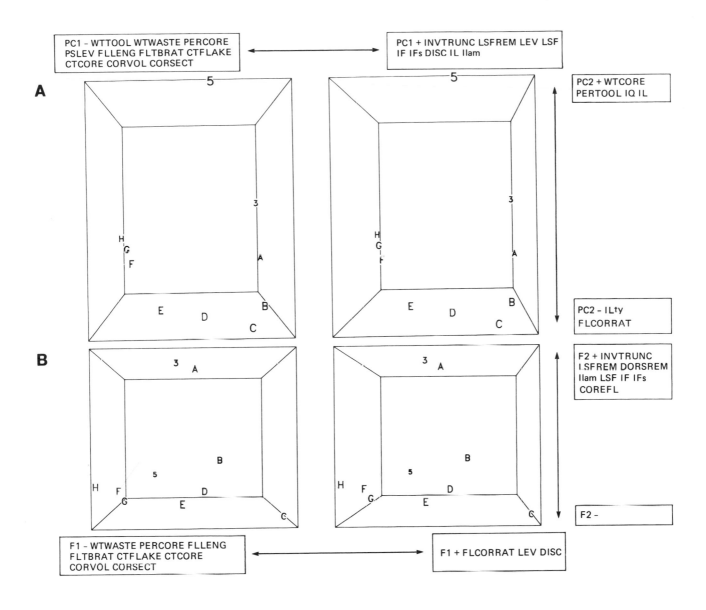

**Fig. 24.6** Dataset B (technology): stereoscopic pairs of scattergrams of (A) principal component and (B) factor scores.

## 24.5 THE LINK WITH RAW MATERIAL

We have seen that the technological and typological data exhibit remarkably similar structure, and that at certain nodal points in the sequence important behavioural shifts occur. Can we explain these? Technology lends itself better than does typology to investigation of possible causal factors, because the reasons for variation are more likely to be identifiable. If change is imposed by the action of some constraint, then the influence of raw material is most likely to be responsible. If the material supply is stable, some other cause must be sought.

The fall in the percentage of flint at La Cotte was mentioned in section 24.2. Do the industrial developments represent an appropriate response to this? Under such circumstances we may envisage varied reactions: more efficient use of the material through more economical primary flaking, longer tool life and a greater degree of compromise in the choice of blanks, or a shift (partial or complete) of artefact manufacture to another more favourable location, so avoiding the need to transport over long distances the material which will eventually be wasted as knapping debris. Taking these as our starting point we may predict the character of the assemblages that would result.

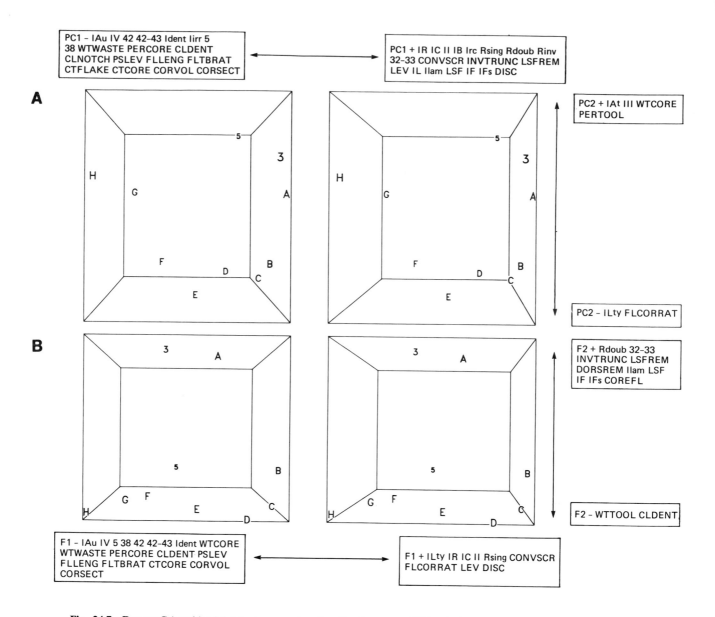

PC1 – IAu IV 42 42–43 Ident Iirr 5 38 WTWASTE PERCORE CLDENT CLNOTCH PSLEV FLLENG FLTBRAT CTFLAKE CTCORE CORVOL CORSECT

PC1 + IR IC II IB Irc Rsing Rdoub Rinv 32–33 CONVSCR INVTRUNC LSFREM LEV IL Ilam LSF IF IFs DISC

PC2 + IAt III WTCORE PERTOOL

PC2 – ILty FLCORRAT

F2 + Rdoub 32–33 INVTRUNC LSFREM DORSREM Ilam LSF IF IFs COREFL

F2 – WTTOOL CLDENT

F1 – IAu IV 5 38 42 42–43 Ident WTCORE WTWASTE PERCORE CLDENT PSLEV FLLENG FLTBRAT CTCORE CORVOL CORSECT

F1 + ILty IR IC II Rsing CONVSCR FLCORRAT LEV DISC

**Fig. 24.7** Dataset C (combined data): stereoscopic pairs of scattergrams of (A) principal component and (B) factor scores.

1. Improved efficiency of primary technique:
   more 'advanced' flaking methods e.g. disc cores;
   a higher flake to core ratio;
   (relatively) thinner flakes.

2. Longer tool life — a 'strategy of desperation':
   greater acceptability of small tools;
   tool rejuvenation;
   more 'multiple' tools;
   smaller cores (and waste);
   re-use of tools and flakes as cores.

3. Importation:
   debris from primary flaking is scarce;
   numerous finished tools;
   waste and cores inconsistent with tool technology;
   tools larger in size than under (2) above.

We may further postulate that once primary technique had been optimised (and this need not require flint shortage as a stimulus) increased tool life would gain in importance (though a less appealing option in view of the need to accept probably less suitable tools for some tasks). Importation represents a very much more significant change, however, as it implies a different perception of the landscape and its resources. In a situation of gradual deterioration the three tactics may therefore be expected to have been exercised in sequence.

Dataset B provided three factors. The first represents primary flaking technique, and the second increased tool life. The third, with the IL and the tool percentage, corresponds broadly to (3) in the above list; in layer 5 the high IL value is inconsistent with the waste and cores (moreover the tools are much larger than in the preceding layers). Scores for the three factors are shown in

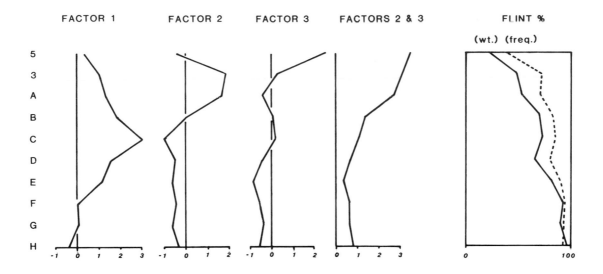

**Fig. 24.8** Changes in the factor scores for technical data (dataset B), compared with the percentage of flint in the assemblage. The combined score for factors 2 and 3 is the Euclidean distance from the point representing the assemblage to a transformed origin (defined by the minima for the two factors), i.e.

$$[(x_2 - \min x_2)^2]^{0.5}$$

chronological order in Fig. 24.8. They yield a succession of maxima following the sequence predicted.

An additional test of the correctness of the model may be made. The decline in factor 1 after layer B is the result of the large quantities of small debris produced by tool rejuvenation and the use of flakes as cores, rather than of lessened efficiency in primary technique, so the history of this factor has to be treated with some caution. But since factors 2 and 3 are rather more exclusive they can be evaluated together against the availability of flint by calculating the distance of each assemblage from the joint minimum of the two factors. The Pearson correlation between this measure and the percentage of flint (by weight) is -.917, which is very highly significant ($p < 0.001$).

We may enquire whether it is really the availability of flint that is the prime mover in this technological development or whether the passage of time itself plays a part; indeed the Spearman coefficient for the depositional order against our combined 2nd and 3rd factors is 0.830. But time can only produce such a result in certain ways:

1. through haphazard change ('random walk');
2. through man's search for technological 'progress' (more output for less effort), the developments taking place within a single tradition or as a result of imported ideas;
3. through the replacement of one human group by another;
4. through the effects of another agency such as climate.

Of these, (1) is clearly insufficient to explain the La Cotte case, and (2), though perhaps acceptable for the early development on factor 1, is inappropriate in the light of the disadvantages probably accompanying factor 2 (the LSF technique was difficult and not always successful, while smaller tools cannot always have been convenient). Equally, the continuing occasional employment in layer 5 of the highly specialised LSF technique provides a case against (3). For at least the upper part of the sequence, therefore, we are left with raw material availability as controlled by climate and sea-level (chapter 22), which themselves show a downward trend during this time. Though the thick points in layer 5 are a new element, perhaps indicating another specialised response to environmental change (see section 26.5.1), there is nothing in the typological evidence to contradict this explanation.

The mechanisms at work in the lower layers, as far as D, are less clear. The development of efficient flaking techniques is widespread at about this date and need not reflect a direct response to changing circumstances at La Cotte (though advantageous). Moreover the typological contrasts are too great for population replacement to be excluded.

229

## 24.6 CONCLUSIONS

The technological changes at La Cotte are dominated by the effects of declining flint supplies, though the use of disc cores in the primary flaking process may have arisen quite independently of this. There is a very great typological shift between layers H and E, with domination passing from notches and denticulates to side-scrapers, and greater stability thereafter. Some of the later variation is itself probably a consequence of the technological modifications. Handaxes appear surprisingly late in the sequence, the first really typical examples being from layer C. Stone and quartz are too little used in the lower layers for it to be possible to consider comparable variation through time, and indeed the three richest quartz series, from the middle and later part of the sequence, are remarkably alike.

Although the assemblages seriate well in three dimensions, it is possible to suggest a number of 'break points'.

After G, with the development of a 'Middle Palaeolithic' character including flaking by the disc core method.

Between D and C, with the appearance of classic Acheulian bifacial tools and the first signs of elaborate resharpening.

Between 3 and 5, with a move to importing finished tools and the use of specialised points.

In addition to the above we may note that there are important technological and typological differences between H and G (regarding the relative importance of denticulates and notches, and the incidence of pseudo-Levallois points), though these are not picked out clearly in the multivariate studies.

# THE FLINT DEBITAGE AND CORES

J. C. C. Hutcheson and P. Callow

## 25.1 INTRODUCTION

Classification of the debitage according to butt type was carried out as part of the initial post-excavation treatment of the La Cotte material. The identification of Levallois flakes and other special categories was performed by Callow as part of the expansion of the tool type-list when the Bordes scheme was adopted; his results, given in chapter 26, are referred to here where appropriate. The debitage was felt to merit a detailed investigation of its qualitative and quantitative characteristics, and was the subject of a B.A. dissertation by Hutcheson (1982). This has been expanded into the more elaborate analysis described here.

Three objectives were defined at the beginning of the study:

Description of the waste flakes and cores from the principal assemblages.

Interpretation in terms of the technology used.

Exploration of the potential of multivariate techniques, applied to attribute data, to identify the most effective variables for describing and summarising variation at both the inter- and intra-assemblage level.

Several points should be stressed at the outset. Firstly, although it might be preferable in a technological study to sample *all* flakes, whether retouched or not, this was not done — partly because of the limited time available, but also because retouch often renders pieces unsuitable for many of the observations. However, *unretouched* Levallois flakes, blades and points, pseudo-Levallois points and *couteaux à dos naturel* were included in the study (the question of whether they were used as tools may equally be applied to other unretouched pieces, of course). The term 'debitage' as employed here should therefore be taken to mean the unretouched flakes, rather than *all* the products of the flaking process or the misleadingly named 'waste' alone. Secondly, the cores have been investigated much less exhaustively, chiefly because the small sample sizes (particularly after breakdown into subtypes) preclude elaborate statistical treatment. Thirdly, the non-flint artefacts were not included in this study, those of 'stone' because of their comparative rarity, those of quartz because of the material's peculiar fracture characteristics. Finally, the

limited extent to which the interpretation can be carried forward on the basis of these data alone is fully recognised. The significance of the results in the light of raw material availability and other potentially important questions is discussed in chapter 22.

The investigation falls into two principal parts:

**Measurement and description.** An essential requirement for a clear understanding of variation in lithic assemblages is "the accurate definition of the actual patterns of similarities and differences among artefact assemblages" (Isaac 1977, 97). In recent years it has become apparent that these patterns can only be delineated satisfactorily by the application of quantitative methods. Accordingly, explicitly defined attributes were recorded for flakes and cores, and summarised by the use of routine descriptive statistics. The results are given in sections 25.2 and 25.3.

**Multivariate analysis.** Bradley (1977, 2), in attempting the analysis of technological processes by the use of experimental replication, regarded the isolation of "those areas of variability that are influenced by basic primary technology" as a prerequisite to an understanding of the factors controlling inter- and intra-assemblage variation. He was able to choose archaeological material particularly appropriate to the questions with which he was concerned, the number of technological options exercised in its manufacture being relatively limited. The far wider range of techniques in use at La Cotte would present formidable problems in the application of this approach, though it should be borne in mind as a project for the future. But we have attempted to profit from the results of Bradley and others when selecting the attributes to be studied, and when interpreting analyses designed to highlight the internal structuring of assemblages or the contrasts between them.

## 25.2 THE UNRETOUCHED FLAKES

### 25.2.1 Definition of Study Units

The unbroken flint debitage from layers H–A and 5, excluding pieces classed *during initial processing* as less than 20 mm long (see section 21.2), was sampled to give something over 100 examples of each butt type for each assemblage, though a lower figure had sometimes to be accepted. Pieces with indeterminable butts were excluded

**Table 25.1** Technological Indices[a]

| Index | Layer | | | | | | | | |
|-------|-------|-----|-----|-----|-----|-----|-----|-----|-----|
| | H | G | F | E | D | C | B | A | 5 |
| IL | 1.58 | 2.36 | 3.02 | 2.43 | 3.08 | 4.31 | 5.15 | 4.07 | 8.15 |
| Ilam (waste only) | 4.77 | 4.66 | 4.13 | 4.28 | 5.92 | 4.10 | 6.67 | 6.66 | 6.11 |
| Pseudo-Levallois points | 3.79 | 1.45 | 1.54 | 0.69 | 0.71 | 0.56 | 0.45 | 0.46 | 0.54 |
| *Couteaux à dos naturel* | 1.89 | 2.39 | 2.00 | 1.23 | 1.78 | 1.50 | 1.62 | 1.81 | 2.86 |
| LSF/TSF | 1.58 | 1.45 | 1.63 | 2.01 | 2.03 | 2.88 | 5.22 | 10.02 | 4.89 |
| For all flakes except 'reduced' tools | | | | | | | | | |
| IF | 33.81 | 35.99 | 38.09 | 42.86 | 43.43 | 47.20 | 47.60 | 55.38 | 54.86 |
| IFs | 16.44 | 21.96 | 24.75 | 22.93 | 27.33 | 31.66 | 37.05 | 42.88 | 42.36 |
| IFss | 10.97 | 16.14 | 17.17 | 15.69 | 21.21 | 25.37 | 28.08 | 36.05 | 31.60 |
| For all flakes | | | | | | | | | |
| IF | 33.56 | 37.33 | 38.86 | 37.80 | 43.22 | 47.79 | 48.75 | 55.18 | 53.93 |
| IFs | 17.06 | 22.07 | 24.91 | 23.12 | 27.57 | 32.60 | 37.75 | 43.00 | 43.37 |
| IFss | 11.38 | 15.98 | 16.21 | 15.55 | 21.20 | 25.38 | 28.68 | 35.73 | 33.26 |

[a]See chapter 23.

from analysis; they do not form a homogeneous group, nor could the full set of butt measurements be obtained for them. The selection procedure was based on computer-generated random numbers, as described in chapter 21. It yielded a total of 4522 flakes for study (the sampling percentage for each butt type varied from layer to layer, as indicated in mf 25.1). A selection of pieces is illustrated in Fig. 25.1.

### 25.2.2 The Attribute List

It was expected that some attributes would eventually prove redundant, but this was a small price to pay for greater comprehensiveness. There is fairly wide agreement as to the importance of such observations as length, breadth and scar count; moreover, these can be defined quite rigorously (though the preferred definitions may vary from writer to writer) and yield fairly repeatable values. However, certain other attributes, notably those based on angles, give less consistent results: this is partly because of difficulties in the mechanics of measurement, and partly because of the difficulty of defining precisely what is to be measured. In the case of edge angles, the modal value of measurements taken at intervals along the piece gives an approximate estimate. Flaking and cone angles, however, are hard to measure when the butt is small, and by no means easy to define in such a way that 'anomalous' cases do not occur too often. Measurements were taken to the nearest mm or degree. The instrument used is specified in the description.

1. **Butt type** (mf 25.2). This had been recorded during initial cataloguing of the material. Six categories were defined.

   *Cortical.* The whole of the butt is formed of cortex or old patina.
   *Plain.* Formed by a single flake scar whose point of impact is almost invariably missing.
   *Dihedral.* Formed by the intersection of two flake scars, almost invariably without preservation of the point of impact on either.

   *Polyhedral.* Formed by three or more scars resulting from flakes struck from the direction of the dorsal surface (occasionally one of these 'facets' may be cortical); however, the impact points of these scars are missing, having been removed by primary flaking when the piece was still part of the core, i.e. they relate to the preparation for an earlier removal.
   *Facetted.* Formed by a number of flake scars (often small) most of which have been struck from the dorsal surface by simple percussion or abrasion with a hammerstone; i.e. from the current margin of the core. Unlike those on polyhedral butts, these scars relate directly to the preparation of the core for the removal of the flake concerned.
   *Punctiform.* The breadth and thickness of the butt are 2 mm or less.

2. **Length** (mf 25.3). The length of the sides of a containing rectangle parallel to the direction of the blow that produced the flake (not necessarily the largest dimension). Measuring board.
3. **Breadth** (mf 25.3). The length of the other sides of the containing rectangle, i.e. perpendicular to the blow. Measuring board.
4. **Thickness** (mf 25.4). The maximum thickness of the flake measured at right angles to the bulbar axis. Callipers.
5. **Butt breadth** (mf 25.5). Maximum distance between left and right extremities of the butt. Callipers.
6. **Butt thickness** (mf 25.6). Maximum distance across the butt between the bulbar and dorsal surfaces. Callipers.
7. **Butt projection** (mf 25.7). Maximum projection of the butt beyond a chord drawn between the ends of the butt on the bulbar surface. Graph paper.
8. **Left edge angle** (mf 25.8). Angle of intersection between the dorsal and ventral surfaces along the left edge of the flake, when the butt is orientated towards the observer with the bulbar surface downward. A mean of three observations. Contact goniometer.
9. **Right edge angle** (mf 25.9). As for the left edge angle.
10. **Flaking angle** (mf 25.10). Angle between the butt and the plane of flattening of the flake, measured in the direction of the blow. Contact goniometer.
11. **Cone angle** (mf 25.11). Angle between the butt and the cone of percussion. Contact goniometer.
12. **Flake scar count** . Number of primary removal scars whose length is at least one fifth of the longest dimension of the piece.
13. **Flake scar orientation** (mf 25.12). Patterns were defined in relation to the longitudinal axis of the flake: parallel unidirectional, perpendicular unidirectional, irregular unidirectional, convergent unidirectional, parallel bidirectional, perpendicular bidirectional, irregular bidirectional, centripetal or radial, divergent unidirectional, no pattern observable.

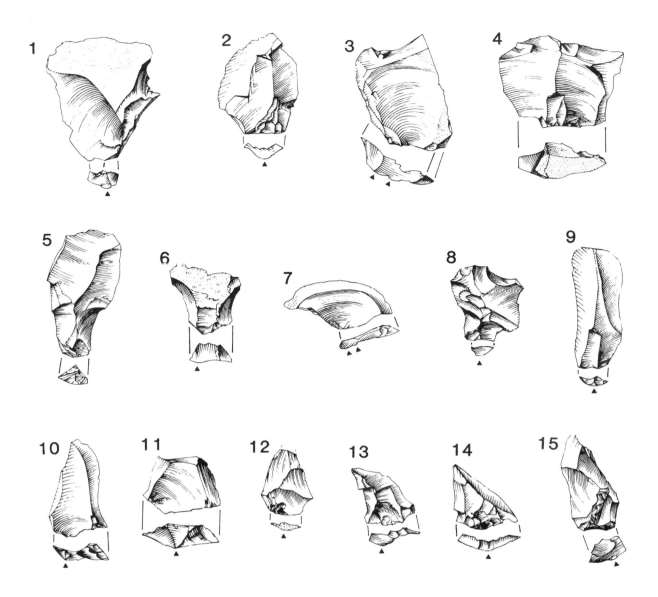

**Fig. 25.1** Examples of waste flakes. They are described in terms of the following attributes: butt, planform shape, preparation pattern, and scar count. 1 Polyhedral, divergent, irreg. unid., 4 (layer E); 2 Plain, medially expanded, parallel unid., 3 (layer A); 3 Polyhedral, parallel, pattern not discernible, 2 (layer E); 4 Cortical, divergent, parallel bidirectional, 8 (layer F); 5 Facetted, elliptic, parallel unid., 3 (layer 5); 6 Plain, hour-glass, parallel unid., 3 (layer A); 7 Facetted, convergent rounded, parallel unid., 2 (layer G); 8 Dihedral, elliptic, centripetal, 6 (layer E); 9 Polyhedral, parallel, parallel unid., 4 (layer E); 10 Facetted, triang. irreg., parallel unid., 3 (layer E); 11 Facetted, convergent straight, irreg. bidirectional, 4 (layer E); 12 Cortical, medially expanded, prep. unid., 4 (layer A); 13 Facetted, triang. irreg., convergent unid., 5 (layer E); 14 Dihedral, triang. reg., irreg. bidirectional, 5 (layer A); 15 Polyhedral, medially expanded, perp. bidirectional, 5 (layer A). Of these, 4, 6, 7 and 11 are transverse. The cortex on 1 and 5 was estimated at 40 and 45% respectively.

**Table 25.2** Debitage: Flake and Butt Measurements[a]

| Layer | Length mean | Length s.d. | Flake Breadth mean | Flake Breadth s.d. | Thickness mean | Thickness s.d. | Breadth mean | Breadth s.d. | Butt Thickness mean | Butt Thickness s.d. | Projection mean | Projection s.d. | Total |
|-------|------|------|------|------|------|------|------|------|------|------|------|------|------|
| 5 | 27.3 | 10.3 | 25.4 | 8.4 | 6.45 | 3.40 | 12.9 | 8.9 | 3.98 | 2.77 | 1.79 | 2.92 | 229 |
| A | 24.6 | 8.7 | 22.0 | 6.4 | 5.62 | 2.50 | 12.4 | 7.2 | 3.78 | 2.34 | 1.23 | 1.79 | 7001 |
| B | 26.1 | 9.9 | 22.8 | 8.0 | 5.97 | 3.12 | 13.2 | 8.3 | 3.96 | 2.58 | 1.88 | 2.52 | 1274 |
| C | 24.5 | 9.7 | 24.4 | 8.8 | 5.82 | 2.99 | 13.3 | 7.8 | 4.04 | 2.72 | 1.70 | 2.18 | 2266 |
| D | 25.9 | 9.2 | 24.3 | 8.6 | 6.41 | 3.13 | 13.9 | 9.1 | 4.32 | 2.52 | 1.88 | 2.36 | 1849 |
| E | 25.8 | 9.8 | 24.8 | 8.3 | 7.34 | 3.64 | 14.6 | 8.9 | 4.99 | 3.12 | 1.60 | 2.45 | 1777 |
| F | 29.9 | 11.6 | 27.4 | 8.7 | 8.41 | 3.71 | 15.8 | 8.2 | 5.68 | 3.22 | 1.55 | 2.60 | 1984 |
| G | 29.3 | 11.6 | 27.4 | 9.1 | 7.98 | 3.98 | 16.1 | 9.2 | 5.47 | 3.43 | 2.36 | 2.72 | 1503 |
| H | 27.7 | 11.4 | 26.2 | 8.4 | 8.02 | 3.80 | 16.3 | 9.7 | 5.81 | 3.47 | 2.63 | 3.26 | 692 |

14. **Cortex percentage**. Percentage of cortex preserved on the dorsal surface of the flake, relative to the total surface area. Estimated by eye, after practice using a set of digitised drawings of handaxes from Olduvai Gorge, Tanzania (Callow, in Leakey in press) for which this information had been computed.

15. **Cortex position** (mf 25.13). The dorsal surface was broken down into six zones and the position of the cortex was recorded on a presence/absence basis: left distal, right distal, left mesial, right mesial, left proximal, right proximal.

16. **Terminal release** (mf 25.14). Divided into five categories: feathered, hinge, step concave, step vertical, plunging.

17. **Planform shape** (mf 25.15). This is based upon the scheme used by Isaac (1977) which depends upon the position of the maximum width of the flake in relation to its longitudinal axis.

*Triangular irregular*. Greatest width at the proximal end (butt end).
*Triangular regular*. Straight sided version of the above.
*Medially expanded*. Greatest width in the mesial (middle third) area of the flake.
*Divergent*. Greatest width in the distal segment of the flake.
*Parallel*. Subequal widths in all three segments of the flake.
*Convergent straight*. Greatest width at the proximal end, with a straight distal edge set roughly at right angles to the longitudinal axis, thus forming a trapezoid.
*Convergent round*. As previous, but with a rounded tip.
*Hour-glass*. Both sides concave (i.e. wide at the proximal end, thinning medially then expanding at the distal end).
*Elliptic*. Greatest width at the distal end (which is pronouncedly round).

18. **Planform orientation**. Recorded as either normal or transverse i.e. end-struck or side-struck.

19. **Planform symmetry**. Observed with the bulbar surface uppermost, the butt resting on a base line, with the flake orientated along the axis of percussion: symmetrical or undiagnostic, left skew, right skew.

In addition to the attributes listed above, several derivative variables (ratios) were calculated using the metrical data: flake breadth/flake length; flake thickness/flake breadth; butt breadth/flake length; butt thickness/butt breadth; butt projection/butt breadth; edge angle difference (relative); mean relative size of primary scars; cross-section index.

The relative edge angle difference was computed by dividing the difference between the angles for the two sides by their sum. To obtain the mean relative size of the primary scars, the percentage area of cortex was deducted from 100, and divided by the scar count to give a measure of the fineness of the dorsal preparation.

The cross-section index, designed to give a more generalised indication of 'thinness' than would a simple thickness/breadth ratio, consists of twice the thickness divided by the sum of the length and breadth.

### 25.2.3 Debitage Characteristics and Time-related Variation

Several of the attributes show very close co-variation and fairly consistent trends, at least until layer A; these are the size attributes length, breadth and thickness. After an initial rise until F, they steadily decrease until the trend is reversed in layer 5. In the case of the butt measurements, however, the decrease in mean values continues in 5. The various facetting indices rise quite steadily through the sequence, with only a very slight downwards turn in layer 5. A similar general trend (from an initially very low level) is apparent in the Levallois index, IL, though here there is a very sharp increase in layer 5. Also (as remarked in chapter 26) the lower levels of the site have a relatively high proportion of *atypical* Levallois pieces.

Interpretation of the size measurements (which exhibit strongly skewed distributions) requires caution. If only the flakes with a minimum length of 40 mm are used, the mean length is remarkably stable (varying between 48.0 and 49.7 mm). The variation in overall mean length reflects shifts in the frequency of these larger pieces as well as in the position of the mode at the lower end of the range (Fig. 25.7).

Other variables showing change through time include: the percentage of blades (defined as having a length/breadth ratio of at least 2) which, while always low, rises from about 4 to about 6; butt projection, which drops sharply after layer G (due partly to a fall in the number of pseudo-Levallois points); scar count, which drops between H and F, and then recovers to reach its highest value in layer 5; cortex (as a percentage of total area), which reaches a maximum in F, falls dramatically until C and then rises again.

It should be stressed that the flaking and cone angle measurements were taken in a manner which yielded

**Table 25.3** Debitage: Flake and Butt Ratios

| Layer | B/L ratio mean | s.d. | T/B ratio mean | s.d. | Cross-section mean | s.d. | Butt T/B mean | s.d. | Total |
|---|---|---|---|---|---|---|---|---|---|
| 5 | 1.01 | 0.39 | 0.258 | 0.113 | 0.242 | 0.096 | 0.345 | 0.184 | 229 |
| A | 0.98 | 0.39 | 0.261 | 0.101 | 0.241 | 0.083 | 0.338 | 0.154 | 7001 |
| B | 0.95 | 0.39 | 0.269 | 0.115 | 0.243 | 0.093 | 0.340 | 0.176 | 1274 |
| C | 1.09 | 0.44 | 0.245 | 0.109 | 0.238 | 0.090 | 0.339 | 0.201 | 2266 |
| D | 1.03 | 0.60 | 0.272 | 0.113 | 0.256 | 0.098 | 0.359 | 0.239 | 1849 |
| E | 1.03 | 0.36 | 0.303 | 0.129 | 0.289 | 0.107 | 0.385 | 0.278 | 1777 |
| F | 1.00 | 0.38 | 0.311 | 0.112 | 0.295 | 0.101 | 0.395 | 0.221 | 1984 |
| G | 1.02 | 0.36 | 0.295 | 0.123 | 0.280 | 0.104 | 0.384 | 0.390 | 1503 |
| H | 1.04 | 0.38 | 0.311 | 0.126 | 0.297 | 0.106 | 0.414 | 0.302 | 692 |

**Table 25.4** Debitage: Edge and Flaking Angles[a]

| Layer | Skewness (left) % | total | Edge angle (left) mean | s.d. | Edge angle (right) mean | s.d. | Edge angle asymmetry mean | s.d. | Flaking angle mean | s.d. | Cone angle mean | s.d. | Total |
|---|---|---|---|---|---|---|---|---|---|---|---|---|---|
| 5 | 52.0 | 57 | 47.5 | 19.4 | 45.6 | 18.4 | -1.85 | 21.33 | 70.5 | 9.3 | 64.4 | 12.4 | 229 |
| A | 52.3 | 2096 | 52.5 | 17.8 | 49.3 | 17.3 | -3.34 | 20.62 | 68.3 | 10.3 | 62.1 | 10.0 | 7001 |
| B | 52.1 | 263 | 49.2 | 16.3 | 46.2 | 16.4 | -3.34 | 19.75 | 68.9 | 10.7 | 64.3 | 10.8 | 1274 |
| C | 57.1 | 295 | 50.0 | 18.4 | 45.7 | 16.0 | -4.16 | 19.82 | 68.4 | 11.2 | 64.7 | 12.9 | 2266 |
| D | 50.2 | 279 | 50.6 | 18.8 | 50.6 | 18.9 | -0.21 | 23.47 | 69.6 | 10.2 | 62.9 | 11.1 | 1849 |
| E | 49.4 | 265 | 54.3 | 18.9 | 50.5 | 17.8 | -3.58 | 20.68 | 70.1 | 9.7 | 62.2 | 11.1 | 1777 |
| F | 43.0 | 319 | 56.7 | 17.5 | 53.3 | 17.6 | -3.27 | 19.84 | 70.1 | 10.8 | 61.3 | 11.6 | 1984 |
| G | 61.7 | 192 | 53.8 | 17.2 | 51.9 | 16.8 | -1.87 | 19.65 | 68.9 | 8.9 | 60.9 | 10.3 | 1503 |
| H | 51.5 | 37 | 53.7 | 17.7 | 51.8 | 19.9 | -2.62 | 21.37 | 67.7 | 10.1 | 63.3 | 12.7 | 692 |

[a]The percentage of left as opposed to right skewed pieces is also given, as being in agreement with the asymmetry of the edge angle measurements (see text).

values typically less than 90°, whereas it is the practice of some workers to record the *complement* of these, which would yield a value of around 120° for a piece with pronounced cone, as against our 60°.

The cone and flaking angles, rather surprisingly, exhibit no clear chronological trends (though the latter is especially low in H, and high in F, E and 5). This is also true, for the most part, of the shape (lending support to Isaac's view that a strict geometrical classification of flakes is of doubtful value); a general predominance of medially expanded, elliptic and divergent forms is discernible, though layer 5 is exceptional in its very high incidence of elliptic pieces. Feathered terminations become more common with the passing of time, and there is also a strong (and predictable) correlation between fluctuations in the number and relative size of scars and the presence and extent of cortex.

An interesting aspect of the edge angle measurements (which themselves show a general reduction from F onwards) is that the left-hand edge is usually less acute than the right. This is consistent with the preference for right-skewed planforms (given that the bulbar surface was upwards when skewness was recorded, and downwards for the angle measurements).

There follows a brief description of the distinctive

**Table 25.5** Debitage: Scar Count and Average Scar Size[a]

| Layer | Scar count mean | s.d. | Scar size (%) mean | s.d. | Total |
|---|---|---|---|---|---|
| 5 | 3.99 | 2.09 | 29.4 | 19.7 | 229 |
| A | 3.41 | 1.58 | 33.2 | 19.4 | 7001 |
| B | 3.38 | 1.55 | 34.7 | 21.3 | 1274 |
| C | 3.46 | 1.64 | 35.4 | 21.8 | 2266 |
| D | 3.17 | 1.55 | 37.4 | 23.2 | 1849 |
| E | 2.99 | 1.53 | 39.8 | 24.4 | 1777 |
| F | 2.92 | 1.63 | 35.9 | 22.2 | 1984 |
| G | 3.17 | 1.73 | 34.6 | 20.4 | 1503 |
| H | 3.78 | 1.93 | 30.1 | 20.1 | 692 |

[a]Expressed as a percentage of the surface area of the flake.

features of the debitage for each assemblage in turn (where general comparisons are drawn, e.g. in the use of the term 'long', these are related to typical values for the whole collection).

**Layer H.** The flakes are of above average size in all three dimensions; butts are large and often projecting. True facetted butts are rare, while dihedral ones are unusually

**Table 25.6**  Debitage: Dorsal Surface Preparation Pattern

| Layer | Unid. para. | Unid. perp. | Unid. irreg. | Unid. converg. | Unid. diverg. | Bid. para. | Bid. perp. | Bid. irreg. | Centrip. radial | Total |
|-------|------|------|------|------|------|------|------|------|------|------|
| 5 | 27.9 | 2.7 | 0.0 | 5.8 | 0.5 | 6.3 | 1.6 | 40.0 | 15.2 | 213 |
| A | 14.7 | 0.3 | 19.4 | 13.8 | 0.0 | 0.8 | 0.6 | 44.0 | 6.4 | 5507 |
| B | 21.1 | 1.0 | 23.5 | 11.5 | 0.6 | 1.6 | 0.0 | 36.3 | 4.2 | 1068 |
| C | 31.6 | 2.1 | 23.1 | 8.1 | 1.8 | 1.1 | 1.9 | 19.9 | 10.4 | 2059 |
| D | 15.6 | 3.4 | 19.7 | 6.3 | 1.2 | 1.4 | 1.2 | 44.0 | 7.1 | 1444 |
| E | 15.5 | 0.8 | 23.3 | 4.6 | 0.6 | 2.7 | 0.9 | 39.8 | 11.8 | 1283 |
| F | 14.7 | 1.6 | 20.2 | 8.4 | 0.1 | 0.9 | 0.4 | 42.9 | 10.8 | 1416 |
| G | 24.4 | 3.3 | 26.0 | 8.7 | 0.3 | 2.1 | 0.6 | 27.8 | 6.9 | 1236 |
| H | 33.6 | 2.7 | 3.8 | 8.5 | 0.5 | 1.8 | 5.1 | 33.8 | 10.1 | 629 |

**Table 25.7**  Debitage and Cores: Incidence and Extent of Cortex

| Layer | Debitage | | | | | | | Cores | |
|-------|------|------|------|------|------|------|------|------|------|
| | Incidence of flakes | | Mean area of cortex (%) for: | | | | | | |
| | with cortex % | with 50% cortex % | all flakes | | cortical flakes only | | | with cortex % | made on flake % |
| | | | mean | s.d. | mean | s.d. | | | |
| 5 | 27.5 | 6.0 | 9.6 | 20.6 | 34.4 | 25.8 | | 63.0 | 3.7 |
| A | 20.0 | 7.2 | 8.2 | 20.9 | 41.1 | 29.0 | | 57.5 | 13.5 |
| B | 17.3 | 6.7 | 6.6 | 17.5 | 38.5 | 23.6 | | 57.1 | 8.2 |
| C | 9.5 | 3.6 | 4.2 | 15.1 | 44.5 | 24.7 | | 52.1 | 6.3 |
| D | 21.2 | 6.0 | 8.3 | 19.1 | 39.1 | 22.8 | | 57.8 | 7.8 |
| E | 21.5 | 7.1 | 8.7 | 21.3 | 40.2 | 28.8 | | 67.0 | 8.9 |
| F | 34.1 | 18.0 | 16.7 | 28.4 | 49.1 | 28.0 | | 65.2 | 8.3 |
| G | 28.9 | 12.7 | 12.8 | 25.0 | 44.2 | 27.7 | | 63.6 | 5.8 |
| H | 27.1 | 10.9 | 11.7 | 24.1 | 43.0 | 27.9 | | 73.7 | 8.8 |

common. The high mean value for butt projection arises from this, and from the presence of extensive polyhedral relict margins on such pieces as pseudo-Levallois points. Levallois flakes are rare, and usually atypical. Edges are steep. Both cortex and stepped terminations are common. The predominant preparation patterns are parallel unidirectional and irregular bidirectional; perpendicular bidirectional flaking (otherwise extremely rare) makes its only noteworthy contribution, at 4.7%.

**Layer G.** Rather similar to H. The preparation scar patterns are most commonly parallel or irregular unidirectional, or irregular bidirectional; as these remain generally the most common until layer 5, the flaking patterns will not be commented on henceforward.

**Layer F.** Similar in many respects to the previous layers, though size attributes are at a maximum and the edges are less acute. Feathered terminations occur more frequently. The incidence of flakes with cortex is extremely high, and a substantially increased proportion have 50% or more cortex on their dorsal surfaces.

**Layer E.** The debitage is strikingly different from that of earlier assemblages, smaller in all respects (including the

butts), with shallower edge angles and a much reduced incidence of cortex. Again many pieces have no discernible preparation pattern. The most common planforms are divergent, medially expanded and elliptic.

**Layer D.** Rather similar in many ways to that of layer E: flake and butt size, planform and preparation pattern. However, the thickness (absolute and relative) is further reduced, as is the incidence of stepped terminations (feathering again increasing). The occurrence of cortex is unchanged, and the incidence of indeterminate preparation pattern remains high (though lower than in the two previous layers).

**Layer C.** Apart from a continuation of the gradual size reduction, the main differences between the debitage of C and D lie in a marked (though temporary) decrease in the frequency of parallel unidirectional flaking and a fall in the percentage of pieces with cortex, to 9.5%, with a mean percentage area of cortex of only 4.2%.

**Layer B.** The debitage shows slightly greater elongation than in other layers (though comparable in overall size, butt size and edge angle to that from layer C). Triangular irregular, divergent and medially expanded shapes

**Table 25.8** Debitage: Coefficients of Correlation (Pearson) between Quantitative Observations

| | | 1 | 2 | 3 | 4 | 5 | 6 | 7 | 8 | 9 | 10 | 11 |
|---|---|---|---|---|---|---|---|---|---|---|---|---|
| 1 | Flake length | | | | | | | | | | | |
| 2 | Flake breadth | **0.50** | | | | | | | | | | |
| 3 | Flake thickness | **0.58** | **0.58** | | | | | | | | | |
| 4 | Butt breadth | **0.17** | **0.60** | **0.42** | | | | | | | | |
| 5 | Butt thickness | **0.33** | **0.48** | **0.70** | **0.66** | | | | | | | |
| 6 | Butt projection | 0.08 | **0.34** | **0.31** | **0.76** | **0.55** | | | | | | |
| 7 | Left edge angle | **0.15** | **0.11** | **0.35** | 0.06 | **0.21** | 0.05 | | | | | |
| 8 | Right edge angle | **0.14** | **0.11** | **0.35** | 0.05 | **0.19** | 0.03 | **0.20** | | | | |
| 9 | Flaking angle | 0.04 | 0.03 | -0.02 | 0.00 | -0.01 | -0.06 | -0.02 | -0.01 | | | |
| 10 | Cone angle | **-0.11** | **-0.11** | **-0.15** | -0.06 | -0.06 | -0.06 | **-0.13** | -0.10 | **0.58** | | |
| 11 | Scar count | **0.13** | 0.03 | **0.11** | -0.06 | -0.03 | -0.03 | 0.09 | **0.12** | **-0.11** | -0.04 | |
| 12 | Area of cortex % | **0.21** | **0.15** | **0.25** | -0.03 | 0.06 | -0.07 | **0.11** | **0.13** | 0.00 | -0.05 | **-0.35** |

[a]In this and subsequent tables, **bold** type indicates correlation significant at the 5% level.

are the most common planforms. Cortex continues to be scarce.

**Layer A.** The flakes are very small and thin; the butt measurements also fall to minimum values. The percentage of cortical pieces, at 20%, is slightly higher than in layer B, but comparable to the same statistic for D and E. Elliptic planforms show a recovery after low values in the two previous layers, matching a decrease in parallel-sided and triangular irregular planforms. However, easily the most common shape is divergent, at 38.5%, followed by medially expanded.

**Layer 5.** There is a clear increase in the size attributes, returning to the levels of the lower layers in the sequence. On the other hand, the butt measurements, though larger on average than in A, are close to those for B. Thus in relative terms the butts are rather small. Levallois flakes (generally retouched as well-defined tools) are more common than in any of the other Saalian assemblages. The edge angles are more acute than in any other layer, and feathered terminations reach a maximum. A particularly important observation, in view of the interesting problems of raw material availability posed by this site, is that the percentage of flakes with cortex returns to the high values found in the early assemblages, but not many of them have as much as 50% of the surface covered by cortex.

### 25.2.4 Multivariate Analysis

The further investigation of the debitage falls into two parts, based on the whole (sampled) collection and on the individual assemblages respectively. Multivariate techniques were employed in order to evaluate the impressions gained during examination of the artefacts, and of the statistics derived from them; also to examine the relationship between variables.

As a further, albeit rough-and-ready, test of the effectiveness of the attributes chosen, discriminant function analysis was performed using the quantitative data to separate the groups of flakes defined by butt type. Using only the butt measurements, reassignment of the flakes according to their scores on the discriminant functions gave agreement of the order of 50–60% with the original classification when the computation was carried out for each layer separately. The percentage match taking all layers together was 43% (the fall is to be expected because of inter-assemblage variation). An improvement of about 5% resulted from inclusion of all continuous variables (including the percentage area of cortex).

Even without the butt data, matches were around 40% for the individual layers. This is well above the value to be expected (for this number of categories) had the distributions been similar irrespective of butt type; therefore there are areas of the multidimensional space which are given over to specific types, though overlap is considerable. The 'best' result, on the other hand, shows that the attribute list, if limited in size, is likely to be of considerable value in technological analysis.

### THE WHOLE COLLECTION

The Pearson (product-moment) correlation matrix was computed for the quantitative variables. The individual cases were weighted as described above, so reconstituting the entire debitage. It must be stressed that this has the disadvantage of overloading the sample with material from the very rich layer A, and the coefficient must be interpreted with this in mind (Table 25.8). Some experimenting was also undertaken with differential weighting of the layers to give constant sample sizes for each; however, in view of the uncertainties regarding the lapse of time between assemblages, and indeed the relationship between geological and archaeological units, this was not pursued.

Several points are worth noting. Firstly, most of the coefficients attain a very high degree of statistical significance (with a less than one in a thousand likelihood

237

**Table 25.9** Debitage: Loadings of the Variables in Table 25.8 on the Principal Components of the Correlation Matrix

| Variable | Factor | | | | |
| | 1 | 2 | 3 | 4 | 5 |
|---|---|---|---|---|---|
| Flake length | 0.48 | −0.28 | 0.35 | 0.06 | −0.31 |
| Flake breadth | **0.70** | −0.02 | 0.09 | −0.00 | −0.34 |
| Flake thickness | **0.86** | −0.32 | 0.32 | 0.04 | 0.17 |
| Butt breadth | **0.80** | 0.46 | −0.35 | 0.01 | −0.06 |
| Butt thickness | **0.78** | 0.14 | −0.04 | 0.01 | 0.18 |
| Butt projection | 0.56 | 0.39 | −0.33 | 0.03 | 0.05 |
| Left edge angle | 0.24 | −0.19 | 0.08 | 0.01 | 0.25 |
| Right edge angle | 0.26 | −0.22 | 0.12 | 0.06 | 0.25 |
| Flaking angle | −0.05 | 0.49 | **0.58** | 0.09 | 0.03 |
| Cone angle | −0.17 | **0.58** | 0.49 | 0.14 | 0.05 |
| Scar count | 0.03 | −0.25 | −0.04 | **0.75** | −0.02 |
| Area of cortex % | 0.19 | −0.17 | 0.23 | −0.52 | 0.01 |
| Eigenvalues | 3.22 | 1.32 | 1.12 | 0.87 | 0.41 |
| % of variance | 46.4 | 19.0 | 16.1 | 12.6 | 5.9 |

**Table 25.10** Debitage: Loadings after Varimax Rotation of the Principal Axes

| Variable | Factor | | | | |
| | 1 | 2 | 3 | 4 | 5 |
|---|---|---|---|---|---|
| Flake length | 0.02 | **0.70** | 0.22 | 0.01 | 0.01 |
| Flake breadth | 0.43 | **0.64** | 0.11 | −0.04 | 0.06 |
| Flake thickness | 0.30 | 0.57 | **0.74** | −0.02 | 0.07 |
| Butt breadth | **0.97** | 0.18 | 0.03 | −0.02 | 0.02 |
| Butt thickness | **0.66** | 0.23 | 0.42 | 0.02 | 0.06 |
| Butt projection | **0.76** | 0.01 | 0.04 | −0.01 | −0.02 |
| Left edge angle | 0.04 | 0.04 | 0.39 | −0.06 | 0.02 |
| Right edge angle | 0.02 | 0.09 | 0.43 | −0.04 | −0.02 |
| Flaking angle | −0.01 | 0.06 | −0.03 | **0.76** | 0.04 |
| Cone angle | −0.01 | −0.08 | −0.13 | **0.78** | −0.03 |
| Scar count | −0.10 | 0.13 | 0.16 | −0.07 | **−0.75** |
| Area of cortex % | −0.07 | 0.20 | 0.19 | −0.05 | 0.56 |

that the relationship between the two variables has arisen by chance) despite the often extremely low values — a very slight relationship is liable to be recognised as significant given the large sample size. Secondly, inspection of the matrix shows some obvious groupings of high values, such as those concerned with flake size and butt size (there is also a clear relationship between these two ratios).

Some of the individual coefficients are instructive. Thus the bigger pieces tend to have a small value for the cone angle (i.e. 'open' flaking) and also a high scar count. Butts tend to be extensive on pieces with few scars, while butt projection is inversely related to the amount of cortex.

Extraction of principal components from the correlation matrix results in a first component which is dominated by the butt measurements, and to a slightly lesser degree by flake size (the loadings on the subsequent components are less strong). In all, five components have eigenvalues greater than 1 — these account for 75.2% of

the original variance (Table 25.9). Varimax factoring of the component matrix was carried out in order to clarify the nature of the variability — this involves rotating the principal component axes to define new axes for which the loadings of the original variables are as close as possible to ±1 or 0 (i.e. perfect or nil correlation, respectively, between the factors and the input data). Three factors proved significant in this case; the resulting matrix is given as Table 25.10. It shows that (apart from the butt and flake size factors 1 and 2), there is a clear axis of 'cross-section' variation based on flake and butt thickness linked to the edge angles (factor 3), while the cone and flaking angles (factor 4) and the percentage of cortex (factor 5) behave rather independently, but not sufficiently so to achieve statistical significance.

## THE ASSEMBLAGES

A data matrix was compiled using summary statistics (i.e. means for the quantitative and percentages for the qualitative attributes). This had the advantage of allowing the introduction of a number of variables which could not be included in factoring based on individual artefacts (being non-ordinal). Also, the small number of units (9) meant that distance matrices could be used as a basis for analysis.

The Pearson correlations (Table 25.11) are quite similar to those for the entire debitage, apart from the much higher values (resulting from the small sample size and the use of mean values — and ignoring within-assemblage variance). However, extraction of the varimax-rotated factors for the same variables as were used in the previous section produces very different second and third factors. The former contrasts the edge angles with cone angle and scar count, while the latter is derived essentially from the flaking angle and, to a lesser degree, butt projection. Factor 1 is thus the gross size component which clearly plays an important part in the characterisation of the debitage throughout the succession, while factor 2 describes a preference for somewhat blunt-edged pieces with few scars and a shallow angle between butt and cone (i.e. 'low-technology' flakes) irrespective of size. Factor 3 is harder to interpret; flaking angle shows little patterning through time and is in any case rather liable to variation caused by hinging or plunging of the flake. On the other hand, it is particularly affected by the presence of a strongly convex bulbar surface, which suggests that it may have important technological implications as discussed below. The scores on factors 1 and 2 are plotted for each assemblage in Fig. 25.2D.

The CLUSTAN program package made it possible to search for potential groupings of assemblages based on the same and other data. Though experiments were carried out with other approaches (such as standardisation of the input values) the most satisfactory solution, in view of the known high correlations, was to use principal component scores in order to devise Euclidean distances between assemblages. Again, the most convenient clustering algorithm proved to be average-linkage,

**Table 25.11**  Debitage: Correlation Matrix based on Assemblage Means

| | | 1 | 2 | 3 | 4 | 5 | 6 | 7 | 8 | 9 | 10 | 11 |
|---|---|---|---|---|---|---|---|---|---|---|---|---|
| 1 | Flake length | | | | | | | | | | | |
| 2 | Flake breadth | **0.87** | | | | | | | | | | |
| 3 | Flake thickness | **0.87** | **0.89** | | | | | | | | | |
| 4 | Butt breadth | **0.77** | **0.85** | **0.95** | | | | | | | | |
| 5 | Butt thickness | **0.77** | **0.84** | **0.97** | **0.98** | | | | | | | |
| 6 | Butt projection | 0.45 | 0.53 | 0.50 | **0.67** | 0.57 | | | | | | |
| 7 | Left edge angle | 0.50 | 0.51 | **0.76** | **0.73** | **0.80** | 0.04 | | | | | |
| 8 | Right edge angle | **0.61** | 0.58 | **0.81** | **0.80** | **0.83** | 0.25 | **0.90** | | | | |
| 9 | Flaking angle | 0.29 | 0.26 | 0.17 | −0.09 | −0.03 | −0.37 | −0.05 | −0.01 | | | |
| 10 | Cone angle | −0.56 | −0.46 | **−0.63** | −0.57 | **−0.61** | 0.02 | **−0.83** | **−0.86** | −0.13 | | |
| 11 | Scar count | −0.18 | −0.15 | −0.31 | −0.30 | −0.31 | 0.31 | **−0.66** | −0.57 | −0.27 | **0.67** | |
| 12 | Area of cortex % | **0.92** | **0.77** | **0.87** | **0.74** | **0.80** | 0.23 | **0.71** | **0.79** | 0.28 | **−0.72** | −0.29 |

**Table 25.12**  Debitage: Loadings on Principal Components based on Assemblage Means

| Variable | Factor 1 | Factor 2 | Factor 3 |
|---|---|---|---|
| Flake length | **0.86** | 0.18 | 0.38 |
| Flake breadth | **0.85** | 0.28 | 0.33 |
| Flake thickness | **0.97** | 0.14 | 0.10 |
| Butt breadth | **0.94** | 0.27 | −0.14 |
| Butt thickness | **0.96** | 0.19 | −0.11 |
| Butt projection | 0.43 | **0.83** | −0.15 |
| Left edge angle | **0.84** | −0.36 | −0.34 |
| Right edge angle | **0.90** | −0.22 | −0.27 |
| Flaking angle | 0.12 | −0.42 | **0.86** |
| Cone angle | **−0.77** | 0.48 | 0.16 |
| Scar count | −0.45 | **0.75** | 0.17 |
| Area of cortex % | **0.90** | −0.06 | 0.24 |
| Eigenvalues | 7.50 | 2.09 | 1.35 |
| % of variance | 62.5 | 17.4 | 11.2 |

**Table 25.13**  Debitage: Loadings on Varimax-related Factors based on Assemblage Means

| Variable | Factor 1 | Factor 2 | Factor 3 |
|---|---|---|---|
| Flake length | **0.90** | 0.20 | 0.24 |
| Flake breadth | **0.93** | 0.14 | 0.14 |
| Flake thickness | **0.90** | 0.40 | 0.01 |
| Butt breadth | **0.87** | 0.38 | −0.26 |
| Butt thickness | **0.86** | 0.44 | −0.19 |
| Butt projection | **0.71** | −0.30 | −0.54 |
| Left edge angle | 0.42 | **0.87** | −0.12 |
| Right edge angle | 0.56 | **0.77** | −0.13 |
| Flaking angle | 0.13 | 0.02 | **0.95** |
| Cone angle | −0.36 | **−0.85** | −0.09 |
| Scar count | 0.04 | **−0.86** | −0.22 |
| Area of cortex % | **0.78** | 0.45 | 0.24 |

**Table 25.14**  Debitage: Assemblage Scores on the Varimax-rotated Factors listed in Table 25.13

| Layer | Factor 1 | Factor 2 | Factor 3 |
|---|---|---|---|
| 5 | 0.35 | −1.87 | 1.47 |
| A | −1.57 | 0.82 | −0.45 |
| B | −0.68 | −0.67 | −0.16 |
| C | −0.84 | −0.66 | −0.55 |
| D | −0.40 | 0.14 | 0.21 |
| E | −0.30 | 0.92 | 0.49 |
| F | 1.00 | 1.28 | 1.19 |
| G | 1.16 | 0.47 | −0.34 |
| H | 1.28 | −0.43 | −1.86 |

groups, and included in a single dataset; the assemblage means for the same variables were used in other experiments. Finally, to the quantitative variables from LENGTH to SCAR COUNT (see the list given in section 25.2.2 above) were added the mean percentage of area of cortex, and a number of percentages obtained from qualitative attributes: the pieces with dorsal cortex, feathered terminations, transverse planforms, cortex on the left edge only, and cortical butts. The first component, as expected, proves consistently to be related to flake and butt size (as well as to other variables) and therefore imposes a strong chronological bias. Even so, this turns out to be more robust than might have been anticipated (Fig. 25.2). Layer 5 is invariably recognised as idiosyncratic, as indeed is H except when the assemblages are broken down according to butt data. F and G are always linked very closely to one another, and less so to H. However, the positions of the stratigraphically intermediate assemblages (C, D and E) are more equivocal; sometimes they join the samples from the upper layers, and sometimes those from below. Much the same picture is given by the scatter plots of principal components scores.

The analysis based on factor scores, referred to above, indicates that one cluster includes the waste from layer A–E, with that from layer A being only weakly entered as a member of the cluster (this is clearly emphasised in the scatter diagram of the first two factors, Fig. 25.2D) while there is also a pairing of F and G.

inasmuch as it was not too prone to either chaining or forcing outliers into clusters.

A wide range of analyses was performed in order to test the stability of the configuration of assemblages. At one extreme, means were calculated for each of the butt

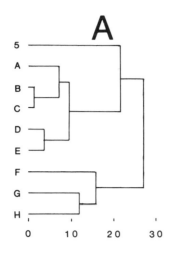

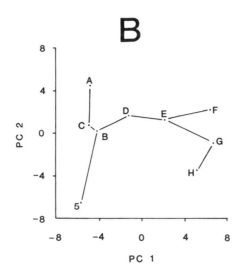

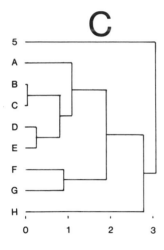

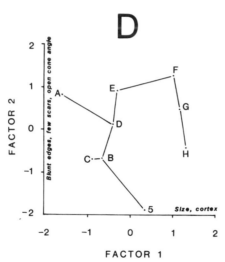

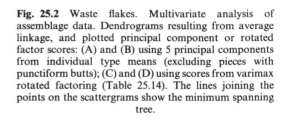

**Fig. 25.2** Waste flakes. Multivariate analysis of assemblage data. Dendrograms resulting from average linkage, and plotted principal component or rotated factor scores: (A) and (B) using 5 principal components from individual type means (excluding pieces with punctiform butts); (C) and (D) using scores from varimax rotated factoring (Table 25.14). The lines joining the points on the scattergrams show the minimum spanning tree.

**Table 25.15** Debitage: Assemblage 'Nearest Neighbours', with corresponding Euclidean Distances[a]

| Layer | Neighbour | | | | | | | | |
|---|---|---|---|---|---|---|---|---|---|
| | 1st | | 2nd | | 3rd | | 4th | | 5th |
| 5 | 1.21 B | | 1.54 D | | 1.75 C | | 2.31 E | | 2.36 G |
| A | 0.57 D | | 0.63 E | | 0.69 C | | 0.78 B | | 1.91 G |
| B | 0.05 C | | 0.22 D | | 0.78 E | | 0.78 A | | 1.18 G |
| C | 0.05 B | | 0.36 D | | 0.69 A | | 0.98 E | | 1.34 G |
| D | 0.18 E | | 0.22 B | | 0.36 C | | 0.57 A | | 0.71 G |
| E | 0.18 D | | 0.58 F | | 0.63 A | | 0.76 G | | 0.78 B |
| F | 0.58 E | | 0.76 G | | 1.05 D | | 2.11 B | | 2.39 A |
| G | 0.71 D | | 0.76 F | | 0.76 E | | 0.79 H | | 1.18 B |
| H | 0.79 G | | 1.57 C | | 1.69 B | | 1.86 D | | 2.47 E |

[a]Calculated from the scores in Table 25.14. The distance is given in units of standard deviation.

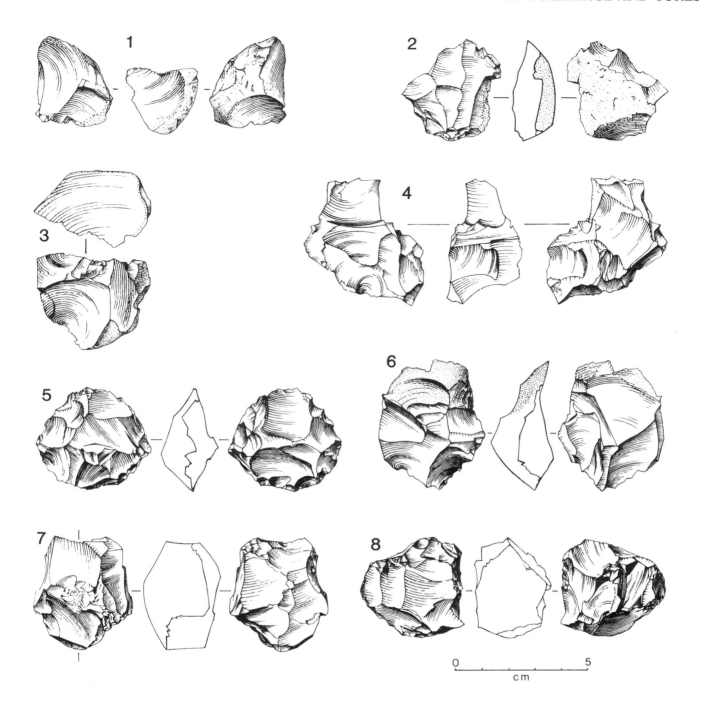

**Fig. 25.3** Examples of cores (layers are indicated in parenthesis). 1 Single platformed bifacial (E); 2 double platformed, irregular (H); 3 double platformed, opposed (F); 4 bipyramidal, irregular (D); 5 bipyramidal, thin, tending to discoidal (A); 6 bipyramidal, on flake (H); 7,8 globular (H, G).

Lastly, layers 5 and H have little in common with other assemblages. The separation between these two is reinforced by extreme scores in the third component. As already suggested, this is at least partly related to the convexity of the bulbar surfaces, which are more rounded in H.

The result of the multivariate analysis, therefore, is to show that although stratigraphically adjacent assemblages may be closely linked, there is little tendency to form stable clusters. However, it is clear that the waste asemblages from layers H and 5 are quite distinctive, both exhibiting higher scar counts (for example) than do those of neighbouring layers. From G–A there is a clear trend towards reduced size, and also a less regular one of developing sophistication (in terms of cone angle, scar count and the sharpness fl of the edges).

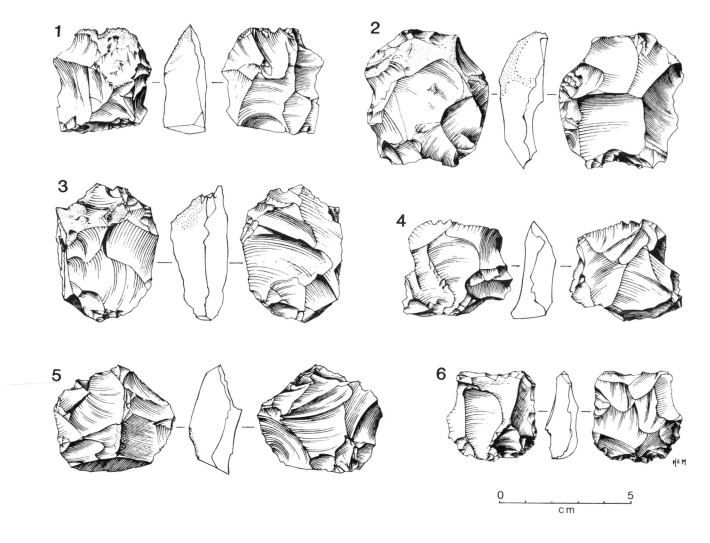

**Fig. 25.4** Examples of cores (layers are indicated in parenthesis). 1–3 Levallois flake cores: 1 proto-Levallois (H); 2 with some marginal retouch (5); 3 slightly offset (A). 4–6 discoidal cores: 4 irregular, with platform preparation (H); 5 regular, with more careful preparation (B); 6 with the lower face worked from one edge (E).

## 25.3  THE CORES

Some of these are illustrated in Figs. 25.3 and 25.4. An initial study was carried out by B. Bauer (1981) as an M.Phil. dissertation at the University of Cambridge. Subsequently, however, it was discovered that even at this time there were still a number of provenances which required correction; also it was felt that some additional observations were called for in the light of progress made in other areas of investigation. Hutcheson therefore undertook a fresh examination of the material. In all, 1091 cores were measured (i.e. a 100% sample).

### 25.3.1  Typology

Choice of a classificatory system suitable for La Cotte cores proved rather difficult, as the Bordes typology is somewhat inadequate in the face of the considerable numbers that fit best into its *divers* category, or which suggest the need for finer subdivision than the scheme allows. On the other hand, it was felt that a general cor-

respondence to the Bordes typology should be preserved. As far as possible this principle was adhered to when the following categories were devised, though the two schemes are not precisely concordant; the approximate Bordes equivalent is given in parenthesis after the name of each class. Isaac's (1977) scheme has been influential in the construction of this type-list, as it was clear that some of the problems raised by the general character of the Olorgesailie material were likely to be encountered at La Cotte.

1.  **Levallois (*Levallois à éclats/pointes/lames*).** Bordes's definitions were employed here (Bordes 1961b, 71–2). The category was subdivided into cores intended for the production of flakes or points — no Levallois blade cores occur at La Cotte despite the presence of their products — and atypical or proto-Levallois cores.
2.  **Disc (*discoïdes*).** True 'Mousterian' disc cores, flattish and with a regular radial pattern of removal scars on one face, and peripheral platform preparation, giving only limited scarring of the other.
3.  **Discoidal (*discoïdes*).** Again rather flat, with more or less radial scar patterns and removals of roughly equal size on both faces.
4.  **Prismatic (*prismatiques, à un plan de frappe*).** The single platform

is formed by a natural fracture or an early flake scar at one end of the core; from much of its circumference parallel-sided flakes or blades were removed.

5. **High backed discoidal (*pyramidaux*).** Rather similar to discoidal cores in the treatment of both sides, but much thicker relative to length and breadth, and with a strongly asymmetrical cross-section. This type may be regarded as intermediate between the pyramidal and bipyramidal categories.

6. **Pyramidal/conical (*pyramidaux*).** From around a single flattish circular platform, flakes (often subtriangular) have been removed so that their scars meet at the apex of a moderately shallow cone.

7. **Bipyramidal/biconical (*pyramidaux*).** The platform is circular, and peripheral; flaking is on both faces, and alternate, making use of one flake scar as the platform for the next.

8. **Polyhedral (*globuleux*).** A multi-platformed core, roughly globular in shape, with flaking carried out in any direction and from any platform capable of yielding a usable flake; often exhausted, and sometimes with arrêtes showing signs of shattering caused by unsuccessful attempts to continue debitage.

9. **Multiple platformed (*globuleux*).** Thinner than the polyhedral cores, but rather similar in other respects (notably in the possession of two or more platforms without evidence of a systematic reduction strategy).

10. **Shapeless or miscellaneous (*informes*).** Formed by a few removals without any obviously preferred orientation.

11. **Single platformed, unifacial (*divers*).** Flakes have been struck on one face only, from a single platform.

12. **Single platformed, bifacial (*divers*).** Flakes have been removed from two faces, from a single platform. This sometimes results in a core looking rather like a chopping tool, though with an edge too obtuse, or irregular, to be useful.

13. **Double platformed, opposed (*divers*).** The two platforms are opposed to each other, on the ends or sides of the cores; from these, flaking has been performed on either the same or adjacent faces

14. **Double platformed, at right angles (*divers*).** The two platforms are perpendicular to each other, more or less adjacent: flaking has been carried out on either the same or adjacent faces.

It should be stressed that many pieces were difficult to classify even with the scheme outlined above, both because of the constraints imposed by adherence in principle to Bordes's guidelines — and indeed to the concept of discrete types — and also because (in order to keep the number of classes within reasonable limits) not all of the categories which could have been defined from the attributes considered have been admitted to the type-list.

## 25.3.2 Attribute Definitions

During measurement the cores were orientated according to an axis passing through the centre of the principal platform and the furthest point from it, in the case of single platformed pieces (where the core had opposed platforms, both were employed to obtain the correct positioning). The second axis was provided by the maximum dimension at right angles to the first, in the case of globular or related forms, and by a perpendicular passing through the core margin where there was a fairly complete peripheral platform. In this way a containing cuboid was defined. The recorded attributes were:

1. **Length.** The length of the first axis as defined above.
2. **Breadth.** The length of the second axis as defined above.
3. **Thickness.** The length of the remaining axis. Note that this may be greater than the two previous measurements under certain cirumstances.
4. **Scar count.** The number of removals whose length is at least one tenth of the largest dimension of the piece.

5. **Cortex.** Presence or absence of cortex.
6. **Primary technique.** In some cases the core was worked down from a large flake; this is recorded.

In addition, two functions were computed from the metrical data:

> **Volume** of the containing cuboid (obtained by multiplying length, breadth and thickness — i.e. not true volume).

> **Cross-section index.** Twice the thickness, divided by length plus breadth. The purpose in using this rather complicated ratio, in preference to simple thickness/breadth, is to counter the distortion caused in the latter by varying elongation.

## 25.3.3 The Results

The relative frequencies, from layer to layer, of the various types (Table 25.16) show some obviously important changes, though the small samples for certain layers — especially 5, but also B and H — mean that caution must be exercised in their interpretation.

Layer H is the only one in which globular cores are found in any quantity, whereas the multiple platformed type, rather variable in its occurrence, is particularly important in layer 5. The percentage of disc cores (which are absent until F) climbs until B and then falls again to a very low level in 5; discoidal cores are less well defined in their distribution, but are at their least frequent in H and G, and then again in 5. High-backed discoidal pieces are not very common, but clearly concentrated in the lower middle part of the sequence, especially in F. The other class which shows a slight (decreasing) trend through time is the single platformed unifacial. Layers A and B exhibit an exceptionally low incidence of miscellaneous cores, while B is the only layer in which Levallois cores seem important in percentage terms — the sample size is low, but the comparatively large number of unmodified Levallois flakes in the same layer (see chapter 26) suggests that the observation in respect of the cores is valid. Both of the double platformed and also the single platformed bifacial types show little systematic variation through the sequence, but the two first, together, are particularly well represented in G. Bipyramidal cores are always present with a high, if fluctuating, percentage; pyramidal and prismatic cores, however, are consistently scarce.

The full type-list has been compressed, in Table 25.17, to give the percentages according to the Bordes scheme (some of the variation referred to above is masked thereby, but the development of the *discoïde* class is particularly clear).

There is a general downward tendency in the size attributes — length, breadth, thickness and of course volume — between layer H and A (Table 25.18). However, B is an exception in respect of length and breadth, but not of volume; its cores are relatively large in plan, but unusually flat. The cross-section index suggests that the decrease in thickness of the cores is appreciably stronger than that in length and breadth. This may be partly explained by the increase in the percentage of discoidal and disc cores (as well as Levallois cores in layer B),

**Table 25.16**   Cores: Type Frequency (%)

| Layer | Leval-lois | Disc | Disc-oidal | Prism-atic | High-backed disc. | Pyr-amidal | Bipyr-amidal | Poly-hedral | Multi-plat. | Irreg-ular | Single plat. unif. | Single plat. bif. | Double plat. opp. | Double plat. per. | Total |
|---|---|---|---|---|---|---|---|---|---|---|---|---|---|---|---|
| 5 | 3.7 | 3.7 | 7.4 | 0.0 | 3.7 | 0.0 | 25.9 | 3.7 | 22.2 | 18.5 | 3.7 | 3.7 | 0.0 | 3.7 | 27 |
| A | 6.9 | 13.8 | 15.5 | 1.0 | 1.2 | 3.4 | 26.3 | 2.9 | 5.2 | 7.9 | 3.2 | 4.7 | 5.9 | 2.2 | 407 |
| B | 14.3 | 22.4 | 10.2 | 0.0 | 2.0 | 2.0 | 24.5 | 2.0 | 0.0 | 8.2 | 4.1 | 4.1 | 4.1 | 2.0 | 49 |
| C | 4.2 | 16.7 | 17.7 | 1.0 | 2.1 | 1.0 | 20.8 | 6.3 | 3.1 | 20.8 | 0.0 | 2.1 | 2.1 | 2.1 | 96 |
| D | 1.1 | 10.0 | 13.3 | 3.3 | 2.2 | 2.2 | 26.7 | 4.4 | 0.0 | 22.2 | 3.3 | 5.6 | 2.2 | 3.3 | 90 |
| E | 2.7 | 8.0 | 10.7 | 5.4 | 2.7 | 1.8 | 15.2 | 5.4 | 10.7 | 21.4 | 4.5 | 6.3 | 0.9 | 4.5 | 112 |
| F | 6.8 | 3.0 | 12.9 | 7.6 | 1.5 | 0.8 | 25.0 | 3.0 | 6.1 | 17.4 | 6.1 | 3.0 | 4.5 | 2.3 | 132 |
| G | 2.5 | 0.0 | 8.3 | 1.7 | 0.0 | 2.5 | 14.9 | 1.7 | 14.1 | 27.3 | 5.0 | 5.0 | 10.7 | 6.6 | 121 |
| H | 8.8 | 0.0 | 7.0 | 3.5 | 3.5 | 1.8 | 19.3 | 15.8 | 5.3 | 14.0 | 8.8 | 8.8 | 3.5 | 0.0 | 57 |

**Table 25.17**   Cores: Bordes Type Frequency (%)

| Layer | Levallois | Discoïde | Prism-atique | Pyramidal | Globuleux | Informe | Divers | Total |
|---|---|---|---|---|---|---|---|---|
| 5 | 3.7 | 11.1 | 3.7 | 25.9 | 25.9 | 18.5 | 11.1 | 27 |
| A | 6.9 | 29.3 | 1.2 | 30.7 | 8.1 | 7.9 | 16.0 | 407 |
| B | 14.3 | 32.6 | 2.0 | 26.5 | 2.0 | 8.2 | 14.3 | 49 |
| C | 4.2 | 34.4 | 2.1 | 22.8 | 9.4 | 20.8 | 6.3 | 96 |
| D | 1.1 | 23.3 | 2.2 | 32.2 | 4.4 | 22.2 | 14.4 | 90 |
| E | 2.7 | 18.7 | 2.7 | 22.4 | 16.1 | 21.4 | 16.2 | 112 |
| F | 6.8 | 15.9 | 1.5 | 33.4 | 9.1 | 17.4 | 15.9 | 132 |
| G | 2.5 | 8.3 | 0.0 | 19.1 | 15.7 | 27.3 | 27.3 | 121 |
| H | 8.8 | 7.0 | 3.5 | 24.6 | 21.1 | 14.0 | 21.1 | 57 |

**Table 25.18**   Cores: Measurements and Scar Counts

| Layer | Length mean | Length s.d. | Breadth mean | Breadth s.d. | Thickness mean | Thickness s.d. | Scar count mean | Scar count s.d. | Volume mean | Volume s.d. | Cross-section mean | Cross-section s.d. | Total |
|---|---|---|---|---|---|---|---|---|---|---|---|---|---|
| 5 | 45.4 | 11.2 | 37.1 | 11.7 | 21.0 | 6.3 | 8.07 | 3.74 | 42.32 | 46.15 | 0.520 | 0.132 | 27 |
| A | 40.9 | 8.5 | 33.7 | 7.3 | 19.2 | 6.4 | 8.25 | 3.71 | 29.39 | 21.36 | 0.517 | 0.143 | 407 |
| B | 45.4 | 8.7 | 36.3 | 8.5 | 18.5 | 6.2 | 6.37 | 3.11 | 32.74 | 21.42 | 0.462 | 0.158 | 49 |
| C | 41.9 | 9.3 | 33.7 | 8.4 | 18.8 | 6.0 | 7.11 | 3.09 | 29.26 | 23.27 | 0.507 | 0.153 | 96 |
| D | 42.8 | 8.5 | 33.9 | 7.4 | 19.7 | 6.3 | 6.98 | 3.14 | 31.02 | 20.57 | 0.520 | 0.147 | 90 |
| E | 41.5 | 7.2 | 32.8 | 6.5 | 20.7 | 6.9 | 7.70 | 3.01 | 30.39 | 19.18 | 0.559 | 0.152 | 112 |
| F | 43.3 | 9.6 | 35.3 | 8.7 | 21.4 | 7.6 | 5.85 | 2.72 | 37.45 | 34.33 | 0.547 | 0.148 | 132 |
| G | 44.8 | 11.3 | 34.8 | 9.2 | 23.0 | 7.2 | 6.33 | 3.21 | 41.53 | 36.40 | 0.587 | 0.145 | 121 |
| H | 45.9 | 11.9 | 36.9 | 10.2 | 23.8 | 8.3 | 8.82 | 3.65 | 49.15 | 47.92 | 0.575 | 0.129 | 57 |

since these are the thinnest of the core types. However, the statistics obtained by breaking down the collection according to layer and type together indicate that the overall reduction in size corresponds to a trend which is independent of typology, and not merely to an increase in the representation of categories characterised by smaller dimensions.

In layer 5, the pattern is reversed and the means of the size attributes return to levels encountered in the lowest layers (it will be recalled that a similar phenomenon was observed in respect of the debitage). This is due in particular to the size of multiple platformed and bipyramidal cores, as well as to the presence of a single large prismatic core.

The cross-section index provides an additional piece of information which is of some interest, as it falls substantially from B to A (which has a value similar to that for 5); the relative thickness rises before the general size increase occurs at the end of the sequence. This is largely explicable in terms of the proportionally thick cross-section of the small disc, discoidal and Levallois cores in A.

The mean scar count behaves in a more complicated manner. It starts high, at 8.8 in H, and falls to 5.8 in F; it then stabilises at around 7 until layer B (6.3), rising again to 8.2 in A and 8.1 in 5. Since polyhedral cores almost always have high counts, the initial value for the mean is easily explained. Another category which plays an important part in the development of this statistic is the disc cores, which are also liable to possess numerous qualifying scars — since the underlying technique allows the removals to be quite invasive without seriously truncating earlier ones. This is less true of Levallois cores, given the definition used here, since many of the scars are truncated by the detachment of the Levallois flake and are thereby left too short to be counted; however, in layer 5 the single Levallois core does in fact achieve an exceptionally high score in this respect, as do the bipyramidal cores from the same layer.

The percentage of cores on which cortex is preserved falls from 73.7 in H to 52.1 in C, subsequently rising to 63.0 in 5. The categories which most commonly possess cortex are single platformed (unifacial and bifacial), polyhedral, and high backed discoidal, while cortex occurs least in the disc cores. However, there is variation from layer to layer as regards its frequency for specific core types. An observation which is of importance as regards the underlying strategy is that, after relative stability during most of the succession, the incidence of cores made on flakes takes a dramatic upwards turn in layer A, to 13.5% (a single piece, i.e. 3.7%, qualifies in layer 5).

### 25.3.4 Multivariate Analysis

The study of the core data has been influenced by both the importance of typological data in the above analysis and the relatively small number of other observations which had been taken. The latter consideration meant that there was little to be gained by an attempt to preserve a piece-by-piece analysis of correlation etc. As a

**Table 25.19** Cores: Principal Component Loadings based on Frequencies

| Type | Factor | | | | |
|---|---|---|---|---|---|
| | 1 | 2 | 3 | 4 | 5 |
| Levallois | -0.42 | **0.58** | 0.37 | -0.41 | -0.22 |
| Disc | **-0.91** | 0.02 | 0.01 | -0.04 | 0.25 |
| Discoidal | **-0.73** | -0.26 | -0.06 | **0.58** | 0.00 |
| Prismatic | 0.39 | 0.10 | 0.06 | **0.66** | -0.55 |
| HB discoidal | 0.16 | **0.73** | **-0.62** | -0.05 | 0.21 |
| Pyramidal | -0.21 | -0.14 | **0.83** | 0.09 | 0.33 |
| Bipyramidal | **-0.62** | 0.25 | -0.26 | -0.13 | -0.36 |
| Polyhedral | 0.39 | **0.75** | 0.03 | 0.30 | 0.30 |
| Multi-platformed | **0.59** | -0.29 | -0.43 | -0.51 | -0.02 |
| Irregular | **0.58** | **-0.62** | -0.26 | 0.32 | 0.12 |
| Single plat., unif. | 0.29 | 0.53 | 0.41 | -0.10 | -0.33 |
| Single plat., bif. | 0.49 | 0.52 | 0.42 | 0.09 | 0.35 |
| Double plat., opp. | **0.87** | -0.46 | **0.79** | -0.18 | -0.16 |
| Eigenvalues | 3.89 | 3.58 | 2.52 | 1.50 | 1.08 |
| % of variance | 27.8 | 25.6 | 18.6 | 10.7 | 7.7 |

**Table 25.20** Cores: Varimax Factor Loadings based on Frequencies

| Type | Factor | | | | |
|---|---|---|---|---|---|
| | 1 | 2 | 3 | 4 | 5 |
| Levallois | **0.92** | 0.10 | 0.06 | 0.05 | -0.10 |
| Disc | 0.39 | -0.01 | -0.28 | **-0.64** | -0.51 |
| Discoidal | -0.04 | 0.02 | -0.38 | **-0.89** | 0.04 |
| Prismatic | -0.15 | -0.02 | 0.12 | -0.08 | **0.93** |
| HB discoidal | 0.17 | **-0.90** | 0.34 | 0.16 | -0.08 |
| Pyramidal | 0.11 | **0.76** | 0.34 | -0.36 | -0.22 |
| Bipyramidal | 0.56 | -0.30 | -0.47 | -0.19 | -0.00 |
| Polyhedral | 0.12 | -0.39 | **0.83** | -0.01 | 0.18 |
| Multi-platformed | -0.42 | -0.19 | -0.18 | **0.79** | -0.15 |
| Irregular | **-0.92** | 0.02 | -0.05 | 0.13 | 0.18 |
| Single plat., unif. | 0.33 | 0.13 | 0.56 | 0.56 | 0.51 |
| Single plat., bif. | **0.91** | 0.07 | **0.93** | 0.21 | 0.09 |
| Double plat., opp. | -0.02 | **0.94** | -0.01 | 0.16 | 0.06 |

**Table 25.21** Cores: Assemblage Scores on the Varimax-rotated Factors, based on Frequencies

| Layer | Factor | | | | |
|---|---|---|---|---|---|
| | 1 | 2 | 3 | 4 | 5 |
| 5 | -0.19 | -1.54 | -0.89 | 1.64 | -0.82 |
| A | 0.80 | 0.92 | -0.14 | -0.73 | -0.58 |
| B | 1.70 | 0.42 | -0.36 | 0.00 | -0.99 |
| C | -0.65 | -0.75 | -0.48 | -1.51 | -0.53 |
| D | -0.59 | -0.21 | 0.16 | -1.03 | 0.23 |
| E | -1.10 | -0.34 | 0.70 | -0.13 | 0.14 |
| F | 0.54 | 0.05 | -1.17 | 0.17 | 2.31 |
| G | -1.26 | 1.92 | -0.05 | 1.06 | -0.26 |

245

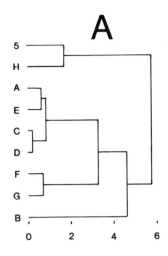

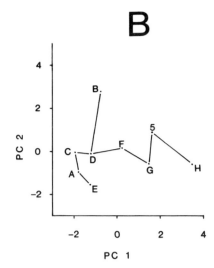

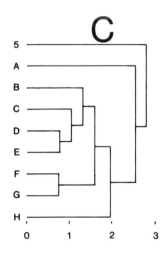

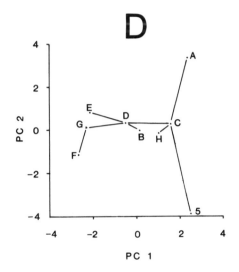

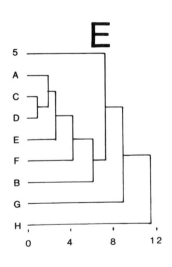

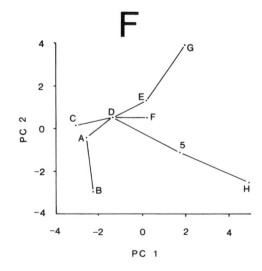

**Fig. 25.5** Cores. Multivariate analysis of assemblage data. Dendrograms resulting from average linkage, and plotted principal component scores: (A) and (B) using 3 principal components, from the assemblage means for attribute data (Table 25.18); (C) and (D) using 5 principal components, from the type percentages (Table 25.16); (E) and (F) using 5 principal components, with the metrical data and type percentages combined.

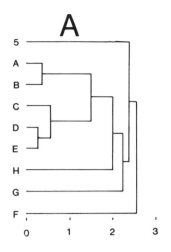

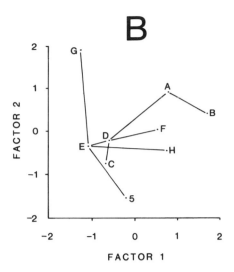

**Fig. 25.6** Cores. Multivariate analysis of assemblage data, using varimax rotated scores (Table 25.21): (A) dendrogram resulting from average linkage; (B) and (C) scores for the first four factors.

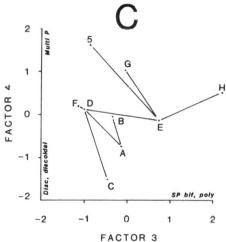

result, the account which follows is concerned only with the summary data for each layer.

Three sets of variables were investigated: (a) The means for the quantitative core attributes; (b) The percentages of the 14 types defined in section 25.3.2; (c) The two previous datasets combined.

As with the waste, the procedure used was to compute principal component scores prior to construction of the Euclidean distance matrices and (average-linkage) hierarchical cluster analysis. In addition, the percentage data was subjected to varimax factoring followed by cluster analysis, as described for the waste in section 25.2.4).

The majority of the resulting dendrograms exhibit considerable chaining — this is somewhat less marked in the diagram based on the attribute means alone. In the case of the percentage data (both principal components and factor scores) there is a nucleus of related assemblages, B–E, with the others only very loosely attached. The factor scores add layer A to this group (Fig. 25.6A); however, in this analysis layer B is connected through A — its next nearest neighbour, D, is very much less similar. In every diagram, cores from layer 5 are seen to be very far removed from those of the immediately underlying layers.

The first principal component for the quantitative core attributes is essentially concerned with size coupled to relatively thick cross-section — thereby placing H and 5 close together — whereas the second component gives a positive result for large but thin pieces with relatively few scars, and so separates B from the rest (Fig. 25.5B summarises graphically some of the comments made in the previous section). The initial four-layer cluster in the accompanying dendogram is thus seen to be based in part on small size and medium-to-thick cross-section. On

**Table 25.22** Cores: Assemblage 'Nearest Neighbours', with corresponding Euclidean Distances

| Layer | \ | Neighbour | | | | | | | | |
|-------|------|-----|------|-----|------|-----|------|-----|------|-----|
| | 1st | | 2nd | | 3rd | | 4th | | 5th | |
| 5 | 1.77 | E | 2.09 | B | 2.21 | C | 2.25 | D | 2.65 | A |
| A | 0.36 | B | 0.81 | D | 1.12 | C | 1.36 | E | 1.71 | G |
| B | 0.36 | A | 1.70 | D | 1.89 | C | 2.09 | 5 | 2.17 | H |
| C | 0.30 | D | 0.82 | E | 1.12 | A | 1.89 | B | 2.21 | 5 |
| D | 0.28 | E | 0.30 | C | 0.81 | A | 1.70 | B | 1.72 | H |
| E | 0.28 | D | 0.82 | C | 1.26 | H | 1.36 | A | 1.46 | G |
| F | 1.77 | D | 2.21 | A | 2.22 | E | 2.61 | B | 2.68 | C |
| G | 1.46 | E | 1.71 | A | 1.93 | D | 2.55 | B | 2.86 | C |
| H | 1.26 | E | 1.72 | D | 2.04 | A | 2.17 | B | 2.91 | C |

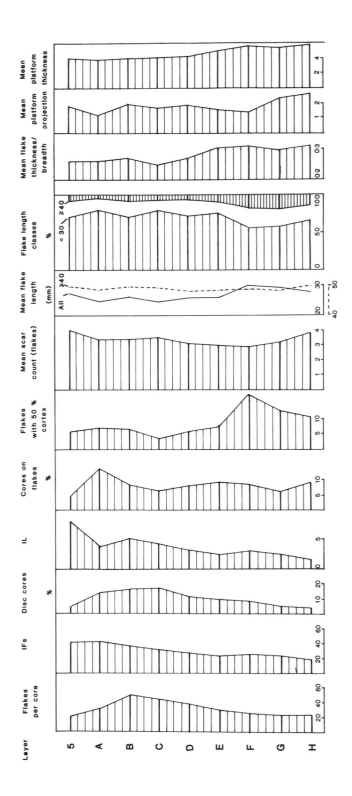

**Fig. 25.7** Technological changes at La Cotte de St. Brelade.

248

the other hand, the grouping of layers A–E on the rotated factor scores is defined chiefly by low within-group variance in factors 2–5, with generally numerous (relative to the collection as a whole) disc and discoidal, but few high-backed discoidal, cores, and middling frequencies of several other types; Levallois cores are extremely variable in their representation.

## 25.4    CONCLUSIONS

The information which may be derived from the study of the debitage and cores, particularly taken in conjunction with other data, includes (a) the character and availability of primary material; (b) the technological processes employed; (c) the extent to which the various stages of artefact manufacture may have taken place at different locations.

At La Cotte, the raw material for the flint artefacts was evidently obtained in the form of beach pebbles, in many cases apparently with a diameter of 100–150 mm (to judge from the surface curvature of cortical pieces). The incidence of cortical flakes is extremely variable, however; the significance of this will become apparent below.

There are very great technological changes as the archaeological succession develops. In layer H, few true facetted (though numerous dihedral) butts occur. The butts are large and thick, tending to an open flaking angle. Levallois technique is very rare, despite the presence of a few proto-Levallois cores; instead, relatively simple core types predominate. Subsequently, the sophistication of the technology increases very greatly, with extensive use of disc and Levallois technique to produce thinner flakes with sharp edges, and to obtain more flakes from each core. However, in layer 5 there is a divergence between the stories told by the flakes and the cores — the former continue to be technologically advanced (with the Levallois index reaching its maximum) while the latter are comparatively simple.

When the above observations, and those relating to size, are taken into account, it is clear that there were important changes in strategy at several points in the sequence.

1.  In the lowest layers, there is quite profligate use of flint.
2.  The exceptionally large quantity of cortical flakes in F (especially those with less than 50% of their dorsal surface scarred by primary flaking), and also the average size of flakes, indicate that decortication of pebbles commonly occurred within the cave at this time.
3.  Later the reduction in flake and core size and in the incidence of cortex, coupled with increasing use of disc or discoidal (or Levallois) cores, reflect both an increase in efficiency and (as is also apparent in the general reduction in tool size) the growing acceptability of smaller blanks.
4.  This trend reaches its culmination in layer A, when all types of artefact are exceptionally small and the knappers have recourse to the treatment of flakes as cores to a far greater extent than before. The dearth of Levallois cores relative to flakes (when absolute counts are used) may perhaps be explained by the extent to which the cores have been worked down. However, restoration of the percentage of cortex in B and A to something approaching its former level, despite the small size of the flakes, cannot be entirely accounted for by technology; thus preparation of the margins of disc and Levallois cores might result in some flakes with distal cortex, but these will be particularly short (liable to have been excluded because of the 20 mm cut-off, and so not represented in the statistics). Rather, the incidence of cortex here is likely to reflect the use of smaller pebbles from the outset.
5.  The industry of layer 5 has an altogether different character, with its high IL and other indications of advanced technology, and its considerable increase in the size of waste and cores and in the percentage of flakes which bear (generally rather limited) residual cortex. This may seem to point to a generous supply of good raw material, worked in the cave. However, the incidence of Levallois flakes is completely at odds with the evidence of the cores, which are mostly of simple type; moreover the amount of debitage present in this layer is far too small for Levallois technique to have been practised there, yet the presence of a relatively large number of cores, with waste, is an indication that knapping took place within the excavated area. The best explanation that can be offered is that Levallois debitage (and probably much of the tool manufacture) took place away from the cave, but that a few pebbles were imported and used to supply sharp flakes as and when required.

249

# 26

# THE FLINT TOOLS

P. Callow

## 26.1 INTRODUCTION

In this chapter the assemblages of Saalian age are described in detail, in stratigraphic order and in terms of the Bordes type-list. Pseudo-Levallois points and naturally backed knives, and the poorly retouched tool types 45–50, are considered separately from the more clearly defined tools of the 'reduced' list (see chapter 23). In section 26.5, after the tools from all the assemblages have been described, a number of specific problems and tool types are considered in some detail. For typological analysis the whole collection was used, but for the more detailed observations described in section 23.2 sampling of the better represented categories was performed. The sample sizes are given in microfiche (mf 26.1).

With the account of each assemblage is provided a set of tables giving the typological counts and indices (Tables 26.1–20). To facilitate cross-reference, tables in which data for all of the assemblages are presented are placed at the end of this chapter (Tables 26.21–28). Cumulative curves based on the Bordes type-list are given in Figs. 26.1–2.

In the description of the side–scrapers, which comprise types 9-29, an account is first given of the noteworthy features of the class as a whole. It is followed by a breakdown into major groups which do not follow the strict sequence of the Bordes list; instead, they are organised more hierarchically:

SIMPLE
|  |  |  |
|---|---|---|
| Single lateral | | 9–11 |
| Transverse | | 22–24 |

MULTIPLE
|  |  |  |
|---|---|---|
| Double separate | | 12–17 |
| Convergent | axial | 18–20 |
| | déjeté | 21 |

OTHER
|  |  |
|---|---|
| On ventral face | 25 |
| With abrupt retouch | 26 |
| With thinned back | 27 |
| With bifacial retouch | 28 |
| With alternate retouch | 29 |

This represents no more than a reordering of the types, whose definitions remain unchanged, to bring the transverse scrapers (which are almost invariably single-edged) into juxtaposition with the single lateral scrapers; it has been done on the grounds that the lateral or transverse position of the retouch seems to be more the result of the interaction between blank technology and functional needs than between function and style — at least at La Cotte. Classification in order of morphological complexity offers the advantage of permitting a more immediate comparison of very similar classes. It should be noted that the terms 'simple' and 'multiple' refer only to pieces with normal (dorsal) retouch. Those assigned to types 25–29 can also be classified on the same morphological criteria, but in the Bordes scheme their special characteristics take precedence over shape. For this reason the equivalent types in the range 9–24 are given for these pieces in Table 26.24.

In describing the assemblages, the writer has adopted the policy of commenting on changes from one series to the next. Some consistent features are also noted. Perhaps the most conspicuous of these is the tendency for certain types to be made on smaller and others on larger blanks than average (e.g., in particular, *raclettes* and double side-scrapers respectively). Truncated pieces and end-scrapers are of course shorter than their original blank length as a result of the location of their retouch. It should also be borne in mind that, because of the practice employed during initial processing of the finds from the 1960s onwards (see section 23.2.1), the length recorded for transverse scrapers (and all pieces on transverse blanks) was taken on an axis which usually ran approximately at right-angles to the percussion axis.

## 26.2 WELL-RETOUCHED TOOLS

### *LAYER H*

The number of pieces making up the reduced total (i.e. well-retouched tools only) is 160, or 12.0% of the unbroken flint material. The real equivalent is 22.3%. Partly retouched types 45–50 are at their maximum combined level, 34.1% of the real total.

In the reduced series the extended denticulate group (consisting of notched and end-notched pieces, denticulates and Tayac points) is overwhelmingly predominant, amounting to 67.5%. Notched pieces alone exceed 30%, including many with large Clactonian notches; occurrences of multiple notches are common. Among the most distinctive artefacts from this layer are some thick pseudo-Levallois points bearing deep Clactonian notches.

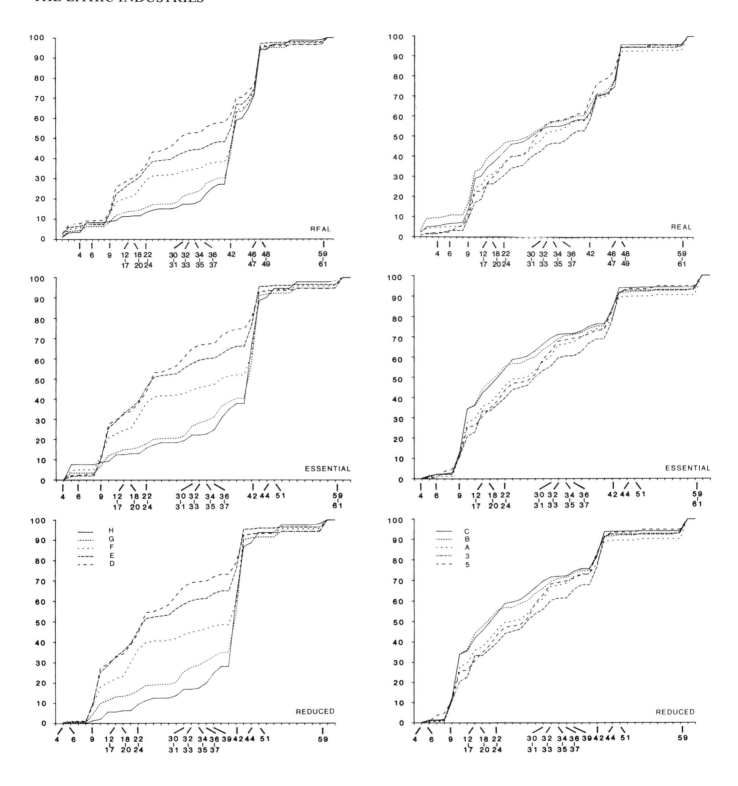

**Fig. 26.1** Cumulative curves for flint tools (layers H to D).

**Fig. 26.2** Cumulative curves for flint tools (layers C to 5).

**Table 26.1**  Layer H. Tool Type Frequencies

| Type | | N | Real | Ess. | Red. |
|---|---|---|---|---|---|
| 1 | Typical Levallois flakes | 3 | 1.01 | – | – |
| 2 | Atypical Levallois flakes | 6 | 2.03 | – | – |
| 3 | Levallois points | 1 | 0.34 | – | – |
| 5 | Pseudo-Levallois points | 14 | 4.73 | 7.57 | – |
| 9 | Single straight side-scrapers | 2 | 0.68 | 1.08 | 1.25 |
| 10 | Single convex side-scrapers | 1 | 0.34 | 0.54 | 0.63 |
| 11 | Single concave side-scrapers | 6 | 2.03 | 3.24 | 3.75 |
| 19 | Convergent convex side-scrapers | 1 | 0.34 | 0.54 | 0.63 |
| 22 | Straight transverse side-scrapers | 2 | 0.68 | 1.08 | 1.25 |
| 23 | Convex transverse side-scrapers | 2 | 0.68 | 1.08 | 1.25 |
| 24 | Concave transverse side-scrapers | 1 | 0.34 | 0.54 | 0.63 |
| 25 | Side-scrapers with inverse retouch | 3 | 1.01 | 1.62 | 1.88 |
| 26 | Abrupt retouched side-scrapers | 2 | 0.68 | 1.08 | 1.25 |
| 29 | Alternate retouched side-scrapers | 2 | 0.68 | 1.08 | 1.25 |
| 30 | Typical end-scrapers | 3 | 1.01 | 1.62 | 1.88 |
| 31 | Atypical end-scrapers | 2 | 0.68 | 1.00 | 1.25 |
| 35 | Atypical borers | 1 | 0.34 | 0.54 | 0.63 |
| 36 | Typical backed knives | 1 | 0.34 | 0.54 | 0.63 |
| 37 | Atypical backed knives | 3 | 1.01 | 1.62 | 1.88 |
| 38 | Naturally backed knives | 11 | 3.72 | 5.95 | – |
| 39 | *Raclettes* | 8 | 2.70 | 4.32 | 5.00 |
| 40 | Truncated pieces | 5 | 1.69 | 2.70 | 3.13 |
| 42 | Notched pieces | 54 | 18.24 | 29.19 | 33.75 |
| 43 | Denticulates | 40 | 13.51 | 21.62 | 25.00 |
| 44 | *Becs burinants alternes* | 3 | 1.01 | 1.62 | 1.88 |
| 45 | Retouched with inverse retouch | 14 | 4.73 | – | – |
| 46-47 | Abrupt and alternate retouch (thick) | 20 | 6.76 | – | – |
| 48-49 | Abrupt and alternate retouch (thin) | 67 | 22.64 | – | – |
| 51 | Tayac points | 8 | 2.70 | 4.32 | 5.00 |
| 54 | End-notched pieces | 6 | 2.03 | 3.24 | 3.75 |
| 61 | Chopping tools | 1 | 0.34 | 0.54 | 0.63 |
| 62 | Miscellaneous | 3 | 1.01 | 1.62 | 1.88 |
| | Bifaces (whole) | 0 | | | |
| | Totals | | 296 | 185 | 160 |

The denticulates are particularly coarsely made, often consisting of a series of large notches. The percentage of Tayac points in this layer, 5.0, may be compared with the next highest value, 0.55% (in layer F). Side-scrapers are scarce (13.8%) and often poorly made. Moreover, there is a considerable blurring of the boundary between these and the notches, as concave-edged pieces are much more common than in other layers.

**Side-scrapers** (22). At 13.8% of the retouched tools, side-scrapers play a minimal part in this assemblage. Their limited representation restricts the usefulness of detailed statistics, but certain features are quite striking. Taking the typological breakdown into major groups, we have:

| | | | |
|---|---|---|---|
| single lateral | 9 | double | 0 |
| convergent | 1 | transverse | 5 |
| other (25–29) | 7 | | |

The relative frequencies for transverse scrapers (22.7% of all side-scrapers) and also for types 25–29 (31.8%) are exceptionally high. However, the two pieces on pseudo-Levallois points have been classified as transverse, since the retouched edge is opposed to the point of detachment (e.g. Fig. 26.4, 2); because of the extent to which the core margin has been retained, it could be argued that they should be regarded as lateral.

Eight pieces have concave edges — the percentage (36.4) is double that for G, and greatly in excess of that for any of the later layers, in which 10% is usual; this is responsible for the exceptionally low mean curvature (0.01 $\pm$ 0.10) of the retouched edges. In a layer dominated by notches and denticulates, there is a real possibility that some of the concave scrapers are functional equivalents of the larger notches (which would reduce IR to only 8.75), though in the case of Fig. 26.4, 1 the large radius of curvature may preclude this. It is therefore of particular interest that scalar/semi-abrupt retouch occurs on only 12 (i.e. 54%) of the scrapers — whereas in every other layer the percentage exceeds 80. Three pieces are in fact somewhat denticulated. A combination of the retouch type and concavity of the edges gives rise to a high mean edge angle, of 67.8°. The mean length, 43.7 $\pm$ 12.3 mm, is actually below average for the tools.

Four of the scrapers (18.1%) are on Levallois blanks (the percentage for all retouched tools is 6.25); the facetting index (IF) is 35.7, lower than usual for the layer.

Simple scrapers

*Single lateral* (9) (Fig. 26.4, 1). Six are concave, including the example illustrated; these have a mean retouched edge length of 29.2 mm, whereas that for the other three is only 17 mm (giving an overall mean of 25.1 mm). The retouch has a mean invasiveness of 2.6 mm. Seven of the nine are worked on the left hand edge. The mean length of the scrapers in this group is only 40.7 mm. All the blanks are simple flakes.

*Transverse* (5) (Fig. 26.4, 2–4). The pieces in this group are in marked contrast to the above, with a retouch invasiveness averaging 7.6 mm, retouched edge length of 34.0 mm and blank length

**Table 26.2**  Layer H. Typological Groups and Indices

| Index | | Real | Ess. | Red. |
|-------|---|------|------|------|
| ILty | Typological Levallois | 3.38 | – | – |
| IR | Side-scrapers | 7.43 | 11.89 | 13.75 |
| IC | Charentian | 2.03 | 3.24 | 3.75 |
| IAu | Acheulian unifacial | 1.35 | 2.16 | 2.50 |
| IAt | Acheulian total | 1.35 | 2.16 | 2.50 |
| IB | Bifaces | 0.00 | 0.00 | 0.00 |
| I | Levallois | 3.38 | 0.00 | 0.00 |
| II | Mousterian | 7.43 | 11.89 | 13.75 |
| III | Upper Palaeolithic | 5.07 | 8.11 | 9.38 |
| IV | Denticulate | 13.51 | 21.62 | 25.00 |
| 5 | Pseudo-Levallois points | 4.73 | 7.57 | – |
| 38 | Naturally backed knives | 3.72 | 5.95 | – |
| I rc | Convergent retouched | 0.34 | 0.54 | 0.63 |
| 6-7 | Mousterian points | 0.00 | 0.00 | 0.00 |
| Rsingle | Side-scrapers, single | 3.04 | 4.86 | 5.63 |
| Rdouble | Side-scrapers, double | 0.00 | 0.00 | 0.00 |
| Rtransv | Side-scrapers, transverse | 1.69 | 2.70 | 3.13 |
| Rinv | Side-scrapers, inverse/alternate | 1.69 | 2.70 | 3.13 |
| 30-31 | End-scrapers | 1.69 | 2.70 | 3.13 |
| 32-33 | Burins | 0.00 | 0.00 | 0.00 |
| 36-37 | Knives | 1.35 | 2.16 | 2.50 |
| 42 | Notched pieces | 18.24 | 29.19 | 33.75 |
| 42-43 | Notched pieces & denticulates | 31.76 | 50.81 | 58.75 |
| Ident | Extended denticulate group | 4.73 | 7.57 | 8.75 |
| Iirr | Irregularly retouched | 34.12 | – | – |
| IQ | Quina | 0.00 | | |
| IQ+½Q | Quina + demi-Quina | 0.00 | | |
| C1 | Clactonian notches | 61.11 | | |

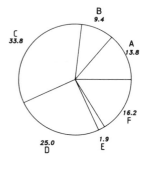

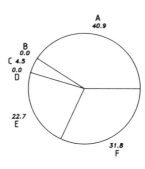

**Fig. 26.3**  Flint tools from layer H. Relative frequencies of the principal tool categories and of the major side-scraper types. Tools: A Group II; B Group II; C Group IV; D notched pieces; E miscellaneous; F other. Side-scrapers: A simple lateral; B simple transverse; C double separate; D convergent axial; E *déjeté*; F other.

of 51.1 mm (the reader is reminded that this last is *not* measured along the axis of percussion — see section 23.2). The first of the illustrated examples is a borderline case, inasmuch as the relict margin is separated from the retouched edge by only a short length of cortex. It is also one of two made on crude pseudo-Levallois points. One piece has a thinned butt.

Multiple scrapers

*Convergent axial* (1). This is biconvex on a Levallois flake 65 mm long with a thinned butt. It is an unusual piece inasmuch as the left-hand edge has denticulated retouch, and the other scalar.

Other side-scrapers

The seven pieces in this group constitute an important element in the side-scraper component in H, and make a major contribution to the incidence of straight and convex working edges (Fig. 24.1). Taken with their relatively invasive retouch (mean 7.0 mm, contrasting with that for the single lateral scrapers) this gives credibility to the concept of scrapers as a tool class in an industry with a quite different emphasis.

*On the ventral surface* (3). One is a double straight scraper with steep retouch on a small (L = 40 mm) plain flake. The other two are concave-edged, one being transverse on a Levallois flake, the other lateral on a cortex-backed flake (Fig. 26.4, 5). This last piece is the most invasively worked (11 mm). The mean edge angle is fairly high (65.5°).

*With abrupt retouch* (2). Both are straight. One is made on a small core-like blank; the other, more dubious, is on a flake only 27 mm long.

*With alternate retouch* (2). These are double, with the longest retouch edge of convex form, and of small size (mean length 40 mm). One, on an atypical Levallois flake, has alternating retouch.

**End-scrapers** (5; 3 typical, 2 atypical). All are simple in form, and on plain flakes (Fig. 26.4, 6 is on a transverse blank, with a deep Clactonian notch adjacent to the end-scraper). One piece has a convexity of 0.25, with a very steep working edge; the others are relatively straight, and could very easily be reclassified as truncated flakes (e.g. Fig. 26.4, 7). There are single examples with left- and right-skewed tips.

**Borer** (1; atypical). So classified only because of slight retouch or utilisation adjacent to the butt, forming a moderately acute angle, this is on a smallish (43.5 mm) flake with plain butt.

**Knives** (4; 1 typical, 3 atypical). The mean length is 49.0 mm, i.e. rather above average for the tools from H. The butts are varied in type. The best example has slightly denticulated, thin semi-abrupt retouch along most of the right hand edge — it could also qualify as a side-scraper. Two of the other pieces are on atypical Levallois blanks.

**Raclettes** (8). As is usual for this class, these are smaller than the average for tools from the same layer (38.0 mm). There is no clear preference for the location of the retouch, but six are straight, the others being convex and concave respectively. Two are inverse; these are also the pieces most likely to be the result of utilisation. One is made on a pseudo-Levallois point, another on a blade-like core rejuvenation flake, and a third on a TSF. Two pieces have facetted butts, the other being plain.

**Truncated pieces** (5). As usual, these are slightly shorter than average (37.4 mm). Two of the truncations are on the proximal end, and all but one (concave) are straight. On a single example the truncation is inverse. The truncation width is small (mean 16 mm). One piece is on a cortex-backed blade.

**Notched pieces** (54). Amounting to 33.7% of the retouched tools, these are, with the denticulates, the characteristic artefacts of layer H (Fig. 26.4, 8–12). Clactonian (single-blow) notches predominate (61%) and are often of considerable size, with a mean width of 21.5 ± 11.7 mm, against 9.8 ± 2.8 for retouched notches. Accordingly, their scars run further across the surface of the blank than in any other layer; it has been observed during analysis that, as a rough approximation, the invasiveness of the removal forming a Clactonian notch is generally about half of the width of the scar in any of the assemblages studied. More surprising, at first, was that the curvature was lower for notches

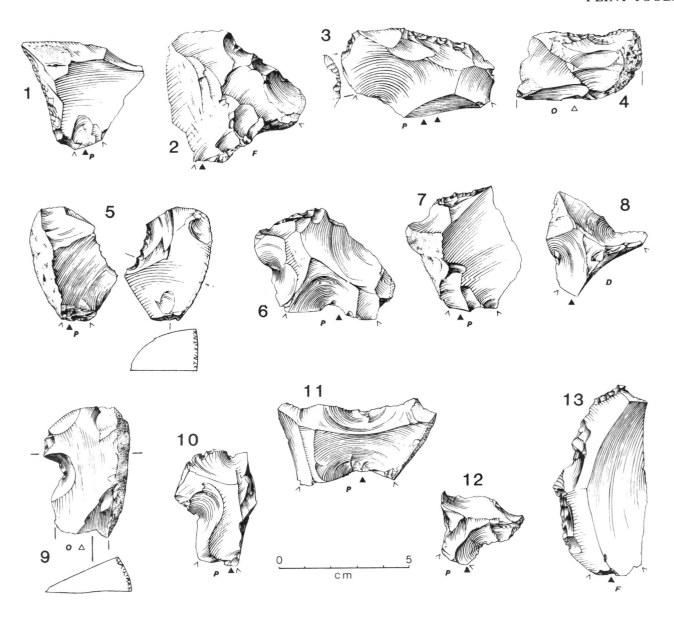

**Fig. 26.4** Flint tools from layer H. 1–5 Side-scrapers: 1 simple lateral; 2–4 transverse; 5 on the ventral surface; 6, 7 end-scrapers; 8–12 Clactonian notches (8 is on a pseudo-Levallois point); 13 denticulate. Information about the butt is given by the following symbols: *C* cortical; *P* plain; *D* dihedral; *X* polyhedral; *F* facetted; *R* removed; *O* broken; *?* indeterminable; ∧ ... ∧ limits of butt; ▲ point of percussion; △ direction of blow (where the point of percussion is missing).

of this type than for those formed by retouch — the only layer in which this was found to be the case. The explanation is that the travel of the blow is restricted in the case of the wider notches by the dorsal ridges of the blank (cf. Fig. 26.4, 8 and 10). As a result, both mean (0.23) and modal curvature are low.

| Curvature | 0.0 | 0.1 | 0.2 | 0.3 | 0.4 | 0.5 |
|---|---|---|---|---|---|---|
| % | 3.1 | 44.4 | 28.7 | 20.6 | 3.1 | |

One very noticeable aspect of this particular layer is that more than 14 out of 33 Clactonian notched pieces have two such removals, and 4 more have three — i.e. more than half are multiple (14 are opposed). In contrast, only 5 out of 21 of the retouched variety are multiple, with 3 opposed. Among the most distinctive artefacts are thick pseudo-Levallois points with one or two notches; in six out of seven cases these are Clactonian (e.g. Fig. 26.4, 8). Retouched notches occur on

the single occurrences of (atypical) Levallois and TSF blanks; two of the three cortex-backed blanks have Clactonian notches. No examples of thinned butts or sides occur on the notched pieces sampled from this or any other layer.

**Denticulates** (40). These are generally well made, and form one of the dominant types of this assemblage, amounting to 25.0% of the reduced total (Figs. 26.4, 13 and 26.5, 1–4). One is made on a Levallois flake and three each on pseudo-Levallois points (all of these being manufactured by Clactonian notches) and cortex-backed flakes (Fig. 26.5, 3 and 4) respectively; one is on a core.

Multiple adjacent Clactonian notches (32.2%) have been more extensively employed for the manufacture of denticulates in this layer than in any other. That the pieces on pseudo-Levallois points were made in this way perhaps indicates a continuum with the notches made on the same type of blank (see above). Coarse denticulate retouch,

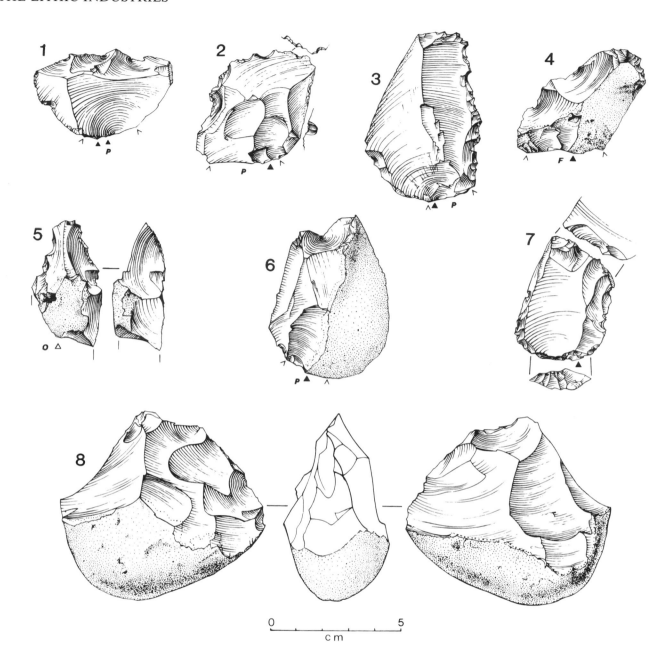

**Fig. 26.5** Flint tools from layer H. 1–4 Denticulates; 5 Tayac point; 6 end-notched flake; 7 miscellaneous (denticulate with distal scaling); 8 chopping tool.

often with hinging of the removals, has been used on 37.2%, and fine retouch on 21.5%. Microdenticulates account for 9.1%, which is a fairly typical figure for the site as a whole. Thus the great majority of the denticulates (69.4%) have coarse working edges.

Morphological classification shows that 87.5% fall into the lateral (60.0%) and transverse (27.5%) subtypes; there are three *déjeté*, one double but no convergent pieces (apart from the Tayac points mentioned below). This is the simplest denticulate subassemblage at La Cotte apart from that of layer E (which is of a quite different character, however). The tools are relatively large, as is usual for this layer: mean length 48.3 ± 11.0 mm, and retouch length 33.9 ± 13.3 mm. The removal scars are deeply invasive (6.9 ± 3.9 mm), as one would expect in view of the importance of Clactonian notches.

**Becs burinants alternes** (3). These may be accidental.

**Tayac points** (8). Of average size (43.8 mm), these are quite a distinctive feature of the assemblage, made in all but one case using Clactonian notches. (Fig. 26.5, 5); the exception has a row of notches opposed to an edge formed by *surélevée* retouch. Three are somewhat *déjeté*. None exhibits fresh facetting of the butt, though two are on pseudo-Levallois points.

**End-notched pieces** (6). They are made on flakes which are only slightly shorter than average (39.3 mm). The largest piece has a cortical back (Fig. 26.5, 6). Only one notch has been made by retouch.

**Table 26.3** Layer G. Tool Type Frequencies

| Type | | N | Real | Ess. | Red. |
|---|---|---|---|---|---|
| 1 | Typical Levallois flakes | 8 | 1.17 | – | – |
| 2 | Atypical Levallois flakes | 19 | 2.77 | – | – |
| 3 | Levallois points | 1 | 0.15 | – | – |
| 4 | Retouched Levallois points | 2 | 0.29 | 0.45 | 0.49 |
| 5 | Pseudo-Levallois points | 13 | 1.90 | 2.94 | – |
| 9 | Single straight side-scrapers | 16 | 2.34 | 3.62 | 3.95 |
| 10 | Single convex side-scrapers | 22 | 3.21 | 4.98 | 5.43 |
| 11 | Single concave side-scrapers | 6 | 0.88 | 1.36 | 1.48 |
| 12 | Double straight side-scrapers | 1 | 0.15 | 0.23 | 0.25 |
| 13 | Double straight-convex side-scrapers | 1 | 0.15 | 0.23 | 0.25 |
| 14 | Double straight-concave side-scrapers | 1 | 0.15 | 0.23 | 0.25 |
| 15 | Double convex side-scrapers | 3 | 0.44 | 0.68 | 0.74 |
| 17 | Double concave-convex side-scrapers | 1 | 0.15 | 0.23 | 0.25 |
| 20 | Convergent concave side-scrapers | 1 | 0.15 | 0.23 | 0.25 |
| 21 | Déjeté scrapers | 4 | 0.58 | 0.90 | 0.99 |
| 22 | Straight transverse side-scrapers | 2 | 0.29 | 0.45 | 0.49 |
| 23 | Convex transverse side-scrapers | 3 | 0.44 | 0.68 | 0.74 |
| 24 | Concave transverse side-scrapers | 3 | 0.44 | 0.68 | 0.74 |
| 25 | Side-scrapers with inverse retouch | 10 | 1.46 | 2.26 | 2.47 |
| 26 | Abrupt retouched side-scrapers | 1 | 0.15 | 0.23 | 0.25 |
| 27 | Side-scrapers with thinned back | 1 | 0.15 | 0.23 | 0.25 |
| 29 | Alternate retouched side-scrapers | 5 | 0.73 | 1.13 | 1.23 |
| 30 | Typical end-scrapers | 7 | 1.02 | 1.58 | 1.73 |
| 31 | Atypical end-scrapers | 14 | 2.04 | 3.17 | 3.46 |
| 32 | Typical burins | 7 | 1.02 | 1.58 | 1.73 |
| 33 | Atypical burins | 1 | 0.15 | 0.23 | 0.25 |
| 35 | Atypical borers | 5 | 0.73 | 1.13 | 1.23 |
| 36 | Typical backed knives | 3 | 0.44 | 0.68 | 0.74 |
| 37 | Atypical backed knives | 5 | 0.73 | 1.13 | 1.23 |
| 38 | Naturally backed knives | 24 | 3.50 | 5.43 | – |
| 39 | Raclettes | 9 | 1.31 | 2.04 | 2.22 |
| 40 | Truncated pieces | 8 | 1.17 | 1.81 | 1.98 |
| 42 | Notched pieces | 77 | 11.24 | 17.42 | 19.01 |
| 43 | Denticulates | 147 | 21.46 | 33.26 | 36.30 |
| 44 | Becs burinants alternes | 3 | 0.44 | 0.68 | 0.74 |
| 45 | Retouched with inverse retouch | 26 | 3.80 | – | – |
| 46–47 | Abrupt and alternate retouch (thick) | 34 | 4.96 | – | – |
| 48–49 | Abrupt and alternate retouch (thin) | 154 | 22.48 | – | – |
| 50 | Bifacial retouch | 1 | 0.15 | – | – |
| 51 | Tayac points | 2 | 0.29 | 0.45 | 0.49 |
| 54 | End-notched pieces | 11 | 1.61 | 2.49 | 2.72 |
| 62 | Miscellaneous | 23 | 3.36 | 5.20 | 5.68 |
| | Bifaces (whole) | 0 | | | |
| | Totals | | 685 | 442 | 405 |

**Chopping tool** (1). This has been made on a flint pebble by means of alternating removals, and weighs 213 gm (Fig. 26.5, 8).

**Miscellaneous** (3). One is a denticulate with scaling off a distal inverse truncation (Fig. 26.5, 7). The others are scrapers; one is made by retouching the butt, and the second is a simple scraper with, again, scaling off an inverse truncation.

## LAYER G

The integrity of the assemblages from layers G and F is subject to an element of uncertainty, in that the admittedly sparse palynological evidence (chapter 11) raises the possibility of an intermediate phase of climatic deterioration, and hence that the two assemblages should be further subdivided. This is impracticable because of the small number of samples available for pollen analysis and the poor microstratigraphic detail at this point in the sequence. The colluviation reported by van Vliet-Lanoë in chapter 9 also leaves open the question of geological mixing. Notwithstanding these caveats, the assemblages possess sufficient distinctive and internally consistent features for it to be likely that each is primarily representative of a single industry. The reduced total for layer G is 405, which at 14% of the total unbroken flint assemblage is slightly higher than for H, as is the equivalent real figure of 23.7%. There is a small reduction in the abundance of types 45–50, to 31.4% of the real total.

Among the well-retouched tools the extended denticu-

**Table 26.4** Layer G. Typological Groups and Indices

| Index | | Real | Ess. | Red. |
|---|---|---|---|---|
| ILty | Typological Levallois | 4.38 | – | – |
| IR | Side-scrapers | 11.82 | 18.33 | 20.00 |
| IC | Charentian | 4.38 | 6.79 | 7.41 |
| IAu | Acheulian unifacial | 1.17 | 1.81 | 1.98 |
| IAt | Acheulian total | 1.17 | 1.81 | 1.98 |
| IB | Bifaces | 0.00 | 0.00 | 0.00 |
| | | | | |
| I | Levallois | 4.38 | 0.45 | 0.49 |
| II | Mousterian | 11.82 | 18.33 | 20.00 |
| III | Upper Palaeolithic | 7.30 | 11.31 | 12.35 |
| IV | Denticulate | 21.46 | 33.26 | 36.30 |
| | | | | |
| 5 | Pseudo-Levallois points | 1.90 | 2.94 | – |
| 38 | Naturally backed knives | 3.50 | 5.43 | – |
| I rc | Convergent retouched | 0.73 | 1.13 | 1.23 |
| 6-7 | Mousterian points | 0.00 | 0.00 | 0.00 |
| Rsingle | Side-scrapers, single | 6.42 | 9.95 | 10.86 |
| Rdouble | Side-scrapers, double | 1.02 | 1.58 | 1.73 |
| Rtransv | Side-scrapers, transverse | 1.17 | 1.81 | 1.98 |
| Rinv | Side-scrapers, inverse/alternate | 2.19 | 3.39 | 3.70 |
| 30- 31 | End-scrapers | 3.07 | 4.75 | 5.19 |
| 32-33 | Burins | 1.17 | 1.81 | 1.98 |
| 36-37 | Knives | 1.17 | 1.81 | 1.98 |
| 42 | Notched pieces | 11.24 | 17.42 | 19.01 |
| 42-43 | Notched pieces & denticulates | 32.70 | 50.68 | 55.31 |
| Ident | Extended denticulate group | 3.07 | 4.75 | 5.19 |
| Iirr | Irregularly retouched | 31.39 | – | – |
| | | | | |
| IQ | Quina | 0.00 | | |
| IQ+½Q | Quina + demi-Quina | 0.00 | | |
| Cl | Clactonian notches | 44.16 | | |

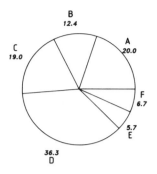
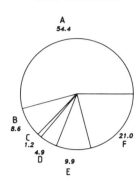

**Fig. 26.6** Flint tools from layer G. Relative frequencies of the principal tool categories and of the major side-scraper types. Tools: A Group II; B Group II; C Group IV; D notched pieces; E miscellaneous; F other. Side-scrapers: A simple lateral; B simple transverse; C double separate; D convergent axial; E *déjeté*; F other.

no Quina or demi-Quina retouch, but a major departure from H is that 85.2% of the scrapers have scalar or semi-abrupt retouch. Edge length and retouch angle are almost identical to those in H; there is a small increase in length, to 45.9 ± 13.0 mm.

Transverse scrapers are of reduced importance, while single lateral pieces account for more than half, and multiple scrapers are present in appreciable quantities (Fig. 26.6). The percentages for each group are

| single lateral | 54.3 | double | 8.6 |
|---|---|---|---|
| convergent | 6.2 | transverse | 9.9 |
| other | 21.0 | | |

Six pieces (14.8%) are on Levallois blanks — slightly above the average for retouched tools — with two more on pseudo-Levallois points and five on cortex-backed blanks. The facetting index (IF = 55.4) is above average for the layer.

Simple scrapers

*Single lateral* (44) (Fig. 26.7, 1, 3, 5 and 6). The low retouch invasiveness (mean 2.9 mm) owes much to the value for straight pieces (2.6). Also the retouched edge is unusually short for straight and concave pieces (23.7 and 23.3 mm, against 36.7 for convex edges). None have notches or thinning, though two have additional retouch on another, adjacent, edge — either for use or to provide a more comfortable grip.

*Transverse* (8). These are scarce but have some distinctive features: in particular, the mean retouch angle (75.3%) and invasiveness (4.5 mm) are high for this layer. One piece is made on the proximal end.

Multiple scrapers

*Separate* (7). The only subtype not represented is biconcave, while the only one of which more than one example occurs is biconvex. Typical Levallois blanks are favoured (3 cases), and the IF is 83.3 — surprisingly high. Also, the mean length (50.7 mm) is nearly 5 mm greater than the average for scrapers from this layer, suggesting that the blanks may have therefore been specially selected (this tends to be true in other layers also). However, the retouch is relatively slight (invasiveness 2.7 mm) and in two instances the longer retouched edge extends over less than 30 mm. Equally, on Fig. 26.7, 4 the secondary work on the principal edge could have served as backing, while that on the other edge may be no more than regularisation (e.g. of the blade of a knife). Thus the status of some of these pieces is uncertain.

late group is still dominant (58.5%), though now denticulates are more common than notches (36.3% against 19.1%). The former have coarse edges, but are more elaborate in morphology; the latter include fewer, and smaller, Clactonian notches. Multi-edged notched pieces are scarce. Side-scrapers are again relatively uncommon (20.0%); their edges are often only lightly retouched. The percentage for Group III ('Upper Palaeolithic' types 30–37) is unusually high, at 10.4. Some of the end-scrapers are particularly fine. Although the relative frequencies of side-scrapers and the extended denticulate group are rather similar to those of layer H, there are important differences of detail which make it clear that these two assemblages cannot have been seriously mixed geologically.

**Retouched Levallois points** (2). Both are of the secondary type. One has light semi-abrupt retouch on one edge, but only a small notch on the other; the second piece (Fig. 26.7, 2) has better secondary work, but also possesses a notch close to the tip. The butt of the former example is dihedral, that of the latter facetted.

**Side-scrapers** (81). Though still relatively scarce (IR$_{red}$ = 20.0), there are many more typical examples in this layer than in layer H. There is a halving, to 18.5%, of the percentage of the concave-edged pieces, whose equivocal position between scrapers and notches caused problems in H (but see, for example, Fig. 26.7, 7). Instead, convex-edged scrapers predominate, at 46.9% (Fig. 26.7, 3 and 5). As a result, the mean of the curvature ratio is 0.05, midway between those for layer H and for the upper layers C to 5. However, the mean invasiveness of the retouch is very low, at 3.2 ± 2.1 mm, apparently because the straighter-edged pieces often have only rather light trimming. There is

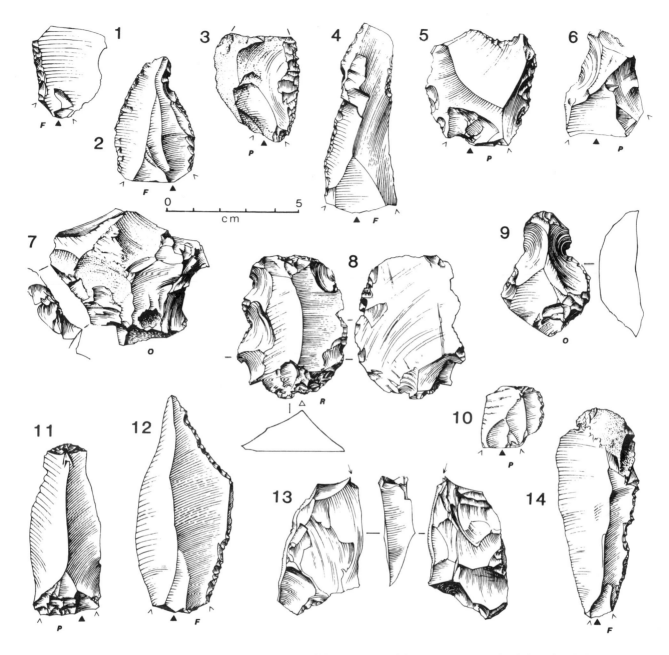

**Fig. 26.7** Flint tools from layer G. 2 Retouched Levallois point; 1, 3–7 side-scrapers: 1, 3, 5, 6 simple lateral; 4 double; 7 on the ventral surface; 8–11 end-scrapers (8 is on a denticulate); 13 burin; 12, 14 knives.

*Convergent axial* (1). This is a small (L = 37.5) concave-convex piece with a narrow tip (15 mm wide at 10 mm from the point).

*Déjeté* (4). In three cases (one an atypical Levallois flake, another double *déjeté*) the longer edge is convex; in the fourth it is concave. One piece has a particularly sharp point.

Other side-scrapers

Table 26.24 shows that these are dominated morphologically by two types which at first sight appear strongly contrasted: single inverse scrapers and double-sided alternate scrapers. The group, which is numerically quite important (21.0% of side-scrapers), may therefore be redefined, at the cost of some approximation, as consisting largely of scrapers with inverse retouch, with or with-

out another (normally) retouched edge. Though not at all extreme, in its invasiveness the retouch is surpassed only by that of the transverse pieces; the worked edges are, however, quite short.

*On the ventral surface* (10). These are all single scrapers, only one being transverse (convex). Of the lateral pieces, four are straight, three concave (Fig. 26.7, 7) and only two convex. Metrically, the edges yield means fairly close to those for the totality of layer G scrapers: retouched edge length 29.4 mm, angle 67.4°, invasiveness 3.7 mm (the *maximum value* is only 8 mm). The blanks include a typical Levallois flake (concave edge) and a cortex-backed flake (straight edge).

*With abrupt retouch* (1). This is straight-edged, on a plain flake.

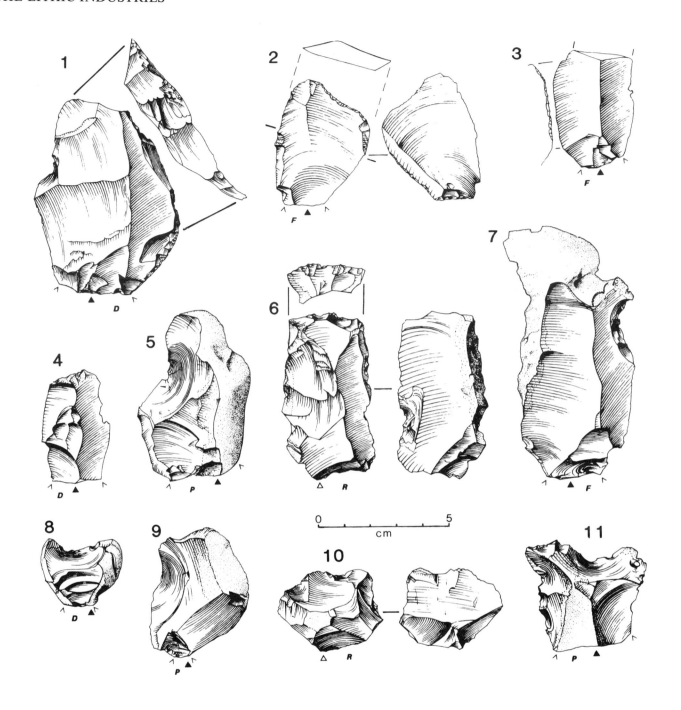

**Fig. 26.8** Flint tools from layer G. 1 Knife; 2–3 *raclettes*; 4, 6 truncated pieces; 5, 7–9 notches; 10, 11 denticulates (11 is a *bec*, or could be regarded as a *déjeté* Tayac point).

*With thinned back* (1). It is otherwise a simple convex scraper.

*With alternate retouch* (5). One is convex-edged (and transverse), two are straight and two concave. The convex piece has alternating working, and the platform has been removed by retouch.

**End-scrapers** (21; 7 typical, 14 atypical). There is a fall in the average length (39.9 mm) of nearly 10 mm relative to layer H, reflecting the presence of a number of rather small pieces (e.g. Fig. 26.7, 10). Nevertheless, there are a number of outstanding pieces which would be quite

at home in an Upper Palaeolithic assemblage — for example, Fig. 26.7, 9 (double-shouldered, with the scraper isolated by Clactonian notches) and 11 (simple, and the only example made on a blade). The end-scraper tips are usually well made (7 have subparallel and 2 have parallel retouch); two are right-skewed and two left-skewed. Six pieces, i.e. 28.6%, also have retouch on a lateral margin (for example Fig. 26.7, 8, on a denticulate). Apart from the double-shouldered piece illustrated, and another with a single shoulder, the end-scrapers are of the 'simple' subtype. None are on transverse flakes; single examples occur of Levallois and cortex-backed blanks.

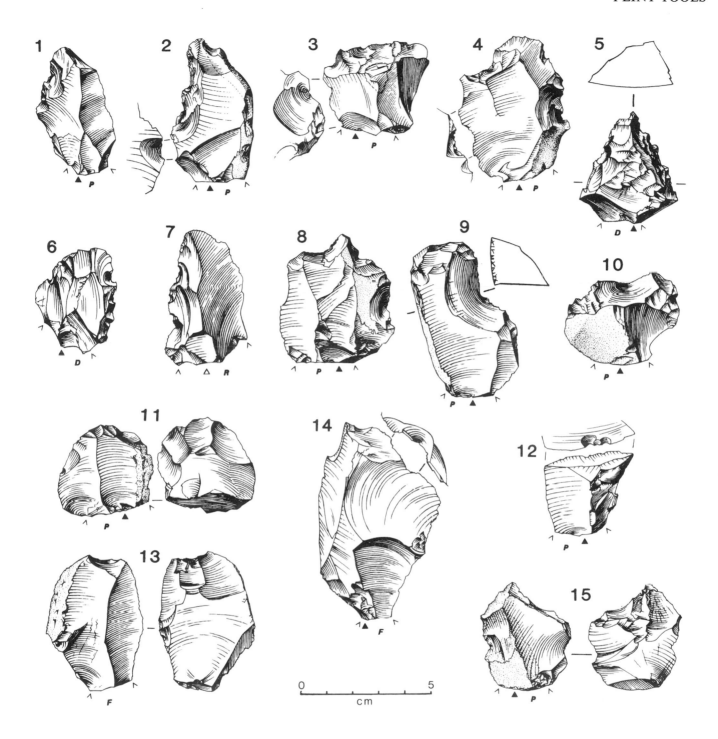

**Fig. 26.9** Flint tools from layer G. 1–4, 6–11 Denticulates (9 and 10 dominated by Clactonian notches); 5 Tayac point; 12 ventrally-retouched flake; 14 end-notched flake; 13, 15 miscellaneous (with scaling).

**Burins** (8; 7 typical, 1 atypical). Six of these are angle burins (four distal, one proximal and the other indeterminate), one dihedral and one *plan*. The dihedral piece is on a corner of the flake, formed by the intersection of transverse and longitudinal removals. Despite the high proportion which for strictly technical reasons have been classified as typical, several of these are questionable and may indeed be the result of spontaneous damage occasioned during debitage or breakage. The tips are exceptionally acute on average (mean $62.6 \pm 7.7°$), and two of the most extreme cases in this respect have very narrow burin scars — only 2 mm wide.

Most are on plain flakes, but there are single examples on a pseudo-Levallois point and a cortex-backed flake. The illustrated piece (Fig. 26.7, 13) is unusual in that a core (probably formerly a plain-butted flake) has been used. One is on an inverse truncation, two on breaks and one on a hinge. Half of them have marginal retouch elsewhere on the piece.

**Borers** (5, all atypical). These are questionable; two are more strictly *becs* than borers.

**Knives** (8; 3 typical, 5 atypical). The three typical examples are all on blades (Fig. 26.7, 12 and 14); the unillustrated piece has an angled back made by thin abrupt retouch along the right-hand edge. The atypical pieces include three with cortical backs extended by retouch, and another on a blade whose backing is provided by a relict margin with a short length of abrupt retouch. The last has a normal truncation which probably served as a finger rest, and an abrupt (primary) edge. The mean length is 59.4 mm. Only the typical examples have facetted butts.

**Raclettes** (9). They are slightly smaller (mean length 40.5) than the average tool size. One example (Fig. 26.8, 2) is on a possible LSF; it is unique in the series from this layer in being both concave and transverse. The others are equally divided between straight and convex.

**Truncated pieces** (8). Two of these have cortical backs (Fig. 26.8, 6), another is on a pseudo-Levallois point and a fourth on an atypical Levallois flake. The mean length is quite low, 40.2 mm. The truncations are straight, apart from a convex and a concave example; their mean width is 18.7 mm.

**Notched pieces** (77). These are less important in relative terms than in layer H (19.0%), but are still one of the dominant components of the assemblage (Fig. 26.8, 5 and 7–9). The Clactonian variety is also rather less common (44%), and includes fewer really large examples. Their mean notch width is 14.9 mm, against 11.3 for the notches made by retouch, which are more invasive than in H.

A difference in the mean curvature between the Clactonian (0.26) and retouched (0.21) notches arises chiefly from a small but extremely incisive example of the former which is probably accidental; the combined mean (0.24) is very slightly higher than for the previous layer. The modal value for all notches falls into a higher interval, but standardisation is improved (the 'odd' piece has been discounted here):

| Curvature | 0.0 | 0.1 | 0.2 | 0.3 | 0.4 |
|---|---|---|---|---|---|
| % | 6.5 | 42.9 | 46.8 | 3.9 | |

Opposed lateral notches are absent (a great contrast to H), though three pieces have removals on more than one edge. Double notches occur in 9% of cases (Clactonian notches are slightly favoured here). The blanks include 3 typical Levallois pieces, 3 pseudo-Levallois and 5 cortex-backed (2 of each having Clactonian notches).

**Denticulates** (147). Layer G, with 36.3%, has the highest frequency of denticulates at the site. Five each are on Levallois flakes, pseudo-Levallois points (four of these being made by adjacent Clactonian notches, as in H), and cortex-backed flakes respectively, with one made on a core. In terms of the edge type also, the denticulates from G resemble those from H: 28.2% Clactonian notches, 35.5% coarse retouch, 31.2% fine retouch, and 5.0% microdenticulation. However, the denticulates of layer G show considerable elaboration of shape, only 63.9% being 'simple' (49.0% lateral, 15.0% transverse), and 6.8% double and 5.4% convergent. The most striking feature, however, is the high incidence of *déjeté* pieces (22.4%) some of which (e.g. Fig.

26.8, 11) could alternatively be described as *becs* or may be thinner equivalents of the Tayac points in the preceding assemblage. There is a slight general size reduction compared with H: mean length $45.7 \pm 11.8$, retouch length $31.4 \pm 9.5$, invasiveness $6.1 \pm 3.3$ mm.

**Becs burinants alternes** (3). One, on a typical Levallois flake, is unusual, formed by the juxtaposition of dorsal removals (from an inverse truncation) adjacent to a slightly denticulated scraper edge. The others are formed by notches next to inverse removals. They average 39 mm in length.

**Tayac points** (2). These are of moderate size (43.0 mm) but symmetrical, thick and quite typical (Fig. 26.9, 5).

**End-notched pieces** (11). They are on ordinary flakes (with one blade) and are of average length (43.0 mm). Only four of the notches have been made by retouch.

**Miscellaneous** (23). The (multiple-response) classification gives the following counts and percentages:

| | n | % | | n | % |
|---|---|---|---|---|---|
| scrapers | 17 | (73.9) | notches | 6 | (26.1) |
| denticulates | 4 | (17.4) | overstruck removals | 1 | (4.3) |
| scaled | 3 | (13.0) | burins | 0 | (0.0) |

Ten pieces (43.5%) are compound, possessing more than one of the above characteristics — half of these combining scraper edges with notches. Only one has dorsal scaling off a (distal) inverse truncation. Two of the scraper-like pieces are on pseudo-Levallois points (there are also two cases of cortex-backed blanks).

## LAYER F

The reduced total, 542, is equivalent to 13.9% of the unbroken flint from this layer; the corresponding real value is 23.6%. These figures, and also the 31.4% of 'real' tools represented by types 45–50, are remarkably close to those for layer G, but in many other respects the two series are very different. Numerically, the extended denticulate group, with a reduced percentage of 46.1, is now only slightly more important than the side-scrapers (40.8%). The notches decline markedly, from 19.0 to 11.3%, and are almost always single. The frequency of denticulates is only a little reduced (32.5%), and coarse edges are still in a clear majority.

The side-scrapers are often extremely well made (contrasting with most from the underlying layers) and include many *déjeté* pieces. Straight retouched edges are more common than in the scraper-dominated series from subsequent layers. Another aspect of the tools from F which serves to distinguish this assemblage from that of layer G is a low percentage of Group III tools.

**Retouched Levallois points** (4). These are of good size (mean 58.5 mm) and of the secondary type; one is a pointed blade. The retouch is very light in every case.

**Mousterian points** (2). One is on a pointed Levallois flake, 67 mm long, with relatively light retouch towards the distal end. The other is much more typical, 46.5 mm long and very sharply pointed. The retouch on the right-hand edge is particularly invasive.

**Side-scrapers** (221). This is the first layer at La Cotte in which side-scrapers may be regarded as the most characteristic element of the assemblage ($IR_{red} = 40.8$), though not yet in an absolute majority. Moreover, they are generally well made, with invasive retouch (mean

**Table 26.5**  Layer F. Tool Type Frequencies

| Type | | N | Real | Ess. | Red. |
|---|---|---|---|---|---|
| 1 | Typical Levallois flakes | 12 | 1.31 | – | – |
| 2 | Atypical Levallois flakes | 34 | 3.71 | – | – |
| 3 | Levallois points | 1 | 0.11 | – | – |
| 4 | Retouched Levallois points | 4 | 0.44 | 0.69 | 0.74 |
| 5 | Pseudo-Levallois points | 23 | 2.51 | 3.95 | – |
| 6 | Mousterian points | 2 | 0.22 | 0.34 | 0.37 |
| 9 | Single straight side-scrapers | 43 | 4.69 | 7.39 | 7.93 |
| 10 | Single convex side-scrapers | 49 | 5.34 | 8.42 | 9.04 |
| 11 | Single concave side-scrapers | 11 | 1.20 | 1.89 | 2.03 |
| 12 | Double straight side-scrapers | 1 | 0.11 | 0.17 | 0.18 |
| 13 | Double straight-convex side-scrapers | 7 | 0.76 | 1.20 | 1.29 |
| 15 | Double convex side-scrapers | 3 | 0.33 | 0.52 | 0.55 |
| 18 | Convergent straight side-scrapers | 3 | 0.33 | 0.52 | 0.55 |
| 19 | Convergent convex side-scrapers | 3 | 0.33 | 0.52 | 0.55 |
| 20 | Convergent concave side-scrapers | 1 | 0.11 | 0.17 | 0.18 |
| 21 | *Déjeté* scrapers | 36 | 3.93 | 6.19 | 6.64 |
| 22 | Straight transverse side-scrapers | 16 | 1.74 | 2.75 | 2.95 |
| 23 | Convex transverse side-scrapers | 5 | 0.55 | 0.86 | 0.92 |
| 24 | Concave transverse side-scrapers | 15 | 1.64 | 2.58 | 2.77 |
| 25 | Side-scrapers with inverse retouch | 17 | 1.85 | 2.92 | 3.14 |
| 26 | Abrupt retouched side-scrapers | 4 | 0.44 | 0.69 | 0.74 |
| 28 | Side-scrapers with bifacial retouch | 2 | 0.22 | 0.34 | 0.37 |
| 29 | Alternate retouched side-scrapers | 5 | 0.55 | 0.86 | 0.92 |
| 30 | Typical end-scrapers | 2 | 0.22 | 0.34 | 0.37 |
| 31 | Atypical end-scrapers | 8 | 0.87 | 1.37 | 1.48 |
| 32 | Typical burins | 3 | 0.33 | 0.52 | 0.55 |
| 33 | Atypical burins | 3 | 0.33 | 0.52 | 0.55 |
| 34 | Typical borers | 2 | 0.22 | 0.34 | 0.37 |
| 35 | Atypical borers | 3 | 0.33 | 0.52 | 0.55 |
| 36 | Typical backed knives | 1 | 0.11 | 0.17 | 0.18 |
| 37 | Atypical backed knives | 4 | 0.44 | 0.69 | 0.74 |
| 38 | Naturally backed knives | 17 | 1.85 | 2.92 | – |
| 39 | *Raclettes* | 7 | 0.76 | 1.20 | 1.29 |
| 40 | Truncated pieces | 3 | 0.33 | 0.52 | 0.55 |
| 42 | Notched pieces | 61 | 6.65 | 10.48 | 11.25 |
| 43 | Denticulates | 176 | 19.19 | 30.24 | 32.47 |
| 44 | *Becs burinants alternes* | 5 | 0.55 | 0.86 | 0.92 |
| 45 | Retouched with inverse retouch | 23 | 2.51 | – | – |
| 46–47 | Abrupt and alternate retouch (thick) | 68 | 7.42 | – | – |
| 48–49 | Abrupt and altetouch (thin) | 197 | 21.48 | – | – |
| 51 | Tayac points | 3 | 0.33 | 0.52 | 0.55 |
| 54 | End-notched pieces | 10 | 1.09 | 1.72 | 1.85 |
| 61 | Chopping tools | 1 | 0.11 | 0.17 | 0.18 |
| 62 | Miscellaneous | 23 | 2.51 | 3.95 | 4.24 |
| Bifaces (whole) | | 0 | | | |
| Totals | | | 917 | 582 | 542 |

5.3 ± 3.1 mm) in marked contrast to those from layer G. The mean edge angle is also appreciably lower, at 59.8 ± 13.4° — a level which is quite consistently maintained in subsequent layers. However, the mean length of the retouched edges is only 31.4 ± 10.7 mm, very like that from the previous layers and substantially lower than for those immediately following (the mean length of the scrapers is 46.2 ± 10.2 mm). Scalar/semi-abrupt retouch is used on 82.3% of the pieces, while at 8.0% subparallel retouch is more important than in any other layer (see for example Figs. 26.11, 12 and 13, and 26.12, 2). The percentages for the major groups are

| | | | |
|---|---|---|---|
| single lateral | 46.6 | double | 5.0 |
| convergent | 19.4 | transverse | 16.2 |
| other | 12.6 | | |

One of the most notable aspects of this assemblage is that no less than 16.3% of the scrapers are of the *déjeté* type — a percentage between two and three times higher than for other layers (Figs. 26.11, 12–14 and 26.12, 1–5). Another unique feature is that straight edges (45.0%) are more common than convex or concave ones. For this reason, and despite a continuation of the trend away from concave edges, the mean curvature is the same as for layer G, i.e. 0.05.

The technology of the blanks very well reflects the pattern for the retouched tools as a whole: 11.8% are Levallois, 4.1% are on pseudo-Levallois points, and 6.3% are cortex-backed; also the IF is 49.1.

Simple scrapers

*Single lateral* (103) (Fig. 26.11, 4–7). The pieces in this category comprise 43 straight, 49 convex and 11 concave, having a mean curvature of 0.06. Three pieces have, as an additional feature, a Clactonian notch, while one has a thinned butt. Eight (i.e. 7.7%) have some retouch on one end, adjacent to the scraper edge. There is a very strong preference for retouching the right-hand edge (65, against 37 on the left edge). The convex scrapers are more steeply retouched (62.4°) than are the straight (54.3°) or concave (55.5°) pieces.

**Table 26.6** Layer F. Typological Groups and Indices

| Index | | Real | Ess. | Red. |
|---|---|---|---|---|
| ILty | Typological Levallois | 5.56 | – | – |
| IR | Side-scrapers | 24.10 | 37.97 | 40.77 |
| IC | Charentian | 9.49 | 14.95 | 16.05 |
| IAu | Acheulian unifacial | 0.55 | 0.86 | 0.92 |
| IAt | Acheulian total | 0.55 | 0.86 | 0.92 |
| IB | Bifaces | 0.00 | 0.00 | 0.00 |
| | | | | |
| I | Levallois | 5.56 | 0.69 | 0.74 |
| II | Mousterian | 24.32 | 38.32 | 41.14 |
| III | Upper Palaeolithic | 3.16 | 4.98 | 5.35 |
| IV | Denticulate | 19.19 | 30.24 | 32.47 |
| | | | | |
| 5 | Pseudo-Levallois points | 2.51 | 3.95 | – |
| 38 | Naturally backed knives | 1.85 | 2.92 | – |
| I rc | Convergent retouched | 4.91 | 7.73 | 8.30 |
| 6–7 | Mousterian points | 0.22 | 0.34 | 0.37 |
| Rsingle | Side-scrapers, single | 11.23 | 17.70 | 19.00 |
| Rdouble | Side-scrapers, double | 1.20 | 1.89 | 2.03 |
| Rtransv | Side-scrapers, transverse | 3.93 | 6.19 | 6.64 |
| Rinv | Side-scrapers, inverse/alternate | 2.40 | 3.78 | 4.06 |
| 30–31 | End-scrapers | 1.09 | 1.72 | 1.85 |
| 32–33 | Burins | 0.65 | 1.03 | 1.11 |
| 36–37 | Knives | 0.55 | 0.86 | 0.92 |
| 42 | Notched pieces | 6.65 | 10.48 | 11.25 |
| 42–43 | Notched pieces & denticulates | 25.85 | 40.72 | 43.73 |
| Ident | Extended denticulate group | 2.07 | 3.26 | 3.51 |
| Iirr | Irregularly retouched | 31.41 | – | – |
| | | | | |
| IQ | Quina | 0.00 | | |
| IQ+½Q | Quina + demi-Quina | 0.90 | | |
| Cl | Clactonian notches | 49.18 | | |

MAJOR TYPES

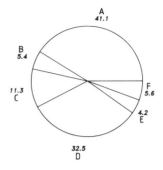

SIDE–SCRAPERS

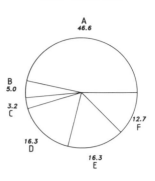

**Fig. 26.10** Flint tools from layer F. Relative frequencies of the principal tool categories and of the major side-scraper types. Tools: A Group II; B Group II; C Group IV; D notched pieces; E miscellaneous; F other. Side-scrapers: A simple lateral; B simple transverse; C double separate; D convergent axial; E *déjeté*; F other.

Other side-scrapers

Taken overall, these are less significant than in most other layers, both as a percentage of the scrapers (12.6%) and metrically. In the latter respect, their retouched edges are very short on average; however, their mean invasiveness is the highest for any of the major groups. A feature which may perhaps be explained by a relative dearth of alternately worked pieces is the rarity of multiple scrapers, other than those of *déjeté* type.

*On the ventral surface* (17). As in the previous layer, all are single-edged; two are transverse. The most remarkable aspect of this series is that only one piece is convex (lateral). Of the others, twelve (or 70.6%) are straight and four concave, one of each shape being transverse. This may be contrasted with the simple lateral scrapers with normal retouch, among which convex pieces are in a majority. This seems to reflect the presence of a category of inverse scraper marked by a short working edge which is slightly steeper than usual and not very invasive. The means for the edge measurements are rather striking: length 20.8 mm, angle 68.3°, invasiveness 3.0 mm (the highest individual value achieved for this last is 8 mm). A typical Levallois flake has been employed in one instance, a cortex-backed piece in another. One piece also has a Clactonian notch.

*With abrupt retouch* (4). One of these (concave) is on a typical Levallois flake (length 45 mm). This, and two of the others (straight and concave), have ordinary abrupt (and not very invasive) retouch. The fourth piece (convex) has retouch of *surélevée* type.

*With bifacial retouch* (2). These are on rather small (mean length 41.5 mm) plain flakes; one is convex, the other a straight *déjeté* scraper apart from its bifacial working.

*With alternate retouch* (5). All but one (convex, with rather weak retouch) are essentially straight-edged. Three, including the convex piece, are *déjeté* — one of the finest of these is on a Levallois flake.

**End-scrapers** (10; 2 typical, 8 atypical). An insignificant component in the assemblage as a whole, these are made on plain blanks, apart from one possibly reworked from a core; one is on a blade, another on a transverse flake. A single piece is double-shouldered and very rounded (curvature 0.55), another is made on the butt, while the rest (all atypical) are simple pieces made on the distal end. Four have parallel or subparallel retouch. One of the tips is left-skewed, and three more are right-skewed.

*Transverse* (36) (Figs. 26.11, 10 and 26.12, 6 and 8). Only five are convex; 16 are straight and 15 concave. Hence the mean curvature is negative (-0.01). The retouch is slightly more invasive than on the single lateral scrapers (5.2 mm compared to 4.6), but the mean retouch angles are very similar (58.7° against 58.3°). The retouched edges are very slightly shorter (27.3 and 29.4 mm respectively).

Multiple scrapers

*Separate* (11). Easily the most important subtype is straight-convex, with seven examples, followed by three biconvex and a single double straight. Thus none have concave edges. They are of relatively good quality (retouch invasiveness 5.8 ± 3.0 mm, with a rather low edge angle, 54.3 ± 10.8°), though the retouched edges are rather short (35.7 ± 9.9 mm). Four are on typical Levallois flakes (e.g. Fig. 26.11, 8). In view of the sample size, an IF of 44.9 is much as expected, but the blanks are relatively long (49.0 mm). One piece (straight) bears a retouched notch, another (straight-convex) has additional retouch on the tip.

*Convergent axial* (7). These include three with straight (e.g. the poorly made Fig. 26.11, 11) and three with convex (e.g. Fig. 26.11, 9) edges. The seventh is classified as concave (in accordance with Bordes's rules) but with one convex edge, and is the only one on a Levallois flake. There is a tendency to pointedness; the mean width at 10 mm from the tip is 15.6 ± 6.8 mm.

*Déjeté* (36). These are an important component of the assemblage; however, they are extremely varied in shape (compare Figs. 26.11, 12–14 and 26.12, 1–5). Twelve are double *déjeté*. As the illustrations suggest, they are generally well made, with invasive secondary work (mean 7.5 ± 3.2 mm). Eight pieces are on Levallois flakes, four on pseudo-Levallois points, and one is cortex-backed. A single piece has a thinned butt.

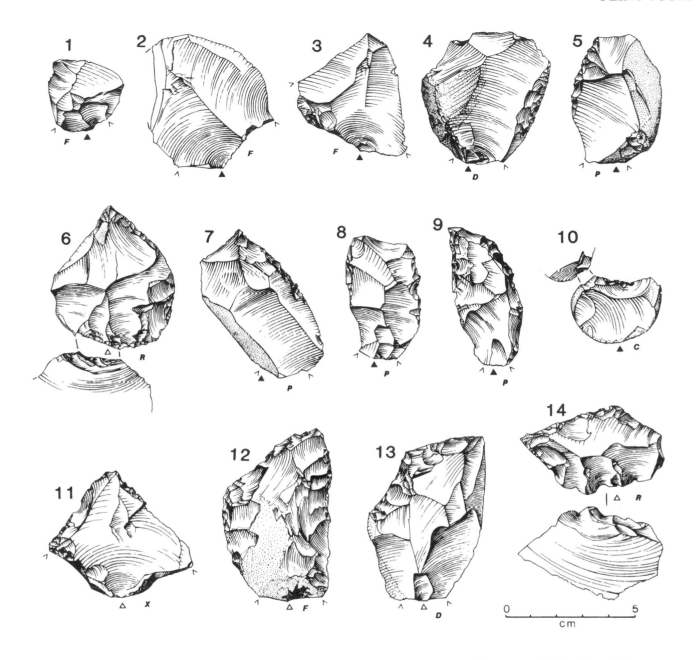

**Fig. 26.11** Flint tools from layer F. 1 Levallois flake; 2, 3 pseudo-Levallois points; 4–14 side-scrapers: 4–7 simple lateral; 8 double; 9, 11 convergent; 10 transverse; 12–14 *déjeté*.

**Burins** (6; 3 typical, 3 atypical). All are of the angle variety, on plain flakes and made by longitudinal burin blows (five being on the distal end, the other proximal). Three are on breaks, the rest on primary surfaces. One piece has multiple burin blows. The mean tip angle (72.8 ± 10.5°) is more typical than was that for layer G. Only two pieces are retouched. The burins are generally very short (mean length 33.4 ± 7.6); however, the use of broken pieces in this and other layers obviously raises questions about the meaning of this statistic.

**Borers** (5; 2 typical, 3 atypical). None of these is entirely convincing. The best has a fine tip formed by alternate semi-abrupt retouch on a corner of a flake 50 mm long; the other piece classed as typical (length 34 mm) is the result of alternate notching. The atypical examples are small (mean length 27 mm), rather obtuse and on the corners of flakes.

**Knives** (5; 1 typical, 4 atypical). One is a fine example on a blade 84

mm long, with a cortical distal end, and has regular abrupt retouch along its right-hand edge. The atypical pieces are shorter (mean length 56 mm) and have retouch on edges which are otherwise cortical (in three cases), or which are steep as a result of the primary flaking. All of the butts are plain.

**Raclettes** (7). Only 39.9 mm long on average, they are made on simple flakes (except one, on a broken blade). There is no clear preference for a particular edge shape or position. One is inverse.

**Truncated pieces** (3). They comprise one concave and two straight truncations (one of the latter being at the proximal end of a cortex-backed flake). The mean length is only 36.3 mm.

**Notched pieces** (61). A further decline reduces the percentage to 11.2; of these, 49% are Clactonian. Only two examples are double (Fig.

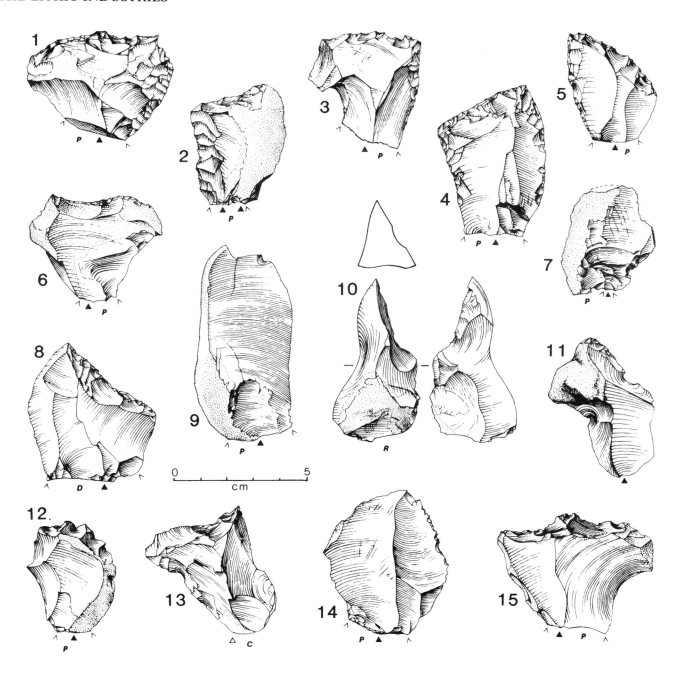

**Fig. 26.12** Flint tools from layer F. 1–6, 8 Side-scrapers: 1–5 *déjeté*; 6, 8 transverse; 9 naturally-backed knife; 7, 10, 11 notches; 12–15 denticulates.

26.12, 11). The mean width is marginally less than before, at 11.8 mm. The curvature is almost identical for the two types of notches; both the overall mean, 0.24, and the form of the distribution are similar to that for G.

| Curvature | 0.0 | 0.1 | 0.2 | 0.3 | 0.4 | 0.5 | 0.6 |
|---|---|---|---|---|---|---|---|
| % | 0.0 | 36.2 | 50.0 | 9.0 | 3.0 | 1.6 | |

However, the invasiveness of the retouched notches, 2.5 mm, is only just over half that for the previous layer. The blanks include 2 Levallois flakes (neither notch being Clactonian) and 2 pseudo-Levallois points (1 Clactonian), 2 cortex-backed (2 Clactonian) and one on a rather curious core — this last piece (Fig. 26.12, 10) is perhaps morphologically equivalent to a Tayac point.

**Denticulates** (176). Again these form one of the predominant types (32.5% in reduced count), and are usually very positive in style (Figs. 26.12, 12–15 and 26.13, 1). Seven (4.8%) are made on Levallois flakes (Fig. 26.12, 14), against 13 on pseudo-Levallois points, while 9 are on cortex-backed blanks (Fig. 26.12, 12) and one on a core. For the first time, none of the pieces on pseudo-Levallois points has been produced by multiple Clactonian notches; instead, 9 are made by coarse denticulate retouch, and the other 4 are microdenticulates. The incidence of multiple Clactonian notches (29.5%) and coarse retouch (40.0%) gives a total of 69.5% coarse edges, in good agreement with the figures for G and H. A further 19.2% are finely denticulated, and 11.3% are microdenticulates.

In their morphology the denticulates of layer F resemble those of layer G, inasmuch as 50.4% are simple lateral and 18.4% are transverse; 7.1% are double and 6.4% convergent, but only 3.5% are *dejeté* (compared with 22.4% in G). It is curious that this last shape should have been favoured for scrapers in F, but is largely absent for denticulates. Though the tools are relatively large (mean length 46.1 ± 11.4 mm, retouch length 33.2 ± 9.7 mm) the removal scars are slightly less invasive than in layer H (mean 6.2 ± 3.9 mm), associated with the rise in the incidence of microdenticulation in F.

**Becs burinants alternes** (5). All are made by means of alternate notches, though only one is a really good example (on the proximal end of a small, thick, convex side-scraper). The mean length is only 36.3 mm.

**Tayac points** (3). One is *dejeté*, on a flake with a cortical back (length 43 mm), and is formed by Clactonian notches; another, also *dejeté*, has two large retouched notches, one of them with denticulations inside the notch, and is 35 mm long (the third edge is a relict margin). The last piece (56.5 mm) has a large Clactonian notch on either side near the base with, towards the tip, subparallel retouch on the left hand edge and steep, slightly denticulate retouch on the other.

**End-notched pieces** (10). Clactonian and retouched notches are present in equal numbers. Four of them may be accidental, however. The average length is 41.9 mm. One example is on a cortex-backed flake.

**Chopping tool** (1). It is quite small (65 x 53 mm; weight 158 gm) on a re-used, cream-patinated unifacial core on a pebble. This has been flaked at one end to give a sinuous edge.

**Miscellaneous** (23). Only three of these (or 13%) are compound pieces according to the multiple-response classification, which gives

|  | n | % |  | n | % |
|---|---|---|---|---|---|
| scrapers | 13 | (56.5) | notches | 2 | (8.7) |
| denticulate | 6 | (26.1) | overstruck removals | 0 | (0.0) |
| scaled | 3 | (13.0) | burins | 0 | (0.0) |

Six pieces have dorsal removals from an inverse truncation or a break (Fig. 26.13, 4).

## LAYER E

According to the thin section analysis described in chapter 9, this is the least disturbed of the archaeological deposits. It was also fairly easy to follow during excavation. The lithic assemblage is thus likely to be less contaminated than most others, though some post-depositional movement of artefacts is inevitable. The reduced tool frequency is the lowest reached at La Cotte, being only 11.5% (398 pieces); the real value falls nearly 5%, to 18.8%. There is only a slight decrease in the real percentage of types 45–50, to 30.5%.

This is the earliest series from La Cotte in which side-scrapers form a clear majority of the well-retouched tools (IR = 55.5). Though they were already important in the previous layer, there are striking differences of detail between the scrapers from the two series. In E the formerly numerous *dejeté* pieces are scarce, and the proportion of transverse scrapers is much reduced. On the other hand, scrapers with inverse or alternate retouch are twice as numerous as before. The denticulates are both fewer (19.6%) and less elaborate in shape; their edges are often produced by fine denticulate retouch. Notches amount to only 10.8% of the reduced total. For the first time atypical Levallois blanks (as opposed to unretouched flakes) are outnumbered by typical ones.

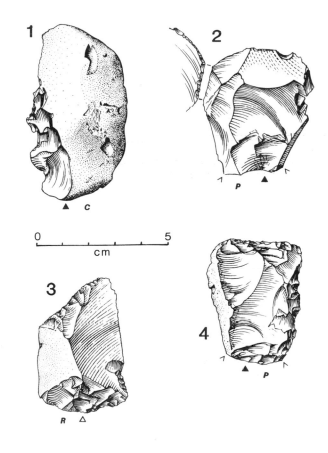

**Fig. 26.13** Flint tools from layer F. 1 Denticulate; 2 retouched on the ventral surface; 3, 4 miscellaneous: 3 is a side-scraper or knife with sharpened butt and 4 a denticulate with scaling from a distal break.

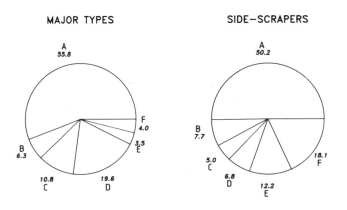

**Fig. 26.14** Flint tools from layer E. Relative frequencies of the principal tool categories and of the major side-scraper types. Tools: A Group II; B Group III; C Group IV; D notched pieces; E miscellaneous; F other. Side-scrapers: A simple lateral; B simple transverse; C double separate; D convergent axial; E *dejeté*; F other.

**Table 26.7** Layer E. Tool Type Frequencies

| Type | | N | Real | Ess. | Red. |
|---|---|---|---|---|---|
| 1 | Typical Levallois flakes | 17 | 2.62 | – | – |
| 2 | Atypical Levallois flakes | 21 | 3.23 | – | – |
| 3 | Levallois points | 2 | 0.31 | – | – |
| 4 | Retouched Levallois points | 2 | 0.31 | 0.49 | 0.50 |
| 5 | Pseudo-Levallois points | 5 | 0.77 | 1.21 | – |
| 6 | Mousterian points | 1 | 0.15 | 0.24 | 0.25 |
| 9 | Single straight side-scrapers | 37 | 5.69 | 8.98 | 9.30 |
| 10 | Single convex side-scrapers | 61 | 9.38 | 14.81 | 15.33 |
| 11 | Single concave side-scrapers | 13 | 2.00 | 3.16 | 3.27 |
| 12 | Double straight side-scrapers | 5 | 0.77 | 1.21 | 1.26 |
| 13 | Double straight-convex side-scrapers | 4 | 0.62 | 0.97 | 1.01 |
| 14 | Double straight-side-scrapers | 2 | 0.31 | 0.49 | 0.50 |
| 15 | Double convex side-scrapers | 3 | 0.46 | 0.73 | 0.75 |
| 16 | Double concave side-scrapers | 1 | 0.15 | 0.24 | 0.25 |
| 17 | Double concave-convex side-scrapers | 2 | 0.31 | 0.49 | 0.50 |
| 18 | Convergent straight side-scrapers | 4 | 0.62 | 0.97 | 1.01 |
| 19 | Convergent convex side-scrapers | 6 | 0.92 | 1.46 | 1.51 |
| 20 | Convergent concave side-scrapers | 1 | 0.15 | 0.24 | 0.25 |
| 21 | *Déjeté* scrapers | 15 | 2.31 | 3.64 | 3.77 |
| 22 | Straight transverse side-scrapers | 15 | 2.31 | 3.64 | 3.77 |
| 23 | Convex transverse side-scrapers | 8 | 1.23 | 1.94 | 2.01 |
| 24 | Concave transverse side-scrapers | 4 | 0.62 | 0.97 | 1.01 |
| 25 | Side-scrapers with inverse retouch | 21 | 3.23 | 5.10 | 5.28 |
| 26 | Abrupt retouched side-scrapers | 3 | 0.46 | 0.73 | 0.75 |
| 27 | Side-scrapers with thinned back | 2 | 0.31 | 0.49 | 0.50 |
| 28 | Side-scrapers with bifacial retouch | 2 | 0.31 | 0.49 | 0.50 |
| 29 | Alternate retouched side-scrapers | 12 | 1.85 | 2.91 | 3.02 |
| 30 | Typical end-scrapers | 2 | 0.31 | 0.49 | 0.50 |
| 31 | Atypical end-scrapers | 7 | 1.08 | 1.70 | 1.76 |
| 32 | Typical burins | 4 | 0.62 | 0.97 | 1.01 |
| 33 | Atypical burins | 3 | 0.46 | 0.73 | 0.75 |
| 34 | Typical borers | 1 | 0.15 | 0.24 | 0.25 |
| 35 | Atypical borers | 2 | 0.31 | 0.49 | 0.50 |
| 37 | Atypical backed knives | 1 | 0.15 | 0.24 | 0.25 |
| 38 | Naturally backed knives | 9 | 1.38 | 2.18 | – |
| 39 | *Raclettes* | 10 | 1.54 | 2.43 | 2.51 |
| 40 | Truncated pieces | 5 | 0.77 | 1.21 | 1.26 |
| 42 | Notched pieces | 43 | 6.62 | 10.44 | 10.80 |
| 43 | Denticulates | 78 | 12.00 | 18.93 | 19.60 |
| 45 | Retouched with inverse retouch | 20 | 3.08 | – | – |
| 46–47 | Abrupt and alternate retouch (thick) | 30 | 4.62 | – | – |
| 48–49 | Abrupt and alternate retouch (thin) | 147 | 22.62 | – | – |
| 50 | Bifacial retouch | 1 | 0.15 | – | – |
| 51 | Tayac points | 2 | 0.31 | 0.49 | 0.50 |
| 54 | End-notched pieces | 2 | 0.31 | 0.49 | 0.50 |
| 62 | Miscellaneous | 14 | 2.15 | 3.40 | 3.52 |
| | Bifaces (whole) | 0 | | | |
| | Totals | | 650 | 412 | 398 |

**Retouched Levallois points** (2). Both are of secondary type (lengths 61 and 49 mm) with zones of scalar retouch.

**Mousterian point** (1). This is not a good example of the type; on a primary Levallois point 43 mm long, it has a slightly rounded tip and is retouched only along the distal half of either edge (abrupt on the left-hand edge, subparallel on the right).

**Side-scrapers** (221). For the first time, more than half of the retouched tools are side-scrapers (IR$_{red}$ = 55.53); this situation persists for the remainder of the sequence at the site. They are usually well made, with invasive retouch (mean invasiveness 5.1 ± 2.9 mm) and longer worked edges than before (34.3 ± 12.7 mm); their mean length is 47.5 ± 10.6 mm. The principal categories show an increase in types 25–29 and in the single scrapers, at the expense of convergent and transverse pieces:

| | | | |
|---|---|---|---|
| single lateral | 50.2 | double | 7.7 |
| convergent | 11.8 | transverse | 12.2 |
| other | 18.1 | | |

*Déjeté* scrapers fall to 6.8% from 16.3% in layer F, and concave transverse scrapers to 1.8% from 6.8%. The edge shape is most commonly convex (48.4%), 39.8% are straight and 11.8% concave; this situation persists in all subsequent assemblages. Six pieces have thinned butts, three are thinned along one edge, and seven have additional (minor) retouch on an end adjacent to the main working edge. Nine also have notches, including five of Clactonian type. The technology of manufacture of the blanks is consistent with that for the other tools. Thus the facetting index, IF, is 35.8 (agreeing with the value for all retouched tools as regards the fall relative to layer F). Of the blanks, 9.5% are Levallois, 2.3% pseudo-Levallois points and 8.1% cortex-backed.

**Table 26.8** Layer E. Typological Groups and Indices

| Index | | Real | Ess. | Red. |
|---|---|---|---|---|
| ILty | Typological Levallois | 6.46 | – | – |
| IR | Side-scrapers | 34.00 | 53.64 | 55.53 |
| IC | Charentian | 13.85 | 21.84 | 22.61 |
| IAu | Acheulian unifacial | 0.15 | 0.24 | 0.25 |
| IAt | Acheulian total | 0.15 | 0.24 | 0.25 |
| IB | Bifaces | 0.00 | 0.00 | 0.00 |
| | | | | |
| I | Levallois | 6.46 | 0.49 | 0.50 |
| II | Mousterian | 34.15 | 53.88 | 55.78 |
| III | Upper Palaeolithic | 3.85 | 6.07 | 6.28 |
| IV | Denticulate | 12.00 | 18.93 | 19.60 |
| | | | | |
| 5 | Pseudo-Levallois points | 0.77 | 1.21 | – |
| 38 | Naturally backed knives | 1.38 | 2.18 | – |
| I rc | Convergent retouched | 4.15 | 6.55 | 6.78 |
| 6-7 | Mousterian points | 0.15 | 0.24 | 0.25 |
| Rsingle | Side-scrapers, single | 17.08 | 26.94 | 27.89 |
| Rdouble | Side-scrapers, double | 2.62 | 4.13 | 4.27 |
| Rtransv | Side-scrapers, transverse | 4.15 | 6.55 | 6.78 |
| Rinv | Side-scrapers, inverse/alternate | 5.08 | 8.01 | 8.29 |
| 30-31 | End-scrapers | 1.38 | 2.18 | 2.26 |
| 32-33 | Burins | 1.08 | 1.70 | 1.76 |
| 36-37 | Knives | 0.15 | 0.24 | 0.25 |
| 42 | Notched pieces | 6.62 | 10.44 | 10.80 |
| 42-43 | Notched pieces & denticulates | 18.62 | 29.37 | 30.40 |
| Ident | Extended denticulate group | 1.69 | 2.67 | 2.76 |
| Iirr | Irregularly retouched | 30.46 | – | – |
| | | | | |
| IQ | Quina | 1.36 | | |
| IQ+½Q | Quina + demi-Quina | 3.17 | | |
| Cl | Clactonian notches | 51.16 | | |

Simple scrapers

*Single lateral* (111) (Fig. 26.15, 1–10). Thirty-seven are straight, 61 convex and 13 concave; the mean curvature is correspondingly high (0.07). The mean edge lengths vary considerably according to shape: straight 29.8, convex 39.8 and concave 21.8 mm. There is little variation in retouch angle and invasiveness, however. Additional retouch occurs on one end in six cases, thinning of the butt in three, and one has a Clactonian notch. There is near equality between left (53) and right (57) edged pieces, whereas in the previous layer the great majority of retouch on scrapers of this group was on the right-hand edge.

*Transverse* (27) (Fig. 26.15, 16–18). In contrast to the distribution for the single lateral scrapers, the most common edge shape is straight (15 pieces), compared to 8 convex and 4 concave. This brings the mean curvature down to 0.03; the retouch is also slightly more invasive.

Multiple scrapers

*Separate* (17). All types are represented, though with a clear preference for straight and convex edges (see Table 26.23) — those with concave edge also tend to have the shortest lengths of retouch. As is usual for this group, Levallois flakes (4, all typical) have been specially favoured, though the length of the blanks and the metrical characteristics of the edges are about average for the layer. A single piece has some thinning along an edge. One has Quina retouch, and another demi-Quina. The distinction between separate and convergent scrapers is not always easy to draw in

this assemblage, as is shown by Fig. 26.15, 11 (separate by virtue of a notch at its tip) and 12.

*Convergent axial* (11). The convex subtype is not only the most common but is represented by larger pieces than are the other subtypes; this pattern persists for the remainder of the Saalian assemblages. Two of these are on Levallois flakes (one of them with a thinned butt). The tips are blunter than before; the mean width at 10 mm from the point is 19.0 ± 4.0 mm, a value more or less maintained until layer 5.

*Déjeté* (15). Six of these are double. The retouch is very invasive (mean 7.9 ± 3.8 mm), more so than for the convergent pieces. The choice of blanks is varied: one is a typical Levallois flake, one a pseudo-Levallois point (Fig. 26.15, 13) and three are flakes with cortical backs.

Other side-scrapers

The frequencies of the 'equivalent' scraper types (Table 26.24) reflect quite closely those for the dorsally worked types 9–24. Thus, more truly than in the earlier layers, a similar result has been achieved by different styles of retouch — indeed, an initial handling of the material suggested this even before statistical analysis. Metrically also, the agreement is very good (apart from the predictable raising of the edge angle caused by the scrapers with abrupt retouch).

*On the ventral face* (21). These are often well made, with quite invasive secondary working (Fig. 26.16, 1 and 3). All are single-edged; only two are transverse. The relative frequencies of the three edge shape categories are more like those of the dorsally-retouched pieces than previously, though the figure for convex edges is still slightly depressed: 10 straight, 9 convex and 2 concave.

The measurements on the retouched edges yield means in keeping with those for the single normal scrapers, apart from a somewhat reduced length. Thus the edge length is 30.4 mm, the angle 62.2° and the invasiveness 4.9 mm. One piece is Levallois, and three are on cortex-backed flakes. There is a single occurrence of a thinned butt.

*With abrupt retouch* (3). On ordinary flakes, two are straight-edged, the other convex. One of the former is extremely small (27 mm), the others 50+ mm and quite invasively worked.

*With thinned back* (2). Both are convex-edged, with invasive retouch (12 mm). One piece is twice the size of the other, with a length of 85 mm (Fig. 26.16, 2); it has subparallel retouch, whereas the other is scaled.

*With bifacial retouch* (2). These are straight-edged, and on ordinary flakes of only average size.

*With alternate retouch* (12). Straight and convex edges total five each, with two concave. Half have alternating retouch. Apart from two Levallois flakes, there is nothing distinctive about the blanks.

**End-scrapers** (9; 2 typical, 7 atypical). One of the typical pieces is ogival (on a Levallois flake), while among the atypical ones are individual examples of the single-shouldered and 'on butt' subtypes. The remainder are simple. Three have retouched edges (Fig. 26.16, 4, on which a good end-scraper tip runs into a side-scraper edge, is on an atypical Levallois flake). Apart from the two pieces already mentioned, the blanks are plain: three are on blades, and three on transverse flakes. One of the tips is skewed to the left and three to the right. Two pieces are manufactured using stepped retouch, and two by subparallel removals. Two of the atypical simple pieces also have Clactonian notches.

**Burins** (7; 4 typical, 3 atypical). Five are angle burins (all on breaks), the others of the *plan* type (one somewhat *déjeté* on a convex normal

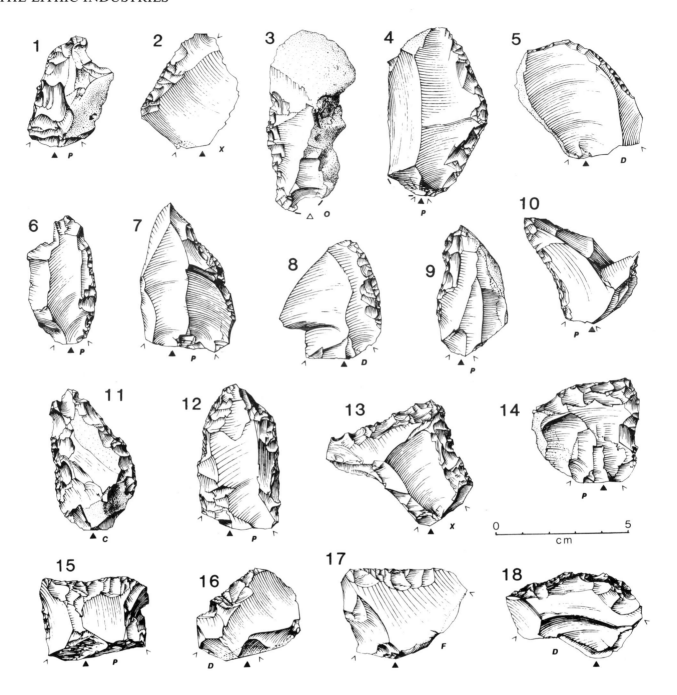

**Fig. 26.15**  Flint tools from layer E. Side-scrapers: 1–10 simple lateral; 11 double; 12 convergent; 13–15 *déjeté*; 16–18 transverse.

truncation). Two of the former are proximal (Fig. 26.16, 5), and one of the latter is struck from a lateral margin; the remainder are on the distal end. In every case the burin is made by a single blow; six have some form of retouch as well. The blanks are ordinary flakes apart from one with a cortical back (one of the angle burins).

**Borers** (3; 1 typical, 2 atypical). The most typical example is formed by the intersection of the retouched butt with a normal retouched edge to give a rather blunt point at one corner. The other two may be the result of accidents. There are no clear indications of use on any of the three, which are all small (mean length 34 mm).

**Knife** (1, atypical). This is on a blank with a partly cortical back, and

is 45 mm long. It has a couple of small removals near the tip, on the cortex, which may have served to regularise the back. The sharp edge exhibits utilisation damage.

**Truncated pieces** (5). With a mean length of 34.6 mm, these comprise two proximal and three distal examples. One truncation (proximal) is inverse. Four of the five are straight, the last being concave. The mean width of the truncations is 18.6 mm; it would be higher but for one which is only 10 mm wide, on a cortex-backed flake.

**Notched pieces** (43). Of these, 51% are Clactonian, including one on a typical (?) Levallois flake (Fig. 26.16, 10) and two cortex-backed. A single example on a pseudo-Levallois point has been manufactured by

270

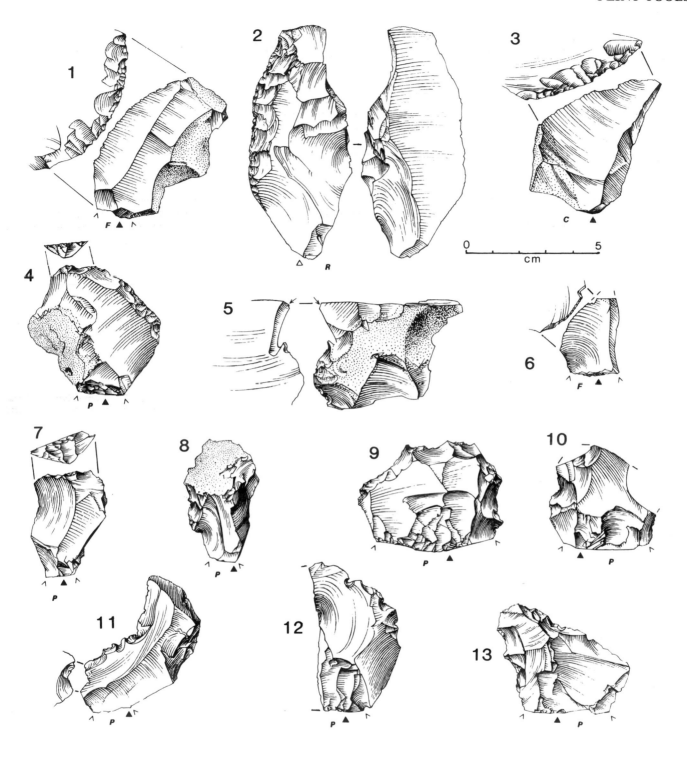

**Fig. 26.16** Flint tools from layer E. 1–3 Side-scrapers: 1, 3 on the ventral surface; 2 with thinned back; 4 end-scraper (on side-scraper); 5 burin; 6 *raclette*; 7 truncated flake; 10 notch; 8, 9, 11, 12 denticulates; 13 Tayac point (slightly *déjeté*).

retouch, as have two of the eight double-notched pieces (two are opposed laterally). Notch width is very similar to that for F, allowing for the relatively small sample, but invasiveness is much greater for both types of notch (combined mean 8.1 mm, retouched alone 4.8). The curvature is not significantly different, but the values are more standardised.

| Curvature | 0.0 | 0.1 | 0.2 | 0.3 | 0.4 | 0.5 | 0.6 |
|-----------|-----|-----|-----|-----|-----|-----|-----|
| % | | 0.0 | 26.5 | 61.7 | 7.8 | 0.0 | 3.9 |

**Denticulates** (78). A reduction in frequency (19.6% in reduced count) is by now apparent — as are some stylistic changes. Only 20.5% are made by means of Clactonian notches, as against 34.4% coarse and 43.8% fine retouch (e.g. Fig. 26.16, 11) and a single microdenticulate. There is a reversion to the simple shapes (88.5%) so common in layer H, with 71.8% lateral and 16.7% transverse pieces. However, as in F, the pseudo-Levallois points employed as blanks for this type have not been used for tools whose edges are formed by adjacent Clactonian notches.

**Table 26.9**  Layer D. Tool Type Frequencies

| Type | | N | Real | Ess. | Red. |
|------|------|-----|------|------|------|
| 1 | Typical Levallois flakes | 22 | 2.91 | – | – |
| 2 | Atypical Levallois flakes | 30 | 3.96 | – | – |
| 3 | Levallois points | 5 | 0.66 | – | – |
| 4 | Retouched Levallois points | 2 | 0.26 | 0.39 | 0.42 |
| 5 | Pseudo-Levallois points | 9 | 1.19 | 1.77 | – |
| 6 | Mousterian points | 1 | 0.13 | 0.20 | 0.21 |
| 7 | Elongated Mousterian points | 1 | 0.13 | 0.20 | 0.21 |
| 8 | *Limaces* | 1 | 0.13 | 0.20 | 0.21 |
| 9 | Single straight side-scrapers | 37 | 4.89 | 7.28 | 7.72 |
| 10 | Single convex side-scrapers | 89 | 11.76 | 17.52 | 18.58 |
| 11 | Single concave side-scrapers | 10 | 1.32 | 1.97 | 2.09 |
| 12 | Double straight side-scrapers | 1 | 0.13 | 0.20 | 0.21 |
| 13 | Double straight-convex side-scrapers | 3 | 0.40 | 0.59 | 0.63 |
| 15 | Double convex side-scrapers | 7 | 0.92 | 1.38 | 1.46 |
| 16 | Double concave side-scrapers | 1 | 0.13 | 0.20 | 0.21 |
| 17 | Double concave-convex side-scrapers | 3 | 0.40 | 0.59 | 0.63 |
| 18 | Convergent straight side-scrapers | 2 | 0.26 | 0.39 | 0.42 |
| 19 | Convergent convex side-scrapers | 7 | 0.92 | 1.38 | 1.46 |
| 21 | *Déjeté* scrapers | 21 | 2.77 | 4.13 | 4.38 |
| 22 | Straight transverse side-scrapers | 12 | 1.59 | 2.36 | 2.51 |
| 23 | Convex transverse side-scrapers | 15 | 1.98 | 2.95 | 3.13 |
| 24 | Concave transverse side-scrapers | 1 | 0.13 | 0.20 | 0.21 |
| 25 | Side-scrapers with inverse retouch | 47 | 6.21 | 9.25 | 9.81 |
| 26 | Abrupt retouched side-scrapers | 2 | 0.26 | 0.39 | 0.42 |
| 27 | Side-scrapers with thinned back | 7 | 0.92 | 1.38 | 1.46 |
| 28 | Side-scrapers with bifacial retouch | 9 | 1.19 | 1.77 | 1.88 |
| 29 | Alternate retouched side-scrapers | 22 | 2.91 | 4.33 | 4.59 |
| 30 | Typical end-scrapers | 9 | 1.19 | 1.77 | 1.88 |
| 31 | Atypical end-scrapers | 12 | 1.59 | 2.36 | 2.51 |
| 32 | Typical burins | 6 | 0.79 | 1.18 | 1.25 |
| 33 | Atypical burins | 3 | 0.40 | 0.59 | 0.63 |
| 35 | Atypical borers | 2 | 0.26 | 0.39 | 0.42 |
| 37 | Atypical backed knives | 2 | 0.26 | 0.39 | 0.42 |
| 38 | Naturally backed knives | 20 | 2.64 | 3.94 | – |
| 39 | *Raclettes* | 10 | 1.32 | 1.97 | 2.09 |
| 40 | Truncated pieces | 6 | 0.79 | 1.18 | 1.25 |
| 42 | Notched pieces | 28 | 3.70 | 5.51 | 5.85 |
| 43 | Denticulates | 65 | 8.59 | 12.80 | 13.57 |
| 44 | *Becs burinants alternes* | 1 | 0.13 | 0.20 | 0.21 |
| 45 | Retouched with inverse retouch | 23 | 3.04 | – | – |
| 46-47 | Abrupt and alternate retouch (thick) | 25 | 3.30 | – | – |
| 48-49 | Abrupt and alternate retouch (thin) | 143 | 18.89 | – | – |
| 50 | Bifacial retouch | 1 | 0.13 | – | – |
| 51 | Tayac points | 2 | 0.26 | 0.39 | 0.42 |
| 54 | End-notched pieces | 5 | 0.66 | 0.98 | 1.04 |
| 61 | Chopping tools | 1 | 0.13 | 0.20 | 0.21 |
| 62 | Miscellaneous | 26 | 3.43 | 5.12 | 5.43 |
| Bifaces (whole) | | 4 | | | |
| Totals | | | 757 | 508 | 479 |

The number of blanks of different types is as follows: six Levallois flakes, six pseudo-Levallois points, seven cortex-backed knives, and three cores. Metrically, the denticulates are almost identical to those from layer F.

**Tayac points** (2). Both are *déjeté* (Fig. 26.16, 13) with very coarse denticulation.

**End-notched pieces** (2). Neither is typical. The smaller (33 mm long) is a cortical flake with a 6 mm wide distal Clactonian notch, but also a larger notch on the right-hand edge, and could be regarded as a *bec*. The other (45 mm long) has a 7 mm wide 'retouched' notch (possibly the result of accidental damage) on the proximal end.

**Miscellaneous** (14). These were characterised as follows:

| | n | % | | n | % |
|---|---|---|---|---|---|
| scrapers | 8 | (57.1) | notches | 1 | (7.1) |
| denticulates | 3 | (21.4) | overstruck removals | 2 | (14.3) |
| scaled | 2 | (14.3) | burins | 1 | (7.1) |

Three are compound: a notched scraper, a scraper with overstrike, and a notched burin; both the overstrikes are marginal, of LSF type — at least superficially.

## *LAYER D*

The 480 tools in the reduced group constitute 13.2% of the total unbroken flint — the real tools amount to 20.8%. The real percentage of types 45–50 is lower than in the previous layer, at 25.2%. The most conspicuous feature of the reduced tool assemblage is the very high (61.7%) side-scraper component. There is an exceptionally high incidence of scrapers with inverse or alternate retouch (14.4% of the reduced tools, or 23.3% of the side-scrapers); the percentages of these types are quite high in the immediately adjacent series, but those for D represents a very considerable increase. The extended denticulate group accounts for only 20.8% of the reduced count. The percentage of denticulates is 13.5, and

**Table 26.10** Layer D. Typological Groups and Indices

| Index | | Real | Ess. | Red. |
|---|---|---|---|---|
| ILty | Typological Levallois | 7.79 | - | - |
| IR | Side-scrapers | 39.10 | 58.27 | 61.80 |
| IC | Charentian | 16.78 | 25.00 | 26.51 |
| IAu | Acheulian unifacial | 0.26 | 0.39 | 0.42 |
| IAt | Acheulian total | 0.79 | 1.17 | 1.24 |
| IB | Bifaces | 0.53 | 0.78 | 0.83 |
| | | | | |
| I | Levallois | 7.79 | 0.39 | 0.42 |
| II | Mousterian | 39.50 | 58.86 | 62.42 |
| III | Upper Palaeolithic | 5.28 | 7.87 | 8.35 |
| IV | Denticulate | 8.59 | 12.80 | 13.57 |
| | | | | |
| 5 | Pseudo-Levallois points | 1.19 | 1.77 | - |
| 38 | Naturally backed knives | 2.64 | 3.94 | - |
| I rc | Convergent retouched | 4.36 | 6.50 | 6.89 |
| 6–7 | Mousterian points | 0.26 | 0.39 | 0.42 |
| Rsingle | Side-scrapers, single | 17.97 | 26.77 | 28.39 |
| Rdouble | Side-scrapers, double | 1.98 | 2.95 | 3.13 |
| Rtransv | Side-scrapers, transverse | 3.70 | 5.51 | 5.85 |
| Rinv | Side-scrapers, inverse/alternate | 9.11 | 13.58 | 14.41 |
| 30–31 | End-scrapers | 2.77 | 4.13 | 4.38 |
| 32–33 | Burins | 1.19 | 1.77 | 1.88 |
| 36–37 | Knives | 0.26 | 0.39 | 0.42 |
| 42 | Notched pieces | 3.70 | 5.51 | 5.85 |
| 42–43 | Notched pieces & denticulates | 12.29 | 18.31 | 19.42 |
| Ident | Extended denticulate group | 2.11 | 3.15 | 3.34 |
| Iirr | Irregularly retouched | 25.36 | - | - |
| | | | | |
| IQ | Quina | 1.01 | | |
| IQ+½Q | Quina + demi-Quina | 2.36 | | |
| C1 | Clactonian notches | 39.29 | | |

of notched pieces 5.8 — the latter little more than half that in the previous layer. 'Upper Palaeolithic' tools (Group III) increase to 7.1%, principally as a result of a greater percentage of end-scrapers. There are four bifaces in the assemblage, but all are nucleiforms.

**Retouched Levallois points** (2). One, 41 mm long, is on a primary Levallois point with dihedral butt, and has semi-abrupt discontinuous retouch as well as a small notch on the right-hand edge. The other is 53 mm long, secondary and blade-like, with a plain butt. It has slight semi-abrupt retouch at the proximal end on one edge.

**Mousterian point** (1). 54 mm long but with a damaged butt, it has a curved longitudinal cross-section and rather irregular edges.

**Elongated Mousterian point** (1). This is 53 mm long, and thin, with the extreme tip missing. Its butt has been destroyed by a large bulbar scar.

**Limace** (1). It is 63 mm long by 29 mm wide, and pointed at both ends (though a small area of the butt remains). The left edge is appreciably more convex than the right.

**Side-scrapers** (296). There is a further climb in the scraper index (IR$_{red}$ = 61.7). Perhaps the most spectacular development is the growth in the combined percentage for types 25–29:

| | | | |
|---|---|---|---|
| single lateral | 45.9 | double | 5.1 |
| convergent | 10.1 | transverse | 9.5 |
| other | 29.4 | | |

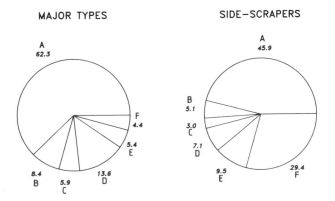

MAJOR TYPES

SIDE-SCRAPERS

A 62.3
F 4.4
E 5.4
D 13.6
C 5.9
B 8.4

A 45.9
B 5.1
C 3.0
D 7.1
E 9.5
F 29.4

**Fig. 26.17** Flint tools from layer D. Relative frequencies of the principal tool categories and of the major side-scraper types. Tools: A Group II; B Group II; C Group IV; D notched pieces; E miscellaneous; F other. Side-scrapers: A simple lateral; B simple transverse; C double separate; D convergent axial; E *déjeté*; F other.

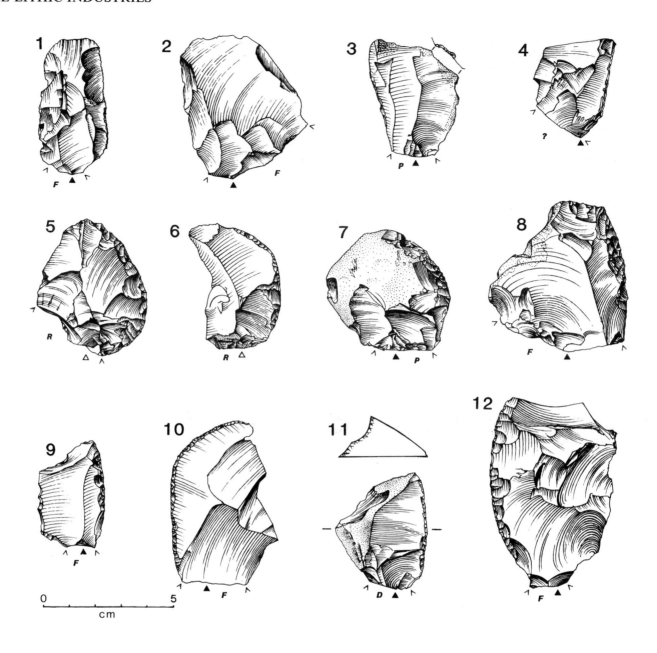

**Fig. 26.18**  Flint tools from layer D. 1–2 Levallois flakes; 3–12 simple lateral side-scrapers.

This is due mainly to an increase in the employment of ventral retouch, either on inverse scrapers (rising to 15.9% from 9.5% in the previous layer) or alternate scrapers (7.4% from 5.4%). Convex-edged scrapers become even more important than before, at 58.7%; there are 32.1% straight and 9.2% concave. The mean curvature is thus slightly higher than for layer E, at 0.08. The edge measurements are virtually unchanged, as is length (47.2 ± 10.9 mm). There is an increase in facetted butts (IF = 44.7, with an even more evident relative increase in IFs from 24.2 to 33.2). Levallois (11.5%) and cortex-backed (8.4%) blanks are very little changed, but there is only one pseudo-Levallois point (0.3%). Two pieces (0.7%) are made on cores.

### Simple scrapers

*Single lateral* (136) (Fig. 26.18, 3–12). Convex scrapers are even more preponderant (89, or 65.4%); there are only 37 straight and 10 concave pieces. Accordingly, the mean curvature is 0.09. The convex scrapers have the longest edges (36.9 mm, against 30.2 and 27.0), as also the lowest retouch angle (55.5°). The concave edges have the least invasive retouch as well as being the shortest. As in F (but not in E) there is a strong preference for retouching the right (63%) rather than the left edge. Clactonian notches are present on two pieces, and eight have additional retouch; there is no thinning of the butt.

*Transverse* (28) (Fig. 26.19, 7 and 10). This group continues to diminish in relative importance among the scrapers, amounting to only 9.5%. Following the usual pattern, the transverse scrapers are smaller than the single lateral pieces (by some 4 mm), but are to all intents and purposes identical to the latter in respect of their edge measurements and curvature. This last feature is in marked contrast to layer E; in D, the distribution of shape tends to follow that for the lateral scrapers, favouring convex edges.

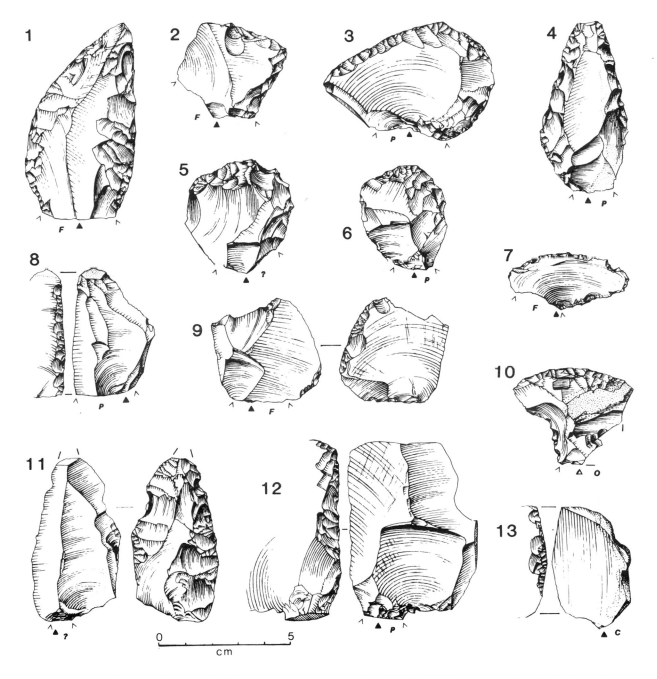

**Fig. 26.19** Flint tools from layer D. Side-scrapers: 1, 4 convergent; 2, 3, 5, 6 *déjeté*; 7, 10 transverse with a Clactonian notch on the left side; 8, 9, 11–13 on the ventral surface.

Multiple scrapers

*Separate* (15). Biconvex scrapers (7) are the clear favourites, though all subtypes except straight-concave are represented. The retouch is generally quite strong (mean invasiveness 6.1 mm) and relatively shallow (53.6°). A single piece (biconvex) has a Clactonian notch. Careful selection of the blanks is apparent; the mean length is 7 mm greater than for the scrapers as a whole (the working edge length being 13 mm greater!). Five pieces, or 33.3%, are Levallois. Also the facetting index is high (54.5), the maximum for the scraper groups from this layer.

*Convergent axial* (9). Seven of these are clearly convex (e.g. Fig. 26.19, 1); the other two have both been classified as straight,

though in the case of Fig. 26.19, 4 this is clearly an oversimplification. The quality of retouch is generally high (invasiveness 8.7 ± 3.1 mm), and the convex pieces in particular are on relatively large blanks (mean length 53 mm), of which two are typical and a third atypical Levallois; two have demi-Quina retouch. One piece has a thinned butt, another some thinning of one edge.

*Déjeté* (21). Once again these are very heterogeneous (Fig. 26.19, 2, 5 and 6). Two are on typical Levallois flakes, and another on a cortex-backed flake. Only two are double — a considerable change from earlier layers. The retouched edges are shorter than for any layer except A (the mean length of the longer edge is 37.8 ± 10.7 mm); the retouch itself is less invasive, at 5.5 mm, than

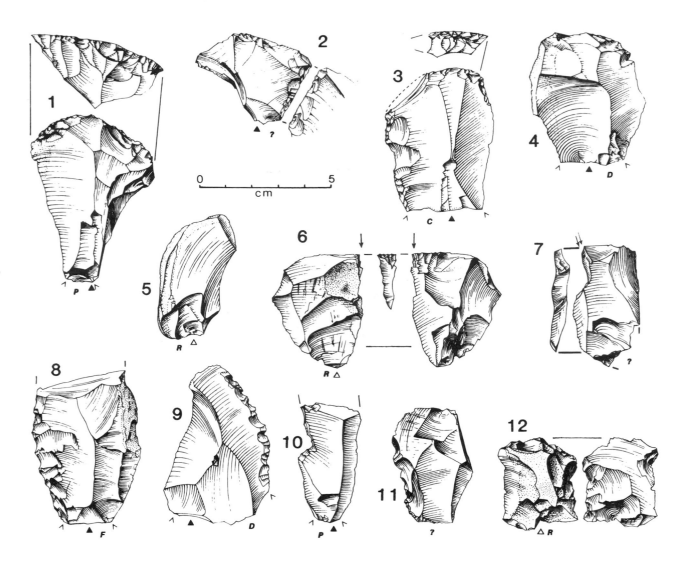

**Fig. 26.20** Flint tools from layer D. 2 Side-scraper with bifacial retouch on an overstruck flake; 1, 3, 4 end-scrapers; 5 naturally-backed knife; 6, 7 burins; 10 notch; 8, 9, 11, 12 denticulates.

for the *déjeté* scrapers of any other layer but G. A single example has demi-Quina retouch. Two have thinned butts.

Other side-scrapers

Reference has already been made to the increase in inverse and alternate scrapers (the contribution of Bordes types 26–28 is less important). Connected with the rise in alternate pieces, in particular, is the addition of appreciable numbers of scrapers of double or convergent form. It was suggested, when these categories were discussed for layer E, that the makers of the scrapers appear to have treated inverse retouch as a viable alternative to normal retouch (i.e. that they used the two techniques to produce somewhat similar edges). For layer D, one can perhaps go further and postulate that the alternate scrapers of this assemblage — and this may apply to others — offer a very satisfactory means of obtaining the equivalent of two single scrapers on one blank, each with the same positional relationship to the butt. In comparison, the more conventional double scraper has the disadvantage that it must be reversed when the second edge is used (giving a different longitudinal section). Lending support to this view is that, in terms of edge length, the inverse/alternate

scrapers resemble the single-edged rather than the double or convergent pieces with normal retouch.

*On the ventral face* (47). Scrapers of this type play quite an important part, so the sample is a good one for analysis. One of the most impressive examples is a convergent convex piece made on a Levallois point by very invasive subparallel flaking (Fig. 26.19, 11); another is *déjeté* and a third is double (straight-convex, but rather lightly retouched). Four are tranverse (two straight and two convex).

Allowing for the presence of multiple pieces, the dominant edges include 19 (40.4%) straight, 20 (42.6%) convex, and 8 (17.0%) concave — thus the straight and concave percentages are higher than might be expected. The working edges are of good size (mean length 33.3 mm), relatively sharp (58.8°) and moderately invasive (4.4 mm, with six pieces reaching double figures).

In choosing the location of retouch the makers have shown a marked preference for the right-hand side of single-edged pieces (where this can be ascertained): 25, against 9 left-hand. Cortex-backed flakes seem also to have been somewhat favoured, being used in eight cases (17.0%). Six of the blanks are Levallois (8.5%, near enough the expected figure for scrapers for this layer). The facetting index IF is slightly lower than that for the totality of retouched tools, at 34.5. Thinning of the butt has occurred in three cases, of the side in four; there is also a solitary case of a Clactonian notch.

*With abrupt retouch* (2). Both are convex, on plain flakes (50+ mm), and moderately invasively retouched (5.0 mm).

*With thinned back* (7). Five are convex, one is straight, and one concave; the butts of two have been removed. There is a single (atypical) Levallois blank.

*With bifacial retouch* (9). These include four with straight edges (Fig. 26.20, 2), four with convex and one with concave; none has extensive bifacial work. Two have thinning of the butt, and two of an edge — the butt has been removed in one case. The blanks include one cortex-backed and two Levallois flakes.

*With alternate retouch* (22). In six cases the secondary work is of alternating type. Convex (10) and straight (9) edges are predominant. Three pieces are Levallois, and another has cortical backing. No thinning has occurred, though one butt has been removed by retouch.

**End-scrapers** (21; 9 typical, 12 atypical). Three pieces (two of them typical) are on Levallois flakes. The remainder are on plain blanks. None are on blades or transverse flakes. Apart from a single example made on the butt, all are of simple type. Nearly half are retouched on one or both lateral margins (Fig. 26.20, 3). In general, the worked tips are both wide (retouched edge length 23.2 mm) and well curved (mean 0.24); good examples are illustrated as Fig. 26.20, 1 and 4. Four are formed by subparallel and one by parallel retouch. Three (all typical) have Clactonian notches. Seven of the tips are skewed (four to the right).

**Burins** (13; 9 typical, 4 atypical). The typical examples are all of the angle variety on plain flakes, struck longitudinally (six distal, one proximal and two indeterminate). Six are made on blanks (Fig. 26.20, 7). In the case of Fig. 26.20, 6 the surface used is possibly the termination of an overstruck flake which has been shaped into a U by repeated burin-blows — in fact three pieces have multiple removals from a single platform. Only two have no additional retouch of any kind.

**Borers** (2, atypical). One, 39 mm long, is formed by the intersection, at the distal end, of a notch and a short length of flat inverse retouch. The extreme tip is sharp, but the overall effect is to produce an obtuse angle. The other is on a broken double side-scraper with thinned butt, and could be spontaneous (associated with the break).

**Knives** (2, atypical). In one case, 39 mm long, a short length of cortical back on the right-hand side has been extended by irregular semi-abrupt retouch — though not as far as the tip, which is pointed. The other edge is mint. The second example, 47 mm in length, is on a rectangular flake with a steep primary right-hand edge which has been further blunted by light abrupt irregular retouch extending around the end of the piece.

**Raclettes** (10). These have a mean length of 39.6 mm, and include three examples on blades. Only two are inverse. Six of the retouched edges are straight, two convex, and two concave; three are on the left side and four on the right, four being transverse.

**Truncated pieces** (6). Their mean length is 39.7 mm, two being on cortex-backed flakes. Five are truncated at the distal end (including one with an inverse truncation). Three are straight, two convex and one concave.

**Notched pieces** (28). This class now plays a rather insignificant part in the assemblage (5.8% in reduced count). The notches include 39% of Clactonian type (e.g. Fig. 26.19, 11). Six pieces have double notches (two being opposed laterally); none is triple. The dimensions of the notches themselves are comparable to those for the previous layer, apart from a lower mean curvature (0.23); the mode is also lower.

| Curvature | 0.0 | 0.1 | 0.2 | 0.3 | 0.4 | 0.5 |
|---|---|---|---|---|---|---|
| % | | 3.6 | 53.6 | 28.6 | 10.7 | 3.6 |

Further statistics would probably be misleading in view of the smallness of the sample.

**Denticulates** (65). These are much less numerous than in the preceding layers (13.6% in reduced count); many of them could be described as side-scrapers with denticulate retouch (e.g. Fig. 26.20, 7 and 8). Five are on Levallois flakes, 6 on pseudo-Levallois points and 7 on cortex-backed flakes. For the first time, more typical than atypical blanks have been used. Also the blanks are flatter than before. Only 16.7% of the denticulates are made by adjacent Clactonian notches, while 38.3% have coarse retouch, giving 55% coarse edges — almost identical to layer E. Fine retouch occurs on 35.0% and 9.9% are microdenticulates. Of the six pieces made on pseudo-Levallois points, four have fine retouch, the others being microdenticulated, thus continuing the trend away from the features which characterised the earliest industries at La Cotte.

The denticulates of layer D establish the principal morphological characteristics which persist throughout the remaining assemblages, despite some minor variations from layer to layer. The 'simple' morphological group now amounts to 63.1% (47.7% lateral, 15.4% transverse) with 7.7% double and 9.2% *déjeté* pieces. As in layer E there are no convergent denticulates (apart from Tayac points). The average length is the lowest yet (43.0 ± 11.1 mm, with mean retouch length 29.9 ± 11.2 mm), and there is a substantial fall in invasiveness (to 5.1 ± 3.1 mm).

**Bec burinant alterne** (1). Though typical in appearance, this could be the result of accidental damage; a lateral normal retouched notch is adjacent to a distal inverse Clactonian notch (the other edge has been retouched as a side-scraper).

**Tayac points** (2). These average 41.5 mm in length. Both are atypical. One of them has one edge consisting of a single large notch, which may be accidental. The other is a curious piece, made on an overstruck flake. After one edge had been denticulated (removing the butt), several further blows were struck on the old core edge; as a result, the tool has a keeled lower surface, and falls midway between the Tayac and Quinson types of point.

**End-notched pieces** (5). All are slightly questionable. The mean length is 39.5 mm.

**Chopping tool** (1). This is small (41 x 37 mm, weight 44 gm) and crudely pointed; it is made on a thick pebble of which about a third of the cortex is preserved. The flaking sequence is rather unsystematic, and the blows on the sides have caused extensive shattering.

**Miscellaneous** (26). They include

| | n | % | | n | % |
|---|---|---|---|---|---|
| scrapers | 22 | (84.6) | notches | 1 | (3.8) |
| denticulates | 6 | (23.1) | overstruck removals | 1 | (3.8) |
| scaled | 1 | (3.8) | burins | 0 | (0.0) |

Seven pieces (or 26.9%) are compounded of more than one of these. Among them are a scraper with an overstruck LSF-type removal from the butt, and four more which are combined with denticulates.

**Bifaces** (4). All are relatively thin nucleiforms (two with residual cortex, in one instance on the butt) and are quite small (50 x 37, 61 x 43, 45 x 28 and 47 x 44 mm respectively).

## LAYER C

The reduced tool total of 646 pieces represents 14.1% of the unbroken flint, and the real tools amount to 20.6%; the real percentage of types 45–50 is 25.1. These values

**Table 26.11** Layer C. Tool Type Frequencies

| Type | | N | Real | Ess. | Red. |
|------|------|---|------|------|------|
| 1 | Typical Levallois flakes | 20 | 2.11 | – | – |
| 2 | Atypical Levallois flakes | 25 | 2.64 | – | – |
| 3 | Levallois points | 3 | 0.32 | – | – |
| 4 | Retouched Levallois points | 2 | 0.21 | 0.30 | 0.31 |
| 5 | Pseudo-Levallois points | 8 | 0.84 | 1.21 | – |
| 6 | Mousterian points | 4 | 0.42 | 0.60 | 0.62 |
| 7 | Elongated Mousterian points | 2 | 0.21 | 0.30 | 0.31 |
| 8 | *Limaces* | 3 | 0.32 | 0.45 | 0.46 |
| 9 | Single straight side-scrapers | 69 | 7.28 | 10.42 | 10.68 |
| 10 | Single convex side-scrapers | 139 | 14.66 | 21.00 | 21.52 |
| 11 | Single concave side-scrapers | 11 | 1.16 | 1.66 | 1.70 |
| 12 | Double straight side-scrapers | 11 | 1.16 | 1.66 | 1.70 |
| 13 | Double straight-convex side-scrapers | 10 | 1.05 | 1.51 | 1.55 |
| 14 | Double straight-concave side-scrapers | 3 | 0.32 | 0.45 | 0.46 |
| 15 | Double convex side-scrapers | 15 | 1.58 | 2.27 | 2.32 |
| 17 | Double concave-convex side-scrapers | 2 | 0.21 | 0.30 | 0.31 |
| 18 | Convergent straight side-scrapers | 4 | 0.42 | 0.60 | 0.62 |
| 19 | Convergent convex side-scrapers | 18 | 1.90 | 2.72 | 2.79 |
| 20 | Convergent concave side-scrapers | 1 | 0.11 | 0.15 | 0.15 |
| 21 | *Déjeté* scrapers | 27 | 2.85 | 4.08 | 4.18 |
| 22 | Straight transverse side-scrapers | 12 | 1.27 | 1.81 | 1.86 |
| 23 | Convex transverse side-scrapers | 14 | 1.48 | 2.11 | 2.17 |
| 24 | Concave transverse side-scrapers | 3 | 0.32 | 0.45 | 0.46 |
| 25 | Side-scrapers with inverse retouch | 30 | 3.16 | 4.53 | 4.64 |
| 26 | Abrupt retouched side-scrapers | 4 | 0.42 | 0.60 | 0.62 |
| 27 | Side-scrapers with thinned back | 7 | 0.74 | 1.06 | 1.08 |
| 28 | Side-scrapers with bifacial retouch | 17 | 1.79 | 2.57 | 2.63 |
| 29 | Alternate retouched side-scrapers | 21 | 2.22 | 3.17 | 3.25 |
| 30 | Typical end-scrapers | 8 | 0.84 | 1.21 | 1.24 |
| 31 | Atypical end-scrapers | 12 | 1.27 | 1.81 | 1.86 |
| 32 | Typical burins | 9 | 0.95 | 1.36 | 1.39 |
| 33 | Atypical burins | 4 | 0.42 | 0.60 | 0.62 |
| 35 | Atypical borers | 2 | 0.21 | 0.30 | 0.31 |
| 37 | Atypical backed knives | 1 | 0.11 | 0.15 | 0.15 |
| 38 | Naturally backed knives | 8 | 0.84 | 1.21 | – |
| 39 | *Raclettes* | 15 | 1.58 | 2.27 | 2.32 |
| 40 | Truncated pieces | 8 | 0.84 | 1.21 | 1.24 |
| 42 | Notched pieces | 37 | 3.90 | 5.59 | 5.73 |
| 43 | Denticulates | 80 | 8.44 | 12.8 | |
| 44 | *Becs burinants alternes* | 1 | 0.11 | 0.15 | 0.15 |
| 45 | Retouched with inverse retouch | 14 | 1.48 | – | – |
| 46-47 | Abrupt and alternate retouch (thick) | 24 | 2.53 | – | – |
| 48-49 | Abrupt and alternate retouch (thin) | 200 | 21.10 | – | – |
| 52 | Notched triangles | 1 | 0.11 | 0.15 | 0.15 |
| 54 | End-notched pieces | 1 | 0.11 | 0.15 | 0.15 |
| 62 | Miscellaneous | 38 | 4.01 | 5.74 | 5.88 |
| | Bifaces (whole) | 6 | | | |
| | Bifaces (broken) | 1 | | | |
| | Totals | | 948 | 662 | 646 |

are little changed from the previous layer.

Side-scrapers reach their maximum frequency, 64.7%, and are generally very well made. They exhibit a strong preponderance of convex edges (60.7%) and are often made on Levallois blanks. The Charentian index IC (i.e. *limaces*, and single lateral convex, all transverse, and bifacially retouched side-scrapers) reaches a peak of 29.1. The percentage of convergent pieces (I rc) is also high, at 9.1 — a value exceeded only in layer 5. Types 25 and 29 now represent only 7.9% of the tools, not much more than half their representation in D. It is in this layer that the practice of edge rejuvenation of scrapers by means of the LSF technique (chapter 29) begins to play a recog-

nisable part, though as yet it is scarce.

The extended denticulate group is little changed from before in frequency (18.3%) or composition, except that the denticulates have more delicate edges. A decline in Group III is due to there being fewer end-scrapers and knives; burins maintain their frequency. This is the first layer in which a biface of 'classic' shape occurs (in this case a cordiform).

**Retouched Levallois points** (2). One has semi-abrupt retouch on the distal half of a primary Levallois point 41 mm long; the other is secondary and blade-like, 53 mm long, with very slight retouch or utilisation. Both butts are plain.

**Table 26.12** Layer C. Typological Groups and Indices

| Index | | Real | Ess. | Red. |
|---|---|---|---|---|
| ILty | Typological Levallois | 5.27 | – | – |
| IR | Side-scrapers | 44.09 | 63.14 | 64.71 |
| IC | Charentian | 19.83 | 28.40 | 29.10 |
| IAu | Acheulian unifacial | 0.11 | 0.15 | 0.15 |
| IAt | Acheulian total | 0.84 | 1.20 | 1.23 |
| IB | Bifaces | 0.73 | 1.05 | 1.07 |
| I | Levallois | 5.27 | 0.30 | 0.31 |
| II | Mousterian | 45.04 | 64.50 | 66.10 |
| III | Upper Palaeolithic | 4.64 | 6.65 | 6.81 |
| IV | Denticulate | 8.44 | 12.08 | 12.38 |
| 5 | Pseudo-Levallois points | 0.84 | 1.21 | – |
| 38 | Naturally backed knives | 0.84 | 1.21 | – |
| I rc | Convergent retouched | 6.22 | 8.91 | 9.13 |
| 6-7 | Mousterian points | 0.63 | 0.91 | 0.93 |
| Rsingle | Side-scrapers, single | 23.10 | 33.08 | 33.90 |
| Rdouble | Side-scrapers, double | 4.32 | 6.19 | 6.35 |
| Rtransv | Side-scrapers, transverse | 3.06 | 4.38 | 4.49 |
| Rinv | Side-scrapers, inverse/alternate | 5.38 | 7.70 | 7.89 |
| 30-31 | End-scrapers | 2.11 | 3.02 | 3.10 |
| 32-33 | Burins | 1.37 | 1.96 | 2.01 |
| 36-37 | Knives | 0.11 | 0.15 | 0.15 |
| 42 | Notched pieces | 3.90 | 5.59 | 5.73 |
| 42-43 | Notched pieces & denticulates | 12.34 | 17.67 | 18.11 |
| Ident | Extended denticulate group | 1.48 | 2.11 | 2.17 |
| Iirr | Irregularly retouched | 25.11 | – | – |
| IQ | Quina | 2.15 | | |
| IQ+½Q | Quina + demi-Quina | 5.02 | | |
| Cl | Clactonian notches | 37.84 | | |

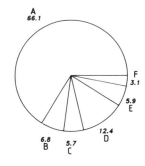
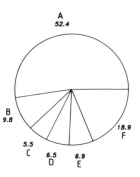

**Fig. 26.21** Flint tools from layer C. Relative frequencies of the principal tool categories and of the major side-scraper types. Tools: A Group II; B Group II; C Group IV; D notched pieces; E miscellaneous; F other. Side-scrapers: A simple lateral; B simple transverse; C double separate; D convergent axial; E *déjeté*; F other.

| | | | |
|---|---|---|---|
| single lateral | 52.4 | double | 9.8 |
| convergent | 12.0 | transverse | 6.9 |
| other | 18.9 | | |

The most obvious development is a further reduction in the frequency of transverse scrapers, and a drop in types 25–29 relative to layer D. More explicitly, those with ventral retouch fall to 7.2% from their peak of 15.9% in the previous layer. Another distinctive aspect of the layer C scrapers, which is repeated in B, is the frequency with which additional treatment of some kind has been applied: thus 9.1% have retouch at one end, adjacent to the working edge (e.g. Fig. 26.22, 7 and 9, both of which have a somewhat pointed tip as a result); thinning of the butt or an edge each occur on 4.5%. Ten pieces (2.4%) have notches, six of them Clactonian.

The percentage of Levallois blanks reaches its highest level (20.6); eight pieces (1.9%) are on pseudo-Levallois points, one on an LSF, and 39 (9.3%) are cortex-backed. In addition, three of the scrapers are on former cores, and one is made on a pebble. The facetting index for scrapers is 11.3 higher than before, at 56.0.

Simple scrapers

*Single lateral* (219) (Fig. 26.22, 3–10, 12 and 13). These are longer than in any layer except 5 (mean 49.9 mm), with high values for invasiveness (4.9 mm) and retouched edge length (37.9 mm); this last is due to the values of 5.7 and 41.2 mm for convex pieces (straight and concave pieces score 3.4 and 32.2, and 3.0 and 32.0 mm respectively). The mean retouch angle is extremely low (52.9°). Mean curvature (0.106) is higher than the average for the layer, matching the changes in representation of the shape categories: 139 are convex, 69 are straight and 11 are concave. Five of the convex scrapers (3.6%) have Quina retouch, and four more have demi-Quina. Twenty-seven (12.3%) have additional retouch at one or other end, 6 have thinned butts and 4 have (Clactonian) notches. A small majority (57%) are worked on the right-hand edge.

*Transverse* (27) (Fig. 26.23, 3 and 4). These comprise 12 straight, 14 convex and 3 concave pieces (mean curvature 0.065). Besides being smaller than the single lateral scrapers, they have steeper edges (mean 61.2°). All are opposed to the butt.

Multiple scrapers

*Separate* (41). Biconvex, and straight and straight-convex scrapers account for almost the whole group (Fig. 26.22, 11, 14 and 15). A single piece has demi-Quina retouch, and — as one might expect

**Mousterian points** (4). With one exception, these are of above average length for the layer (mean 50.9 mm), but of variable quality. The best and largest, on a typical Levallois blank, is slightly skewed, with scalar retouch. It has a plain butt (those of the others are facetted). The smallest piece is also the least characteristic, with irregular light semi-abrupt retouch on both edges.

**Elongated Mousterian points** (2). These are rather small for pieces in this category (48 and 37 mm, though the latter figure is too low because of tip damage). Both are on typical Levallois blanks and have weak semi-abrupt retouch on the left-hand edge. The other edge is worked with predominantly scalar removals, though the smaller piece also has some subparallel scars.

**Limaces** (3). These have a mean length of 63 mm, i.e. well above average for tools from layer C. Two are broad, but the other is blade-like in its proportions. The largest piece is markedly asymmetrical in plan, on a very thick cortical flake, and is double-pointed. The smallest of the three has rounded ends. All are shaped principally by scalar retouch, though some removals are subparallel on the largest example, and demi-Quina on the smallest.

**Side-scrapers** (418) These reach a peak in this layer, as 64.7% of the well-retouched tools. Also, with the exception of those from layer 5 (which is itself very different from the other series), they achieve the highest mean values for length, 49.3 ± 12.3 mm, retouched edge length, 38.1 ± 14.1 mm, and invasiveness, 5.7 ± 3.7 mm. The mean curvature rises to 0.10 ± 0.09, at which value it stays for the remaining assemblages. The frequency of concave-edged pieces is at a minimum for the site, at 5.5%, with a further rise in convex edges to 60.7%. The percentages resulting from a breakdown into major groups are

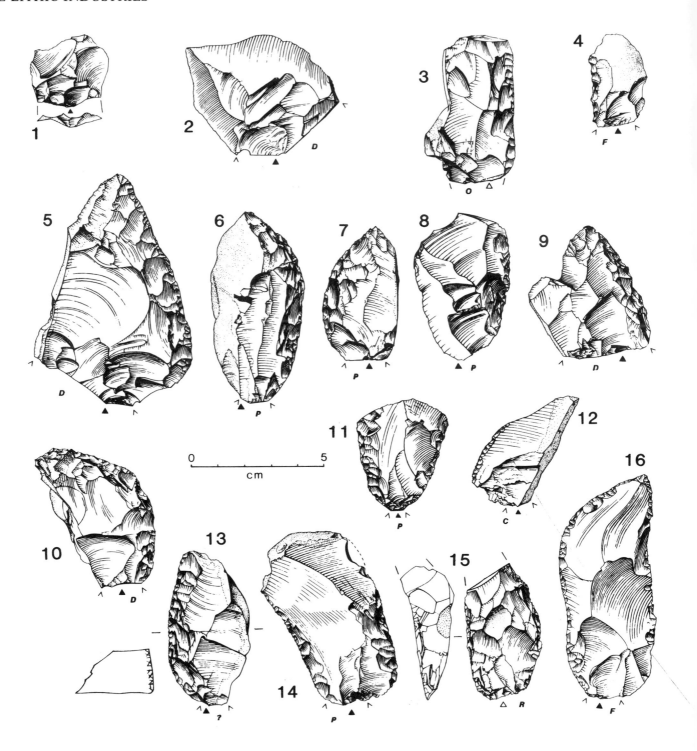

**Fig. 26.22** Flint tools from layer C. 1 Atypical Levallois flake; 2 pseudo-Levallois point; 3–16 side-scrapers: 3-10, 12, 13 simple lateral; 11, 14 and 15 double; 16 convergent.

from the sample size — other rare features are present: four cases of thinning of the butt and one of the side, two of retouch at the tip, and a single Clactonian notch. The secondary work is invasive (mean 6.6 mm) and extensive (length 40.8 mm), though much less so than in layer D. This last observation must be related to a diminution in the size of the blanks, though they are still very often Levallois (11 typical, 4 atypical). Rather unusually for this group, the facetting index is relatively low (45.0).

*Convergent axial* (23). As usual, the convex variety includes the best as well as the largest examples (mean invasiveness 7.4 mm, length 62 mm). Eleven of the blanks are typical Levallois, and another atypical. Two of the convex pieces have thinned butts, and two some thinning of an edge (on the piece illustrated in Fig. 26.23, 5 this has been done by shallow flaking before application of the normal retouch). There are two examples of Quina retouch, and three of demi-Quina.

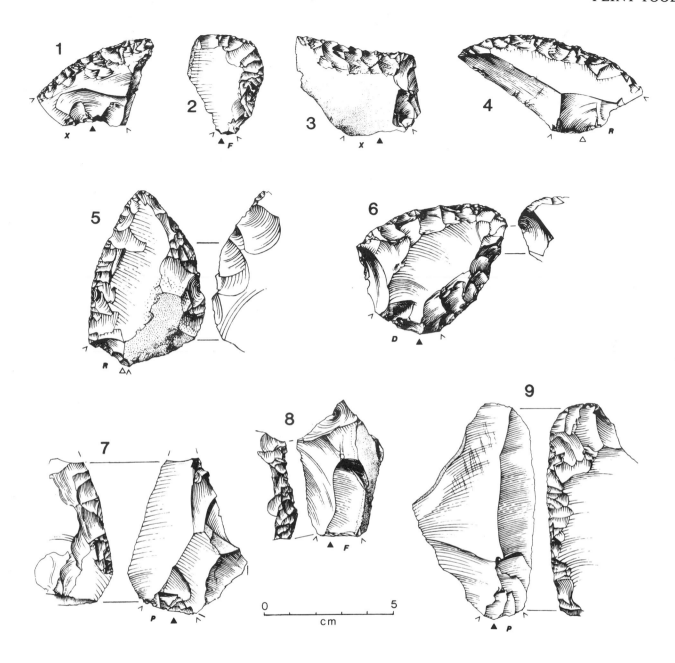

**Fig. 26.23** Flint tools from layer C. Side-scrapers: 5 convergent, with some thinning of one edge; 1, 2, 6 *déjeté*; 3, 4 transverse; 7–9 on the ventral surface.

*Déjeté* (27). For the first time, these are strongly dominated by convex-edged pieces, numbering 19, against 8 straight (Fig. 26.23, 1, 2 and 6); previously this majority has been slight. The retouch is very invasive (mean 7.4 ± 3.2 mm). The blanks include only two Levallois flakes (both typical), and a cortex-backed flake, in marked contrast to those used for the convergent axial scrapers. The butt has been removed by retouch in one case.

Other side-scrapers

Again, it is the alternate scrapers in particular that provide the equivalents of double or convergent pieces (Table 26.24). The high incidence of simple straight forms seems to be a fairly recurrent feature of the inverse scrapers from the site, and may imply no more than that the technique is well suited to the production of such edges. The metrical data exhibit moderate values.

*On the ventral face* (30). Apart from a single *déjeté* piece, these are all single-edged; four are transverse (2 straight, 2 concave). Of the lateral examples, seventeen are straight, six convex (Fig. 26.23, 9, on which the retouch extends round the distal end) and two concave. Thus the edge shape distribution is very different from that of the scrapers with dorsal retouch. The mean retouched edge length is appreciably lower than in layer D (28.8 compared to 33.3 mm); the retouch is also a little steeper (61.6°), but the invasiveness is effectively unchanged at 4.5 mm. The preponderance in favour of right-hand edges is only 13:10, in contrast to the extreme situation in D.

The facetting index, IF, is around the expected value, at 55.0 (the butt has been removed by retouch in a single example); however, the three Levallois blanks are fewer than normal for the layer. Two pieces (6.7%) have cortical backs (Fig. 26.23, 8). Additional features include a single case of a thinned side, two of

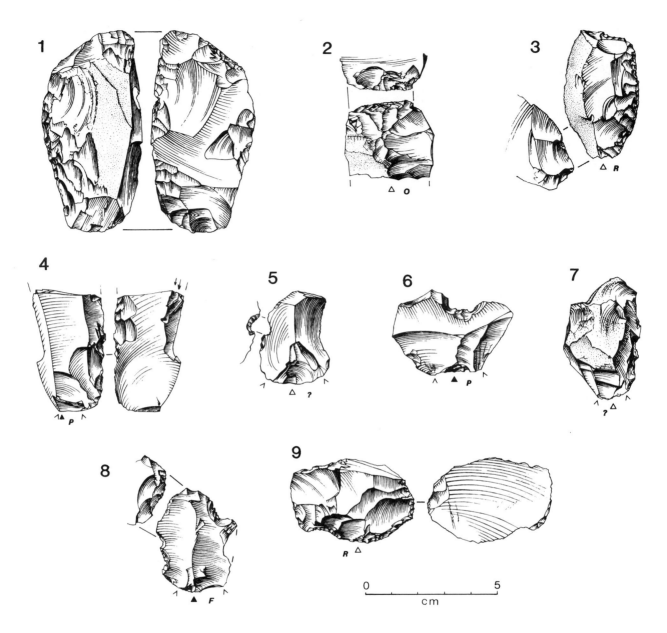

**Fig. 26.24** Flint tools from layer C. 1–2 Side-scrapers with bifacial retouch (alternatively, 2 could be classified as miscellaneous, with scaling); 3 end-scraper (on side-scraper with thinned butt); 4 angle-burin on a break; 5, 6 notches; 7, 8 denticulates; 9 miscellaneous (transverse side-scraper with scaling from inverse truncation).

retouch at an end, and one of a Clactonian notch.

*With abrupt retouch* (4). Three are straight-edged, and the other (the largest) is convex; all have quite invasive retouch (mean 6 mm). The shortest piece (46 mm) has a cortical back.

*With thinned back* (7). Two are straight, four convex, and one is concave. Thinning has extended to the removal of the butt in one instance. As a rule the retouch is very invasive ( $\geqslant 10$ mm in six cases); on three pieces it is subparallel. One example is on a typical Levallois flake.

*With bifacial retouch* (17). One of these is fully bifacial rather than merely bifacially retouched (Fig. 26.24, 1); another is rather core-like. Thirteen are convex, two straight, and two concave. The retouched edge angle (mean 66.7°) is rather open. The blanks include 4 typical Levallois and 3 cortex-backed flakes, one core and one pebble.

*With alternate retouch* (21). Twelve are convex-edged, the other nine straight. Alternating retouch occurs on seven. There is one case of butt removal by retouch, and another of thinning of the bulbar surface from the butt end. A single example occurs of an additional (retouched) notch.

Six pieces are on Levallois blanks, and one on a cortex-backed flake.

**End-scrapers** (20; 8 typical, 12 typical). Two are on Levallois flakes, and one on a cortex-backed flake (Fig. 26.24, 3); two of the blanks are blades, and one is a transverse flake. The illustrated piece combines an end-scraper with a side-scraper whose butt has been thinned by removals from one edge; in all, 13 pieces have retouched lateral margins. The mean length is relatively high, at 48.2 ± 13.2 mm, but none of the other measurements show unusual mean values. Six of the worked tips are skewed to the right, and two to the left. As usual, scalar/semi-abrupt retouch predominates, but six have subparallel and one each stepped and *raclette*-type working. Sixteen are of the simple variety, one denticulated, one shouldered (single), and two on the butt.

**Burins** (13; 9 typical, 4 atypical). Nine are angle burins, the rest of the *plan* type. They are longitudinal apart from two transverse pieces. One burin has been struck from the proximal end of its blank. Eight pieces are made on breaks, and three more on truncations. The blanks include three Levallois flakes (all of them used for typical burins). Eight examples possess some kind of marginal trimming.

**Borers** (2, atypical). One is created by the intersection of the edges of a *déjeté* scraper made with predominantly inverse retouch. The other is double, with borers formed on each end of a ventrally retouched edge.

**Knife** (1, atypical). This is on a broad, typical Levallois flake, 67 mm long. It has a semi-circular back made by semi-abrupt retouch, and is thus on the borderline between the knife and side-scraper classes.

**Raclettes** (15). Averaging only 35.8 mm in length, these include two pieces on blades and three on Levallois flakes. Six have inverse retouch. Other characteristics of the retouched edges are: eight straight, four convex, and two concave; four on the left-hand edge, three on the right, and eight transverse.

**Truncated pieces** (8). The mean length is only 39.7 mm, with a mean truncation width of 14.2 mm. All but one of the truncations are on the distal end, and three are oblique. One is convex, the rest being straight. As is not unusual for this type, there is a possibility that in three cases the truncation is the result of a knapping accident rather than being intentional.

**Notched pieces** (37). Clactonian notches are again in a clear minority (37.8%); nevertheless they occur on the only pseudo-Levallois point present, and on one of the two cortex-backed pieces. The single typical Levallois flake used for this purpose has a retouched notch. Nine pieces have double notches (five in lateral opposition). Metrically, the notches themselves are virtually identical to those from the previous layer (for which reason the curvature has not been tabulated) except that the Clactonian notches are somewhat deeper than the retouched ones, while (unusually) the latter are wider.

**Denticulates** (80). There is a further slight decrease in the frequency of this type, to 12.4% of the reduced tool total, as well as a marked reduction in the incidence of coarse edges, to 36.7% (as opposed to 55.0% in D); Clactonian notches are used in only 10.7% whereas 53.1% have fine retouch. The blanks include seven Levallois flakes, four pseudo-Levallois points, and six cortex-backed flakes. Metrically and morphologically there is a strong resemblance to the denticulates from layer D, even if in E there is a rather high percentage (15.0%) of ventrally worked pieces.

**Bec burinant alterne** (1). This has been produced by a slight invasive removal adjacent to a shallow normal notch at the distal end of a short (41 mm) flake.

**Notched triangle** (1). An inverse retouched notch has been made on a tapering mesial flake fragment.

**End-notched flake** (1). The only possible example of this type has a small Clactonian notch which could well be accidental.

**Miscellaneous** (38). They include 19 pieces (i.e. 50%) which fit into more than one of the following categories:

|  | n | % |  | n | % |
|---|---|---|---|---|---|
| scrapers | 23 | (60.5) | notches | 9 | (23.7) |
| denticulates | 13 | (34.2) | overstruck removals | 3 | (7.9) |
| scaled | 2 | (5.3) | burins | 2 | (5.3) |

Eleven have dorsal thinning/scaling — that in Fig. 26.24, 9 executed from an inverse truncation. One of the three overstruck secondary removals is of LSF type (all three pieces are, in other respects, side-

scrapers). No fewer than eleven (28.9%) of the blanks are Levallois; such a high percentage is not repeated until layer 5.

**Bifaces** (6). Apart from one nucleiform 82 mm long, these are very small, in the range 36–44 mm. Two of the 'partial' bifaces (pieces on flakes which have not been fully worked on the ventral face) have residual cortex. *Méplats* occur on two of the butts.

## *LAYER B*

There is a small but definite increase in the frequency of tools: the reduced count, 426, is 15.7% of the unbroken flint, whereas in C the corresponding figure was 14.1%, and in E it was as low as 11.5%. The real total is equivalent to 23.5%, a very much bigger increase. On the other hand, the real percentage of types 45–50 continues its decline, to 22.9.

Many typological similarities exist between the assemblages from B and the underlying C — for example, IR (62.2), IC (27.9), I rc (8.2) and the extended denticulate group (18.1%). Convex retouched edges predominate on the side-scrapers, as in C. On the other hand, burins are much commoner than before (3.2%) and include dihedral examples; this marks the onset of a development which reaches its peak in layer A. The general reduction of size is likewise part of a trend. Peculiar to B, however, is the presence of a relatively large number of unmodified Levallois blanks (ILty = 9.4).

**Retouched Levallois point** (1). This is of the secondary type, with thin alternating retouch on the right hand side near the tip.

**Mousterian points** (5). Their mean length is 49.5 mm. Four are slightly *déjeté* (Fig. 26.26, 1). One of them has demi-Quina retouch, and also ventral thinning at the tip and butt (prior to the normal retouch). Another is unusual in that only one edge has been completely retouched; the other has been left unmodified except near the butt (the tip is extremely finely pointed). Two pieces have facetted butts.

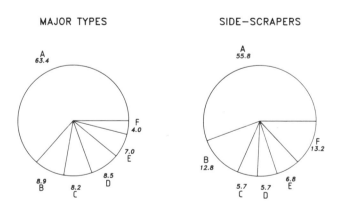

**Fig. 26.25** Flint tools from layer B. Relative frequencies of the principal tool categories and of the major side-scraper types. Tools: A Group II; B Group II; C Group IV; D notched pieces; E miscellaneous; F other. Side-scrapers: A simple lateral; B simple transverse; C double separate; D convergent axial; E *déjeté*; F other.

THE LITHIC INDUSTRIES

**Table 26.13** Layer B. Tool Type Frequencies

| Type | | N | Real | Ess. | Red. |
|---|---|---|---|---|---|
| 1 | Typical Levallois flakes | 21 | 3.29 | – | – |
| 2 | Atypical Levallois flakes | 36 | 5.63 | – | – |
| 3 | Levallois points | 2 | 0.31 | – | – |
| 4 | Retouched Levallois points | 1 | 0.16 | 0.23 | 0.23 |
| 5 | Pseudo-Levallois points | 4 | 0.63 | 0.92 | – |
| 6 | Mousterian points | 5 | 0.78 | 1.15 | 1.17 |
| 9 | Single straight side-scrapers | 44 | 6.89 | 10.14 | 10.33 |
| 10 | Single convex side-srapers | 95 | 14.87 | 21.89 | 22.30 |
| 11 | Single concave side-scrapers | 9 | 1.41 | 2.07 | 2.11 |
| 12 | Double straight side-scrapers | 8 | 1.25 | 1.84 | 1.88 |
| 13 | Double straight-convex side-scrapers | 13 | 2.03 | 3.00 | 3.05 |
| 15 | Double convex side-scrapers | 10 | 1.56 | 2.30 | 2.35 |
| 16 | Double concave side-scrapers | 1 | 0.16 | 0.23 | 0.23 |
| 17 | Double concave-convex side-scrapers | 2 | 0.31 | 0.46 | 0.47 |
| 18 | Convergent straight side-scrapers | 2 | 0.31 | 0.46 | 0.47 |
| 19 | Convergent convex side-scrapers | 9 | 1.41 | 2.07 | 2.11 |
| 20 | Convergent concave side-scrapers | 4 | 0.63 | 0.92 | 0.94 |
| 21 | Déjeté scrapers | 15 | 2.35 | 3.46 | 3.52 |
| 22 | Straight transverse side-scrapers | 5 | 0.78 | 1.15 | 1.17 |
| 23 | Convex transverse side-scrapers | 11 | 1.72 | 2.53 | 2.58 |
| 24 | Concave transverse side-scrapers | 2 | 0.31 | 0.46 | 0.47 |
| 25 | Side-scrapers with inverse retouch | 6 | 0.94 | 1.38 | 1.41 |
| 27 | Side-scrapers with thinned back | 8 | 1.25 | 1.84 | 1.88 |
| 28 | Side-scrapers with bifacial retouch | 6 | 0.94 | 1.38 | 1.41 |
| 29 | Alternate retouched side-scrapers | 15 | 2.35 | 3.46 | 3.52 |
| 30 | Typical end-scrapers | 2 | 0.31 | 0.46 | 0.47 |
| 31 | Atypical end-scrapers | 8 | 1.25 | 1.84 | 1.88 |
| 32 | Typical burins | 10 | 1.56 | 2.30 | 2.35 |
| 33 | Atypical burins | 4 | 0.63 | 0.92 | 0.94 |
| 34 | Typical borers | 2 | 0.31 | 0.46 | 0.47 |
| 35 | Atypical borers | 5 | 0.78 | 1.15 | 1.17 |
| 37 | Atypical backed knives | 2 | 0.31 | 0.46 | 0.47 |
| 38 | Naturally backed knives | 4 | 0.63 | 0.92 | – |
| 39 | Raclettes | 9 | 1.41 | 2.07 | 2.11 |
| 40 | Truncated pieces | 5 | 0.78 | 1.15 | 1.17 |
| 42 | Notched pieces | 35 | 5.48 | 8.06 | 8.22 |
| 43 | Denticulates | 36 | 5.63 | 8.29 | 8.45 |
| 44 | Becs burinants alternes | 1 | 0.16 | 0.23 | 0.23 |
| 45 | Retouched with inverse retouch | 15 | 2.35 | – | – |
| 46-47 | Abrupt and alternate retouch (thick) | 28 | 4.38 | – | – |
| 48-49 | Abrupt and alternate retouch (thin) | 103 | 16.12 | – | – |
| 51 | Tayac points | 2 | 0.31 | 0.46 | 0.47 |
| 54 | End-notched pieces | 4 | 0.63 | 0.92 | 0.94 |
| 62 | Miscellaneous | 30 | 4.69 | 6.91 | 7.04 |
| | Bifaces (whole) | 3 | | | |
| | Totals | | 639 | 434 | 426 |

**Side-scrapers** (265). After reaching a peak in layer C, the percentage of side-scrapers now begins a slight decline ($IR_{red}$ = 62.2). There is also some reduction of size towards the minimum in A (mean length 46.9 ± 10.8 mm); similar developments occur in respect of mean retouched edge length and invasiveness, but the retouch angle remains shallow, at 56.2°. Though there are more convex-edged pieces than in any other layer (straight 28.9%, convex 64.6%, concave 6.5% respectively) the curvature is unchanged, at 0.10 ± 0.09. The principal group percentages are

| single lateral | 55.8 | double | 12.8 |
|---|---|---|---|
| convergent | 11.3 | transverse | 6.8 |
| other | 13.2 | | |

The frequencies of major scraper classes are very similar to those for layer C; the chief differences are the increases in single lateral side-scrapers to 55.8%, due to the maximum representation of convex scrapers, and in double scrapers to 12.8% (straight-convex pieces rise

from 2.4 to 4.9%). Minor additional features include thinned butts (6.8%), thinned sides (4.5%), retouched ends (8.1%) and notches (only 1.1% Clactonian and 1.6% retouched). Levallois blanks (51 cases, or 19.2%) have been employed slightly more often than for the totality of the retouched tools; there are also 1.9% pseudo-Levallois points, 1.9% LSFs, and 8.3% cortex-backed pieces. One scraper is made on a tranchet flake. The facetting index, IF, is 62.8 for the scrapers.

Simple scrapers

*Single lateral* (148) (Fig. 26.26, 2–10 and 14). Forty-four are straight, 95 convex and 9 concave. The mean curvature, at 0.093, is somewhat lower than in C. As for the length of the retouched edge, the pattern which has existed since layer E continues: the convex scrapers have a very much higher mean (38.6 mm) than the others (31.6 for straight scrapers, 25.6 for concave). The concave scrapers are more steeply retouched than the others by about 6°, and the working is appreciably less invasive (cf. Fig. 26.26, 7

284

**Table 26.14** Layer B. Typological Groups and Indices

| Index | | Real | Ess. | Red. |
|---|---|---|---|---|
| ILty | Typological Levallois | 9.39 | – | – |
| IR | Side-scrapers | 41.47 | 61.06 | 62.21 |
| IC | Charentian | 18.62 | 27.42 | 27.93 |
| IAu | Acheulian unifacial | 0.31 | 0.46 | 0.47 |
| IAt | Acheulian total | 0.78 | 1.14 | 1.17 |
| IB | Bifaces | 0.47 | 0.69 | 0.70 |
| I | Levallois | 9.39 | 0.23 | 0.23 |
| II | Mousterian | 42.25 | 62.21 | 63.38 |
| III | Upper Palaeolithic | 5.95 | 8.76 | 8.92 |
| IV | Denticulate | 5.63 | 8.29 | 8.45 |
| 5 | Pseudo-Levallois points | 0.63 | 0.92 | – |
| 38 | Naturally backed knives | 0.63 | 0.92 | – |
| I rc | Conretouched | 5.48 | 8.06 | 8.22 |
| 6-7 | Mousterian points | 0.78 | 1.15 | 1.17 |
| Rsingle | Side-scrapers, single | 23.16 | 34.10 | 34.74 |
| Rdouble | Side-scrapers, double | 5.32 | 7.83 | 7.98 |
| Rtransv | Side-scrapers, transverse | 2.82 | 4.15 | 4.23 |
| Rinv | Side-scrapers, inverse/alternate | 3.29 | 4.84 | 4.93 |
| 30-31 | End-scrapers | 1.56 | 2.30 | 2.35 |
| 32-33 | Burins | 2.19 | 3.23 | 3.29 |
| 36-37 | Knives | 0.31 | 0.46 | 0.47 |
| 42 | Notched pieces | 5.48 | 8.06 | 8.22 |
| 42-43 | Notched pieces & denticulates | 11.11 | 16.36 | 16.67 |
| Ident | Extended denticulate group | 3.13 | 4.61 | 4.69 |
| Iirr | Irregularly retouched | 22.85 | – | – |
| IQ | Quina | 1.13 | | |
| IQ+½Q | Quina + demi-Quina | 4.53 | | |
| Cl | Clactonian notches | 54.29 | | |

so classified according to the rules of precedence defined by Bordes (1961, 27), have a *convex* longer edge, for instance Fig. 26.26, 12. This is also true of Fig. 26.26, 11, which was classified as straight for the same reason. Eight pieces are made on Levallois blanks. A single example has Quina retouch, while three have demi-Quina.

*Déjeté* (15). As in C, convex edges (10) predominate (e.g. Fig. 26.22, 13) and the retouch is distinctly invasive (7.6 ± 3.0 mm). Typical Levallois flakes are unusually common as blanks (6); there is also a pseudo-Levallois point. No fewer than eight pieces are double *déjeté*. Two examples have lost their butts as a result of retouch.

Other side-scrapers

The relatively high proportion of alternate scrapers weights the morphological classification heavily in favour of double and more especially *déjeté* forms (Table 26.24). The retouch is more invasive than usual for the layer, though the worked edge length for the group is relatively low.

*On the ventral face* (6). These are too rare for the statistics to mean a great deal (though it is perhaps worth noting that the low edge angle, 53°, matches that for the scrapers with normal retouch). One piece is *déjeté*, and another transverse. One is on an LSF.

*With thinned back* (8). Three of these are straight (e.g. Fig. 26.27, 4), four are convex and one concave. Removal of the platform has taken place in three instances. The blanks include one Levallois flake and one cortex-backed flake. The retouch is generally very invasive (10+ mm in four cases).

*With bifacial retouch* (6). None of these is fully bifacially worked, though one example has extensive biting removals as if used as a core. Five have convex edges, the sixth being straight. The blanks include a Levallois flake and a pseudo-Levallois point.

*With alternate retouch* (15). Three are straight-edged, ten convex and two concave. In two instances the retouch is of alternating type. The mean retouch angle is low (56.1°), typical for this layer. Four of the blanks are Levallois, and one is cortex-backed. This last (Fig. 26.27, 5) is one of no less than 5 pieces of this class which are *déjeté* in form. Two pieces have thinning of the butt, and two more of an edge; the butt has been removed by retouch in one instance.

**End-scrapers** (10; 2 typical, 8 atypical). Two of these are on typical Levallois flakes: two of the others are on blades. Four pieces have marginal retouch, e.g. Fig. 26.27, 7; Fig. 26.27, 6 is a well made end-scraper on a double side-scraper with thinned butt. Six are simple, one carinated, one shouldered (double), and two on the butt. Of these last, one is skewed to the left. Parallel or subparallel retouch occurs in 5 instances. The mean edge angle is rather low (67.9°), and the retouch is less invasive than usual (4.7 mm). These figures are consistent with the use of relatively flat blanks.

**Burins** (14; 10 typical, 4 atypical). The small rise in frequency marks the beginning of the increasing importance of this class. It is therefore of some interest that the first dihedral burins (apart from a solitary occurrence in G) are encountered in this layer. Four are of this type, eight are on an angle, and two are of the *plan* variety. Only one of the dihedral burins is axial; the other three are on a corner of the blank, and formed by transverse and longitudinal blows. All but one of the other pieces have been worked longitudinally (the exception is transverse). Multiple blows have been applied in two instances. Nine pieces are on breaks, and two on normal truncations (straight and concave respectively). The blanks are quite varied: two typical Levallois flakes, an LSF, a cortex-backed flake, and a core. Eleven of them have retouched edges.

**Borers** (7; 2 typical, 5 atypical). Only one has a really sharp, well-defined tip, made by the intersection of two normal denticulated edges

which also possesses opposed Clactonian notches at the distal end).

Eleven pieces (no less than 7.5%) have thinned butts, and 17 (11.2%) carry additional retouch at one end (for example Fig. 26.26, 9). Three pieces have Clactonian notches. In this layer, there is the unusual feature of a marginal (and statistically insignificant) preference for retouching the left-hand edge (74, against 71) instead of a definite preference for the right in the others.

*Transverse* (18) (Fig. 26.27, 2). The sample, when divided into three shape categories, is too small to yield very useful descriptive statistics. However, it is worth noting that the relationship between retouch length and edge shape is in good agreement with that for the single lateral scrapers. In fact the means of the various edge measurements are almost identical for the two groups, except that the lateral pieces have more invasive retouch.

Multiple scrapers

*Separate* (34). The subtype frequencies are very similar to those for layer C, though with straight pieces leading. A quite exceptional number (47.1%) are based on Levallois blanks, and the facetting index is 72.0. The blanks and the retouched edges are little larger than average, however. One piece is on a pseudo-Levallois point. Three pieces have thinned butts, one some thinning of the edge, and as many as five have retouched ends. The retouch is rather less invasive (5.1 mm) than for the scrapers as a whole, and much less so than for the double scrapers of layer C.

*Convergent axial* (15). These are rather smaller than those of layer C (mean length 47.8 ± 11.2 mm, as against 55.2 ± 16.5). The convex pieces are also less invasively retouched than might be expected (5.0 ± 2.6 mm). Three of the concave examples, though

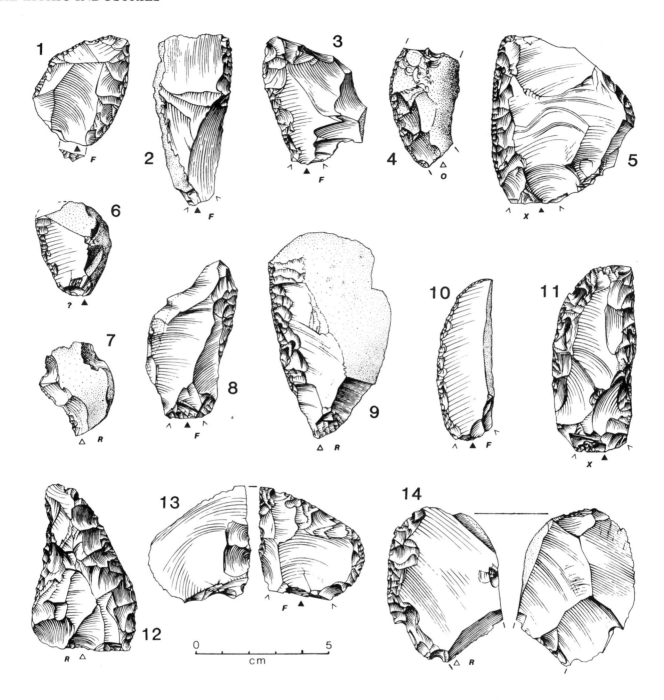

**Fig. 26.26** Flint tools from layer B. 1 Mousterian point (slightly *déjeté*); 2–14 side-scrapers: 2–10, 14 simple lateral (7 also possesses two notches, while the bulb has been removed from 14); 11, 12 convergent; 13 *déjeté*.

at a corner. Three of the atypical pieces are not very convincing as tools. There is only one facetted butt (on a typical borer). The mean length is 37.4 mm.

**Knives** (2, atypical). Both are cortex-backed. One (46 mm long) has inverse retouch along most of its back. The other (71 mm long) is strictly a thick *couteau à dos naturel*. However, it is terminated distally by an inverse truncation which had been employed for the removal of an LSF from the otherwise unretouched sharp edge.

**Raclettes** (9). Small in size (mean length 37.5 mm), they are varied in respect of shape and location of the retouched edges; three are inverse. Three examples are on Levallois blanks (two of them atypical).

**Truncated pieces** (5). One is on a blade, and another on an atypical Levallois flake. One 'truncation' may perhaps be no more than the result of spontaneous retouch when the flake was detached from the core. All of the truncations are straight, with a mean width of 14 mm. The mean length of the pieces themselves is 37.5 mm.

**Notched pieces** (35). Somewhat idiosyncratically for this stage of the sequence, the percentage of Clactonian notches rises to 54. The difference in width between the two categories is slight, giving a mean of 11.7 mm. The retouch is very shallow, however (mean 2.4 mm). Both types have a mean curvature of 0.25: the combined frequencies are

| Curvature | 0.0 | 0.1 | 0.2 | 0.3 | 0.4 | 0.5 |
|---|---|---|---|---|---|---|
| % | | 0.0 | 36.9 | 43.6 | 15.6 | 3.9 |

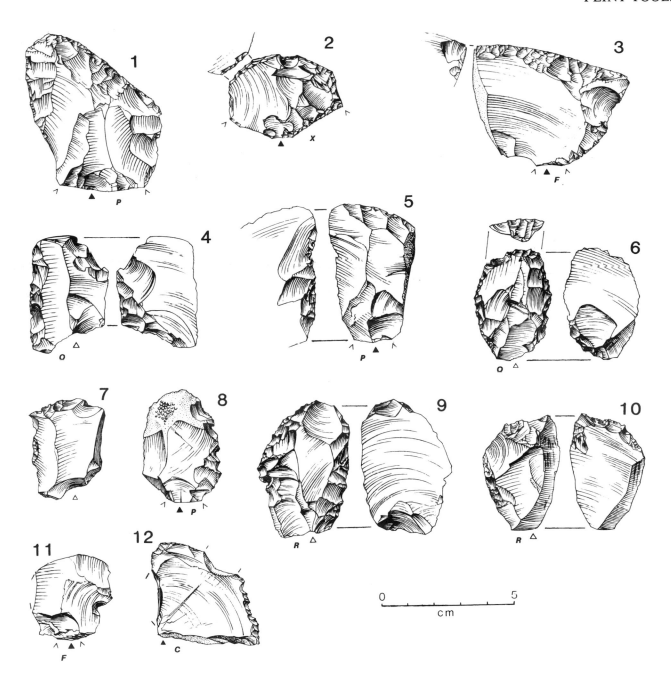

**Fig. 26.27** Flint tools from layer B. 1–5 Side-scrapers: 1, 3 *déjeté*; 2 transverse; 4 with thinned back; 5 with alternate retouch; 6, 7 end-scrapers; 8, 12 denticulates; 11 notch (Clactonian); 9–10 miscellaneous — essentially side-scrapers (10, on a LSF blank, has itself been modified by a LSF removal to which further ventral retouch has been applied).

Of three pieces made on Levallois flakes, two have Clactonian notches, as does the only example on a pseudo-Levallois point. This also applies to one of the two cortex-backed pieces (the other being made by retouch). Five examples are double, and one triple — of these, four are laterally opposed.

**Denticulates** (36). These account for only 8.3% of the reduced assemblage. Technologically, metrically and morphologically they are indistinguishable from those of layers C and D. Even allowing for the small sample, however, it is perhaps surprising that in a layer in which Levallois technique is fairly prevalent there should be only one denticulate made on such a blank.

**Bec burinant alterne** (1). It is on the proximal end of a side-scraper, and may be accidental.

**Tayac points** (2). One is typical though slightly *déjeté*, and has been shaped by mean of multiple Clactonian notches. The other has a point defined by at least two large Clactonian notches (the piece is slightly damaged). The mean length is 45 mm.

**End-notched pieces** (4). One has Clactonian notches at both ends. There is a single example with a (shallow) retouched notch. The mean length is 41 mm.

**Miscellaneous** (30). Broken down typologically, they include

|            | n  | %      |                    | n | %      |
|------------|----|--------|--------------------|---|--------|
| scrapers   | 25 | (83.3) | notches            | 6 | (20.0) |
| denticulates | 3 | (10.0) | overstruck removals | 4 | (13.3) |
| scaled     | 2  | (6.7)  | burins             | 1 | (3.3)  |

Of these, ten (i.e. one third) are multiple — especially scrapers combined with notches or overstriking. Fig. 26.27, 10 is a good example of

a piece which, by virtue of its complex history, is now effectively unclassifiable. In this assemblage many of the 'miscellaneous' tools owe their attribution to this category to the special removals from one or both ends: LSF removals or dorsal thinning/scaling from a break or truncation.

**Bifaces** (3). Of the two nucleiforms (43 x 39 and 43 x 31 mm) one is twisted and has some relict cortex and a rather blunt butt. The butt of the other piece is sharper and has a *méplat* — as does the third, 'miscellaneous', handaxe (51 x 41 mm).

**Table 26.15**  Layer A. Tool Type Frequencies

| Type |  | *N* | Real | Ess. | Red. |
|------|---|-----|------|------|------|
| 1 | Typical Levallois flakes | 87 | 2.44 | – | – |
| 2 | Atypical Levallois flakes | 53 | 1.49 | – | – |
| 3 | Levallois points | 14 | 0.39 | – | – |
| 4 | Retouched Levallois points | 7 | 0.20 | 0.27 | 0.28 |
| 5 | Pseudo-Levallois points | 14 | 0.39 | 0.54 | – |
| 6 | Mousterian points | 4 | 0.11 | 0.15 | 0.16 |
| 7 | Elongated Mousterian points | 4 | 0.11 | 0.15 | 0.16 |
| 8 | *Limaces* | 2 | 0.06 | 0.08 | 0.08 |
| 9 | Single straight side-scrapers | 278 | 7.81 | 10.77 | 11.05 |
| 10 | Single convex side-scrapers | 402 | 11.29 | 15.57 | 15.98 |
| 11 | Single concave side-scrapers | 47 | 1.32 | 1.82 | 1.87 |
| 12 | Double straight side-scrapers | 28 | 0.79 | 1.08 | 1.11 |
| 13 | Double straight-convex side-scrapers | 41 | 1.15 | 1.59 | 1.63 |
| 14 | Double straight-concave side-scrapers | 12 | 0.34 | 0.46 | 0.48 |
| 15 | Double convex side-scrapers | 61 | 1.71 | 2.36 | 2.42 |
| 16 | Double concave side-scrapers | 4 | 0.11 | 0.15 | 0.16 |
| 17 | Double concave-convex side-scrapers | 15 | 0.42 | 0.58 | 0.60 |
| 18 | Convergent straight side-scrapers | 6 | 0.17 | 0.23 | 0.24 |
| 19 | Convergent convex side-scrapers | 35 | 0.98 | 1.36 | 1.39 |
| 21 | *Déjeté* scrapers | 99 | 2.78 | 3.83 | 3.93 |
| 22 | Straight transverse side-scrapers | 39 | 1.10 | 1.51 | 1.55 |
| 23 | Convex transverse side-scrapers | 71 | 1.99 | 2.75 | 2.82 |
| 24 | Concave transverse side-scrapers | 8 | 0.22 | 0.31 | 0.32 |
| 25 | Side-scrapers with inverse retouch | 85 | 2.39 | 3.29 | 3.38 |
| 26 | Abrupt retouched side-scrapers | 12 | 0.34 | 0.46 | 0.48 |
| 27 | Side-scrapers with thinned back | 19 | 0.53 | 0.74 | 0.76 |
| 28 | Side-scrapers with bifacial retouch | 76 | 2.13 | 2.94 | 3.02 |
| 29 | Alternate retouched side-scrapers | 116 | 3.26 | 4.49 | 4.61 |
| 30 | Typical end-scrapers | 28 | 0.79 | 1.08 | 1.11 |
| 31 | Atypical end-scrapers | 38 | 1.07 | 1.47 | 1.51 |
| 32 | Typical burins | 91 | 2.56 | 35.2 | 36.2 |
| 33 | Atypical burins | 58 | 1.63 | 2.25 | 2.31 |
| 34 | Typical borers | 2 | 0.06 | 0.08 | 0.08 |
| 35 | Atypical borers | 10 | 0.28 | 0.39 | 0.40 |
| 36 | Typical backed knives | 10 | 0.28 | 0.39 | 0.40 |
| 37 | Atypical backed knives | 17 | 0.48 | 0.66 | 0.68 |
| 38 | Naturally backed knives | 52 | 1.46 | 2.01 | – |
| 39 | *Raclettes* | 81 | 2.27 | 3.14 | 3.22 |
| 40 | Truncated pieces | 31 | 0.87 | 1.20 | 1.23 |
| 41 | Mousterian tranchet | 1 | 0.03 | 0.04 | 0.04 |
| 42 | Notched pieces | 121 | 3.40 | 4.69 | 4.81 |
| 43 | Denticulates | 278 | 7.81 | 10.77 | 11.05 |
| 44 | *Becs burinants alternes* | 11 | 0.31 | 0.43 | 0.44 |
| 45 | Retouched with inverse retouch | 54 | 1.52 | – | – |
| 46-47 | Abrupt and alternate retouch (thick) | 248 | 6.96 | – | – |
| 48-49 | Abrupt and alternate retouch (thin) | 520 | 14.60 | – | – |
| 50 | Bifacial retouch | 3 | 0.08 | – | – |
| 51 | Tayac points | 4 | 0.11 | 0.15 | 0.16 |
| 52 | Notched triangles | 1 | 0.03 | 0.04 | 0.04 |
| 54 | End-notched pieces | 16 | 0.45 | 0.62 | 0.64 |
| 56 | *Rabots* (push planes) | 2 | 0.06 | 0.08 | 0.08 |
| 62 | Miscellaneous | 245 | 6.88 | 9.49 | 9.74 |
| | Bifaces (whole) | 66 | | | |
| | Bifaces (broken) | 4 | | | |
| | Totals | | 3561 | 2582 | 2516 |

## LAYER A

By far the richest assemblage from La Cotte, this has 2516 tools in the reduced count (16.3% of the unbroken flints; the percentage for types 1 to 63 is 23.1). The poorly retouched pieces (types 45–50) amount to 24.0% of the real total, a figure very similar to those for the underlying layers.

The size of the collection from this layer is due not only to the density of the archaeological material in the deposit, but also to the unit's considerable thickness (Fig. 6.4); unfortunately post-depositional chemical changes coupled with complex local erosional features made it impossible to subdivide the layer further over an area large enough to provide usable samples. Consequently the possibility that there were cultural changes in the course of the deposition of the unit cannot be excluded; indeed, to judge from the material from the base of layer 3 (see below) there appears to have been a strengthening through time in some of the industry's most distinctive features. Taken as a whole, however, the layer A series is remarkably consistent in character.

The side-scraper index $IR_{red}$ is very high (57.8), though lower than for the preceding assemblages; this group includes many pieces with inverse, bifacial or alternate retouch. The percentage represented by the extended denticulate group (notched and end-notched pieces, denticulates and Tayac points) is only 16.6, and that of notches only 4.8. Denticulates, at 11.1%, are slightly more common than in layer B. But the most remarkable change in the typological frequencies is undoubtedly that for Group III ('Upper Palaeolithic' tools) to 10.1%, due to the contribution by burins (5.9%).

The biface index (3.1) is the highest recorded at the site, though low when compared to values at some other sites of comparable age. Two points which should be borne in mind are that materials other than flint were used in the manufacture of handaxes at La Cotte and are not reflected in this statistic (see chapters 27 and 28), and that by no means all are of the classic Acheulian shapes. Nucleiform, partial and miscellaneous bifaces account for nearly half.

The most unusual feature of the layer A artefacts, though one which already appears to a lesser extent in earlier layers, is the very frequent use of the specialised technique for the rejuvenation of tool edges (mainly those of side-scrapers) which is referred to throughout this monograph as LSF technique; it is discussed in detail in chapters 29 and 30. Technologically it may be linked with two other types of treatment which are particularly common in this assemblage: the formation of (usually inverse) truncations, and the use of these or other suitable surfaces as platforms for the removal of small flakes from the dorsal surface of the piece (scaling). These are considered in more detail in section 26.5.6. The degree to which such modifications of the standard tool types have taken place has a considerable influence upon the frequency of the 'miscellaneous' category, number 62 in the type-list, which is unusually high in layer A (9.7%).

**Retouched Levallois points** (7). These have a mean length of 48.9 mm; all but one are on secondary points. In general the retouch is light, usually semi-abrupt, and limited in extent.

**Table 26.16**  Layer A. Typological Groups and Indices

| Index | | Real | Ess. | Red. |
|---|---|---|---|---|
| ILty | Typological Levallois | 4.52 | – | – |
| IR | Side-scrapers | 40.83 | 56.31 | 57.79 |
| IC | Charentian | 16.79 | 23.16 | 23.77 |
| IAu | Acheulian unifacial | 0.76 | 1.05 | 1.07 |
| IAt | Acheulian total | 2.67 | 3.66 | 3.75 |
| IB | Bifaces | 1.93 | 2.64 | 2.71 |
| I | Levallois | 4.52 | 0.27 | 0.28 |
| II | Mousterian | 41.11 | 56.70 | 58.19 |
| III | Upper Palaeolithic | 8.00 | 11.04 | 11.33 |
| IV | Denticulate | 7.81 | 10.77 | 11.05 |
| 5 | Pseudo-Levallois points | 0.39 | 0.54 | – |
| 38 | Naturally backed knives | 1.46 | 2.01 | – |
| I rc | Convergent retouched | 4.21 | 5.81 | 5.96 |
| 6–7 | Mousterian points | 0.22 | 0.31 | 0.32 |
| Rsingle | Side-scrapers, single | 20.42 | 28.16 | 28.90 |
| Rdouble | Side-scrapers, double | 4.52 | 6.24 | 6.40 |
| Rtransv | Side-scrapers, transverse | 3.31 | 4.57 | 4.69 |
| Rinv | Side-scrapers, inverse/alternate | 5.64 | 7.78 | 7.99 |
| 30–31 | End-scrapers | 1.85 | 2.56 | 2.62 |
| 32–33 | Burins | 4.18 | 5.77 | 5.92 |
| 36–37 | Knives | 0.76 | 1.05 | 1.07 |
| 42 | Notched pieces | 3.40 | 4.69 | 4.81 |
| 42–43 | Notched pieces & denticulates | 11.20 | 15.45 | 15.86 |
| Ident | Extended denticulate group | 4.75 | 6.55 | 6.72 |
| Iirr | Irregularly retouched | 23.17 | – | – |
| IQ | Quina | 0.62 | | |
| IQ+½Q | Quina + demi-Quina | 1.44 | | |
| Cl | Clactonian notches | 41.32 | | |

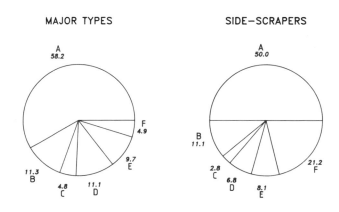

MAJOR TYPES · SIDE–SCRAPERS

**Fig. 26.28**  Flint tools from layer A. Relative frequencies of the principal tool categories and of the major side-scraper types. Tools: A Group II; B Group II; C Group IV; D notched pieces; E miscellaneous; F other. Side-scrapers: A simple lateral; B simple transverse; C double separate; D convergent axial; E *déjeté*; F other.

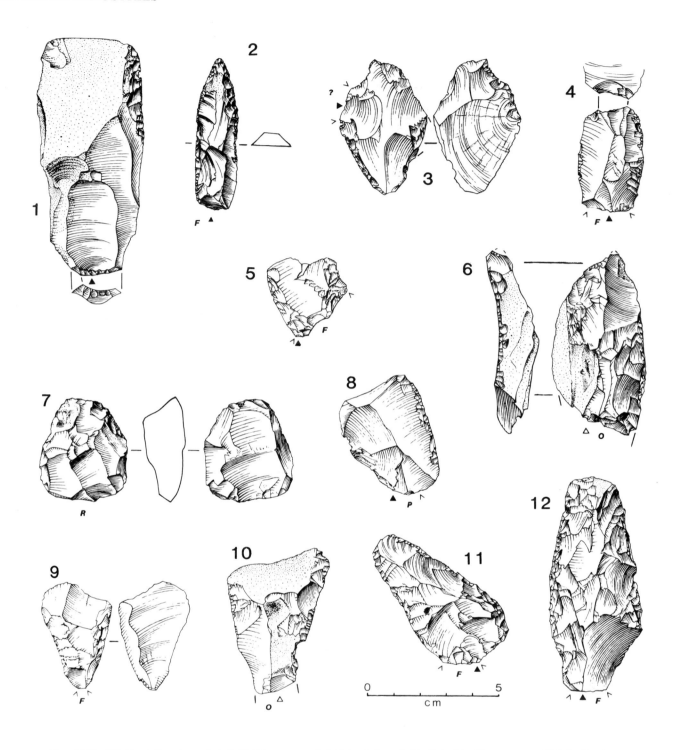

**Fig. 26.29** Flint tools from layer A. 1, 3–12 Side-scrapers: 1, 3–9 simple lateral; 10, 12 double; 11 convergent (slightly *déjeté*). 2 Thin elongated Mousterian point. 5 and 9 are made on LSF blanks, 3 on a handaxe tranchet flake, and 6 on a core. 4 exhibits thinning of the distal end from an inverse truncation, while 8 has a distal LSF removal.

**Mousterian points** (4). These are all flat, and have a mean length of 59.4 mm. Two are relatively broad, but the others only just fail to qualify as elongated Mousterian points (allowing for the damaged tip of one of them). Two exhibit thinning of the butt; one of these is the only finely pointed example (but it is also broad and slightly skewed). In two cases the retouch on one of the edges is very light. The shortest piece has a very small distal inverse truncation which has apparently

been used for detachment of an LSF (the scar was retouched afterwards). The two largest pieces are on Levallois blanks.

**Elongated Mousterian points** (4). Their mean length is 64.3 mm; all are comparatively thin. The longest example is a Levallois blade which has been retouched only at the distal end; another has rather irregular secondary working, which is inverse on part of one of the edges. The

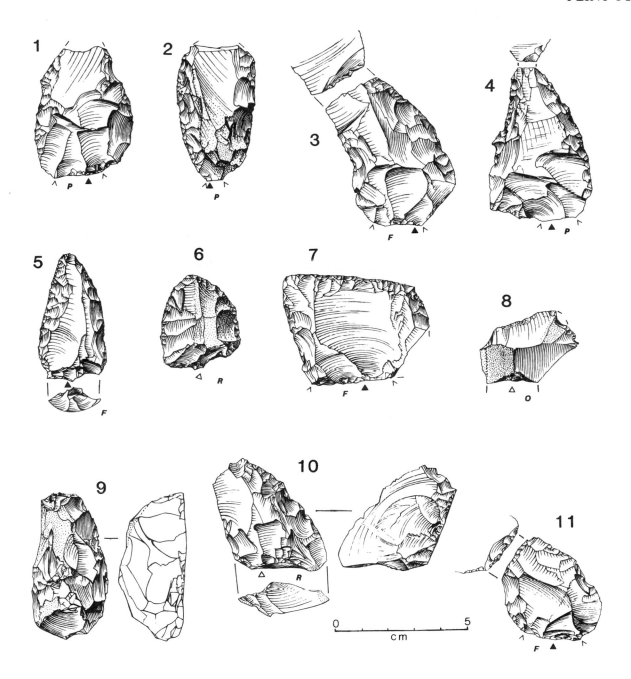

**Fig. 26.30** Flint tools from layer A. Side-scrapers: 1–3 double; 4, 5 convergent; 6, 7 *déjeté*; 8 transverse; 9 with abrupt (*surélevée*) retouch; 10, 11 with alternate retouch. An attempt at a distal LSF removal has been made on 3.

most striking piece is a very elegant and finely pointed example, 57.5 mm long and only 16 mm wide, made by carefully retouching a Levallois blade (Fig. 26.29, 2).

**Limaces** (2). Both are short (mean length 45.8 mm) and unpointed. One, with demi-Quina retouch, is biconvex and has rounded ends. It has a distal truncation from which dorsal scaling has taken place. The other is straight convex, much thinner distally than near the butt, and has normal truncations at either end; the distal one is oblique.

**Side-scrapers** (1454). This very important series of scrapers (57.8% in the reduced count) offers an unusual opportunity for detailed analysis

of even the rarer types. Only three categories (double concave, convergent straight and transverse concave) are represented in single-figure quantities, and only one (convergent concave) is altogether absent. A complication is introduced by techniques devised for resharpening and other purposes; indeed, the percentage of tools formally classified as scrapers would be higher but for this, which has forced many pieces into the 'miscellaneous' category, type 62. In this, and also in their small size, the side-scrapers and former side-scrapers of layer A are extremely distinctive.

Among the major typological groups, the most conspicuous change is a rise in the importance of types 25–29 (21.1%, as opposed to 13.2% in layer B), reflecting substantial increases in the incidence of scrapers with inverse, bifacial or alternate retouch.

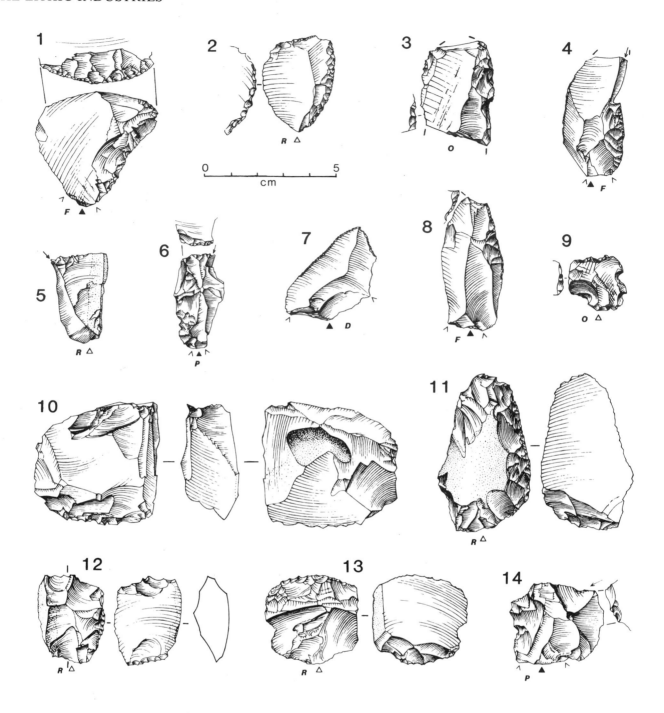

**Fig. 26.31** Flint tools from layer A. 1–3 Side-scrapers with alternate retouch; 4–6 burins (though 6 could be an attempt at LSF removal with scaling); 7 *raclette*; 8 truncation on side-scraper; 9 notch (multiple); 10 denticulated *rabot* (push-plane) on a core; 11–14 miscellaneous: 11–13 are essentially side-scrapers (12 with dorsal scaling from inverse truncations at either end), and 14 is composite (a denticulate with one scraper edge modified by LSF removal).

| | | | |
|---|---|---|---|
| simple | 48 | ogival | 2 |
| carinated | 3 | shouldered (single) | 4 |
| shouldered (double) | 2 | on butt | 4 |
| double | 2 | | |

Concave-edged pieces remain at the low level which has persisted since layer C; however, there is a decline in convex edges, from 64.6% in B to 56.2% in A, and a corresponding increase in straight edges to 37.2%. Although the curvature, 0.10, is the same as in layer B the

standard deviation is 0.11 (the highest for side-scraper curvatures at the site) as opposed to 0.09.

The smallness of the pieces (mean length 41.7 $\pm$ 9.7 mm) is matched by a reduction in edge length, to 31.4 $\pm$ 10.7 mm. Although the invasiveness is also less than in the previous layers (5.2 $\pm$ 3.1 mm, compared to 5.4 in B and 5.7 in C) this is not in proportion to the size change; in relative terms the scrapers are more heavily worked. Thinning of the butt occurs on 94 pieces (i.e. 6.5%) — a slight decrease in comparison with the previous layer. Thinned sides (2.9%) and retouch of the tip (5.4%) are somewhat less usual. 1.7% of the scrapers have Clactonian notches, and 0.7% retouched notches. Levallois blanks (16.7%) are slightly scarcer than in B and C; 24 pieces (1.7%) are on pseudo-Levallois points. Cortex-backed blanks, at 8.9%, are at a level typical of the site as a whole. However, 38 pieces (2.6%) are made on sharpening flakes (almost exclusively LSFs) — e.g. Fig. 26.29, 9. The minor blank categories are represented by 12 scrapers on cores (Fig. 26.29, 6), one on a handaxe thinning flake and three on handaxe tranchet flakes (e.g. Fig. 26.29, 3).

### Simple scrapers

*Single lateral* (727) (Fig. 26.29, 1 and 3–9). Of these, 278 (38.2%) are straight, 402 (55.3%) convex, and 47 (6.5%) concave, giving a mean curvature of 0.098. The differences between these categories are slight as regards the length of the working edge (29.1, 31.1 and 29.4 mm respectively). However, the pattern is unusual for retouch angle (54.3, 61.0 and 61.8°) and invasiveness (4.2, 5.1, 5.4 mm). Thus the straight scrapers are distinctly 'sharp', but less invasively worked than the others. Fifty-nine pieces (8.0%) have retouch at an end, and 30 (4.2%) have thinned butts. These are high values, albeit lower than in the preceding assemblage. Thirteen (1.8%) have Clactonian notches. A strong preference for retouching the right hand edge is apparent (430, or 60.6%, against 279).

*Transverse* (118) (Fig. 26.30, 8). Thirty-nine (33.1%) are straight, 71 (60.2%) convex and 8 (6.8%) concave — in quite good agreement with the figures for single lateral scrapers (curvature 0.094). As a group, they are virtually identical to the lateral scrapers as regards edge length, but the edges of the straight and concave pieces are appreciably shorter. The mean edge angle is very similar, but the retouch is somewhat less invasive on average (4.2 mm). Nine examples are made on the butt.

### Multiple scrapers

*Separate* (161). Biconvex pieces (e.g. Fig. 26.30, 1 and 2) are in a clear majority, followed by straight-convex and straight; straight-concave and convex-concave (Fig. 26.30, 3, with an attempted LSF removal) are quite common however (Table 26.23). The biconcave pieces have particularly short retouched edges (mean 24.3 ± 4.2, compared to 33.3 ± 9.9 mm for the whole group).

The members of this group are only marginally larger than the average for the scrapers, with a mean length of 42.5 mm (much the lowest value for this group). Though the IF is only 56.8 (rather below expectation), no less than 31.1% of the double separate are on Levallois blanks, almost all typical. Two are on cores, one on a pseudo-Levallois point, and a further example once possessed a cortical back. The butts have often been thinned (22 cases); five pieces have some thinning of an edge. Ten more have additional retouch at one end. The mean retouch angle (62.8°) and invasiveness (5.0 mm) are very close to those for all layer A scrapers.

*Convergent axial* (41). No fewer than 35 of these are convex, on blanks 7 mm longer than the average for scrapers from A. Eight are on Levallois blanks (six typical), one on a tranchet flake, and one on a cortex-backed flake. A single example has a thinned butt, and another a Clactonian notch. The scraper illustrated as Fig. 26.30, 4 is somewhat unusual in having an irregular, rather denticulated left-hand edge and also in respect of its tip, which has been thinned dorsally, from an inverse truncation, after the retouch which shaped the piece. Fig. 26.30, 5 is much more steeply retouched on one edge than the other. Two pieces have Quina retouch, and another demi-Quina. Despite the small size of these scrapers (reflecting the general situation for this layer), the retouch is strongly invasive (mean 7.2 ± 3.9 mm). Five pieces have had their butts removed by retouch.

*Déjeté* (99). Straight edges are relatively common, at 41.1%. The blanks are only slightly longer than average for layer A scrapers.

They include 23 Levallois (only 1 typical), 4 pseudo-Levallois points, one LSF, a handaxe thinning flake, a tranchet flake, two cortex-backed flakes and a core. Thirteen pieces have thinned butts, four Clactonian notches. Double *déjeté* scrapers (27) are present in moderate numbers.

### Other side-scrapers

As in layer B, alternate scrapers are more common than those with inverse retouch; bifacially retouched pieces are also important. Consequently, double and convergent forms are particularly numerous (Table 26.24). The retouch is somewhat invasive (though well exceeded in this respect by that of the convergent scrapers *sensu stricto*); the length of the working edge matches closely that of the simple lateral and transverse scrapers.

*On the ventral face* (85). Two are *déjeté*, one double, eight transverse (straight) and the remainder single lateral. As in layer C, straight scrapers are in a clear majority (57.1%, compared to 29.8% convex and 13.1% concave). Pieces with retouch on the right-hand side of the bulbar surface are in a small majority (37:31). Two have thinned butts, and four thinned sides. The butt has been removed in four cases.

The retouched edges are substantially shorter and steeper than for the normally-worked scrapers (23.7 ± 8.0 mm, and 64.8 ± 13.1°). The retouch is slightly less invasive than expected (4.3 ± 2.7 mm). The blanks include 11 Levallois pieces of which 10 are typical. There are also 2 pseudo-Levallois points, 8 cortex-backed flakes, and a single LSF.

*With abrupt retouch* (12). Six are straight, five convex and one concave; the retouched edge is generally short (mean 27 mm). One piece has *surélevée* retouch. Apart from a single example on a core, these are on plain flakes, one of them with its butt removed by retouch.

*With thinned back* (19). Eight are straight, nine convex and two concave; only one has retouch. The butt has been removed in four instances (with thinning of the butt in three). Five of the blanks (26%) are Levallois, and one is cortex-backed. The retouch is somewhat denticulated in three cases; the length of the working edge is in good agreement (mean 32.1 mm) with the average for scrapers from this layer (the retouch angle is 60.3°).

*With bifacial retouch* (76). The distribution of principal edge shape reflects very well that for the totality of scrapers for layer A: straight 33.6%, convex 60.0%, concave 6.5%. Four pieces are fully bifacial. In size and character of the edge, as well as in the choice of blanks, this series is fairly typical of layer A.

*With alternate retouch* (116). Again, the shape of the longer edge is distributed in the usual way: 35.2 straight, 62.6 convex and 2.2 concave. Alternating retouch has been used on 27.6% of the pieces.

It is important to remember that, in this layer in particular, the practice of forming inverse truncations on an existing scraper (for a variety of possible reasons, as discussed in section 26.5.6) results in many pieces which are difficult to classify — this is especially true in respect of alternate scrapers (cf. for example Fig. 26.30, 11). Three of the blanks are typical Levallois flakes, and another is cortex-backed; eight pieces in all have had their butts removed.

**End-scrapers** (66; 28 typical, 38 atypical). Despite the low percentage (2.6% in reduced count) the sample is large enough for some of the rarer types to be represented:

| | | | |
|---|---|---|---|
| simple | 48 | ogival | 2 |
| carinated | 3 | shouldered (single) | 4 |
| shouldered (double) | 2 | on butt | 4 |
| double | 2 | | |

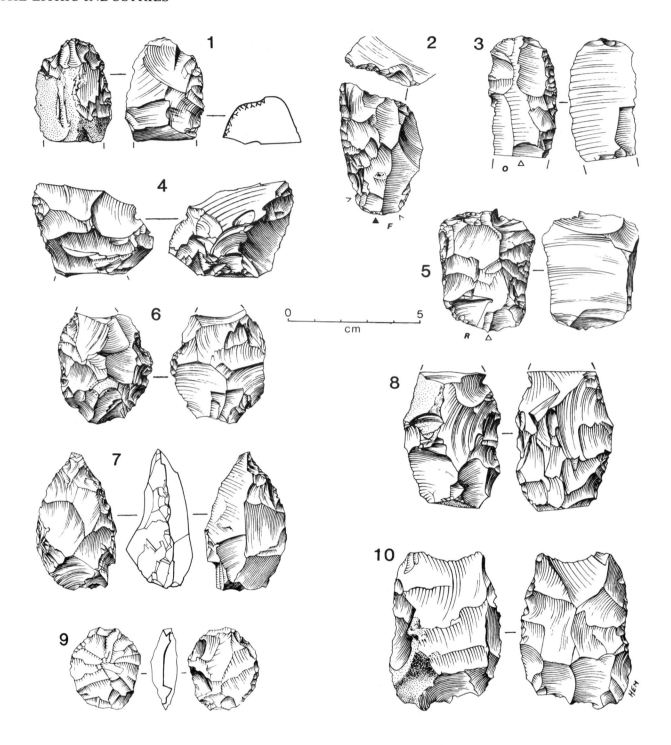

**Fig. 26.32** Flint tools from layer A. 1–5 Miscellaneous: 1 end- and side-scraper with thinned bulbar surface; 2 side-scraper with extensive scaling; 3 side-scraper with distal scaling and attempted LSF removal; 4 former side-scraper with thinned butt, and opposed LSF removals on the principal working edge (formerly retouched?); 5 side-scraper with proximal and distal scaling, and thinning of the opposite edge; 6–10 handaxes: 6–8 amygdaloids; 9 discoidal; 10 cleaver with tranchet finish.

The retouch is equally varied:

| | | | |
|---|---|---|---|
| scalar/semi-abrupt | 30 | subparallel | 21 |
| parallel | 6 | step | 1 |
| denticulate | 4 | abrupt | 4 |

Six pieces have thinned butts, and two are thinned along one of the long sides. Three end-scrapers also have notches (two of them Clact-

onian). Twenty-two have one or both of the lateral margins retouched. Eleven of the tips are skewed to the left and eleven to the right.

Five are on Levallois blanks, one on an LSF, and two are cortex-backed. There are five on blades, and only two on transverse flakes. None of the metrical observations relating to the working tip gives particularly extreme values for the mean; e.g. curvature (0.20), retouch angle (73.7°) and invasiveness (5.8 mm); however, the low mean length (37.5 mm) is typical of this layer generally.

294

**Burins** (149; 91 typical, 58 atypical). This class achieves its greatest importance in this layer, amounting to 5.9% of the reduced tools — an unusually high percentage for an industry of Middle Palaeolithic type. The subtypes represented are angle (longitudinal or transverse) (116), dihedral (12), mixed (1), *plan* (2), double at opposite ends (5), and double at the same end (13). Double burins on the same end are always longitudinal angle burins; the opposed ones include a transverse example, however. Six of the single angle burins are transverse. In all, seven pieces are skewed rather than strictly longitudinal or transverse — four of them are dihedral. The dihedral pieces also include four which are axial, and four made by transverse and longitudinal blows at a corner (six of them have multiple removals from one surface, whereas such treatment has been applied to only 20% of the burins as a whole).

Breaks have been used to provide the platforms for the burin blows in 72 cases (Fig. 26.31, 4). Of the truncations used as platforms for burin removal, fifteen are normal and fifteen inverse (e.g. Fig. 26.31, 6). One of each type is concave; the convex truncations include six normal but only one inverse. A number of other pieces bear inverse truncations which have not been used in burin manufacture; some of these may simply be unstruck, rather than intended for other purposes.

Marginal retouch of some kind is present on 60% of the blanks, but only seven pieces were definitely retouched after the burin blow(s). The blanks include 26 Levallois (3 atypical), 2 LSFs, 13 cortex-backed (Fig. 26.31, 5, which is one of 12 pieces which have been overstruck) and a single core. The facetting index (IF) is 66.0 — though this is based on only 53 determinable butts (partly as a result of the use of broken flakes as blanks).

**Borers** (12; 2 typical, 10 atypical). These average only 33.7 mm in length. The largest, which is one of the typical examples, is on a core; the presumed borer is formed by two retouched notches which truncate some earlier inverse removal scars. The other typical piece has a large inverse notch on the distal end, with bifacial removals defining projections at either corner. Among the remainder, three are pointed splinters. One TSF has two projecting corners which have been emphasised by single notches.

**Knives** (27; 10 typical, 17 atypical). The mean length is 50.5 mm; the typical examples are smaller than the atypical ones by nearly 5 mm, on average. The numbers of blades are 5 and 12 respectively. In all, nine pieces are on Levallois blanks, and eight are cortex-backed (though others have more restricted areas of cortex). Two have inverse backing. There are also four pieces with distal inverse truncations of Kostienki type (see section 26.5.6 below), including one with an LSF scar partly removing a zone of retouch on the thinner edge. The typical knives include several thin pieces backed with abrupt or semi-abrupt removals only 1 mm long. The atypical ones include eight examples with retouch on, or extending, zones of cortex.

**Raclettes** (81). These are only 35.4 mm long, on average. Eight, or 9.9%, are multiple; of these, two are straight-convex and the rest have straight edges only (e.g. Fig. 26.31, 7). There is a fairly even balance between retouching of the left and right sides; 18.5% are transverse. The shape and location of the retouched edges are distributed as follows: 37 straight, 27 convex, 9 concave, 2 straight-convex, and 6 straight-straight; 28 left, 30 right, 15 transverse, and 8 multiple. The retouch is inverse in 24.7% of cases. Nine pieces are made on blades. The blanks include 7 typical Levallois, 4 atypical Levallois, 3 pseudo-Levallois points, 4 TSF/LSFs, and one cortex backed.

**Truncated pieces** (31). In only one case does the length exceed 50 mm, the mean being 34.5 mm. Eight of the truncations are proximal, six oblique, and six inverse (in one case with some dorsal scaling). The great majority are straight (24, against three convex and four concave); their mean width is 15.9 mm. The blanks include five Levallois flakes or blades (Fig. 26.31, 8), one LSF and three flakes with cortical backs.

**Mousterian tranchet** (1). The tranchet edge appears to have suffered heavy utilisation (though *concassage* cannot be entirely ruled out); the butt has been retouched, and then thinned with parallel ventral removals.

**Notched pieces** (121). Layer A has the lowest percentage of notches in the site (4.8%); 41% are Clactonian. Only seven pieces have multiple notches (one is triple, the rest double); in three cases these are opposite to each other, on the lateral margins (for example Fig. 26.31, 9, though this particular piece may also qualify as a denticulate). The retouched notches are unusually small (9.9 mm mean width), with only moderately invasive retouch (3.5 mm); the Clactonian notches (12.0, 6.8 mm) are virtually identical to those of B and C. The blanks in both cases are rather small (mean length 38.3 mm), as is usual for this layer. The curvature is very slightly more marked than in B (mean 0.26).

| Curvature | 0.0 | 0.1 | 0.2 | 0.3 | 0.4 | 0.5 | 0.6 |
|---|---|---|---|---|---|---|---|
| % | 0.0 | 27.1 | 49.2 | 18.2 | 4.7 | 0.8 | |

The choice of blanks is interesting. Six, or 4.9%, are Levallois (only one with a Clactonian notch); since the Levallois index (IL) for retouched tools is 15.93, notched pieces have certainly not been favoured in this respect. On the other hand, 6 pieces are on pseudo-Levallois points — three times the expected number (two have Clactonian notches). Six more (including 3 Clactonian) are on cortex-backed blanks. Single examples occur of a tranchet flake and a core. Seven pieces (3 Clactonian) are made on sharpening flakes (SFs).

**Denticulates** (278). As in the other assemblages from layer C onwards, these play a very limited role (11.0% in reduced count). Levallois blanks are relatively common (18.4%, against 5.4% for cortex-backed flakes); none of the denticulates are made on pseudo-Levallois points or LSFs. While tools of this kind are obviously rather poor candidates for edge rejuvenation by the LSF technique, it is interesting that the blanks resulting from modification of other tools were not employed as blanks in the manufacture of denticulates.

As also in layer B and C, Clactonian notches (8.0%) are a trivial element in the fabrication of the edges. Coarse retouch accounts for 33.0%, fine for 51.4% and microdenticulation for 7.7%; thus once again the percentage of coarse edges is low, at 41.0%. Morphologically, the denticulates of layer A are notable for the comparative rarity of simple lateral pieces (35.4%) and the abundance of transverse (21.3%). The small size of the pieces reflects trends apparent in other artefacts from layer A: mean length 39.1 ± 8.6 mm, retouch length 28.0 ± 8.7 mm, with invasiveness 5.4 ± 3.1 mm.

**Becs burinants alternes** (11). Most are formed by pairs of alternate notches. One example, however, has an inverse truncation on which one corner has been used as a platform for a few dorsal removals, giving a burin-like angle; another is the result of inverse trimming of the tip of a *déjeté* scraper. As a rule the blanks are small (mean length 34.6 mm).

**Tayac points** (4). All are relatively broad, and crudely pointed; their mean length is 46.0 mm. The butt exhibits some thinning in two cases. The degree of denticulation of the edges varies; on one piece, *surelevée* retouch has been used along one edge.

**Notched triangle** (1). A wedge-shaped flake fragment bears a Clactonian notch 15 mm wide.

**End-notched pieces** (16). As usual, these are on relatively small flakes (mean length 33.9 mm), two of which are Levallois. Nine of the notches are single-blow (Clactonian) and two are at the proximal end; one of the latter has the largest of the retouched notches, 19 mm wide. The largest Clactonian notch (21 mm) is on a pseudo-Levallois point with retouch on one of the other edges. The mean width of the notches is 9.9 mm.

**Rabots** (push-planes) (2). Both are on cores. One (Fig. 26.31, 10) has a steep, convex denticulated edge. The other is smaller (35 x 34 mm) and very high-backed, with invasive parallel removals at an angle of 88° to their platform forming a convex working edge.

**Miscellaneous** (245). In view of the extensive and intensive resharpening of tools which occurred in this layer, it is perhaps not altogether surprising that it should possess the highest incidence of unclassifiable and complex tools. The examples illustrated in Figs. 26.31, 11–14 and 26.32, 1–5 give an idea of the wide range of the artefacts placed in this

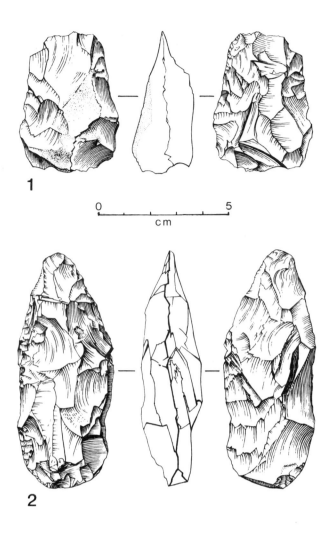

0           5
cm

**Fig. 26.33** Flint tools from layer A. 1 Cleaver; 2 thick foliate pick.

'last-resort' category. Many undoubtedly functioned as side-scrapers for at least part of their history, but the significance of other features is unclear, while in some cases rejuvenation has removed an edge completely and a new function is not apparent. A typological breakdown gives

|  | n | % |  | n | % |
|---|---|---|---|---|---|
| scrapers | 150 | (61.1) | notches | 30 | (12.2) |
| denticulates | 38 | (15.6) | overstruck removals | 42 | (17.1) |
| scaled | 27 | (11.0) | burins | 7 | (2.9) |

More than one of these categories applies in 81 cases (33.2%). Also, 'special' removals — LSF, burin or dorsal thinning — off inverse truncations etc. occur on 162 pieces (66.4%).

**Bifaces** (66). The biface index $IB_{red}$, though the highest for flint from La Cotte, is only 2.71, but the sample of this typological group is a large one notwithstanding, when compared to most others from cave sites of comparable age. A selection of pieces is illustrated in Figs. 26.32, 6–10, and 26.33. The small size of the bifaces is quite striking; their mean length is $45.5 \pm 11.1$ mm.

It will be seen from Table 26.26 that nucleiform handaxes are the most common type, forming one-third of the series, and are followed by amygdaloids (27.3%) and discoidal handaxes (10.6%). A wide range of other forms is present, in small numbers. Flake-cleavers are entirely

lacking in flint, but there are four bifacial cleavers (Figs. 26.32, 10 and 26.33, 1). When the series is plotted on a diagram of the type used by Bordes (1961b, Fig. 7), almost all the pieces are found to lie in the cordiform and ovate bands (Fig. 26.45), while on that designed by Roe (1981, Fig. 5:15) two-thirds fall in his ovate section (see mf 26.3).

Four pieces are markedly twisted in profile, and two more slightly so. Five have tranchet finish at the tip. As for the butts, only three are heavily cortical, and twelve more have localised cortex (at the corners in eight instances).

## LAYER 3

Though a considerable number of artefacts was recovered from the base of this layer, the bulk of these pieces are unlikely to be related to the formation of the bone heap. In the small section visible in 1981 near the west wall of the cave, small lenses of layer A deposits were seen to be intercalated with the overlying loess. They result from the erosion of projecting areas of the very uneven surface of layer A while the loess was being laid down. The series of artefacts assigned to 3 is therefore primarily of interest as probably indicating the character of the layer A industry in its very latest form.

296

**Table 26.17** Layer 3. Tool Type Frequencies

| Type | | N | Real | Ess. | Red. |
|---|---|---|---|---|---|
| 1 | Typical Levallois flakes | 1 | 0.78 | – | – |
| 2 | Atypical Levallois flakes | 1 | 0.78 | – | – |
| 5 | Pseudo-Levallois points | 1 | 0.78 | 1.04 | – |
| 6 | Mousterian points | 1 | 0.78 | 1.04 | 1.08 |
| 9 | Single straight side-scrapers | 9 | 6.98 | 9.38 | 9.68 |
| 10 | Single convex side-scrapers | 9 | 6.98 | 9.38 | 9.68 |
| 11 | Single concave side-scrapers | 2 | 1.55 | 2.08 | 2.15 |
| 12 | Double straight side-scrapers | 3 | 2.33 | 3.13 | 3.23 |
| 13 | Double straight-convex side-scrapers | 3 | 2.33 | 3.13 | 3.23 |
| 15 | Double convex side-scrapers | 2 | 1.55 | 2.08 | 2.15 |
| 17 | Double concave-convex side-scrapers | 2 | 1.55 | 2.08 | 2.15 |
| 21 | Déjeté scrapers | 3 | 2.33 | 3.13 | 3.23 |
| 22 | Straight transverse side-scrapers | 1 | 0.78 | 1.04 | 1.08 |
| 23 | Convex transverse side-scrapers | 2 | 1.55 | 2.08 | 2.15 |
| 25 | Side-scrapers with inverse retouch | 4 | 3.10 | 4.17 | 4.30 |
| 26 | Abrupt retouched side-scrapers | 1 | 0.78 | 1.04 | 1.08 |
| 27 | Side-scrapers with thinned back | 1 | 0.78 | 1.04 | 1.08 |
| 28 | Side-scrapers with bifacial retouch | 4 | 3.10 | 4.17 | 4.30 |
| 29 | Alternate retouched side-scrapers | 3 | 2.33 | 3.13 | 3.23 |
| 31 | Atypical end-scrapers | 2 | 1.55 | 2.08 | 2.15 |
| 33 | Atypical burins | 4 | 3.10 | 4.17 | 4.30 |
| 34 | Typical borers | 1 | 0.78 | 1.04 | 1.08 |
| 38 | Naturally backed knives | 2 | 1.55 | 2.08 | – |
| 39 | Raclettes | 4 | 3.10 | 4.17 | 4.30 |
| 40 | Truncated pieces | 2 | 1.55 | 2.08 | 2.15 |
| 42 | Notched pieces | 8 | 6.20 | 8.33 | 8.60 |
| 43 | Denticulates | 14 | 10.85 | 14.58 | 15.05 |
| 44 | Becs burinants alternes | 1 | 0.78 | 1.04 | 1.08 |
| 45 | Retouched with inverse retouch | 1 | 0.78 | – | – |
| 46–47 | Abrupt and alternate retouch (thick) | 9 | 6.98 | – | – |
| 48–49 | Abrupt and alternate retouch (thin) | 21 | 16.28 | – | – |
| 61 | Chopping tools | 1 | 0.78 | 1.04 | 1.08 |
| 62 | Miscellaneous | 6 | 4.65 | 6.25 | 6.45 |
| | Bifaces (whole) | 1 | | | |
| | Totals | 129 | | 96 | 93 |

Because the assemblage is small, the various types are not described in detail here. In most respects it is very similar to that from A. However, there is a slight further reduction in the size of the tools. Also the peculiar technological features of the tools from A all reach their maximum representation: 11.2% inverse truncations, 7.5% LSF removals and 7.5% dorsal scaling (moreover the percentage of SF blanks in the entire flint assemblage peaks here, to 7.8%). These figures suggest that the industry of layer A may have undergone an evolution with the passage of time, becoming increasingly specialised.

## LAYER 5

This is the only series in which the artefacts often show signs of light patination, or of physical alteration as a result of soil movement; their condition varies from absolutely fresh to very lightly concassé with polished surfaces. An indication that this need have little or no chronological significance is provided by the different patinas exhibited by the refitted burins illustrated in Fig. 26.37. As explained in chapter 21, material from the base of layer 6.1 is included with that from 5 because the available provenance data are insufficient to distinguish two groups; it is likely that the former is partly the result of geological reworking of a small number of pieces from the underlying deposit, as in layer 3. In any case, it is not possible to establish which of the artefacts may have been used in the butchering of the animals whose remains form the bone heap at the base of 6.1.

The percentage of tools is 35.1 (based on the reduced count, 273). This is twice as high as in other layers; the figure for types 1–63 is 43.6%. Despite the effects of soil movement, only 17.0% of the flints are of types 45–50 (the lowest percentage for any of the series).

The tools are once again dominated by side-scrapers (54.2%), of which most are convex. This is the only layer in which Quina retouch has been much used; the value of IQ is 9.4, rising to 17.7 with demi-Quina retouch included. The extended denticulate group (19.8%) is more important numerically than in any other layer since D, largely as a result of an increase in the representation of notched pieces, to 9.9% (the great majority of these are of Clactonian type). There are many 'Upper Palaeolithic' tools (10.6%); as in layer A, burins form

**Table 26.18**   Layer 3. Typological Groups and Indices

| Index | | Real | Ess. | Red. |
|---|---|---|---|---|
| ILty | Typological Levallois | 1.55 | – | – |
| IR | Side-scrapers | 37.98 | 51.04 | 52.69 |
| IC | Charentian | 12.40 | 16.67 | 17.20 |
| IAu | Acheulian unifacial | 0.00 | 0.00 | 0.00 |
| IAt | Acheulian total | 0.77 | 1.03 | 1.06 |
| IB | Bifaces | 0.77 | 1.03 | 1.06 |
| I | Levallois | 1.55 | 0.00 | 0.00 |
| II | Mousterian | 38.76 | 52.08 | 53.76 |
| III | Upper Palaeolithic | 6.98 | 9.37 | 9.68 |
| IV | Denticulate | 10.85 | 14.58 | 15.05 |
| 5 | Pseudo-Levallois points | 0.78 | 1.04 | – |
| 38 | Naturally backed knives | 1.55 | 2.08 | – |
| I rc | Convergent retouched | 3.10 | 4.17 | 4.30 |
| 6-7 | Mousterian points | 0.78 | 1.04 | 1.08 |
| Rsingle | Side-scrapers, single | 15.50 | 20.83 | 21.51 |
| Rdouble | Side-scrapers, double | 7.75 | 10.42 | 10.75 |
| Rtransv | Side-scrapers, transverse | 2.33 | 3.12 | 3.23 |
| Rinv | Side-scrapers, inverse/alternate | 5.43 | 7.29 | 7.53 |
| 30-31 | End-scrapers | 1.55 | 2.08 | 2.15 |
| 32-33 | Burins | 3.10 | 4.17 | 4.30 |
| 36-37 | Knives | 0.00 | 0.00 | 0.00 |
| 42 | Notched pieces | 6.20 | 8.33 | 8.60 |
| 42-43 | Notched pieces & denticulates | 17.05 | 22.92 | 23.66 |
| Ident | Extended denticulate group | 3.10 | 4.17 | 4.30 |
| Iirr | Irregularly retouched | 24.03 | – | – |
| IQ | Quina | 2.04 | | |
| IQ+½Q | Quina + demi-Quina | 4.08 | | |
| Cl | Clactonian notches | 37.50 | | |

the most important class (5.1%) but in layer 5 end-scrapers are almost as common as these.

One of the most interesting components of the assemblage is a series of thick elongated Mousterian points, several of which exhibit damaged or imperfectly repaired tips; their significance is discussed further in section 26.5.1). It must be stressed that these very distinctive tools were found *in situ* only in layer 5. Consequently the discovery by Burdo of a similar piece (Fig. 26.41) at a much lower level, in the slumped deposits to the south of the fossil cliff, provides confirmation of our interpretation of the stratigraphic position of layer 5 relative to the Eemian marine deposits.

Flint bifaces are scarce ($IB_{red} = 1.80$); once again, though, the presence of several examples on other materials must be taken into account (chapter 28).

The tools from layer 5 are about 10 mm longer on average than those from 3 and A (Table 26.28). The increase is even greater in the case of the side-scrapers. The Levallois index IL, for all of the artefacts, is 8.2, high by the standards of La Cotte, but the *typological* Levallois index ILty is only 1.8. Although Levallois blanks have been used in the manufacture of 17.7% of the retouched tools, hardly any have been left unmodified.

**Retouched Levallois point** (1). This is a thin, somewhat skewed example of the secondary type, 48 mm long and with a facetted butt. The right-hand edge has been modified by a series of flat, trapezoidal scars.

**Mousterian point** (1). This piece (Fig. 26.35, 4; catalogue no. 68/876) just fails to qualify as an elongated Mousterian point (69.5 x 36 mm). Nevertheless, it should probably be regarded as at the shorter end of shape distribution of that category, given both the characteristics of the other points from layer 5, as described below, and the real possibility that the piece has been shortened by rejuvenation. The tip has both ventral thinning (antedating normal retouch) and a burin-like (post-retouch) scar which is characteristic of impact damage often suffered by projectile points. The piece is fairly fresh.

**Elongated Mousterian points** (8). Forming one of the most distinctive components of the assemblage, they include some quite superb examples of the flint-knapper's art. Because of their importance, they will be described individually. The last item given for each piece is its catalogue number.

1.  63 x 30 mm; thin; finely pointed on a blank which retains some cortex towards the butt on the left-hand side (on this edge, the otherwise normal scalar retouch changes to inverse in the area of the cortex); butt damaged; condition: absolutely fresh. 69/482.

2.  78 x 27 mm; thick; cortex-backed on the right-hand side, which is retouched only near the tip; butt broken off; Quina retouch; a narrow bladelet scar along the central ridge at the point may be impact damage; condition: slightly lustrous. 66/193.

3.  71 x 28.5 mm; moderately thick; laurel-leaf shape; Quina retouch;

**Table 26.19**  Layer 5. Tool Type Frequencies

| Type | | N | Real | Ess. | Red. |
|---|---|---|---|---|---|
| 1 | Typical Levallois flakes | 3 | 0.89 | – | – |
| 2 | Atypical Levallois flakes | 1 | 0.30 | – | – |
| 3 | Levallois points | 2 | 0.59 | – | – |
| 4 | Retouched Levallois points | 1 | 0.30 | 0.36 | 0.37 |
| 6 | Mousterian points | 1 | 0.30 | 0.36 | 0.37 |
| 7 | Elongated Mousterian points | 8 | 2.37 | 2.91 | 2.93 |
| 8 | *Limaces* | 3 | 0.89 | 1.09 | 1.10 |
| 9 | Single straight side-scrapers | 17 | 5.03 | 6.18 | 6.23 |
| 10 | Single convex side-scrapers | 40 | 11.83 | 14.55 | 14.65 |
| 11 | Single concave side-scrapers | 1 | 0.30 | 0.36 | 0.37 |
| 12 | Double straight side-scrapers | 3 | 0.89 | 1.09 | 1.10 |
| 13 | Double straight-convex side-scrapers | 6 | 1.78 | 2.18 | 2.20 |
| 15 | Double convex side-scrapers | 2 | 0.59 | 0.73 | 0.73 |
| 16 | Double concave side-scrapers | 1 | 0.30 | 0.36 | 0.37 |
| 17 | Double concave-convex side-scrapers | 1 | 0.30 | 0.36 | 0.37 |
| 18 | Convergent straight side-scrapers | 4 | 1.18 | 1.45 | 1.47 |
| 19 | Convergent convex side-scrapers | 7 | 2.07 | 2.55 | 2.56 |
| 21 | *Déjeté* scrapers | 10 | 2.96 | 3.64 | 3.66 |
| 22 | Straight transverse side-scrapers | 4 | 1.18 | 1.45 | 1.47 |
| 23 | Convex transverse side-scrapers | 10 | 2.96 | 3.64 | 3.66 |
| 25 | Side-scrapers with inverse retouch | 10 | 2.96 | 3.64 | 3.66 |
| 26 | Abrupt retouched side-scrapers | 1 | 0.30 | 0.36 | 0.37 |
| 27 | Side-scrapers with thinned back | 2 | 0.59 | 0.73 | 0.73 |
| 28 | Side-scrapers with bifacial retouch | 9 | 2.66 | 3.27 | 3.30 |
| 29 | Alternate retouched side-scrapers | 20 | 5.92 | 7.27 | 7.33 |
| 30 | Typical end-scrapers | 3 | 0.89 | 1.09 | 1.10 |
| 31 | Atypical end-scrapers | 8 | 2.37 | 2.91 | 2.93 |
| 32 | Typical burins | 11 | 3.25 | 4.00 | 4.03 |
| 33 | Atypical burins | 3 | 0.89 | 1.09 | 1.10 |
| 35 | Atypical borers | 2 | 0.59 | 0.73 | 0.73 |
| 37 | Atypical bives | 2 | 0.59 | 0.73 | 0.73 |
| 38 | Naturally backed knives | 2 | 0.59 | 0.73 | – |
| 39 | *Raclettes* | 4 | 1.18 | 1.45 | 1.47 |
| 40 | Truncated pieces | 5 | 1.48 | 1.82 | 1.83 |
| 42 | Notched pieces | 27 | 7.99 | 9.82 | 9.89 |
| 43 | Denticulates | 23 | 6.80 | 8.36 | 8.42 |
| 44 | *Becs burinants alternes* | 6 | 1.78 | 2.18 | 2.20 |
| 45 | Retouched with inverse retouch | 5 | 1.48 | – | – |
| 46-47 | Abrupt and alternate retouch (thick) | 14 | 4.14 | – | – |
| 48-49 | Abrupt and alternate retouch (thin) | 37 | 10.95 | – | – |
| 50 | Bifacial retouch | 1 | 0.30 | – | – |
| 54 | End-notched pieces | 4 | 1.18 | 1.45 | 1.47 |
| 62 | Miscellaneous | 14 | 4.14 | 5.09 | 5.13 |
| Bifaces (whole) | | 7 | | | |
| Totals | | | 338 | 275 | 273 |

the tip may have suffered a lateral fracture; dihedral butt; condition: fresh. 69/610.

4. 46.5 x 21.5 mm; thin, on a Levallois blank; finely pointed, with demi-Quina retouch; apparently the tip of a larger piece (hence the small size), the break having been repaired; condition: slight abrasion of the edges. 72/159.

5. (Fig. 26.38). 89 x 37 mm; thick; quite sharply pointed; local Quina retouch, otherwise scalar; flat inverse flaking has preceded the normal retouch at the tip (perhaps to straighten out the bulbar surface if it originally plunged slightly); plain butt; condition: slightly lustrous. 68/891.

6. 73.5 x 30 mm; very thick and irregular; Quina retouch; the edges are parallel, except near the tip, where they turn sharply to form a point (very thick, possibly as a result of a repair after damage); broken butt (pre-retouch); condition: glossy, worn. 72/149.

7. (Fig. 26.35, 2). 90.5 x 41 mm; thick; straight-sided, with Quina retouch except at the tip; missing butt; large multiple bulbar scar (alternatively, this may represent deliberate thinning); condition: fairly fresh. 66/414.

8. 63 x 26 mm; on a thick cortical flake; sharply pointed; Quina retouch on the left hand side, subparallel on the other; two small burin-like scars at the tip may result from impact damage; butt removed; condition: slightly lustrous. 67/274.

Unfortunately the freshest pieces were coated with aluminium in order to carry out microwear studies in the 1960s, when cleaning techniques were less effective than those of today. Consequently they could not be included in the investigation described in chapter 30.

**Limaces** (3). One piece (Fig. 26.35, 5) is rather broad and flat, with Quina or somewhat parallel scalar retouch locally, according to the nature of the edge. Another, also with Quina retouch and much thicker, has a truncation at one end (Fig. 26.35, 6); it is not impossible that this was formerly an elongated Mousterian point, however. Both of the above have plain butts. The third example is very thin, on a Levallois blank with a markedly concave bulbar surface. It is sharply pointed at both ends (the butt has been removed by retouch); however, the central part of one edge has been left unmodified, presumably because it was already the required shape. The mean length is 71 mm.

**Side-scrapers** (149). For all that there is a slight further drop in frequency, to 54.2% of the reduced total, the scrapers from this layer include some of the most impressive from the site, being generally both large and well made. Fourteen pieces have Quina retouch, and 8 have demi-Quina (very greatly in excess of the levels encountered elsewhere at La Cotte). The percentages for the principal groups show a marked fall in simple scrapers, and increases in convergent scrapers and types 25–29.

| | | | |
|---|---|---|---|
| single lateral | 38.9 | double | 8.7 |
| convergent | 14.1 | transverse | 9.4 |
| other | 28.9 | | |

In more detail, alternate scrapers increase to 13.5% (easily a maximum for the site). The greater importance of convergent (straight and convex) pieces may of course be linked to the incidence of points in the layer. Convex edges reach their peak frequency, at 69.4% — a considerable increase from layer A. 29.2% of the scrapers are essentially straight; and only one (1.4%) is concave! As a result, the curvature is both high and relatively stable (0.10 ± 0.08).

In accordance with the increased size of the scrapers (mean length 53.4 ± 16.5 mm), the retouched edge length, and indeed the invasiveness, are well in excess of the values encountered in other layers (44.0 ± 18.5 and 6.45 ± 4.07 mm respectively). The mean retouch angle is also high (60.7 ± 12.4°), contrasting strongly with the values for layers A to C. Even allowing for the smaller sample, thinning of the butt or sides is rarer than usual (one piece, or 1.4%, each). Two (both simple scrapers) have retouch at one end (Fig. 26.36, 1). Five scrapers also have notches (four of them Clactonian). The Levallois Index for the scrapers (18.3) is almost identical to that for the tools as a whole. Single examples occur of scrapers on pseudo-Levallois points, LSFs and cores. The use of cortex-backed blanks (10.8%) is somewhat more frequent than expected, however.

Simple scrapers

*Single lateral* (58) (Figs. 26.35, 7 and 8, and 26.36, 1). The mean retouch length and angle are close to the average for this layer's scrapers; on the other hand, invasiveness is slightly less than expected (5.9 against 6.5 mm). Seventeen out of 58 are straight, 40 convex, and 1 concave — giving a mean curvature of 0.103. Two pieces have Clactonian notches, and two are retouched on the tip. Convex scrapers possess very much longer working edges (48.2 mm) than the others, and their retouch is also very invasive (6.5

**Table 26.20** Layer 5. Typological Groups and Indices

| Index | | Real | Ess. | Red. |
|---|---|---|---|---|
| ILty | Typological Levallois | 2.07 | – | – |
| IR | Side-scrapers | 43.79 | 53.82 | 54.21 |
| IC | Charentian | 19.53 | 24.00 | 24.18 |
| IAu | Acheulian unifacial | 0.59 | 0.73 | 0.73 |
| IAt | Acheulian total | 2.61 | 3.19 | 3.21 |
| IB | Bifaces | 2.03 | 2.48 | 2.50 |
| I | Levallois | 2.07 | 0.36 | 0.37 |
| II | Mousterian | 47.34 | 58.18 | 58.61 |
| III | Upper Palaeolithic | 10.06 | 12.36 | 12.45 |
| IV | Denticulate | 6.80 | 8.36 | 8.42 |
| 5 | Pseudo-Levallois points | 0.00 | 0.00 | – |
| 38 | Naturally backed knives | 0.59 | 0.73 | – |
| I rc | Convergent retouched | 9.76 | 12.00 | 12.09 |
| 6-7 | Mousterian points | 2.66 | 3.27 | 3.30 |
| Rsingle | Side-scrapers, single | 17.16 | 21.09 | 21.25 |
| Rdouble | Side-scrapers, double | 3.85 | 4.73 | 4.76 |
| Rtransv | Side-scrapers, transverse | 4.14 | 5.09 | 5.13 |
| Rinv | Side-scrapers, inverse/alternate | 8.88 | 10.91 | 10.99 |
| 30-31 | End-scrapers | 3.25 | 4.00 | 4.03 |
| 32-33 | Burins | 4.14 | 5.09 | 5.13 |
| 36-37 | Knives | 0.59 | 0.73 | 0.73 |
| 42 | Notched pieces | 7.99 | 9.82 | 9.89 |
| 42-43 | Notched pieces & denticulates | 14.79 | 18.18 | 18.32 |
| Ident | Extended denticulate group | 5.33 | 6.55 | 6.59 |
| Iirr | Irregularly retouched | 16.86 | – | – |
| IQ | Quina | 9.40 | | |
| IQ+½Q | Quina + demi-Quina | 14.77 | | |
| C1 | Clactonian notches | 18.52 | | |

mm). Invasiveness is much lower for the straight pieces. The edge angles are very similar for straight and convex scrapers, but that for the single concave piece is extremely shallow (38°). Once again, there is a marked bias towards pieces retouched on the right hand edge (35, or 62.5%, compared to 21).

*Transverse* (14). These are generally smaller and less invasively retouched than the single lateral scrapers. In shape distribution (4:10:0), as also curvature (mean 0.101), they are very similar. All are distal apart from one (with a thinned butt) which is double.

Multiple scrapers

*Separate* (13). The six straight-convex pieces (e.g. Fig. 26.35, 9) are much the most common in this group (the only subtype missing is straight-concave). In general, these scrapers are well made, often with extremely invasive retouch (7.9 ± 3.5 mm) on blanks which are 11 mm longer than those for the entire scraper series for this layer. Seven (i.e. 53.9%) are on Levallois blanks; however, the facetting index is very low, at 25.0 (only 8 butts are determinable, however). Two pieces have Quina retouch and one demi-Quina. In addition, there are three with thinned butts, one with some thinning on one side, and one with a Clactonian notch. The piece illustrated in Fig. 26.35, 9, with a retouched tip, is rather irregularly worked on the left-hand edge.

*Convergent axial* (11). These are in clear contrast to pieces similarly classified from the immediately preceding layers, not only in respect of their size (62.8 mm for the four straight examples, and 57.4 for the others, which are convex), but also of the invasiveness of their retouch (mean 9.0 mm) and its steepness (65.1°). Three have Quina retouch, and two more demi-Quina. Another possibly important development is that they are rather more pointed (particularly the straight scrapers): the mean width

MAJOR TYPES                    SIDE—SCRAPERS

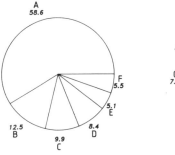
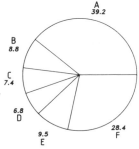

**Fig. 26.34** Flint tools from layer 5. Relative frequencies of the principal tool categories and of the major side-scraper types. Tools: A Group II; B Group II; C Group IV; D notched pieces; E miscellaneous; F other. Side-scrapers: A simple lateral; B simple transverse; C double separate; D convergent axial; E *déjeté*; F other.

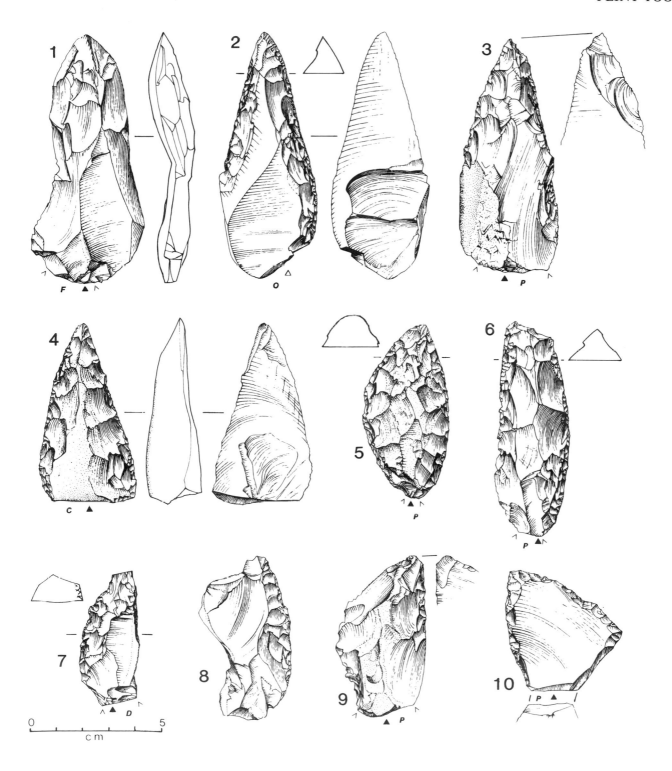

**Fig. 26.35** Flint tools from layer 5. 1 Levallois point (slightly overstruck); 2–3 elongated Mousterian points; 4 Mousterian point; 5, 6 *limaces* (the tip of 6 has been truncated); 7–10 side-scrapers: 7, 8 simple; 9 double; 10 *déjeté*.

at 10 mm from the tip is 17.0 ± 4.7 mm. The tips are also much thicker than usual, notably on the more elongated pieces (the mean thickness/breadth ratio *at the same position* is 0.52, as opposed to around 0.36 for earlier layers). In fact, the possibility cannot be excluded that some of these represent the 'blunter' fringe of a continuum including the Mousterian points already described. Three of the blanks are Levallois (all typical).

*Déjeté* (10). These are much smaller than the convergent axial scrapers (mean length 44.9 ± 13.3 mm) with much less invasive retouch (5.7 ± 3.5 mm), as in Fig. 26.35, 10. Three are on typical Levallois blanks. A single piece has thinning of the butt, and two some thinning of an edge. There is also one with a retouched notch. The tips are noticeably more pointed than in previous layers.

301

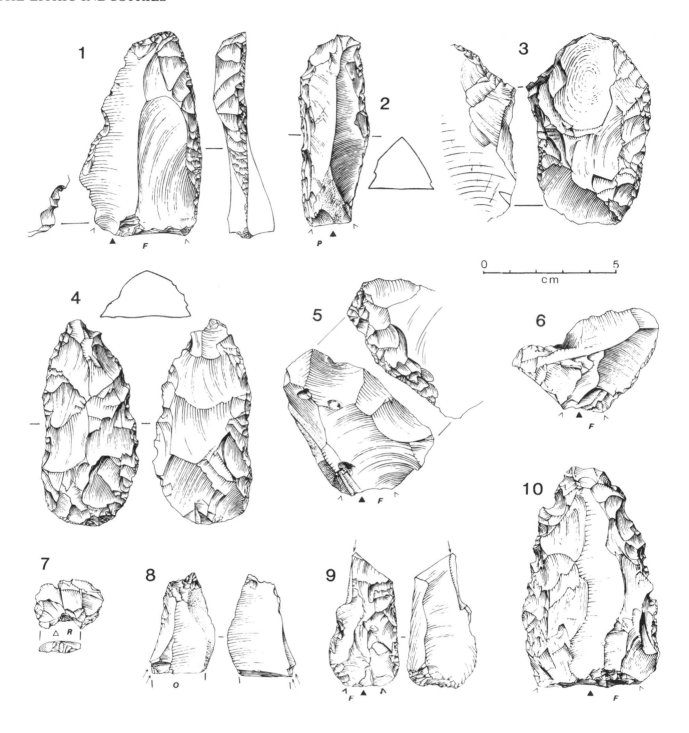

**Fig. 26.36** Flint tools from layer 5. 1–5 Side-scrapers: 1 simple lateral; 2 double; 3 with thinned back; 4 with bifacial retouch; 5 on the ventral surface; 6, 7 notches; 8, 9 burins; 10 denticulate. 7 is on a pseudo-Levallois point, and 9 on a LSF.

*Other side-scrapers*

Clearly dominated by alternate, inverse and bifacially retouched pieces, this very important group substantially raises the overall representation of the double and convergent forms (Table 26.23). Notwithstanding this, the retouched edges are generally short by the standards of the assemblage, at 39.0 mm.

*On the ventral face* (11). Eight are straight, two convex and one concave. Apart from two pieces (one double straight, one double transverse straight with the butt removed by retouch) they are of single lateral type. The retouched edges are very short when compared with the dorsally retouched scrapers (29.6 compared to 44.0 mm); the secondary working is also less invasive (4.7 against 6.45 mm). Two pieces have notches (one being Clactonian). Single examples are made on a cortex-backed, typical and atypical Levallois blanks.

*With abrupt retouch* (1). This is straight, on the right-hand edge of a plain flake 54 mm long (though the retouched edge length is only 35 mm). The retouch is not very invasive (4 mm). In addition, it possesses a Clactonian notch.

*With thinned back* (2). Both are convex, on simple blanks (Fig. 26.36, 3).

*With bifacial retouch* (9). Two of these are fully bifacial, as illustrated in Fig. 26.36, 4. The dominant edges are straight in three cases, and convex in the other six. The secondary working is generally invasive ( ⩾10 mm on all but two pieces). For a single example a typical Levallois flake was used, and for another a core; however, the extent of the retouch may have obscured the true character of some of the other blanks. In two instances the butt has been thinned. The tool illustrated in Fig. 29.3, with conjoining LSF, is strongly reminiscent of the *prondniks* or bifacial knives well known from central European sites. The mean length is high (63.6 mm).

*With alternate retouch* (20). Eleven are straight-edged, eight convex and one concave. Alternating retouch occurs in a slight majority of pieces (11). The blanks are smaller on average (50.3 mm) than those of the bifacially retouched pieces, and include three which are typical Levallois and have cortical backs. There are single occurrences of thinning of the butt, additional retouch, and Clactonian and retouched notches.

**End-scrapers** (11; 3 are typical, 8 atypical). These are generally steeply retouched (mean 81.7°), often with strong curvature (mean 0.24). Six have retouch along the edges, one piece has abrupt retouch, one is slightly denticulated (though not sufficiently to be placed in the denticulate subtype) and a third has subparallel retouch; the remainder are scaled or semi-abrupt. Two tips are skewed to the left, two to the right. Except for one typical end-scraper on a Levallois flake, they are made on plain flakes (including 3 transverse) and blades (2). Six are simple, one ogival, two shouldered (single), and two on the butt.

along one edge, giving an appearance somewhere between a busked burin and a thick end-scraper. No fewer than thirteen pieces have marginal retouch. Three of the burins are overstruck (five have multiple blows on the same platform). The blanks include 4 typical Levallois, 2 LSFs, and one with a cortical back.

**Borers** (2, both atypical). The larger of the two (55 mm) is also the more suspect, on the corner of a broken abrupt scraper. The other (only 39 mm) is on a proximal corner of a flake, and is made by normal retouch on the butt (which has removed the impact point) and lateral inverse retouch.

**Knives** (2, atypical). Both are 73 mm long, on blades. One is Levallois; the other has a steep primary back with only a few small removals.

**Raclettes** (4). They are on very small simple blanks, including one blade, with a mean length of only 24.4 mm. Three are retouched on the right hand edge; the fourth is transverse.

**Truncated pieces** (5). Only one piece is of any size (62 mm long); the mean is only 34 mm. All are distal and straight, and one, the largest is oblique and on a blade. Another is on a Levallois flake with retouched lateral margins. The width of the truncations averages only 14 mm.

**Notched pieces** (27). Only 18.5% are of Clactonian type — the lowest figure for the site. Generally speaking the blanks are small by the standards of this layer (mean length 39.1 mm). Nine have double notches (eight made by retouch). The notches themselves are a couple of mm wider than in A (mean width 12.6 mm), with marginally less curvature (mean 0.25).

| Curvature | 0.0 | 0.1 | 0.2 | 0.3 | 0.4 | 0.5 | 0.6 |
|---|---|---|---|---|---|---|---|
| % | | 7.4 | 22.2 | 37.0 | 29.6 | 0.0 | 3.7 |

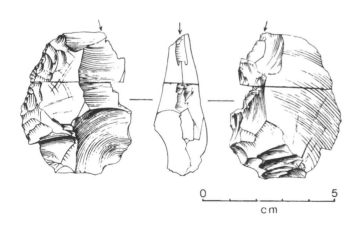

**Fig. 26.37** Flint tools from layer 5. Refitted burins, made on two halves of the same broken flake (one is much more strongly patinated than the other).

**Burins** (14; 11 typical, 3 atypical). Though their percentage (5.1) would be considered high in other industries of this age, it represents a slight decline from the level encountered in layer A.

Twelve pieces are single, longitudinally struck angle burins — e.g. Fig. 26.36, 8 (proximal, on a broken flake with cortical back) and 9 (distal, on the hinged termination of a very large LSF). In all, three of this type are on hinges and nine are on breaks, the last being on a truncation. Nine are distal. A pair of these proved to be conjoinable; it seems that when the original piece broke, the proximal half was itself turned into a burin. The considerable difference in patination is a pointer to the geological complexity of the layer (both pieces were too surface-altered for microwear traces to be preserved). Apart from the simple angle burins, there is a solitary 'mixed' piece (i.e. dihedral opposite angle) and one which has multiple parallel spalls which converge

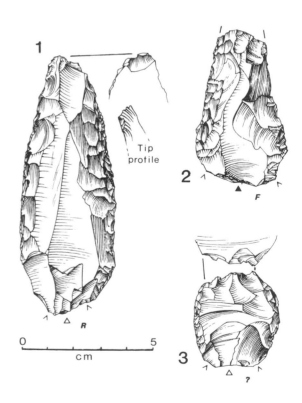

**Fig. 26.38** Flint tools from layer 5. Miscellaneous: 1 elongated Mousterian point with modified tip and butt; 2 broken side-scraper or point with later scaling at the tip; 3 side-scraper with thinning/scaling from a platform prepared on the distal end.

In so small a sample it is not surprising that some of the rarer types of blank are absent. The only special categories consist of 3 Levallois (11.1%, of which one has a Clactonian notch) and a single cortex-backed piece (retouched type).

**Denticulates** (23). Despite a small sample size in this instance (the percentage, for the reduced type list, is 8.4) there is generally good agreement with the results from layer A. The percentages of the various types of edge are: Clactonian notches 17.4, coarse retouch 34.8, fine retouch 43.5, and microdenticulate 4.3. Those for shape are: lateral 43.5, transverse 21.7, double 4.3, convergent 8.7, and *déjeté* 17.4.

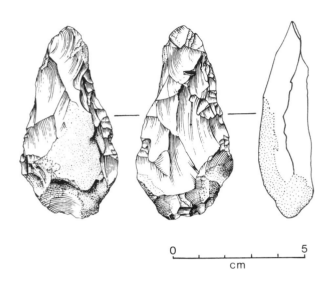

0       5
cm

**Fig. 26.39** Ficron from layer 5.

Four pieces (17.4%) are on pseudo-Levallois points, 30.4% are Levallois, and 26.1% are on cortex-backed flakes, leaving only 26.1% on plain flakes. There is also a very great increase in size, to a mean length of 47.7 $\pm$ 16.3 mm, and a retouch length of 35.1 $\pm$ 18.4. Invasiveness rises sharply as a result of a number of extreme pieces, and is second only to that for layer H, at 6.6 $\pm$ 5.3 mm.

**Becs burinants alternes** (6). Only two of these pieces are entirely satisfactory typologically, and none is particularly sharply pointed. They are appreciably smaller (37.8 mm) than the average tool size for layer 5.

**End-notched pieces** (4). Their mean length is only 39 mm; however, this is partly because one is on the end of a broken flake, and another on a broken side-scraper. Only one of the notches has been formed by retouch.

**Miscellaneous** (14). They include only the following categories:

|  | n | % |  | n | % |
|---|---|---|---|---|---|
| scrapers | 13 | (92.9) | notches | 2 | (14.3) |
| denticulates | 6 | (14.9) |  |  |  |

Half are compound, being scrapers combined with notches or, in five cases, denticulates. Six have LSF or other removals off breaks or truncations; good examples are illustrated as Figs. 26.38, 1–3; the first of these obviously began as one of the thick points characteristic of the assemblage, while the last has a curious denticulated distal end on what would otherwise have been a double convex scraper.

**Bifaces** (7). The pieces are larger than those from layer A (the mean length is 56.6 $\pm$ 15.9 mm). The reduced biface index, 2.50, is very similar to that for A (2.71). Two pieces deserve special mention in view of the morphological contrast that they represent. One is the small lanceolate or ficron illustrated as Fig. 26.39. The other (Fig. 26.40), a cordiform with a slight S-twist, was found by Burdo and has not been included in the statistics, therefore; there is no question, from its co-ordinates, but that it came from layer 5. The other pieces in the series

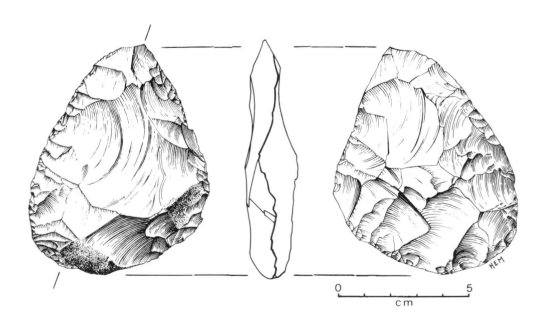

0       5
cm

**Fig. 26.40** Twisted cordiform handaxe from layer 5 (Burdo excavations). The stratigraphic attribution is based on its coordinates, viz. 23.7 m below datum, 7.0 m into the north ravine (measuring from the southern entrance), 3.0 m from the west wall.

comprise an amygdaloid, a bifacial cleaver, two nucleiforms, one partial biface and one 'miscellaneous' biface. One of the specimens from the McBurney excavations is twisted, and another slightly twisted; none have tranchet finish. Cortex on the butt is more common than in A: two pieces are heavily cortical, and three more have cortical corners.

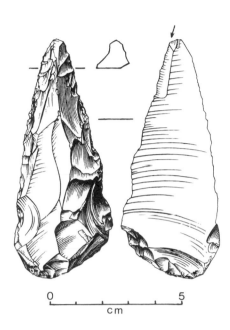

**Fig. 26.41** A thick elongated Mousterian point from the Burdo excavations. Probably displaced from layer 5, this was found in layer 7.2 (i.e. the collapsed deposits above the '8 m' beach) about 2 m south of the fossil cliff. Comparable pieces (cf. Figs. 26.35, 2–4) are found *in situ* only in layer 5, which supports a pre-Eem date for the latter. A burin blow (inverse) which may be an impact scar has removed part of the retouch along the right-hand edge.

*LAYERS 6.1 (UPPER PART) TO 9*

Eight tools (in real count) were recovered from 6.2 and the upper part of 6.1, and eleven from layer 9. Apart from one notched piece from layer 9, the reduced tools comprise only side-scrapers. Given that the raw material frequencies are the same as for layer 5, and that the artefacts are physically damaged and distributed throughout geliflucted or cryoturbated deposits, there is no reason to suppose we have here anything other than derived pieces.

Quite apart from the above, the lens of charcoal stratified within the beach deposits, which Lanoë has suggested may provide evidence of a human visit to the site at a time of lowered sea-level during the Last Interglacial complex (chapter 9), is reworked; the artefacts found in and around it need not have been directly associated with the former hearth. The overburden of slumped material (layer 7.2) was quite rich archaeologically. For the most part it was excavated by Burdo; the finds have not been studied in detail but a cursory examination suggests that most or all of them must have

been derived from relatively high up in the Saalian layers. They include the thick Mousterian point illustrated in Fig. 26.41 whose significance has already been discussed. Layer 8.1 seems also to have been rich in artefacts of the same general character, including the handaxe illustrated in mf 26.2, 2. A more unusual find, stratigraphically speaking, is the handaxe shown in mf 26.2, 1. It was discovered at the top of layer 8.3 (Burdo's [3B]) where it was presumably washed from a higher deposit. A post-Saalian date cannot be excluded, but it is perfectly in keeping with the late Saalian industries.

## 26.3 UNMODIFIED 'TOOLS'

Under this heading are included types 1–3, 5 and 38, i.e. typical and atypical Levallois flakes, Levallois and pseudo-Levallois points, and naturally backed knives (*couteaux à dos naturel*). Some of the pieces in these categories exhibit utilisation damage, or even small areas of retouch (probably intended to regularise the edge, but insufficient to justify attribution to one of the other types in the Bordes list). The frequencies are generally low (Table 26.21). In particular, no series from La Cotte (including that from the Weichselian layers described in Appendix B) is dominated by Levallois flakes; the highest ILty is only 9.2, for layer B, falling to 1.8 for layer 5.

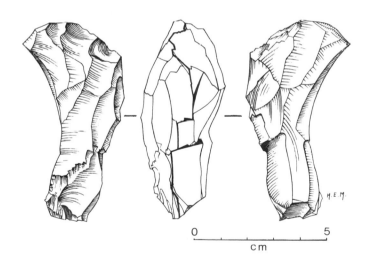

**Fig. 26.42** Overstrike from a Levallois blade core, from layer A.

**Levallois flakes and points.** Within the unmodified Levallois group consisting of types 1–3, and notwithstanding the small sample sizes for layers H, G and 5 (which render their statistics unreliable), some striking technological variation occurs (Table 26.21). Thus layer A exhibits a very high incidence of Levallois blades and blade-like points (25.3%). In a few cases these may be no more than LSFs which have failed because directed too axially to remove the edge of a tool; the old retouch scars, if invasive, will have been truncated by the LSF to give the appearance of partly centripetal preparation. That true Levallois blade technique was practised, despite the absence of the cores themselves, is clearly demonstrated by the existence of pieces such as that illustrated in Fig. 26.42, which is an overstrike from a double-platformed Levallois core.

**Table 26.21**  Unmodified Levallois Pieces (Types 1–3): Technological Indices

| Index | | | | | Layer | | | | |
|-------|------|------|------|------|------|------|------|------|------|
|       | H    | G    | F    | E    | D    | C    | B    | A    | 5    |
| IF    | 55.55 | 62.50 | 70.73 | 45.71 | 58.82 | 48.76 | 80.39 | 75.87 | 66.67 |
| IFs   | 44.44 | 37.50 | 56.10 | 28.57 | 43.14 | 45.45 | 68.63 | 68.28 | 50.00 |
| IFss  | 33.33 | 33.33 | 41.46 | 22.86 | 39.22 | 36.36 | 60.78 | 62.07 | 33.33 |
| Ilam  | 20.00 | 14.29 | 8.51 | 0.00 | 14.04 | 16.67 | 11.86 | 25.32 | 33.33 |
| Sample | 10  | 28   | 47   | 40   | 57   | 48   | 59   | 154  | 6    |

The behaviour of the facetting indices, particularly IFss, is interesting. There is a very clear fall in layer E, to the surprisingly low value of 22.9, while in B and A the values are in excess of 60.

Though the samples are too small for the results for individual layers to be statistically significant, the distributions of the criteria causing pieces to be put into the atypical class are noteworthy. Leaving aside the minority categories such as the presence of an excessively large relict margin (i.e. 'éclats débordants' as described by Beyries and Boëda 1983) the most important are thickness and cortex.

|          | H–D | C–5 |
|----------|-----|-----|
| Thickness | 13  | 35  |
| Cortex    | 32  | 29  |

The $\chi^2$ value for these figures, 6.13, is significant at the $P = 0.02$ level. So the increase, in the upper layers, in pieces regarded as atypical because of their thickness at the expense of those with cortex is not due to sampling error.

Only five of the Levallois points (four from layer A, the other from C) are primary, i.e. single-ridged; the others are secondary and bear a scar resulting from the removal of a point of primary type from the same platform. Those from the lowest layers are poorly made and may be fortuitous.

**Pseudo-Levallois points** (Figs. 26.11, 2–3 and 26.22, 2). These are rarely points in the strict morphological sense. Though most are triangular in shape, polyhedral examples occur in all layers in which the type is represented. Layer D is exceptional in that triangular pieces are in a minority. Many of the best examples are from the lower layers, and appear to have been struck from globular or bipyramidal cores, and are often quite thick. Some of those from higher up in the sequence are likely to have been produced from disc cores.

The individual points are of course asymmetrical about the axis of percussion. A record was kept of their orientations, but the deviation from a 50:50 ratio was significant in only one layer. This is H (in which pseudo-Levallois points are particularly well represented). Only 2 out of 14 had their points to the right when positioned with the butt towards the observer and the ventral surface downwards; according to the binomial distribution this is likely to represent a real bias in the population, but in the absence of examples which can be refitted to their cores the technological implications of this observation remain uncertain.

**Cortex-backed, or naturally backed, knives** (Figs. 26.12, 9 and 26.20, 5). In 127 cases (87.0%) the 'back' is formed of pebble cortex, and on a further 13 (8.9%) it consists of a patinated and sometimes stained flake surface. In six more it is unclear whether it is of pebble or nodular cortex (generally because the surface is essentially concave and thus protected from battering). In no instance is it possible to say with certainty that the piece was struck from an unrolled nodule. The angle between the back and the ventral surface, and the position (left or right) of the cortex, vary erratically from layer to layer. The knives from layer A are characterised by a high percentage of blades (38.5); the samples for the other layers are too small for much reliance to be placed on them.

## 26.4  IRREGULARLY RETOUCHED TOOLS

Types 45–50 always present problems during the use of the Bordes typology as they include not only the pieces which exhibit a short length of obviously deliberate marginal retouch, but also those which are probably merely 'retouched' by utilisation or soil movement. As yet no satisfactory means exists for distinguishing these effects on large numbers of artefacts. To draw sound functional conclusions about the examples from La Cotte would require a major programme of research, far beyond the scope of the present book. Nevertheless a few observations may be made.

1. In the course of the sequence there is a persistent decline in their frequency from 34.1% of the real count in layer H to 16.9% in layer 5. Bordes (1972, 52) has remarked that "a layer without any disturbance will usually have less 'utilised' flakes than one which has been slightly cryoturbated". Soil movement and colluviation may go some way towards explaining high frequencies in the lower layers of La Cotte, though in general the extreme freshness of the associated tools and the lack of strictions suggest otherwise. In the assemblage from layer 5, however, many artefacts show unequivocal effects of soil movement, yet the percentage of types 45–50 is low. It is therefore more likely that this group consists chiefly of pieces whose edges have been damaged by human rather than natural agencies, and variation in frequency may be due to differences in the amount of trampling or in the use of weakly retouched or unmodified flakes.

2. As a rule the pieces in this group are smaller (by around 3 mm on average, in the case of the length) than are the formal tools; nevertheless they are considerably larger than the waste. This suggests that they have been deliberately selected, and lends support to the view that they have been used by man.

It is unlikely that we have here the totality of the pieces which have been utilised — those lacking macroscopic traces will have been classified as waste. But they serve as a forceful reminder of the probable importance of 'casual' tools in processing activities.

## 26.5 A DISCUSSION OF SOME ASPECTS OF THE FLINT TOOLS

### 26.5.1 Mousterian Points and Elongated Mousterian Points

A very distinct, stratigraphically related dichotomy was observed for these categories. In layer A and below, the points are generally quite thin, and vary considerably in their relative elongation. In layer 5, on the other hand, they are thick, elongated, and are not uncommonly made by means of Quina or demi-Quina retouch. They often exhibit signs of damage or repair at the tip.

The contrast between the two groups is shown very clearly by breadth and thickness measurements taken at 10 mm from the tip. The mean T/B ratio at this position is 0.37 for pieces from layer A, and 0.72 for those from 5. The distributions for this attribute overlap very little (Fig. 26.43). The ratio is unlikely to have been much affected by any small error in choosing the position for measurement, which may be the result of damage, as the cross-section is quite consistent near the extremity. It should be noted that the mean of the ratio in the case of the layer 5 points is very close indeed to 0.707, corresponding to the proportions of an equilateral triangle (an observation whose implications are considered below).

In any discussion of this tool class, perhaps the first question which comes to mind is whether indeed we are dealing with weapons or merely pointed scrapers (or some other type of tool). The elongated pieces from layer 5 would appear to be quite suitable as borers, for example, but they show no macroscopic signs of use in this way; their surfaces are not in sufficiently good con-

**Table 26.22** Major Tool Categories (Percentage of 'Reduced' Tools)[a]

| Layer | II (6–29) | III (30–37) | IV (43) | Notches (42) | Misc. (62) | Other |
|-------|-----------|-------------|---------|--------------|------------|-------|
| 5 | 58.6 | 10.6 | 11.4 | 8.4 | 5.1 | 5.9 |
| 3 | 53.8 | 7.5 | 8.6 | 15.1 | 6.5 | 8.5 |
| A | 50.2 | 10.1 | 5.5 | 11.1 | 9.7 | 5.9 |
| B | 63.4 | 7.8 | 9.2 | 8.9 | 7.0 | 3.7 |
| C | 66.1 | 5.6 | 5.9 | 12.5 | 6.0 | 3.9 |
| D | 62.4 | 7.1 | 6.9 | 14.0 | 5.4 | 4.2 |
| E | 55.8 | 5.0 | 11.9 | 20.1 | 3.5 | 3.7 |
| F | 41.4 | 4.8 | 13.1 | 33.0 | 4.2 | 3.5 |
| G | 20.0 | 10.4 | 21.7 | 36.8 | 5.7 | 5.4 |
| H | 13.8 | 6.3 | 37.5 | 30.0 | 1.9 | 10.5 |

[a]The column headings II, III and IV refer to typological groups (see section 23.1).

## Points

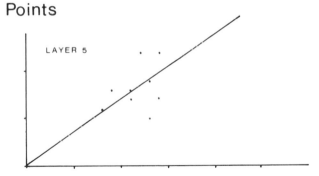

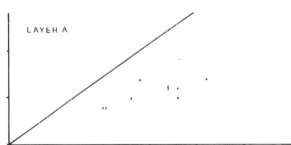

## Convergent scrapers

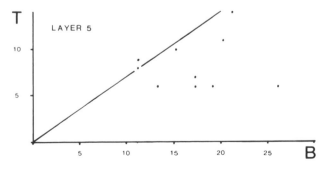

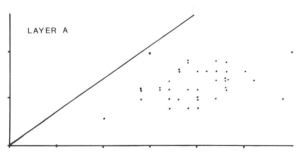

**Fig. 26.43** Mousterian points and convergent side-scrapers from layers 5 and A: thickness and breadth at 10 mm from the tip.

dition to permit high resolution study. And even if it *had* been possible to identify use-wear traces indicating employment as scrapers, this need imply no more than that a discarded weapon-head had been re-used. It is possible to make a more positive statement than this, however. The damage to the tips of a number of points from layer 5 accords very well with the types of impact fracture observed by Bergman and Newcomer (1983) during experiments on flint arrowheads:

| | |
|---|---|
| *Bend fracture*: | Artefact no. 72/159; repaired after break. |
| *Flute (dorsal)*: | 66/193. |
| *Burin-like*: | 68/876; possibly with an earlier ventral flute fracture as well (Fig. 26.35, 4). |
| | 67/274. |

Also:

| | |
|---|---|
| *Repaired tip*: | 68/891 (Fig. 26.35, 3). |
| *Lateral snap*: | 69/610; this is not a typical impact fracture, however. |

Thus four (or possibly six) out of nine pieces exhibit impact damage. The pieces illustrated as Figs. 26.35, 6 and 26.38, 1, which can no longer be classified as points, may have suffered similarly before modification. The point found by Burdo in layer 7.2 also exhibits a single burin-like scar at the tip (Fig. 26.41). It is therefore interesting to recall Bergman and Newcomer's comment (1983, 239) that Mousterian points "*could* have been projectile points" although "breakage attributable to impact is, in our experience, very rare on such points".

The points from layer A, on the other hand, lack evidence of impact damage. Being much thinner, though, they may have been more liable to suffer bend fractures or more extreme shattering; if so, the smaller debris may be among the retouched fragments which could not be examined in detail during this study. Therefore the function of these points (and their relationship to convergent scrapers) has yet to be established.

Assuming that the earlier pieces are indeed weapon heads, the shift towards thicker, more elongated points may be explained in three ways.

Technologically, as a consequence of employing a debitage process with a tendency to produce thick flakes and blades.

Functionally, reflecting a change of hunting objectives, tactics or technology.

Stylistically, implying no more than a culturally determined preference for points of a different shape.

The technological alternative can probably be ruled out, despite increased tool thickness. Taking side-scrapers as an example, estimated *absolute* thickness (see chapter 23) is greater in 5 than in A (12.8 as against 10.5 mm), but this reflects a general size change; the estimated *relative* thickness is almost identical for the two series (0.310 and 0.305). Therefore the cross-section of the points is not the result of technological constraints, and must reflect a morphological preference.

The question of function *versus* style is less easily resolved, particularly as there is much ethnographic and historical evidence testifying to the diversity of shape of projectile or spear points whose functions are very similar. Nevertheless, the differences between the La Cotte specimens are of a kind that would certainly affect their functional qualities. In particular, an equilateral triangle is one of the strongest possible cross-section shapes.

In order to obtain information about penetrative ability, it is necessary to turn to studies of metal weapon heads (e.g. Pope 1974; Barker 1975). Blade-like points are inefficient against an armoured target — into which category may perhaps be placed *M. primigenius* and *C. antiquitatis* with their thick hair and hides. Against plate armour, wood, etc. a strong, relatively blunt tip is effective, but a very sharp point would presumably be necessary to penetrate a dense outer covering of hair (by analogy with the bodkin type of point which is the most successful against chain mail). The flint points from layer 5 at La Cotte may therefore represent the best compromise for dealing with giant herbivores, inasmuch as they are both strong and sharp. Experiments to test this are planned for the future; ethnographic evidence is of little value here, as living elephants and rhinos are virtually hairless. If the hypothesis is correct the change of design is likely to reflect purpose rather than style, as the earlier points would be much less effective against well-protected animals.

### 26.5.2 Side-scrapers

These are numerically important for the first time in layer F, and are predominant thereafter, undergoing considerable typological development in the course of the sequence; the changes have been dealt with in some detail in the descriptive section above. The most interesting aspects of the variation (see also Fig. 26.44) are summarised here.

1. Size changes broadly parallel those for other artefact types.
2. The relative frequencies of the edge shape categories shift towards a strong predominance of convex scrapers.
3. There is a rise in the *curvature* of convex scrapers, followed by a decline in layer 5.
4. The retouched edges are particularly obtuse in G and H, and particularly acute in B and C.
5. The invasiveness of retouch usually averages just over 5 mm. The exceptions are layers G (with mar-

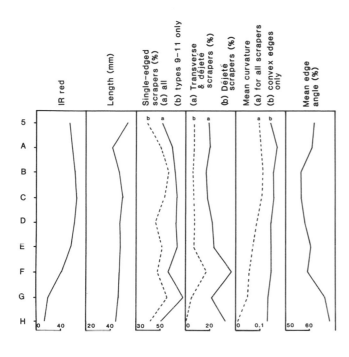

**Fig. 26.44**  Side-scrapers: typological and other changes through time.

ginal retouch predominant) and 5 (6.5 mm). The latter has a particularly high proportion of pieces whose retouch scars exceed 10 mm in length (Fig. 26.44).

6. The incidence of Quina retouch is generally low, but rises sharply in layer 5.

7. There is a sudden and temporary increase in *déjeté* scraper frequency in layer F.

8. The percentage of inverse and alternate scrapers of good quality peaks in D, and again in 5 (the high percentages in G and H are due for the most part to less typical pieces).

9. Transverse scrapers are rather scarce except in the lower layers (but the percentage in G is also low).

To take the side-scraper series layer by layer, those from each of H, G and F have distinctive features which preclude a simple sequence of development, insofar as they exhibit considerable typological and metrical variability. Those from E and D are very much more alike. There is some consistency of characteristics from C to A (or 3, representing the terminal state of the layer A industry) but also a clear evolution; as described elsewhere, the morphological complexity of the pieces grows as the practice of resharpening becomes more common. Much of the observed variation, such as a slight decline in invasiveness and an increase in the curvature of convex scrapers in A, is probably causally linked to size reduction. On a small convex scraper, rounding of the ends of the retouched edge, to prevent damage when working skins, may contribute substantially to the curvature, whereas the effect is less marked for large pieces.

In layer 5 the scrapers are larger, and retouched more invasively. The most striking development, though, is the increase in Quina (and demi-Quina) retouch. The lower curvature of the convex scrapers is likely to be partly due to the size change; nevertheless this series of side-scrapers might appear totally unrelated to that of layer A were it not for continued resharpening (albeit at a reduced frequency) by the LSF technique and the use of the other practices referred to in section 26.5.6.

From an economic viewpoint, one of the potentially most interesting developments in the La Cotte side-scrapers is the shift in favour of convex-edged pieces. Its possible significance is discussed in section 31.5.

### 26.5.3 Burins

Though these are very common in the later assemblages (e.g. 5.9% of the reduced tools from layer A) their purpose is uncertain. In the Upper Palaeolithic (and in many later stone-using cultures) burins were certainly employed for engraving and splitting bone and antler in the course of artistic and tool-making activities. Earlier, such processes appear to have taken place on a more limited scale; bone tools are rare, as are burins them-

**Table 26.23**  Principal Side-scraper Categories (as a Percentage of Side-scrapers only)

| Layer | Type | | | | | | Sample |
| | Single lateral (9-11) | Double lateral (12-17) | Converg. axial (18-20) | *Déjeté* (21) | Transv. (22-24) | Other side-scr. (25-29) | |
|---|---|---|---|---|---|---|---|
| 5 | 39.2 | 8.8 | 7.4 | 6.8 | 9.5 | 28.4 | 148 |
| 3 | 40.8 | 20.4 | 0.0 | 6.1 | 6.1 | 26.5 | 49 |
| A | 50.0 | 11.1 | 2.8 | 6.8 | 8.1 | 21.2 | 1454 |
| B | 55.8 | 12.8 | 5.7 | 5.7 | 6.8 | 13.2 | 265 |
| C | 52.4 | 9.8 | 5.5 | 6.5 | 6.9 | 18.9 | 418 |
| D | 45.9 | 5.1 | 3.0 | 7.1 | 9.5 | 29.4 | 296 |
| E | 50.2 | 7.7 | 5.0 | 6.8 | 12.2 | 18.1 | 221 |
| F | 46.6 | 5.0 | 3.2 | 16.3 | 16.3 | 12.7 | 221 |
| G | 54.3 | 8.6 | 1.2 | 4.9 | 9.9 | 21.0 | 81 |
| H | 40.9 | 0.0 | 4.5 | 0.0 | 22.7 | 31.8 | 22 |

**Table 26.24** Scraper Types 25–29: Reclassified according to Morphology (Percentages)

| Layer | Limace | Single lateral | Double lateral | Type Converg. axial | Déjeté | Transv. | Sample |
|---|---|---|---|---|---|---|---|
| | (8) | (9–11) | (12–17) | (18–20) | (21) | (22–24) | |
| 5 | 2.3 | 44.2 | 25.6 | 18.6 | 7.0 | 2.3 | 43 |
| 3 | 0.0 | 46.2 | 30.8 | 7.7 | 15.4 | 0.0 | 13 |
| A | 0.0 | 46.1 | 21.8 | 9.7 | 19.2 | 3.2 | 308 |
| B | 0.0 | 45.7 | 17.1 | 5.7 | 22.9 | 8.6 | 35 |
| C | 0.0 | 57.0 | 17.7 | 5.1 | 12.7 | 7.6 | 79 |
| D | 0.0 | 57.5 | 19.5 | 5.7 | 12.6 | 4.6 | 87 |
| E | 0.0 | 72.5 | 5.0 | 2.5 | 7.5 | 12.5 | 40 |
| F | 0.0 | 64.3 | 3.6 | 0.0 | 14.3 | 17.9 | 28 |
| G | 0.0 | 58.8 | 5.9 | 5.9 | 17.6 | 11.8 | 17 |
| H | 0.0 | 28.6 | 42.9 | 0.0 | 0.0 | 28.6 | 7 |

selves in the majority of pre-Upper Palaeolithic industries. Burins have been used for other purposes (boring, whittling, etc.), and flakes of suitable shape are also known through microwear studies to have been used as graving tools (Keeley 1980, 104). The La Cotte burins examined so far have yielded no such microscopic traces, however (chapter 30).

That some of the La Cotte burin scars were caused accidentally is possible in view of the high percentage on broken flakes and blades, but this can hardly explain their overall frequency. A few may have resulted from failed attempts at LSF removals, but examples on retouched truncations are very rare. A deliberate policy of manufacture is attested by the presence of a number of very carefully made pieces (excellent dihedral and angle burins, and also burins and spalls with evidence of multiple removals). It is worth bearing in mind that the burin itself may not have been the part of the tool which was brought into contact with the material being worked. Semenov (1964, 108–9) describes whittling knives from Kostienki IV which have at one end a dihedral burin which was fitted into a haft. An angle burin facet might also, in some circumstances, serve as a form of backing. Although it is not yet possible to draw firm conclusions about the purpose of the La Cotte burins,

therefore, their high incidence in the upper part of the Saalian deposits points to an interesting area of investigation for the future and underlines the uncertain status of such tools in the Lower and Middle Palaeolithic.

### 26.5.4 Notched Pieces and Denticulates

Both types are very well represented in the lower layers; the borderline between them is not easily drawn, as the denticulation tends to be coarse and is often formed by multiple adjacent Clactonian notches. In the scraper-dominated assemblages, on the other hand, the denticulation is finer (possibly reflecting a different function). The notches (both Clactonian and retouched) are particularly wide and deep in H; much smaller in G, they undergo a further gradual but irregular reduction in size thereafter. Throughout the Saalian assemblages the mean length of the notched pieces is consistently less than that of the denticulates.

### 26.5.5 Handaxes

These only occur in the scraper-dominated series. The first 'classic' example (discoidal) is from C, though there are a few nucleiforms in D. Only in A (Fig. 26.45) and 5

**Table 26.25** Denticulates: Reclassified according to the Typology for Scrapers (Percentages)

| Layer | Single lateral | Double lateral | Converg. axial | Type Déjeté | Transv. | Other side-scr. | End-scr. | Sample |
|---|---|---|---|---|---|---|---|---|
| | (9–11) | (12–17) | (18–20) | (21) | (22–24) | (25–29) | (30–31) | |
| 5 | 43.5 | 4.3 | 8.7 | 17.4 | 21.7 | 4.3 | 0.0 | 23 |
| 3 | 50.0 | 7.1 | 0.0 | 21.4 | 7.1 | 14.3 | 0.0 | 14 |
| A | 35.3 | 5.0 | 6.8 | 9.4 | 21.2 | 18.0 | 4.3 | 278 |
| B | 54.3 | 11.4 | 5.7 | 5.7 | 11.4 | 8.6 | 2.9 | 36 |
| C | 45.1 | 4.9 | 3.7 | 8.5 | 17.1 | 20.7 | 0.0 | 80 |
| D | 48.4 | 9.4 | 0.0 | 9.4 | 15.6 | 17.2 | 0.0 | 65 |
| E | 70.3 | 3.1 | 0.0 | 1.6 | 15.6 | 9.4 | 0.0 | 78 |
| F | 50.0 | 7.1 | 6.4 | 3.6 | 18.6 | 14.3 | 0.0 | 176 |
| G | 48.3 | 6.7 | 5.4 | 22.1 | 14.8 | 2.7 | 0.0 | 147 |
| H | 59.0 | 2.6 | 0.0 | 7.7 | 28.2 | 2.6 | 0.0 | 40 |

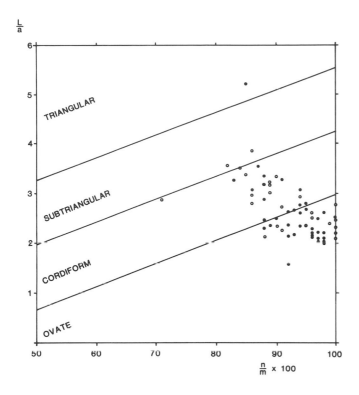

**Fig. 26.45** Handaxe shape diagram (flint pieces only) from layer A.

are handaxes at all numerous; even so, they form a tiny percentage of the tools from either layer. Most of the classic Acheulian forms (especially the amygdaloids) are represented, but the nucleiform type is common (Table 26.26), as is also the case in some of the French cave series of similar age, such as Combe Grenal. Flat triangular and subtriangular pieces of the kind found in the M.T.A. in N. France and elsewhere are completely absent.

### 26.5.6 Truncations, Scaling and Resharpening

Some truncations (mainly of the normal type) are probably no more than rather straight end-scrapers, or were intended to provide a rest for the forefinger when one of the edges of the tool was used for cutting; these have generally been classed as type 40 ('truncated pieces'). Nevertheless:

1. Effectively beginning in layer D and rising to 11.2% of the real tool total in layer 3 with a subsequent decline to 4.2% in layer 5 (Fig. 26.46) there is a series of inverse truncations, on various classes of tool, for which the above explanation seems inappropriate.
2. In almost all the layers, well over half of these exhibit some degree of dorsal scaling (generally not very invasive); the exceptions are layers E (in which only one piece has an inverse truncation) and A (47.0%).

**Table 26.26**  Biface types: Frequency[a]

| Type | Layer | | | | | |
|------|---|---|---|---|---|---|
| | D | C | B | A | 3 | 5 |
| Lanceolate | 0 | 0 | 0 | 2 | 0 | 0 |
| Ficron | 0 | 0 | 0 | 1 | 0 | 1 |
| Cordiform | 0 | 0 | 0 | 2 | 0 | 0 |
| Ovate | 0 | 0 | 0 | 1 | 0 | 0 |
| Amygdaloid | 0 | 0 | 0 | 18 | 0 | 1 |
| Discoidal | 0 | 1 | 0 | 7 | 0 | 0 |
| Limande | 0 | 0 | 0 | 2 | 0 | 0 |
| Cleaver (bifacial) | 0 | 0 | 0 | 4 | 0 | 1 |
| Nucleiform | 4 | 2 | 2 | 22 | 1 | 2 |
| Miscellaneous | 0 | 0 | 1 | 2 | 0 | 1 |
| Partial | 0 | 3 | 0 | 4 | 0 | 1 |
| Pick | 0 | 0 | 0 | 1 | 0 | 0 |
| Determinable | 4 | 6 | 3 | 66 | 1 | 7 |
| Broken | 0 | 1 | 0 | 4 | 0 | 0 |
| Total | 4 | 7 | 3 | 70 | 1 | 7 |
| 'Classic' forms | 0 | 1 | 0 | 37 | 0 | 3 |

[a]'Classic' bifaces exclude the last four types listed here.

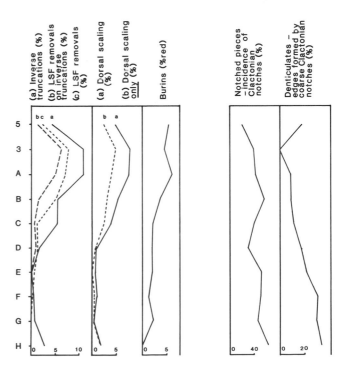

**Fig. 26.46** Specialised technological features: changes through time.

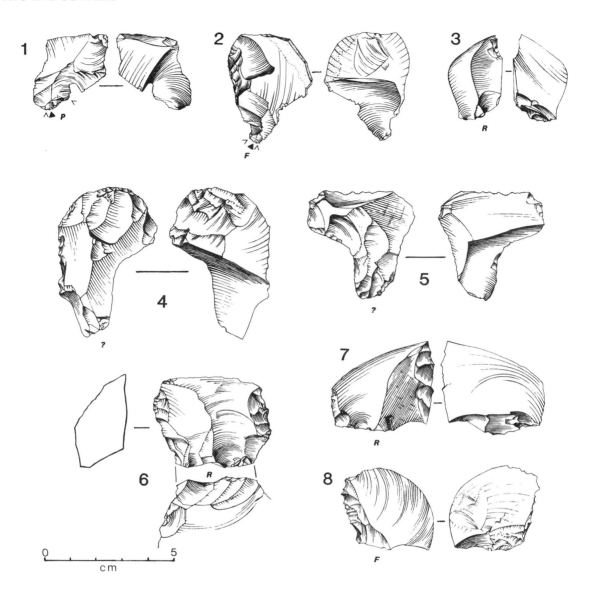

**Fig. 26.47** Examples of pieces exhibiting interesting technological features. 1 Overstruck (plunging) TSF; 2 plunging and misdirected LSF; 3 tool spoilt by plunging LSF removal; 4 and 5 plunging flakes produced during dorsal scaling; 6–8 tools spoilt by plunging flakes during dorsal scaling (lateral in the case of 8).

3. Except in the lowest layers there is a steady rise in the percentage of the inverse truncations which have been used as platforms for the LSF edge rejuvenation technique, to a maximum of 53.3% in layer 3. This parallels the developing importance of the technique as expressed in the frequencies of the LSF removal scars relative to the tools, and of the characteristic waste products relative to the whole assemblage; see chapter 29 for a full discussion. These rejuvenated pieces quite commonly exhibit scaling from the same truncation or truncations.

4. In a very few cases the inverse truncation has been used as the platform for the removal of a burin; in layer A the percentage of inverse truncations so treated is only 7.1. The use of normal truncations for this purpose is very much rarer, as a fracture surface is the more usual in practice.

5. A few inverse truncations, in the layers in which these are present in more than negligible quantity, have been left without further modification; in layer A it is 11.9%.

6. Most of the inverse truncations are on the distal end (Table 26.27). The comparative rarity of those on the proximal end is no doubt due to the suitability of the butt itself as a platform for further removals (in such cases, dorsal scaling in particular is less likely to be recognised).

From the above it is clear that, although the production of inverse truncations is partly linked to resharpening (and to a much lesser degree to the production of burins), in very many cases the practice must have served some other purpose. A close parallel is provided by the Upper Palaeolithic Kostienki knife, first identified

**Table 26.27**  Tools: Special Technological Features (Percentages)

| Layer | Inv. trunc. | LSF removal | Dorsal scaling | LSF rem. on inv. trunc. | Dorsal scaling only | Sample | For inverse truncated pieces only LSF removal | Dorsal scaling | Burin removal | Sample |
|---|---|---|---|---|---|---|---|---|---|---|
| 5 | 4.2 | 2.2 | 4.5 | 1.1 | 2.2 | 338 | 26.7 | 60.0 | 6.7 | 15 |
| 3 | 11.2 | 7.5 | 7.5 | 6.0 | 4.5 | 129 | 53.3 | 60.0 | 6.7 | 15 |
| A | 10.7 | 6.8 | 7.2 | 5.0 | 3.9 | 3561 | 47.1 | 47.0 | 7.1 | 410 |
| B | 5.3 | 2.3 | 5.1 | 1.5 | 2.9 | 639 | 33.3 | 61.1 | 2.8 | 36 |
| C | 3.2 | 1.2 | 3.6 | 0.5 | 2.3 | 948 | 15.6 | 78.1 | 3.1 | 32 |
| D | 1.5 | 1.3 | 0.9 | 0.9 | 0.6 | 757 | 25.0 | 50.0 | 0.0 | 12 |
| E | 0.1 | 0.6 | 0.5 | 0.1 | 0.0 | 650 | 100.0 | 50.0 | 0.0 | 1 |
| F | 0.4 | 0.1 | 0.8 | 0.0 | 0.3 | 917 | 0.0 | 75.0 | 0.0 | 4 |
| G | 0.6 | 0.0 | 0.4 | 0.0 | 0.4 | 685 | 0.0 | 75.0 | 25.0 | 4 |
| H | 2.6 | 0.0 | 1.6 | 0.0 | 1.6 | 296 | 0.0 | 62.5 | 12.5 | 8 |

**Table 26.28**  Tools: Mean Lengths for Principal Classes

| Layer | Levallois flakes (1–2) | Side-scrapers (9–29) | End-scrapers (30–31) | Burins (32–33) | Knives (36–37) | Notched pieces (42) | Denticulates (43) | All (real) | All (reduced) |
|---|---|---|---|---|---|---|---|---|---|
| 5 | 47.8 | 53.4 | 49.6 | 45.6 | 72.7 | 39.1 | 47.7 | 47.7 | 50.3 |
| 3 | 40.3 | 39.7 | 51.8 | 44.4 | – | 33.8 | 38.9 | 40.1 | 39.5 |
| A | 44.4 | 41.7 | 37.5 | 37.2 | 50.5 | 38.3 | 39.1 | 39.6 | 40.4 |
| B | 39.7 | 46.9 | 41.6 | 42.1 | 58.5 | 39.8 | 42.2 | 42.6 | 45.1 |
| C | 42.1 | 49.3 | 48.2 | 42.0 | 53.5 | 40.6 | 43.0 | 44.2 | 47.1 |
| D | 42.3 | 47.2 | 41.3 | 37.6 | 43.0 | 38.0 | 43.0 | 42.7 | 45.3 |
| E | 40.9 | 47.5 | 40.9 | 44.1 | 45.5 | 39.8 | 47.2 | 42.9 | 45.6 |
| F | 43.3 | 46.2 | 45.6 | 33.5 | 61.6 | 43.0 | 46.1 | 43.5 | 45.3 |
| G | 48.8 | 45.9 | 39.9 | 47.2 | 59.4 | 43.9 | 45.7 | 43.9 | 45.3 |
| H | 48.3 | 43.7 | 49.2 | – | 49.0 | 43.6 | 48.3 | 43.8 | 44.6 |

in the Ukraine but now known to occur in the Gravettian of central Europe; many of the pieces illustrated by Otte (1980) can be matched almost exactly at La Cotte. It is worth recalling that Kostienki knives have been reported in deposits of Riss III age, more or less contemporary with those of La Cotte, at La Chaise de Vouthon (Debénath 1976b, 935; 1980). They are also present in the Mousterian of Bisitun (Iran) and of the Levant (Dibble 1984). Otte concluded from his study that the truncations may have served two purposes: (1) thinning of an extremity prior to hafting; (2) preparation of a chisel (or wedge) which, when used, produced scaling which might eventually mask the evidence of the initial stage of manufacture. The second of these brings to mind another parallel worth considering, this time for those examples from La Cotte which have *overstruck* dorsal scaling either from the butt or from an inverse truncation (e.g. Figs. 26.47, 6 and 7). Similar pieces found in the Howieson's Poort layers at Klasies River Mouth in South Africa (Singer and Wymer 1982, 104) have been described as a special unifacial form of *outil ecaillé*.

The La Cotte specimens show no sign of utilisation on the 'chisel' end, nor do the more typical pieces with restricted scaling exhibit the crushing which might be expected if they had been used violently. Of Otte's two suggestions, therefore, the former seems the more likely here. Although the percentage of inverse truncations bearing dorsal scaling, without LSF or burin scars as well, is at its lowest in layers A and 3, this is chiefly because resharpening is most common in these layers; the overall frequency of such pieces is actually at a maximum (3.9 and 4.5% of the real tool total). If hafting is the correct explanation for them, this (like resharpening by the LSF technique) is closely linked to variations in raw material frequency and artefact size. The practice of hafting (particularly when applied to the scrapers) would certainly have facilitated the use of smaller tools.

A third explanation of the character of the La Cotte pieces is provided by analogy with the work of Newcomer and Hivernel-Guerre (1974), who suggested that in the Upper Capsian of Gamble's Cave, Kenya, artefacts of the kind illustrated in Fig. 26.47 were the result of using flakes *as cores* (see their Fig. 1). Despite a time difference of the order of two hundred millenia the similarities between the Gamble's Cave and La Cotte specimens are remarkable. In the case of the former site obsidian was quarried elsewhere, suggesting that some economy of material was desirable, as at La Cotte. In the light of the model proposed by Rolland (1986) to explain certain aspects of Middle Palaeolithic variability, in which the 'Charentian' is seen as characterised by maximal exploitation of raw material (see chapter 32), it is interesting that one of the pieces illustrated by de Lumley-Woodyear (1971, Fig. 312) from the 'proto-

Ferrassie' site of La Baume les Peyrards, dated to the end of Riss III, should be identical to the overstrikes from La Cotte.

Whichever of the above proves to be the correct explanation for the inverse truncations and overstrikes — and a considerable amount of work remains to be done on these problems — it is clear that in the second half of the Saalian sequence at La Cotte a set of closely related techniques was devised, apparently to permit greater economy in the use of flint. These techniques achieve their greatest frequency in the assemblages from **A and** 3, when artefact size is at a minimum.

# 27

# ARTEFACTS MADE OF QUARTZ

F. Hivernel

## 27.1 INTRODUCTION

### 27.1.1 Problems Linked with the Study of Quartz as a Raw Material

The artefacts are made of several different qualities of quartz. A minority are made of a fine-grained variety with flaking properties almost indistinguishable from those of flint. The majority are of vein quartz, which in structure is either sugar-like or transected by marked cleavage planes. From a typological perspective this presents several difficulties. In particular, it is often impossible to orientate a piece whose ventral surface follows a cleavage plane, and difficult to identify burin removals. Looking at the light through some pieces, one can clearly see fracture lines that could result in burin-like removals either intentionally or by accident. As for the sugar-like quartz, it is quite likely that some of the discontinuous retouch and Clactonian notches were accidentally produced. A recent experimental study using various types of Australian quartz (Dickson 1977) demonstrated not only the unpredictable mode of fracture but also the difficulty of discerning conchoidal features on flakes of all but the finest variety. Therefore the degree of confidence attached to the qualitative and quantitative attributes of these artefacts will be lower than for flint.

### 27.1.2 Methodology

Of the 11929 pieces of quartz from the Saalian layers H to 5 the great majority have been classed as 'waste' (Table 27.1). This includes large numbers of splinters and chunks as well as true flakes. In view of the problems the material presents these were not included in the further analysis. Such technological and typological

**Table 27.1** Quartz Tools, Cores and Waste[a]

| Layer | Tools | | Cores | | Waste | | Total |
|-------|-------|------|-------|-----|-------|------|-------|
| | n | % | n | % | n | % | |
| 5 | 169 | 12.3 | 52 | 3.8 | 1150 | 83.9 | 1371 |
| A | 464 | 6.3 | 166 | 2.3 | 6678 | 91.4 | 7308 |
| B | 46 | 7.6 | 12 | 2.0 | 551 | 90.5 | 609 |
| C | 28 | 5.7 | 10 | 2.0 | 452 | 92.2 | 490 |
| D | 99 | 8.2 | 32 | 2.7 | 1071 | 89.1 | 1202 |
| E | 55 | 9.7 | 14 | 2.5 | 496 | 87.8 | 565 |
| F | 20 | 10.3 | 6 | 3.1 | 169 | 86.7 | 195 |
| G | 17 | 14.8 | 2 | 1.7 | 96 | 83.5 | 115 |
| H | 6 | 8.1 | 2 | 2.7 | 66 | 89.2 | 74 |
| Sample | 904 | 7.6 | 296 | 2.5 | 10729 | 89.8 | 11929 |

[a]Tools here comprise Bordes's types 1–63, handaxes and broken tools.

**Table 27.2** Quartz Cores

| Type | | | | | Layer | | | | | Total |
|------|------|-------|------|------|------|------|------|------|------|-------|
| | H | G | F | E | D | C | B | A | 5 | |
| Discoidal | 1 | – | 3 | 1 | 8 | – | 2 | 32 | 16 | 63 |
| % | 50.0 | – | 75.0 | 7.1 | 25.0 | – | 16.7 | 19.3 | 30.8 | 21.4 |
| Globular | – | – | – | 7 | 3 | 3 | 1 | 26 | 6 | 46 |
| % | – | – | – | 50.0 | 9.4 | 30.0 | 8.3 | 15.7 | 11.5 | 16.7 |
| Prismatic | – | – | – | – | – | – | – | 5 | – | 5 |
| % | – | – | – | – | – | – | – | 3.0 | – | 1.7 |
| Pyramidal | – | – | – | – | 1 | 1 | – | 2 | 1 | 5 |
| % | – | – | – | – | 3.1 | 10.0 | – | 1.2 | 1.9 | 1.7 |
| Miscellaneous | – | – | – | – | – | – | – | 11 | 3 | 14 |
| % | – | – | – | – | – | – | – | 6.6 | 5.8 | 4.8 |
| Shapeless | 1 | 2 | 1 | 6 | 20 | 6 | 9 | 90 | 26 | 161 |
| % | 50.0 | 100.0 | 25.0 | 42.9 | 62.5 | 60.0 | 75.0 | 54.2 | 50.0 | 54.8 |
| Sample | 2 | 2 | 4 | 14 | 32 | 10 | 12 | 166 | 52 | 294 |

**Table 27.3**  Quartz Tool Totals and Percentages[a]

| Index | | | | | Layer | | | | |
|---|---|---|---|---|---|---|---|---|---|
| | H | G | F | E | D | C | B | A | 5 |
| Real | 4 | 16 | 20 | 53 | 97 | 28 | 45 | 421 | 165 |
| % | 6.8 | 13.9 | 10.3 | 9.4 | 8.1 | 5.7 | 7.4 | 5.8 | 12.0 |
| Essential | 4 | 16 | 14 | 42 | 84 | 22 | 40 | 343 | 127 |
| % | 6.8 | 13.9 | 7.2 | 7.4 | 7.0 | 4.5 | 6.6 | 4.7 | 9.3 |
| Reduced | 4 | 16 | 14 | 42 | 84 | 22 | 40 | 343 | 126 |
| % | 6.8 | 13.9 | 7.2 | 7.4 | 7.0 | 4.5 | 6.6 | 4.7 | 9.2 |

[a]The percentages are based on the quartz artefacts only.

**Table 27.4**  Major Tool Types: Distribution by Layer (%)[a]

| Type | | | | | Layer | | | | |
|---|---|---|---|---|---|---|---|---|---|
| | H | G | F | E | D | C | B | A | 5 |
| Side-scrapers | – | – | 10.0 | 9.4 | 9.3 | 17.9 | 13.3 | 14.7 | 15.2 |
| End-scrapers | 1[a] | 12.5 | 5.0 | 11.2 | 17.5 | 10.7 | 15.6 | 13.4 | 10.4 |
| Burins | – | – | – | 9.4 | 2.1 | 10.7 | 8.9 | 5.5 | 1.2 |
| Borers | – | – | – | 3.8 | 1.0 | 3.6 | 2.2 | 2.1 | 1.8 |
| *Raclettes* | – | 18.8 | – | 3.8 | 2.1 | – | – | 1.7 | 1.2 |
| Truncated pieces | – | – | – | 3.8 | 8.2 | – | 4.4 | 3.6 | 4.2 |
| Notched pieces | 2[a] | 25.0 | 15.0 | 26.4 | 27.8 | 21.4 | 15.5 | 18.8 | 18.3 |
| Denticulates | – | 31.3 | 20.0 | 3.8 | 12.4 | 14.3 | 17.8 | 16.1 | 15.2 |
| Irreg. retouched pieces | – | – | 30.0 | 20.8 | 13.4 | 21.4 | 11.1 | 18.5 | 22.6 |
| Sample (tools) | 4 | 16 | 20 | 53 | 96 | 28 | 45 | 421 | 163 |
| Handaxes | 1[a] | – | – | – | 2[a] | – | – | 8[a] | 3[a] |

[a]Raw counts are given for the very poor assemblage from layer H and for the handaxes.

**Table 27.5**  Quartz Side-scrapers: Distribution by Layer

| Type | | | | Layer | | | | | Total | % |
|---|---|---|---|---|---|---|---|---|---|---|
| | | F | E | D | C | B | A | 5 | | |
| 9 | Single straight | 1 | – | – | – | 1 | 4 | 2 | 8 | 7.1 |
| 10 | " convex | – | – | 4 | 1 | 1 | 15 | 7 | 28 | 24.8 |
| 11 | " concave | – | 1 | 1 | 1 | 1 | 5 | 1 | 10 | 8.8 |
| 13 | Double/convex | – | – | – | – | – | 1 | 2 | 3 | 2.7 |
| 17 | " concave/convex | – | – | 1 | – | – | – | – | 1 | 0.9 |
| 19 | Convergent convex | – | – | – | – | – | – | 1 | 1 | 0.9 |
| 21 | *Déjeté* | 1 | – | – | – | – | 1 | – | 2 | 1.8 |
| 22 | Transverse straight | – | – | – | 1 | – | 1 | – | 2 | 1.8 |
| 23 | " convex | – | 1 | – | – | 1 | 5 | 3 | 10 | 8.8 |
| 24 | " concave | – | – | – | – | – | 1 | – | 1 | 0.9 |
| 25 | With ventral retouch | – | 2 | 1 | 1 | 1 | 13 | 5 | 23 | 20.4 |
| 26 | With abrupt retouch | – | – | 2 | 1 | 1 | 13 | 4 | 21 | 18.6 |
| 27 | With thinned back | – | – | – | – | – | 2 | – | 2 | 1.8 |
| 29 | With alternate retouch | – | – | – | – | – | 1 | – | 1 | 0.9 |

observations as can be made are therefore based on the cores and the tools. They have been classified according to the Bordes typology (1961). Measurements are of length (L) and breadth (B) in millimetres, and of weight (W) in grammes, and, for the bifaces, thickness (T) in millimetres. The curvature (C) of the working edges of side-scrapers and end-scrapers and of notches was measured according to the criteria used by Girard (1978; this volume, section 23.2.2), and the values given here are of the mean and the standard deviation.

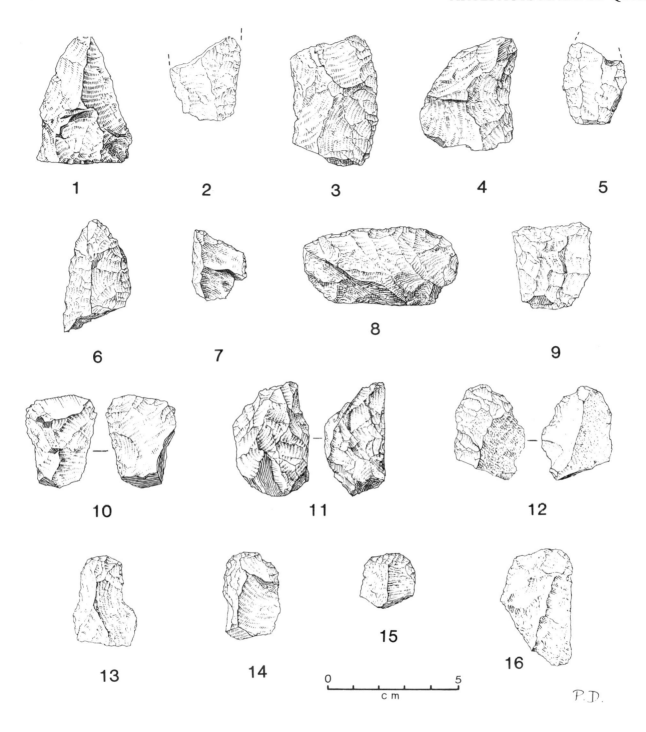

**Fig. 27.1** Quartz tools. 1 Levallois point; 2–12 side-scrapers: 2 single straight; 3 single convex; 4 single concave; 5 double bi-convex; 6 convergent bi-convex; 7 *déjeté*; 8 transverse convex; 9 transverse concave; 10 on ventral surface; 11 with abrupt retouch; 12 with thinned back; 13–16 typical end-scrapers.

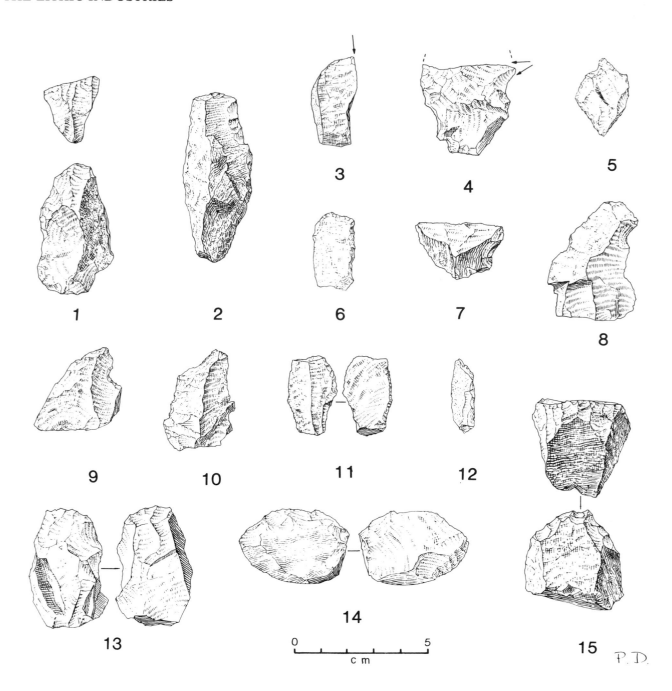

**Fig. 27.2** Quartz tools. 1 and 2 Typical end-scrapers; 3 and 4 typical burins; 5 awl; 6 *raclette*; 7 *tranchet*; 8 and 9 notched pieces; 10 denticulate; 11 retouched on the ventral surface; 12 thin abrupt retouch; 13 *hachoir*; 14 chopper; 15 *rabot*.

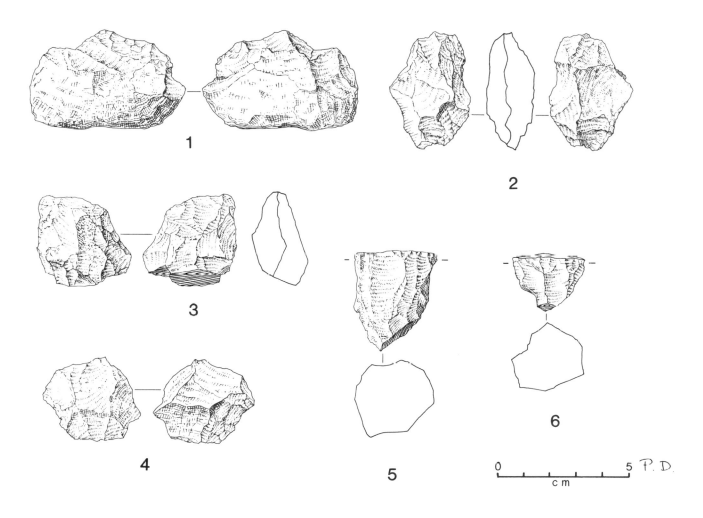

**Fig. 27.3** Quartz tools and cores. 1 Chopping tool; 2 core-like biface; 3 partial biface; 4 discoidal core; 5 prismatic core; 6 pyramidal core.

## 27.2 GENERAL CHARACTER OF THE INDUSTRIES

The frequencies of the various core types are given in Table 27.2. Allowing for sampling problems in the poorer layers, the distribution is quite uniform. Shapeless (*informe*) cores predominate, followed by discoidal and globular; Levallois cores are totally absent, as might be expected from the intractability of the raw material.

A selection of the tools is illustrated in Figs. 27.1, 27.2 and 27.3 to give a general picture of the material. They are ordered typologically and come mainly from layer A. The paucity of tools in most layers (only layers D, A, and 5 include most of the principal types of the Bordes list) restricts the possibility of interpreting variations in terms of cultural difference or activity facies. For example, while the side-scraper category is well represented in most layers, it is totally absent in layers H and G, where, however, end-scrapers occur. Notched and lightly retouched pieces, on the other hand, are well represented throughout.

Side-scrapers of types 12, 14–16, 18, 20 and 28 do not occur at all. Of the other types (Table 27.5) the most common is the single convex (24.1%), followed by the ventrally retouched (20.7%), the abruptly retouched (18.9%) and the transverse with convex edge (8.6%). Single greatly outnumber double side-scrapers in all layers. End-scrapers are much more numerous in the quartz than in the flint assemblages, although they diminish with time as side-scrapers increase, and are responsible for the anomalously high 'Upper Palaeolithic' index (Table 27.7).

Handaxes (bifaces) are present in the three richest layers (D, A and 5, described below), although in very small numbers. In addition, one was found in layer H (which has only four tools), but in view of its 'core-like' character, and the absence of 'true' handaxes among the flint artefacts from the lower layers, its authenticity is in some doubt. The one example of Levallois technique is a Levallois point, from layer E (Fig. 27.1, 1).

319

**Table 27.6**  Quartz Tools from Layers D, A and 5: Typological Counts

| Type | | N | Real | D Ess. | Red. | N | Real | A Ess. | Red. | N | Real | 5 Ess. | Red. |
|---|---|---|---|---|---|---|---|---|---|---|---|---|---|
| 9 | Single straight side-scrapers | 0 | 0.00 | 0.00 | 0.00 | 4 | 0.95 | 1.17 | 1.17 | 2 | 1.23 | 1.57 | 1.59 |
| 10 | Single convex side-scrapers | 4 | 4.12 | 4.76 | 4.76 | 15 | 3.56 | 4.37 | 4.37 | 7 | 4.29 | 5.51 | 5.56 |
| 11 | Single concave side-scrapers | 1 | 1.03 | 1.19 | 1.19 | 5 | 1.19 | 1.46 | 1.46 | 1 | 0.61 | 0.79 | 0.79 |
| 13 | Double straight-convex side-scrapers | 0 | 0.00 | 0.00 | 0.00 | 1 | 0.24 | 0.29 | 0.29 | 2 | 1.23 | 1.57 | 1.59 |
| 17 | Double concave-convex side-scrapers | 1 | 1.03 | 1.19 | 1.19 | 0 | 0.00 | 0.00 | 0.00 | 0 | 0.00 | 0.00 | 0.00 |
| 19 | Convergent convex side-scrapers | 0 | 0.00 | 0.00 | 0.00 | 0 | 0.00 | 0.00 | 0.00 | 1 | 0.61 | 0.79 | 0.79 |
| 21 | *Déjeté* (offset) scrapers | 0 | 0.00 | 0.00 | 0.00 | 1 | 0.24 | 0.29 | 0.29 | 0 | 0.00 | 0.00 | 0.00 |
| 22 | Straight transverse side-scrapers | 0 | 0.00 | 0.00 | 0.00 | 1 | 0.24 | 0.29 | 0.29 | 0 | 0.00 | 0.00 | 0.00 |
| 23 | Convex transverse side-scrapers | 0 | 0.00 | 0.00 | 0.00 | 5 | 1.19 | 1.46 | 1.46 | 3 | 1.84 | 2.36 | 2.38 |
| 24 | Concave transverse side-scrapers | 0 | 0.00 | 0.00 | 0.00 | 1 | 0.24 | 0.29 | 0.29 | 0 | 0.00 | 0.00 | 0.00 |
| 25 | Side-scrapers on ventral face | 1 | 1.03 | 1.19 | 1.19 | 13 | 3.09 | 3.79 | 3.79 | 5 | 3.07 | 3.94 | 3.97 |
| 26 | Abrupt retouched side-scrapers | 2 | 2.06 | 2.38 | 2.38 | 13 | 3.09 | 3.79 | 3.79 | 4 | 2.45 | 3.15 | 3.17 |
| 27 | Side-scrapers with thinned back | 0 | 0.00 | 0.00 | 0.00 | 2 | 0.48 | 0.58 | 0.58 | 0 | 0.00 | 0.00 | 0.00 |
| 29 | Alternate retouched side-scrapers | 0 | 0.00 | 0.00 | 0.00 | 1 | 0.24 | 0.29 | 0.29 | 0 | 0.00 | 0.00 | 0.00 |
| 30 | Typical end-scrapers | 11 | 11.34 | 13.10 | 13.10 | 45 | 10.69 | 13.12 | 13.12 | 8 | 4.91 | 6.30 | 6.35 |
| 31 | Atypical end-scrapers | 6 | 6.19 | 7.14 | 7.14 | 12 | 2.85 | 3.50 | 3.50 | 9 | 5.52 | 7.09 | 7.14 |
| 32 | Typical burins | 2 | 2.06 | 2.38 | 2.38 | 11 | 2.61 | 3.21 | 3.21 | 2 | 1.23 | 1.57 | 1.59 |
| 33 | Atypical burins | 0 | 0.00 | 0.00 | 0.00 | 12 | 2.85 | 3.50 | 3.50 | 0 | 0.00 | 0.00 | 0.00 |
| 34 | Typical borers | 1 | 1.03 | 1.19 | 1.19 | 7 | 1.66 | 2.04 | 2.04 | 2 | 1.23 | 1.57 | 1.59 |
| 35 | Atypical borers | 0 | 0.00 | 0.00 | 0.00 | 2 | 0.48 | 0.58 | 0.58 | 1 | 0.61 | 0.79 | 0.79 |
| 38 | Naturally backed knives | 0 | 0.00 | 0.00 | – | 0 | 0.00 | 0.00 | – | 1 | 0.61 | 0.79 | – |
| 39 | *Raclettes* | 2 | 2.06 | 2.38 | 2.38 | 7 | 1.66 | 2.04 | 2.04 | 2 | 1.23 | 1.57 | 1.59 |
| 40 | Truncated pieces | 8 | 8.25 | 9.52 | 9.52 | 15 | 3.56 | 4.37 | 4.37 | 7 | 4.29 | 5.51 | 5.56 |
| 41 | Mousterian tranchet | 0 | 0.00 | 0.00 | 0.00 | 0 | 0.00 | 0.00 | 0.00 | 1 | 0.61 | 0.79 | 0.79 |
| 42 | Notched pieces | 27 | 27.84 | 32.14 | 32.14 | 79 | 18.76 | 23.03 | 23.03 | 30 | 18.40 | 23.62 | 23.81 |
| 43 | Denticulates | 12 | 12.37 | 14.29 | 14.29 | 68 | 16.15 | 19.83 | 19.83 | 25 | 15.34 | 19.69 | 19.84 |
| 44 | *Becs burinants alternes* | 0 | 0.00 | 0.00 | 0.00 | 1 | 0.24 | 0.29 | 0.29 | 2 | 1.23 | 1.57 | 1.59 |
| 45 | Retouched on ventral face | 0 | 0.00 | – | – | 10 | 2.38 | – | – | 4 | 2.45 | – | – |
| 46-47 | Abrupt and alternate retouch (thick) | 3 | 3.09 | – | – | 0 | 0.00 | – | – | 11 | 6.75 | – | – |
| 48-49 | Abrupt and alternate retouch (thin) | 10 | 10.31 | – | – | 67 | 15.91 | – | – | 21 | 12.88 | – | – |
| 50 | Bifacial retouch | 0 | 0.00 | – | – | 1 | 0.24 | – | – | 0 | 0.00 | – | – |
| 55 | *Hachoirs* | 1 | 1.03 | 1.19 | 1.19 | 2 | 0.48 | 0.58 | 0.58 | 4 | 2.45 | 3.15 | 3.17 |
| 56 | *Rabots* (push-planes) | 0 | 0.00 | 0.00 | 0.00 | 4 | 0.95 | 1.17 | 1.17 | 4 | 2.45 | 3.15 | 3.17 |
| 59 | Choppers | 0 | 0.00 | 0.00 | 0.00 | 2 | 0.48 | 0.58 | 0.58 | 0 | 0.00 | 0.00 | 0.00 |
| 61 | Chopping tools | 0 | 0.00 | 0.00 | 0.00 | 1 | 0.24 | 0.29 | 0.29 | 0 | 0.00 | 0.00 | 0.00 |
| 62 | Miscellaneous | 5 | 5.15 | 5.95 | 5.95 | 13 | 3.09 | 3.79 | 3.79 | 4 | 2.45 | 3.15 | 3.17 |
| Bifaces | | 2 | | | | 8 | | | | 3 | | | |
| Totals | | | 97 | 84 | 84 | | 421 | 343 | 343 | | 163 | 127 | 126 |

## 27.3  THE TOOLS FROM THE MAJOR ASSEMBLAGES

Only the tools from layers that yielded ninety or more — D, A and 5 — are described here. Their metrical attributes, and those of the tools from the poorer layers, are more fully tabulated in microfiche (mf 27.1–27.10). The frequencies of tools by layer on which the three sets of Bordes indices are computed are given in Table 27.3 and the cumulative frequency curves for layers D, A and 5 in Fig. 27.4. The distributions by layer of tools according to Bordes's 'real' index are shown in Table 27.4, and of the different side-scraper types in Table 27.5. In accordance with the practice adopted elsewhere in this volume, the curvature of concave side-scrapers is recorded as negative.

## *LAYER D* (96)

**Side-scrapers** (9). Of the five single lateral scrapers four are convex (C = 0.21 ± 0.06), with scalar retouch, and one concave (C = 0.08), with stepped retouch associated with a distal truncation. The one double scraper has scalar retouch on the concave (C = -0.08) and stepped retouch on the convex (C = 0.09) edge. Other side-scrapers (3) comprise one with thin scalar retouch on the ventral surface, and two with stepped abrupt retouch. All are convex (C = 0.13 ± 0.05). The edge curvature for all side-scrapers is 0.12 ± 0.12.

**End-scrapers** (17). Eleven are typical, six atypical. In either case the majority (10 altogether) are made on raw rather than retouched flakes, but three of the typical are on retouched flakes. There are also two of the carinated, one of the denticulate and one of the nosed type. Eight are axial; seven are offset, six to the left and one to the right; two could not be orientated. The mean curvature of the working edge is 0.25 ± 0.05.

**Table 27.7**  Quartz Tools from Layers D, A and 5: Typological Indices

| Index | | Real | D<br>Ess. | Red. | Real | A<br>Ess. | Red. | Real | 5<br>Ess. | Red. |
|---|---|---|---|---|---|---|---|---|---|---|
| ILty | Typological Levallois | 0.00 | – | – | 0.00 | – | – | 0.00 | – | – |
| IR | Side-scrapers | 9.28 | 10.71 | 10.71 | 14.73 | 18.08 | 18.08 | 15.34 | 19.69 | 19.84 |
| IC | Charentian | 4.12 | 4.76 | 4.76 | 5.23 | 6.41 | 6.41 | 6.13 | 7.87 | 7.94 |
| IAu | Acheulian unifacial | 0.00 | 0.00 | 0.00 | 0.00 | 0.00 | 0.00 | 0.00 | 0.00 | 0.00 |
| IAt | Acheulian total | 2.02 | 2.33 | 2.33 | 1.86 | 2.28 | 2.28 | 1.81 | 2.31 | 2.33 |
| IB | Bifaces | 2.02 | 2.33 | 2.33 | 1.86 | 2.28 | 2.28 | 1.81 | 2.31 | 2.33 |
| I | Levallois | 0.00 | 0.00 | 0.00 | 0.00 | 0.00 | 0.00 | 0.00 | 0.00 | 0.00 |
| II | Mousterian | 9.28 | 10.71 | 10.71 | 14.73 | 18.08 | 18.08 | 15.34 | 19.69 | 19.84 |
| III | Upper Palaeolithic | 28.87 | 33.33 | 33.33 | 24.70 | 30.32 | 30.32 | 17.79 | 22.83 | 23.02 |
| IV | Denticulate | 12.37 | 14.29 | 14.29 | 16.15 | 19.83 | 19.83 | 15.34 | 19.69 | 19.84 |
| 5 | Pseudo-Levallois points | 0.00 | 0.00 | – | 0.00 | 0.00 | – | 0.00 | 0.00 | – |
| 38 | Naturally backed knives | 0.00 | 0.00 | – | 0.00 | 0.00 | – | 0.61 | 0.79 | – |
| I rc | Convergent retouched | 0.00 | 0.00 | 0.00 | 0.24 | 0.29 | 0.29 | 0.61 | 0.79 | 0.79 |
| 6-7 | Mousterian points | 0.00 | 0.00 | 0.00 | 0.00 | 0.00 | 0.00 | 0.00 | 0.00 | 0.00 |
| Rsingle | Side-scrapers, single | 5.15 | 5.95 | 5.95 | 5.70 | 7.00 | 7.00 | 6.13 | 7.87 | 7.94 |
| Rdouble | Side-scrapers, double | 1.03 | 1.19 | 1.19 | 0.24 | 0.29 | 0.29 | 1.23 | 1.57 | 1.59 |
| Rtransv | Side-scrapers, transv. | 0.00 | 0.00 | 0.00 | 1.66 | 2.04 | 2.04 | 1.84 | 2.36 | 2.38 |
| Rinv | Side-scrapers, inverse/alternate | 1.03 | 1.19 | 1.19 | 3.33 | 4.08 | 4.08 | 3.07 | 3.94 | 3.97 |
| 30-31 | End-scrapers | 17.53 | 20.24 | 20.24 | 13.54 | 16.62 | 16.62 | 10.43 | 13.39 | 13.49 |
| 32-33 | Burins | 2.06 | 2.38 | 2.38 | 5.46 | 6.71 | 6.71 | 1.23 | 1.57 | 1.59 |
| 42 | Notched pieces | 27.84 | 32.14 | 32.14 | 18.76 | 23.03 | 23.03 | 18.40 | 23.62 | 23.81 |
| 42-43 | Notched pieces & denticulates | 40.21 | 46.43 | 46.43 | 34.92 | 42.86 | 42.86 | 33.74 | 43.31 | 43.65 |
| Ident | Extended denticulate group | 40.21 | 46.43 | 46.43 | 34.92 | 42.86 | 42.86 | 33.74 | 43.31 | 43.65 |
| Iirr | Irregularly retouched | 13.40 | – | – | 18.53 | – | – | 22.09 | – | – |

**Burins** (2). Both are typical. One is an axial dihedral burin, the other an angle burin on a break.

The one **borer** (typical) is formed by a notch on the left and retouch on the right edge.

**Raclettes** (2). One has semi-abrupt distal retouch; the other is retouched on the left edge.

**Truncated pieces** (8). The truncation is direct in all cases, convex in seven, concave in one. It is axial on five, oblique (and to the left) on three.

**Notched pieces** (27). Retouched notches (19) are more common than Clactonian, fifteen of them direct and four inverse; thirteen are on one or other lateral edge and six are distal. Four pieces have direct Clactonian notches, and one inverse. On three others the notches are multiple or mixed. On two of these they are alternating (i.e. direct and inverse on the same edge); in one case the direct notch is Clactonian and the inverse retouched; in the other both notches are Clactonian. The third piece has an inverse notch on each edge, one retouched and the other Clactonian. The mean curvature of both retouched and Clactonian notches is 0.19 ± 0.02.

**Denticulates** (12). On four the denticulation is made by retouched notches, one direct and three inverse (one on a transverse edge). Two are microdenticulates, one direct and one inverse. The six Clactonian notches are equally divided between direct and inverse.

**Irregularly retouched pieces** (13) comprise three with thick abrupt retouch, inverse on one, more or less continuous round the edges of the two others, and ten with thin abrupt retouch, which is direct in eight cases.

One **hachoir**, with a retouched or utilised cutting edge, was found. Five fragments of flakes with scraper-like retouch could not be orientated and were classed as **miscellaneous**.

**Bifaces** (2). Both are partial, one whole and one incomplete. The dimensions of the former are L = 81, B = 41, T = 45.

## *LAYER A* (421)

**Side-scrapers** (62). Of the 24 single lateral scrapers, 4 are straight, 15 convex and 5 concave. The retouch is scalar on one of the straight, invasive on another, and stepped on the other two, as it is on all the concave. It is mainly scalar on all the convex pieces. Five of the latter and one concave piece are cortically backed. The curvature of the convex edges is 0.20 ± 0.08, of the concave 0.09 ± 0.04, and of all combined 0.11 ± 0.14. The transverse scrapers (7) comprise five with convex, one with straight and one with concave edges. Retouch is scalar on three of the convex and on the straight, stepped on two of

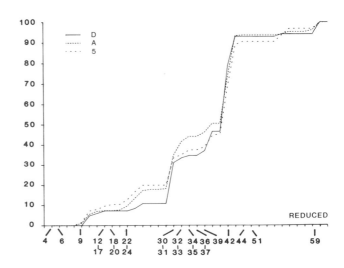

**Fig. 27.4**  Layers D, A and 5: cumulative curves (reduced) for quartz tools.

321

the convex and on the concave. The curvature of the convex edges is 0.18 ± 0.03. For all transverse side-scrapers combined it is 0.11 ± 0.11. Both double scrapers are biconvex, with a curvature of 0.14 ± 0.03. One is *déjeté*, with scalar retouch. On the other the retouch is stepped.

There are 29 other side-scrapers. The inversely and abruptly retouched types are well represented with 13 pieces each; two of the former are cortically backed. All the edges are convex; the curvature of the inverse examples is 0.14 ± 0.04, of the abrupt 0.17 ± 0.06. There are two pieces with thinned back and one with alternating retouch; one of the former and seven of the inversely retouched pieces show scalar retouch. On the others, and on all the abruptly retouched, it is stepped. The mean curvature for the side-scrapers in this group is 0.16 ± 0.05. For all types of side-scraper in layer A it is 0.13 ± 0.10.

**End-scrapers** (57; 45 typical, 12 atypical). Nineteen (40.4%) are made on retouched flakes, a further twelve (21.0%) on raw flakes. There are also nine (15.8%) thumbnail, eight (14.0%) carinated, two nosed, two shouldered, and one denticulated. Most (35) are axial, against 14 offset to the left and 7 to the right. The curvature (0.25 ± 0.06) is a little higher than for the flint end-scrapers (0.20) in this layer, while in both materials the length is appreciably less than previously, that of the quartz scrapers falling to 31.1 mm from 43.4 in layer D.

**Burins** (23; 11 typical, 12 atypical). All are angle burins. Six of the typical and three of the atypical are made on a convex truncation, four of each on a break, and one typical and two atypical on the ventral surface. Three of the atypical are dihedral.

**Borers** (9; 7 typical, 2 atypical). Five of the typical are made on a truncation associated with retouch on the ventral surface; one is made by alternating and one by direct retouch. Four are axial. The two atypical forms are offset, one made by direct retouch, the other by a notch.

**Raclettes** (7). Thin direct abrupt retouch can be seen on five of these, sometimes discontinuous; on two the retouch is ventral.

**Truncated pieces** (15). Fourteen are made by direct retouch; of these, five are convex, four concave, two straight and one is convex-concave. The two others are oblique, one to the right and one to the left. The inverse truncation is oblique and to the right.

**Notched pieces** (79). The retouched notches are in the great majority (81.0%) and are mainly direct (64), as are 14 of the 15 Clactonian. The retouched notches are rather less curved (0.21 ± 0.06) than the latter (0.25); for all together, it is 0.21 ± 0.06.

**Denticulates** (68). The majority (50) are made by retouched notches, direct on 44, inverse on 5, and alternate on one. Twelve are worked on the transverse edge. The denticulation made by Clactonian notches is direct on seven pieces. Eleven are microdenticulates, also made by direct retouch, which is in most cases discontinuous.

**Irregularly retouched pieces** (78). There is a marked increase in these, above all in those with thin abrupt retouch (55). There are ten ventrally worked pieces with semi-abrupt, often discontinuous, retouch, and twelve with alternate semi-abrupt retouch. They are made on widely different types of blanks, no one type being preferred. There is one bifacially retouched piece.

The rest of the assemblage consists of a **bec burinant alterne** made by two notches on the left; two **hachoirs**, one made on a flake shaped by flat direct and the other by alternating retouch on the transverse edge; four **rabots**, all made on thick flakes, with semi-abrupt fronts from 14 to 28 mm in height made by subparallel retouch, their curvatures ranging from 0.11 to 0.40; two small **choppers** on cortical flakes, and a **chopping tool** on a cobble. The 13 assigned to the **miscellaneous** category include eight side-scrapers that could not be orientated, two *becs* with indeterminate retouch, one *pièce esquillée*, a composite denticulate borer, and a disc with invasive retouch.

There are 34 broken pieces, 27 of which have semi-abrupt, side-scraper retouch; also two pieces with bifacial and two with inverse retouch.

**Bifaces** (8). Five are partial, two 'core-like' and one miscellaneous. They range in length from 42 to 94 mm, in breadth from 28 to 54 mm, and in thickness from 21 to 37 mm.

## LAYER 5 (163)

**Side-scrapers** (25). Seven of the ten single lateral scrapers are convex, with mixed scalar and stepped retouch, two are straight, one with scalar, the other with stepped retouch, and one is concave, with stepped retouch. For the convex edges the curvature is 0.13 ± 0.04; for all combined (concave: -0.14) it is 0.09 ± 0.08. The three transverse scrapers are convex, with mixed scalar and stepped retouch and an edge curvature of 0.14 ± 0.06. The three double scrapers are convex, one convergent with scalar retouch, the others with stepped retouch. Their edge curvature is 0.13 ± 0.02.

The other side-scrapers (9) consist of five ventrally worked pieces, four convex and one concave, on three of which the retouch is scalar, on two stepped, and four abruptly retouched pieces, all convex. The convex edges together have a curvature of 0.15 ± 0.06; with the concave (-0.05) the mean is 0.13 ± 0.09. The combined curvature for all side-scrapers in layer 5 is 0.12 ± 0.07.

**End-scrapers** (17; 8 typical, 9 atypical). Five are made on retouched flakes, three on raw flakes; there are also four of the nosed, one of the carinated and one of the shouldered type. All are axial. The higher mean curvature (0.30 ± 0.08) is due to the unusually high value for the atypical scrapers (0.34 ± 0.09); the mean for the typical alone (0.25 ± 0.04) is the same as that for both together in the other two layers.

**Burins** (2). Both are typical angle burins on a break.

**Borers** (3). Two are typical (both axial), one made by a direct retouched notch on the left and a direct semi-abrupt notch on the right, the other by a Clactonian notch and semi-abrupt direct retouch. The atypical borer is made by a direct retouched notch opposed to the straight edge of the flake.

**Raclettes** (2). The retouch is abrupt and discontinuous on both.

**Truncated pieces** (7). All are distal. Six are direct and one is inverse. Four of the former are straight; two are oblique, both to the left, and have retouch on the right edge also, which is abrupt and inverse on one piece, direct on the other. Two are concave, as is the only inverse truncation.

**Notched pieces** (30). Only two retouched (which are in the great majority, 24) and one of the four Clactonian notches are inverse; the rest are direct. A further two pieces have two adjacent notches on the same edge, both retouched on one, one Clactonian and one retouched on the other. The curvature for both types of notch together is 0.23 ± 0.06 (retouched only: 0.22 ± 0.06; Clactonian only: 0.25 ± 0.08).

**Denticulates** (25). None are made by Clactonian notches. Five are microdenticulates, three made by direct and two by inverse retouch. Of the twenty made by retouched notches, thirteen are direct and seven inverse.

**Irregularly retouched pieces** (36). As a proportion of the assemblage these continue to rise, now to 22.6%. As in layer A, this is due largely to the greater numbers of thin abruptly retouched pieces (21). On eighteen of these the working is direct, on three inverse. There are also eleven thick abruptly retouched pieces, made on a wide variety of blanks. The often discontinuous retouch is direct on ten — on both edges in two cases — and inverse on one. A further four pieces show semi-abrupt retouch on the ventral surface, on the transverse edge of one, and on both edges of another.

Among other types present in very small numbers, two appear for the first time. One is a naturally backed knife made on a large pointed flake with traces of use on the cutting edge; the other is a **Mousterian** *tranchet* resembling a transverse arrow head, with both sides obliquely truncated and showing traces of wear on its transverse edge. There are also two **becs burinants alternes** — each with a broken tip — made by adjacent notches, direct in one case, alternating in the other; four **hachoirs**, two made by direct and two by inverse retouch, and four **rabots**, the fronts of which are very abrupt and made by subparallel retouch; their curvature values range from 0.13 to 0.38 and their heights from 14 to 24 mm.

Of the **miscellaneous pieces** (4), two show side-scraper retouch, but could not be orientated; one is a *pièce esquillée*; one is a point made by direct semi-abrupt retouch on the left edge and a large retouched notch on the right.

All three **bifaces** are 'partial'. They range in length from 7 to 13 mm, in breadth from 3.4 to 7.2 mm and in thickness from 2.3 to 5.4 mm.

## 27.4 DISCUSSION

### 27.4.1 Comparative Frequencies of Tool Types

The types that are well enough represented for any possibly significant trend of change through time to be detected are side-scrapers and end-scrapers, notches and denticulates, and irregularly retouched pieces (Table 27.4). The wider spectrum of tool types in layer A could be due either to cultural factors or to the greater number of artefacts it contains. Side-scrapers rise as end-scrapers fall, though not quite reciprocally. After an initial increase from 9.3% in D to 14.7% in A, they rise by only 0.5%, to 15.2, in layer 5, whereas end-scrapers start at 17.5%, decrease less steeply in A to 13.4%, but then fall to 10.4% in 5.

Rather similarly, denticulates rise as notched pieces fall, except that, comparatively, their rise (from 12.4% in D to 15.2 in layer 5) is smaller and the fall in notched pieces greater (from 27.8 to 18.4%).

The irregularly retouched pieces rise the most steadily — from 13.4% in layer D to 18.5% in layer A and to 22.6% in layer 5. But on these pieces the retouch is so often discontinuous that it is not always possible to be sure that it was not naturally or accidentally caused.

Although burins are always too few for much importance to be attached to fluctuations in their relative frequency, it is worth noting that, after rising from 2.1%

in layer D to 5.5% in the more diversified and numerous assemblage of layer A, they fall to 1.2% in layer 5, although in this layer the percentage of quartz tools to waste rises to 12.3% (from 6.3% in A) and the proportion of all quartz to all flint (40%, from 18 in A) is at its highest.

Handaxes occur in about the same proportion to other tools in all three layers: two in D, eight in A and three in 5. But as all are of the 'core-like', 'partial', or 'miscellaneous' types they are of little value for comparative purposes.

### 27.4.2 Metrical and Technological Features

**Size variation.** The mean length, breadth and weight of the grouped cores and tools are given in Table 27.8. The values for individual types are given in microfiche (mf 27.5, 8 and 10).

There is a fall in size in both cores and tools from layer D to A, and thereafter a rise in layer 5. Whereas in the lower layers globular cores are the largest, in layer 5 there is a great increase in the size of discoidal and shapeless cores, although the large standard deviations bear witness to the great variation within categories.

As a group, the tools vary in much the same way as the cores. In layer A they are very much smaller than in either of the other two layers and in layer 5 generally larger. But, as with the cores, the increase in length and breadth in layer 5 is outstripped by the increase in weight, indicating the same trend towards greater thickness that is observed in the flint tools. Of the more numerous types in these layers, only side-scrapers depart at all significantly from this pattern, the changes in size and weight from layer to layer being less regular and less pronounced. In layer A they are only 1.9 mm shorter and 0.2 mm narrower than in layer D, and yet a little heavier, even — by 1.3 g. In layer 5 they are no longer than in D, and only 1 mm broader, but 8.5 g heavier. End-scrapers conform quite closely on the other hand, with a marked diminution in size in layer A particularly in length (43 mm in D, 31 in A, 42 in 5) and to a lesser

**Table 27.8** Overall Dimensions of Quartz Tools and Cores in Layers D, A and 5

| Measurement | Layer D | A | 5 |
|---|---|---|---|
| **TOOLS** | | | |
| Length | 40.00 | 35.06 | 41.64 |
| Breadth | 26.56 | 24.41 | 28.87 |
| Weight | 17.80 | 13.85 | 32.42 |
| Sample | 96 | 421 | 163 |
| **CORES** | | | |
| Length | 45.31 | 40.77 | 59.61 |
| Breadth | 34.94 | 32.60 | 44.71 |
| Weight | 48.30 | 43.87 | 85.90 |
| Sample | 32 | 165 | 51 |

extent in breadth (28 mm in D, 23.5 in A, 27 in 5). They are a little shorter and narrower in 5 than in D, but, again, considerably heavier (21 g in D, 11 in A, 27 in 5).

**Butts.** The great majority of these are plain (Table 27.9). Facetting is poorly represented in all layers. Its slight rise in layer A may echo its much greater increase among the flint artefacts but is negligible in comparison and serves rather to illustrate the essential difference in the treatment of the two materials. Dihedral butts rise a little more steeply from D to A. The fall in cortical butts in layer 5 might indicate some change of habit in the collecting or knapping of quartz, such as occurs in the flint, but on its own is too small to be of much interest.

**Curvature.** The only detectable trend is in the rising curvature of notches (0.19 in D, 0.21 in A and 0.23 in 5), coinciding in A with a fall in the size of their blanks and in 5 with a rise. The curvature of end-scrapers, on the other hand, shows no change in A (despite a considerable reduction in the size of the pieces), and rises in layer 5 (to 0.30 from 0.25) when the blanks are again consistently larger.

**Retouch.** On the side-scrapers, retouch is either scalar or stepped, with a slight preponderance of the latter in layer A. Among the notched and denticulated pieces (combined) there is a steady rise of retouched notching at the expense of the Clactonian type, from 62.2% in D to 83.8% in A and 90.4% in 5. Taking all types of tools into account, there is also a rise in the proportion of direct to inverse retouch through these layers, from about twice as much in D to nearly three times as much in A and 5.

The principal differences between the quartz artefacts and those of flint (and stone) are described in subsection 24.3.3.

Table 27.9  Distribution of Butt Types on Quartz Artefacts, by Layer (%)

| Butt type | Layer | | |
|---|---|---|---|
| | D | A | 5 |
| Cortical[a] | 9.6 | 9.6 | 5.3 |
| Plain | 80.8 | 75.9 | 81.6 |
| Facetted | 2.7 | 4.0 | 3.9 |
| Dihedral | 6.8 | 10.4 | 9.2 |

[a]Here, 'cortex' is used to mean the contact zone between the quartz vein and the containing granite, and does not imply surface alteration.

# ARTEFACTS MADE OF OTHER LITHIC MATERIALS

P. Callow

## 28.1 INTRODUCTION

The artefacts made of flint and quartz have been described in the immediately preceding chapters. Because of their relatively small numbers, those remaining are treated here collectively under the heading 'stone', although several different types of hard rock with different flaking properties were used. As described in chapter 22, these include the locally obtainable basic igneous rocks (henceforward referred to as dolerite irrespective of their exact identity), and siltstones, quartzites and sandstones of local or exotic origin. Granite appears to have been flaked only very occasionally; no formal tools have been identified. The relative frequencies of the various raw materials are listed in Tables 22.3 and 22.4.

The only assemblage which has a large enough sample of stone artefacts to give reliable statistics is that from layer A, which is therefore examined in detail here. The much smaller sample from layer 5 is also described and illustrated because it possesses features of particular interest. The other layers yielded too few stone artefacts to be worth describing fully, though they are referred to in the general discussion. Statistical data are given in mf 28.1. The Bordes typology has been used throughout, and for the cleavers reference is made to Tixier's (1956) also.

It should be borne in mind that the raw materials vary not only petrologically but also in the form and manner in which they were obtained — notably as beach pebbles, thermally shattered fragments, or direct debitage from outcrops. Detailed analysis of raw material preferences in respect of particular tool types has not been attempted as, even for layer A, the samples are too small for the number of categories involved. For much

the same reasons, an exhaustive study of the debitage and cores was not undertaken.

## 28.2 TECHNOLOGY

No attempt has been made to compute the blade index (Ilam) because it would be misleading, though blades do occur (e.g. Fig. 28.3,2). Some pieces have been too badly corroded to be orientated with certainty; also the likelihood of successful blade removal is in part a function of the raw material itself — in some types strong cleavage planes are present. Because of chemical damage suffered by the basic rocks in particular, only two facetting indices (IF and IFs) have been computed; the distinction between facetted and polyhedral butts cannot be made on many of the pieces.

### 28.2.1 Layer A

The technological indices for the stone component of this assemblage (and of layer 5 also) are given in Table 28.1. Of the 2245 whole flakes and blades, only 53 are Levallois (IL = 2.4); ten more are pseudo-Levallois points. There are only four *couteaux à dos naturel*. (These figures exceed the *typological* counts in Table 28.3 because they include the blanks used for retouched tools). Sharpening flakes, with 67 examples (3.0%), are fairly well represented, but much less so than in the flint component of this series; most of them are of *grès lustré*, which seems to have been treated as flint. The facetting indices are moderately high (IF = 42.2, IFs = 24.2), though much lower than for the flint (cf. Table 25.2).

The cores (Table 28.2) are dominated to a striking degree by the *discoïde* type, which accounts for 65% of the whole pieces — more than twice their frequency among the flint cores. Though no Levallois cores were found in layer A itself, it is worth noting that two tortoise cores (of length 101 and 60.5 mm respectively) occurred at the base of layer 3, and may have been reworked from A. All types of raw material are present, dolerite and sandstone being the most common. With a mean length of 64 as against 41 mm, the stone cores are very much larger than their flint counterparts.

The number of flakes and flake tools per core (including core tools), based on whole pieces, is 56.1. This value is consistent with the use of disc technique; it is even *too* high in view of the nature of some of the materials, which are rather too coarse-grained for it to be worth

**Table 28.1** Stone Artefacts : Technological Indices[a]

| Layer | IL | IF | IFs | Total for IL | Total for IF(s) |
|---|---|---|---|---|---|
| 5 | 4.20 | 39.85 | 22.56 | 286 | 133 |
| A | 2.36 | 42.20 | 24.17 | 2245 | 1349 |
| B | 1.59 | 42.86 | 23.36 | 251 | 126 |
| C | 1.29 | 37.43 | 20.47 | 621 | 342 |
| D | 3.75 | 31.85 | 10.83 | 240 | 157 |
| E | 2.48 | 40.00 | 11.25 | 121 | 80 |
| F | 1.89 | 24.14 | 6.90 | 53 | 29 |
| G | 0.00 | 8.89 | 4.44 | 45 | 45 |
| H | 0.00 | – | – | 24 | 10 |

[a]Whole pieces only.

working down the cores to the same extent as those of flint. It is therefore probable that a number of the flakes were manufactured elsewhere. The larger Levallois flakes and the flake cleavers, for example, must have been struck from cores whose size would favour manufacture at the source of raw material, and which would certainly have provided other blanks worth carrying away.

**Table 28.2** Stone Cores: Frequencies

| Type | | | | Layer | | | | | |
| | H | G | F | E | D | C | B | A | 5 |
|------|---|---|---|---|---|---|---|---|---|
| *Discoïde* | 0 | 0 | 0 | 0 | 0 | 0 | 3 | 15 | 0 |
| Globular | 0 | 0 | 0 | 0 | 1 | 0 | 0 | 2 | 0 |
| Prismatic | 0 | 0 | 0 | 0 | 1 | 0 | 0 | 0 | 0 |
| Pyramidal | 0 | 0 | 0 | 0 | 0 | 0 | 0 | 1 | 1 |
| Miscellaneous | 0 | 1 | 0 | 0 | 0 | 0 | 0 | 4 | 3 |
| Shapeless | 0 | 0 | 0 | 1 | 0 | 0 | 0 | 1 | 2 |
| Total | 0 | 1 | 0 | 1 | 2 | 0 | 3 | 23 | 6 |
| Broken | 0 | 1 | 0 | 1 | 1 | 3 | 0 | 7 | 0 |
| Total (including broken) | 0 | 2 | 0 | 2 | 3 | 3 | 3 | 30 | 6 |

### 28.2.2 The Other Layers

The sample size is very low in the early part of the sequence, and little reliance can be placed on the data for layer H in particular. However, there is a generally rising trend in the facetting indices, from extremely low levels in G (IF = 8.9, IFs = 4.4) to fairly consistent values of around 40 and 20–30 from C upwards (Fig. 28.9). Analysis of the contingency tables based on the raw counts for the butts shows that these differences are statistically significant. On the other hand, the technological Levallois index IL is consistently very low, and such fluctuations as occur in its value are all within the limits of sampling error.

Only four whole cores were found below layer B. None of them are Levallois or *discoïde* — types found only in layers 3, A and B. The absence of *discoïde* cores in layer 5 is striking in view of their importance in A, but parallels their decline in the flint; despite the small number of pieces the difference is great enough to exclude sampling error as an explanation.

**Table 28.3** Stone Tools from Layers A and 5: Typological Counts

| Type | | N | Real | A Ess. | Red. | N | Real | 5 Ess. | Red. |
|------|------|---|------|--------|------|---|------|--------|------|
| 1 | Typical Levallois flakes | 21 | 7.84 | – | – | 4 | 6.56 | – | – |
| 2 | Atypical Levallois flakes | 7 | 2.61 | – | – | 1 | 1.64 | – | – |
| 3 | Levallois points | 2 | 0.75 | – | – | 0 | 0.00 | – | – |
| 5 | Pseudo-Levallois points | 7 | 2.61 | 4.83 | – | 1 | 1.64 | 3.23 | – |
| 6 | Mousterian points | 1 | 0.37 | 0.69 | 0.73 | 0 | 0.00 | 0.00 | 0.00 |
| 9 | Single straight side-scrapers | 5 | 1.87 | 3.45 | 3.65 | 1 | 1.64 | 3.23 | 3.33 |
| 10 | Single convex side-scrapers | 19 | 7.09 | 13.10 | 13.87 | 1 | 1.64 | 3.23 | 3.33 |
| 11 | Single concave side-scrapers | 1 | 0.37 | 0.69 | 0.73 | 0 | 0.00 | 0.00 | 0.00 |
| 12 | Double straight side-scrapers | 0 | 0.00 | 0.00 | 0.00 | 1 | 1.64 | 3.23 | 3.33 |
| 18 | Convergent straight side-scrapers | 1 | 0.37 | 0.69 | 0.73 | 1 | 1.64 | 3.23 | 3.33 |
| 19 | Convergent convex side-scrapers | 1 | 0.37 | 0.69 | 0.73 | 0 | 0.00 | 0.00 | 0.00 |
| 21 | *Déjeté* (offset) scrapers | 4 | 1.49 | 2.76 | 2.92 | 0 | 0.00 | 0.00 | 0.00 |
| 22 | Straight transverse side-scrapers | 1 | 0.37 | 0.69 | 0.73 | 1 | 1.64 | 3.23 | 3.33 |
| 23 | Convex transverse side-scrapers | 6 | 2.24 | 4.14 | 4.38 | 1 | 1.64 | 3.23 | 3.33 |
| 25 | Side-scrapers on ventral face | 7 | 2.61 | 4.83 | 5.11 | 3 | 4.92 | 9.68 | 10.00 |
| 27 | Side-scrapers with thinned back | 1 | 0.37 | 0.69 | 0.73 | 0 | 0.00 | 0.00 | 0.00 |
| 28 | Side-scrapers with bifacial retouch | 5 | 1.87 | 3.45 | 3.65 | 2 | 3.28 | 6.45 | 6.67 |
| 29 | Alternate retouched side-scrapers | 10 | 3.73 | 6.90 | 7.30 | 3 | 4.92 | 9.68 | 10.00 |
| 30 | Typical end-scrapers | 1 | 0.37 | 0.69 | 0.73 | 0 | 0.00 | 0.00 | 0.00 |
| 31 | Atypical end-scrapers | 1 | 0.37 | 0.69 | 0.73 | 0 | 0.00 | 0.00 | 0.00 |
| 32 | Typical burins | 5 | 1.87 | 3.45 | 3.65 | 0 | 0.00 | 0.00 | 0.00 |
| 38 | Naturally backed knives | 1 | 0.37 | 0.69 | – | 0 | 0.00 | 0.00 | – |
| 39 | *Raclettes* | 6 | 2.24 | 4.14 | 4.38 | 1 | 1.64 | 3.23 | 3.33 |
| 40 | Truncated pieces | 0 | 0.00 | 0.00 | 0.00 | 1 | 1.64 | 3.23 | 3.33 |
| 42 | Notched pieces | 11 | 4.10 | 7.59 | 8.03 | 7 | 11.48 | 22.58 | 23.33 |
| 43 | Denticulates | 24 | 8.96 | 16.55 | 17.52 | 4 | 6.56 | 12.90 | 13.33 |
| 44 | *Becs burinants alternes* | 1 | 0.37 | 0.69 | 0.73 | 0 | 0.00 | 0.00 | 0.00 |
| 45 | Retouched on ventral face | 7 | 2.61 | – | – | 2 | 3.28 | – | – |
| 46–47 | Abrupt and alternate retouch (thick) | 59 | 22.01 | – | – | 17 | 27.87 | – | – |
| 48–49 | Abrupt and alternate retouch (thin) | 27 | 10.07 | – | – | 6 | 9.84 | – | – |
| 54 | End-notched pieces | 2 | 0.75 | 1.38 | 1.46 | 0 | 0.00 | 0.00 | 0.00 |
| 59 | Choppers | 2 | 0.75 | 1.38 | 1.46 | 1 | 1.64 | 3.23 | 3.33 |
| 61 | Chopping tools | 4 | 1.49 | 2.76 | 2.92 | 0 | 0.00 | 0.00 | 0.00 |
| 62 | Miscellaneous | 18 | 6.72 | 12.41 | 13.14 | 2 | 3.28 | 6.45 | 6.67 |
| | Bifaces | 20 | | | | 4 | | | |
| | Totals | | 268 | 145 | 137 | | 61 | 31 | 30 |

326

**Table 28.4**  Stone Tools from Layers A and 5: Typological Indices

| Index | | A | | | 5 | | |
|---|---|---|---|---|---|---|---|
| | | Real | Ess. | Red. | Real | Ess. | Red. |
| ILty | Typological Levallois | 11.19 | – | – | 8.20 | – | – |
| IR | Side-scrapers | 22.76 | 42.07 | 44.53 | 22.95 | 45.16 | 46.67 |
| IC | Charentian | 11.57 | 21.38 | 22.63 | 8.20 | 16.13 | 16.67 |
| IAu | Acheulian unifacial | 0.00 | 0.00 | 0.00 | 0.00 | 0.00 | 0.00 |
| IAt | Acheulian total | 6.94 | 12.12 | 12.74 | 6.15 | 11.43 | 11.76 |
| IB | Bifaces | 6.94 | 12.12 | 12.74 | 6.15 | 11.43 | 11.76 |
| I | Levallois | 11.19 | 0.00 | 0.00 | 8.20 | 0.00 | 0.00 |
| II | Mousterian | 23.13 | 42.76 | 45.26 | 22.95 | 45.16 | 46.67 |
| III | Upper Palaeolithic | 2.61 | 4.83 | 5.11 | 1.64 | 3.23 | 3.33 |
| IV | Denticulate | 8.96 | 16.55 | 17.52 | 6.56 | 12.90 | 13.33 |
| 5 | Pseudo-Levallois points | 2.61 | 4.83 | – | 1.64 | 3.23 | – |
| 38 | Naturally backed knives | 0.37 | 0.69 | – | 0.00 | 0.00 | – |
| I rc | Convergent retouched | 2.61 | 4.83 | 5.11 | 1.64 | 3.23 | 3.33 |
| 6-7 | Mousterian points | 0.37 | 0.69 | 0.73 | 0.00 | 0.00 | 0.00 |
| Rsingle | Side-scrapers, single | 9.33 | 17.24 | 18.25 | 3.28 | 6.45 | 6.67 |
| Rdouble | Side-scrapers, double | 0.00 | 0.00 | 0.00 | 1.64 | 3.23 | 3.33 |
| Rtransv | Side-scrapers, transverse | 2.61 | 4.83 | 5.11 | 3.28 | 6.45 | 6.67 |
| Rinv | Side-scrapers, inverse/alternate | 6.34 | 11.72 | 12.41 | 9.84 | 19.35 | 20.00 |
| 30-31 | End-scrapers | 0.75 | 1.38 | 1.46 | 0.00 | 0.00 | 0.00 |
| 32-33 | Burins | 1.87 | 3.45 | 3.65 | 0.00 | 0.00 | 0.00 |
| 42 | Notched pieces | 4.10 | 7.59 | 8.03 | 11.48 | 22.58 | 23.33 |
| 42-43 | Notched pieces & denticulates | 13.06 | 24.14 | 25.55 | 18.03 | 35.48 | 36.67 |
| Ident | Extended denticulate group | 13.81 | 25.52 | 27.01 | 18.03 | 35.48 | 36.67 |
| Iirr | Irregularly retouched | 34.70 | – | – | 40.98 | – | – |
| IQ | Quina | 0.00 | | | 14.29 | | |
| IQ+½Q | Quina + demi-Quina | 1.64 | | | 14.29 | | |

## 28.3  TYPOLOGY

### 28.3.1  Layer A

The typological counts and other statistics are given in Tables 28.3 and 28.4, using the three sets of indices — Real, Essential and Reduced — as defined in chapter 23. The values of ILty and IB are high by the standards of La Cotte, albeit low when compared with series of approximately the same age from many other sites. In the essential and reduced counts, side-scrapers dominate the series, though denticulates — especially when grouped with notches and related pieces — are also important. The 'Upper Palaeolithic' group, III, is poorly represented, and consists chiefly of burins.

In the selection of blanks for the well-retouched tools there is a slight preference for Levallois pieces (15, or 10.9%), which are more numerous than among the stone artefacts as a whole for this layer (IL = 2.36); on the other hand, only one of the sharpening flakes has been used as a blank.

#### UNMODIFIED 'TOOLS'

**Levallois flakes** (21 typical, 7 atypical). These are not particularly large, with a mean length of 58.5 mm. Only four pieces exceed 80 mm, the biggest being that illustrated in Fig. 28.2, 1; three, all less than 40 mm long, are of *grès lustré*. Of the seventeen determinable butts, seven are facetted, three more are dihedral, and the rest plain.

**Levallois points** (2). Poorly made, they are probably accidental products of disc core reduction.

**Pseudo-Levallois points** (7). These are all 'technical' examples, none appearing to be deliberately manufactured as points.

#### IRREGULARLY RETOUCHED TOOLS

The 93 pieces in categories 45–50 are very important in percentage terms (34.8% of the real total), but it should be remembered that the materials under discussion are particularly liable to edge damage, both accidentally and during use. Only nine pieces are of *grès lustré*, the material best suited to preserve convincing evidence of deliberate working.

#### WELL-RETOUCHED TOOLS

**Mousterian point** (1). This is made on a *grès lustré* LSF; not very sharply pointed, it is 55 mm long, with scalar retouch.

**Side-scrapers** (61). No fewer than 28 of these, mostly of the more elaborately worked types, are made of *grès lustré*. The representation of major typological groups is

| | | | |
|---|---|---|---|
| single lateral | 25 | double | 0 |
| convergent | 6 | transverse | 7 |
| other | 23 | | |

Four of the convergent pieces are *déjeté* (Fig. 28.2, 4). Single convex scrapers are much the most common individual type (Figs. 28.2, 2 and 28.3, 1); in all, the principal retouched edges are convex in 47 cases (77%) while only two are concave. There is one example of demi-Quina retouch. The single scrapers are appreciably bigger (mean length 65.7 mm) than the scrapers of other categories — the overall mean length is 58.5 mm. Nine of the blanks are Levallois (two atypical); the others include a pseudo-Levallois point, two cortex-backed flakes and a core.

**End-scrapers** (2). One is typical, the other atypical. Both are of medium-grained quartzite, of simple form on flakes. The typical example is on a small (48 mm) Levallois flake; the other, only 30.5 mm long, has short abrupt retouch along part of the distal end only.

**Burins** (5). All are typical. Four are of *grès lustré*: one of these is of mixed type, the others angle burins on breaks (Fig. 28.3, 3). The fifth, of dolerite, is another angle burin (Fig. 28.3, 2).

**Raclettes** (6). Morphologically, these are extremely varied. In one instance, inverse retouch on a convex edge may have served as backing rather than as the working part of the tool. Four pieces are made of *grès lustré*.

**Notched pieces** (11). The notches are made by retouch on three of these and are Clactonian on the other eight (Fig. 28.3, 4). They are shallow, and relatively narrow (10–19 mm, apart from two in the 20–29 and two in the 30-39 mm width intervals). Five pieces are of dolerite, including the four with the largest notches; one of these is on a tranchet flake.

**Denticulates** (24). Only three are of dolerite, and four of *grès lustré*. Those made with Clactonian notches are in a majority (13); six more are made by coarse and five by fine retouch; there are no microdenticulates, which is not altogether surprising in view of the materials used. Ten have normal, six alternate and eight bifacial working; none are inverse. The blanks include two typical Levallois flakes and one tranchet flake. Additional features are rare: one example of a thinned edge, two of retouch at one end, and a single isolated Clactonian notch.

**Bec burinant alterne** (1). This is on a flake from a quartzite pebble. The piece has been truncated by a single blow to the dorsal surface. A notch has been made next to this, and the truncation itself has been used as a platform for dorsal removals.

**End-notched pieces** (2). One notch is made by retouch; the other is Clactonian (and possibly accidental).

**Choppers** (2). Both are of dolerite. One is 71 x 63.5 mm, weighing 209 g, and the other 62 x 44 mm, weighing 109 g.

**Chopping tools** (4). Two are extensively flaked, on dolerite, and measure 95 and 60.5 mm (weighing 403 and 152 g respectively). A third is atypical, inasmuch as it is on a fragment of a large dolerite flake (85 mm and 172 g). The last is of sandstone, with alternating flaking around the periphery except where two intersecting cleavage planes form a butt.

**Miscellaneous** (18). No fewer than twelve of these are made of *grès lustré*. This is consistent with their typological character, which is complex but favours scraper edges (11). Four pieces are scaled, four denticulated (Fig. 28.3, 5), and three notched. Two have LSF removal scars.

**Bifaces** (20). Amygdaloid handaxes are the most common variety, represented by seven examples; two are slightly twisted. Cleavers of either dolerite or quartzite are also important, with three bifacially worked (e.g. Fig. 28.3, 6) and four on flakes (such as the divergent Tixier type V piece in Fig. 28.4, from which a tranchet flake has been removed to straighten the working edge).

The single occurrences of the nucleiform and miscellaneous types are of *grès lustré*; the latter is amygdaloid in shape but one edge consists largely of a thick crust of cortex, so that the piece may have served as a *biface-racloir*. There is also a crude pick on a thick elongated cortical flake of dolerite, worked mostly on the ventral surface but with normal removals at each end. The partly bifacial category, represented by three pieces, is very heterogeneous. One example is a massive (486 g) dolerite flake, predominantly ventrally worked and broken at the tip. At the other extreme (16 g) is a *grès lustré* twisted amygdaloid, one edge of which is entirely primary. The third is also small, and is roughly discoidal on an overstruck siltstone (?) flake. The handaxes made of *grès lustré* are generally much smaller than average (mean length 52 as opposed to 86 mm), which accords with the mode of occurrence of this raw material (chapter 22).

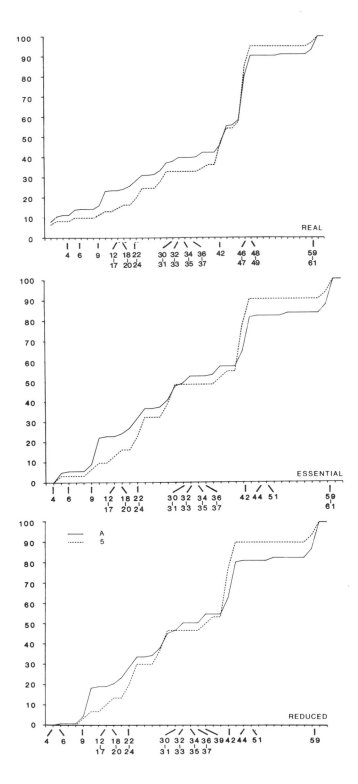

**Fig. 28.1** Layers A and 5: cumulative curves for the stone tools.

Two of the most remarkable pieces from the whole excavation may well be from A, but could not be included in the statistics for the layer as they are from the 1961 excavations and were originally assigned to layer B (see section 21.1). Both are of quartzite. One is an amygdaloid; the other is a large side-struck flake cleaver of Tixier's type II (Figs. 28.5 and 28.6).

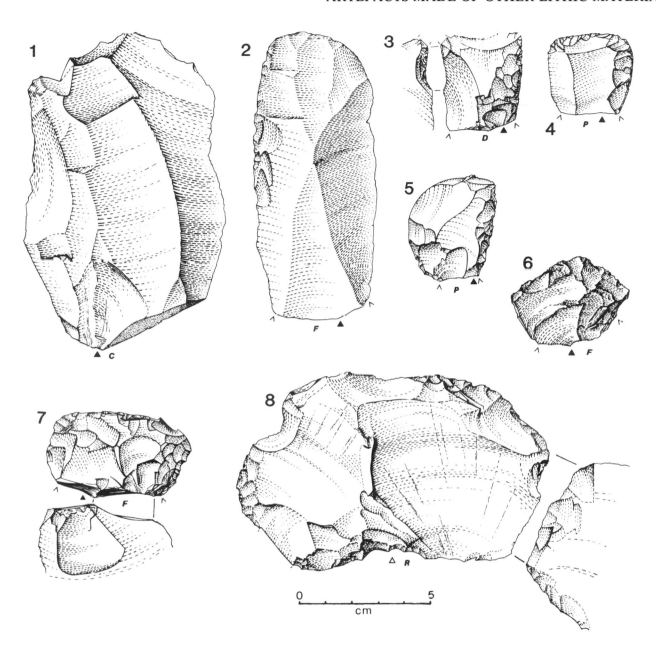

**Fig. 28.2** Stone tools from layer A (except 5, which is from layer C). 1 Levallois flake. 2–8 side-scrapers: 2, 3, 5 single; 4 *déjeté*; 6–8 transverse. Raw materials: 1 and 8 sandstone; 2 basic igneous (probably dolerite); 5 siltstone; 3, 4, 6, 7 *grès lustré*. The conventions used for the butts are those described in the caption for Fig. 26.4.

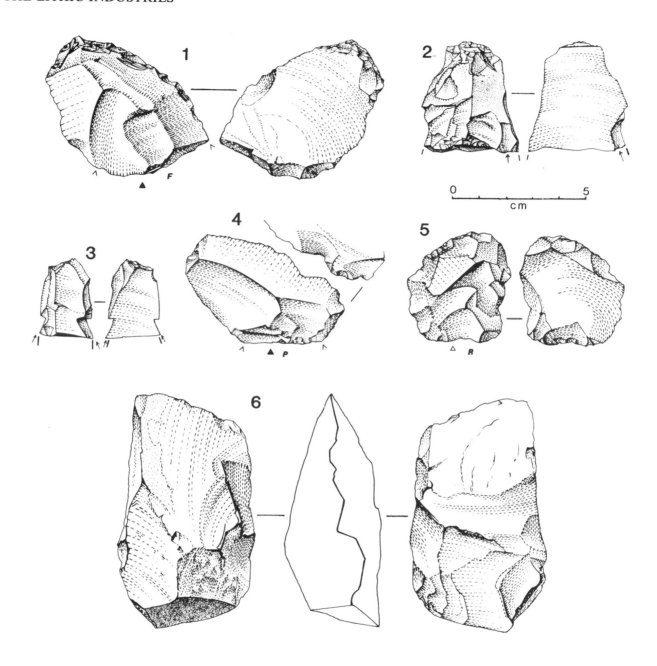

**Fig. 28.3** Stone tools from layer A. 1 single side-scraper; 2, 3 burins (on break); 4 notched piece (Clactonian); 5 miscellaneous (denticulate with scaling); 6 bifacial cleaver. Raw materials: 2, 5, 6 basic igneous (probably dolerite); 4 quartzite; 1, 3 *grès lustré*.

330

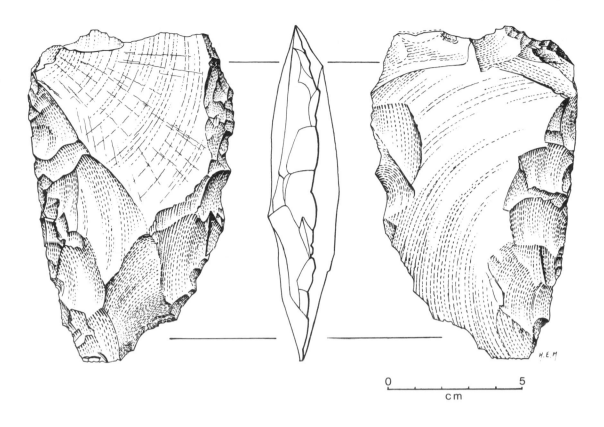

**Fig. 28.4**  Layer A. Flake cleaver (quartzite): Tixier type V.

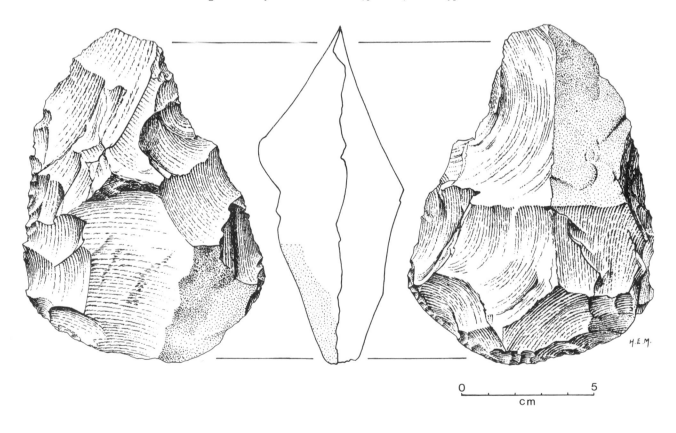

**Fig. 28.5**  1961 excavations (spit 20). Amygdaloid handaxe (quartzite). Originally assigned to layer B, but probably from layer A.

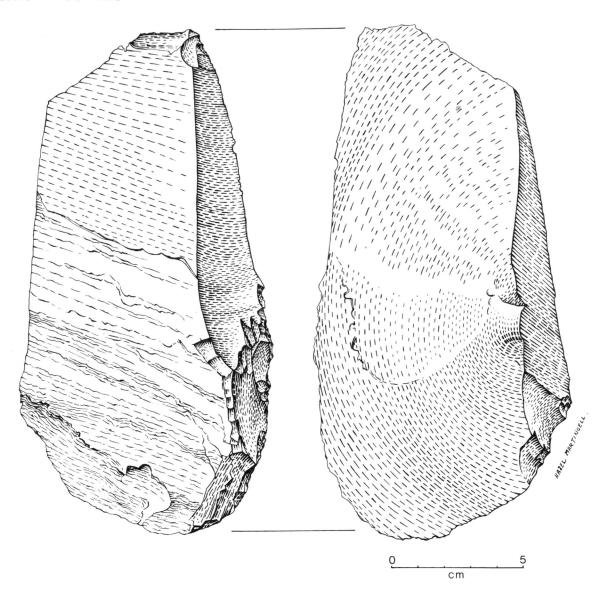

**Fig. 28.6** 1961 excavations (spit 20). Flake cleaver (quartzite). Originally assigned to layer B, but probably from layer A.

**Table 28.5**  Stone Bifaces: Frequencies

| Type | Layer | | | | |
|------|---|---|---|---|---|
| | C | B | A | 3 | 5 |
| Ovate | 1 | 0 | 0 | 0 | 0 |
| Amygdaloid | 1 | 0 | 7 | 0 | 2 |
| Cleaver (bifacial) | 0 | 0 | 3 | 0 | 0 |
| Flake-cleaver | 1 | 1 | 4 | 1 | 0 |
| Nucleiform | 1 | 0 | 1 | 0 | 0 |
| Miscellaneous | 0 | 0 | 1 | 0 | 0 |
| Partly bifacial | 0 | 0 | 3 | 0 | 2 |
| Pick | 0 | 0 | 1 | 0 | 0 |
| Total | 4 | 1 | 20 | 1 | 4 |
| Broken | 0 | 1 | 0 | 0 | 0 |
| Total (including broken) | 4 | 2 | 20 | 1 | 4 |

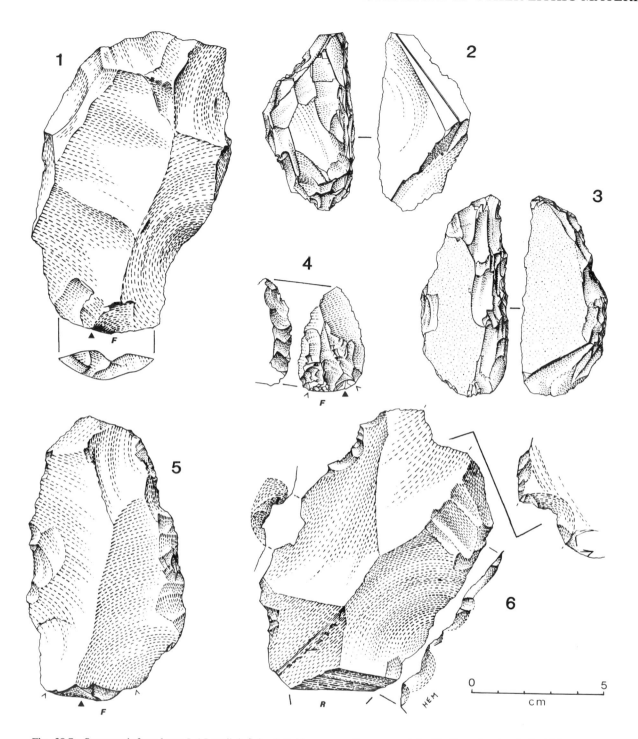

**Fig. 28.7** Stone tools from layer 5. 1 Levallois flake. 2–5 side-scrapers: 2 convergent; 3 with alternate retouch; 4 with inverse retouch; 5 double, on Levallois flake; 6 notched piece (multiple, Clactonian, on Levallois flake). Raw materials: 1 sandstone; 2, 3 siltstone; 4 *grès lustré*; 5, 6 quartzite.

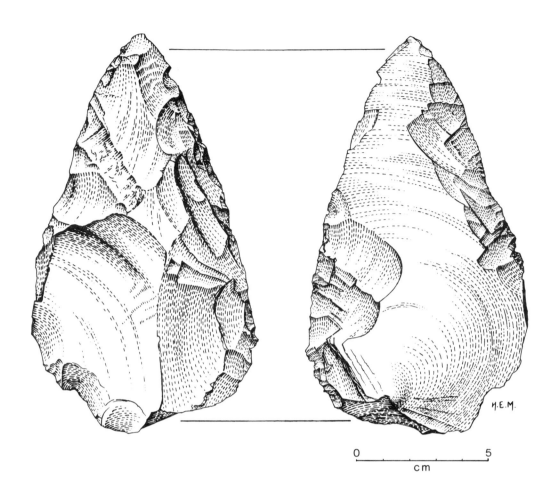

**Fig. 28.8**  Partially bifacial amygdaloid handaxe (quartzite) from layer 5.

### 28.3.2  Layer 5

For the typological counts and indices see Tables 28.3 and 28.4. Although the sample size is small there is a general resemblance to the figures for layer A.

#### UNMODIFIED 'TOOLS'

**Levallois flakes** (4 typical, 1 atypical). These are very much larger than their counterparts in layer A, with a mean length of 74.5 mm (80 mm is exceeded in three cases); the two smallest pieces are of quartzite. The three determinable butts are facetted.

**Pseudo-Levallois point** (1). This is small (28 mm) but relatively elongated, and of quartzite. Not a very typical example, it cannot be taken to indicate on-site debitage of this material using disc or Levallois technique.

#### IRREGULARLY RETOUCHED TOOLS

The 35 pieces (41%) in this category are subject to the same reservations as those in layer A.

#### WELL-RETOUCHED TOOLS

**Side-scrapers** (14). The subgroup consisting of types 25–29 is predominant:

| | | | |
|---|---|---|---|
| single lateral | 2 | double | 1 |
| convergent | 1 | transverse | 2 |
| other | 8 | | |

Straight and convex edges are equally represented; there are no examples of concave edges. Whereas in layer A the favoured material for the more complex scrapers is *grès lustré*, in layer 5 siltstone has been used (for seven pieces). The two *grès lustré* scrapers are, respectively, single convex and inversely retouched (Fig. 28.7, 4).

Two pieces, both of siltstone, have Quina retouch (Figs. 28.7, 2 and 3; another siltstone piece has a possible LSF removal. One scraper has a Clactonian notch, and another is retouched at one end. Three are made on Levallois blanks (Fig. 28.7, 5). With a mean length of 72.6 ±25.0 mm, the layer 5 scrapers are considerably larger than those of layer A (58.5±26.3 mm).

**Raclette** (1). This is straight-edged, on the left hand margin of a quartzite Levallois flake.

**Truncation** (1). Being on a sandstone flake and poorly retouched, this is rather questionable.

**Notched pieces** (7). Three are Clactonian, and four made by retouch. One is on a Levallois flake (Fig. 28.7, 6). Two are of siltstone, two of sandstone and three of quartzite.

**Denticulates** (4). These are single examples on dolerite, siltstone, sandstone and quartzite (this last on a core). One piece is denticulated by adjacent Clactonian notches, one by coarse retouch and the others by fine retouch. Three are worked on only one edge, the fourth on both.

**Chopper** (1). This is made on a flat (dolerite?) pebble, with rather invasive removals around about three quarters of its periphery.

**Miscellaneous** (2). One has flat inverse retouch around the tip of a large, pointed sandstone flake. The other is double notched on the tip of a broken siltstone flake, possibly retouched as a scraper after the break occurred.

**Bifaces** (4). There is a short amygdaloid handaxe of dolerite, 80 mm long, and another — possibly a thick ovate — whose tip has been removed by a tranchet blow. The latter, of sandstone, is 83 mm in length. The other two pieces, both of quartzite, are only partly bifacial and are on flakes. One is 74 mm long and is cleaver tipped, with concave sides. The other is the magnificent, finely pointed elongated cordiform illustrated in Fig. 28.8, and is 141 mm in length.

### 28.3.3  The Other Layers

The type-lists and indices are given in mf 28.2. There are very few well-retouched stone tools in the lowest layers: a scraper with inverse retouch and an inverse chopper in H, a 'miscellaneous' piece in G (with inverse removals on a thick flake), and a convex single scraper in F. Five of the six in E are side-scrapers, but only one, with inverse retouch, of the ten in D (though two classified as miscellaneous have scraper-like edges). In layer C, however, 17 of the 22 tools included in the reduced count are side-scrapers (Fig. 28.2, 5), giving an $IR_{red}$ of 77.3. A $\chi^2$ test shows that this is significantly higher than in layer A; seven of them are made of *grès lustré*. On the other hand, the scraper index for A falls within the sampling error of that for the much smaller series from layer B, for which $IR_{red}$ is 28.6.

Bifaces are lacking from layer D and below. Those from layer C are varied: a small but fine untwisted ovate and a nucleiform, both of *grès lustré*, a small and crudely worked siltstone amygdaloid, and a dolerite flake cleaver (Tixier type I). From layer B there is only a dolerite flake cleaver (Tixier type II); a similar piece came from the base of layer 3.

### 28.4  HAMMERSTONES

Hammerstones were found in all layers except H (see mf 28.2), though the single example from D is broken. Layer A provides the only sample of any size, with 15 whole pieces; these are very small, with a mean weight of 89 g as opposed to 352 g for all the whole examples from layers C–G combined. This is in keeping with the suggestion that knapping activity in the upper layers was concentrated on tool finishing and resharpening rather than on primary flaking (see chapter 24), and that the

largest stone blanks may have been manufactured elsewhere (see section 28.2.1). Another feature of the upper layers is the relative abundance of hammerstone fragments. Newcomer (in Green 1984), describing experimental knapping of 'hard' rocks such as occur at the broadly contemporary site of Pontnewydd, north Wales, draws attention to the frequency with which hammerstones are shattered during debitage of such materials; flint, on the other hand, tends to cause abrasion more commonly than breakage. It is therefore not surprising that there should be evidence of major damage to the hammerstones in the later Saalian layers, in which the hard rocks were exploited more often than before: in A, the ratio of fragments to complete examples is over three to one, though it is quite possible that some of the former are shattered cores on pebbles rather than hammerstones.

### 28.5  DISCUSSION

In layer A, which has provided the largest sample, the size and frequency of the Levallois blanks (retouched or unretouched) are entirely consistent with the high percentage of disc cores. But as such cores are liable to give rise to 'technical' Levallois flakes in some numbers it is suggested that knapping of the minor raw materials was conducted without employment of much true Levallois technique. It is not necessary to suppose that these disc cores represent the final stage of reduction of former Levallois cores. Some of the larger Levallois flakes may have been imported.

Why were so many Levallois flakes abandoned, unretouched, in layer A, especially given that their mean length is identical to that of the stone side-scrapers from

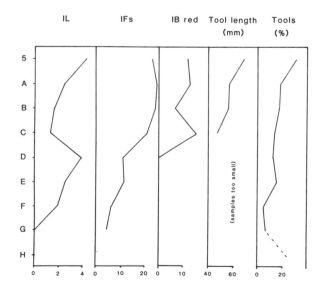

**Fig. 28.9**  Stone artefact variation through time.

this layer (Fig. 28.9)? An answer is suggested by the 23 tools (including the 'irregularly retouched' types 45–50) made on Levallois blanks. With a mean length of only 48 mm, these are characterised by a high incidence of *grès lustré* (used for 12 pieces, or 52%); it is clear that this material was also particularly favoured for the more elaborate flake tools. However, it occurs in only 11% of the *unmodified* Levallois flakes. In deciding suitability for retouch, therefore, the raw material was at least as important as size. The unmodified pieces were probably rejected as blanks for scrapers and other tools for which retouch quality or edge durability was important, because too soft or coarse-grained; they were perhaps reserved for tasks for which unretouched edges are appropriate.

The great increase in the size of the stone artefacts in layer 5 is partly due to the rarity of *grès lustré*, as during the earlier occupations it had been available in the form of pebbles of restricted dimensions which, moreover, were usually worked down to cores of very small size. Two of the six stone cores from layer 5 are of this material, however, leaving the other rock types poorly represented in view of their importance in other artefact categories. This lends further support to the suggestion that at this time much of the stone debitage was carried out elsewhere.

# SPECIALIZED RESHARPENING TECHNIQUES AND EVIDENCE OF HANDEDNESS

J. M. Cornford

## 29.1 INTRODUCTION

An especially interesting feature of the lithic material at La Cotte is the growing use of a type of *coup du tranchet* technique to modify or resharpen the edges of existing flint tools, apparently as a response to diminishing supplies of flint. It was through its numerous by-products that it was recognized first. These are two types of flake, one struck longitudinally, the other laterally, which will be called 'long sharpening flakes' and 'transverse sharpening flakes', abbreviated commonly to 'LSFs' and 'TSFs'.

The LSFs are particularly idiosyncratic, and have not as yet been recorded from any other European palaeolithic site. They were first noted by Burdo in his site diary for 1954 as "the small blades worked on the flaking side which puzzled so much in London and Cambridge last November". When he showed these to Harper Kelley and François Bordes in Paris they explained them to him as "retrimmings of damaged larger tools". From this is it clear that they recognised the marginal retouch to be prior, not subsequent, to the flakes' detachment, but there is nothing to suggest that they regarded them as a common or well known component of palaeolithic assemblages elsewhere, though the affinities of other artefacts from La Cotte were particularly discussed. Nor, as a class, had they become familiar to Bordes twenty-six years later when, in 1980, he examined further examples.

McBurney had picked out a few exceptionally arresting examples from among the finds of his first two seasons at La Cotte. He remarked on their technological similarity to a small number of waste flakes from one of the Mousterian levels at the Haua Fteah (McBurney 1967a, 108 and Fig. V.3, 5–8). But it was not until their recovery in greater numbers during the last two seasons of excavation that their identity as a distinct type began to be appreciated. Thereafter many more were discovered while processing the material in the laboratory.

The most conspicuous and distinctive feature of a long sharpening flake is the narrow, well-defined facet lying alongside its bulbar surface, parallel to its longitudinal axis and at a highly obtuse angle to it (Figs. 29.1, 4–13, 29.2, 6–7 and 29.7). This facet is a marginal strip of the ventral surface of the tool from which the flake was struck; the intersection of its outer edge with the dorsal surface formed, before removal, the edge of the parent tool. The relationship between flake and tool is illustrated diagrammatically in Figs. 29.2, 29.7 and 29.11, and by the two pieces in Fig. 29.3 — the only two

that were found to conjoin (see section 29.5) — although the flake here is atypically thick and narrow. Ideally, the technique creates a new edge of the greatest possible length and sharpness on the parent tool.

The essential asymmetry of the mode of production, as mirrored in these by-products, made it possible to observe a strong bias to one side in the relative position of the new and old bulbar surfaces. This led to the speculation that the bias might be due to the makers' use of a preferred hand (see section 29.6). Transverse sharpening flakes, in comparison, are short and broad (Figs. 29.1, 1–3 and 29.13). They are struck from the edge of a retouched tool by a blow more or less at right angles to it and delivered on the ventral surface, which provides their characteristic plain butt. In their technological evolution they, too, add to the evidence for handedness.

## 29.2 METHOD OF ANALYSIS

The specimens had been classified initially into three broad groups: burin spalls, long sharpening flakes and transverse sharpening flakes, the LSFs being at the same time subdivided by butt type into four subgroups: facetted, dihedral, plain and-indeterminable. Unbroken specimens were measured and described according to the criteria listed below. Because of the far greater numbers in layer A, samples of 20% of those with facetted, 50% of those with plain and 33% of those with indeterminable butts were taken. The values were weighted accordingly in the final analysis. The relative frequencies by layer for the unbroken and broken specimens in each group are given in Fig. 29.4, and for the four subgroups of LSFs (unbroken) in Fig. 29.5.

### 29.2.1 Metrical Attributes

*ANGLES*

These were measured by goniometer. They are explained diagrammatically in Figs. 29.6 and 29.7 and discussed in section 29.3. (Burin spalls were measured for removal angle only.)

Dorsal angle: the angle at which the facet of the old bulbar surface meets the plane of the *unmodified* dorsal surface. This gives an estimate of the angle on the parent edge at the penultimate stage.

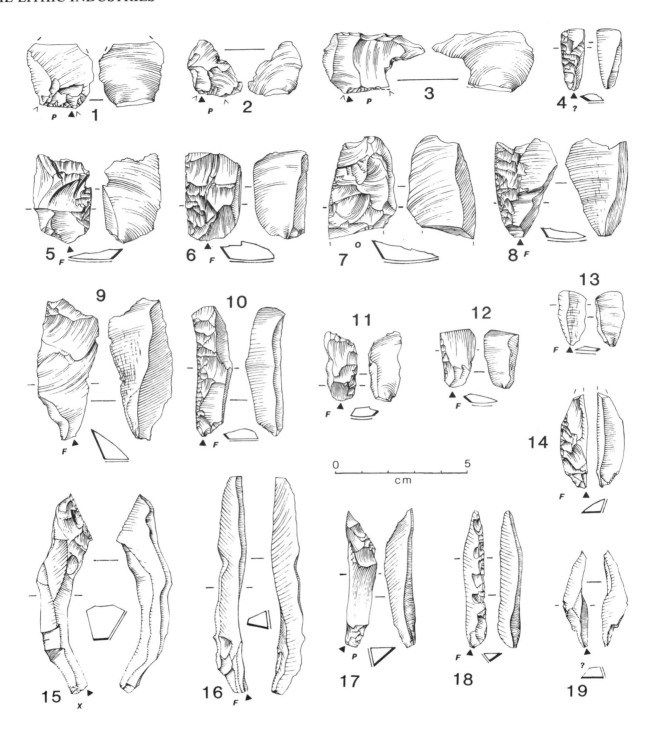

**Fig. 29.1** Transverse sharpening flakes, long sharpening flakes and burin spalls. 1–3 TSFs; 4–13 LSFs: the facet of the old bulbar surface is to the right of the new bulbar surface in 4, 6, 7 (broken), 8, 9, 10, 12, and to the left of it in 5, 11, 13; 14–19 burin spalls. 4 and 5 are from layer B, and 16 and 17 from layer 3. The rest are from layer A. Key to butt symbols: ▲ point of percussion; ∧ ... ∧ limits of butt; *P* plain; *F* facetted; *X* polyhedral; *O* broken; *?* indeterminable. Key to cross section conventions: thick line, old bulbar suface; double line, new bulbar surface; thin line, dorsal surface.

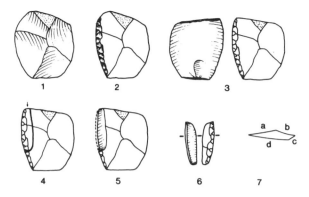

**Fig. 29.2** The stages preceding and following the removal of a long sharpening flake from the parent tool. In (1)–(5) the dorsal surface of the parent is shown, with the ventral face also in (3): (1) before and (2) after modification of an edge; (3) after preparation of a platform at the distal end; (4) with the flake to be removed shown in outline; (5) with the scar remaining after detachment of the flake — the broken line indicates the strip removed from the ventral surface. (6) The LSF (butt end pointing downwards): on the left, the new bulbar surface with the facet of the parent ventral surface to its left and, on the right, the dorsal surface with the retouch from the parent edge; (7) in cross section (x4) where (a) is the dorsal surface (b) the retouched edge (c) the old bulbar surface, from the parent ventral surface, and (d) the new bulbar surface.

Retouch angle: the angle of retouch at the extreme edge of the dorsal surface where it meets the facet of the old bulbar surface, as registered at the first point of resistance on the goniometer. Consistently higher than the values for the whole retouch plane (as measured and given in chapters 26 and 30), but too imprecise, for mechanical reasons, to be used for comparative purposes.

Removal angle: the angle between the newly struck and the old bulbar surface. The complement of this gives the angle of the rejuvenated edge on the parent tool.

## DIMENSIONS

Length, breadth, thickness; length/breadth ratio (L/B); breadth/thickness ratio (B/T).

Length and breadth (invasiveness) of retouch on primary edge.

Length and breadth of old bulbar facet — the 're-moval length' and 'removal breadth'.

### 29.2.2 Descriptive Attributes

**Flake termination**: (a) hinge (b) feather.

**Direction of asymmetry**. Determined by the relative positions of the new and old bulbar surfaces (bulbar surface facing upwards, bulb towards the observer). TSFs oriented with butt to the right of the bulbar surface. (a) left — when the facet of the old bulbar surface is to the left of the new bulbar surface as in Figs. 29.1, 5, 11, 13 and 29.2, 6; (b) right — when the facet of the old bulbar surface is to the right of the new bulbar surface (Fig. 29.1, 4, 6–10, 12); (c) TSFs only: with point of impact (i) central — within middle third, (ii) distal — at far end, (iii) proximal — at near end of the butt.

**Edge type**. (a) Mint (b) utilized or damaged (c) retouched.

**Retouch type**, if applicable. (a) Scalar: flat, invasive retouch, usually associated with a sharp acute-angled edge; (b) nibble: fine, small scars invading less than 2 mm of the edge, at a steep angle to the bulbar surface — possibly due to utilization; (c) scalar and nibble: two tiers of retouch combining (a) and (b); (d) denticulate; (e) stepped; (f) Quina; (g) mixed, miscellaneous or indeterminable.

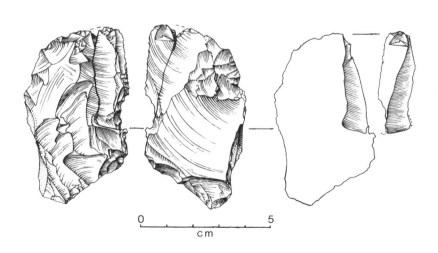

0       5
c m

**Fig. 29.3** Refitted tool and long sharpening flake from layer 5. The tool is a bifacial scraper (possibly a *prondnik*; see section 29.5).

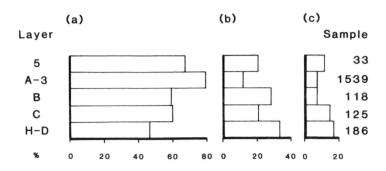

**Fig. 29.4**  Relative frequencies of (a) LSFs, (b) TSFs, and (c) burin spalls (whole and broken specimens).

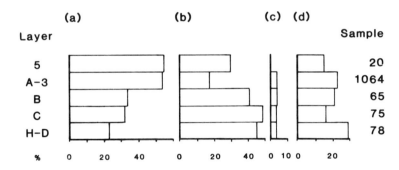

**Fig. 29.5**  Relative frequencies of butt types on LSFs: (a) facetted, (b) plain, (c) dihedral, (d) indeterminable (whole specimens only).

**Scar of earlier removal** (a) Single (b) multiple (c) burin spall.

These attributes were analysed by group and by layer and the results compared with other aspects of the assemblage to identify any significant patterns of association or covariation. Canonical discriminant analysis (Doran and Hodson 1975) based on the values of the quantitative attributes was used to test the validity of the initial subjective classification and to detect any unsuspected similarities or differences between or within groups.

### 29.3  MORPHOLOGICAL DEFINITION AND DESCRIPTION OF TYPES

As a preliminary to further analysis it was necessary to establish the validity of the long sharpening flakes as a typological entity in their own right and, in particular, to isolate the attributes that distinguish them from burin

spalls. In the basic mode of production the two are obviously alike: both are detached by a blow struck vertically down the edge of a tool or blank and both remove some or all of that edge. Beyond this, they begin to diverge.

The difference is evident in the different properties of the edges created by each type of removal. The object, or the result, of removing a burin spall is the production of a burin — a tool with an established typological identity, whereas the removal of a LSF does not so much make a tool (as generally defined) as unmake an existing one, by removing some or all of the retouch that characterized it. The removal of a burin spall leaves a narrow, two-edged facet more or less at right angles to the dorsal and ventral surfaces of the parent piece, of which, traditionally, only the chisel-ended tip is intended for use. Despite some evidence from recent wear studies (Vaughan 1986) and ethnographic observations (Hayden 1977) to challenge this view of their function, it remains unlikely that the edges of burin facets were primarily designed for cutting, in the slicing sense at least; nor are they readily amenable to further modification by lateral

**Table 29.1**  Metrical Attributes differentiating Main Groups[a]

| Layer | Length mm | Breadth mm | Thickness mm | L/B | B/T | Removal angle | Sample |
|---|---|---|---|---|---|---|---|
| **A. Burin spalls** | | | | | | | |
| A–3 | 32.5 ± 9.4 | 7.4 ± 2.5 | 7.8 ± 3.1 | 4.7 ± 1.7 | 1.04 ± 0.49 | 81.0 ± 15.8 | 72 |
| C–B | 35.5 ± 9.3 | 7.1 ± 2.5 | 7.6 ± 2.4 | 5.4 ± 1.9 | 0.95 ± 0.23 | 83.7 ± 17.2 | 17 |
| H–D | 30.9 ± 9.1 | 7.2 ± 2.9 | 8.2 ± 4.6 | 4.8 ± 1.5 | 0.98 ± 0.29 | 74.0 ± 19.5 | 15 |
| **B. Long sharpening flakes** | | | | | | | |
| A–3 | 29.4 ± 7.9 | 16.8 ± 5.3 | 4.2 ± 1.6 | 1.9 ± 0.6 | 4.5 ± 2.5 | 125.2 ± 11.5 | 1064 |
| C–B | 30.2 ± 6.8 | 18.2 ± 4.7 | 5.2 ± 1.9 | 1.7 ± 0.4 | 3.9 ± 1.5 | 127.0 ± 10.5 | 140 |
| H–D | 30.3 ± 9.2 | 20.7 ± 7.4 | 5.9 ± 3.6 | 1.5 ± 0.4 | 3.9 ± 1.1 | 127.8 ± 12.5 | 78 |
| **C. LSFs with facetted butts** | | | | | | | |
| A–3 | 30.4 ± 9.0 | 16.8 ± 5.1 | 4.1 ± 1.7 | 1.9 ± 0.7 | 4.8 ± 2.4 | 127.3 ± 10.2 | 576 |
| C–B | 32.6 ± 8.1 | 18.8 ± 4.6 | 5.1 ± 1.8 | 1.8 ± 0.4 | 4.2 ± 1.8 | 129.3 ± 12.4 | 46 |
| H D | 30.7 ± 9.8 | 21.4 ± 8.1 | 5.8 ± 4.5 | 1.5 ± 0.4 | 4.2 ± 1.3 | 129.4 ± 14.7 | 18 |
| **D. LSFs with plain butts** | | | | | | | |
| A–3 | 28.4 ± 6.8 | 16.8 ± 5.6 | 4.3 ± 1.5 | 1.8 ± 0.6 | 4.3 ± 1.8 | 123.2 ± 13.4 | 174 |
| C–B | 27.8 ± 5.6 | 17.7 ± 4.8 | 5.4 ± 1.9 | 1.6 ± 0.4 | 3.7 ± 1.3 | 125.0 ± 8.9 | 64 |
| H–D | 29.9 ± 8.7 | 20.0 ± 6.8 | 6.1 ± 2.8 | 1.5 ± 0.3 | 3.6 ± 1.1 | 126.3 ± 10.2 | 35 |
| **E. Transverse sharpening flakes** | | | | | | | |
| A–3 | 23.1 ± 7.7 | 20.2 ± 5.8 | 3.2 ± 1.2 | 1.2 ± 0.5 | 7.0 ± 2.9 | 137.8 ± 11.0 | 151 |
| C–B | 22.0 ± 7.6 | 21.1 ± 6.1 | 3.2 ± 1.2 | 1.1 ± 0.5 | 7.2 ± 2.9 | 126.1 ± 9.5 | 63 |
| H–D | 18.7 ± 4.7 | 23.8 ± 6.8 | 4.9 ± 1.9 | 0.9 ± 0.4 | 5.3 ± 1.8 | 123.1 ± 6.9 | 67 |

[a]Means and standard deviations for whole pieces only.

retouch, because of the steepness of the angle at each edge (mean 99.0°). The removal of a LSF, by contrast, leaves an acute-angled edge of about 52°, well enough suited either to immediate cutting use (see section 29.5) or to further retouch. On these grounds alone it seems unlikely that the two types belong to a single technological continuum, or that LSFs are merely the aberrant by-products of faulty burin technique, particularly in view of the consistency of their design in far greater numbers.

Morphological and metrical attributes endorse this conclusion more explicitly (Table 29.1). Burin spalls are narrow and thick in relation to their bulbar surface (length/breadth ratio 4.7; breadth/thickness ratio 1.0). In cross section, with the new bulbar surface for base, they approximate to an isosceles triangle (Figs. 29.1, 15–19 and 29.6) and they are detached from the parent at a steep angle clustering round 90°. LSFs, on the other hand, are broad and flat, their L/B (1.7) and B/T (4.6) ratios almost the inverse of the burin spalls'. In cross section they resemble a scalene triangle, or asymmetrical wedge, and they are detached from the parent piece at an obtuse angle between 122° and 128° (Figs. 29.1, 4–13 and 29.7). Using these values, canonical discriminant analysis indicated consistency of the initial classification according to the criteria used: only three of the 106 burin spalls and three of the 560 sampled long sharpening flakes were less compatible with their allotted group than with the other.

Divergences in the descriptive attributes are less marked, but not insignificant. In termination, for instance, burin spalls are less prone to hinge fracture, owing perhaps to different technological constraints

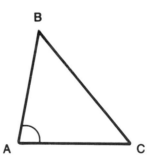

**Fig. 29.6**  Diagram of a burin spall in cross section, based on the means for L, B, T and removal angle (all layers): BC dorsal surface; AB old bulbar surface; AC new bulbar surface; BAC removal angle.

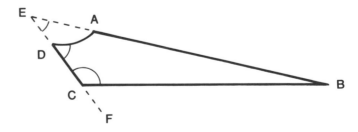

**Fig. 29.7**  Diagram of a long sharpening flake in cross section, based on the means for L, B, T, dorsal angle and removal angle of LSFs in layer A. AB dorsal surface; AD retouch; AE projected line of previous dorsal surface; CD old bulbar surface; BC new bulbar surface; BEC dorsal angle, giving an estimate of the previous edge angle; BCD removal angle; BCF the angle on the new edge created by the LSF removal (it is the complement of BCD).

**Table 29.2**  Flake Termination (layer A)[a]

| Type | Hinge % | Feather % | Sample |
|------|---------|-----------|--------|
| Burin spalls | 12.6 | 87.4 | 57 |
| LSFs | 30.2 | 69.8 | 851 |

[a]Where determinable.

(Table 29.2 and section 29.5). The greater degree of skill and forethought entailed in the successful operation of the sharpening technique is reflected in the consistently higher incidence and steady rise of facetting on LSF butts — to 55.9% in layer A, when for burin spalls it is only 26.6%. There is no significant difference between them in asymmetry or in type of retouch, where present, though rather more burin spalls (15.0%) than LSFs (about 8.0%) came from edges that had not been re-touched at all.

The transverse sharpening flakes were identified (later, in the laboratory) by the primary retouch on the dorsal surface convergent with the butt — itself always plain. The morphological differences that distinguish them from LSFs are immediately apparent (Fig. 29.1, 1–3 and Fig. 29.13), and are borne out metrically by their mean length (21.1 mm), mean breadth (21.9 mm) and L/B ratio (1.05) (Table 29.1E; Figs. 29.1, 1–3 and 29.13). The inversion of the values for length and breadth should be explained: to ensure comparability, the definition of 'length' as the axis of the removal blow on the bulbar surface was used for both LSFs and TSFs. Thus measured, the breadth of TSFs sometimes exceeds the length, particularly in the lower layers; by layer C, the normal relationship is restored. Other major changes in the TSFs between the lower and upper layers call the linearity of their evolution into question, but their original defining traits — plain butts and transverse removal — persist throughout. They share with the LSFs the function of rejuvenating an edge, but whether the purpose of removal is the same in each case is open to doubt (see section 29.5). Certainly it is different in effect: an edge from which a succession of TSFs had been removed would be uneven, or at best undulating, unlike the clean, continuous edge produced by a single longitudinal removal.

**Table 29.3**  Left/right Asymmetry of LSFs with Facetted and with Plain Butts

| Butt type | Left % | Right % | Sample |
|-----------|--------|---------|--------|
| Layer A | | | |
|   Facetted butts | 16.5 | 83.5 | 565 |
|   Plain butts | 28.6 | 71.4 | 174 |
| Lower layers | | | |
|   Facetted butts | 20.2 | 79.8 | 64 |
|   Plain butts | 39.3 | 60.7 | 99 |

As the analysis proceeded, some differences between the two main subgroups of LSFs emerged (Table 29.1, C and D). Those with plain butts would have left a rather less acute angle on the new edge (56.5°) than would those with facetted butts (52.5°), as reflected in their less obtuse removal angle — 123.5° against 127.5°. Their total length, also, is rather less (28.4 against 30.4 mm). But in terms of function this is hardly significant, as it is the *removal* length, not the total length, that corresponds to the length of new edge created. In this, the two types are almost identical — 22.9 and 23.2 mm respectively. In asymmetry, those with plain butts are less strongly biased to the right, although the difference becomes less pronounced with time (Table 29.3).

No significant difference in type of retouch was found between them (therefore all LSFs are combined in the diagram (Fig. 29.14) showing retouch distribution by layer). Rather more of those with facetted butts — 25.0% against 15.0% — show the scar of a previous removal on the dorsal surface. This may reflect the role of facetting in a chain of resharpening activity wherein the need to strike a greater length from an edge could best be met by more careful platform preparation. The reversal of their relative frequencies is in itself a strongly discriminatory feature. In the lower layers (and excluding the other two subgroups) there are 39.3% with facetted butts and 60.7% with plain; by layer A, the former have risen to 76.5% and the latter fallen to 23.5% (Fig. 29.8).

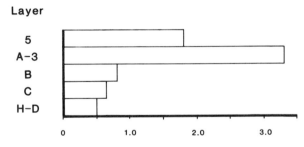

**Fig. 29.8**  Ratio of LSFs with facetted butts to LSFs with plain butts.

The LSFs with dihedral butts are too few (2.8% of the total) and those with indeterminable butts too poorly defined — by this criterion — to justify more detailed study. They are, however, included in the figures for overall frequencies (as are broken specimens) and for the distribution of asymmetry and of retouch by layer.

## 29.4  PLACE IN THE INDUSTRIAL SEQUENCE

Although always a small proportion of them numerically, the sharpening flakes echo the response throughout the assemblages to diminishing supplies of flint. They belong to the overall linear pattern of covariance that can be observed from the lowest layers — until its disruption in layer 5 — and, in their steeply rising rate of increase, throw a particular light on it (Fig. 29.9).

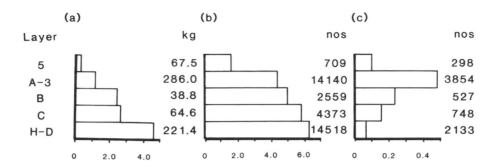

**Fig. 29.9** Covariation between the total amount of flint relative to other raw materials (by weight) and the frequencies of flint waste, tools and sharpening flakes. (a) Ratio of flint to other raw materials; (b) ratio of flint waste to flint tools; (c) ratio of sharpening flakes (LSFs and TSFs) to flint tools.

In the lower layers, H to D, where nearly 90% of the artefacts are of flint, they arc too few to be separately tabulated. Even when the totals are combined, they would not warrant closer analysis if it were not for their later increase and evolution, especially in relation to retouched flint tools. In these lower layers, also, flint tools are at their lowest in relation to general flint waste. It is the subsequent differential rates, of increase in sharpening flakes to tools on the one hand and of decrease in general waste to tools on the other, that is of specific interest. The first appreciable rise in sharpening flakes, in layer C, coincides with a drop both in the proportion of all flint to other raw materials and in the ratio of general flint waste to flint tools. As this trend continues, until its culmination in layer A, the concomitant rise in the ratio of sharpening flakes to flint tools is increasingly steep.

The changes in relative frequency are accompanied by a reduction in the size of retouched tools, from a mean weight of about 17.0 g in the lower layers to 12.3 g in A. The intensification of flaking activity implicit in this process might be expected to raise, rather than lower, the proportion of general waste to tools. Instead, it is the by-products of the specialized sharpening techniques that accrue. At the same time, the excess of LSFs over TSFs rises sharply: previously between one and a quarter and three times more numerous, they become, in layer A, over six times more so (Fig. 29.4). The change is due to the increase in the ratio of LSFs to tools (from 0.15 in B to 0.45 in A) rather than to the slight decrease in that of TSFs to tools (from 0.08 in B to 0.06 in A). The decrease in TSFs could be explained partly by a mechanical factor: the rise in relative thickness (breadth/ thickness ratio) that would accompany the progressive reduction in tool size would render the removal of transverse flakes from the steeper edge ever more difficult. The greater increase in LSFs could reflect a growing shortage of fresh blanks with suitably long edges for cutting use and the consequent need to resort to existing retouched tools for the making of these. The shortage of blanks could in turn be explained by the conversion of greater numbers of those still available into retouched tools, as the higher proportion of tools to other flint seems to indicate; this, on the other hand, should have been offset by an increase in the number of smaller waste flakes, if knapping took place on the spot. An evaluation of the roles of the large and of the small waste flakes (i.e. blanks and debitage respectively) might help to resolve the paradox, but is handicapped by the exclusion from the present analysis of the very flakes — those smaller than 20 mm — that would have been produced in the greatest numbers. The alternative explanation, that finished tools were being brought in from elsewhere, is difficult to reconcile with the the economies to which the sharpening flakes and the tools themselves bear witness.

A further point of comparison between the LSFs and the general waste is the differential incidence and rise of facetted butts within each group (Fig. 29.10). They are always more numerous among the LSFs, in layer A amounting to 75.3% of those with determinable butts. The lower frequency and more gradual rise (itself of interest) of facetting in the general waste reflect, by comparison, the greater care needed in the execution of the more specialized technique and the role of platform preparation in accomplishing a predesigned end.

In layer 5 the picture changes. No longer are sharpening flakes an integral feature of the response to flint shortage. While total flint drops even more sharply than

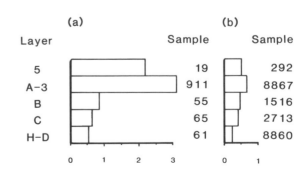

**Fig. 29.10** Ratio of facetted to other (determinable) butts: (a) LSFs (b) general flint waste.

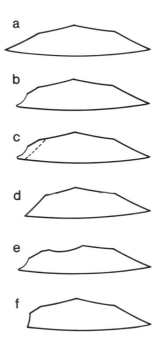

**Fig. 29.11** Diagram of a parent piece in cross section to illustrate LSF removal (a–d) and other types of longitudinal removal (e, f). (a) Before and (b) after modification of the edge; (c) showing the location of the 'future' LSF, its bulbar surface in broken outline; (d) after detachment of the LSF; (e) after detachment of a flake from the dorsal surface; (f) after detachment of a burin spall.

before, to 22.8% of all raw materials (by weight), and the ratio of flint waste to flint tools falls from 4.4 to 1.6, there is no corresponding rise in sharpening flakes. On the contrary, their ratio to retouched tools drops from 0.45 in A to 0.10. This new pattern of inter-relationship is reflected throughout the assemblage, particularly in the attributes of the retouched tools (chapter 26), and points to a fundamental change in behaviour.

### 29.5.1 Technique and Purpose

The morphological attributes of the long and the transverse sharpening flakes have been defined and described in section 29.3. It remains to examine the process of manufacture and discuss its purpose.

To produce an edge of the optimum sharpness and length the vertical blow for removing a LSF must be more finely directed and angled than that required for burin spall removal because of the complexity of the planes to be manipulated. Faulty coordination or orientation in any of these will lead to a defective result. The holding hand must simultaneously adjust both the angle of tilt and the degree of rotation. If the blow is too outwardly directed the new edge will be too short; if too inwardly, it may break out abruptly in a hinge fracture, which would impede effective use (Figs. 29.1, 6, 12 and 29.3). According to which edge is the target, too much or too little rotation will result either in the removal of a

flake from the dorsal surface, leaving the edge unchanged (Fig. 29.11e), or of a burin spall, leaving a burin facet rather than a cutting edge (Fig. 29.11f). The angle of removal from the parent bulbar surface is of particular importance. The greater its obtuseness, the acuter the angle and the greater the sharpness of the new edge (Figs. 29.7, 29.11, a–d). Competence in achieving this increases as the frequency of LSFs with facetted butts rises: in layer A their mean removal angle is 127.3°, whereas for LSFs with other types of butt and in earlier layers (combined) it is 122°.

The preference, in 78.5% of cases, for removing LSFs from the dorsal rather than the ventral surface can be explained by the cleaner edge thereby produced. Thus, it is formed by the meeting of two continuous, unbroken planes: the bulbar surface of the tool and the new (negative) scar on its dorsal surface left by the removal of the flake. If removed from the ventral surface, one of these planes would be formed by the tool's dorsal surface, ridged and furrowed by truncated retouch or primary scar facets. The relationship between the location of removal and the preference, in 63 (79.0%) cases, for striking the flake from the edge of the tool nearest, or proximal to, the grip, rather than from that distal to it, is

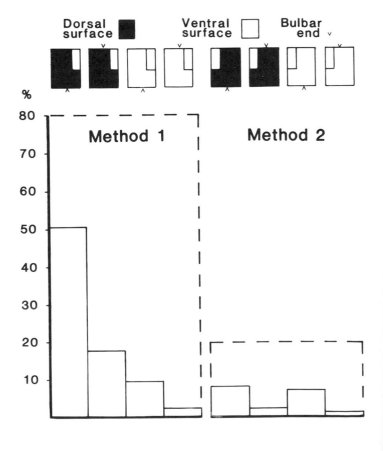

**Fig. 29.12** Distribution of LSF scar position on parent tools, relative to their dorsal and ventral surfaces and distal and proximal ends. The four positions are schematically represented above as if each piece were held in the left hand and struck with the right, and then turned over to reveal the new scar. Method 1: removal from the same edge as the grip; Method 2: removal from the edge opposite the grip (n = 76).

**(a)**      **(b)**

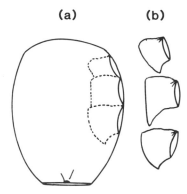

**Fig. 29.13** Successive overlapping TSF removals, reconstructed from the patterning of scars on TSFs. (a) Bulbar surface of a tool showing the platforms for 'future' TSFs (plain line) and corresponding removals from dorsal surface (broken line), less overlap; (b) the flakes detached (note the point of percussion).

shown in Fig. 29.12. The bearing of the factor of handedness on this distribution is discussed in section 29.6.

Striking from this edge is in accordance with the assumption, based on observation and experiment, that in using direct percussion techniques 'the tendency was to strike flakes down and away from the knapper' (Bradley and Sampson 1986). To some extent it was endorsed also by a short experiment in replication carried out by Peter Jones, research assistant to Dr. Mary Leakey at Olduvai, during a morning's visit to Cambridge. Flake tools were manufactured and flakes removed from their retouched edges by vertically directed blows, after platform preparation by abrasion. Eight fairly close replicas were produced, seven of which were struck from the proximal edge (as defined above). Attempting to strike from the other edge resulted more often in the detachment of a burin spall or of an abruptly hinging flake from the dorsal surface. But even the 'successful' removals fell far short of the prototypes in obtuseness of removal angle (116° and 127° respectively), and thus, correspondingly, in the sharpness of the new edge.

No systematic attempt has yet been made to replicate the transverse sharpening flakes — in their later, more complex form at least. The technological peculiarities of these are of particular interest. Until layer C, almost all TSFs are short, broad and thick, the product of a blow at right angles to the tool's edge, the point of whose impact lies correspondingly within the central area of the butt. Each such removal would leave a deep Clactonian notch on the parent edge, as reflected in their thickness (4.9 mm). In layer C (when also the first important rise in LSFs occurs) their thickness falls to 3.2 mm; their length/breadth and breadth/thickness ratios move from 0.86 and 5.3 to 1.2 and 7.0 respectively, and the point of impact with increasing frequency lies at the extreme end or corner of the butt, at the point of intersection with the dorsal surface (Fig. 29.1, 2–3). In this process the move is almost exclusively in one direction (see Fig. 29.18 and section 29.6). The more refined technique re-

moves a series of flat, overlapping, scale-like flakes down the length of a retouched edge by a succession of obliquely angled blows (Fig. 29.13). The shallow diffuse bulb and the butt's overhanging lip are features often associated with the use of a soft hammer. The angle of the butt to the bulbar surface becomes progressively more obtuse, from 123° in layers H–D and 126° in C–B to 138° in layer A. Thus the edge created by each such removal would be even sharper than that left by the removal of LSFs with facetted butts in layer A (removal angle 127.5°), but very short, corresponding to the length of the flake's butt (mean 10 mm). A series of such would leave an undulating, scalloped edge, less useful for cutting than one created by a LSF removal. No scars of TSF removals were detected on the tools; all trace of them may have been obscured or obliterated by subsequent lateral retouch.

The view, based on their technological features, that LSFs were not themselves intended for use is corroborated by microwear study (chapter 30). The 55 that have been retouched along the new edge (and have therefore been classified as tools) amount to only 4.3% of the total (1357, if they are included); their greater mean length (41, against 30 mm for the rest) would have made them, fortuitously, more eligible for further retouch and use. To the naked eye, none of the rest appear to have been even utilized on the new edge. Only one of those microscopically examined bore traces of wear. Such traces as were found on primary edges could more reasonably be attributed to use before detachment from the parent tool.

Further clues to the purpose of the technique might be sought in the state of the primary edges. The majority of these (70%) are retouched, but a substantial minority (22%) are utilized or damaged only. The types of retouch on the flakes (Fig. 29.14) might be expected to agree broadly with those on the tools, in distribution of frequency from layer to layer. A significant disagreement, on the other hand, could point to a tendency to favour removing one sort of retouched edge above another. In the lower layers, a broad correspondence is indeed apparent in the prevalence of coarse stepped and denticulate retouch on tools and flakes alike, though the 'scale and nibble' category is already high for the LSFs; (later it becomes by far their most common). As this category is not among the accepted designations by which the retouch on the tools was classified, the opportunity for further direct comparison is limited. But it could be informative that the need to nominate such an unorthodox category became compelling in the course of examining the LSFs as a group, because of its distinctiveness and very frequent occurrence (Fig. 29.1, 8 and 10 are examples of it). The 'nibble' component may well be due to use, and thus in itself help to provide an explanation for the flakes' detachment, though these edges are not very much less keen than those with scalar retouch only. In any case, however, either type of edge could equally well have been sharpened or rejuvenated by lateral retouch while still part of the parent tool, if the re-creation of a similar edge for a similar function

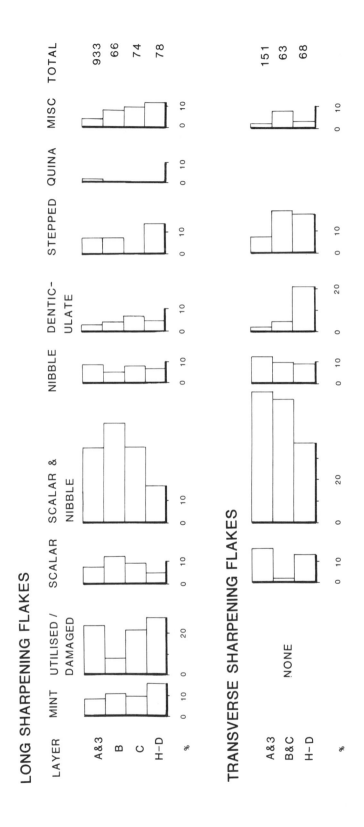

**Fig. 29.14** Distribution of retouch on (a) LSFs (all) and (b) TSFs.

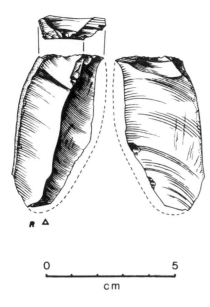

0 _____ 5

cm

**Fig. 29.15** Atypical knife (on *couteau à dos naturel*), resharpened by LSF technique. Its history includes the following stages: (a) removal of a blade with cortical back from a core with opposed platforms; (b) loss of proximal end through breakage; (c) fabrication of an inverse truncation on the break; (d) rejuvenation of the right hand margin (with loss of the butt) by means of a LSF removal (it is not known whether the rejuvenated edge had previously been retouched). From the A/3 interface.

were intended. Only the edges with stepped retouch (about 8%) are of too steep an angle for this to be effected easily.

The flakes with damaged or utilized edges have, of course, no equivalents among the retouched tools. They would have been struck off the utilized edge of a blank, in their case probably for continuance rather than change of function. Those with mint-fresh edges are more difficult to explain, but one can surmise that they are the product of an abortive attempt to remove damage or blunting further down a parent edge. Their shorter length (20 mm, against 30 mm) lends some support to this. That length was a primary object of LSF removal is indicated by the presence of short removal scars on the dorsal surface of 22% of the LSFs themselves. The mean length of these is only 14 mm, and their edges are never utilized or damaged. They would seem to record the failure of a first attempt to create a serviceable length of new edge which was then followed by the second, successful, attempt that produced the LSF on the back of which their scar remains.

The discussion so far has focussed on the by-products of the technique rather than its products. This is because it was through them that it was recognized in the first place, because their far greater numbers provide a better statistical sample for analysis, and because they mirror the technological event more faithfully, having remained virtually unchanged since the moment they were made.

The identification of the products came later, and was more difficult. It was geared initially to the hope and expectation of conjoining at least some of the flakes and their original pieces, with a view to learning more about the mechanics and purpose of the technique. All the retouched tools were systematically re-examined with this object. A total of 105 removal scars were identified, but only the 79 (53 from layer A) that were longer than 20 mm were eligible for matching, as all flakes (of whatever kind) below that size had been excluded at the outset (see chapter 21). The overall ratio of flakes to scars is 19.5 in layer A, 17.0 in the other layers (together). The remarkable paucity of parent pieces, compared with flakes, may be attributable to extensive further working having rendered many of them unrecognizable; moreover, while a single piece may have had a succession of LSFs detached from it, it will bear the scar (or scars) of the most recent only. In a few instances (e.g. Fig. 29.15), the whole length of a parent edge had been successfully removed, making their detection more difficult. Few, however, are likely to have escaped identification in the course of successive increasingly critical rounds of scrutiny, whether of the retouched tools or the waste, because of evidence of platform preparation associated with a negative cone.

It was quickly found that no matches could be made between any of the parents and the flake candidates from the lower layers. The far greater numbers of each in A required more careful organization to optimize the chances of success. All LSFs and parent pieces were sorted into staggered groups by length and scar length at 2.5 mm intervals. In the event, however, no matches were found between any of these either. The sole exception, illustrated in Fig. 29.3, had been effected by pure chance during the initial sorting of material long before. The two pieces come from layer 5 and were found some 2.5 m apart.

This negative result was quite unexpected and seems to require explanation. Two factors are proposed as contributory to it. The first is that the scar surviving on a piece after several LSFs have been removed from it will be beyond the size range of any smaller previous removals — and indeed the excess of flakes to scars is greatest in the two lowest size bands. But this cannot explain why none of the surviving scars could be matched with any of the flakes of corresponding size that were available. Here the factor of mobility comes into play. A tool may be used first where it was sharpened but then carried away for use elsewhere, either to another part of the living floor within the cave or outside and far beyond it. Equally, it may have been sharpened elsewhere before being finally dropped in apparent association with flakes that had never belonged to it. This factor will operate all the more strongly where, as at La Cotte, the excavation of the living floor is incomplete. It is estimated, for example, that only about a third of layer A has been uncovered; the chances of finding fitting pieces are therefore correspondingly reduced. The force of the mobility factor is demonstrated by the fact that not one of the parent pieces (105 in all) was made from any of the twelve rare and highly distinctive types of flint that had been designated in the laboratory, although nineteen LSFs were made from one or other of these.

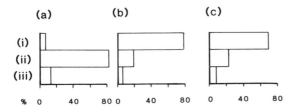

**Fig. 29.16** State of (a) the new edge, (b) the remaining part of the original edge on 76 parent pieces, and (c) LSF edges: (i) retouched, (ii) utilized, and (iii) mint.

Study of the parent edges themselves was more rewarding. Comparison of the state of the remaining length of old edge with that of the new edge created by LSF removal showed that at least 58 out of 76 (72.2%) edges had originally been retouched, whereas the new edge had been retouched in only 5 (6.4%) cases. Utilization, however, was clearly visible on 62 (81.5%) of the new edges; the remaining 10% were apparently mint (Fig. 29.16). How many more new longitudinal edges may in fact have been retouched, but too invasively for any trace of them to remain, it is of course impossible to say. But the survival of this fairly substantial number which had *not* been retouched, but had clearly been put to use, is consistent with the view that it was the cutting property of a new edge that was the object of removing the old one. Microscopic examination of the only two parent edges not too heavily patinated for study also indicates a backward and forward action, as used in cutting (chapter 30). One bore traces of wood and the other of antler polish.

The angles on these edges were 44° and 47°. The angle of a cutting edge is important to its effectiveness. The mean for the 53 parent edges in layer A is 49°; the estimate of the mean edge angle for the many more, unknown, pieces from which the LSFs had been removed, derived from the complement of the LSFs' removal angle (Fig. 29.7), is 52° — remarkably close, given the manipulative difficulties of measuring between the various planes. Indeed, as measured, the angles on our only two refitted pieces (the parent's 65, the flake's 119°) add up to *184°*.

The mean edge angle of 51° (reached by combining the parent edge angles with those derived from the flakes) is considerably less acute than the optimum for cutting suggested by experimental studies and ethnographic observation. This is something between 25° and 40°, while for scraper use an angle between 46° and 67° is most effective (Semenov 1964; Wilmsen 1968; Gould and Koster 1971; Hayden 1977). At La Cotte, where existing flint was reworked to the limit (tools in A to a length and breadth of about 41 mm x 28 mm, for instance) the feasibility of creating an 'ideally' sharp cutting edge would be steadily reduced. A sharper edge at an earlier stage in the process can be gauged from the value of the dorsal angle (Fig. 29.7) which gives an approximate estimate of the parent edge angle at one stage further

back, before the current 'round' of retouch preserved on a LSF's primary edge. The mean for this — 44.0° — is significantly closer to the suggested optimum.

### 29.5.2 Parallels and Purpose

Besides the *coup du tranchet* technique already well known from removal scars on biface edges and their corresponding waste products, there is a longitudinally struck variant which closely resembles that represented by the LSFs at La Cotte. Removal scars typical of it have been observed, particularly, as a common but not essential feature of a small class of tools known as *prondniks*. First singled out from sites in Poland near the river of that name, these have since been identified in Middle Palaeolithic assemblages from Germany, central France and Belgium (Desbrosse, Kozlowski and Zuate y Zuber 1976). Their attributes have been variously defined, but common to all are a broad, thick base and a pointed distal end, a curved back on one side opposite a straight edge on the other, and bifacial working irrespective of whether they are made on flakes or cores. A *coup du tranchet* removal is not integral to their design, and where it is present the fact that it truncates a line of retouch down the tool's edge has not excited comment. None of their essential features is in fact commonly characteristic of the parent pieces from La Cotte (nor of those that can be inferred from the LSFs, only 6.0% of which bear traces of bifacial retouch), with the notable exception of the piece shown in Fig. 29.3. Furthermore, the great majority are on unifacially worked flakes. For comparative purposes, however, the real disparity between La Cotte and other European sites is the absence from these of any record of the corresponding by-products. They may (indeed must) exist, but have escaped notice — perhaps because of relatively poor representation in an assemblage. If they now come to light, it will be of great interest to see if any correlation with environmental pressures can be observed in their occurrence, as appears to be the case at La Cotte.

Otherwise, the nearest parallels known so far come from prehistoric Indian sites in North America. From two in Texas (Shafer 1970 and 1971) both the transverse and the longitudinal type of flake have been recovered, as well as parent pieces with longitudinal removal scars. At Piney Creek, in Wyoming (Frison 1970), transverse flakes with features strikingly reminiscent of those from the upper layers at La Cotte were found. The edge angles on the parent scars and the removal angles on the longitudinal flakes from the Texan sites were measured in the same way as those from La Cotte, but the total values for these were combined with those of a different type of removal flake so that no direct comparison is possible. The flakes are referred to as 'burin spalls' — and none of those illustrated approach the flatness of LSFs — but the purpose of their removal is judged to be the creation of fresh, sharp edges on 'dulled unifaces'. Apparently there was no shortage of good raw material here but, as with La Cotte, no flakes and tools could be conjoined. Here, too, the excavated area was only a part of the estimated whole. At Piney Creek, a buffalo drive kill site, several 'transverse sharpening flakes', some of the overlapping variety described earlier in this chapter, were refitted either to each other or to their parent edges. Moreover the distance between the find spots of the flakes and the parent pieces, and the change in function that they attest, tend to support both the explanations proposed for the failure to refit pieces from La Cotte. Here, where the whole activity area was excavated, large numbers of small transverse flakes removed from retouched edges — but very few large flakes or retouched tools — were found in the immediate area where the buffalo had fallen. Some 30 m away, in a secondary butchery area, the tools from which these flakes had been removed were found in concentration, some of them further modified by scraper retouch.

The importance of raw and utilized edges in the interpretation of function and activity is becoming increasingly recognized. Palaeontological evidence from kill sites of elephant and mammoth (Clark and Haynes 1969 and their references) in Africa, Spain, Germany, Siberia and Hungary has already drawn attention to the absence or scarcity of retouched tools or large flakes (e.g. cleavers) in the immediated vicinity of carcases in the primary kill area. In the experimental field, some workers (Walker 1978) have found raw stone (flint?) edges to be the most effective for butchering animals in the middle size range (sheep, goats etc.). Others (for example Jones 1980 and 1982) have found such edges too brittle and their butts too thin and tiring to hold for more than very short periods of use. Jones therefore questions the validity of concluding that small unretouched waste flakes were preferred for butchery simply because they are almost the only artefacts to be found at some kill sites. The answer he suggests is that the larger 'business' tools were taken away for use elsewhere and their by-products left behind. But this does not dispose of the evidence for the use of raw edges at La Cotte, where the artefacts in question are not small waste flakes but relatively substantial previously retouched tools.

In any case, it is clear that typological systems and interpretations of tool function can no longer be based on the evidence of retouched tools alone. The material from La Cotte can contribute to the re-evaluation by showing that unmodified edges were not only used but deliberately manufactured for use.

## 29.6  THE EVIDENCE FOR HANDEDNESS

Before discussing the wider implications of our findings, it should be stressed that the presence and direction of asymmetry in the long sharpening flakes is quite unmistakable. It is not a matter of degree or opinion, or an arbitrary point in a continuum, but is established at once by the position of the new ventral surface relative to the facet of the old one adjacent to it. Although this facet was originally part of a ventral surface, bulbar features are barely discernible on it compared with those on the new ventral surface, alongside which it appears smooth

and lustrous. Unfortunately, except where this facet is to the left of the new ventral surface, as in Fig. 29.1, 5, 11 and 13, this distinctive contrast tends to be lost or disguised in illustration because of the convention of drawing specimens with the light coming from the left.

At an early stage it was noticed that this facet lay far more often to the right (as in Fig. 29.1, 4, 6–10, 12). On analysis this proved to be the case for between 71 and 84% of the sample (varying according to butt type and layer). Such a preponderance can hardly be due to chance, or to the random contingencies of flint-knapping. It seems reasonable to suppose that it might be associated with the makers' use of a predominantly preferred hand, especially as it approximates to the preponderance of right-handers to left-handers today. The incidence of left-handedness in modern populations has been variously estimated, by different criteria and in different cultural contexts, at between 5 and 20%. A recent survey (of 1100 British children) put it at 11.2% (McManus 1981).

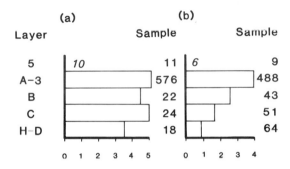

**Fig. 29.17**  Ratio of right to left asymmetry of LSFs: (a) with facetted butts; (b) with other types of butt, including indeterminable.

Our sample consisted of 1590 unbroken specimens (1302 long and 288 transverse sharpening flakes). The attribution of sidedness was based solely on the mode of production. The incidence of right asymmetry (Fig. 29.17) is highest for the long sharpening flakes with facetted butts in layer A. This underlines the association of manual operations requiring a high degree of precision and control with the use of the most 'dextrous' hand, as the evolution of the TSFs also illustrates — though rather differently. In the lower layers, where most of the TSFs are detached by a blow struck roughly at right angles to the tool's edge, the point of impact falls correspondingly within the central area of the flake's butt. In the higher layers, increasingly, a more refined technique to create a much smoother, sharper edge was used. The flakes were removed with an obliquely directed blow which leaves its point of impact at the very end or corner of the butt (Figs. 29.1–3 and 29.13). As the process of shift in the position of the point of impact from mid-butt advances, it is almost exclusively in one direction, or to one end rather than the other, that it 'moves' — that is to say, to the end distal to the observer, with the

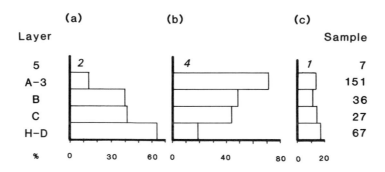

Fig. 29.18 Position of the point of impact of the removal blow on TSF butts, with the butt oriented to the right of the bulbar surface: (a) central, (b) distal to, and (c) proximal to the observer. (Unlike LSFs, TSFs can have been struck only from the edge opposite the grip).

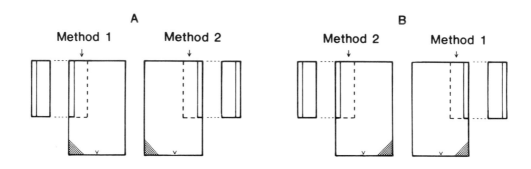

Fig. 29.19 Diagram illustrating the relationship of the sidedness of the asymmetry on an LSF to the edge of the parent tool from which it is struck relative to the grip and to the hand in which it is held. With bulbar surface facing, the stipple shows the position of the grip on the parent piece held in the left hand (pair A) and in the right hand (pair B). Method 1: the flake is removed from the edge on the same side as the grip; Method 2: from the edge opposite the grip. Note that the removed strip of the parent bulbar surface is on the same side of the flake in A/1 as in B/2, and in A/2 as in B/1.

butt to the right of the bulbar surface — until, in layer A, this is so in 84% of cases (Fig. 29.18), thus matching the percentages of right-asymmetrical LSFs (with facetted butts) in this layer.

It might be argued that the preponderance of right asymmetrical over left asymmetrical flakes in our sample does not justify the inference that the makers were right-handed and left-handed in the same proportion. Certainly a flake struck off by the right hand from the near edge of a tool will be identical, in the sense of asymmetry, to a flake struck off by the left hand from the far edge (Fig. 29.19). But the chances of striking from one edge rather than the other are equal in either case. Similarly (although the sample is small) the evidence from the 76 parent pieces of a preference for striking LSFs from the edge on the same side as the grip (Method 1) rather than from the other (Method 2) in 80% of cases (Fig. 29.12) is likely also to reflect the distribution of handedness in the population. The argument is systematically demonstrated in Appendix D and Fig. D.1. It disposes of the anticipated objection that there is no way of knowing

whether the asymmetry of a flake is due to its removal with one hand by one method or with the other hand by the other. The alternative inference that the data would stand, that it was *left*-handers that made up the majority, would be admissible only if it could be supposed that the whole trend of handedness had become reversed during the last 200,000 years.

If these findings and the interpretation put on them are valid they could have a bearing on the antiquity and evolution of human handedness, and even of speech. Hand preference is thought to be a typically human characteristic. In other mammals, including primates, preference does occur in the individual but it is divided equally between left and right in the population (Passingham 1982). The association of both handedness and speech with cerebral dominance is well known, although the precise nature of the association and the order of precedence within it are not fully understood. The cerebral hemisphere responsible for speech is almost invariably on the left in right-handed people and damage to it results in impairment of speech and of function in the right hand. It is not known when or by what stages this triad of human evolution emerged but there is some evidence that one or more of its components was present from a very early date.

Insubstantial indications, for handedness and for speech, reach back some 1.5 to 2.0 million years to the australopithecines of Africa. That a majority of these were right-handed was deduced by Dart (1949) from club injuries thought to have been inflicted by them on the skulls of forty baboons and six other hominids. Seventeen blows were delivered laterally. Of the eleven to the facial area, nine are on the left; the six to the back of the skull, in the parietal region, are all on the right. More conjectural, but based on more extensive observations, is the evidence from concentrations of broken food bones and stone artefacts of hominid societies in which language-like communication would have important functions (Isaac 1975; 1982). The habit of bringing meat back to a home base for sharing is not characteristic of other primates, and would tend to require exchanges of information beyond the range of primate call systems. From Neanderthal burials much later (c. 50–75 kya), traces of funerary ritual have been interpreted as evidence for conceptual thought, for which speech is held to be prerequisite (Solecki 1975; Leroi-Gourhan 1975).

Patterns of asymmetry similar to those present in the brain of modern man (and some higher apes) have been observed on endocasts of seven fossil hominid skulls. The three Neanderthals (Spy I, Spy II and Djebel Irhoud) and the *Homo erectus* specimen (Salé) studied by Holloway (1981), and the Neanderthal from Chapelle-aux-Saints and two *Homo erectus* skulls (Pithecanthropus I and II) summarized and discussed in Galaburda, LeMay, Kemper and Geschwind (1978) show these. In all but one — Pithecanthropus I — they are on the left, as in right-handed man today.

For the comparatively recent stage of human evolution marked by *Homo sapiens sapiens*, it is often alleged that the Upper Palaeolithic cave paintings of western Europe demonstrate a preponderance of right-handedness in the population by the 'vast' (one writer) majority of left-facing animal profiles they portray. This simply is not the case. Founded on the natural propensity of right-handers and left-handers in modern conditions to draw profiles facing to the left and right respectively, the assertion originates in a misreading of Wilson's largely conjectural remarks (1885 and 1891) suggesting that the paintings *might* provide a clue to handedness, since, of the thirteen that were then known to him, eight faced to the left and five to the right. The many more such paintings discovered since have not borne out his surmise, however. In fact, when some 300 paintings — 100 or so from Lascaux (Windels 1948) and 200 of those figured by Breuil — are considered, the proportion facing each way is almost exactly equal; on the other hand, 124 out of 137 of the hand outlines on the walls of Gargas cave are of the left hand (Breuil 1952). But, in any case, if the correlation between speech, handedness and cerebral dominance holds good, a preponderance of right-handers among Upper Palaeolithic groups might be assumed, since the existence of speech in the full sense by this time (15–20 kya) can hardly be doubted.

It might be expected that, above all else, the tools actually made and handled by early man would yield information of particular relevance to the matter, since they are available in large numbers over the whole time span in question. Until recently, however, apart from occasional perceptive but unsystematic observations (e.g. Wilson 1885), work in this field has concentrated on the microscopic study of edges and surfaces with a view to detecting directions of use. The conclusions that can be drawn from these are limited by uncertainty about the mode of handling and holding in the first place. Tools involving a boring action could prove more informative, since the mode of holding is not in dispute and it appears that right-handers rotate a boring tool clockwise and left-handers anti-clockwise, where any degree of force is required. Convincing traces of anti-clockwise (as well as clockwise) rotation have been found on borers from Meer II in Belgium (Cahen and Keeley 1980), but so far in too small numbers for statistical purposes, and again from a site probably too recent (c. 9 kya) to be relevant to the origins of handedness or speech.

The material from La Cotte belongs to a stage of evolution intermediate or transitional between *Homo erectus* and Neanderthal man. Unlike much of the evidence outlined above, the findings are unambiguous, drawn from a large enough sample to be statistically acceptable, and grounded not on any hypothetical use but on the act of manufacture itself. New criteria for determining handedness based on experiments in tool making are emerging from the work of Bradley and Sampson (1986) and Toth (1985). It is to be hoped that these will prove generally applicable and that the identification of asymmetrical features will become a routine procedure in the study of stone tools.

## NOTE

The possibility that stone tools would yield evidence bearing on the handedness of early man was a particular interest of Charles McBurney's. He had begun to approach the question through edge wear study and by seeking to find a clue to the mode of holding and position of grip in the course of handling specimens in the laboratory. Sadly, however, awareness of bias in the asymmetry of the sharpening flakes was only just beginning to dawn at the time of his death, and its possible significance had not been realized.

We should like to thank Dr. Bernard Campbell for his initial encouragement from the anthropological viewpoint, and also Dr. C.B. Stringer, of the British Museum (Natural History) and Drs. J. Harriman, J. Harris and J. Musgrove, of Bristol University, for their interest and help.

30

# MICROSCOPIC USE-WEAR TRACES

H. Frame

## 30.1 INTRODUCTION

Microwear analysis is a comparatively new technique for determining the function of stone (usually, but by no means only, flint) artefacts from a microscopic examination of their surface and working edges (Anderson 1980, Anderson-Gerfaud 1981; Keeley 1980; Odell 1977). It has long been hoped that such traces of use might be found on archaeological specimens (e.g. Curwen 1930) but it was not until the published results of S.A. Semenov's work were translated into English in the 1960s (Semenov 1964) that the reality of such techniques reached a wider audience. The years since then have been a period of considerable advance, in the West at least, in both our use and our understanding of the method. This progress stems from an early dichotomy, based on the magnification ranges employed. The aim — to assign specific functions to specific artefacts — is the same in each case, but the approaches differ. The low magnification method for the most part employs a binocular microscope capable of magnifications up to x80, and is concerned with recording the extent and orientation of damage scars on the edge and striations on the surface of an artefact. The high magnification method, on the other hand, uses a stereomicroscope with a magnification capability usually in the region of x50 to x500, sometimes more, and concentrates principally on the location and differentiation of particular 'polishes' on the tool surface. The two methods overlap insofar as that the examination of striations and edge damage is a part of the high magnification approach also: the extent and orientation of striations play an important part in the process of analysis, but the characterisation of edge damage takes a very secondary place in the high magnification technique, whereas it is fundamental to the low magnification method.

It has to be admitted that similar types of traces may be produced by agencies other than use. For example, edge damage and striations can be the result of post-excavation handling and storage. Any but the most controlled cleaning processes and the most careful storage conditions could damage the surface and especially the more fragile edge of an artefact. Equally, such damage could be the result of soil movements during or after deposition, or the trampling of the cave floor by the occupants. Moreover, some technological effects of tool manufacture can create pitfalls for the unwary (Keeley 1974); what may seem to be the typical pattern of edge damage scars resulting from use may in reality be the result of the method of tool manufacture such as the

preparatory 'scrubbing' of an edge to provide a suitable series of platforms for pressure flaking.

The removal of a flake from the edge of a tool, or the formation of striations on its surface, is the work of an instant, but polishes are known to take longer to develop. It is highly unlikely that an artefact in a deposit would both come into contact with and be repeatedly rubbed against something such as a tree root for long enough for a polish visible under the microscope to form. The polish which develops when a stone object comes into contact with a stone artefact is quite distinctive and unlikely to be mistaken for any use-wear polish, nor would it necessarily be related to the working edge, as would be expected of true use-wear polish.

It would perhaps be more accurate today to call the x200 plus magnification work medium range in view of the recent advances made using the much higher magnifications possible with the scanning electron microscope (Anderson 1980; Anderson-Gerfaud 1981) which have added considerably to our knowledge of how polishes may be formed. Anderson-Gerfaud has shown that under some circumstances the part of the tool in contact with the material being worked 'dissolves' to an extent dependent on such factors as the relative coarseness of the tool's raw material and the substance being worked, the length of time the tool is in contact with the worked substance, and the amount of water present in the material. This very high magnification work has also shown that some recognisable plant and animal residues such as grass phytoliths and bone mineral may be found actually adhering to or embedded in the surface of the artefact. The importance of this in terms of the information it could give on the environment of a site need hardly be emphasised. There are, however, practical drawbacks to the S.E.M. approach. The equipment is not always as readily available as a more conventional stereomicroscope of medium magnification range, its cost being beyond the reach of most departments; (it is interesting to note (Odell and Odell-Vereecken 1980) that this same argument has been directed at the 'medium range' technique by those using low power equipment). Furthermore, the size of the objects that can be viewed is limited by the size of the S.E.M. view chamber. Anderson-Gerfaud was fortunate in having access to an S.E.M. with which objects as large as Mousterian hand-axes could be analysed.

The analysis of the La Cotte artefacts follows the medium range approach using a Leitz Metallux II microscope, fitted with an incident light attachment capable of producing both perpendicular and swinging oblique

light. Although the microscope has a possible magnification of x50 to x1000, most of the analysis is carried out
using x100 to x320 magnifications. Each artefact is
scanned over its entire surface at x100 to locate possible
areas of polish which are then examined in detail at x200
to identify the type of use-wear present, increasing the
magnification to x320 as an aid to the interpretation and
differentiation of the polishes.

## 30.2  ARTEFACT PREPARATION

Prior to examination, artefacts are cleaned to remove
any soil particles which might have survived post-
excavation cleaning, and also surface grease (the result
of handling) which tends to mask any traces of polish.
The cleaning process used is similar to that described by
Keeley (1980, 10–11) but with some modifications.
Artefacts are suspended in the solvents in an ultrasonic
cleaning tank so as to minimise the possibility of 'glass
polish' resulting from contact with the sides of the glass
beaker containing the solvents. The process ends with
cleaning in distilled water, as it was found that impurities in the methylated spirits could be deposited on the
flint surface and cause possible confusion in the analysis:
surfaces with this deposit could look at first glance quite
similar to the traces which are associated with processing
meat. So far it has seldom proved necessary to remove
mineral deposits from the surface of any artefact by im-
mersion in warm HCL. Pieces of unusual interest are
given this additional treatment if necessary. A few artefacts were coated with a deposit which I have been unable to remove.

## 30.3  COMPARATIVE MATERIAL

Fundamental to the microwear analysis of any assemblage is the compilation of a reference collection of
pieces with experimentally produced use traces to which
those discerned on the archaeological artefacts can be
compared (Anderson-Gerfaud 1981; Keeley 1980). To be
confident of a functional interpretation, the analyst
should be able to reproduce what he sees on the archaeological artefact. Obviously it would be preferable to
conduct the experiments on pieces made of the same raw
material as the archaeological specimens being examined.
The experimental programme for the study of the La
Cotte artefacts has proved difficult because of the lack of
readily available similar raw material (the possible origin
of the flint used by the inhabitants of La Cotte de St.
Brelade is discussed in chapter 22). However, two possible sources of raw material have been tapped: recently
collected beach pebbles from Jersey, and a few wholly
unstratified waste flakes and fragments from the excavations. When using excavated specimens as a source of
raw material for experiments there is always a danger
that they may already be patinated, stained or otherwise
altered. A simple way of avoiding this hazard is to re-

flake the pieces prior to the experiment to ensure that
fresh unaltered surfaces are in contact with the substance
being worked. Experimental tools are examined under
the microscope immediately before the experiment is carried out to check that the surface is fresh. This unmodified surface is photographed so that it can be compared
later with the same surface after the experiment has been
conducted. The result of these experiments using flint
from the Jersey area has shown that there is no appreciable difference between the features of these experimentally produced polishes and those which were formed by
similar work on implements made of the more readily
available flint from sources in the south of England. This
conclusion agrees with the findings of other analysts
using medium magnification (Anderson 1980; Keeley
1980).

## 30.4  SELECTION OF PIECES FOR STUDY

In view of the limited time available for the investigation, it was decided that the principal objective of the
work should be to obtain a better understanding of the
artefacts known as long sharpening flakes, or LSFs
(chapter 29), which are particularly numerous in layer A
and are such a distinctive feature of the La Cotte industries. Whatever resources could be spared from this
would be devoted to examining other types of artefact
for evidence of use. The process of choosing pieces for
study is not a matter of taking a random sample, nor
can it be. It is generally agreed that artefacts must be in
extremely fresh condition if use polishes are to be discernible under the microscope. Patination affects the surface of flint in ways which are not yet fully understood
but which without doubt make the detection of polish
virtually impossible. This then places an immediate limitation on the proportion of artefacts which will be in a
condition suitable for study.

At the outset a sample of 249 LSFs was available,
consisting essentially of pieces which had been washed
and classified by a certain date. Of these, almost all of
which were from layer A, 74 were rejected after macroscopic examination either because they were made of a
coarse grained raw material which is highly reflective
and does not yield worthwhile results (if any), or because
they were visibly patinated or stained. Fortunately there
is a considerable number of artefacts from La Cotte de
St. Brelade in a suitable state of freshness for microwear
analysis. The method of selection adopted subsequently,
for other categories of artefact, was to choose the most
obviously fresh and least damaged examples, largely
irrespective of provenance. Even so, many pieces proved
on microscopic examination to bear traces of patination
and staining which were not visible to the naked eye.
The extent of the problem is shown in Table 30.1: 'initial
sample' gives the number of artefacts which appeared to
be in sufficiently fresh condition for wear traces, if present, to be visible, and 'final sample' gives the number of
artefacts that eventually proved to be so. In other words,
41% of the pieces examined were in such a condition

**Table 30.1**  Polishes on All Classes of Objects Studied

| Type | | Initial sample | Patin- ated | Final sample | Type of polish | | | | | | |
|---|---|---|---|---|---|---|---|---|---|---|---|
| | | | | | Wood | Bone | Antler | Meat | Hide | Unid. | None |
| 1-2 | Levallois flakes | 1 | - | 1 | - | - | - | - | - | - | 1 |
| 4 | Ret. Lev. points | 1 | - | 1 | - | - | - | - | 1 | - | - |
| 6-7 | Mousterian points | 2 | 2 | 0 | - | - | - | - | - | - | - |
| 9-29 | Side-scrapers | 42 | 25 | 17 | 4 | - | - | - | 8 | 1 | 4 |
| 30-31 | End-scrapers | 11 | 8 | 3 | - | - | 1 | - | 2 | - | - |
| 32-33 | Burins | 30 | 11 | 19 | 1 | - | 1 | 1 | 3 | 1 | 12 |
| 36-38 | Knives | 15 | 11 | 4 | 2 | - | - | 1 | 1 | - | - |
| 39 | *Raclettes* | 4 | - | 4 | 1 | - | - | - | 2 | - | 1 |
| 42 | Notches | 18 | 5 | 13 | 2 | - | - | - | 3 | 2 | 6 |
| 43 | Denticulates | 10 | 2 | 8 | 1 | - | - | - | 3 | 1 | 3 |
| 44 | Retouched becs | 1 | - | 1 | 1 | - | - | - | - | - | - |
| 46-49 | Retouched/damaged | 8 | 5 | 3 | - | - | - | - | 2 | - | 1 |
| 62 | Miscellaneous | 27 | 10 | 9 | 4 | - | 1 | - | 3 | - | 1 |
| Broken tools | | 1 | - | 1 | - | - | - | - | 1 | - | - |
| LSFs | | 175 | 58 | 117 | 23 | 2 | - | 4 | 37 | 10 | 41 |
| Burin spalls | | 18 | 7 | 11 | 1 | - | - | - | 4 | 2 | 4 |
| Total | | 367 | 152 | 212 | 40 | 2 | 3 | 6 | 70 | 17 | 74 |

that no conclusions could be reached about their function.

Patination need not necessarily mean that no functional interpretation is possible. On a few of the pieces examined, polish appears to be on top of patination or staining, suggesting use of the tool after the surface had been altered. It need hardly be said that the alteration which can be seen below the use traces has completely obliterated any polishes which might be underneath it, so that one can only guess from the presence of retouch that the artefact was used before. The use of excavated fragments as raw material is invaluable in experiments to ascertain whether polish can be seen lying on top of surface alteration. These fragments were cleaned and then examined to be sure that their surfaces were altered but bore no polish-like traces. They were then used experimentally and subsequent examination showed that use polish did indeed form on top of the altered surface and that such polishes were distinguishable from the surface. Preliminary results are encouraging and further experimental work will, hopefully, lead to greater confidence in differentiating the various use polishes, which may have interesting implications for the re-use of material at La Cotte de St. Brelade.

## 30.5  RESULTS OF ANALYSIS

### 30.5.1  Long Sharpening Flakes

A full discussion of these artefacts can be found in chapter 29. My aim here is to show what additional light microwear analysis can hope to throw on the function and, for that matter, on the history of each piece studied.

The long sharpening flakes (LSFs) had been divided for study according to butt type: Table 30.2 shows the results of my analysis according to these categories. Ad-

mittedly the numbers in each class, with the exception of the first, are small, but they are nonetheless sufficient to show that such a division is a technological and not a functional one. The only major discrepancy is the seeming lack of evidence for wood working in the case of the plain-butted pieces. If these were functionally distinct from those with facetted butts, one might expect to find a difference in the mean angle of the working edge, but this is not the case (Table 30.3); nor is there any marked difference in the shape of their working edges (Table 30.4). The plain-butted LSFs are comparable to the burin spalls (about 50% of which have plain butts) in showing a preference for hide over wood working, though only 11 of the latter class were studied.

A question to be considered is why such a large number of the LSFs studied bear no traces of use-wear polish, although it is obvious from technological considerations that they were struck from a pre-existing tool. Of those which were in a suitable state of freshness for such traces to be visible, 35% showed none. It is unwise to argue from negative evidence, but such a percentage is too high to dismiss. I consider that there are three possible explanations for this lack of polish (which could apply equally, of course, to all the other artefacts analysed).

1. The tool was never used. This is always a possibility to be borne in mind, but is unlikely in view of the fact that the vast majority of the edges from which LSFs were detached had been retouched.

2. The function of the tool was such, or the material on which it was used was so abrasive, that all traces of polish were removed as quickly as they formed. However, it might be expected that slight traces of polish would remain at some points along the working edge.

**Table 30.2**   Polishes on LSFs

| Butt type | Initial sample | Patin-ated | Final sample | Type of polish | | | | | | |
|---|---|---|---|---|---|---|---|---|---|---|
| | | | | Wood | Bone | Antler | Meat | Hide | Unid. | None |
| Facetted | 97 | 32 | 65 | 17 | 2 | – | 2 | 19 | 7 | 18 |
| Dihedral | 2 | – | 2 | – | – | – | – | – | – | 2 |
| Plain | 34 | 9 | 25 | 1 | – | – | – | 11 | – | 13 |
| Unreadable | 26 | 9 | 17 | 4 | – | – | – | 6 | 2 | 5 |
| Missing | 16 | 8 | 8 | 1 | – | – | 2 | 1 | 1 | 3 |
| Total | 175 | 58 | 117 | 23 | 2 | – | 4 | 37 | 10 | 41 |

**Table 30.3**   Edge Angle of LSFs with Polish

| Butt type | All pieces | Type of polish | | | | |
|---|---|---|---|---|---|---|
| | | Wood | Bone | Meat | Hide | Unid. |
| Facetted | 58 | 60 | 59 | 62 | 57 | 56 |
| Plain | 59 | 60 | – | – | 59 | – |
| Unreadable | 56 | 53 | – | – | 66 | 52 |
| Missing | 74 | 83 | – | 62 | 78 | 82 |

**Table 30.4**   Working Edge Shapes of the LSFs with Polish

| Butt type | Working edge shape | Total | Type of polish | | | | |
|---|---|---|---|---|---|---|---|
| | | | Wood | Bone | Meat | Hide | Unid. |
| Facetted | Concave | 11 | 7 | – | – | 2 | 2 |
| | Straight | 17 | 5 | 1 | 2 | 6 | 3 |
| | Convex | 17 | 5 | 1 | – | 10 | 1 |
| | Unclassifiable | 2 | – | – | – | 1 | 1 |
| Plain | Concave | 4 | – | – | – | 4 | – |
| | Straight | 2 | – | – | – | 2 | – |
| | Convex | 5 | 1 | – | – | 4 | – |
| | Unclassifiable | 1 | – | – | – | 1 | – |
| Unreadable | Concave | 3 | – | – | – | 3 | – |
| | Straight | 6 | 4 | – | – | 1 | 1 |
| | Convex | 1 | – | – | – | 1 | – |
| | Unclassifiable | 2 | – | – | – | 1 | 1 |
| Broken | Concave | 1 | 1 | – | – | – | – |
| | Straight | 3 | – | – | 1 | 1 | 1 |
| | Convex | – | – | – | – | – | – |
| | Unclassifiable | 1 | – | – | 1 | – | – |

3.  The tool was used for an insufficient length of time to allow the formation of polishes which are visible under the microscope. There has been little published discussion on the subject of how long polishes take to form, but my own observations, especially of working hide, lead me to think that under some conditions it may take a considerable time.

It is this last suggestion that I would put forward as the most likely to account for the large number of long sharpening flakes, many of them with perfectly sound retouched edges, that were removed from *seemingly* unused tools. It may be that after a short period of use for one purpose — but before polishes had time to form — the need for a different type of working edge for another

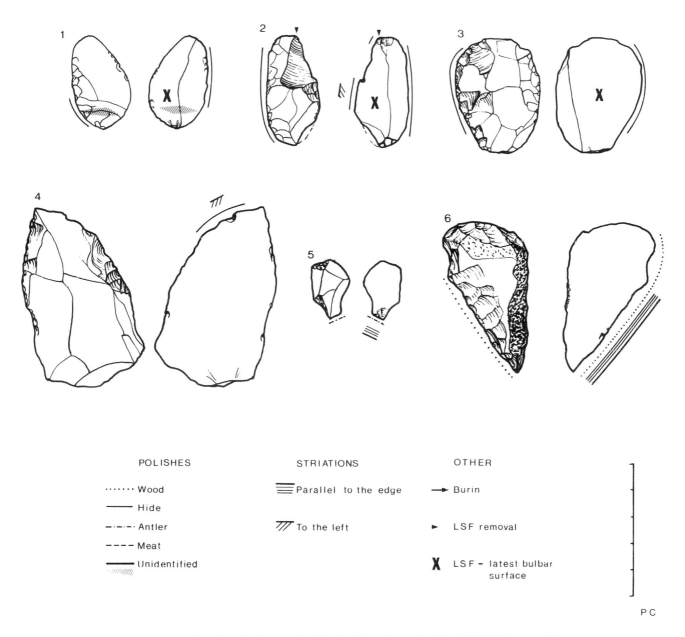

**Fig. 30.1** Wear traces on artefacts from La Cotte de St. Brelade. 1 Long sharpening flake (LSF); 2, 3 convex side-scrapers on LSF blanks; 4 retouched Levallois point; 5 'end-scraper' used as a burin; 6 thick convergent straight scraper (the polish is less marked on the retouched surface). Scars drawn with infilling are secondary (i.e. retouch), but removals predating the detachment of the blank are left open in the case of pieces made on LSFs. All are from layer A except 6, which is from layer 5.

purpose arose, and the existing edge, though still service-able for the first function, was removed to obtain one better suited to the new task (with or without further retouch).

This interpretation assumes that LSFs are a by-product of the operation. But they could also be primary products, if the parent pieces were treated as cores, and used to produce flakes with retouch along one edge and a fresh, natural edge opposite. If this were the case then one would hope to find traces of use along the new unre-touched edge. Out of the entire sample there is only one possible example of this, and that is on an atypical LSF: considerable traces of dry hide polish were found along the new edge, which had sustained some damage. But if

this were the principal reason for removing LSFs one would expect to find many more examples of this kind. It would seem rather to be a case of re-using a suitable flake when it happened to be at hand. This view is sup-ported by the fact that only 4.3% of LSFs are retouched along the new edge.

In the few instances where two different forms of use polish are found on the old working edge it is probable that both occurred when the flake was still attached to the parent piece. Re-use of LSFs subsequent to removal could be positively asserted only if traces of hafting or utilisation were found on the new ventral surface; haft-ing traces on any other part of the artefact (i.e. old sur-faces) would be ambiguous, as they could relate to haft-

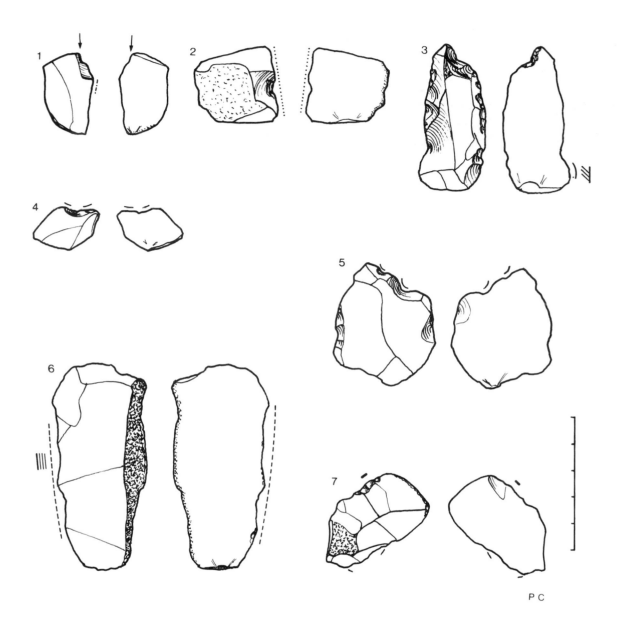

**Fig. 30.2** Wear traces on artefacts from La Cotte de St. Brelade (for key, see Fig. 30.1). 1 Burin; 2, 4, 5 notches; 3 miscellaneous (side-scraper opposed to a large Clactonian notch), 6 naturally backed knife; 7 denticulate. 1 is from layer A, 7 from C, 3 and 5 from G, 2 and 4 from H, and 6 from the G/H interface.

ing of the parent piece, or of the LSF itself, or both. Though possible binding marks were found on one LSF, 78/34031 (Fig. 30.1, 1), the evidence is not conclusive. Traces of dry hide polish (Fig. 30.3B) were found along the retouched edge on both the (old) ventral and dorsal surfaces accompanied by a characteristic rounding of the edge, the traces being most intense in the lower third of the artefact. The new working edge opposite the one described above displayed much fainter traces of polish. On both faces small patches of an as yet unidentified and poorly developed polish were found, again in the lower third portion of the flake. One possible explanation of this arrangement of polishes is that roughly one third of this LSF was in some form of hide binding, but the suggestion is, as yet, a tentative one.

And what of the parent pieces? If LSFs are for the most part the by-product of a process of rejuvenation or renewal, then evidence of re-use would be likely to be found on the parent pieces. Unfortunately, of the 17 parent pieces that I was able to examine, 13 proved to be patinated. Two of the unpatinated pieces did display evidence of re-use after the LSF removal, however. From one of these, 78/37232, the whole working edge has been removed with one blow. Along the entire length of the new working edge, but extending no more than two or three microns back from it, faint traces of antler polish could be seen. A considerable number of damage scars cut into the polish indicating the abrasive nature of the task performed. The second example of a re-used parent piece, 78/40488, was itself originally a

**Fig. 30.3**  (A) Long sharpening flake (78/51170) showing a narrow line of wood polish close to the edge of the old bulbar surface (i.e. antedating detachment of the LSF from its parent); bands of polish can be seen running at right angles to the edge. (B) Long sharpening flake (78/34031, drawn as Fig. 30.1, 1) with hide polish along the edge of the bulbar face (antedating detachment of the LSF) cut into by damage scars. (C) Convex side-scraper (77/14204, drawn as Fig. 30.1, 3) with an area of intense hide polish on the bulbar surface. (D) Naturally backed knife (78/38169, drawn as Fig. 30.2, 6) with possible meat polish on an *arrête* between two damage scars, on the dorsal surface. The scale bar is 1 mm long.

LSF (Fig. 30.1, 2). Traces of well-developed wood polish were found along both its old and new retouched working edges, indicating not only that the initial LSF was re-used but also, by the presence of polish on the rejuvenated dorsal surface, that it continued to be used after the removal of a small LSF from its edge. At the tip, on the ventral surface, a small area of very flat polish, similar to Keeley's 'hammerstone smear', was noted. The sequence of events that can be reconstructed from this piece is thus:

1.  Manufacture and use of a trimmed tool;
2.  Removal from it of a LSF — the piece in question;
3.  Retouch of the LSF's newly formed edge (opposite old scraper edge); use as trimmed tool;
4.  Removal of a small LSF from its dorsal surface, creating a short length of new edge;
5.  Continued use (as in 3, above), now involving the new edge also.

The final aspect of LSFs considered here is the matter of the handedness of the knapper. Only technological features can suggest the handedness of the tool *maker* but the handedness of the tool *user* may occasionally be revealed by microwear analysis (Keeley 1980). Orientation of striations and bands of polish are the principal indicators. Very few of the LSFs examined did have striations or bands of polish visible on the surface, and of the 21 possible 'left-handed' (on technological grounds) LSFs from the initial sample, only four bore striations. On one example these were orientated at right angles to the edge, and could be attributed equally, therefore, to the use of either hand. Nor indeed can striations orientated at an angle to the working edge be taken as an indicator of handedness, as might at first be thought; rotation of the tool through 180 degrees during use could result in striations being orientated in opposite directions relative to the working edge. However, from the small number of specimens bearing evidence of direc-

**Table 30.5** Striation Direction on the LSFs Studied[a]

| Observation | Striation Direction | | | | | Total |
|---|---|---|---|---|---|---|
| | (1) ⫻ | (2) ⫼ | (3) ⫧ | (4) ≡ | (5) ⋈ | |
| LSF symmetry | | | | | | |
| Right | 1 | 6 | 4 | 3 | 4 | 18 |
| Left | 2 | 1 | 0 | 0 | 1 | 4 |
| Polish | | | | | | |
| Wood | 1 | 5 | 0 | 1 | 2 | 9 |
| Hide | 2 | 2 | 0 | 1 | 2 | 7 |
| Unidentified | 0 | 0 | 4 | 1 | 1 | 6 |
| Total | 3 | 7 | 4 | 3 | 5 | 22 |

[a] The striations are classed as follows: (1) oblique left, (2) perpendicular, (3) oblique right, (4) parallel and (5) multidirectional. The terms left and right are chosen to reflect the direction of movement, bearing in mind that striations form on the underside of the piece during use.

tion of use (Table 30.5) there does seem to be a preference, in the case of 'right-asymmetrical' LSFs, for motion tending towards the right, rather than to the left. It is interesting to note that the reverse is true of the two 'left-asymmetrical' LSFs with obliquely oriented striations.

The striations caused by the use of a tool in a circular, boring motion are the only ones that can be taken as certain evidence of handedness of the tool user; and the very shape and character of a LSF militates against its use in this way. Thus the benefit of finding striations and bands of polish on the surface of archaeological specimens is generally not in determining the handedness of the tool user, but rather in giving an indication of how the tool was used. I have assumed that striations running parallel to the working edge imply movement of the tool forwards and backwards across the material being worked, as in cutting or sawing. Where striations run perpendicular to the working edge, and occur on both the ventral and the dorsal faces, I have assumed also that the tool was used in a movement towards and away from the material, as in chopping. Where perpendicular striations occur predominantly on only one face, I would consider the tool to have been used in a planing motion. Striations which run at an angle to the edge imply an oblique motion relative to the tool user, as in whittling. These are the premises used during the analyses in making inferences about the way artefacts were used.

### 30.5.2 Formal Tools

Although the tools were taken from various layers and the LSFs almost exclusively from A, the similarity of the results for the edges of both is remarkable. In each group almost the same percentage (34.5% and 35.0%) appeared unused. Traces of the same types of polish were found in similar proportions, hide polish being equally preponderant on both (56.9% and 56.1%), followed by wood polish almost equally again (31.4% and 34.8%).

**Retouched Levallois points** (Bordes type 4). One example of this type has been examined (Fig. 30.1,4). Faint traces of polish, similar in many respects to hide polish, were found on both the dorsal and ventral surfaces near the tip. Some linear bands of polish, whose position is marked on the illustration, were noted on the ventral surface, running parallel to the right edge. This evidence suggests that the artefact was used with a slicing or sawing motion.

**Side-scrapers** (Bordes type 9–29). Of the seventeen which bore identifiable use traces, eight were of hide, which is not an entirely surprising result. Two of these artefacts bore striations which ran at right angles to the retouched edge. Figure 30.3C illustrates an area of intense hide polish on the ventral surface of a side-scraper (Fig. 30.1, 3). Three scrapers appear to have been used for wood working: one has been discussed above in connection with LSF removal (Fig. 30.1, 2) while the other two bore only very faint traces of wood polish close to the retouched edge on the ventral face. On one (Fig. 30.1, 6) wood polish occurs on both the ventral and dorsal surfaces, but striations could be seen only on the ventral surface and running parallel to the edge. This suggests a cutting or sawing motion, leaning on the ventral face, and causing considerable edge damage, which could remove much of the polish traces with it; it may explain also the extreme narrowness of the band of polish. It is easy to see how four scrapers were found to have no traces of polish; it is quite possible that the edge damage associated with working wood could sometimes remove all traces of polish at the edge margins.

**End-scrapers** (Bordes type 30–31). One piece, 78/6232, which had been classified as an atypical end-scraper (made by truncating one corner of the distal end), proved to bear polish similar to that which results from working antler, but at the proximal end. Here, a notch had been made adjacent to the broken butt (Fig. 30.1, 5). The use traces take the form of quite distinctive bands of polish running parallel to the working edge (and perpendicular to the 'burin facet' formed by the notch), occurring on both surfaces but more strongly on the ventral face. This would suggest a cutting motion. The working edge is abraded.

Of the other two end-scrapers on which signs of use could be detected, both bear faint traces of hide polish, but on only one were striations noted, running parallel to the working edge (again suggesting use as a cutting tool). This interpretation is not without parallel: Pfeiffer (quoted by Semenov 1964, 85) considered end-scrapers to be cutting tools but believed that they would have been hafted. However, not one of the end-scrapers in the sample from La Cotte showed indications of hafting.

**Burins** (Bordes types 32–33). Of the seven which showed recognisable use traces, none could be called typical on demonstrably functional grounds. They have been termed 'burins' only on the basis of technological or morphological features; that is, having at one end of the piece an angle which is formed by the meeting of two facets (or one facet and the natural edge of the artefact). But on none of them did the polish traces occur on the burin facet or on the tip, which argues against their traditionally assigned function as engraving tools. One example, 78/29229 (Fig. 30.2, 1), had a patch of polish similar to that produced by working antler on the ventral surface below the 'burin facet'. In both examples with polish this was located along the edges and bore no relation to the 'burin facet'. In two cases, striations were noted at right angles to the working edge. From the sample examined, therefore, it appears that the burin facets were manufactured after the pieces had already been utilised, and that they were not directly employed in processing.

**Knives** (Bordes types 36–38). It is unfortunate that the word 'knife' is used to describe this class of tool (although there is no satisfactory alternative), for it must conjure up in the minds of most readers some preconceived idea of how the tool was used. By studying the microwear traces on all the edges of such artefacts it should be possible to confirm or refute these notions. It is therefore particularly regrettable that so many of the knives taken for analysis proved to be coated with some form of staining which could not be removed chemically. On only one, a naturally backed knife, were there any recognisable striations (Fig. 30.2, 6). Traces of meat polish were found close to the edge on both the ventral and dorsal surface (Fig. 30.6), with bands of polish running parallel to the working edge, confirming the notion of a cutting action which is implied in the term 'knife'.

**Notched pieces** (Bordes type 42). It is interesting to note that a third of those examined bore no trace of use: some of the possible reasons for this have been mentioned above; moreover, a notch may be fortuitous rather than deliberately manufactured.

The position of polish on the edge of the artefacts in relation to the notch varies considerably. In two examples, 78/25551 and 78/41523 (Fig. 30.2, 4 and 5), the polish is confined to one half of the notch only and to the ventral surface. In both cases the polish resembled that of hide and was more intense on the outer edge of the notch, decreasing in extent towards the centre. It will be noted that the traces occurred on the same 'half' of the notch: this might at first be considered as a possible indicator of the handedness of the tool user, but neither specimen bore any striations or bands of polish which would have assisted in this interpretation. Not all the notches examined had use traces in such restricted areas. On one artefact, 78/33017, I found traces of wood polish on the edge round the entire extent of the notch; in comparison with the two previous examples, the polish was more intense and consistent on the ventral surface. It is interesting to note that this notch was much larger than those with the hide polish traces, being 24 mm in diameter. It may be that this, as well as the dissimilar polish traces, points to a difference in tool function and in the raw material worked. In these three examples it was the notched portion of the working edge that was used. There is one example, 78/45783, where this is not the case (Fig. 30.2, 2): traces of wood polish were found on both faces, occurring most intensively in the lower portion and showing a marked decrease towards the notch. Here the presence of the notch seems irrelevant; it may be that not all notches necessarily serve the same function. The analysis of many more instances of this tool type and more experimental tools with which to compare them is required.

On two of the pieces there were traces of a polish which so far I have been unable to identify with any degree of confidence. It is quite distinctive, being very flat and bright, frequently with deep cracks on the surface. In many ways it resembles what Keeley calls 'hammerstone smear' (Keeley 1980, 28); I have not yet been able to replicate it, but its presence is worth noting.

**Denticulates** (Bordes type 43). Although half the initial sample (10) of this tool class bore use-wear traces, the results are rather inconclusive. The traces of polish in one case are so faint that I cannot identify them, and in two cases the identification of hide polish is tentative. On one example, 78/15765 (Fig. 30.2, 7), the polish was found to be restricted to the edge of a small protrusion: this suggests that what is visible may be the remnant of a more extensive area of polish which has been removed through the abrasive nature of the tool use.

**Pieces with abrupt or alternate retouch, thick or thin** (Bordes types 46–49). The traces found on these artefacts were faint and inconclusive. On only one artefact did the polish extend any distance along the edge of the tool, and then only very close to the edge, extending back from it only a few microns. The considerable damage which could be seen along this working edge suggested that polish could have been removed almost as quickly as it formed.

**Miscellaneous** (Bordes type 62). There is little to be said of the tools assigned to this category. The three artefacts on which traces of wood polish were noted have been discussed above with the LSFs. One of the remaining specimens, 78/50469, could possibly be classed as a notched piece; traces of hide polish were noted on its ventral surface only (Fig. 30.2, 3): again, it is interesting to note, the polish was confined to one side of the 'notch' only. In addition, striations were recorded running at an angle to the edge, the direction of these being indicated on the drawing.

## 30.6 CONCLUDING REMARKS

This is not the first time that microscopic analysis has been conducted on artefacts from La Cotte, and it is to be hoped that further work will be carried out in the future. Professor McBurney was greatly interested in the work of Semenov and the potential of microwear (McBurney 1978). He was probably the first person in Britain to initiate a similar programme, using relatively low magnification and based largely on his own excavated material; in this he was assisted by Mary Pohl and Gillian Farnsworth, but unfortunately the investigation did not reach publication. I have examined a few of the La Cotte pieces from this study, but sadly the aluminium coating which was at that time considered necessary has completely masked any traces of polish.

The findings I have presented here are based on a limited number of artefacts from La Cotte. Because it was not feasible to examine a random sample, there is no way of telling how representative the results may be, and it would be unwise to generalise, from them alone, as to the full range of function of a tool type, for instance. The observations have, however, given useful information about the nature of the materials worked, and contribute to the existing body of data about the manner and direction of use of specific types of tool.

And for the future? This study shows that wear traces have survived on some of the pieces from La Cotte, though the further back one goes in time, the more complicated the interpretation of such traces becomes. I do not expect that microscopic vegetable remains such as phytoliths would have survived on the surface of any of the artefacts, but the examination of a sample of pieces using the higher magnifications and greater visibility of a scanning electron microscope would greatly increase the range of information that could be gleaned from each artefact.

*ACKNOWLEDGEMENTS*

I have received much advice, on aspects of the excavations and on the choice of artefacts for study, from members of the La Cotte Project, and I would like to express my thanks to them. I am especially grateful to P. Callow, who generously allowed me to remove for (often prolonged) analysis a large number of interesting and sometimes important pieces, despite the complications which this caused during studies of other aspects of the material.

# PART V

## DISCUSSION

# 31

# PLEISTOCENE LANDSCAPES AND THE PALAEOLITHIC ECONOMY

## P. Callow

In chapter 7, environmental data were employed to reconstruct the biostratigraphic sequence at La Cotte de St. Brelade. Here, on the other hand, we are concerned rather with the nature of the territory available during the various periods of human occupation, and of man's exploitation of these landscapes. For the location of places mentioned in the text see Fig. 1.2.

## 31.1 THE ENVIRONMENT OF THE CHANNEL ISLANDS AREA DURING THE PLEISTOCENE

The relationship between climate, sea-level, geomorphological processes, and fauna and flora is a complicated one; nevertheless, even at the risk of over-simplification it is sufficient for present purposes to consider three generalised types of environmental conditions which can be reconstructed primarily on the basis of evidence from La Cotte and the surrounding region, while drawing on data of more widespread origin. The headings used are:

1. Interglacial. Despite a temperate climate, we are chiefly concerned with periods when the sea-level is low enough to allow human occupation of Jersey without the use of boats or rafts, and without risking a crossing on foot from the mainland at extreme low tide. A higher sea-level would of course result in a landscape very like that of today.
2. Initial or terminal interglacial, or interstadial. Here a rather lower sea-level is probably accompanied by geological evidence of frost action; alternatively, relatively mild conditions may have been quite short-lived.
3. Stadial, with a very low sea-level and a rigorous climate.

The corresponding changes postulated for the shape and position of the coastline are illustrated in Figs. 31.1 and 31.2. These are very approximate, as are the nominal values for sea-level given in the sections following. Mary (1982) has suggested that a plastic deformation of isostatic origin occurred along the whole Atlantic coast of Europe during the Holocene, with a rate of change in the Brittany area of the order of 1 m per ky since 6000 B.P. A more elaborate programme of modelling on a world scale by Marcus and Newman (1983) has produced isobase maps which are in generally good agreement with Mary's results, and which show a very much greater deformation at 12,500 B.P. The two last-mentioned

authors are pessimistic about the use of eustatic sea-level change alone in the construction of a model of hominid migration, and it follows that the same reservations must apply to extrapolation of site territories onto the continental shelf during earlier periods. However, the most extreme crustal displacement relative to the modern level arose from the maximum extent of the ice-sheets and moreover exhibits a lag effect. Deformation in the period following an interglacial in our area is therefore unlikely to have been severe until continental glaciation was well established. The maps given in this chapter are not claimed to be accurate in detail, being based entirely on eustasy, but are believed to give a fair representation of the topographical changes (with the possible exception of that based on the -100 m isobath, which is subject to greater uncertainty).

Indirect local evidence may be used, with caution, to indicate the probable relationship between sea-level and climate in the area. Firstly, occupation of La Cotte de St. Brelade during Stage II indicates that temperate conditions could occur without preventing human occupation, i.e. when the sea-level was of the order of 20 m below today's. The climatic optimum may have corresponded to a higher level still, in which case the site would have been abandoned during the pedogenetic episode affecting layer E (this cannot be determined because the A horizon is missing). Secondly, the abandonment of the site during periods of extremely severe climate and the total absence of Palaeolithic finds in the other Channel Islands suggest that a fall to below -40 m would have been accompanied by a climatic deterioration to the point at which the area ceased to be attractive, or even habitable, with the technology then available; a parallel is provided by the non-occupation of N. France during much of the Weichselian Upper Pleniglacial. Thirdly, Lautridou's proposal (1984) that the rejuvenation of sources of Normandy loess in the course of a glaciation was the result of estuarine deposition at about -20 or -30 m is consistent with the occupation of La Cotte during some of the interstadials.

### 31.1.1 Interglacial
Sea-level c. -20 m or above: La Cotte Stage II

Jersey was a rocky plateau situated at the end of a peninsula extending westwards from the area of Normandy north of Coutances. It was bounded to north and south by the estuaries of the Ay and the Sienne, the former occupying the Ruau Channel. A second peninsula extended south of the Sienne from Granville, and was ter-

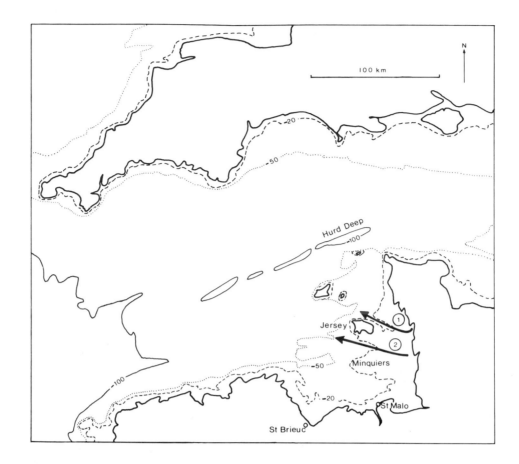

**Fig. 31.1** Sea-bed contours in the Western Approaches (in metres below sea-level). The arrows indicate the palaeovalleys of the rivers Ay (1) and Sienne (2).

minated by what is now the Minquiers reef. The neck of land joining Jersey to the continent was low-lying, only a few metres above the then high tide level. Elsewhere there was only a narrow low-lying strip between the higher ground and the coast; in places this would have been no more than 1–2 km wide.

Taking the palynological data from La Cotte (chapter 11) with that for Flandrian peats (Keen 1981) it is clear that there were extensive marshy areas in the coastal strip. Alder carr was probably a very important feature of the landscape, and ground with better drainage supported deciduous woodland. As today, some parts of the plateau were heathland (heathers and bracken). A maritime component (sea buckthorn, sea plantain, thrift) is apparent from the pollen diagram, and is in accordance with the proximity of the coast. Modern analogy also indicates that it is likely that some parts of the island would have received considerable amounts of blown sand. It has been very tentatively suggested by Keen (1975) that the iron-cemented sand around St. Peter's Church may be equivalent to the 8 m beach; if earlier deposits of this nature still exist, they have not yet been identified or dated.

Faunal evidence relating to these conditions is almost entirely lacking at La Cotte, because so little of the bone refuse from layers H to E is identifiable (chapter 13). Rhinoceros is recorded, and it may be assumed that red deer was present (since it was isolated on the island during the Eemian, as is demonstrated by the finds from Belle Hougue, and would certainly have found the fairly open/mosaic woodland habitat to its liking during the interglacials). As reconstructed here, conditions would have been less favourable for steppe herbivores, and it is unlikely that the herd size for any species would have approached the levels possible under the conditions described next.

### 31.1.2 Initial or Terminal Interglacial or Interstadial
Sea-level c. -30 m: La Cotte Stage III occupation deposits

The area of low relief around Jersey was very much more substantial than during fully interglacial conditions. The peninsulas referred to above were absorbed into a single land mass in a westwards extension of Normandy, and the Gulf of St. Malo was effectively reduced to a 'Gulf of St. Brieuc'. The gentleness of the topography suggests that the inter-tidal zone may have been very wide — perhaps as much as 10 km in places — though its precise extent is hard to estimate because of uncertainties about the former tidal range and about possible erosion or deposition. During interstadials when the sea was relatively low it may have been possible to reach the other Channel Islands between tides, though the estuaries of the Normandy rivers may have rendered this too hazardous. The Ay and the Sienne are unlikely to have been sufficiently vigorous to occupy much of the broad valleys available to them, and their shallow braided channels cannot have greatly impeded animals or men.

The pollen spectra for the appropriate layers at La Cotte show that the plain was occupied by rich grassland with abundant herbs, and with sedges growing in the depressed areas. Trees and shrubs provide little of the pollen, and were probably localised in their distribution (for instance forming gallery woodland along the streams). That the sea was still quite close is clear from pollen of maritime species.

The corresponding fauna is well documented at La Cotte; large herbivores were abundant, probably in substantial herds, and included mammoth, woolly rhinoceros, horse, giant deer, red deer and bison/aurochs. Reindeer were also hunted, but may not always have been available.

### 31.1.3  Stadial
Sea-level below -40 m: La Cotte Stage III sterile deposits (?)

The English Channel was almost entirely dry land, apart from a narrow lake, some 100 km long, to the north of Alderney (the Hurd Deep). The coastline was usually at least 80 km from La Cotte — as much as 200 km when the ice-sheets reached their maximum extent — and once again the inter-tidal zone may have been very wide (the mean gradient between the -50 and -100 m isobaths is of the order of 0.2%).

The vegetation was of steppe-tundra type: grasses (and sedges in marshy areas) and some herbs; the trees and shrubs included birch and willow (probably dwarf species) and some pine. It is not known which large mammals were present at times of glacial maximum, though it is probably safe to assume that they would have included mammoth, woolly rhinoceros and reindeer. Conditions were evidently favourable for 'arctic' rodents (chapter 14); Chaline and Brochet's temperature curve based on the present-day environmental preferences of the species represented at La Cotte (Fig. 14.5) is in fairly good agreement with a model produced for central England by Williams (1975), who obtained values between -25° and +10°, though the former analysis suggests a greater degree of continentality.

## 31.2  SITE EXPLOITATION TERRITORIES AND SITE CATCHMENTS

The similarity between the greater part of the *in situ* finds from La Cotte à la Chèvre (Appendix C) and La Cotte de St. Brelade layer H, and the presence of an LSF at the former site, suggest parallel occupation, at least for part of the Saalian. It is therefore legitimate to consider the economic potential of both sites (hereafter referred to as CC and CSB) in combination as well as individually. Moreover, even though there may be further sites as yet undiscovered, the topography of Jersey is such that their territories will have had much in common with that of one or other of the two known caves.

### 31.2.1  Principles

The methodology of 'site catchment analysis' (SCA) was developed from the late 1960s onwards by Claudio Vita-Finzi, Eric Higgs and their followers (Vita-Finzi *et al.* 1970, Higgs *et al.* 1967). In the course of its evolution the concepts of *site catchment* (SC) and *site exploitation territory* (SET) have acquired separate identities: the former "embraces the terrain covered by occasional forays in search of raw material for tools and other purposes", while the latter is "the territory habitually exploited from a single site" (Higgs and Vita-Finzi 1972, 30). Much of the early work in this field, though conceptually important, can now be seen to rest on simplistic extrapolations from the present-day terrain and the behaviour of its occupants; more recently there have been efforts to overcome these weaknesses, and techniques of considerable utility have been devised (see Bailey and Davidson 1983 for a discussion).

As Bailey and Davidson point out, the incompleteness of archaeological data (e.g. the often poor survival rate of plant remains) means that an analysis of the *potential* of an area may be essential to an assessment of the possible role of some elements in a prehistoric economy. The definition of a SET on the basis of 'reasonable' assumptions about how far people are prepared to go for their regular requirements provides a starting point for such an investigation. Any 'residuals' — materials occurring at the site but not available in the SET — must then be the subject of a further investigation in order to establish the total catchment area. The assumptions underlying such a model are obvious: availability is not a synonym for exploitation; goods move centripetally towards a particular locality, so the idea of collecting 'on the run' between short-lived encampments (for example when following a moving herd) is excluded; and so on. Nevertheless, if the sites under consideration are indeed major foci of activity, and the demands made of the model are not too severe, it may at the very least demonstrate the untenability of a hypothesis and generate others for evaluation.

In the case of Lower and Middle Palaeolithic sites such as those on Jersey, we have to accept that there are severe limitations to the accuracy of the analysis — especially when our concern is with conditions pertaining to an area of several hundred square km, and not merely to the micro-environment of the site itself. A spring, a rock shelter or some other topographic feature which was of importance to the ancient inhabitants in their day-to-day activities may have disappeared completely. For instance it may be possible to make a general statement about the nature of the vegetation, but not a detailed estimate of game trails through marshy areas.

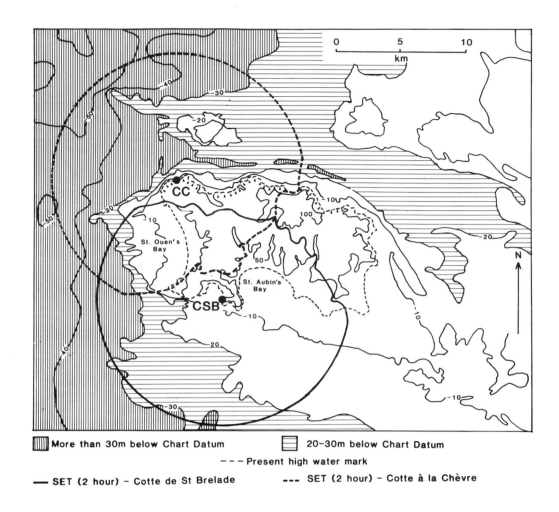

**Fig. 31.2** Sea bed contours around Jersey, and the 2 hour site exploitation territories of the two caves known to have been occupied during the Old Stone Age (note the overlapping SETs, and the small size of that for La Cotte à la Chèvre during periods of high sea-level). Chart datum (used for the isobaths) is 5.88 m below mean sea-level. For a tidal range similar to that of today's, therefore, each isobath roughly approximates the low tide line for a corresponding sea-level fall, and the one above is equivalent to high tide. Thus the unshaded area corresponds to the dry land during a regression to -30 m. Contours based on British Admiralty Chart 3655 (Crown Copyright) by permission of The Controller of Her Majesty's Stationery Office, and of the Hydrographer of the Navy.

Under these circumstances, a result accurate to within an order of magnitude is an acceptable goal.

In order to estimate the SETs of the two sites on Jersey, Naismith's formula has been used, following the example of Bailey and Davidson (1983, 94): it is assumed that to walk 1 km on flat ground takes 12 minutes, while each variation in altitude of 100 m adds 10 minutes. In the interest of simplicity, the impediments offered by streams, marshland, vegetation etc. have been ignored, so the SETs are probably somewhat distorted. Ideally, the accuracy of the estimates in respect of modern conditions should have been verified by walking along a number of transects, but the built-up or agricultural nature of much of the Jersey terrain (as well as the height of today's sea-level) would render such an exercise futile. In accordance with widespread practice, a two hour limit (each way) has been assumed as defining the travelling time which might be regarded as acceptable on a regular basis. The SETs defined in this way for CSB and CC are mapped in Fig. 31.2.

### 31.2.2 The SETs for the Jersey Sites

**La Cotte de St. Brelade (CSB).** This territory covers the southern half of the island as far east as St. Clement's Church, and so includes almost all of the major valleys of Jersey. In the west it extends to the northern end of St. Ouen's Bay — this, and St. Brelade's and St. Aubin's Bays, would certainly have allowed rapid movement whenever the ground was not too marshy or wooded. Water is freely available; indeed, the source nearest the site was probably only about 100 m from the north entrance. The southern half of the SET must have been severely restricted by the coast-line when the sea-level was at about -30 m or above.

An important feature of this SET is that it provides a number of excellent look-out points which would have permitted the identification of target animals at very great distances on the coastal plain. La Cotte Point itself gives a good view towards the south and west (an important consideration during periods of lowered sea-level); a more extensive area can be covered from Noirmont, while from La Moye almost the whole of the seaward part of the territory is visible.

**La Cotte à la Chèvre (CC).** The SET available during periods of high sea-level is situated almost entirely in the bottom right-hand quadrant of the theoretical territorial maximum. To the east, it extends approximately to La Belle Hougue along the modern coast; on the plateau, changes in relief limit its radius to some extent. Moreover the height and steepness of the cliffs along the north side of the island divide the area into two zones which would have been quite distinct both ecologically and in their exploitation. To the south only the area beyond St. Aubin and that around La Moye lie outside the SET, but the whole of St. Aubin's Bay, most of the valleys, and the eastern half of the island are excluded.

Because the principal drainage of Jersey is towards the south, and the catchment areas of the few streams in the extreme north are small, water was probably less readily available in the immediate vicinity of CC than around CSB, particularly when the climate was slightly continental. At times of lowered sea-level, therefore, the river along the northern side of Jersey, originating in Normandy, may have been important as a source. But when the sea was higher (though well below the present level) the same submerged valley system brought the sea very close to the site.

Once again, the site itself, with the cliff top above, is admirably placed to provide lines of sight over the coastal area. Better still, Grosnez Point is only a few hundred metres away, and gives a field of view of well over 200°. The remaining low-lying areas of the SET are covered from Plémont Point and Grande Etaquerel, both within a thirty minute walk.

### 31.2.3 Inter-site Relationships

**The completeness of the record.** Neither of the sites is close to the eastern side of Jersey and the area which would have formed the link with the Continent. Systematic exploitation of the whole region may therefore have required at least one other site, ideally in the Trinity or St. Martin's parishes, though none has yet been discovered. During interglacials, coverage of the landward end of the peninsula could have been provided by a single site in what is now Normandy.

**Overlap.** The two SETs are far from mutually exclusive; at times of high sea-level the percentage area shared by the sites was considerable (indeed, that exclusive to CC was always very limited in extent). This observation may be developed into a generalisation concerning any two sites situated on the north and south coasts at about the same longitude — an important point if other, undiscovered, caves were ever occupied. In the event that both La Cottes were used by the same human group, they are unlikely to have had identical functions. CSB is evidently the more important site, given the richness of its industries and the diversity of activities which took place there, and also its size and aspect; CC is too small to

have given shelter to more than a handful of people, and, though entirely roofed over, faces north and is at times even more inhospitable.

If the above reasoning is correct, CSB would have served as a centre for the exploitation of the larger area, whereas CC (and perhaps other sites) would have provided a means of extending hunting and other activities to the north of the island; the latter site may have been used not for regular occupation but as a temporary shelter when expediency or bad weather demanded it. At times of lowered sea-level (and less equable climate) the extended land mass must have permitted more far-ranging expeditions of greater duration. The overlap of SETs on Jersey itself may then have had less significance, but the nature of CC again seems to preclude its use as a 'home base'. A major change of role is therefore unlikely, except perhaps as a shelter linked to a long-range 'lookout post' for herds approaching from that side.

## 31.3 RESOURCES IN THE LA COTTE DE ST. BRELADE AREA

### 31.3.1 Mammalian Resources

The only identifiable faunal remains from the lower layers are fragments of rhinoceros teeth from F and E; their species cannot be determined (chapter 13). Bone debris is abundant, however, and presumably derives from a wider range of prey; traces of hide polish were observed on some of the artefacts (chapter 30).

If interglacial conditions were as proposed in section 31.1, medium-sized mammals may have provided a fairly stable year-round resource, though subject to demographic fluctuations. The environment would have been ideal for red deer in small herds of very limited mobility. Studies on the island of Rhum, off the west coast of Scotland, indicate that in the course of its lifetime a hind normally ranges over an area of only a few hundred hectares, though the males leave the herd when old enough (Dr. S. Albon, personal communication); there is no reason to suppose movement between winter and summer grazing. New genes are provided by the arrival of stags from other herds (in the case of Jersey, probably from the continent). A degree of territorial stability may be expected of other types of deer and of pigs; smaller mammals and other fauna would have contributed very greatly to the exploitable biomass, though liable to considerable cyclical variation in abundance (see Winterhalder 1981 for a fuller discussion of similar questions, though for a boreal forest environment).

The fauna of the layers after E is dominated by herd animals which would have been almost entirely migratory, implying very great variation in the local biomass. This is in keeping with the exposure of extensive tracts of land as the sea retreated, and with the development of a rich, well-watered grassland. There is no seasonality data for CSB, and it is not possible to establish where the site lay in relation to the animals' principal grazing areas, which would have been separated by many,

369

possibly hundreds, of kilometres (compare Sturdy 1975). If migration routes ran east-west, Jersey would lie near the western extreme of the range so long as the sea-level fall was moderate; if north-south, across the main drainage, CSB may have provided an opportunity for intercepting the herds on the move. Whatever the details, the annual pattern of availability of meat must have been very different from that outlined earlier. Predictability may have been a problem, as if the site lay beside a migration route, rather than in the pastures at either end, local and temporary circumstances (such as high river levels after heavy precipitation) may occasionally have caused diversion of the herds, with serious consequences for hunters relying on them.

Apart from meat, the animals whose remains were found at La Cotte must have provided valuable raw materials — at the very least, sinews and skins. Dressing of the latter is indicated by the occurrence of hide polish upon many of the tools (chapter 30). The possibility that mammoth bone was used as a fuel is discussed in chapter 20.

### 31.3.2 Marine Resources

No marine (or freshwater) shells, or bones of fish or marine mammals, have been recorded in the Pleistocene deposits at CSB; even if they were formerly present, the acidity of the deposits would not favour the preservation of fish or shellfish remains. On the other hand, it is known that exploitation of such resources took place during the Last Interglacial Complex and before, in South Africa, at Klasies River Mouth (Singer and Wymer 1982) and on a more limited scale in southern France, at Terra Amata, Lazaret, and Orgnac (Desse and Desse 1976); at the last-named site, trout and other freshwater fish were found. It is therefore worth considering the economic possibilities for the prehistoric inhabitants of Jersey.

During the later Saalian occupations (from layer D or perhaps C onwards) the sea, and major rivers, would have been many km distant; thus shellfish in particular (because of their high weight/food value ratio) are unlikely to have been transported to the site. Their use, if it occurred, cannot have been functionally related to the occupation of La Cotte, as it implies removal of the group to a point outside the SET in order to benefit from the resource. Even during the periods of relatively high sea-level compatible with human occupation, the position of the coast-line would have left a shelf 1–2 km wide in front of the cave. In the event that the inhabitants of CSB were in the habit of exploiting marine resources, it is likely that their middens would have been situated on this shelf, and so destroyed by the sea during subsequent transgressions. For all the richness of Jersey's marine resources today, therefore, the question of their former exploitation is unanswerable.

### 31.3.3 Plant Resources

As is usual for sites of this age — though exceptions have been claimed on the basis of seeds at Saint-Estève-Janson, Lunel-Viel and Terra Amata (see Boone and Renault-Miskovsky 1976) — no evidence of vegetable foodstuffs has been recovered from CSB. Nevertheless, they must have been readily available when the site was occupied under temperate or boreal conditions. The same is true of bedding material; compare Worthington Smith's (1894, 292) suggestion about the masses of *Osmunda regalis* fern at the Acheulian site of Stoke Newington in north London.

Processing of wood or other vegetable matter with the aid of stone tools is indicated by the use-wear traces reported in chapter 30. The nature of this activity is unclear, but it is a reasonable supposition that some of the lithic artefacts were mounted in wooden or bone hafts, e.g. the elongated points in layer 5 (note also the suggestion by Frame, in subsection 30.5.1, that traces of polish on one of the tools from layer A may be the result of hafting). Finally, the presence of charcoal in several layers, though never abundant, points to the use of wood as a fuel, either on its own or in conjunction with bone (chapters 12 and 20).

Detailed reconstruction of the vegetation in the SET for CSB is not possible. However, the soils of the area indicate that a wide range of plant types must always have been available. Thus the main drainage of the island, running through St. Aubin's Bay, ensures a rich supply of alluvium provided enough of the foreshore is exposed; thin acid soils on the headlands and cliffs contrast with the loess deposits on the plateau, and presumably on the expanded coastal zone. Despite changes in the presence and abundance of species arising from climatic change or soil impoverishment, therefore, CSB must always have been well placed for plant foraging.

### 31.3.4 Raw Materials for Tool-making

The availability of rocks suitable for lithic tools is discussed in general terms in chapter 22. Application of the SET concept to the procurement of exotic pebbles is informative, and gives a rough indication of the point at which tactical changes may have been required. For the various sea-levels the probable position (relative to CSB) of the high tide mark, and hence, approximately, of the storm beaches containing the pebbles, is as follows:

-20 m    Across the entrance of St. Brelade's Bay, perhaps only a few hundred metres from the site.

-30 m    Off Pt. La Moye, and also due south of La Cotte Pt., at c. 3 km.

-40 m    At a minimum of 6 km to the west, and probably 10 km to the south.

Lower    At least 10 km distant.

Thus at one extreme there were times when good raw material was easily obtained (despite the problems accompanying rising sea-level as outlined in chapter 22), to the extent that during layer F even the decortication of pebbles was commonly performed within the site itself. At the other, when the sea-level approached -40 m, the active beaches must have been near the limit of the SET, or outside it if the tidal range was much less than today's.

## 31.4 THE BONE HEAP INCIDENTS AND MEAT PROCUREMENT STRATEGIES

### 31.4.1 Evidence and Hypotheses

Leaving aside layer E and below, in which the rarity of animal bone is probably the result of complex chemical processes, the Saalian deposits containing faunal remains fall into two classes:

1. With dense concentrations of small bone splinters (often burnt) but few large bones; a wide range of species; association with numerous artefacts;
2. With (mostly) large bones, usually broken; few small splinters; very limited species representation (essentially mammoth and rhino); artefacts scarce and/or at the base of the layer containing the bones.

The first (represented by layers 5, A, C and D) is characteristic of human occupation debris in a repeatedly and intensively occupied site, while the range of lithic material, and of the microwear traces, suggests a number of different processing activities. The second class of deposits (6.1, 3 and possibly B) is less familiar in caves and shelters occupied by man. In an open situation it would immediately suggest a kill/scavenging butchery site (once an entirely natural origin has been discounted, as argued in chapter 18 and 19); given the topography of the headland, and the presence of cut marks on some of the bones, the same explanation suggests itself, with the fissure system providing the means of trapping animals which would otherwise be killed with difficulty.

Thus if both types of deposit demonstrate successful procurement of meat, for the most part by hunting herd animals, when considered in more detail they appear to indicate very different patterns of behaviour. In the dense occupation layers, the range of species suggests that as a rule any suitable prey might be taken during hunting expeditions — the site serving as a base camp to which game was brought. On the other hand, the use of CSB as a natural 'pitfall' for large quantities of megafauna might be expected to interfere with the day-to-day activities at the bottom of the ravines, at least locally. At first sight the two types of behaviour appear incompatible. A further element in the discussion is the stratigraphic relationship between the two types of deposit; on two (or possibly three) occasions a rich occupation layer with small debris is followed by a bone heap embedded in loess devoid of hearths — in the case of layers 3 and 6.1 the considerable overburden of loess is sterile of anthropogenic material immediately above the bones. The concentrations of large herbivore remains are therefore evidence of the last human activity at the site before it was abandoned for a period.

The combination of events represented by variation in the nature of the anthropogenic material and the accumulation, or non-accumulation, of loess suggests two very different behavioural hypotheses.

**Hypothesis A.** Just before the site was abandoned for long enough to permit deposition of up to a metre of loessic material (or more if allowance is made for possible later erosion) there was a considerable shift in hunting strategy at CSB, from one of individual kills to one of mass trapping.

**Hypothesis B.** Both types of meat procurement were available to, and practised by, the hominids using CSB; however, only the special circumstances pertaining to abandonment of the site permitted the survival of evidence of on-the-spot kills of large herbivores.

The essential difference between the models is that the former attempts to explain the evidence in terms of transient changes in the occupants' methods of food procurement, the latter in terms of preservation.

### 31.4.2 Hypothesis A

This postulates a cycle (home-base/kill site/abandonment) repeated three times, if that of layers C to B is included (Fig. 31.3). One sub-model, A1, explains the change of use as arising from a tactical shift by the occupying group, while A2 requires replacement of one group by another with a different approach to hunting.

The second sub-model can be dismissed as inherently improbable, as it implies that the 'kill site' groups in each case made only a single visit to the site before abandoning it; if they were better adapted to new environmental conditions than were their predecessors, they ought to have left evidence of more prolonged activity.

Had the period of time represented by layers B to 6.1 been very short, with the loess accumulations the result of occasional very severe dust-storms, sub-model A1 would be quite attractive, as one could consider the possibility that the conditions before abandonment (perhaps reflecting seasonal weather patterns) favoured drives of large animals into the ravines, after the site had been used for routine activities while the hunters were waiting for their opportunity. However, the vegetational changes that took place between layers 3 and 6 (chapter 11) are incompatible with such an interpretation. The probability that the same sequence would have been repeated at very much longer intervals is much lower, requiring as it does the probably three times repeated coincidence of identical behavioural and environmental shifts.

The case for independence of the two tactics in fact appears less strong when the contents of the 'non-bone heap' layers are considered in detail. The same species are present, so a change in the fauna cannot have provoked the new behaviour. There are firm indications that at least during the layer A occupation very large game was hunted — a single molar of a virtually full-grown mammoth; adult and old adult rhino molars (some are upper dentition, and may imply the presence of complete skulls in the ravines at this time). The head may have supplied food items which are often considered delicacies — cheeks, brain, tongue and perhaps even the pulp from

# DISCUSSION

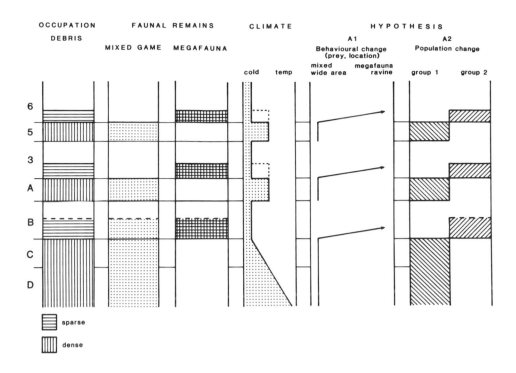

**Fig. 31.3** Hypotheses A1 and A2, rejected as an explanation for the bone heaps in the upper Saalian layers (see text).

the roots of teeth (see Hill 1983). Nevertheless it would probably have been economically attractive to extract this meat at the site of the kill itself in the case of very large animals. The presence of these few remains therefore suggests that large animals were being taken locally, perhaps in the same manner as in the layers with bone heaps. The mammoth teeth from the rich culture layers are mostly those of very young animals which could have been killed more easily than adults (though scavenging is also a possible explanation for their presence, given the high natural mortality of lower age-groups), and which would not be too difficult to transport.

Thus both variants of model A appear improbable.

### 31.4.3 Hypothesis B

This, on the other hand, assumes that the hunters employed a consistent and broad-based strategy throughout, and that the abandonment of the site and deposition of loess permitted the preservation of evidence which in the course of regular occupation would be destroyed (Fig. 31.4). However, if this explanation is the correct one, some mechanism must have existed to account for the disappearance, from the denser archaeological layers, of most of the megafauna and particularly the larger body parts. It is also necessary to account for the rarity of the remains of animals other than rhino and mammoth in the bone heap layers.

Bearing in mind the narrowness of the north ravine — only about 6 m at this point — the presence of several elephant carcasses, even dismembered, must have rendered impossible the day-to-day activities of the oc-

cupants. Elephant and rhino bones (unlike those of most other animals, or the abalone shells in the Tasmanian example cited below) individually constitute very significant obstacles, especially when concentrated in a narrow cave. Stacking the remains against one of the walls can have been at best a temporary solution, whether it was done in order to provide a concentrated store of meat or merely to clear enough space for butchery to continue. Only about 40% of an elephant carcase is likely to have been regarded as fit to eat (see Hanks 1979, 78). Fluids, intestines and stomach contents total roughly a quarter of the body weight, adding further to the practical difficulties of dismemberment in a confined space. Although removal of the habitation area to another part of the headland may have taken place, if use of the site were to return to normal once the meat had been consumed then either the floor level would have had to rise to cover the bones — demanding exceptionally rapid sedimentation — or else man would have had to rely upon his own efforts to restore the site to a suitable condition.

Hypothesis B includes a provision for occasional clearance of medium-to-large bone debris from occupied areas (the sweepings being burnt as fuel or eventually ejected from the site). This would have been specially desirable after large kills, but probably also as a result of excessive accumulation of bone over a longer period. Such a practice would result in a strong bias towards preservation of unidentifiable splinters, and bones or teeth (whole or broken) whose size and shape rendered them liable to be trodden into the ground surface, though the trampling process itself is likely to have reduced the mesh size of this 'depositional filter' by compacting the cave floor. Factors influencing the degree to which sorting of this kind is likely to take place at a site include the permeability of the occupation surface (Gifford 1978), the ease with which rubbish can be deposited elsewhere, and the purpose for which the site was used. How important is refuse disposal for successful continued occupation? Studies of modern camp-sites confirm the effectiveness of such constraints (Schiffer 1978). Even before ethno-archaeological studies of these matters were far advanced (see Scott's remarks in chapter 18), such behaviour was being recognised in archaeological sites where conditions of preservation are better than at CSB:

> During the Mousterian occupation of one of the galleries of the Grotte du Renne at Arcy-sur-Cure (Yonne, France), the faunal remains were swept to the sides of the cave, leaving the central and anterior areas relatively clear except for lithic artefacts and a few bone fragments. A similar distribution was noted in the Grotte de l'Hyène, another of the Arcy caves (de Lumley and Boone 1976, 654).

> At Rocky Cape (Tasmania) the South Cave, abandoned c. 6700 years ago, large quantities of abalone shells were piled around the sides of the chamber, yet there were hardly any in the ashy central area (Jones, Rh. 1971).

### 31.4.4  Conclusions

The scarcity of remains of smaller animals in the bone heaps is also to be expected, for two reasons. There is no reason to suppose that other species would have accompanied the megafauna at the time of the principal kills. Also, when the butchery of the latter took place, any such debris would have been lying on or even in the floor, already processed. Much less three-dimensional than the rhino and mammoth remains, most of it must have been very rapidly trampled into the ground during this period of intensive activity, and so incorporated into what would appear during excavation to be the underlying layer.

On Hypothesis B, non-return of the hominids to the site is sufficient to explain the existence of the bone heaps. Though limited clearance may have occurred at any time, if the occupation were normally seasonal or sporadic rather than continuous any major accumulations would almost certainly have had to be disposed of at the beginning of the next occupation (by which time much of the residual meat would have rotted away). If the elimination of refuse was indeed one of the first jobs to be tackled on returning to CSB, the initial state of the site would be precisely that indicated in the archaeological record, with the centre of the cave effectively cleared of large remains during butchery, massive accumulations of bone against the wall, and the lithic and other small debris of the previous occupation scattered on or in the surface of the denser deposits underneath, or embedded in recent local mudflows.

The occurrence of a few remains of adult rhino and mammoth in layer A has already been mentioned. These

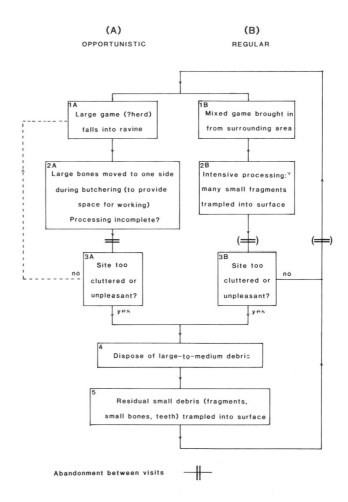

**Fig. 31.4** Interpretation of faunal exploitation strategies at La Cotte during the upper Saalian layers, according to the preferred hypothesis, B.

are splinters or body parts small enough to become trampled into the surface quite rapidly, and would not need to be ejected with the larger debris. Mammoth post-cranial remains are rare in layers D, C, A and 5, but allowance must be made for their fragile structure (and vulnerability to chemical destruction) and the possibility of their use as fuel. The contents of these layers are therefore in accordance with the predictions of Hypothesis B.

Hypothesis B — i.e. a strategy combining wide-ranging hunting of many different species with occasional, and probably opportunistic, large game kills in the ravines — is capable of accounting for all of the evidence, whereas the variants of Hypothesis A are less robust because of their complexity, requiring repeated behavioural changes, and they also deal less satisfactorily with the surviving large faunal elements. The bone heaps should not be taken as indicating a specialised way of life, involving very high possible returns at the expense of very high risks, but as the exploitation of a local asset to supplement what may otherwise have been low-to-medium risk tactics once the herds were in the vicinity. Prolonged abandonment of the site, leading to the preservation of the bone-heaps almost as left by the occupants, may have been due to a general deterioration of climate, or to the onset of severe loess-laden winds.

The evidence from La Cotte de St. Brelade appears to contradict the conclusion of Guérin and Faure (1983) that it is unlikely that early man hunted rhinoceros, though he probably killed young or injured animals (and also scavenged). In this context, it is worth recalling that at Mont-Dol, near the Brittany coast directly opposite Jersey, large Weichselian accumulations of mammoth, horse and rhinoceros remains were associated with Mousterian artefacts at the foot of a cliff (Sirodot 1873); the site may have functioned in much the same way as La Cotte, though the early date of its discovery, and subsequent destruction, makes this impossible to verify.

## 31.5 LITHIC TOOLS AND ECONOMIC INFERENCES

Leaving aside the constraints upon tool size and morphology which may have been imposed by the fluctuating availability of raw materials, certain aspects of the lithic industries have wider implications for the activities of the hominids.

1. At the base of the sequence the predominant flint tool types are flakes with large notches, and denticulates; later, convex side-scrapers are by far the most important group. Moreover, in the first layer in which scrapers play a numerically significant part (F), they are chiefly straight or concave; not until layer D, when a climatic deterioration is apparent, do the side-scrapers include more than 50% that are convex.

   In the time available it was not possible to arrange for a comprehensive study of the use-wear traces on different types of artefact from different layers. Such work as has been done (chapter 30) suggests that notches may be involved in the working of several materials. On the other hand, the only *large* notch which provided evidence of polish proved to have been used on wood. Many authors (e.g. Coles and Higgs 1975, 65; Brézillon 1969, 91) have suggested that notched pieces may have been used as spokeshaves on wood, bone and antler, though articles made of these last are found only rarely before the Upper Palaeolithic. They are entirely unsuitable for dressing whole skins, though they might be employed in the manufacture of leather thongs. Keeley (1980, 111) concluded that at Clacton "the 'preferences' for wood scraping seem to be either straight or concave edges, while those for hide scraping are for convex and straight edges". Binford and Binford (1966), among others, have also proposed that denticulates may have played a part in the processing of vegetable materials.

   The changes in frequency of the different kinds of retouched tool are at least not inconsistent with a reduction, through time, in the amount of woodworking at the site and a corresponding increase in the preparation of hides, changes which very roughly parallel the environmental sequence outlined earlier. However, in the absence of more use-wear data, especially for the unretouched artefacts, it is not possible to develop this argument in greater detail.

2. Artefacts suitable for use as pointed weapons are lacking or very rare in the oldest assemblages. Layer C is the first in which retouched points (types 4, 6 and 7) exceed 1% of the reduced tools. While weapon point frequency is one which, more than most, must be treated with caution (because of the special complications of loss, discard, replacement etc. associated with the function of these artefacts) there is a *prima facie* case for believing that such spears as may have been in use during much of Stage II were not tipped with flint. The use, during the Old Stone Age, of simple spears made by sharpening a wooden pole is recorded at Clacton (Oakley *et al.* 1977) and Lehringen (Movius 1950); at Clacton the associated lithic industry is technologically and typologically not so very different from that of our layer H. If the flint points were indeed used for tipping spears during the later occupation, this practice may have offered two advantages: a more effective wound because of extra tissue damage caused by the sharp edges of the flint, and a slight reduction in the risk of damage to the shaft, which in a relatively open environment may have been more valuable than a stone spear-head.

   The points found in the last of the Saalian occupation deposits (layer 5) are very different from earlier examples, having a thick cross-section which approximates an equilateral triangle and provides great strength. This may be an adaptation appropriate to their use in conjunction with the hunting of megafauna (see section 26.3).

3. The reduced tool count (from which are excluded unmodified and slightly modified pieces) is under 20% of the flint assemblages in layers H to A, though a slow but statistically significant increase is apparent (the Spearman correlation coefficient is -0.769). The trend is steady from layer E onwards, and is consistent with the conclusion drawn elsewhere that there was more efficient use of the available flint as supplies dwindled. In layer 5, however, the percentage of tools more than doubles, to 35.1, and there is a considerable increase in their size, despite a dramatic reduction in the amount of flint (as opposed to other material) brought into the site. The knapping waste from this layer is insufficient to account for the finished tools. This is interpreted as indicating that artefacts were being manufactured elsewhere, for the most part (see chapter 24). Thus most of the points referred to above are likely to have been discarded at the site when spears were being repaired, but their replacements were probably brought in ready made rather than being manufactured locally. To manufacture a wide range of artefacts at one locality, and then to carry them around — and perhaps even cache them elsewhere —

**Table 31.1**  Summary of Resource Availability in the SET of La Cotte de St. Brelade

| Resources | Stratigraphic stages | |
| --- | --- | --- |
| | Stage II | End of Stage II & Stage III interstadials |
| Water | Abundant | Available |
| Mammalian | Fairly stable, abundant | Strongly cyclical (seasonal), and with periods of very great abundance. Not always predictable. Probably capable of supporting large human bands for part of the year, but poor for the rest |
| Vegetable | Abundant | More localised |
| Marine | Available, but uncertain whether exploited | None in SET |
| Lithic material | Generally abundant, accessible | Less accessible, sometimes flint readily available only at margin of or outside SET |

together with spares, for use and resharpening as needed (curation and maintenance, to employ the terminology of Binford 1973 and 1977), is a type of behaviour particularly appropriate to exploitation of a large territory whose raw material sources are irregularly and possibly sparsely distributed. That curation seems best developed at the time of deposition of layer 5 lends support to the view that at that time CSB was used by bands which were more mobile than before.

## 31.6  TERRITORIAL ADEQUACY, BAND SIZE AND ECONOMIC STRATEGY

Economic data derived from the Cambridge excavations has been reviewed above, with extrapolation from other sites of broadly comparable age. Introduction of the concept of the SET has served to underline the availablity near La Cotte of a wide range of resources, despite the reduced accessibility of some of these at certain periods (Table 31.1).

Bettinger (1980) has summarised the main points of a general hunter-gatherer model based on a large number of ethnographic studies, among which are

a low population density (.05–.001 individuals per km$^2$);

operation at around 30–70% of the potential carrying capacity of the environment;

demographic arrangement at two organisational levels — "the *maximum band*, or mating network

(connubium) consisting of about 500 individuals, and the *minimum band*, or local group, consisting of about 25 individuals" (Bettinger 1980, 192–3). In practice, band size may vary considerably according to hunting requirements (see for example Heffley 1981, 141).

How well does the prehistoric occupation of CSB fit Bettinger's model, bearing in mind the incompleteness of the data? It is legitimate to enquire whether the carrying capacity of the area of the proposed site exploitation territory is in keeping with the suggested size of the minimum band, either throughout the year or on a seasonal basis. Because the range of resources exploited may have been well below the total available, and because no indications exist of the extent to which plant foods may have been used, a 'worst-case' approach is adopted here in assessing site potential.

The interglacial situation lends itself quite well to investigation because of easily defined limits on the SET and indeed on probable site catchment. Formulated explicitly, the hypothesis to be evaluated is that the meat available within the SET would be insufficient to maintain a band of 25 people. We shall take an extreme case in supposing that the occupants restricted themselves to hunting the only species recorded from Pleistocene interglacial deposits on the island, i.e. *Cervus elaphus*.

Of course it is not suggested that they lived exclusively on deer during much of Stage II (and it has not been overlooked that *no* deer remains have been identified in the layers in question). However, it may be seen from the calculations which follow that even a group employing an unnecessarily selective strategy (ignoring all foods but one) would come close to calorific suffici-

375

ency. Shortage of meat alone need not have imposed mobility upon the band (the influence of vegetable food availability cannot be assessed).

We shall employ figures used by Clark (1975) in his Star Carr study (190 kg kill weight, 60% meat yield, and annual culling of one animal in six), and the lowest density of red deer recorded by Darling (1969), i.e. one animal per 40.5 hectares. For the SET corresponding to a 30 m fall in sea level (area c. 17,000 ha) the results are: deer population 420, cull 70, meat weight 7741 kg. This gives a daily productivity of 21.1 kg, a little lower than the 21.7 kg used by Clark as the requirement for a band of four families (20 people). So despite a very pessimistic set of assumptions, a single resource might well be capable of filling almost all the needs of such a group (or 80% of those of a 25 person band), especially if CC or other satellites were used to extend the area of exploitation.

When the sea-level was somewhat lower, giving place to steppe, the expanded SET probably had a much greater total carrying capacity during the growing season. Even in the much harsher environment of present-day Greenland, Sturdy (1972) estimated that one reindeer per 40 ha was a reasonable figure for most types of grazing; the presence of giant deer suggests a very much richer grassland and many times the Greenland biomass per unit area. Hunting animals in large herds encourages formation of human groups larger than the minimum band, and increased mobility. Whether or not the presence of megafauna also contributed to pressures towards enlarged groups would have depended upon

tactical requirements, but also upon the efficiency with which the carcasses were exploited.

## 31.7 CONCLUSION

The interplay of climate, landscape and economy was extremely elaborate in the Channel Islands area during the Middle and Upper Pleistocene, as Fig. 31.5 shows. The Jersey sites were particularly sensitive to climatic variation of even limited magnitude, inasmuch as it could lead to severe modification of their territories and resources.

The different environments of La Cotte de St. Brelade must have imposed very different constraints upon the occupants. During fully temperate periods, small bands of limited mobility may have been best suited to the terrain; an enlarged territory and more open landscape will have favoured occasional larger aggregations and greater movement. In the latter case, the combination of mobility and distancing of sources of good raw material from the site appear eventually to have encouraged an important shift in practices of tool manufacture and use. The peculiar topography of La Cotte de St. Brelade also allowed it to be used as a kill site as well as for more domestic activities. Perhaps its most important lesson is the demonstration of early man's adaptability in the face of different challenges and opportunities.

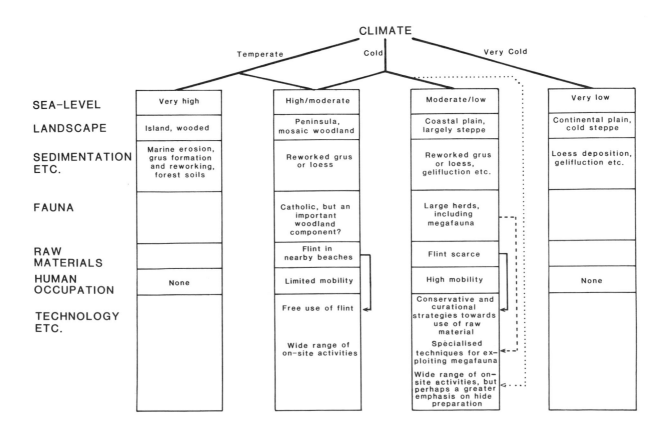

**Fig. 31.5** Relationships between environment and human activity during the Palaeolithic occupation of Jersey.

# THE LA COTTE INDUSTRIES AND THE EUROPEAN LOWER AND MIDDLE PALAEOLITHIC

P. Callow

## 32.1  INTRODUCTION

The dating and character of the La Cotte industries have been considered in earlier chapters. To complete the discussion, two further questions must be asked. What parallels, if any, can be found elsewhere in Europe? And what are the implications of the new data for models attempting to explain industrial variation?

In recent years the industries of the European late Middle and early Upper Pleistocene have attracted increasing attention as new discoveries and modern excavations have supplemented, and partly supplanted, the considerable volume of information obtained in the past, some of which was of indifferent quality and which generally exhibited a bias in favour of the flint-rich areas of Britain and north France. Integration of local and regional records into a wider framework has scarcely begun, and investigations of raw material and site function (which, as we have seen, play so important a part at La Cotte) have largely taken second place to taxonomy. It may be expected that the complications of interpreting Saalian industries will be not dissimilar to those so keenly debated for the early Weichselian (Bordes 1981). It was apparent from the concluding discussion of the 1980 Haifa conference (Ronen 1982) that the nature and even the existence of a 'Lower to Middle Palaeolithic transition' had yet to be universally accepted. Since then a rapid increase in the number of chronometrically dated sites (and the re-evaluation of stratigraphic data) has led to growing agreement that the use of the term 'Middle Palaeolithic' should be extended back in time to well before the last interglacial complex (see, for example, various papers in Tuffreau in press).

It is now clear that by the early Saalian, at latest, there had been an important change in the character of stone industries in western Europe. Prior to this, most assemblages were dominated by core tools (handaxes or chopper/chopping tools) sometimes accompanied by small numbers of tools made on flakes. But although core tools persisted in many series during the period with which we are concerned they usually played a subsidiary role to large numbers of often well-made flake tools. At about the same time, more advanced methods were often adopted to provide the blanks, notably Levallois technique in its several forms (Bordes 1980) and the use of disc cores. The dates of their first occurrence have been pushed back by recent research. Though true tortoise-cores certainly occur around the onset of Saale, the manufacture of Levallois blades and points (as opposed to flakes) and the practice of detaching usable flakes from a disc by progressive centripetal reduction (which allows very efficient use of raw material) were once thought to be of relatively late date — appearing at the end of Saale; these are now unequivocally recorded from the middle part of this period, e.g. at Etaples (Tuffreau and Zuate y Zuber 1975), Biache-St.-Vaast (Tuffreau 1982a) and La Cotte itself. A conspicuous feature of the industries based on these knapping strategies is the high value of the indices of facetted butts (IF and more especially IFs). It is these developments and their timing that raise the question of a possible major transition; an extreme view of its significance is offered by Bosinski (1982, 167), who sees it as accompanying a better mastery of environment permitting the first occupation of Central Europe during cold periods, and (contributing to the closing discussion in Ronen 1982, 324) equates it in importance with that from the Middle to Upper Palaeolithic. The dating of these events is summarised in section 32.4.1 below.

Apart from the patchy distribution of sites, in both time and space, the problems facing any attempt at synthesis of the late Middle and early Upper Pleistocene archaeology of Europe include: dating evidence often of questionable reliability; non-comparability of published artefact data from different sites; uneven quality of the older collections; often inadequate information about probable site function, environment and raw material sources, for instance.

Dating is critical, of course, not only for the interpretation of any observed variation but also for the initial definition of an appropriate set of assemblages for consideration (in this chapter, as usual, Alpine and British glacial/interglacial terms have been converted to those of northern Europe). In fact great controversy has arisen as to the position within the Middle Pleistocene of several key sites, even. One of the best known is the Caune de l'Arago cave in the eastern Pyrenees (de Lumley et al. 1979; Bada and Helfman 1976), where conflicting biostratigraphic and chronometric data have been used to support either an early Saale or an Elster age (see Cook et al. 1982 for further discussion). Moreover the age estimates for some of the classic sites in northern France and Britain have been revised in recent years. Thus the fluviatile deposits at Cagny-La Garenne, which were assigned to the beginning of Saale by Bordes (1956), are now regarded as late Elster (e.g. Bourdier et al. 1974b; Tuffreau et al. 1982b)). The Middle Gravels at Swanscombe, long regarded as Holsteinian (i.e. isotope stage 11), are appreciably younger, probably stage 9 (Kerney 1971; Callow 1976; Bridgland 1980; Hubbard 1982).

The problem of data comparability has two distinct components. One is the lack of a universally recognised descriptive scheme; in France and Belgium the Bordes (1961b) typology is now generally employed, but a number of different approaches are to be found in the works of German and British authors. The other is the inconsistency of the detail in which results are reported; when the Bordes typology is used the raw counts should be given in addition to the derived indices, yet it is quite common practice for only selected figures to be provided (this is especially true for some of the more important southern French sites). Information about older collections is often incomplete because the typology was in its infancy at the time of publication (Bourgon 1957 for the Perigord; Bordes 1954 for the Seine valley). Bias during collection is likely to have resulted in over-representation of certain types (particularly handaxes, but also the better-made scrapers), so it has not been possible to make use of material from many well-known sites, especially in Britain, Belgium and northern France.

The frequent dearth of information supplementing bald statements about typology and technology is especially restricting. Long ago Bordes (1950), dismissing the 'Levalloisian culture' proposed by Breuil and Koslowski (1931–4), pointed out that Levallois technique appeared to be related to the ready availability of good flint, an observation later refined by the definition of the term 'Levallois facies' to denote an abundance of unretouched Levallois blanks (often indicative of factory sites). More recently, Rolland (1981) has suggested that during the Weichselian much of the assemblage variation apparent in the Mousterian of southern France arises from interrelated changes in environment and social morphology; he has since extended the same model to southern French industries of greater antiquity (Rolland 1986). This theme is taken up again later in the present chapter.

## 32.2 EXISTING REGIONAL MODELS FOR THE LATE MIDDLE PLEISTOCENE INDUSTRIES

It is not necessary for us to consider explicitly industries older than the Saalian, though it is as well to recall that some authors believe that the industrial partitions proposed for this period reflect deeper rooted traditions. Nor shall we discuss the much more detailed information available for the Weichselian (see Appendix B).

The area of direct relevance to La Cotte may be broken down into three major regions.

1. Northern France, Belgium and the southern part of Great Britain: predominantly open sites; raw material essentially good quality flint from the Chalk. See Roe (1981, Fig. 5:1–4) for a distribution map of Lower and Middle Palaeolithic findspots in Britain, showing that these are largely restricted to southern and eastern England. Areas of crystalline rocks include Wales, Brittany and the Vosges.
2. Southern France, as far north as the Charente: numerous caves and rock-shelters, as well as open sites (very few of the latter have been excavated, however); raw material type, quality and availability vary greatly.
3. Germany and central Europe: predominantly open sites, in the north (caves and rock-shelters further south have generally yielded rather poor assemblages); raw materials are varied.

From the regional summaries which follow it is apparent that three aspects of the assemblages have consistently been regarded by European prehistorians as particularly important: Levallois technique, handaxes, and the relative abundances of side-scrapers as against notched

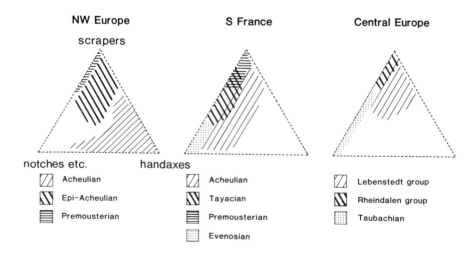

**Fig. 32.1** Major components of the regional classificatory schemes described in section 32.2. Despite the triangular format (more usually employed for precise abundance data) the representations have been somewhat schematised in the interests of clarity.

pieces and denticulates. The significance accorded to other features varies from region to region. The structure of each of the schemes in terms of these features is set out in Fig. 32.1, and the location of sites mentioned in the text is shown in Fig. 32.2.

Two points should be stressed here. Firstly, within each region the absence of full quantitative data for many of the sites makes it impossible to substantiate the validity of the subdivisions (quite apart from the artificial influence of modern political boundaries which determine the method of description employed!). Secondly, the principal motive for introducing these schemes is to provide a terminology permitting comparison of the La Cotte material with substantial numbers of assemblages, as this would not be possible without some prior grouping.

### 32.2.1 Northern France, Belgium, and Britain

Numerous instances of handaxe-rich **Acheulian** assemblages have been attributed to the Saalian. Though many of these owe their present composition to the selectivity of collectors in the days when gravel and brickearth were dug by hand, a number of excavated or carefully collected series show that the percentage of handaxes was indeed sometimes very high: e.g. Cuxton (Tester 1965), Swanscombe Barnfield Pit Middle Gravels (Wymer 1968, 334 ff.), Vimy (Sommé and Tuffreau 1976). In this area the handaxes are usually of classic form (ovate, *limande*, elongated cordiform, lanceolate, etc.). Discussion of possible temporal and regional subdivisions of the Acheulian has concentrated on variation in handaxe morphology (e.g. Waechter 1973; Wymer 1968 and 1974; Callow 1974 and 1976); a fresh review of the whole problem is urgently needed in view of recent revisions in the dating of terrace sequences.

In some assemblages, such as Bapaume-Les Osiers (series B), handaxes are scarce; Tuffreau (1982 and 1984) has suggested the name **Epi-Acheulian** for these. Many other assemblages lack handaxes altogether: e.g. Etaples-La Bagarre (Tuffreau and Zuate y Zuber 1975), the lustrous series from Beaumetz-les-Loges (Hurtrelle *et al.* 1972), Biache-St.-Vaast (Piningre 1978; Tuffreau 1978 and 1982; Tuffreau *et al.* 1982a), Champvoisy (Tuffreau 1982, 143). In Britain, High Lodge (Marr *et al.* 1921) provides the clearest example of an industry of this type in view of the stratigraphic separation of handaxe and scraper series demonstrated by recent excavation (Anon. 1968). A case can be made that these new finds, and also some of the older ones referred to by Breuil, mark an important development in the European Old Stone Age, inasmuch as they are indistinguishable, typologically and technologically, from the assemblages of the 'Mousterian technocomplex' of the last glaciation; they are sometimes called **Premousterian**. Side-scrapers are usually a fairly important component; in the Biache and Beaumetz assemblages IR$_{ess}$ is 47.32 and 42.96 respectively, while at Champvoisy the same index reaches the very high figure of 73.66. It should be noted, however, that side-scrapers occur in considerable numbers in some of the handaxe-rich series such as Vimy (IR$_{ess}$ = 45.30) and Hoxne

(upper industry), where Wymer's (1983) figures give an IR$_{ess}$ of around 70.

One recently excavated site of particular interest is Mesvin IV, in Belgium (Cahen 1984, 144), with uranium-series dates — admittedly on bone — of around 250 kya (Gilot 1984, following analyses by Szabo). Though lacking typical Acheulian handaxes it possesses biface-scrapers and *prondniks*, some with lateral tranchets or LSF removals (see chapter 29), leading Cahen to draw parallels with the Micoquian of central Europe (normally dated to the Weichselian).

Levallois technique was widely used, but there are important exceptions: in north France at St. Acheul (*Atelier de Commont*), Mareuil (*série blanche*) and Beaumetz-les-Loges (*série jaune*), and in England at Hoxne (upper industry) and Swanscombe (Barnfield Pit Middle Gravels), for example. But in most cases such assemblages are likely to be relatively early, say isotope stage 9 (see references already cited).

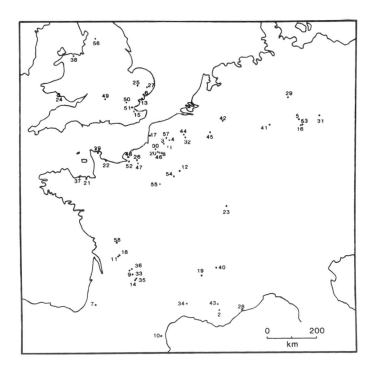

**Fig. 32.2** Map of sites mentioned in the text. Key: 1 Bapaume, 2 La Baume Bonne, 3 Beaumetz-les-Loges, 4 Biache, 5 Bilzingsleben, 6 Bobbitshole, 7 Bouheben, 8 Cagny, 9 Cantalouette, 10 Caune de l'Arago, 11 La Chaise, 12 Champvoisy, 13 Clacton, 14 Combe-Grenal, 15 Cuxton, 16 Ehringsdorf, 17 Etaples, 18 Fontéchevade, 19 Fouillousse terrace, 20 Grâce, 21 Grainfollet, 22 Grandcamp-les-Bains, 23 La Grande Pile, 24 Gower peninsula, 25 High Lodge, 26 Houppeville, 27 Hoxne, 28 Lazaret, 29 Lebenstedt, 30 Mareuil, 31 Markkleeberg, 32 Mesvin, 33 La Micoque, 34 Orgnac, 35 Pech de l'Azé, 36 Les Pendus, 37 Piégu, 38 Pontnewydd, 39 Port Pignot, 40 Prélétang, 41 Reutersruh, 42 Rheindalen, 43 Rigabe, 44 Rissori, 45 Rocourt, 46 St. Acheul, 47 St. Pierre-lès-Elbeuf, 48 St. Romain de Colbosc, 49 Stanton Harcourt, 50 Stoke Newington, 51 Swanscombe, 52 Tancarville, 53 Taubach, 54 Le Tillet, 55 Verrières-le-Buisson, 56 Victoria Cave, 57 Vimy, 58 Voulème.

# DISCUSSION

## 32.2.2 Southern France

If we exclude surface-collected sites on river terraces, it seems that handaxes are generally scarcer than in the north; by local standards, the IB$_{ess}$ of 12.23 for the open site of Cantalouette (Guichard 1965) is high. 'Classic' forms are rarer, and nucleiform, partial, backed and other poorly characterised types are often numerically important. Flake-cleavers occur in some series from the southwest, e.g. Pech de l'Azé, Cantalouette, Les Pendus (see Laville 1982; Guichard 1976), though they are absent from the late Saalian layers of Combe-Grenal. The technological indices IL, Ilam, IF and IFs generally take low values. These and other features have resulted in the use of the term **Acheuléen méridionale** (Southern Acheulian) to distinguish this facies from that of the north; the flake-cleavers led Bordes (1971) to suggest links with Spain. Its northeasternmost extension is perhaps in the Rhône valley, on the Fouillousse terrace, where surface finds of flake-cleavers, handaxes, chopping-tools and especially choppers are abundant, but it is possible that these are substantially earlier than the other finds considered here (Brochier 1976, 876–7).

Elsewhere in the south the **Upper Acheulian** is generally characterised by numerous and well-made flake tools, and scarce handaxes; de Lumley (1976c, 834) describes an IB$_{ess}$ of 5.3 for Lazaret as quite high. Nevertheless, some open sites have yielded higher frequencies: in the north of the region under discussion, at Les Vignes, Voulème, the biface index is 23.40 (Texier 1972).

As in the north, assemblages without handaxes (or in which handaxes are exceedingly rare) are well represented. De Lumley (1975) has proposed the name **Evenosian** for those with rare side-scrapers but numerous notched, truncated and 'Upper Palaeolithic' tools (chopping-tools and polyhedric balls are also common). On the other hand, assemblages with numerous well-made side-scrapers (often with Quina retouch), Tayac points and some *limaces* or *proto-limaces* are termed **Tayacian** (the differences are tabulated in full by de Lumley 1976, 847). The use of this name in such a context is unfortunate, as it was originally coined by Breuil (1932) with a very different meaning — closer, indeed, to that of the Evenosian!

Yet another type of industry without handaxes is described by de Lumley (1975, 797; de Lumley-Woodyear 1969 and 1971) as **Premousterian**; IL and ILty are high, as too are IR and I rc, and there is some Quina retouch. Examples are recorded from Rigabe and Baume des Peyrards; Orgnac III 1 has been assigned to the same group (Combier 1967 and 1976). Such industries, which often possess Mousterian-style disc cores, have often been regarded as ancestral to the Typical Mousterian of the last glaciation, as has the Tayacian to the Charentian (e.g. de Lumley-Woodyear 1971; Le Tensorer 1978; Rigaud and Texier 1981).

Whereas the Tayacian and Premousterian occur in caves and rock-shelters, assemblages attributed to the Evenosian are found chiefly in open sites (a possible exception is the rather problematical series from Fontechevade layer E). The Upper Acheulian is known from sites of both types.

## 32.2.3 Germany

A number of sites situated in the north European plain in East and West Germany are of particular interest, but their evaluation is complicated by poorly standardised nomenclature. A form of Upper Acheulian occurs in open sites, thinning out to the east (Bosinski 1976). Handaxes generally form a small proportion of the assemblage (flake-cleavers are occasionally present); Levallois technique has been extensively employed and includes blade production. For examples see Markkleeberg (Baumann *et al.* 1983), Reutersruh (Luttropp and Bosinski 1971), etc.

Bosinski (1982), in suggesting that the Saalian industries of northwest Germany should be redefined as Middle rather than Lower Palaeolithic, has named such assemblages the **Lebenstedt Group**. In addition to the features already mentioned he lists a relatively low frequency of retouched pieces, predominance of side-scrapers (mainly single and double), and the presence of bifacial scrapers resembling leaf-points and of rare unifacial points. Another type of assemblage, termed by Bosinski the **Rheindalen Group** and likewise attributed to the Saalian, is represented by finds from layer B3 at the type-site (Thieme *et al.* 1981). Handaxes are absent, Levallois technique is rarer and less carefully performed, and the percentage of retouched tools is relatively high. The side-scrapers are much thicker than in the Lebenstedt Group, and often possess Quina retouch. Points are a very common feature. In East Germany the assemblage from Ehringsdorf (Behm-Blancke 1960; Feustel 1983) is broadly similar typologically to that from Rheindalen, but includes a number of leaf-points; furthermore, Levallois technique is effectively lacking.

The material from Taubach (Behm-Blancke 1960) and Bilzingsleben (Mania 1976; Burdukiewicz *et al.* 1979; Mania *et al.* 1980; Mai *et al.* 1983) is reported to be of totally different character, made on pebbles and of almost microlithic proportions. It includes numerous notched pieces and denticulates, as well as Tayac points. Levallois technique is absent. Valoch (1976 and 1982) has suggested that the last interglacial industry from layer 11 of Kůlna cave in Czechoslovakia belongs to the same group, the **Taubachian**, seen as extending as far back in time as Elster at Vértesszöllös. Recently obtained uranium-series dates place this last site at the isotope stage 7/6 transition, however (Schwarcz and Latham 1984), bringing it more into line with the others. For Kůlna he quotes several of the Bordes indices (IR$_{ess}$ is 52.78). Comparable data has yet to be published for the other sites, though Valoch indicates that at Taubach notched pieces and denticulates outnumber side-scrapers. The thick loess deposits at Achenheim, which though situated in France (Alsace) perhaps belongs more naturally to this region, have provided a small series of Ehringsdorf-like pieces in the middle Older Loess, which dates to an intermediate phase of Saale, perhaps isotope stage 8 (Wernert 1957; Thévenin 1976; Heim *et al.* 1982; Sainty and Thévenin 1978).

**Table 32.1**  Middle Pleistocene Assemblages Used in the Multivariate Analysis[a]

| Symbol | Label | Assemblage | Region | Site type | Source |
|---|---|---|---|---|---|
| None | H | La Cotte de St. Brelade H | Jersey | C | |
| | G | La Cotte de St. Brelade G | | | |
| | F | La Cotte de St. Brelade F | | | |
| | E | La Cotte de St. Brelade E | | | |
| | D | La Cotte de St. Brelade D | | | |
| | C | La Cotte de St. Brelade C | | | |
| | B | La Cotte de St. Brelade B | | | |
| | A | La Cotte de St. Brelade A | | | |
| | 3 | La Cotte de St. Brelade 3 | | | |
| | 5 | La Cotte de St. Brelade 5 | | | |
| | X | La Cotte à la Chèvre | Jersey | C | |
| □ | 1 | La Chaise, Abri Suard 52 | France: Poitou-Charentes | C | Debénath 1976a |
| | 2 | La Chaise, Abri Suard 51 | | | |
| | 3 | La Chaise, Abri Suard V–VI | | | |
| | 4 | La Chaise, Abri Suard IV | | | |
| | 5 | La Chaise, Abri Suard II | | | |
| △ | 1 | Combe-Grenal 60 | France: Aquitaine | C | Bordes (pers. comm.) |
| | 2 | Combe Grenal 59 | | | |
| | 3 | Combe Grenal 58 | | | |
| | 4 | Combe Grenal 57 | | | |
| | 5 | Combe-Grenal 56 | | | |
| ◇ | 1 | Orgnac 5–7 | France: Rhône-Alpes | C | Combier 1967 |
| | 2 | Orgnac 4 | | | |
| | 3 | Orgnac 2–3 | | | |
| | 4 | Orgnac 1 | | | |
| ○ | 1 | Pech de l'Azé IIB 9 | France: Aquitaine | C | Bordes (pers. comm.) |
| | 2 | Pech de l'Azé IIB 6–8 | | | |
| ⌂ | 1 | Bapaume A | France: Nord | O | Tuffreau 1971 |
| | 2 | Bapaume | | | |
| | 3 | Beaumetz-les-Loges | France: Nord | O | Hurtrelle et al. 1972 |
| | 4 | Bouheben | France: Aquitaine | O | Bordes (pers. comm.) |
| | 5 | Cantalouette | France: Aquitaine | O | Guichard 1965 |
| | 6 | Grainfollet | France: Bretagne | C | Monnier 1980b |
| | 7 | Houppeville s. rousse | France: Haute-Normandie | O | Bordes (pers. comm.) |
| | 8 | Piégu | France: Bretagne | O | Monnier 1980b |
| | 9 | Prélétang | France: Rhône-Alpes | C | Bordes (pers. comm.) |
| | 10 | Rissori | Belgium | O | Adam & Tuffreau 1972 |
| | 11 | Le Tillet s. blanche | France: Région parisienne | O | Bordes (pers. comm.) |
| | 12 | Le Tillet s. grise | | | |
| | 13 | Verrières-le-Buisson 3 | France: Région parisienne | O | Daniel et al. 1973 |
| | 14 | Les Vignes | France: Poitou-Charentes | O | Texier 1972 |

[a]The symbols are those employed in Figs. 32.3 and 32.4. The site type is given as C (cave or shelter) or O (open).

## 32.3  TYPOLOGICAL AND TECHNOLOGICAL COMPARISONS

The search for parallels for the La Cotte assemblages was performed in two stages: multivariate analysis of the series for which more or less complete Bordes data are available, and evaluation of the La Cotte material in the light of the taxonomic schemes described in section 32.2. Because of frequent repetition, La Cotte de St. Brelade is referred to as CSB and La Cotte à la Chèvre as CC for the remainder of this chapter.

### 32.3.1  Multivariate Analysis

Besides CSB and CC, 30 assemblages have been included in the study; they are listed in Table 32.1. The requirement that the Bordes scheme should have been employed means that they are all French apart from Rissori in Belgium; even so, sites in the Midi (especially those assigned to the Tayacian and Evenosian) and the handaxe-

# DISCUSSION

**Table 32.2** Correlation Coefficients between Typological and Technological Indices and Varimax-rotated Factors[a]

| Index | Factor | | | | | | |
|---|---|---|---|---|---|---|---|
| | 1 | 2 | 3 | 4 | 5 | 6 | 7 |
| ILty | 0.07 | 0.81 | 0.02 | 0.27 | -0.02 | 0.31 | 0.06 |
| IR | 0.95 | -0.05 | -0.21 | 0.00 | 0.07 | 0.06 | 0.04 |
| IC | 0.77 | -0.07 | -0.35 | 0.08 | -0.33 | -0.11 | 0.25 |
| IAu | -0.09 | 0.54 | 0.20 | 0.53 | 0.20 | 0.30 | 0.28 |
| IAt | 0.04 | 0.58 | 0.26 | 0.11 | 0.70 | -0.10 | 0.21 |
| IB | 0.08 | 0.47 | 0.23 | -0.05 | 0.75 | -0.23 | 0.18 |
| I | 0.27 | 0.81 | -0.00 | 0.03 | 0.08 | -0.15 | -0.24 |
| II | 0.95 | 0.13 | -0.21 | 0.08 | 0.03 | -0.00 | -0.03 |
| III | -0.15 | 0.21 | 0.89 | -0.02 | 0.16 | 0.17 | 0.10 |
| IV | -0.81 | -0.28 | -0.30 | -0.01 | -0.04 | -0.02 | 0.05 |
| 5 | -0.13 | 0.34 | -0.01 | -0.15 | -0.78 | -0.13 | 0.20 |
| 38 | 0.14 | 0.02 | 0.00 | -0.15 | -0.09 | 0.79 | -0.21 |
| I rc | 0.43 | 0.68 | 0.01 | 0.06 | 0.19 | -0.22 | -0.35 |
| 6-7 | 0.29 | 0.73 | -0.08 | 0.29 | -0.16 | -0.23 | -0.27 |
| Rsingle | 0.71 | -0.16 | -0.52 | 0.08 | -0.20 | 0.23 | 0.08 |
| Rdouble | 0.47 | 0.25 | 0.02 | 0.34 | 0.11 | 0.24 | -0.29 |
| Rtransv | 0.12 | -0.18 | 0.06 | -0.14 | 0.03 | -0.19 | 0.81 |
| Rinv | 0.38 | -0.60 | 0.05 | -0.13 | -0.03 | -0.12 | 0.22 |
| 42 | -0.86 | -0.10 | -0.04 | -0.23 | -0.11 | -0.02 | -0.02 |
| 42-43 | -0.93 | -0.23 | -0.21 | -0.12 | -0.08 | -0.03 | 0.02 |
| Ident | -0.93 | -0.22 | -0.18 | -0.15 | -0.13 | -0.01 | 0.04 |
| Iirr | -0.48 | -0.63 | -0.12 | -0.02 | -0.10 | -0.28 | -0.12 |
| 30-31 | -0.14 | -0.14 | 0.70 | -0.51 | 0.03 | 0.04 | -0.12 |
| 32-33 | 0.34 | -0.25 | 0.69 | 0.06 | 0.01 | -0.35 | 0.09 |
| IL | 0.14 | 0.82 | 0.07 | 0.47 | 0.06 | 0.13 | 0.07 |
| Ilam | 0.17 | 0.38 | 0.51 | 0.22 | 0.24 | 0.43 | 0.08 |
| IFs | 0.24 | 0.25 | -0.08 | 0.85 | 0.12 | 0.00 | -0.18 |
| IF | 0.20 | 0.23 | -0.09 | 0.86 | -0.01 | -0.24 | -0.11 |

[a]Assemblage scores are plotted in Fig. 32.4.

rich facies of the Acheulian in northern France are under-represented. Most are of Saalian age, but several Eemian and very early Weichselian series attributed to older traditions have been included; e.g. Orgnac III 1 (Premousterian), and Houppeville *série rousse* and Grainfollet (Micoquian). The sites of Combe-Grenal, Orgnac III, and the Abri Suard at La Chaise are particularly important in that they have yielded multiple stratified assemblages whose internal variation invites comparison with that at CSB.

The variables have already been listed in section 24.4, in connection with the analysis of CSB on its own (see the description of dataset A). For the most part they are based on the reduced tool count, but it was found necessary to omit the Quina index (IQ), since this was not available for about half of the assemblages, and also the percentage of sharpening flakes, of course.

Seven principal components have eigenvalues greater than unity and may therefore be regarded as potentially of some consequence, but only the first three are equivalent in weight to two or more of the input variables. PC1 opposes [notched pieces, denticulates etc. + irregularly retouched pieces] and [IR, II, I rc, Rdouble, I, IL, ILty, and to a lesser extent IF and IFs]. PC2 consists of [Rsingle and Rinv] against [IAu and, more weakly, IB], while PC3 concerns [III, end-scrapers and burins]. Between them they account for 64.3% of the variance.

Varimax rotation results in a set of axes whose relationship to the input variables is of great interest. Factor 1 (see Table 32.2) is the classic 'Mousterian' axis, contrasting side-scrapers (especially the simpler forms) and the Charentian index against denticulates and notches. Factor 2 is based on the typological and technological Levallois indices, $I_{red}$ (retouched Levallois points), points and convergent scrapers. Factor 3 is concerned purely with facetting; Factor 4 with Group III, end-scrapers and burins. Factor 5 opposes bifaces and pseudo-Levallois points, and Factors 6 and 7 relate to *couteaux à dos naturel* and transverse side-scrapers respectively.

The principal component score distributions (Fig. 32.3) show that the northern French open-air series (and also Prélétang) are well separated from the others; Prélétang is further isolated by its very low score on PC3, however. The remaining assemblages are dispersed in a loose ellipsoid pattern. Conspicuous features are:

1. the proximity of CC to the early CSB series;
2. the position of CSB, ranged diagonally on the left hand side of the distribution;
3. the reversal of direction of the trend in PC1 for CSB from layer D onwards, and the lack of consistent trends at the other multi-layer sites;
4. the gross similarity between CSB G and H and certain of the Southern Acheulian series (though Pech de l'Azé IIB 9 stands out on PC3), despite the presence of handaxes in the latter;
5. similarities between the later CSB series and those from the 'Riss III' deposits at Combe-Grenal in particular, and also those from other southern French caves.

The most obvious result of rotating factors is to realign CSB along Factor 1 (Fig. 32.4); moreover the separation of 'Levallois-ness' and 'platform preparation' as Factors 2 and 3 results in much greater separation between the various cave series (some of the Abri Suard assemblages stand out because of their high facetting indices, contrasting with the two earliest from Orgnac). Factors 4 and 5 are very similar to PC2 and PC3.

### 32.3.2 Similarities and Dissimilarities

The regionally based subdivisions of the Lower and early Middle Palaeolithic of Europe, described in section 32.2, yield parallels whose 'goodness of fit' to the CSB assemblages varies considerably. Except where otherwise stated the features mentioned indicate points of divergence from CSB.

**LAYERS H AND G** (no handaxes; denticulates and notched pieces dominant). Despite the differences be-

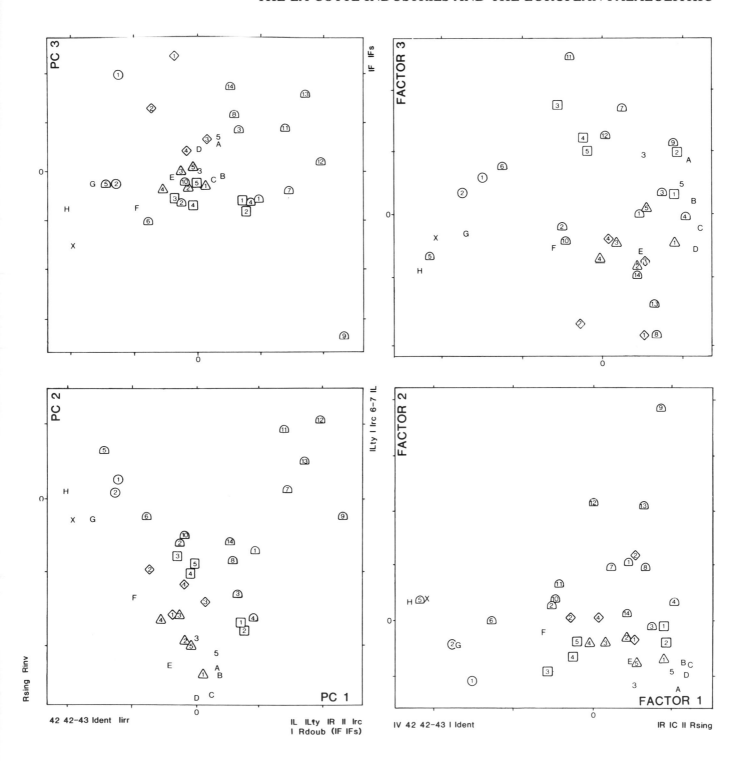

**Fig. 32.3** Late Middle Pleistocene assemblages. Scatter plots of principal component scores (for a key see Table 32.1).

**Fig. 32.4** Late Middle Pleistocene assemblages. Scatter plots of varimax-rotated factor scores.

tween these assemblages, especially in the relative abundances of denticulates and notches, they will be considered together for comparative purposes.

1. Acheulian of northern France. In the early Saalian cailloutis at Cagny-L'Epinette (Tuffreau *et al.* 1982b)

denticulates and notches are common and the debitage is non-Levallois, the tools being manufactured on casual flakes (many of which are clearly the by-products of handaxe roughing-out).

2. Southern Acheulian. Handaxes are present (or even numerous); in Pech de l'Azé IIB, Upper Palaeolithic

## DISCUSSION

tools are common. In fact the late series from Combe-Grenal are altogether different, with much higher percentages of end-scrapers.

3. Evenosian. There are numerous chopping tools and 'Upper Palaeolithic' tools; pseudo-Levallois points are rare.
4. Taubachian. There are considerable technological differences: pebbles are employed more directly, and pseudo-Levallois points have not been reported (raw material differences may be responsible). At some Taubachian sites side-scrapers are rather numerous.

Thus CSB H–G (and CC) are not easily matched. They may be thought to qualify as Middle Palaeolithic by virtue of the overwhelming predominance of tools made on flakes. Yet their technology is primitive, if competent. Several interpretations are therefore possible. Perhaps they represent an early equivalent of the Denticulate Mousterian, a functionally distinct variant of the Premousterian. Or, if the absence of handaxes is seen as related to site function, a link with the Acheulian cannot be ruled out (but the best parallels would suggest a relatively early date). Finally, they may be a technologically evolved form of the Clactonian (Wymer 1974), known from southern Britain during Holstein and perhaps Elster.

The last suggestion in particular is intriguing in view of the residual uncertainties about the age of the early CSB (and CC) assemblages. That from layer H appears to resemble the British Clactonian more closely than do most of the European post-Holstein series for which similar claims have been made, usually on no better grounds than hard-hammer debitage. But this line of enquiry cannot be pursued further until the British evidence has been systematically re-examined; Ohel (1979a), in the latest review of the subject, apparently treats Clactonian notches as accidental (see Ohel 1979b), while Collins's data (1969) are partly superseded by the results of later excavations at Swanscombe and Clacton.

**LAYERS F–D** (no handaxes; side-scrapers dominant; Quina retouch absent). Layer D has been included here because, as in the earlier assemblages, 'classic' bifaces are lacking; the implications of the few nucleiform handaxes are uncertain.

1. Premousterian of northern France. These industries make heavy use of Levallois technique, and often include a high percentage of convergent side-scrapers.
2. Premousterian of southern France. Differences include: at Baume des Peyrards and Rigabe, a high percentage of unmodified Levallois pieces and convergent scrapers, and the presence of Quina retouch; in Orgnac III 1, a high IL and ILty, again, and an IQ of 8.09; IR is also at the lower end of the range for CSB F–D; in La Micoque 4, a very low IR.
3. Upper Acheulian of Germany. Rheindalen Group: thick side-scrapers, often with Quina retouch, and

numerous points; Lebenstedt Group: presence of bifaces, extensive use of Levallois technique.

Typologically, CSB F–D fit very well into the handaxe-less Premousterian industries if allowance is made for technological variation and the possible morphological constraints which this would impose.

**LAYERS C–A** (scarce handaxes; side-scrapers dominant; scarce Quina retouch).

1. Epi-Acheulian of northern France. The most notable difference is in respect of Levallois technique.
2. Acheulian of southern England. Few series from this area can be used for comparative purposes; the upper industry from Hoxne, a recently re-excavated site, at least matches CSB in the abundance of side-scrapers, but its primary technology is simpler.
3. Upper Acheulian of southern France; Southern Acheulian. There are some quite close parallels, e.g. with Combe-Grenal, Lazaret, Orgnac III, La Chaise-Abri Suard (though some of these are more strongly Levallois). The occurrence of flake cleavers at CSB is interesting in view of their presence at some sites in this region.
4. Upper Acheulian of northern Germany (Lebenstedt Group). It has abundant unretouched Levallois flakes.

The technological peculiarities of the CSB series (resharpening, scaling, burin preparation, and so on) should perhaps be ignored here as being a local response to changing raw material availability, despite very occasional minor occurrences of some of these features elsewhere, e.g. at La Chaise (Debénath 1976a and 1976b) and Mesvin IV. In other respects CSB C–A are very typical of the Epi-Acheulian or Upper Acheulian.

**LAYER 5** (scarce handaxes; side-scrapers dominant; thick elongated points; Quina retouch).

1. Industries of southern England. High Lodge resembles CSB 5 in the quality of its scrapers, but these are not associated with handaxes. If allowance is made for possible over-representation of handaxes in the collections, the 'floor' at Stoke Newington (Smith 1894; Wymer 1968, 299) may provide a parallel, but points are absent; there are also considerable stratigraphic problems. CSB 5 differs from the Hoxne upper industry in possessing points and in the high proportion of retouched pieces made on Levallois blanks.
2. Epi-Acheulian of northern France. This differs in the scarcity of Quina retouch and general abundance of unmodified Levallois flakes. Thick elongated Mousterian points are lacking.
3. Tayacian. Most similar to CSB 5 is La Baume Bonne (de Lumley 1976a; de Lumley-Woodyear 1971), with an essential biface index of 1.16 and

numerous thick scrapers (IR$_{ess}$ 50.1), sometimes with Quina retouch. Though points are quite scarce, most are elongated and have Quina retouch, recalling those from CSB. *Surélevée* retouch is extensively used, however, giving rise to *proto-* (rather than true) *limaces*. The tools are rarely made on Levallois blanks, in contrast to those from CSB.

4. Upper Acheulian in Germany. Rheindalen B3 and Ehringsdorf are very like CSB 5 as regards the flake tools but lack typical Acheulian handaxes.

While CSB 5 can certainly be accommodated in the scraper-dominated industries of western Europe, it differs in important respects from all of the examples cited above. It appears to be a specialised facies of the Upper Acheulian with typical bifaces, as well as flake tools (including points) of 'Charentian' or central European character.

### 32.3.3 La Cotte as a Succession

Apart from the basal layers of CSB (as well as CC), which are technologically simple, the assemblages are all highly 'evolved'. In the light of evidence from other long stratified sequences one feature is particularly noteworthy: the occurrence of early Middle Palaeolithic series without handaxes *before* those with handaxes of typical Acheulian type. This supplies conclusive support for the impression gained from single-occupation sites and less securely dated caves that the presence of handaxes has no chronological significance during the second half of Saale.

## 32.4 TOWARDS A SYNTHESIS

It is not realistic to discuss a site of CSB's methodological importance merely in terms of concepts which are rapidly becoming outmoded. Certainly it represents a substantial addition to our basic data for the archaeology of the Saalian. But, by virtue of the unusual clarity with which environmental stimuli and human responses can be identified and linked, it also brings into focus the processual factors which are increasingly being used as a basis for questioning traditional models, but for which only indifferent 'test beds' have hitherto been available. Therefore, after a brief review of time-related industrial trends, the lessons of La Cotte are brought together with suggestions made by different authors at one time or another, in an exploration of possible avenues for future development.

### 32.4.1 The Chronological Dimension

Dating problems make it impossible to construct a very precise scheme even for a comparatively restricted area; also the uncertainty attaching to the age of such sites as Caune de l'Arago could, when resolved, force a drastic revision or at least suggest that the model is too simple. Nevertheless, if we concentrate on the sites at which high quality raw material was most readily available some degree of temporal patterning emerges.

1. During the early part of the Saalian (until perhaps isotope stage 9) there is a continuation of the 'classic' Acheulian in which handaxes play a major part (e.g. Cagny-L'Epinette, Swanscombe).

2. At about this point two developments occur (or at any rate become discernible). Handaxes when present usually occupy a numerically less important position in the assemblage, their place being taken by flake tools which are typologically more clearly defined than before. And the classic Levallois technique, based on tortoise-cores designed to produce a single large flake, is very heavily used at some sites. Examples (with or without handaxes) include the upper industry at Hoxne, Markkleeberg, perhaps the *Atelier de Commont* at St. Acheul (where handaxes may be over-represented in the old collections), Baker's Hole at Northfleet, etc.

3. By isotope stage 7 various alternative approaches to prepared-core technique are in use, and minor technological tricks are becoming quite common as at CSB, La Chaise and Mesvin IV (chronometric dates for these last are given by Blackwell *et al.* 1983 and Aitken *et al.* in press). Unless bifaces of typical Acheulian form are present it is no longer possible to differentiate clearly between the assemblages of this age and those of the Weichselian, as the flake tools are typologically as well as technologically advanced.

To the above we may add two further observations. Firstly, for most of the earlier part of the period under discussion there is no evidence of regular occupation of the north European plain under conditions of extreme cold; it seems that hominid distribution was essentially southerly whenever the climate was very cold, with northwards expansion during interglacials and interstadials being reversed by the next severe cooling. The first occurrences of sites in some numbers in this area and under such conditions appear to date to isotope stage 6: Achenheim '*sol 74*' and Rheindalen B3, for example.

Secondly, within the Saalian as a whole we find the full range of variation on the 'denticulates/notches to side-scrapers' scale. Marked domination by the former is the exception; there is a strong case for placing the most unequivocal French examples — Cagny-L'Epinette (Tuffreau *et al.* 1982b) and Pech de l'Azé II (Bordes 1972) — relatively early in Saale. Schwarcz and Blackwell (1983) have obtained U-series ages consistent with stage 7 for calcite fragments embedded in deposits of the latter site, suggesting that the important implementiferous layers 8 and 9 are older still.

### 32.4.2 Other Elements

In the discussion of phenetic comparisons in section 32.3 no systematic attempt was made to look for underlying causes of industrial variation, though in some instances attention was drawn to raw material and other constraints.

Reference has already been made to Rolland's suggestion (1986) that many assemblages owe much of their character to the patterns of raw material procurement and tool re-use which he sees as partly linked to group mobility; Tavoso (1984) has also discussed the importance of site function and location in determining the constituents of an assemblage. And Villa (1981) goes further, dismissing so-called 'culture provinces' in France during the Middle Pleistocene as merely the product of raw material differences. La Cotte itself provides strong evidence of the validity of the determinist case (see chapter 24). Though Hodder (1982) and others have demonstrated the importance of the symbolic content of human behaviour, this can only operate within such latitude as remains after the operation of economic and technological constraints.

Even if the data are as yet inadequate to allow the construction of a comprehensive model of European material culture and its environmental background, it is possible to identify some of its component elements.

The role of the handaxe is likely to have been complex. Keeley (1980, 161), while arguing that some types must have had a limited functional range, admits that others (the less elaborate forms) could have been used for many different purposes and may have been designed for use away from the home base. Quite apart from this 'Swiss army knife' aspect, handaxes were quite frequently employed as cores for the removal of a few flakes when necessary (there are numerous examples from northern France); moreover it may be supposed that once the technique of thinning had been mastered it would have been possible to strike off several razor-edged flakes without greatly affecting the piece's suitability for other tasks. It is not without interest here that Isaac and Harris (1980) have remarked that at Lake Turkana in Kenya handaxes occur only *outside* the sites marked by occupation debris. This observation has to be treated with caution, though, as deflation of a later surface could also have been responsible; the period in question is of course some million years older than La Cotte.

Some of the variation in the technological indices will have been related to raw material availability, which must also have influenced the use of alternative relatively advanced styles of debitage of varying efficiency (assuming that these formed part of the knappers' repertoire). The classic Levallois technique for producing flakes by centripetal preparation (i.e. using tortoise-cores) is extravagant compared to the disc-core technique. The latter, and that directed at the production of blades, are methods for the continuous manufacture of blanks and incur very little waste (the term 'Levallois blade' is a misnomer when applied to parallel longitudinal preparation, as the underlying concept is entirely 'non-Levallois' compared to that employed at, say, Baker's Hole or Montières). We may also note the observation by Boëda (in press) that at Biache-St.-Vaast, where the principal occupations date to around 175 kya (Aitken in press), the Levallois tortoise-core technique had been modified to yield two or three good flakes for each round of preparation. A question which will clearly bear further investigation is why, at any particular site, one of

these more economical techniques should have been favoured at the expense of the rest.

Even when Levallois technique was employed, the relative abundance of unretouched and retouched Levallois blanks will have been linked to both raw material availability and site function; thus high ILty values occur not only at some factory sites but where a supply of blanks has been imported, as at Prélétang (which being at very high altitude was probably visited by hunters who brought with them everything they were likely to need).

Much the same considerations apply to retouched tools, whose overall representation is related not only to the activities which took place at the site but to the assemblage formation processes of importation, resharpening and discard. Moreover, the weighting of these three behavioural parameters varies according to the tool category; a higher percentage of the total output of side-scrapers may have been abandoned at heavily occupied sites than would have been the case for weapon points, for example.

The choice between two alternative 'types' capable of fulfilling a single need may sometimes have been governed by raw material through technological limitations. Perhaps the most conspicuous example is provided by flake- and bifacial cleavers; the former are more easily made from large slabs of (often tough) rock, the latter from nodules or pebbles of moderate size and fine grain, usually flint (for a fuller review of this topic see Villa 1981). Of more general concern is the durability of working edges, and their need for and ability to take retouch. Jones (1979) has discussed the consequences of this for the bifaces from Olduvai Gorge in Tanzania, but it is equally pertinent for industries based on light-duty flake-tools.

Binford and Binford's (1966) much discussed suggestion that variations in Middle Palaeolithic assemblage composition may be seen in terms of tool-kits for different activities may be considered for the Saalian industries. The existence of multi-purpose artefacts (see especially Beyries 1984) and the assemblage formation processes already mentioned imply that for this purpose simple factorial models (particularly of the varimax type) are unlikely to cope entirely satisfactorily with the archaeological data; not the least of the difficulties is that certain types will have been present in several kits. But as suggested in section 31.5, some tools are indeed better suited by their shape to certain tasks and materials. Assessment of the relative contribution of the tool-kit principle to inter-assemblage variation will not be possible until a number of assemblages with first class environmental data have been examined for microwear, with very much larger samples than have been studied so far.

In the previous chapter attention was drawn to the very great changes in the character of the Channel Islands area in the course of major climatic oscillations. Spatially patterned contrasts of comparable magnitude to the eyes of stone age hunters would have existed between the lowlands stretching from England eastwards into Asia and the more broken relief of southern France. At a single moment in time the economics of living in these two areas may have been very different, as evidenced (for example) by the greater abundance of proboscidean remains in the former, and have led to differences in tool inventories and perhaps in social behaviour as well. If man was sometimes obliged to withdraw from northern Europe because of the cold, we might expect to find clearer cultural discontinuities here than in the permanently occupied zone; the availability of more and better dated sites in the future may provide the means of testing this and also the hypothesis that apparent regional differences are environmentally rather than culturally determined.

### 32.4.3  How Far Can the Data Take Us?

The considerations listed in the previous section make it evident that even a single-phylum model is capable of generating a good deal of assemblage variability, some of which will exhibit regional patterning (Fig. 32.5). In particular, they call into question the significance of the principal variables upon which the classificatory schemes of section 32.2 are largely based. Should we then reject altogether the possibility of independent and perhaps local traditions? The answer is unlikely to be found

through the approach introduced by Bordes, based on comparison of typological or technological indices designed to express the most commonly encountered forms of variability. As we have seen, these are chiefly informative about aspects of the assemblages influenced, at least in part, by external constraints (though Bordes himself recognised the desirability of defining additional indices to cope with new questions).

The principle to which we must turn is one which was introduced in chapter 24, in an intra-site context: that only when it can be shown that the same choice has repeatedly been made between viable alternatives is there a *prima facie* case for considering an explanation in terms of cultural continuity between a group of assemblages — always admitting the possibility of behavioural convergence. It is rare, rather than common, behaviour that is usually the more informative in this context. In fact we are only on firm ground if the assemblages in question come from a single multi-layer site (with a reasonably long occupation) or several sites within a clearly defined area. This approach is based on a regression model: elimination of explanatory factors leaving residuals, followed by examination of the latter in order to identify non-random behaviour (specifically, the localisation of

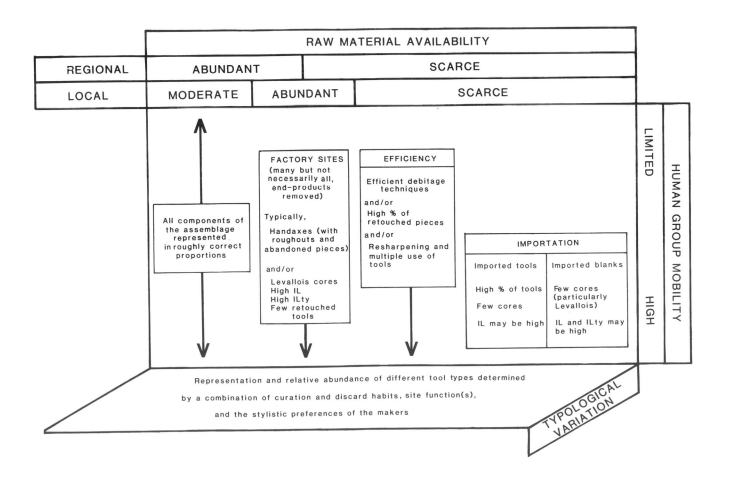

**Fig. 32.5**  Influences contributing to assemblage variability.

particular attribute states in time and/or space).

Development of an appropriate methodology does not guarantee that a problem can be solved, however, as the available data may be insufficient. Thus in a study of the possible relationship between two variables a correlation coefficient of 0.2, say, does not entitle one to reject the null hypothesis of random association between variables for a sample of five pairs of observations, but becomes significant for 5000 observations — yet the pattern of variation has not itself changed. To take an archaeological example (Callow in press), the bifaces from Olduvai are very strongly influenced in form and technology by raw material, size and chronological trends; some of the variation remaining after these factors have been allowed for is consistent with the presence of two industrial traditions (Acheulian and Developed Oldowan) suggested by other aspects of the assemblages, but the sample for the second of these is too small for significance tests to exclude a single-industry model *on the basis of the bifaces alone.*

Turning once more to the late Lower and early Middle Palaeolithic, the scarcity and imperfections of our data mean that at the moment we cannot make much progress even with the first stage of the analysis — the study of independent variables and their effects. But we should not be too hasty in expecting these to yield a complete explanation of variation between assemblages. One prerequisite for statistical analysis on a larger scale is the publication of the numerous important sites excavated in the 1960s and 1970s but never described in a usable form. In a few years' time we may hope to see the fruits of the recent extensive programmes of multi-disciplinary work on open-air sites, notably that directed by Tuffreau in northern France. At present we can only point to a few examples in which stylistic preferences appear to go beyond the individual knapper, and which certainly justify further investigation.

1. Unusually high incidences of certain types of retouch for no very apparent reason, such as scraper edges made by inverse retouch during part of the La Cotte sequence, or of *surélevée* retouch during the Tayacian (given that there are other ways of producing steep edges);
2. Spatial and chronological structuring of some aspects of biface morphology in the flint-rich areas (Callow 1976);
3. Alternative approaches to tool rejuvenation;
4. Alternative 'advanced' flaking techniques.

If in the light of further evidence these or other aspects of the industries are confirmed as exhibiting marked patterning, we may be forced to admit the existence of distinct traditions. Villa (1981, 32) was correct to question

the unconsidered extension into the Middle Pleistocene of models of social organisation which are derived from studies of more recent periods (though the demonstration of handedness on the part of the La Cotte hominids shows that hominisation of the brain, and behaviour, had by then proceeded further than was proven at the time when she was writing). Nevertheless there are risks inherent in over-zealous application of Occam's razor (see Clarke 1968, 79, for a formulation of a Converse-Occam axiom to counter this). It is difficult to justify more than one evolving technocomplex in our area during the Saalian, and we have rejected many of the criteria conventionally used for its subdivision on cultural lines, but wholly to deny the possibility of such subdivisions (whether short- or long-lived) would be to assume a better knowledge of the period than we presently possess.

## 32.5  CONCLUSIONS

The assemblages from La Cotte span much of the typological range recorded in western and central Europe during the Saalian. Apart from those from the two earliest layers, for which close parallels are not easily found, they fit very readily into an early Middle Palaeolithic technocomplex which is already well known from other sites. But allowance has to be made for special features linked to raw material problems; even so, the superposition of assemblages with bifaces over those without them, but in other ways closely resembling each other, is an unusual feature.

The addition of the La Cotte data greatly increases the number of assemblages capable of inclusion in statistical analyses, and this allows the principal axes of variation to be identified more clearly. Nevertheless the absence of a single widely-adopted typology means that we cannot yet systematically compare assemblages over the whole area, nor are there sufficient supporting environmental and chronometric data. Consequently, although existing proposals for regional industrial groupings are suspect, as arising very largely from external factors, it is not possible to arrive at a satisfactory replacement model. Too heavy a reliance on the determinist approach is undesirable, but we can identify some of the component parts of such a model and suggest topics which call for further investigation. Until more of the back-log of earlier excavations is published, together with those recently carried out with newer questions in mind, the most stimulating advances in this field of archaeological research will probably be the improvement of age estimates and consequent shifts in our perceptions of space/time relationships.

# 33

# CONCLUDING DISCUSSION

### P. Callow

La Cotte de St. Brelade is of much greater interest than the archaeological finds alone would suggest. Certainly the richness of its industries and faunal remains place it high in the ranking of Middle and early Upper Pleistocene sites. And although depositional processes have not as a rule favoured the preservation of distinct occupation horizons the bone heaps do provide rare evidence of momentary events at the site. But because of the range of environmental information that can be placed within a chronostratigraphic framework, and the interplay of marine and terrestrial responses to climatic change, La Cotte is also of importance in the wider compass of Pleistocene research. The most significant conclusions reached during the investigations are as follows:

1. The deposits, with a cumulative thickness of some 40 m before excavation, span a quarter of a million years or more, and provide a fairly detailed record of environmental change from isotope stage 7 onwards. Their formation occurred in the course of two important interglacial complexes before the Flandrian, and of the intervening periods of periglacial conditions; during the second of the interglacial complexes marine erosion was responsible for the destruction of part of the contents of the ravines (as has also occurred during the Flandrian). The most important archaeological deposits were formed in relatively mild periods within the Saalian and in the early/middle Weichselian.
2. An apparently extremely complex history of sedimentation and redeposition can be resolved in a cyclical model determined by the interaction of climate and topography, as described in section 6.5; on this are superimposed the processes of soil formation and marine erosion.
3. Not only are the marine and terrestrial events recorded at the site identifiable with those recognised elsewhere, but they themselves provide important additional information about Pleistocene sea-levels. Thus the existence of a +18 m transgression during or subsequent to that part of isotope stage 7 represented at La Cotte is rendered implausible both by the lack of the expected geomorphological evidence and by the very close resemblance between the artefacts from layer H and those from La Cotte à la Chèvre, where such an event *is* recorded. And the younger beach deposits at La Cotte suggest that the regression from the isotope substage 5e maximum was more complex than is usually apparent elsewhere (chapter 7).

4. Human occupation of the site took place only under certain conditions. It required a sea-level low enough to allow access to Jersey from the mainland, and also a not too severe climate; the latter consideration may have been responsible for the absence of Palaeolithic finds on the other islands. The extent and character of the territory exploitable from La

**Fig. 33.1**   La Cotte after a century of research: the north ravine during re-excavation of the deep sounding in 1982, with retaining walls protecting the principal sections (compare Fig. 1.10). The photograph is of particular interest in showing a dark stain, due to water, on the upper concrete face; this closely corresponds to the position of shallow erosion gullies cutting at a number of levels through the deposits on the west side of the site (described in section 6.6.3).

Cotte de St. Brelade and La Cotte à la Chèvre would have varied considerably during the different occupations (chapter 31).

5. Though La Cotte de St. Brelade was usually the focus of a wide range of day-to-day human activities, it also served, during isotope stage 6, for trapping and butchering mammoth and woolly rhinoceros. On two (or possibly three) occasions the failure of the inhabitants to return to the site at the onset of a period of loess deposition allowed the preservation of substantial remains of these animals, in contrast to the small splinters of bone that characterise the dense archaeological layers arising from repeated occupation (chapters 18 and 31).

6. The lithic assemblages are extremely varied (chapters 24 and 32). The earliest is reminiscent of the Clactonian (though it would be unwise to assume any direct link with this British industrial type), but those that follow are strongly Middle Palaeolithic in character, dominated by well-made side-scrapers. Typical handaxes of Acheulian type occur only in the upper part of the Saalian sequence. The Weichselian layers excavated early this century contain a Typical Mousterian very similar to that found on the Breton mainland at Mont-Dol (Appendix B).

7. A particularly interesting aspect of the artefacts is the successive adoption of alternative strategies for the procurement and use of flint when this material was in short supply; they include intensive resharpening and the importation of finished tools (chapters 24 and 31).

8. One of the resharpening techniques provides the clearest behavioural evidence of lateralisation of brain function so far recorded during the Middle Pleistocene, inasmuch as it shows that one hand was

favoured over the other in a ratio of about four to one (chapter 29).

Because of restrictions on funding and on the time available for the preparation of this report many aspects of the Cambridge excavations remain to be exploited. These include detailed palaeontological comparisons with fauna from other sites, refitting of artefacts, and details of artefact distribution within individual layers (bearing in mind the certainty of some degree of geological disturbance, of course). The variation in the colour of flint from different layers (Fig. 22.2) was not recognised until a late stage of the investigation, and the possibilities offered by the identification of the contained fossils have yet to be realised; even though the raw material was obtained as pebbles rather than from outcrops, it may eventually be possible to integrate this and other information into a model of the development of the Channel bed during the Middle Pleistocene. Also, the whole collection of pieces measuring less than 20 mm awaits analysis; their relevance to some of the questions raised during study of the rest of the material has become increasingly apparent. This list is far from exhaustive, but should serve to indicate the immense potential that remains. With the completion of this book the team that has produced it will be disbanded; it is hoped that others will be encouraged to pursue the investigations further.

The site archive is to be lodged with the Société Jersiaise, and arrangements are being made for providing copies of the principal computer files on magnetic tape. Sediment samples are available, but continued study of the *in situ* deposits will be limited by the walling-up of the principal sections, which was carried out by the Société with the intention that they should remain undisturbed until well into the next century.

# APPENDICES

# APPENDIX A

# CORRECTION OF PROVENANCE DATA AND DEFINITION OF ASSEMBLAGES

## P. Callow and P. J. Rose

## A.1   INTRODUCTION

As a result of the accelerating pace of work during the last few years of excavations at La Cotte, and a shortage of the necessary skilled labour, it was only after digging had ceased that a thorough analysis of the stratigraphic observations and artefact records was performed. Computerisation of the artefact provenance data was not completed until early 1980, when some of the geomorphological problems were already becoming apparent. An extensive revision was then undertaken, based on coordination (and, where necessary, reinterpretation) of information from four sources:

1. Field drawings (plans and sections) and photographs.
2. Field notebooks: from 1976, notes were kept for each metre square, which was subdivided into 0.2 x 0.2 m subsquares and excavated in 'spits' (in principle about 0.05 m deep). The observations generally included sketched sections, as well as information about depths and layer changes.
3. Finds inventories, which in the earlier years related to individual discoveries, and later listed the spit coordinates (one provenance could be represented by several bags of finds). These were updated at the end of each day's work, from the information on finds bag labels. At this stage the bags were assigned serial numbers (which therefore were in only very approximate stratigraphic order for each metre square and subsquare).
4. The site diary, which was used as a repository for miscellaneous stratigraphic observations, as well as for recording datum changes, known mistakes, etc.

Once recording and transcription errors had been corrected it was possible to put into effect the changes in layer definitions called for by the re-evaluation of the stratigraphy, and so to define the assemblages into which the finds were eventually divided for study.

The problems encountered at La Cotte are not unusual on other excavations (though certainly exacerbated by the backlog of finds processing and, more importantly, of stratigraphic analysis and interpretation); that is not to say that they are always formally identified, of course. On the other hand, because their resolution took the form of a one-off operation after a good deal of post-excavation work had already been done (rather than running in parallel with the excavation), and because the quantity of material involved called for substantial effort, it was necessary to devise a strategy which took maximum advantage of the power of the computer in order to speed up both the analysis and the implementation of the results to the collections. This was therefore given immediate priority by one of us (Callow) when he renewed his active participation in the investigation in 1979 — until the collections were in sufficiently good order there could be no question of carrying out the specialist study for which a Science Research Council grant had already been awarded (to have continued with the existing policy, of punching artefact provenance and descriptive data together, would have delayed the checking and correction process still further). The approach used is described in some detail as being of general interest to anyone concerned with the mechanics of excavation and post-excavation management, and also to explain the very considerable divergence between field and laboratory records which will be encountered by those studying La Cotte in the future.

## A.2   CATEGORIES OF ERROR

### A.2.1   Measurement and Recording in the Field

On occasion, finds were attributed to the wrong metre square (relatively easily discovered and corrected). Sometimes, also, measurements were taken from the wrong edge of the square, or were transposed so that the north coordinate became the east, and vice versa. Because measurements were being taken from an origin within the excavated area (i.e. north and south, east and west) there were some failures to record the direction of measurements — implying the default values, north and east. Occasionally when another excavator took over a metre square (particularly at the beginning of a new season) he or she did not realise that a layer change had already been observed.

### A.2.2   Transcription

The procedure in the field was to compile the finds inventories from the information on the bag labels (this was done at the end of the day). Mistakes could occur at this stage, or again when data were being entered into the computer (in the latter case, few escaped detection when the printouts were routinely checked against the inventory sheets).

### A.2.3   Stratigraphic Interpretation

The similar appearance of some of the layers, and also lateral facies variation within a single layer, made identification difficult where exposures were restricted; moreover, some of the depositional complexities of the site were not anticipated during excavation. For example, the distinctive yellowish colour of B was thought to render it suitable as a marker horizon for differentiating between A and C; in areas where B was absent, therefore, finds from the upper part of C (and even D) tended to be attributed to A. When sections were being drawn, variations in colour due to the presence of manganese, charcoal and burnt bone (which could be very localised) sometimes caused confusion; this was compounded when the work of interpreting and drawing adjacent sections was carried out by different people at different times.

## A.3   PRELIMINARY INVESTIGATIONS

Several approaches were followed:

Examination of field drawings and notes with a view to producing composite and isometric sections, based on direct comparison and analysis of Harris matrices (Harris 1979). These suggested areas in which the identification of layers had been inconsistent.

The provenance data for finds bags from each of the 0.2 m subsquares (and for such pieces as had been measured individually, reduced to the same system), were sorted according to depth. The computer was then asked to list the bag numbers in each subsquare, flagging possible inconsistencies of the following types (see Fig. A.1):

Layers out of sequence (noting whether the order of the bag numbers also gave an improbable result, which often indicated an error in the coordinates as opposed to the layer attribution).

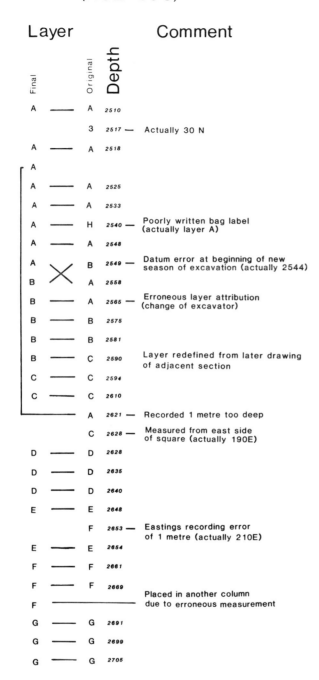

### (110E 30S)

| Layer | | | Comment |
|---|---|---|---|

**Final** — **Original** — **Depth**

| Final | | Original | Depth | Comment |
|---|---|---|---|---|
| A | —— | A | 2510 | |
| | | 3 | 2517 — | Actually 30 N |
| A | —— | A | 2518 | |
| A | | | | |
| A | —— | A | 2525 | |
| A | —— | A | 2533 | |
| A | —— | H | 2540 — | Poorly written bag label (actually layer A) |
| A | —— | A | 2548 | |
| A | ✕ | B | 2549 | Datum error at beginning of new season of excavation (actually 2544) |
| B | | A | 2558 | |
| B | —— | A | 2565 — | Erroneous layer attribution (change of excavator) |
| B | —— | B | 2575 | |
| B | —— | B | 2581 | |
| B | —— | C | 2590 | Layer redefined from later drawing of adjacent section |
| C | —— | C | 2594 | |
| C | —— | C | 2610 | |
| | | A | 2621 — | Recorded 1 metre too deep |
| | | C | 2628 — | Measured from east side of square (actually 190E) |
| D | —— | D | 2628 | |
| D | —— | D | 2635 | |
| D | —— | D | 2640 | |
| E | —— | E | 2648 | |
| | | F | 2653 — | Eastings recording error of 1 metre (actually 210E) |
| E | —— | E | 2654 | |
| F | —— | F | 2661 | |
| F | —— | F | 2669 | |
| F | ———————— | | | Placed in another column due to erroneous measurement |
| G | —— | G | 2691 | |
| G | —— | G | 2699 | |
| G | —— | G | 2705 | |

**Fig. A.1** Typical recording errors. The provenance data for the spits from a 0.2 x 0.2 m subsquare have been sorted by depth; the layer attributions are shown before and after correction (while not uncharacteristic of some parts of the site, this example is an artificial one compiled to include as a wide range of different errors as possible).

Duplication — harmless where two or more bags were genuinely one (rich) spit, but sometimes due to incorrect coordinates for one bag.

An exceptionally large gap (>0.5 m) between two consecutive spits in the same column (usually the result of a coordinate error).

A somewhat smaller or larger gap than usual (<0.03 or >0.08 m). Though often due merely to irregularities in the thickness of the spits, this could also be caused by an error in the coordinates, especially in the horizontal plane.

In addition, some of the more extreme discrepancies were highlighted by scattergrams of the coordinate data plotted in vertical 'slices' 0.2 m thick. In this way, any provenances falling outside the limits of the excavation became apparent.

Once this phase was complete, the results of the provenance analysis were examined. Discrepancies were checked in the following order:

Those which were impossible or highly improbable, implying a major inaccuracy in the coordinates. Most of these were dealt with by ascertaining which parts of the site were being worked at the appropriate date, and then checking the field documentation. From a knowledge of the most common types of recording error (see A.2.1 above) it was often possible to suggest a number of alternative values for the true coordinates. If it was found that any of these corresponded to those of a spit that had been dug on the appropriate day, and for which no other finds bag could be traced, it could be assumed with fair confidence that this was the source of the artefacts. Further verification was sought by comparison of handwriting and individual stylistic features present on the bag labels. If a number of different solutions offered themselves the matter was put on one side until the remaining unallocated spit bags had been checked.

Inconsistencies in the order of layers. During this phase of the investigation, those resulting from provenance errors were the principal target. Comparison with the original bag labels isolated a number of these as having occurred during transcription, and in other cases handwriting permitted identification of the excavator, and hence the correct square (as described in the previous paragraph).

Inconsistencies in respect of depth, but not layer (possibly due to misrecording of the horizontal coordinates, though some unnoticed shifts of depth datum points had occurred).

Elimination was an iterative process, the checking program being run on four occasions altogether. At first, minor problems over layer attribution were ignored as being perhaps due to difficulties of interpretation, leaving us free to concentrate on the coordinates. Eventually most of the residual problems could be sorted out by matching the suspect spits with gaps elsewhere on the site (for instance where the wrong square had been indicated). As the depth error flags became less numerous, the interpretation problems grew in prominence. Where layer inconsistencies occurred in the vicinity of a section, and the notebook information was explicit, resolution was a fairly straightforward matter. However, this left a residue of spits in the central part of each metre squares which could only be dealt with by interpolation.

Mistakes made during section drawing were resolved by cross-checking Harris matrices against the principal field drawings and the notebooks for the metre squares, bearing in mind new developments in our knowledge of the geological processes active in the site. We also had to take into account changes of opinion that had occurred as more extensive exposures became available. It was necessary to weigh the observations in three dimensions made by individuals who had actually excavated the deposits against those made later — sometimes many years later — in two dimensions during redrawing of sections. In the former case, perception was restricted according to what was visible at the time, while in the latter the unrecorded details on which earlier reasoning partly depended might have been forgotten or overlooked. The target was the simplest possible interpretation of the stratigraphy which was consistent with the depositional and erosional mechanisms.

In the course of this operation, a detailed analysis was made of the extensive photographic record.

The final stage called for the production of a series of computer-generated three-dimensional views of the interfaces, based on the provenance data in its intermediate state. Comparison of the mapping of the top of one layer with that of the base of the layer above (performed partly by eye, partly with the aid of the computer) showed where their surfaces displayed marked irregularities — expressed as pinnacles or 'holes'. Smoothing these out by reassignment was then quite straightforward; the same drawings proved to be valuable in aiding interpretation of the geomorphology (Fig. 6.12).

A small number of spits were known to represent interfaces; in other instances a sound decision about provenance could not be reached. The contents of these were omitted from the artefact analyses. Similarly, in the north east corner of the site (excavated in 1977) no discrimination was possible between the base of A, B, and the upper part of C. Apart from these, about 15% of provenances proved to contain an correctable error of some sort (even if only a sideways shift by 0.2 m, or one subsquare).

## A.4 APPLICATION TO THE COLLECTIONS

A step of the greatest importance for the project's development had been taken when McBurney decided to proceed with classification of the artefacts, by raw material and by typology, ahead of the resolution of outstanding stratigraphic questions. As a result, the contents of the finds bags had been individually numbered and then dispersed. The exact provenance of each piece could then be established only by reference to the master catalogue, since the artefact number did not itself encapsulate the spit number or other pertinent information. As it turned out, roughly one in six of the 'in situ' artefacts excavated in the 1977–1978 seasons (which yielded about three-quarters of the total) proved to need either reassignment to a different layer from that originally indicated, or outright rejection as being effectively unstratified. Because of the policy outlined above, these artefacts had to be found individually, rather than grouped together in their spit bags. Consequently about 100,000 separate handlings were required for the whole collection, as opposed to a few thousand had the work been done earlier. The former number may seem surprising, but on average it takes $n/2$ trials to find a particular piece in a bag containing $n$ of them; moreover, as described below it was necessary to adopt slightly inefficient methods.

One other point is worth stressing, since it may not be immediately obvious. Whether or not the artefacts had previously been sorted into provisional layers made negligible difference to the amount of handling required, because of the unexpectedly high percentage of attributions. Thus in the case of the 1977 material more than two-thirds was initially stored as 'Mainly Layer 2' (where '2' refers to our A), and it was hoped that effort would be saved with this group by treating A as the residual attribution and only looking for pieces from other layers. In the event, only 39% of the pieces from this year could be confidently assigned to A, and every category had to be thoroughly searched, to the extent that all the artefacts in it were sorted into numerical order before extraction. Again, the first 25,000 pieces excavated in 1978 were processed with their provenances ignored completely, as McBurney had realised that the latter were not particularly reliable. By July 1979 about 30,000 remained. The cheapest policy was clearly to abandon the processing altogether until the assemblages could be properly defined; however, this appeared impossible to implement because of the likely damage to team morale, because essential labour would be available only for the coming six months, and because of the time constraints imposed by fixed-term staff contracts. At that time it was not appreciated that resolution of the stratigraphy itself would prove a very lengthy operation. Callow therefore reintroduced the policy of making use of the original attribution (in the hope of saving at least a little time). As it turned out, the percentage of reassignments meant that no gains resulted. It was the initial dispersal of finds with the same provenance that proved to be the critical factor.

The only practicable course (short of returning the finds to their original bags and starting again) was to use the computer to produce lists ordered according to the existing storage arrangements, with flags to indicate pieces which had to be transferred. It was mentioned above that error resolution was carried out iteratively, and in fact three 'passes' were required to clean up the data. To have waited until this had been completed before handling the finds would have imposed even greater delays on the programme of work, so at each of these three stages such layer changes as proved necessary were effected for the artefacts. Although this incurred additional expense, in labour and computing, the reclassification was completed much sooner than would otherwise have been the case. All the same, it is worth stressing that the total cost of all operations described in this Appendix (excluding washing, marking and the original sorting of finds) exceeded the budget for the three last — and most expensive — years of excavation. With the benefit of hindsight, therefore, it is clear that, by designing excavation and post-excavation procedures to take particular account of human error, and also by careful training of the staff who implement them, immense sums can be saved in a project of the size of La Cotte.

## A.5 CONCLUDING OBSERVATIONS

It is worth bearing in mind that without the aid of a computer many of the inconsistencies might never have come to light. An important consequence of the partial mixing of the contents of archaeological layers (whether in prehistory or during excavation) is that, by increasing the resemblance of each assemblage to the next, it actually favours interpretation in terms of gradual rather than abrupt change — the implications are rather alarming in view of the extent to which we still rely, for our longer sequences, on the results of cave excavations conducted many years ago.

Another influential consideration in arriving at the decision to make a considerable investment in the process of error correction was that, apart from the dangers for our own research programme in working with contaminated assemblages (over and above a certain amount of unavoidable geological mixing), the discovery of many such errors during future study by other workers would probably have undermined the credibility (and value) of the collection in the longer term. There could be no question of simply excluding all questionable spits, as the contents of erroneous and correct ones were thoroughly mixed together when the first sorting was attempted — and thus were indistinguishable. Because of this, such a policy would in fact have required the rejection of 40–60% of the finds (depending on the strictness of the criteria employed), and virtually eliminated layers B–H from consideration.

Bordes (1971, 46–48) has described how analysis of the finds from the first year of work at Combe-Grenal appeared to show an unusual and surprising association between artefact types in one of the assemblages. This led, during subsequent fieldwork, to more careful investigation which showed that mixing had occurred due to imprecise excavation of a complex interface (as at La Cotte, too much reliance had been placed on the colour of the sediments, and too little on their texture and on looking for evidence of erosion). In dealing with the La Cotte finds our situation was very different, as the excavations had already been completed. It must be stressed that at no time were the typology and technology of the artefacts taken into account during the correction process; such an approach would have been unacceptable because it would have involved an element of circular reasoning.

## NOTE

Much of the preliminary work of cataloguing and collating the field drawings was performed by A. Berquist. Also, in addition to the authors, R. Burnell, B. Byrd and L. M. Evans were involved in various aspects of the operation described above.

# APPENDIX B

# ARTEFACTS FROM THE WEICHSELIAN DEPOSITS

P. Callow

## B.1    INTRODUCTION

Although all but a handful of the finds from the industries of the last glaciation were recovered before the Cambridge involvement with La Cotte began, they are reconsidered here for the sake of completeness, and also because they add an extra dimension to the evidence of the earlier layers. In general the provenance of individual pieces is unknown. This circumstance arises not only from the early date of the excavations and the techniques of recovery and recording then in use but also from the dispersal of the collections and from sometimes over-hasty curatorial decisions, which have resulted in the loss of much contextual information. The gravest problem, however, is posed by the disappearance of much of the original field documentation, since this makes it impossible to subdivide the collections into the stratigraphic units which are known to exist.

As mentioned in earlier chapters, and argued in detail in Appendix E (in microfiche), the artefacts came from the lower part of the loessic head (our layer 11); the upper part was archaeologically sterile. However, Marett (1916, 112ff.) refers to a thin sterile sandy layer separating an upper and a lower industry. An old sample of the 'hearth' material exhibits clay illuviation which may perhaps correspond to the phase of soil formation observed in 1982 in the north of the site (our 'episode 39' of chapter 7, this volume).

The artefacts from the Weichselian layers come from several phases of excavation:

1.  1881–1911. Work by private individuals and the Société Jersiaise, in the north ravine.
2.  1913–1920. Directed by R.R. Marett, with funding from the British Association, in the north ravine.
3.  1936–1940. C. Burdo and R.R. Marett, in the west and south ravines. The first volume of Burdo's manuscript field diaries covers this period, and refers to sections and other documents which seem to have been lost.
4.  1950–1956. The initial stages of Burdo's excavations, at a new and lower level, cut through the Weichselian deposits on the north side of the west ravine. Again, it is clear from the diaries that important documents are missing; there is a specific reference to an inventory of finds, and manuscript sections are entirely lacking.
5.  1961–1978. The Cambridge excavations yielded only a few flakes from a loessic deposit resting on the uppermost 'peat' near the east wall of the north ravine.

Burdo (1960, 31) gave the following figures for the artefacts found during the various stages of excavation:

|   | | | |
|---|---|---|---|
| 1. | pre-1910 | Unknown | |
| | 1910 | 100 + | (waste not counted) |
| | 1911 | 60 + | (waste not counted) |
| 2. | 1914-1915 | 15924 | |
| | 1916 | 1112 | |
| | 1917 | 297 | |
| 3. | 1936-1940 | 251 | |
| 4. | 1950-1951 | c.3900 | |

Although this suggests that perhaps 22,000 pieces were recovered, he also points out that very large quantities of waste were discarded or overlooked in the earlier phases of the work (for that matter, his own excavation methods certainly resulted in some losses, to judge from the recovery rates obtained in the Saalian layers by his and the Cambridge excavations respectively). The biases introduced into the collections during excavation are, however, rather trivial when their subsequent fate is considered.

The majority of the initial finds must be supposed to have found their way into private hands. The dispersal continued officially when the more formal excavations began; thus in 1911 batches of material were sent to the Oxford University Museum and the British Museum; similarly, in 1916 letters from Marett to Nicolle list 110 pieces to be sent to Oxford and Cambridge, and a further 120 to the British Museum. The proportion of the total finds represented in these gifts is small. Unfortunately, however, this is not the end of the tale. The data listed above would lead one to expect that over 16,000 pieces ought still to be preserved in the collections of the Société Jersiaise, even without the finds from the 1936–1940 and 1950–1951 excavations; in the event only about a quarter appear to have survived. The collections have long been stored in a wide range of boxes and trays, and material from all periods of the excavation has been mixed; moreover, much of it is unmarked (this appears to date from the British Association period) and therefore vulnerable. The 1936 and later material is easily identified, since until the beginning of 1952 Burdo marked everything with a three letter code (referring to an inventory which is now missing). Subsequently he wrote coordinates on larger pieces, leaving many smaller pieces unlabelled; these were sent by him with the rest of his Saalian collection directly to Cambridge, and there is thus no danger of admixture with Marett's finds.

Excluding the finds made by Burdo in the 1930s and 1950s, the total Weichselian material surviving at the Jersey Museum amounts to 3621 pieces, of which 3158 are more than 2 cm in length. It is on this collection that the analyses given below are based. At some stage in its history a manuscript partial inventory was made to permit formal accessioning; including the faunal and human remains from the same excavations, 1515 entries were made in all, the objects themselves being numbered in ink. A large proportion of the smaller flints was thus left unmarked, though it is unclear whether it was before this time that most of the losses occurred. If not, it must be supposed that the missing material is essentially small waste which was not thought to be worth marking. The amount of material scattered in museums in England was felt to be too small to justify the time required for its study.

For the present investigation it was decided that the Burdo material should not be used. The relatively few finds in 1936–1940 (obtained from a very large volume of deposit) came from a steeply sloping talus and were likely to have been far from their original context. A somewhat similar problem applies to Burdo's 1950–1951 finds, made before the discovery of the lower layers; the steep dip of the deposits in the west ravine raises important questions about the integrity of the series, and in particular the possibility of partial derivation from earlier layers. In the absence of detailed field documentation, examination of these issues would have been too time-consuming, given that the primary concern of this volume is with the earlier industries.

The fauna from the Weichselian layers is listed in Appendix F (in microfiche).

**Table B.1**  Inventory of Finds from the pre-1936 Excavations Stored at the Museum of La Société Jersiaise, Jersey

| Type | Flint | Quartz | Quartzite | Siltstone | Sandstone | Dolerite | Granite | Total |
|---|---|---|---|---|---|---|---|---|
| Flakes | 2349 | 10 | 13 | 7 | 0 | 1 | 3 | 2383 |
| Tools | 872 | 0 | 3 | 3 | 0 | 1 | 1 | 880 |
| Cores | 137 | 0 | 0 | 1 | 0 | 0 | 0 | 138 |
| Hammerstones/cobbles | 3 | 1 | 1 | 1 | 0 | 1 | 35 | 42 |
| Debris | 168 | 1 | 0 | 2 | 1 | 0 | 6 | 178 |
| Total | 3529 | 12 | 17 | 14 | 1 | 3 | 45 | 3621 |

[a]All pieces have been counted, whole or broken, and irrespective of size.

**Table B.2**  The Weichselian Industry: Technology[a]

| Index | Real | Essential | Reduced |
|---|---|---|---|
| **Tools** | | | |
| IL | 33.37 | 36.82 | 39.06 |
| Ilam | 8.15 | 8.14 | 6.77 |
| IF | 72.70 | 73.37 | 72.31 |
| IFs | 67.73 | 68.27 | 67.69 |
| IFss | 62.41 | 62.61 | 62.77 |
| **Waste** | | | |
| IL | 0.00 | 6.30 | 6.34 |
| Ilam | 4.13 | 4.67 | 4.95 |
| IF | 46.21 | 49.72 | 50.61 |
| IFs | 39.53 | 43.30 | 44.13 |
| IFss | 31.13 | 35.43 | 36.15 |
| **All** | | | |
| IL | 10.96 | – | – |
| Ilam | 5.18 | – | – |
| IF | 54.55 | – | – |
| IFs | 48.41 | – | – |
| IFss | 40.98 | – | – |

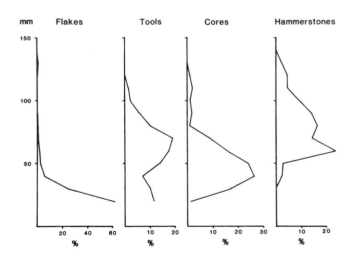

**Fig. B.1**  Weichselian layers. Length distribution for principal categories.

[a]Excluding pieces less than 20 mm long. In this table the term 'waste' refers to all flakes not treated as tools in a given column.

**Table B.3**  The Weichselian Industry: Butt Type, by Shape and Technology[a]

| Shape | Plain | Dihedral | Facetted | Polyhedral | Butt type Cortical | Punctiform | Removed | Missing | Total |
|---|---|---|---|---|---|---|---|---|---|
| **Non-Levallois** | | | | | | | | | |
| Flake | 584 | 92 | 471 | 117 | 122 | 9 | 51 | 823 | 2269 |
| Blade | 35 | 0 | 49 | 2 | 3 | 1 | 1 | 27 | 118 |
| Point | 9 | 3 | 20 | 3 | 6 | 0 | 0 | 3 | 44 |
| | 628 | 95 | 540 | 122 | 131 | 10 | 52 | 853 | 2431 |
| **Levallois** | | | | | | | | | |
| Flake | 32 | 11 | 160 | 9 | 6 | 0 | 32 | 33 | 283 |
| Blade | 4 | 1 | 19 | 1 | 0 | 0 | 0 | 2 | 27 |
| Point | 0 | 1 | 13 | 0 | 0 | 0 | 1 | 2 | 17 |
| | 36 | 13 | 192 | 10 | 6 | 0 | 33 | 37 | 327 |

[a]Excluding pieces less than 20 mm long.

**Table B.4**  The Weichselian Industry: Length [a]

| Type | Mean | Standard deviation | N |
|------|------|-----|---|
| Waste flakes | 26.4 | 11.2 | 1357 |
| Tools ('real') | 57.4 | 22.3 | 694 |
| Handaxes | 65.0 | 7.1 | 2 |
| Cores | 51.5 | 20.2 | 116 |
| Hammerstones and manuports (all materials) | 80.0 | 20.2 | 42 |

[a]Excluding pieces less than 20 mm long.

**Table B.5**  The Weichselian Industry: Core Typology

| Type | N | % | % whole only |
|------|---|---|------|
| Levallois – flake | 16 | 11.7 | 13.9 |
| Levallois – blade | 3 | 2.2 | 2.6 |
| Levallois – point | 1 | 0.7 | 0.9 |
| Discoidal | 65 | 47.4 | 56.5 |
| Globular | 5 | 3.6 | 4.3 |
| Prismatic | 3 | 2.2 | 2.6 |
| Pyramidal | 2 | 1.5 | 1.7 |
| Miscellaneous | 16 | 11.7 | 13.9 |
| Shapeless | 4 | 2.9 | 3.5 |
| Broken (undiagnosable) | 22 | 16.1 | – |
| Total | 137 | | |

## B.2    THE LITHIC INDUSTRY

The frequencies of the principal categories of lithic material are listed in Table B.1. However, for the purpose of describing the industry, all pieces less than 20 mm long (whole or broken) have been excluded, to provide some degree of comparability with finds from the older deposits. Where shape is to be considered (e.g. in calculating the frequency of blades) all broken pieces have been omitted. Only the artefacts made of flint have been used in the computation of typological and technological indices, based on the extended Bordes scheme as described in chapter 23.

Marett believed that there were two somewhat different industries represented (the lower being coarser in appearance, and with fewer fine points and scrapers), but it must be stressed that because individual provenance information was lacking there could be no question of separating the material stratigraphically, and there is a real possibility that the series described here is a mixture (but see section B.3 below).

### B.2.1    Raw Materials

In total contrast to the final Saalian industries, and paralleling much more closely those of the earliest layers at the site (Fig. 22.1), flint has been used almost exclusively: 98.6% when hammerstones and manuports are excluded. Once again, beach pebbles appear to have been the main source of supply. This is discussed in section 22.5.

### B.2.2    Technology

Even when fragments are included, the number of flakes (and flake tools) per core (23.4) serves to emphasise that bias has been exercised in the collection or conservation of the material, since ordinarily a much higher figure might be expected for such an industry. A Levallois index (IL) of 11.0 may therefore be a slight overestimate. For the same reason, there is a possibility of minor under-representation of, for instance, cortical platforms, which may have occurred more often among the less carefully collected waste.

As with the earlier series, three facetting indices, rather than two, have been computed in view of the inclusion of polyhedral platforms in the classification (see chapter 23). The values thus obtained (IF 54.6, IFs 48.4, IFss 41.0) are consistent with flake production based primarily on Levallois and disc core techniques. The facetting indices for the Levallois pieces on their own are of course much higher (IF 83.3, IFs 77.9, IFss 74.0). The blade index (Ilam), calculated for complete pieces only, is quite low, at 5.2. It is worth mentioning, however, that many of the blades are Levallois (18.6%); this is even more marked in the case of the points, though most should be regarded as pointed flakes rather than true Levallois points. Selection of blanks for tool making (with a very strong preference for Levallois flakes) resulted in higher values of the facetting indices for retouched pieces (Tables B.2–3).

The difference between tools and waste might have been even more marked were it not that 10.5% of the former (using the real count) have lost their platforms through retouch. Obviously part of the same process of blank selection is reflected in the high Levallois index for well retouched pieces (39.1). The tools (including unmodified Levallois flakes) are on average more than twice the length of the waste flakes, even allowing for the bias introduced by truncating the distribution at 20 mm.

The cores are small, with a mean length of 51.5 mm, which is actually less than that of the flake tools (Fig. B.1 and Table B.4). They are strongly dominated by the discoidal type. Of the Levallois cores, those for the production of flakes are the most common, though typical examples of the other types are present (Fig. B.4, 3). While some Levallois cores are likely to have yielded several Levallois flakes, and more of the latter may have been produced from the discs, there are far too few Levallois cores to explain the number of Levallois flakes. It is probable therefore that at least a few of the disc cores represent the final stage of reduction of the Levallois flake cores ('tortoise cores'), a view supported by their appearance in several cases.

The inventory includes a number of (chiefly granite) cobbles averaging 80 mm in their maximum dimension. Very few of the 311 referred to by Marett et al. (1916, 293) have survived, however. At least three, of flint, were presumably brought in as raw material but were never used. On the other hand, he comments that a concentration of them was found in the northern part of the site, and amounted to 37.5% of the lithic material in this area. Here, unretouched flakes were much scarcer than further south, but bone refuse was abundant. Despite the uncertainties arising from possible local dissolution of the bone, Marett's suggestion that these particular cobbles may have been used not so much as hammerstones for flint-knapping as for some culinary purpose is not implausible. One activity to which they would have been suited is the fragmentation of animal bone, though the abundant sharp-edged pieces of cave granite would surely have done just as well; alternatively they could have been used in the preparation of pemmican or of plant foods.

The importance of tool resharpening, as opposed to manufacture, in some of the earlier assemblages (chapter 29) raises the question of the extent to which such behaviour occurred in this series also. Only one good example (broken) of a long sharpening flake was recorded during analysis, together with two much more doubtful ones, and a solitary 'parent piece'. There are also five transverse sharpening flakes. These figures are so low that there can be no question of a deliberate policy of resharpening — at least by the techniques used previously — and are consistent with the view that raw material availability was no longer a serious problem. Only one piece has been classed as a burin spall; this matches the rarity of burins themselves.

### B.2.3    Typology

The early excavations in the deposits of the last glaciation yielded a rich assemblage of tools (Tables B.6 and B.7). Including the handaxes, the sample studied numbers 788, with 470 well retouched pieces. Sidescrapers are by far the most important element, and there is a reasonably large component of denticulates and notches, but the 'Upper Palaeolithic' group of types (end-scrapers, burins, awls and backed

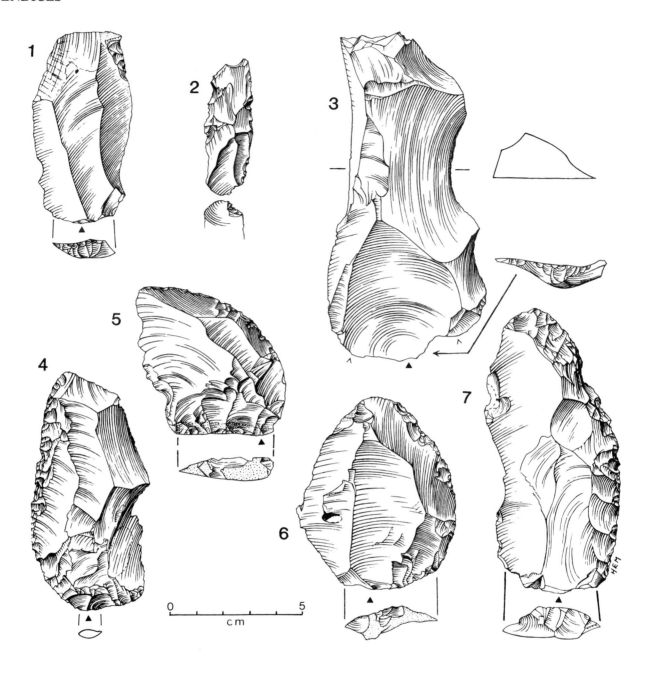

**Fig. B.2** Weichselian layers (Marett and earlier excavations). 1, 2 Typical Levallois flake and blade (both prepared by centripetal flaking); 3 atypical Levallois flake; 4–7 single convex side-scrapers.

**Table B.6**  The Weichselian Industry: Tool Typology

| Type | | N | Real | Ess. | Red. |
|---|---|---|---|---|---|
| 1 | Typical Levallois flakes | 77 | 9.80 | – | – |
| 2 | Atypical Levallois flakes | 38 | 4.83 | – | – |
| 3 | Levallois points | 2 | 0.25 | – | – |
| 4 | Retouched Levallois points | 3 | 0.38 | 0.60 | 0.64 |
| 5 | Pseudo-Levallois points | 15 | 1.91 | 3.01 | – |
| 6 | Mousterian points | 16 | 2.04 | 3.21 | 3.42 |
| 7 | Elongated Mousterian points | 2 | 0.25 | 0.40 | 0.43 |
| 8 | *Limaces* | 2 | 0.25 | 0.40 | 0.43 |
| 9 | Single straight side-scrapers | 32 | 4.07 | 6.41 | 6.84 |
| 10 | Single convex side-scrapers | 89 | 11.32 | 17.84 | 19.02 |
| 11 | Single concave side-scrapers | 15 | 1.91 | 3.01 | 3.21 |
| 12 | Double straight side-scrapers | 4 | 0.51 | 0.80 | 0.85 |
| 13 | Double straight-convex side-scrapers | 6 | 0.76 | 1.20 | 1.28 |
| 14 | Double straight-concave side-scrapers | 2 | 0.25 | 0.40 | 0.43 |
| 15 | Double convex side-scrapers | 21 | 2.67 | 4.21 | 4.49 |
| 17 | Double concave-convex side-scrapers | 5 | 0.64 | 1.00 | 1.07 |
| 18 | Convergent straight side-scrapers | 6 | 0.76 | 1.20 | 1.28 |
| 19 | Convergent convex side-scrapers | 25 | 3.18 | 5.01 | 5.34 |
| 20 | Convergent concave side-scrapers | 3 | 0.38 | 0.60 | 0.64 |
| 21 | *Déjeté* (offset) scrapers | 31 | 3.94 | 6.21 | 6.62 |
| 22 | Straight transverse side-scrapers | 5 | 0.64 | 1.00 | 1.07 |
| 23 | Convex transverse side-scrapers | 4 | 0.51 | 0.80 | 0.85 |
| 25 | Side-scrapers on ventral face | 7 | 0.89 | 1.40 | 1.50 |
| 26 | Abrupt retouched side-scrapers | 4 | 0.51 | 0.80 | 0.85 |
| 27 | Side-scrapers with thinned back | 9 | 1.15 | 1.80 | 1.92 |
| 28 | Side-scrapers with bifacial retouch | 6 | 0.76 | 1.20 | 1.28 |
| 29 | Alternate retouched side-scrapers | 8 | 1.02 | 1.60 | 1.71 |
| 31 | Atypical end-scrapers | 6 | 0.76 | 1.20 | 1.28 |
| 32 | Typical burins | 2 | 0.25 | 0.40 | 0.43 |
| 33 | Atypical burins | 1 | 0.13 | 0.20 | 0.21 |
| 34 | Typical borers | 2 | 0.25 | 0.40 | 0.43 |
| 37 | Atypical backed knives | 4 | 0.51 | 0.80 | 0.85 |
| 38 | Naturally backed knives | 16 | 2.04 | 3.21 | – |
| 39 | *Raclettes* | 7 | 0.89 | 1.40 | 1.50 |
| 40 | Truncated pieces | 6 | 0.76 | 1.20 | 1.28 |
| 42 | Notched pieces | 39 | 4.96 | 7.82 | 8.33 |
| 43 | Denticulates | 55 | 7.00 | 11.02 | 11.75 |
| 44 | *Becs burinants alternes* | 3 | 0.38 | 0.60 | 0.64 |
| 45 | Retouched on ventral face | 9 | 1.15 | – | – |
| 46-47 | Abrupt and alternate retouch (thick) | 40 | 5.09 | – | – |
| 48-49 | Abrupt and alternate retouch (thin) | 119 | 15.14 | – | – |
| 50 | Bifacial retouch | 2 | 0.25 | – | – |
| 54 | End-notched pieces | 3 | 0.38 | 0.60 | 0.64 |
| 61 | Chopping tools | 1 | 0.13 | 0.20 | 0.21 |
| 62 | Miscellaneous | 32 | 4.07 | 6.41 | 6.84 |
| 63 | Bifacial leaf-shaped points | 2 | 0.25 | 0.40 | 0.43 |
| Handaxes | | 2 | | | |
| Totals | | | 786 | 499 | 468 |

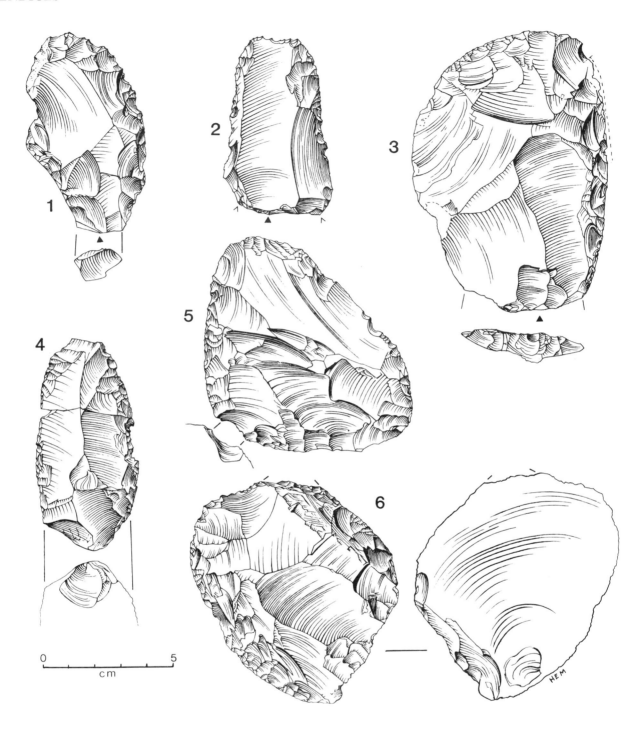

**Fig. B.3** Weichselian layers (Marett and earlier excavations). Side-scrapers: 1 single; 2, 4, 6 double (6 also has thinning along one edge); 3 convergent (tending to *déjeté*); 5 *déjeté*.

**Table B.7** The Weichselian Industry: Typological Groups and Indices

| Index | | Real | Ess. | Red. |
|---|---|---|---|---|
| ILty | Typological Levallois | 15.27 | – | – |
| IR | Side-scrapers | 35.88 | 56.51 | 60.26 |
| IC | Charentian | 13.49 | 21.24 | 22.65 |
| IAu | Acheulian unifacial | 0.51 | 0.80 | 0.85 |
| IAt | Acheulian total | 0.76 | 1.20 | 1.28 |
| IB | Bifaces | 0.25 | 0.40 | 0.43 |
| I | Levallois | 15.27 | 0.60 | 0.64 |
| II | Mousterian | 38.42 | 60.52 | 64.53 |
| III | Upper Palaeolithic | 2.67 | 4.21 | 4.49 |
| IV | Denticulate | 7.00 | 11.02 | 11.75 |
| 5 | Pseudo-Levallois points | 1.91 | 3.01 | – |
| 38 | Naturally backed knives | 2.04 | 3.21 | – |
| I rc | Convergent retouched | 10.81 | 17.03 | 18.16 |
| 6-7 | Mousterian points | 2.29 | 3.61 | 3.85 |
| Rsingle | Side-scrapers, single | 17.30 | 27.25 | 29.06 |
| Rdouble | Side-scrapers, double | 4.83 | 7.62 | 8.12 |
| Rtransv | Side-scrapers, transverse | 1.15 | 1.80 | 1.92 |
| Rinv | Side-scrapers, inverse/alternate | 1.91 | 3.01 | 3.21 |
| 30-31 | End-scrapers | 0.76 | 1.20 | 1.26 |
| 32-33 | Burins | 0.38 | 0.60 | 0.64 |
| 42 | Notched pieces | 4.96 | 7.82 | 8.33 |
| 42-43 | Notched pieces & denticulates | 11.96 | 18.84 | 20.09 |
| Ident | Extended denticulate group | 12.34 | 19.44 | 20.73 |
| Iirr | 'Irregularly retouched' | 21.63 | – | – |
| IQ | Quina | 3.18 | | |
| IQ+½Q | Quina + demi-Quina | 7.42 | | |

knives) is very poorly represented. Because the assemblage is a simple one, and rather similar to many industries using Levallois technique at the same date, only a summary description of the tools is given here.

Untransformed Levallois pieces are quite common by the standards of La Cotte (but not by comparison with many other contemporary sites) and include appreciable numbers of blades and points (Fig. B.2, 1–3). Typical Levallois pieces are in a clear majority. However, the mean length for the group as a whole, at 53 mm, is considerably less than that for most of the commoner retouched tool classes, such as side-scrapers and Mousterian points. It may therefore be argued that this stock of Levallois pieces represents not so much an abandoned supply of usable blanks as, at least in part, the material put on one side as less attractive for tool-making.

The Mousterian points are rarely of the elongated variety, and are usually flat in cross-section, in total contrast to those from the Saalian layer 5. They are fairly large (mean length 76 mm) and mostly made on Levallois blanks.

The side-scrapers (Figs. B.2, 4 to B.4, 1) are well made and of good size (66 mm). Single lateral scrapers are easily the most numerous, followed by convergent (including *déjeté*) and double separate pieces. Transverse scrapers are extremely rare. The percentages of the major typological groupings are:

| | | | |
|---|---|---|---|
| single | 48.1 | double | 13.4 |
| convergent | 23.3 | transverse | 3.2 |
| other | 12.0 | | |

*Déjeté* scrapers are relatively important, forming 11.0% of the class as a whole. As in most industries from La Cotte, the double separate and convergent pieces are generally slightly longer than the others; Levallois blanks have also been strongly favoured for their manufacture (59.5% and 41.5%). One of the double straight scrapers appears to have an LSF removal of the type recorded from earlier layers (chapter 29).

The 'Upper Palaeolithic' group (types 30–37) is almost entirely represented by atypical pieces, usually on small blanks (though the knives, atypical or cortex-backed, average over 70 mm in length). Notched pieces and denticulates, while numerically important, are variable in quality and on comparatively small flakes.

As usual, the miscellaneous category includes a range of multiple tools and others which are hard to classify. Thus there is a convex side-scraper with opposed bifacial denticulated edge; also a discoidal scraper with some bifacial retouch. There are also two pieces which could perhaps be classed as foliates.

The handaxes include one untwisted ovate; the final retouch of the two edges was applied to opposite faces, in contrast to the treatment of the other handaxe from the series (Fig. B.5). This second piece is plano-convex in conception, if not strictly so in terms of the shape of its cross-section: the scars on one face are truncated by the final removals on the other. Morphologically, it is an elongated and rather thin amygdaloid. This rather distinctive tool is quite closely paralleled by some of the 'slipper-shaped' handaxes from Wolvercote in Oxfordshire, as well as from a number of Hampshire sites. Roe (1981, 118–128) has argued that such industries — including those on the continent with rather similar handaxes — seem to be datable to a period ranging from a little before to a little after the last interglacial. It is therefore of some importance that Marett states explicitly (1916, 95) that the La Cotte piece came from the very bottom of the implementiferous deposits excavated by him and his colleagues; it must therefore have been made very soon after the cave was reoccupied during the Weichselian.

### B.2.4 Artefacts Made of Materials Other than Flint

There is a single (possible) core fragment in siltstone. No formal tools exist in quartz or sandstone and there is only a single retouched fragment in dolerite. The tool inventory thus consists of:

- 1 convergent convex side-scraper (quartzite)
- 1 notched piece (siltstone)
- 1 denticulate (siltstone)
- 1 *bec burinant alterne* (siltstone)
- 2 partially retouched pieces (quartzite)
- 1 chopping tool (granite)

The part played by non-flint rocks was clearly negligible.

### B.3 DISCUSSION

#### B.3.1 Broadly Comparable Industries from Neighbouring Areas

In its general character, the series described above is consistent with what is known of the industries of the first part of the last glaciation in western Europe. However, the 'Mousterian technocomplex' (which extends over most of the Old World) encompasses a great deal of variation. Thus in his last review of the question before his death, the late François Bordes (1981) listed the following variants of the Mousterian in southwest France, where large numbers of assemblages are available for study:

1. **Mousterian of Acheulian Tradition**, with two variants (though some transitional series are known). Subtype B is later than A.

   A. Numerous (up to 40%) cordiform or triangular handaxes; fairly numerous (20–40%) side-scrapers; 15–20% denticulates; some points and backed knives.
   B. Rare handaxes (up to 5%); very few side-scrapers (4–10%); numerous and often elongated backed knives (up to 20%); fairly numerous denticulates.

2. **Mousterian of Quina type**, with very many (50–80%) side-scrapers, including thick pieces with stepped ('Quina') retouch, transverse scrapers and plano-convex bifacial scrapers; few denticulates; non-Levallois debitage.

3. **Mousterian of Ferrassie type** with, once again, a very high percentage of side-scrapers, but fewer transverse scrapers and only a moderate incidence of Quina retouch.

403

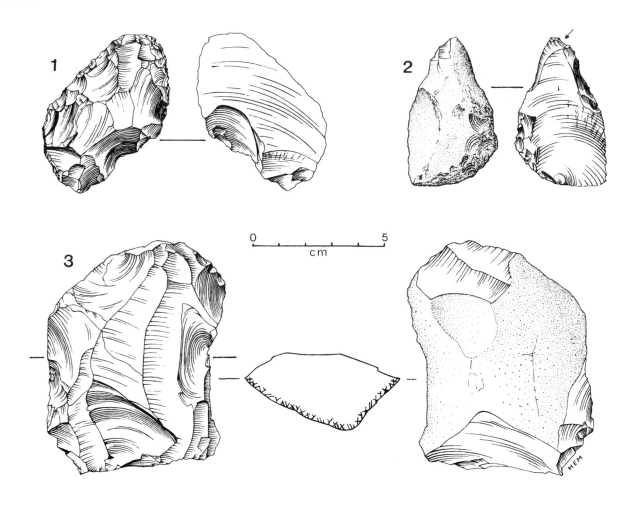

**Fig. B.4** Weichselian layers (Marett and earlier excavations). 1 Side-scraper with thinned back; 2 burin; 3 Levallois blade core.

4. **Denticulate Mousterian**, with fewer scrapers (often of poor quality); no backed knives; rare *nucléiform* handaxes; very numerous (occasionally as much as 80%) denticulates and notched pieces.

5. **Typical Mousterian**. Handaxes and backed knives rare or absent; moderate but variable numbers of denticulates. Two facies are now recognised: one with side-scrapers in large quantities (often over 50%), the other with these in only moderate frequencies.

6. **Asinipodian**. Known only from Pech de l'Azé IV in the Dordogne, this is characterised by specialised techniques of debitage, and few retouched tools except notched pieces.

7. **Vasconian**. This is characterised by flake cleavers, and is restricted to the Basque region.

The Asinipodian and the Vasconian do not concern us here because of their local distribution. Apart from the Quina and Ferrassie Mousterian, the other categories include examples with both high and low incidences of Levallois technique (which is usually regarded as largely a function of raw material availability, and cuts across a classification based on the preferences for different types of retouched tools). The various models which have been offered to account for the existence of the assemblage types will not be discussed here; however, it is worth noting that statistically discrete groups of assemblages do exist, and cannot be explained away as arbitrary subdivisions of a continuum of variation (Callow and Webb 1977 and 1981, with references).

In northern France (Bordes 1981; Tuffreau 1979) the Typical (both subtypes) and Acheulian Tradition Mousterian are well represented, though some assemblages possess distinctive features such as a high percentage of backed knives. There is a single reported example of the Ferrassie type, from Busigny, Nord (Tuffreau 1971). Until recently, only one Denticulate site was recorded; however, examples are now known in the Cotentin (see below). Quina Mousterian is completely lacking (though it is present even further north, in the caves of Belgium).

The industries from the areas closest to the Channel Islands (i.e. Normandy and Brittany) are of particular interest, given that La Cotte was probably only one of a number of sites exploited by human groups with a fairly large territorial range. The Breton Weichselian sites also have relevance for the earlier series from La Cotte because of the use made, at many of them, of local hard rocks.

Until quite recently, although a few occurrences were known in Lower Normandy, it was primarily Upper Normandy (and the Seine valley, especially around Rouen) which attracted most attention (Bordes 1954; Tuffreau 1976; recent discoveries are summarised by Fosse 1980). This was due to both the archaeological richness and the

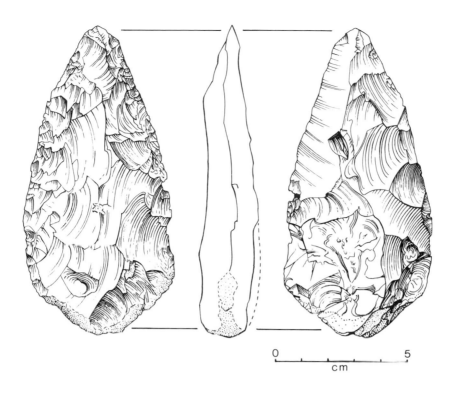

**Fig. B.5** Weichselian layers (Marett excavations). Elongated cordiform handaxe of Wolvercote type, slightly fire-damaged.

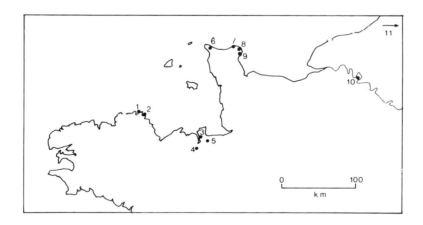

**Fig. B.6** Location of sites mentioned in the text. 1 Roche-Tonnerre; 2 Pte. de la Trinité; 3 Grainfollet; 4 Bois-du-Rocher; 5 Mont-Dol; 6 St. Germain-des-Vaux; 7 Gouberville; 8 Montfarville; 9 St. Vaast-la-Hougue; 10 Oissel; 11 Busigny.

geological interest of many of the localities, since they are situated in an area where the loess is well developed. The usefulness of comparing these industries with the series from La Cotte, however, is restricted by the very different circumstances of raw material procurement. Large flint nodules are easily obtained from the Chalk; consequently in the Seine valley a very high incidence of Levallois technique (and unmodified Levallois flakes) tends to be the rule.

For our purpose, Lower Normandy is of more direct interest, in view of both proximity and raw materials. The area around Caen has not been very productive; on the other hand, the Cotentin peninsula (Manche) has yielded many new and important sites in the last few years. The most significant new discoveries are coastal, at the tip of the Cotentin (see Fig. B.6). A Denticulate Mousterian employing Levallois technique on flint pebbles occurs in the top of, or more usually immediately above, beach deposits assigned to the Eemian: Montfarville-Landemer; Port Racine at St. Germain-des-Vaux; the Fort at St. Vaast-la-Hougue; Gouberville-Gattemare. At St. Vaast-la-Hougue the overlying sands contain a Typical Mousterian of Levallois facies (for a summary, see Fosse 1980; also Lautridou *et al.* 1982). Unfortunately there is as yet a dearth of statistical data for these sites.

405

# APPENDICES

In Brittany the picture is somewhat complicated by the extensive use of local raw materials in some assemblages. Monnier (1980b) lists the following variants for the Weichselian industries:

1. **Micoquian/Late Acheulian**.

   Grainfollet: flint; rare handaxes of Micoquian and lanceolate type; moderate percentages of side-scrapers and denticulates. According to Monnier (1980b, 35–37), the industry overlies an Eemian palaeosol. Under a rock shelter.

   Tréissény: mostly flint; numerous small amygdaloid, ovate and cordiform bifaces; numerous side-scrapers, but very few denticulates. On an Eemian beach.

2. **Mousterian**.

   With rare bifaces, e.g. Pointe de la Trinité (Nord, mostly on tuf; Sud, mostly on flint); Roche Tonnerre (mostly flint). Little Levallois debitage; generally abundant side-scrapers. Possibly related to humiferous palaeosols (Brörup?). All sites lie in front of marine cliffs. Affiliation: early Typical Mousterian?

   Mousterian of Acheulian Tradition: e.g. Bois-du-Rocher, Clos-Rouge, Kervouster, Arcouest. *Grès lustré* has been employed almost exclusively at the first three of these, and microgranite and flint at Arcouest. Numerous bifaces (principally amygdaloid, ovate and cordiform) and some flake-cleavers; side-scrapers common; percentage of denticulates moderate. Open air sites; poorly dated.

   Without bifaces (attributed to the Middle Weichselian).

   Mont-Dol: Typical Mousterian of Levallois facies, with numerous side-scrapers, on flint. At the foot of a cliff.

   Gouréva: Typical Mousterian, rich in denticulates, on flint and dolerite. Under a rock shelter.

Thus the industries of the area of France nearest La Cotte are varied in their typology and technology. However, it should be noted that the Mousterian of Acheulian Tradition is represented by open air factory sites exploiting local raw materials of good quality (*grès lustré* and microgranite). Such sites are as yet entirely unknown in the Channel Islands.

## B.3.2 Cultural Affinities of the La Cotte Weichselian Material

The principal features of the material described in this chapter include: (1) extensive use of Levallois blanks; (2) rarity of unmodified Levallois pieces (and Levallois cores); (3) numerous discoidal cores; (4) high facetting indices; (5) low frequencies of pseudo-Levallois points and *couteaux à dos naturel*; (6) an overwhelming preponderance of side-scrapers; (7) a moderate percentage of denticulates; (8) Extremely rare bifaces.

If the collection from the early excavations at La Cotte is treated as a single assemblage, it may be assigned to the **Typical Mousterian, rich in side-scrapers**; it is not of Levallois facies, since unmodified Levallois pieces are uncommon. However, Levallois technique has been employed to provide many of the blanks required for tool-making; Quina retouch has been very little used, as befits the thinness of the pieces. The importance of Levallois technique also serves effectively to distinguish this industry from those antedating the last glaciation at the same site, despite some similarities in the typological indices (e.g. the percentage of side-scrapers) and cumulative curves (Fig. B.7). In particular, the contrast with the last of the Saalian assemblages (layer 5), with its thick elongated points and rather high Quina index, is very strong.

To consider technology first, La Cotte's high facetting indices are not matched in the Breton Typical Mousterian series, but are more in keeping with those for the other sites listed in Table B.8. On the other hand, its typological Levallois index is exceptionally low; and while the

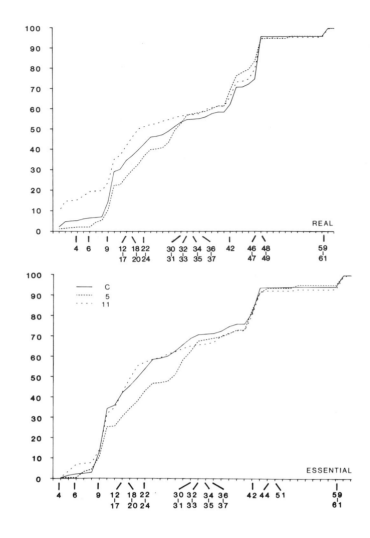

**Fig. B.7** Cumulative frequency curves for tools from the Weichselian and upper Saalian layers at La Cotte de St. Brelade.

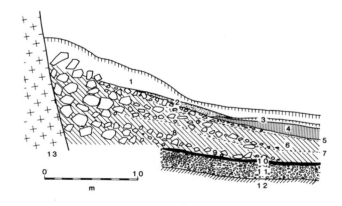

**Fig. B.8** Mont-Dol. Schematic section oriented NE/SW, as obtained by Sirodot; after Monnier (1980a). 1 Recent talus; 2 and 3 granitic sand; 4 carbonated loamy clay; 5 head with blocks of granite; 6 loam with sand and clay (solifluction deposit); 7 head with blocks; 8 as 6; 9 head with blocks of granite and most of the industry and fauna; 10 humified soil; 11 marine gravel; 12 schisty substrate; 13 granite cliff.

percentage of Levallois technique used overall (IL) is matched at Roche Tonnerre and Pointe de la Trinité (Sud), at La Cotte it owes much to the 33.4% of the retouched pieces for which Levallois blanks have been employed. As far as can be estimated from the data available, a comparable proportion of Levallois blanks was used at Mont-Dol. It may be recalled, in passing, that the industry from the Ruquier Pit at Oissel, near Rouen, is altogether exceptional, and Bordes (1981, 80) comments that the IL of 80 can only be explained if Levallois flakes were imported, as it is impossible to obtain such a value by knapping on the spot. The percentage of blades (Ilam) at La Cotte is at the lower end of the range.

Typologically, La Cotte is relatively well endowed with denticulates and notches, though as in most other Typical Mousterian assemblages (except Oissel) side-scrapers form over 50% of the tools — but are appreciably less common than at Busigny, assigned to the Ferrassie Mousterian. 'Upper Palaeolithic' tool types are even rarer than is usual elsewhere.

One of the closest sites to the Channel Islands is Mont-Dol (Fig. B.8). It is worth considering in more detail, since it has a very similar industry (Fig. B.9) and, moreover, shares with La Cotte's pre-Eemian layers the distinction of possessing large quantities of megafauna, including cranial material. It lies at the foot of a cliff at the southern end of a small granite hill which rises 60 m from the low-lying Marais de Dol (the surrounding area was submerged during the Flandrian transgression). Excavated in 1872 by Sirodot (1873), and later by Vayson de Pradenne (1929), it was dismissed by Giot and Philippot (1946) as having been reworked. However, Monnier (1980a) has pointed out that very few of the bones are rolled, and that the industry is extremely fresh (confirmed by this author). Thus it is likely that only the edge of the site was disturbed by the Flandrian sea, and that the effects of solifluction during the last glaciation were minimal.

The lowest layer reached by Sirodot at Mont-Dol was a marine gravel whose top was at about 1.5 m below the present high tide level. Probably developed on a thin layer of *limon* there was a blackish soil; analysis by Monnier of one of Sirodot's samples showed the presence of bone splinters, wood charcoal and flint debris. However, the principal discoveries were evidently at the base of the overlying head, which was initially loessic but became sandy in its upper part. This head included three layers which were particularly rich in granite blocks. There then followed what is very likely an older littoral sand deposit reworked by the wind, and containing derived marine as well as freshwater and terrestrial shells (Monnier 1980a, 358–365).

Allowing for the oversimplification which probably occurred as a result of the early date of the excavations, the sequence is not unlike the upper part of that at La Cotte. Of course, since the Mont-Dol site directly abuts a granite cliff, there was no collapse of more ancient sediments onto the beach, which is presumed to date to the last interglacial. If the soil was indeed preceded by a certain amount of loess deposition the parallel becomes very close indeed.

It has already been suggested that the Mont-Dol and La Cotte industries have many features in common, and may be assigned to the same group within the Mousterian. However, differences do occur, such as the somewhat greater percentages at La Cotte of denticulates and of tools with convergent edges, and the presence of a very few handaxes. The contrast is more marked in the technology (unretouched Levallois flakes, *couteaux à dos naturel*); moreover, at Mont-Dol Levallois cores are very numerous, and discoidal cores are scarce. Mont-Dol is evidently a site at which all stages of Levallois debitage were carried out without any problem of raw material availability, which may explain the technological differences. In view of their typological characteristics it seems likely that the two sites were occupied by human groups with very similar material culture.

The discussion so far has skirted the possibility of two distinct industries at La Cotte, separated by a sterile layer. This was raised by Marett (1916, 112ff.), who commented that the higher levels contained the finest tools, as well as more blades, while near the base the workmanship was poorer. Moreover, the Wolvercote-type handaxe referred to in section B.2.3 is known to have come from the bottom of the implementiferous bed. For want of more exact provenance information, few firm conclusions can be drawn. However, the value for IR is such that one can be quite confident that the upper part at least must have contained a Typical Mousterian rich in side-scrapers; even allowing for dilution of the scrapers and Levallois flake percentages by ad-

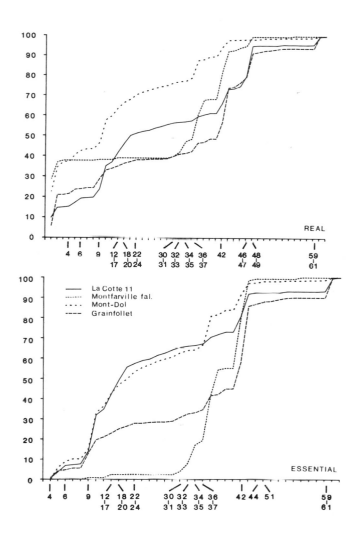

**Fig. B.9** Cumulative frequency curves for tools from La Cotte de St. Brelade (Weichselian layers) and some Breton and Norman sites of approximately similar age.

mixture with a non-Levallois assemblage, the scarcity of Levallois cores rules out a Ferrassie Mousterian. Mixture of such an industry with a Micoquian series rather like that from Grainfollet might explain some of the differences between La Cotte and Mont-Dol. Equally, it is worth recalling the denticulates and notches of the earliest industry at the top of the Eemian beach at St. Vaast-la-Hougue, Montfarville, etc. (and the rarity of handaxes at La Cotte). However, the conclusion seems unavoidable that the typological differences between the series above and below the sterile sand cannot have been great (unless the latter was very much less numerous than the former), as otherwise the collection would fit less readily into the Typical Mousterian. Marett's view (1916, 114) is illuminating; of the upper industry he says: "Altogether, the workmanship, while adhering in the main to the original patterns, renders them in a less ponderous and more graceful way. There is an evolution, but it is an intensive evolution; it is the same industry, but perfected". It is unlikely that he could have reached such a conclusion if the lower industry had contained significant quantities of notches and denticulates.

**Table B.8**  Comparison of the Weichselian Industry of La Cotte de St. Brelade with Others in Northern France. T. Typical; F. Ferrassie; D. Denticulate; Mic. Micoquian[a]

| Index or type | La Cotte | Mont-Dol (T.) | Roche Tonnerre (T.) | Trinité Sud (T.) | Grain-follet falaise (Mic.) | Mont-farville ser. 2 (D.) | Oissel Ruquier (T.) | Bapaume Chateau d'eau (T.) | Busigny (F.) |
|---|---|---|---|---|---|---|---|---|---|
| IL | 11.0 | 27.9 | 17.8 | 10.7 | 12.7 | 78.4 | 80.8 | 40.7 | 38.1 |
| IF | 54.6 | 33.5 | 31.5 | 11.1 | 57.6 | 51.2 | 71.8 | 37.4 | 58.5 |
| IFs | 48.4 | 25.1 | 25.9 | 17.8 | 41.7 | 35.8 | 57.8 | 32.3 | 48.5 |
| Ilam | 5.2 | 9.3 | 4.2 | 9.5 | 9.3 | 4.3 | 25.8 | 14.6 | 4.8 |
| ILty | 14.9 | 37.2 | 19.8 | 32.2 | 26.4 | 43.2 | 61.9 | 43.4 | 32.7 |
| IR | 56.5 | 50.5 | 51.8 | 50.8 | 28.5 | 2.3 | 32.5 | 54.0 | 71.0 |
| IC | 21.2 | 21.5 | 19.8 | 26.2 | 12.5 | 0.0 | 6.9 | 17.7 | 27.6 |
| I rc | 17.0 | 11.0 | 11.9 | 6.0 | 2.8 | 0.0 | 7.4 | 16.9 | 17.9 |
| IAu | 0.8 | 2.6 | 1.2 | 0.5 | 2.5 | 2.32 | 7.4 | 2.4 | 0.0 |
| IB | 0.4 | 0.0 | 2.4 | 2.2 | 1.5 | 0.0 | 0.5 | 0.8 | 0.7 |
| Group III | 3.0 | 6.1 | 8.6 | 7.1 | 10.7 | 17.1 | 7.9 | 6.4 | 4.8 |
| Group IV | 11.0 | 5.5 | 4.0 | 4.8 | 22.7 | 18.4 | 8.9 | 6.4 | 10.3 |
| Moust. points | 3.6 | 4.2 | 2.5 | 1.1 | 0.4 | 0.0 | 2.9 | 4.8 | 0.7 |
| Side-scr., single | 27.3 | 26.3 | 28.4 | 30.0 | 17.6 | 0.8 | 20.1 | 26.6 | 36.5 |
| Side-scr., double | 7.6 | 6.4 | 3.7 | 5.4 | 1.7 | 1.5 | 4.4 | 10.5 | 9.7 |
| Side-scr., transverse | 1.8 | 3.7 | 2.5 | 3.3 | 2.6 | 0.0 | 0.5 | 0.8 | 3.4 |
| Nat. backed knives | 3.2 | 14.0 | 1.2 | 12.6 | 5.9 | 21.7 | 9.8 | 5.6 | 2.8 |
| Notches | 7.8 | 7.7 | 4.9 | 10.9 | 12.7 | 24.8 | 7.9 | 8.9 | 7.6 |

[a]After Monnier (1980a and 1980b) and Michel (1971). The typological indices (IR etc.) are of the 'essential' type.

# APPENDIX C

# LA COTTE A LA CHEVRE*

P. Callow

## C.1 HISTORY AND DESCRIPTION

Discovered in 1861, this north-facing cave in the cliffs at the northeast corner of Jersey (Fig. C.1) was excavated by Dancaster and Sinel in 1881. Among the finds was the only handaxe known from the site; for many years it was in the collection of antiquities owned by the Lukis family of Guernsey, but it has recently been returned to Jersey and is on display in the museum at La Hougue Bie. In 1911, further investigations by Marett (1912), and Nicolle and Sinel (Sinel 1912) showed that not all of the deposits had been removed during the earlier work. A particularly important result of the renewed excavations was the demonstration that, despite its altitude, La Cotte à la Chèvre had been formed as a sea-cave The site has proved vulnerable to unauthorised digging, and when McBurney carried out fresh fieldwork in 1962, 1964 and 1968 it became apparent that most of the surviving sediments had been disturbed. In spite of this, small series of artefacts were obtained from deposits believed to be *in situ* (McBurney 1963).

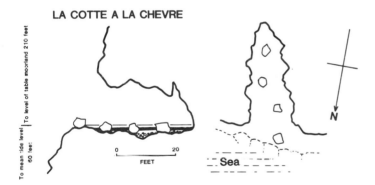

**LA COTTE A LA CHEVRE**

**Fig. C.1** La Cotte à la Chèvre. Plan and section (from Marett 1912, 463).

In its present form the cave is some 9 m deep by 3 wide; it is possible that during the Pleistocene the entrance was several metres further north than it is today. Excavation has effectively removed the archaeological deposits, and the most striking features visible are natural ones, namely large sea-worn boulders and rounding of the cave interior (Fig. C.2) which took place during a period when the sea was active some 18 metres above the present mean sea level.

The recorded stratigraphy is simple — the granite boulders and pebbles of the beach were lying in the lower part of a bed of fine sand, overlain by 'white earth' (actually a sandy silt loam), and finally a decalcified (?) yellowish, slightly sandy loess. The white deposit was considered by McBurney (following analysis by Dr. M. Tamplin) to have been the result of soil-forming processes acting on the sand; it seems to have been the principal source of artefacts and associated hearth remains during the early excavations. Finds were recorded by McBurney in all three layers.

## C.2 INDUSTRY

The writer has recently examined the artefacts recovered by McBurney from the deposits he judged to be undisturbed. Only the 1964 season produced stratified finds in quantity; these have been used as the basis for the statistics given here. Some pieces attributed with rather less confidence to *in situ* basal sand have been included, as without them the sample from this unit is too small to be informative. Very few artefacts were made of raw materials other than flint (Table C.1). The samples are small, but all three are very clearly dominated by denticulates and notches (often of the single-blow 'Clactonian' type), which account for 59 of the combined total of 79 tools (excluding utilised and unmodified pieces). It may be significant that the best-made side-

**Fig. C.2** La Cotte à La Chèvre. Rounded boulders lying in a 'pothole' scoured out in the floor of the cave by the 18 m sea, just inside the entrance.

*Also known as La Cotte de St. Ouen.

409

# APPENDICES

**Table C.1** La Cotte à la Chèvre: Artefacts from the 1964 excavations

| Material/Type | Sand | White earth | Loess |
|---|---|---|---|
| **Flint** | | | |
| Tools | 36 | 47 | 50 |
| Waste flakes | 70 | 219 | 155 |
| Cores | 3 | 6 | 4 |
| Total | 109 | 272 | 209 |
| Quartz – total | 0 | 1 | 1 |
| Stone  – total | 0 | 6 | 6 |

scrapers (including an exceptionally fine *déjeté* one) are from the loess. As for the technology used, this seems to have been based largely on simple globular, discoidal and 'shapeless' cores in the two lower layers. In 1964 the loess yielded one good Levallois point core (two Levallois tortoise-cores found in 1962 are without definite provenance). Two rather questionable Levallois points came from the white earth, but apart from these all the Levallois flakes with a well-defined context are 'atypical'. The facetting indices are rather low: for the three layers in turn, 36.4, 29.9 and 40.7% (falling to 25.8, 19.3 and 26.6 when dihedral platforms are excluded). The only other find of note was a single very fine 'long sharpening flake' of the type particularly characteristic of layer A at La Cotte de St. Brelade.

## C.3    CONCLUSIONS

The only element in the stratigraphy of La Cotte à la Chèvre which gives any indication of age is the 18 m beach. This is certainly too high to be attributable to the Eemian interglacial (see chapter 3). A transgression to such an altitude is unlikely to have occurred during formation of the deposits now exposed in La Cotte de St. Brelade, where there is no evidence of marine erosion of either sediments or walls at this level; supposing that exfoliation of the rock walls had removed the smoothed surfaces, fragments would surely have been identified in the deposits. Indeed, unless the accumulated deposits in the ravines were laterally very thick indeed, a transgression to 18 m would probably have emptied the site. A strong case can therefore be made that the raised beach at La Cotte à la Chèvre antedates the earliest layers yet known at the other site.

The question of dating may be linked to another: are we dealing here with one industry or several, perhaps representing a considerable span of time? Reference to the sequence at La Cotte de St. Brelade suggests that the majority of this material is indistinguishable from the layer H industry there apart from a higher ratio of denticulates to notched pieces (see also Figs. 32.3 and 32.4). The exceptions — side-scrapers of high quality, and the Levallois point core — were found in the loess. The LSF might appear to pose a problem as the type is totally alien to layer H at La Cotte de St. Brelade. This example comes from a bag of finds *provisionally* attributed to the basal sand, but with the caveat that this might be erroneous. In 1964 McBurney had not yet recognised LSFs as a class, much less studied the sequence at La Cotte de St. Brelade and appreciated their limited stratigraphic range. Consequently the discovery of this single piece from La Cotte à la Chèvre

**Table C.2**   La Cotte à la Chèvre: Flint Tools from the 1964 excavations

| Type | | Sand | White earth | Loess |
|---|---|---|---|---|
| 2 | Atypical Levallois flakes | 0 | 4 | 1 |
| 3 | Levallois points | 0 | 2 | 0 |
| 5 | Pseudo-Levallois points | 1 | 2 | 5 |
| 9 | Single straight side-scrapers | 0 | 1 | 3 |
| 10 | Single convex side-scrapers | 0 | 1 | 1 |
| 11 | Single concave side-scrapers | 0 | 0 | 1 |
| 21 | *Déjeté* side-scrapers | 0 | 0 | 1 |
| 23 | Convex transverse side-scrapers | 0 | 1 | 0 |
| 24 | Concave transverse side-scrapers | 0 | 0 | 1 |
| 27 | Side-scrapers with thinned back | 0 | 1 | 0 |
| 29 | Alternate retouched side-scrapers | 0 | 0 | 1 |
| 32 | Typical burins | 1 | 0 | 0 |
| 33 | Atypical burins | 0 | 1 | 0 |
| 38 | Naturally backed knives | 0 | 0 | 1 |
| 40 | Truncated blades and flakes | 0 | 0 | 1 |
| 42 | Notched pieces | 5 | 4 | 9 |
| 43 | Denticulates | 11 | 9 | 11 |
| 45 | Retouched on ventral face | 5 | 1 | 3 |
| 46–47 | Abrupt and alternate retouch (thick) | 9 | 8 | 2 |
| 48–49 | Abrupt and alternate retouch (thin) | 4 | 9 | 5 |
| 51 | Tayac points | 0 | 0 | 1 |
| 54 | End-notched pieces | 0 | 1 | 0 |
| 61 | Chopping tools | 0 | 1 | 0 |
| 62 | Miscellaneous | 0 | 0 | 2 |
| Total (real) | | 36 | 46 | 49 |
| Total (essential) | | 18 | 32 | 38 |
| Total (reduced) | | 17 | 30 | 32 |
| Broken | | 0 | 1 | 1 |

**Table C.3** La Cotte à la Chèvre (1964 excavations): Technology and Typology

| Index | Sand | White earth | Loess |
|---|---|---|---|
| Technological indices | | | |
| IL | 0 | 2.53 | 0.98 |
| IF | 36.39 | 29.86 | 40.67 |
| IFs | 25.78 | 19.26 | 26.61 |
| Ilam | 1.89 | 0.38 | 0.98 |
| Typological indices and groups ('real') | | | |
| ILty | 0 | 13.04 | 2.04 |
| IR | 0 | 8.70 | 16.33 |
| IAu | 0 | 0 | 0 |
| IB | 0 | 0 | 0 |
| I | 0 | 13.04 | 2.04 |
| II | 0 | 8.70 | 16.33 |
| III | 2.78 | 2.17 | 0 |
| IV | 30.56 | 19.56 | 22.45 |
| 42-43 | 44.44 | 39.13 | 40.82 |
| Typological indices and groups ('essential') | | | |
| ILty | – | – | – |
| IR | 0 | 18.18 | 21.05 |
| Au | 0 | 0 | 0 |
| IB | 0 | 0 | 0 |
| I | 0 | 0 | 0 |
| II | 0 | 18.18 | 21.05 |
| III | 5.55 | 4.55 | 0 |
| IV | 61.11 | 28.13 | 28.05 |
| 42-43 | 88.88 | 59.09 | 52.63 |

would not have struck him as necessitating a special investigation of its context even supposing that he had seen it at the time. It would thus be unwise to regard it as genuinely a part of a less advanced industry, at odds with the evidence from the other site.

The most economical explanation for the stratigraphical and archaeological data is that the first — and major — period of occupation of La Cotte à la Chèvre some time after the retreat of the sea corresponds to the base of the sequence at La Cotte de St. Brelade. There were later visits to the site by users of different technological traditions — these being responsible for the Levallois debitage, the LSF, and the handaxe. The northern aspect of the cave would have favoured occupation (as opposed to passing visits) only during relatively temperate periods, if at all. And there is a distinct possibility that solifluction, by disturbing and partly removing deposits during successive cold episodes, has been responsible for collapsing a long sequence (perhaps equivalent to the whole of that at the other cave) into a slight depth of deposits, with some degree of geological mixing of the industries.

**Table C.4** La Cotte à la Chèvre (1964 excavations): Cores

| | Sand | White earth | Loess |
|---|---|---|---|
| Levallois - point | 0 | 0 | 1 |
| Discoidal | 1 | 0 | 0 |
| Globular | 0 | 1 | 2 |
| Pyramidal | 1 | 0 | 0 |
| Shapeless | 0 | 1 | 1 |
| Miscellaneous | 1 | 4 | 0 |
| Total | 3 | 6 | 4 |

# APPENDIX D

# ESTIMATION OF RIGHT-HAND PREFERENCE AMONG THE LA COTTE FLINT-KNAPPERS

## P. Callow and J. M. Cornford

1. In section 29.6 attention was drawn to the asymmetrical character of the LSFs; this observation was discussed with regard to its implication that one hand was favoured more than the other by the prehistoric flint-knappers. This Appendix sets out to demonstrate that, notwithstanding the possible employment of more than one orientation of the parent piece during resharpening, the deviation from equality in the ratio between the two forms' occurrence actually represents a minimum estimate for such lateralisation. This point is an important one as, although technological arguments have occasionally been used elsewhere to support comparable claims, the evidence from La Cotte is of particular interest for the extreme value taken by the ratio. The reasoning is set out here, deliberately in simple terms, in the hope that as many readers as possible will follow it through. (Thanks are due to Professor H. B. Barlow and Mr. C. Chapman for comments and advice.)

2. For the entire population (i.e. ignoring sampling problems), let

$k$ = proportion of knappers that are right-handed;
$m$ = proportion of LSFs made by 'Method 1' — i.e. by removing the flake from the edge on the same side as the grip;
$a$ = proportion of LSFs that are right asymmetrical.

Then the proportion of left-handed knappers is $1-k$, and so on. Assuming that right-asymmetrical LSFs may be made by right-handed knappers using Method 1, or by left-handed knappers using Method 2 (i.e. removing the flake from the edge on the side opposite the grip):

$$a = km+(1-k)(1-m)$$

$$= 1-k+m(2k-1)$$

i.e.
$$k = \frac{a-1+m}{2m-1} \qquad (1)$$

This may be rewritten as

$$(m-0.5)(k-0.5) = 0.5(a-0.5) \qquad (2)$$

which is the equation of a set of rectangular hyperbolae with asymptotes $m = 0.5$ and $k = 0.5$. We may also note that if $a > 0.5$ the hyperbolae are monotone decreasing, and lie in the 1st and 3rd quadrants (as defined by the asymptotes); on the other hand, if $a < 0.5$ they are monotone increasing, in the 2nd and 4th quadrants.

In the case of the La Cotte LSFs we need only consider the case $a > 0.5$. Moreover, because of the limits 0 and 1 on the values of $k$, $m$ and $a$ (which are proportions), the asymptotes cannot in fact be reached. Instead, we find by substitution in equation 2 that, balancing the signs on the two sides, if $a > 0.5$ then in the first quadrant

$$a \leqslant m \leqslant 1 \text{ and } a \leqslant k \leqslant 1 \qquad (3) (4)$$

and in the third quadrant

$$0 \leqslant m (1-a) \text{ and } 0 \leqslant k \leqslant (1-a) \qquad (5) (6)$$

As it is extremely unlikely that a very substantial majority of the La Cotte knappers would have been left-handed (and, incidentally, used Method 2) — which would imply a complete reversal of the modern pattern — we need only consider the first quadrant, and conclude from 3 and 4 that, sampling error apart, $a$ provides a minimum estimate for the incidence of both right-handed knappers and Method 1.

3. Evidently sampling error cannot be ignored. The standard error of the proportion $a$ is estimated by

$$\sigma_a = \sqrt{\frac{a(1-a)}{n}} \qquad (7)$$

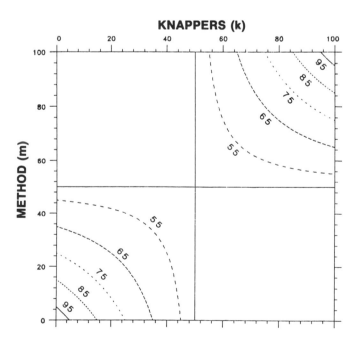

**Fig. D.1** Graph of the relationship between flaking method and knapper's handedness producing different percentages of 'right-handed' LSFs in the range 55–95.

413

where $n$ is the sample size. For an estimate to be made with 95% confidence, from equations 3 and 7, the proportion of right-handed knappers is

$$k \geqslant a \pm 1.96 \sqrt{\frac{a(1-a)}{n}}$$

If $K$ and $A$ are percentages equivalent to $k$ and $a$ this becomes

$$K \geqslant A \pm 1.96 \sqrt{\frac{A(100-A)}{n}} \qquad (8)$$

4. The above calculations assume that the objects included in the sample of LSFs have been correctly identified — i.e. that no 'look-alikes' made by different methods and for different reasons have been included. However, provided that any such pieces are not themselves more 'hand-specific' than the LSFs (which would imply an even higher degree of technological specialisation), they could only result in a tendency of $a$ towards 50% (i.e. randomness).

5. Applying equation 8 to the data for LSFs with facetted butts yields the results shown in Table D.1.

**Table D.1**   Lateralisation: Minimum Estimates based on LSFs

| Layer | Right asymmetrical (%) | Sample size | Standard error (%) | Limits (%) |
|-------|------------------------|-------------|--------------------|------------|
| 5 | 91.6 | 11 | 8.36 | 75.2–100.0 |
| 3 | 63.6 | 11 | 14.51 | 35.2–92.0 |
| A | 83.5 | 103 | 3.66 | 76.3–90.7 |
| B | 81.8 | 22 | 8.23 | 65.7–98.0 |
| C | 81.8 | 22 | 8.23 | 65.7–98.0 |
| D–H | 77.3 | 18 | 9.87 | 58.0–96.6 |

# BIBLIOGRAPHY

Adam, A., and Tuffreau, A. 1972. Le gisement paléolithique ancien du Rissori à Masnuy-Saint-Jean (Hainaut, Belgique). *Bull. Soc. Préhist. Fr.* 70: 293–310.

Agadjanian, A. 1976. Die Entwicklung der Lemmings der zentralen und östlichen Paläarktis im Pleistozän. *Mitt. Bayer Staatssamml. Paläont. Hist. Geol.* 16: 53–64.

Agadjanian, A., and Koenigswald, W. von. 1977. Merkmalsverschiebung an den oberen Molaren von *Dicrostonyx* (Rodentia, Mammalia) im Jungquartär. *N. Jahrb. Geol. Paläont. Abh. Stuttgart* 153(1): 33–49.

Aitken, M.J. 1978. Radon loss evaluation by alpha counting. In

Aitken, M.J., and Mejdahl V. (eds.) *A Specialist Seminar on Thermoluminescence Dating*, pp. 104–14. PACT (J. of the European Study Group on Physical, Chemical and Mathematical Techniques Applied to Archaeology) 2.

Aitken, M.J., and Bowman, S.G.E. 1975. Thermoluminescent dating: assessment of alpha particle contribution. *Archaeometry* 17(1): 132–8.

Aitken, M.J., Huxtable, J., and Debenham, N.C. In press. Thermoluminescence dates in the palaeolithic: burnt flint, stalagmitic calcite, and sediment. In Tuffreau, A. (ed.) *Chronostratigraphie et Faciès Culturels du Palaéolithique Inferieur et Moyen dans l'Europe du Nord-Ouest*. Papers given at Congr. Préhist. France (Lille, 1984). To appear as *Supp. Bull. Assoc. Fr. Et. Quat.*

Alduc, D., Auffret, J.P., Carpentier, G., Lautridou, J.P., Lefèbvfre, D., and Porcher, M. 1979. Nouvelles données sur le Pléistocène de la basse vallée de la Seine et son prolongement sous-marin en Manche orientale. *Bull. Inf. Géol. Bassin de Paris* 16(2): 27–34.

Andersen, S.T. 1961. Vegetation and its environment in Denmark in the early Weichselian glacial. *Dan. Geol. Unders.* 75.

Anderson, P.A. 1980. A testimony of prehistoric tasks: diagnostic residues on stone tool working edges. *World Archaeol.* 12: 181–94.

Anderson-Gerfaud, P.A. 1981. *Contribution Méthodologique à l'Analyse des Microtraces d'Utilisation sur les Outils Préhistoriques*. Doctoral thesis (3ème cycle), Université de Bordeaux I.

Andersson, C.J. 1856. *Lake Ngami; or Explorations and Discoveries during Four Years Wanderings in the Wilds of South Western Africa*. Hurst and Blackett, London.

Andrews, J.T. 1983. Short ice age 230,000 years ago? *Nature, Lond.* 303: 21–2.

Andrews, J.T., Bowen, D.Q., and Kidson, C. 1979. Amino-acid ratios and the correlation of raised beach deposits in south-west England and Wales. *Nature, Lond.* 281: 556–8.

Angel, J.L., and Coon, C.S. 1954. La Cotte de St. Brelade II: present status. *Man* 76: 53–5.

Anon. 1968. High Lodge palaeolithic industry. *Nature, Lond.* 220: 1065–6.

Archipov, S.A. 1976. Research into the teeth of extinct elephants: procedural recommendations. *Akad. Nauk USSR Sib. Otd. Inst. Geol. Geofiz.*

Auffret, J.P., and Alduc, D. 1977. Ensembles sedimentaires et formes d'érosion du Quaternaire sous-marin de la Manche orientale. *Bull. Assoc. Fr. Et. Quat.* 14(4): 71–5.

Avery, B.W., and Bascomb, C.L. 1974. *Soil Survey Laboratory Methods*. Soil Survey Techn. Monogr. 6. Harpenden.

Azoury, I., and Hodson, F.R. 1973. Comparing palaeolithic assemblages. 'Ksar Akil, a case study. *World Archaeol.* 4(3): 292–306.

Bada, J.L., and Helfman, P.M. 1976. Application of amino acid racemisation dating in paleoanthropology and archaeology. *UISPP* 9 (Nice). Colloque 1: 39–62.

Baden-Powell, D.F.W. 1931. Notes on raised beach mollusca from the Isle of Portland. *Proc. Malacol. Soc., Lond.* 19: 67–76.

Bailey, G.N., and Callow, P. 1986. *Stone Age Prehistory. Studies in Memory of Charles McBurney*. Cambridge University Press, Cambridge.

Bailey, G.N., and Davidson, I. 1983. Site exploitation territories and topography: two case studies from Palaeolithic Spain. *J. Archaeol. Sci.* 10: 87–115.

Bailiff, I.K., and Aitken, M.J. 1980. Use of thermoluminescence dosimetry for evaluation of internal beta dose rate in archaeological dating. *Nucl. Instrum. Meth.* 173: 423–9.

Bar-Yosef, O., and Vandermeersch, B. 1969. The stratigraphic and cultural problems of the passage from Middle to Upper Palaeolithic in Palestinian caves. In Bordes, F. (ed.) *L'Origine de l'homme moderne*, pp. 221–5. UNESCO, Paris.

Barker, P.C. 1975. An exact method of describing metal weapon points. *Computer Applic. in Archaeol.* 3: 3–8.

Bates, C.D., Coxon, P., and Gibbard, P.L. 1978. A new method for the preparation of clay-rich sediment samples for palynological investigation. *New Phytol.* 81: 459–63.

Bauer, B.S. 1981. *Core Reduction at the La Cotte Site*. Unpublished M.Phil. dissertation, University of Cambridge.

Baumann, W., Mania, D., Toepfer, V., and Eissmann, L. 1983. *Markkleeberg*. VEB Deutscher Verlag der Wissenschaften, Berlin.

Bay-Petersen, J. 1975. *Pre Neolithic Faunal Exploitation in Southern France and Denmark*. Unpubl. Ph.D. thesis, University of Cambridge.

Behm-Blancke, G. 1960. Altsteinzeitliche Rastplätze im Travertingebiet von Taubach, Weimar, Ehringsdorf. *Alt-Thüringen* 4.

Behrensmeyer, A.K. 1978. Taphonomic and ecologic information from bone weathering. *Paleobiology* 4: 150–62.

Bell, W.T. 1976. The assessment of radiation dose-rate for thermoluminescence dating. *Archaeometry* 18(1): 107–10.

Belyaeva, E.I., Gromova, V.I., and Yanovskaya, N.M. 1968. Order of Perissodactyla. In Gromova, V.I. (ed.) *Fundamentals of Palaeontology* XIII. Monson, Jerusalem.

Bere, R.M. 1966. *The African Elephant*. Golden Press, New York.

Berggren, W.A., Burckle, L.H., Cita, M.B., Cooke, H.B.S., Funnell, B.M., Gartner, S., Hays, J.D., Kennett, J.P., Opdyke, N.D., Pastouret, L., Shackleton, N.J., and Takayanagi, Y. 1980. Towards a Quaternary time scale. *Quat. Res.* 13: 277–302.

Bergman, C., and Newcomer, M.H. 1983. Flint arrowhead breakage: examples from 'Ksar Akil, Lebanon. *J. Field Archaeol.* 10(2): 238–43.

Bettinger, R.L. 1980. Explanatory/predictive models in hunter-gatherer adaptation. In Schiffer, M.B (ed.) *Advances in Archaeological Method and Theory*, vol. 3, pp. 189–255. Academic Press, New York.

Beyries, S. 1984. *Approche Fonctionelle de la Variabilité des Faciès du Moustérien*. Doctoral thesis, Univ. de Paris X.

Beyries, S., and Boëda, E. 1983. Etude technologique et traces d'utilisation des "éclats débordants" de Corbehem (Pas de Calais). *Bull. Soc. Préhist. Fr.* 80(9): 275–9.

Binford, L.R. 1973. Interassemblage variability — the Mousterian and the 'functional' argument. In Renfrew, A.C. (ed.) *The Explanation of Culture Change. Models in Prehistory*, pp. 227–254. Duckworth, London.

1977. Forty seven trips. A case study in the character of archaeological formation processes. In Wright, R.V.S. (ed.) *Stone Tools as Cultural Markers*, pp. 24–36. Australian Institute for Aboriginal Studies, Canberra.

1981. *Bones: Ancient Men and Modern Myths*. Academic Press, New York.

Binford, L.R., and Bertram, J.B. 1977. Bone frequencies — and attritional processes. In Binford, L.R. (ed.) *For Theory Building in Archaeology*, pp. 77–153. Academic Press, New York.

Binford, S.R., and Binford, L.R. 1966. A preliminary analysis of functional variability in the Mousterian of Levallois facies. *Am. Anthrop.* 68(2): 238–95.

Biquand, D., and Lautridou, J.P. 1979. Détermination de la polarité magnétique des loess et sables pléistocènes de Haute-

# BIBLIOGRAPHY

Normandie: premiers résultats. *Bull. Assoc. Fr. Et. Quat.* 16(1–2): 75–81.

Birks, H.J.B., and Birks, H.H. 1980. *Quaternary Palaeoecology.* Arnold, London.

Bischoff, J.L., and Rosenbauer, R.J. 1981. Uranium-series dating of human skeletal remains from the Del Mar and Sunnyvale sites, California. *Science, N.Y.* 213: 1003–5.

Blackwell, B., Schwarcz, H.P., and Debénath, A. 1983. Absolute dating of hominids and Palaeolithic artifacts of the cave of la Chaise-de-Vouthon (Charente), France. *J. Archaeol. Sci.* 10: 493–513.

Blount, A.M. 1974. The crystal structure of crandallite. *Am. Miner.* 59: 41–7.

Blumberg, J.E., Hylander, W.L., and Goepp, R.A. 1971. Taurodontism: a biometric study. *Am. J. Phys. Anthrop.* 34: 243–56.

Boëda, E. In press. Le débitage levallois de Biache-Saint-Vaast (Pas de Calais): première étude technologique. In Tuffreau, A. (ed.) *Chronostratigraphie et Faciès Culturels du Paléolithique Inférieur et Moyen dans l'Europe du Nord-Ouest.* Papers given at Congr. Préhist. France (Lille, 1984). To appear as *Supp. Bull. Assoc. Fr. Et. Quat.*

Bonifay, M.F. 1961. Les Rhinocéros à Narines Cloisonnées de l'Aven de Coulon (Gard). *Bull. Mus. d'Anthrop. Préhist. de Monaco* 8: 135–75.

Boone, Y., and Renault-Miskovsky, J. 1976. La cueillette. In Lumley, H. de (ed.) *La Préhistoire Française,* pp. 684–7. CNRS, Paris.

Bordes, F. 1950. L'évolution buissonnante des industries en Europe occidentale. *L'Anthropologie* 54: 393–420.

— 1954. Les limons quaternaires du Bassin de la Seine. *Arch. Inst. Paléont. Hum.,* 26.

— 1956. Some observations on the Pleistocene succession in the Somme valley. *Proc. Prehist. Soc.* 22: 1–5.

— 1961a. Mousterian Cultures in France. *Science, N.Y.* 134: 803–10.

— 1961b. *Typologie du Paléolithique Ancien et Moyen.* Delmas, Bordeaux.

— 1967. Considérations sur la typologie et les techniques dans le Paléolithique. *Quartär* 18: 25–55.

— 1969. Traitement thermique du silex au Solutréen. *Bull. Soc. Préhist. Fr.* 66: 197.

— 1971. Observations sur l'Acheuléen des grottes de la Dordogne. *Munibe* 22(1): 5–23.

— 1972. *A Tale of Two Caves.* Harper and Row, New York.

— 1980. Le débitage levallois et ses variantes. *Bull. Soc. Préhist. Fr.* 77: 45–9.

— 1981. Vingt-cinq ans après: le complexe moustérien revisité. *Bull. Soc. Préhist. Fr.* 78(3): 77–87.

Bordes, F., Debénath, A., Kervazo, B., Laville, H., Le Tensorer, J.M., Texier, J.P., and Thibault, Cl. 1980. Les dépôts quaternaires en Aquitaine. In Chaline, J. (ed.) *Problèmes de Stratigraphie Quaternaire en France et dans les Pays Limitrophes,* pp. 250–67. Bull. Assoc. Fr. Et. Quat. (supplément).

Borsuk-Bialynicka, M. 1973. Studies on the Pleistocene rhinoceros *Coelodonta antiquitatis* (Blumenbach). *Palaeontologica Polonica* 29: 1–94.

Bosinski, G. 1976. L'Acheuléen en Europe centrale du nord. *UISPP 9* (Nice). Colloque 10: 52–64.

— 1982. The transition Lower/Middle Palaeolithic in north western Germany. In Ronen, A. (ed.) *The Transition from Lower to Middle Palaeolithic and the Origin of Modern Man,* pp. 166–75. British Archaeological Reports, International Series 151.

Bourdier, F., Chaline, J., Munaut, A.V., and Puisségur, J.J. 1974a. La très haute nappe alluviale de la Somme. *Bull. Assoc. Fr. Et. Quat.* 10(3–4): 137–43.

— 1974b. Le complexe mindélien: II — La moyenne terrasse de l'Avre. *Bull. Assoc. Fr. Et. Quat.* 10(3–4): 168–80.

Bourgon, M. 1957. Les industries moustériennes et pré-moustériennes du Périgord. *Mem. Arch. Inst. Pal. Hum.* 27.

Bowen, D.Q. 1973. The Pleistocene succession of the Irish Sea. *Proc. Geol. Assoc., Lond.* 84: 249–72.

— 1978. *Quaternary Geology.* Pergamon, Oxford.

Bradley, B.A. 1977. *Experimental Lithic Technology with Special Reference to the Middle Palaeolithic.* Unpublished Ph.D dissertation, University of Cambridge.

Bradley, B.A., and Sampson, G.R. 1986. Analysis by replication of an Acheulian artefact assemblage from Caddington, England. In

Bailey, G.N., and Callow, P. (eds.). *Stone Age Prehistory. Studies in Memory of Charles McBurney.* Cambridge University Press, Cambridge.

Brain, C.K. 1967. Hottentot food remains and their meaning in the interpretation of fossil bone assemblages. *Sci. Pap. Namib Desert Res. Stn.* 32: 1–11.

— 1969. The contribution of Namib Desert Hottentots to an understanding of Australopithecine bone accumulations. *Sci. Pap. Namib Desert Res. Stn.* 39: 13–22.

— 1981. *The Hunters or the Hunted? An Introduction to African Cave Taphonomy.* University of Chicago Press, Chicago.

Brebion, P., Buge, E., Fily, G., Lauriat, A., Margarel, J.W., and Pareyn, C.L. 1973. Le Quaternaire ancien de St. Nicolas de Pierrepont et St. Sauveur de Pierrepont (Manche). *Bull. Soc. Linn. Normandie* 104: 70–106.

Breuil, H. 1952. *Four Hundred Centuries of Cave Art.* Centre d'Etudes et de Documentation préhistoriques, Montignac.

Breuil, H., and Koslowski, H. 1931–4. Etudes de stratigraphie paléolithique dans le Nord de la France, la Belgique et l'Angleterre. *L'Anthropologie* 41: 471–88, 42: 27–47 and 44: 249–90.

Brézillon, M. 1969. *Dictionnaire de la Préhistoire.* Librarie Larousse, Paris.

Bridgland, D.R. 1980. A reappraisal of Pleistocene stratigraphy in north Kent and eastern Essex, and new evidence concerning former courses of the Thames and Medway. *Quat. Newsl.* 32: 15–24.

Broadbent, N.D., and Knutsson, K. 1975. An experimental analysis of quartz scrapers. Results and applications. *Fornvännen* 70(3–4): 113–28.

Brochet, G. 1981. *Etude comparative des faunes d'Arcy-sur-Cure, la Cotte de St. Brelade et de Gigny à l'aide d'une méthode d'écologie quantitative.* D.E.A. thesis, U.E.R. Sciences de la Terre, Dijon.

Broecker, W.S., Thurber, D.L., Goddard, J., Ku, T.L., Matthews, R.K., and Mesolella, K.J. 1968. Milankovich hypothesis supported by precise dating of coral reefs and deep-sea sediments. *Science, N.Y.* 159: 297–300.

Brochier, J.E. 1976. Les civilisations du Paléolithique inférieur dans le Basse-Isère. In Lumley, H. de (ed.) *Le Préhistoire Française,* pp. 875–8. CNRS, Paris.

Brunnacker, K. 1978. *Geowissenschaftliche Untersuchungen in Gönnersdorf.* Römish-Germanische Kommission des Deutschen Archaologischen Instituts, Franz Steiner-Verlag GMBH, Weisbaden.

Burdo, C. 1960. *La Cotte de Saint Brelade, Jersey, British Channel Islands: Excavation of a pre-Mousterian Horizon 1950–1958.* Société Jersiaise, St. Helier.

Burdukiewicz, J., Mania, D., Kocoń, A., and Weber, T. 1979. Die Silexartefakte von Bilzingsleben. Zu ihrer morphologischen Analyse. *Ethnogr.-Archäol. Z.* 20: 682–703.

Bustin, R.M., and Mathews, W.H. 1979. Selective weathering of granitic clasts. *Can. J. Earth Sci.* 16(2): 215–23.

Butzer, K.W. 1982. *Archaeology as Human Ecology.* Cambridge University Press, Cambridge.

Cahen, D. 1984. Paléolithique inférieur et moyen en Belgique. In Cahen, D., and Haesaerts, P. (eds.) *Peuples Chasseurs de la Belgique Préhistorique dans leur Cadre Naturel,* pp. 133–55. Inst. roy. Sci. nat. Belgique, Bruxelles.

Cahen, D., and Haesaerts, P. 1984. *Peuples Chasseurs de la Belgique Préhistorique dans leur Cadre Naturel.* Inst. roy. Sci. nat. Belgique, Bruxelles.

Cahen, D., and Keeley, L.H. 1980. Not less than two, not more than one. *World Archaeol.* 12: 166–80.

Cahen, D., Keeley L.H., and van Noten, F.L. 1979. Stone tools, toolkits and human behaviour in prehistory. *Curr. Archaeol.* 20 (4): 661–83.

Callow, P. 1974. Le Paléolithique inférieur et moyen de la Grande-Bretagne et de la France septentrionale (Nord, bassin parisien). *Septentrion* 4: 61–70.

— 1976. *The Lower and Middle Palaeolithic of Britain and Adjacent Areas of Europe.* Unpublished Ph.D. dissertation, University of Cambridge.

— In press. The Olduvai bifaces: technology and raw materials. In Leakey, M.D. (ed.) *Olduvai Gorge* Vol 5.

Callow, P., and Webb, R.E. 1977. Structure in the southwest French

Mousterian. *Computer Applic. in Archaeol.* 4: 69–76.

1981. The application of multivariate statistical techniques to Middle Palaeolithic assemblages from southwestern France. *Rev. d'Archéometrie* 5: 129–38.

Campbell, J.B. 1977. *The Upper Palaeolithic in Britain.* Oxford University Press, Oxford.

Catt, J.A. 1977. Loess and cover sands. In Shotton, F.W. (ed.), *British Quaternary Studies — Recent Advances.* Clarendon Press, Oxford.

Chaline, J. 1972. *Les rongeurs du Pléistocène moyen et supérieur de France.* Cahiers de Paléontologie, CNRS, Paris.

1976. Les Rongeurs. In Lumley, H. de (ed.) *La Préhistoire Française,* pp. 420–4. CNRS, Paris.

Chaline, J., and Jerz, H. 1983. Proposition de création d'un étage Würmien par la sous-commission de stratigraphie du Quaternaire européen de l'INQUA. *Bull. Assoc. Fr. Et. Quat.* 30(4): 149–52.

Chappell, J. 1983. A revised sea-level record for the last 300,000 years from Papua, New Guinea. *Search* 14(3–4): 99–101.

Chmielewski, W., and Kubiak, H. 1962. The find of mammoth bones at Skaratki in the Lowicz District. *Folia Quat.* 10: 1–29.

Clark, J.D., and Haynes, C.V. Jr. 1969. An elephant butchery site at Mwanganda's Village, Karonga, and its relevance for palaeolithic archaeology. *World Archaeol.* 1: 390–411.

Clark, J.D., and Wilkinson, L.P. 1986. Charles Brian Montagu McBurney (1914–1979): an appreciation. In Bailey, G.N., and Callow, P. (eds.). *Stone Age Prehistory. Studies in Memory of Charles McBurney,* pp. 7–25. Cambridge University Press, Cambridge.

Clark, J.G.D. 1972. Star Carr: a case study. *Addison-Wesley Modular Publications* 10.

1977. *World Prehistory in New Perspective* (3rd edn.). Cambridge University Press, Cambridge.

Clarke, D.L. 1968. *Analytical Archaeology.* Methuen, London.

Clet-Pellerin, M. 1983. *Le Plio-Pléistocène en Normandie. Apports de la Palynologie.* Doctoral thesis (3ème cycle), Université de Caen.

Coe, M. 1980. The role of modern ecological studies in the reconstruction of paleoenvironments in Sub-Saharan Africa. In Behrensmeyer, A.K., and Hill, A.P. (eds.) *Fossils in the Making: Vertebrate Taphonomy and Paleoecology.* University of Chicago Press, Chicago.

Coles, J.M., and Higgs, E.S. 1975. *The Archaeology of Early Man.* Penguin Books, Harmondsworth.

Collcutt, S.N. 1979. The analysis of Quaternary cave sediments. *World Archaeol.* 10(3): 290–301.

Collenette, A. 1916. The Pleistocene Period in Guernsey. *Trans. Guernsey Soc. Nat. Sci.* 7: 331–408.

Collins, D. 1969. Culture traditions and environment of early man. *Current Anthrop.* 10(4): 267–316.

Combier, J. 1967. Le paléolithique de l'Ardèche dans son cadre paléoclimatique. *Mém. Inst. Préhist. Bordeaux* 4.

1976. Gisement acheuléen d'Orgnac III. *UISPP* 9 (Nice). Livret-guide de l'excursion A8: 217–26.

Cook, J., Stringer, C.B., Currant, A.P., Schwarcz, H.P., and Wintle, A.G. 1982. A review of the chronology of the European Middle Pleistocene hominid record. *Yearbook. Phys. Anthrop.* 25: 19–65.

Coope, G.R., Jones, R.L., and Keen, D.H. 1980. The palaeoecology and age of peat at Fliquet Bay, Jersey, Channel Islands. *J. Biogeogr.* 7: 187–95.

Coope, G.R., Jones, R.L., Keen, D.H., and Whaton, P.V. 1985. The flora and fauna of Late Pleistocene Deposits in St. Aubin's Bay, Jersey, Channel Islands. *Proc. Geol. Ass.* 96(3), in press.

Coope, G.R., Dickson, J.H., Jones, R.L., and Keen, D.H. 1986 The flora and fauna of Late Pleistocene deposits on the Cotentin peninsula, Normandy. *Phil. Trans. Roy. Soc. Lond.,* series B, in press.

Courty, M.A. 1982. *Etude géologique de sites archéologiques holocènes. Définition de processus sédimentaires et post-sédimentaires, caractérisation de l'impact anthropique. Essai de méthodologie.* Doctoral thesis (3ème cycle), Univ. de Bordeaux.

Coutard, J.P., Giresse, J.P., and Lautridou, J.P. 1973. Les loess du Nord-Ouest du Bocage Normand. *Bull. Soc. Linn. Normandie* 103: 58–68

Coutard, J.P., Helluin, M., Lautridou, J.P., Ozouf, J.C., and Pellerin, J. 1970. Les loess de la campagne de Caen. *Bull. Cent. de Géo-morphol. du CNRS, Caen* 8: 1–30.

Coutard, J.P., Helluin, M., Lautridou, J.P., Ozouf, J.C., Pellerin, J. and Clet, M. 1979. Dynamique et stratigraphie des heads de la Hague (Basse-Normandie). *Bull. Cent. de Géomorphol. du CNRS, Caen* 24: 131–51.

Coutard, J.P., Helluin, M., Ozouf, J.C., and Pellerin, J. 1981. Le Quaternaire main et continental du Cap Levi et de ses abords (Fermanville, Manche). *Bull. Soc. Linn. Normandie* 108: 7–22.

Coutard, J.P., Lautridou, J.P., Lefèbvre, D., and Clet, M. 1979. Les bas-niveaux marins Eemiens et pré-Eemiens de Grandcamp-les-Bains. *Bull. Soc. Linn. Normandie* 107: 11–20.

Cox, A. 1969. Geomagnetic reversals. *Science, N.Y.* 163: 237–45.

Crader, D.C. 1983. Recent single-carcass bone scatters and the problem of "butchery" sites in the archaeological record. In Clutton-Brock, J., and Grigson, C. (eds.) *Animals and Archaeology: 1. Hunters and their Prey.* British Archaeological Reports S163, Oxford.

Cumming, R.G. 1850. *A Hunter's Life in South Africa.* Murray, London.

Curtis, G.H., and Evernden, J.F. 1962. Age of basalt underlying Bed I, Olduvai. *Nature, Lond.* 194: 610–2.

Curwen, E.C. 1930. Prehistoric flint sickles. *Antiquity* 4: 179–86.

Dalrymple, B.J. 1957. The Pleistocene deposits of Penfold's Pit, Slindon, Sussex and their chronology. *Proc. Geol. Ass., Lond.* 68: 294–303.

Daniel, M., Daniel, R., Degros, J., and Vinot, A. 1973. Les gisements préhistoriques du Bois de Verrières-le-Buisson (Essonne). I. Le site palaéolithique du Terrier. *Gallia Préhist.* 16: 63–104.

Darling, F. 1969. *A Herd of Red Deer.* Oxford University Press, Oxford.

Dart, R.A. 1949. The predatory implemental technique of Australopithecus. *Am. J. Phys. Anthrop.* n.s.7(1): 1–38.

Davidson, D.A., Jones, R.L., and Renfrew, C. 1976. Palaeoenvironmental reconstruction and evaluation: a case study from Orkney. *Trans. Inst. Br. Geogr.* 1: 346–61.

Davies, K.H. 1983. Amino-acid analysis of Pleistocene marine molluscs from the Gower Peninsula. *Nature, Lond.* 302: 137–9.

Davies, K.H., and Keen, D.H. 1985. The age of Pleistocene marine deposits at Portland, Dorset. *Proc. Geol. Ass.* 96(3), in press.

Dawson, A.G. 1980. Shore erosion by frost. An example from the Scottish Lateglacial. In Lowe, J.H., Gray, J.M., and Robinson, J.E. (eds.) *Studies in the Lateglacial of North-West Europe.* Pergamon Press, Oxford.

Debénath, A. 1976a. Les civilisations du Paléolithique inférieur en Charente. In Lumley, H. de (ed.) *La Préhistoire Française,* pp. 930–5. CNRS, Paris.

1976b. Les gisements préhistoriques de la Chaise-de-Vouthon. *UISPP* 9 (Nice). Livret-guide de l'excursion A4: 141–5.

1980. Quelques particularités techniques et typologiques des industries de La Chaise-de-Vouthon (Charente). *Congr. nat. Soc. sav.* 105: 239–247.

Deraniyagala, B. 1955. *Some Extinct Elephants, their Relatives and Two Living Species.* Ceylon National Museum.

Desbrosse, J., Kozlowski, J.K. and Zuate y Zuber, J. 1976. Prondniks de France et d'Europe Centrale. *L'Anthropologie* 80: 3.

Desse, G., and Desse, J. 1976. La pêche. In Lumley, H. de (ed.) *La Préhistoire Française,* pp. 697–702. CNRS, Paris.

Dibble, H.L. 1984. The Mousterian industry from Bisitun Cave (Iran). *Paléorient* 10(2): 23–34.

Dickson, F.P. 1977. Quartz flaking. In Wright, R.V.S. (ed.) *Stone Tools as Cultural Markers,* pp. 97–103. Humanities Press, New Jersey.

Dimbleby, G.W. 1957. Pollen analysis of terrestrial soils. *New Phytol.* 56: 12–28.

Dines, H.G., and Edmunds, F.H. 1929. The Geology of the country around Aldershot and Guildford. *Mem. Geol. Surv. U.K..* HMSO, London.

Dodge, R.E., Fairbanks, R.G., Benninger, L.K., and Maurrasse, F. 1983. Pleistocene sea levels from raised coral reefs of Haiti. *Science, N.Y.* 219: 1423–5.

Doran, J.E., and Hodson, F.R. 1975. *Mathematics and Computers in Archaeology.* Edinburgh University Press, Edinburgh.

Douglas-Hamilton, F., and Douglas-Hamilton, O. 1975. *Among the Elephants.* Viking Press, London.

Duchadeau-Kervazo, C., and Kervazo, B. 1983. Confrontation de plusieurs types de courbes climatiques et corrélations avec quelques dépôts du Quaternaire récent. *Bull. Ass. Fr. Et. Quat.* 20(1): 25–38.

Eberl, B. 1930. *Die Eiszeitenfolge in nördlichen Alpenvorlande.* Benno Filzer, Augsburg.

Elhai, H. 1963. *La Normandie occidentale entre la Seine et le Golfe normand-breton, étude morphologique.* Bière, Bordeaux.

Emiliani, C. 1955. Pleistocene temperatures. *J. Geol.* 63: 538–578.

Everard, C.E. 1977. Valley direction and geomorphological evolution in west Cornwall, England. *Occas. Pap. Geog. Queen Mary Coll., Univ. Lond.* 10.

Evernden, J.F., and Curtis, G.H. 1965. The potassium/argon dating of Late Cenozoic rocks in East Africa and Italy. *Current Anthrop.* 6: 354–60.

Faegri, K., and Iversen, J. 1975. *Textbook of Pollen Analysis* (3rd edn.). Munskgaard, Copenhagen.

Fairbridge, R.W. 1972. Climatology of a glacial cycle. *Quat. Res.* 2: 283–302.

Feustel, R. 1983. Zur zeitlichen und kulturellen Stellung des Paläolithikums von Weimar-Ehringsdorf. *Alt-Thüringen* 19: 16–42.

Fink, J., and Kukla, G.J. 1977. Pleistocene climates in Central Europe: at least 17 interglacials after the Olduvai event. *Quat. Res.* 7: 363–71.

Fleming, S. 1976. *Dating in Archaeology. A Guide to Scientific Techniques.* Dent, London.

1979. *Thermoluminescence Techniques in Archaeology.* Clarendon Press, Oxford.

Flicoteaux, R., Nahon, D., and Paquet, H. 1977. Genèse des phosphates à partir de sediments argilo-phosphatés du Tertiaire de Lam-Lam (Sénégal). Suite minéralogique. Permanences et changements de structure. *Sci. Geol. Bull.* 30(3): 153–74.

Florschütz, F., Menéndez-Amor, J., and Wijmstra, T.A. 1971. Palynology of a thick Quaternary succession in southern Spain. *Palaeogeog., Paalaeoclim., Palaeoecol.* 10: 233–64.

Formazov, A.A. 1975. Developments in Palaeolithic archaeology in the USSR. In Bruce-Mitford, R. (ed.) *Recent Archaeological Excavations in Europe.* Routledge and Kegan Paul, London.

Fosse, G. 1980. Eléments récents sur le Paléolithique de Normandie. In Bigot, F., Fosse, G., Lautridou, J.P., Tuffreau, A., and Verron, G. *Préhistoire de la Normandie*, pp. 83–91. Direction des Antiquités Préhistoriques de Basse-Normandie, Caen.

1982. Position stratigraphique et paléoenvironnement du Paléolithique ancien et moyen de Normandie. *Bull. Assoc. Fr. Et. Quat.* 19(2–3): 83–92.

Frayer, D.W. 1978. Evolution of the dentition in Upper Palaeolithic and Mesolithic Europe. *Univ. Kans. Publ. Anthrop.* 10: 1–201.

Freeman, L.G. 1976 Acheulian sites and stratigraphy in Iberia and the Maghreb. In Butzer, K.W., and Isaac, G.Ll. (eds.) *After the Australopithecines: Stratigraphy, Ecology and Culture Change in the Middle Pleistocene*, pp. 661–743. Mouton, The Hague.

Frenzel, B. 1964. Sur pollenanalyse von lössen. *Eiszeit. und Gegen.* 15: 5–39.

Frison, G.C. 1968. A functional analysis of chipped stone tools. *Am. Antiq.* 33: 149–55.

Frison, G.C. (ed.) 1974. *The Casper Site: A Hell Gap Bison Kill on the High Plains.* Academic Press, New York.

Frison, G.C., Wilson, M., and Wilson, D. 1976. Fossil bison and artifacts from an early altithermal period arroyo trap in Wyoming. *Am. Antiq.* 41(1): 28–57.

Galaburda, A.M., LeMay, M., Kemper, T.L., and Geschwind, N. 1978. Right-left asymmetries in the brain. *Science, N.Y.* 199: 852–6.

Garrod, D. 1951. A transitional industry from the base of the Upper Palaeolithic in Lebanon and Syria. *Jl. R. Anthrop. Soc.* 81: 121–9

Gascoyne, M., Currant, A.P., and Lord, T.C. 1981. Ipswichian fauna of Victoria Cave and the marine palaeoclimatic record. *Nature, Lond.* 294: 652–4.

George, P.K. 1973. The Quaternary on Guernsey. *Trans. Soc. Guernesiaise* 19: 161–85.

Gifford, D.P. 1978. Ethnoarchaeological observations of natural processes affecting cultural materials. In Gould, R.A. (ed.) *Explanations in Ethnoarchaeology*, pp. 77–101. Univ. of New Mexico Press, Albuquerque.

Giot, P.R., and Philippot, A. 1946. Nouvelles interprétations sur la géologie du Mont-Dol (Ille-et-Vilaine). *Comptes-rendus séances Soc. Géol. France* 1946: 295–6.

Gilot, E. 1984. Datations radiometriques. In Cahen, D., and Haesaerts, P. (eds.) *Peuples Chasseurs de la Belgique Préhistorique dans leur Cadre Naturel*, pp. 111–25. Inst. roy. Sci. nat. Belgique, Bruxelles.

Girard, C. 1978. Les industries moustériennes de la Grotte de l'Hyène à Arcy-sur-Cure (Yonne). *Gallia Préhistoire*, 11e suppl. CRNS, Paris.

Giresse, P. 1980. The Maastrichtian phosphate sequence of the Congo. In Bentor, Y.K. (ed.) *Marine Phosphorites — Geochemistry, Occurrence, Genesis*, pp. 193–205. Soc. Econ. Pal. Mineral., Special Pub. 29.

Giresse, P., Hommeril, P., and Lamboy, M. 1972. Résultats préliminaires d'une campagne de sismique réflexion dans le golfe normand-breton. *Mém. Bur. Rech. Géol. Min.* 79: 193–201.

Goddard, J. 1970. Age criteria and vital statistics of a black rhinoceros population. *E. Afr. Wildl. J.* 8: 105–121.

Godwin, H., and Godwin, M.E. 1952. Pollen analyses from peat on the shore and coastal plain of Jersey, Channel Islands. *Bull. ann. Soc. Jersiaise* 15(4): 457–62.

Gould, R.A. 1980. *Living Archaeology.* Cambridge University Press, Cambridge.

Gould, R.A., and Koster, D.A. 1971. The lithic assemblage of the Western Desert Aborigines of Australia. *Am. Antiq.* 36: 149–69.

Gowlett, J.A.J., Harris, J.W.K., Walton, D., and Wood, B.A. 1981. Early archaeological sites, hominid remains and traces of fire from Chesowanja, Kenya. *Nature, Lond.* 294: 125–9.

Grahmann, R. 1955. The Lower Palaeolithic site of Markkleeberg and other comparable localities near Leipzig. *Trans. Am. Phil. Soc.* 45: 509–687.

Graindor, M.J., and Roblot, M.M. 1976. *Géologie Sous-marine de la Baie de Mont Saint-Michel et de ses Abords.* Electricité de France, Grenoble.

Green, H.S. 1981. The first Welshman: excavations at Pontnewydd. *Antiquity* 55: 184–95.

1984. *Pontnewydd Cave. A Lower Palaeolithic Hominid Site in Wales. The First Report.* Alan Sutton Publishing Ltd., Gloucester.

Green, H.S., *et al.* 1981. Pontnewydd Cave in Wales — a new Middle Pleistocene hominid site. *Nature, Lond.* 294: 707–13.

Guérin, C. 1976. Les Périssodactyles: rhinocérotidés. In Lumley, H. de (ed.) *La Préhistoire Française*, pp. 405–8. CNRS, Paris.

1978. Les nouveaux restes de *Rhinocéros* (Mammalia, Perissodactyla) recueillis dans la formation Pléistocène d'Achenheim (Bas-Rhin). *Rech. Géogr. Strasbourg* 7: 115–21.

Guérin, C., and Faure, M. 1983. Les hommes du Paléolithique européen ont-ils chassé le rhinoceros? In Poplin, F. (ed.) *La Faune et l'Homme Préhistorique. Mém. Soc. Préhist. Fr.* 16.

Guillet, B., *et al.* 1981. Dispersion et migration de minéraux argileux dans les podzols. Contribution des composés organiques associés, leur rôle sur les formes et l'état de l'aluminium. *Colloq. Int. du CNRS; Migrations organo-minérales dans les sols temperés (Nancy, 1979).* 303: pp. 43–56.

Guichard, G. 1976. Les civilisations du Paléolithique inférieur en Périgord. In Lumley, H. de (ed.) *Le Préhistoire Française*, pp. 909–28. CNRS, Paris.

Guichard, J. 1965. Un faciès original de l'Acheuléen: Cantalouette (commune de Creysse, Dordogne). *L'Anthropologie* 69: 413–64.

Haesaerts, P., and Van Vliet, B. 1981. Phénomènes périglaciaires et sols fossiles observés à Maisières-Canal, Harmignies et à Rocourt, lors des excursions des 21 et 22 Septembre 1978. *Biul. Perygl.* 28: 291–325.

Hamilton, D., and Smith, A.J. 1972. Origin and sedimentary history of the Hurd Deep. *Mém. Bur. Rech. Géol. Min.* 79: 59–78.

Hanks, J. 1979. *The Struggle for Survival. The Elephant Problem.* Mayflower, New York.

Harland, W.B., Cox, A.V., Llewellyn, P.G., Picklon, C.A.G., Smith, A.G., and Walters, R. 1982. *A Geologic Time Scale.* Cambridge University Press, Cambridge.

Harmon, R.S., Glazek, J., and Nowak, K. 1980. $^{230}$Th/$^{234}$U dating of travertine from the Bilzingsleben archaeological site. *Nature,*

Lond. 284: 132–5.

Harris, E.C. 1979. *Principles of Archaeological Stratigraphy*. Academic Press, London.

Harris, W.C. 1838. *Narrative of an Expedition into Southern Africa during the Years 1836 and 1837*. American Mission Press, Bombay.

Hayden, B. 1977. Stone tool functions in the Western Desert. In Wright, R.V.S. (ed.) *Stone Tools as Cultural Markers*, pp. 178–88. Humanities Press, New Jersey.

Haynes, G. 1983. Frequencies of spiral and green-bone fractures on ungulate limb bones in modern surface assemblages. *Am. Antiq.* 48(1): 102–14.

Hays, J.D., and Berggren, W.A. 1971. Quaternary bounderies and correlations. In Funnell, B.M., and Riedel, W.R. (eds.) *Micropalaeontology of the Oceans*, pp. 669–91. Cambridge University Press, Cambridge.

Hays, J.D., Imbrie, J., and Shackleton, N.J. 1976. Variations in the earth's orbit: pacemakers of the ice ages. *Science, N.Y.* 194: 1121–32.

Hedberg, H.D. (ed.) 1976. In *International Stratigraphic Guide: A Guide to Stratigraphic Classifications, Terminology and Procedure*. Wiley, New York.

Hedges, R.E.M., and McLellan, M. 1976. On the cation exchange capacity of fired clays and its effect on the chemical and radiometric analysis of pottery. *Archaeometry* 18(2): 203–7.

Heffley, S. 1981. The relationship between Northern Athapaskan settlement patterns and resource distribution: an application of Horn's model. In Winterhalder, B., and Smith, E.A. (eds.) *Hunter-Gatherer Foraging Strategies. Ethnographic and Archeological Analysis*, pp. 126–47. University of Chicago Press, Chicago.

Heim, J., Lautridou, J.P., Maucorps, J., Puisségur, J.J., Sommé, J., and Thévenin, A. 1982. Achenheim: une séquence-type des loess du Pléistocène moyen et supérieur. *Bull. Assoc. Fr. Et. Quat.* 19(2–3): 147–59.

Heintz, A., and Garutt, V.E. 1965. Determination of the absolute age of the fossil remains of Mammoth and Woolly Rhinoceros from the permafrost in Siberia by the help of radiocarbon (C14). *Norsk. Geol. Tidsskr.* 45: 73–79.

Henson, F.A. 1947. The granites of southwest Jersey. *Geol. Mag.* 84(5): 273–80.

Higgs, E.S., and Vita-Finzi, C. 1972. Prehistoric economies: a territorial approach. In Higgs, E.S. (ed.) *Papers in Economic Prehistory*, pp. 27–36. Cambridge University Press, London.

Higgs, E.S., Vita-Finzi, C., Harris, D.R., and Fagg, A.E. 1967. The climate, environment and industries of stone age Greece, part III. *Proc. Prehist. Soc.* 33: 1–29.

Hill, A.P. 1979. Butchery and natural disarticulations: an investigatory technique. *Am. Antiq.* 44: 739–44.

1983. Hippopotamus butchery by *Homo erectus* at Olduvai. *J. Archaeol. Sci.* 10: 135–7

Hinton, M.A.C. 1926. *Monograph of the Voles and Lemmings* (Microtinae) *Living and Extinct*. Vol. I (vol. II not published). British Museum (Nat. Hist.), London.

Hodder, I. 1982. *Symbols in Action*. Cambridge University Press, Cambridge.

Hokr, Z. 1951. Methoda kvantitativniho stanoveni klimatu ve ctvrtohorech podle ssavcich spolecenstev. *Vestnic U.U.G.* 18: 209–19.

Holloway, R.L. 1981. Volumetric asymmetry determination on recent hominid endocasts: Spy I and II, Djebel Irhoud, and the Salé *Homo erectus* specimens, with some notes on Neandertal brain size. *Am. J. Phys. Anthrop.* 55: 385–93.

Hooijer, D.A. 1955. Fossil Proboscidea from the Malay Archipelago and the Punjab. *Zool. Verh.* 28.

Howell, F.G., Cole, G.H., Kleindienst, M.R., Szabo, B.J., and Oakley, K.P. 1972. Uranium-series dating of bones from the Isimila prehistoric site, Tanzania. *Nature, Lond.* 237: 51–2.

Hubbard, R.N.L.B. 1982. The environmental evidence from Swanscombe and its implications for Palaeolithic archaeology. In Leach, P.L. (ed.) *Archaeology in Kent to A.D. 1500*, pp. 3–7. Council for British Archaeology Res. Rept. 48, London.

Hurtrelle, J., Monchy, E., and Tuffreau, A. 1972. Le gisement paléolithique ancien de Beaumetz-les-Loges (Pas-de-Calais). *Ann. Soc. Géol. Nord* 92: 147–53.

Hutcheson, J.C.C. 1982. *Preliminary Statistical Analysis of the Lithic Debitage from the Wolstonian Levels at La Cotte de St. Brelade*. Unpublished B.A. dissertation, University of London.

Huxtable, J. 1981. Light bleaching of archaeological flint samples. *Anc. TL* 16: 2–4.

Huxtable, J., and Jacobi, R.M. 1982. Thermoluminescence dating of burned flints from a British mesolithic site: Longmoor Inclosure, East Hampshire. *Archaeometry* 24(2): 164–9.

Imbrie, J., Hays, J.D., Martinson, D.G., MacIntyre, A., Mix, A., Morley, J.J., Pisias, N.G., Prell, W.L., and Shackleton, N.J. 1984. The orbital theory of Pleistocene climate: support from a revised chronology of the marine $\delta^{18}O$ record. In Berger, A.L., and Imbrie, J., (eds.) *Milankovich and Climate*, Part 1, pp. 269–305. Reidel, Dordrecht.

Isaac, G.Ll. 1971. The diet of early man: aspects of archaeological evidence from Lower and Middle Pleistocene sites in Africa. *World Archaeol.* 2(3): 278–99.

1976. Stages of cultural elaboration in the Pleistocene: possible archaeological indicators of the development of language capabilities. *Ann. N.Y. Acad. Sci.* 280: 275–88.

1977. *Olorgesailie*. Chicago University Press, Chicago.

1983. Aspects of human evolution. In Bendall, D.S. (ed.) *Evolution from Molecules to Man*, pp. 509–43. Cambridge University Press, Cambridge.

Isaac, G.Ll., and Harris, J.W.K. 1980. A method for determining the characteristics of artefacts between sites in the Upper Member of the Koobi Fora Formation, East Lake Turkana. *Proc. 8th Pan-Afr. Congr. Prehist. and Quat. Stud.* (Nairobi, 1977) 19–22.

Isherwood, D., and Street, A. 1976. Biotite-induced grussification of the Boulder Creek Granodiorite, Boulder County, Colorado. *Bull. Geol. Soc. Am.* 87: 366–70.

Ivanovich, M., Rae, A.M.B., and Wilkins, M.A. 1984. Brief report on dating the *in situ* stalagmitic floor found in the East Passage in 1982. In Green, H.S. (ed.) *Pontnewydd Cave. A Lower Palaeolithic Hominid Site in Wales*. National Museum of Wales, Cardiff.

Iversen, J. 1944. *Viscum, Hedera* and *Ilex* as climatic indicators. *Geol. Foren. Stock. Forh.* 66: 463–83.

Jamagne, M., Lautridou, J.P., and Sommé, J. 1981. Préliminaire à une synthèse sur les variations sédimentologiques des loess de la France du Nord-Ouest dans leur cadre stratigraphique et paléographique. *Bull. Soc. Géol. Fr.* 23(2): 143–7.

Jaspers, M.T., and Witkop, C.J.Jr. 1980. Taurodontism, an isolated trait associated with syndromes and X-chromosomal aneuploidy. *Am. J. Hum. Genet.* 32: 396–413.

Jones, P.R. 1979. Effects of raw materials on biface manufacture. *Science, N.Y.* 204: 835–6.

1980. Experimental butchery with modern stone tools and its relevance for palaeolithic archaeology. *World Archaeol.* 12: 153–165.

1982. Experimental implement manufacture and use. In *The Emergence of Man*, pp. 189–95. *Trans. R. Soc., London* B292.

Jones, R.L., and Cundill, P.R. 1978. *Introduction to Pollen Analysis*. Geo-Abstracts, Norwich.

Jones, Rhys. 1971. *Rocky Cape and the problem of the Tasmanians*. Unpublished Ph.D. dissertation, Australian National University, Canberra.

Kahlke, H.D. 1975. The macro-faunas of continental Europe during the Middle Pleistocene: stratigraphic sequence and problems of intercorrelation. In Butzer, K.W., and Isaac, G.Ll. (eds.) *After the Australopithecines: Stratigraphy, Ecology and Culture Change in the Middle Pleistocene*, pp. 309–74. Mouton, The Hague.

Keeley, L.H. 1974. The methodology of microwear analysis: a comment on Nance. *Am. Antiq.* 39: 126–8.

1980. *Experimental Determination of Stone Tool Use: a Microwear Analysis*. Chicago University Press, Chicago.

Keen, D.H. 1975. *Some aspects of the Pleistocene succession in areas adjoining the English Channel*. Unpublished Ph.D. thesis, University of London.

1978. The Pleistocene deposits of the Channel Islands. *Rep. Inst. Geol. Sci.* 78/26.

1980. Report on a short field meeting in Guernsey and Jersey, 24th–28th September 1979. *Quat. Newsl.* 30: 26–30.

1981. The Holocene deposits of the Channel Islands. *Rep. Inst.*

*Geol. Sci.* 81/10.

1982a. Late Pleistocene land mollusca in the Channel Islands. *J. Conchol., Lond.* 36: 57–61.

1982b. Depositional sequence, age and palaeoenvironment of raised beaches and head in the Channel Islands and Central Channel. *Bull. Assoc. Fr. Et. Quat.*, 19(1): 3–12.

1985. Pleistocene deposits and mollusca from Portland, Dorset. *Geol. Mag.* 122(2): 181–186.

Keen, D.H., Harmon, R.S., and Andrews, J.T. 1981. U series and amino-acid dates from Jersey. *Nature, Lond.* 289: 162–4.

Kehoe, T.F. 1973. *The Gull Lake Site: a Prehistoric Bison Drive Site in Southwestern Saskatchewan*. Pub. Anthrop. and Hist., 1. Milwaukee Public Museum.

Keith, A. 1913. Problems relating to the teeth of the earlier forms of prehistoric man. *Proc. R. Soc. Med.* 6: 103–24.

Keith, A., and Knowles, F.H.S. 1912a. A description of teeth of Palaeolithic man from Jersey. *J. Anat.* 46: 12–27.

1912b. A description of teeth of Palaeolithic man from Jersey. *Bull. ann. Soc. Jersiaise* 37: 223–40.

Kerney, M.P. 1971. Interglacial deposits in Barnfield Pit, Swanscombe and their molluscan faunas. *J. Geol. Soc. Lond.* 127: 69–86.

Klein, R.G. 1975. Paleoanthropological implications of the non-archeological bone assemblage from Swartklip 1, southwestern Cape Province, South Africa. *Quat. Res.* 5: 275–88.

1976. The mammalian fauna of the Klasies River Mouth sites, Southern Cape Province, South Africa. *S. Afr. Archaeol. Bull.* 31: 75–98.

1978. Stone Age predation on large African bovids. *J. Archaeol. Sci.* 5: 195–217.

1980a. The interpretation of mammalian faunas from Stone Age archeological sites, with special reference to sites in the southern Cape Province, South Africa. In Behrensmeyer, A.K., and Hill, A.P. (eds.) *Fossils in the Making: Vertebrate Taphonomy and Paleoecology*, pp. 223–46. University of Chicago Press, Chicago and London.

1980b. Environmental and ecological implications of large mammals from Upper Pleistocene and Holocene sites in southern Africa. *Ann. S. Afr. Mus.* 81(7): 223–83.

1981. Ungulate mortality and sedimentary facies in the Late Tertiary Varswater Formation, Langebaanweg, South Africa. *Ann. S. Afr. Mus.* 84(6): 233–54.

1983. The Stone Age prehistory of southern Africa. *Ann. Rev. Anthropol.* 12: 25–48.

Kozlowski, J.K., and Kubiak, H. 1972. Late Palaeolithic dwellings made of mammoth bones in South Poland. *Nature, Lond.* 237: 463–4.

Kretzoi, M., and Vertés, L. 1965. Upper Biharian (Intermindel) pebble-industry occupation site in western Hungary. *Current Anthrop.* 6(1): 74–87.

Kruuk, H. 1972. *The Spotted Hyaena*. University of Chicago Press, Chicago.

Kubiak, H. 1976. Hutten aus Mammutknochen. *Unscham*, Frankfurt a. M. 77: 116–17.

1980. The skulls of *Mammuthus primigenius* (Blumenbach) from Debica and Bzianka near Rzeszow, S. Poland. *Folia Quat.* 51: 31–45.

1982. Morphological characters of the mammoth: an adaptation to the arctic-steppe environment. In Hopkins, D.M., Matthews, J.V., Jr., Schweger, C.E., and Young, S.B. (eds.) *Paleoecology of Beringia*, pp. 281–9. Academic Press, New York.

Kukla, G.J. 1975. Loess stratigraphy of Central Europe. In Butzer, K.W., and Isaac, G.Ll. (eds.) *After the Australopithecines: Stratigraphy, Ecology and Culture Change in the Middle Pleistocene*, pp. 98–188. Mouton, The Hague.

1977. Pleistocene land-sea correlation. *Earth Sci. Rev.* 13: 307–74.

Kukla, G.J., and Kocí, A. 1972. End of the Last Interglacial in the loess record. *Quat. Res.* 2: 374–83.

Kuntz, G., Lautridou, J.P., Cavelier, C., and Clet, M. 1979. Le Plio-Quaternaire de Haute-Normandie. *Bull. Inf. Géol. Bassin de Paris* 16: 93–126.

Kurtén, B. 1968. *Pleistocene Mammals of Europe*. Weidenfeld and Nicolson, London.

1972. *The Ice Age*. Hart-Davis, London.

Lalou, C., and Hoang, C.T. 1979. Les méthodes de datation par les descendants de l'uranium. *Bull. Assoc. Fr. Et. Quat.* 16(1–2): 3–14.

Laloy, J. 1980. Recherche d'une méthode pour l'exploitation des témoins de combustion préhistorique. *Cah. du Cent. de Rech. Préhist., Univ. de Paris I*, 7.

Lamb, H.H. 1977. *Climate: Present, Past and Future. Vol. 2: Climatic History and the Future*. Methuen, London.

Laplace, G. 1964. *Essai de Typologie Systématique*. Univ. de Ferrara.

Lautridou, J.P. 1968. Les loess de Saint-Romain et de Mesnil-Esnard (Pays de Caux). *Bull. Cent. de Géomorphol. du CNRS, Caen* 2: 1–55.

1979. Granulométrie des sédiments éoliens périglaciaires; comparaison avec la fraction fine obtenue par gélifraction expérimentale. 7ème R.A.S.T., Lyon. *Soc. Géol. Fr. Ed. Paris.* 280.

1980. Stratigraphie du Quaternaire de Normandie et du Bassin Parisien. In Chaline, J. (ed.) *Problèmes de Stratigraphie Quaternaire en France et dans les Pays Limitrophes*, pp. 180–91. Bull. Assoc. Fr. Et. Quat. (supplément).

1984. *Le Cycle Périglaciaire Pléistocène en Europe du Nord-Ouest et plus particulièrement en Normandie*. Thèse Doctorat d'Etat, Univ. de Caen.

Lautridou, J.P., and Giresse, P. 1981. Genèse et signification paléoclimatique des limons à doublets. *Biul. Perygl.* 28: 149–61.

Lautridou, J.P., *et al.* 1982. The Quaternary of Normandy. Guidebook of the Q.R.A. Normandy meeting, May, 1982. *Bull. Cent. de Géomorphol. du CNRS, Caen* 26.

Lautridou, J.P., Kuntz, G., Clozier, L., and Giot, D. 1976. Les loess des cartes géologiques au 1/50.000ème Doudeville, Saint-Saens, Yvetot (Pays de Caux). *Bull. Bur. Rech. Géol. Min., Orléans* (2),1: 25–37.

Lautridou, J.P., Masson, M., Paepe, R., Puisségur, J.J., and Verron, G. 1974. Loess, nappes alluviales et tuf de Saint-Pierre-les-Elbeuf, près de Rouen; les terrasses de la Seine de Muids à Caudebec. *Bull. Assoc. Fr. Et. Quat.* 10(3–4): 193–201.

Lautridou, J.P., and Puisségur, J.J. 1977. Données nouvelles sur le microfaunes malacologiques et sur les rongeurs du Pléistocène continental de la Basse-Seine. *Bull. Soc. Géol. Normandie et Amis du Mus. du Havre* 64: 119–27.

Laville, H. 1976. Les remplissages de grottes et abris sous roche dans le Sud-Ouest. In Lumley, H. de (ed.) *La Préhistoire Française*, pp. 250–70. CNRS, Paris.

1982. On the transition from 'Lower' to 'Middle' Palaeolithic in southwest France. In Ronen, A. (ed.) *The Transition from Lower to Middle Palaeolithic and the Origin of Modern Man*, pp. 131–5. British Archaeological Reports, International Series 151.

Laville, H., Raynal, J.-P., and Texier, J.-P. 1984. Interglaciaire ... ou déjà glaciaire? *Bull. Soc. Préhist. Fr.* 81(1): 8–11.

Laville, H., Rigaud, J.P., and Sackett, J. 1980. *Rockshelters of the Périgord*. Academic Press, New York.

Laws, R.M. 1966. Age criteria for the African elephant *(Loxodonta a. africana)*. *E. Afr. Wildl. J.* 4: 1–37.

Le Tensorer, J.M. 1978. Le Moustérien type Quina et son évolution dans le sud de la France. *Bull. Soc. Préhist. Fr.* 75(3): 141–9.

Leakey, M.D. 1971. *Olduvai Gorge* Vol. 3. Cambridge University Press, Cambridge.

In preparation. *Olduvai Gorge* Vol. 5.

Lefèbvre, D. 1978. *Contribution à l'étude géochimique et minéralogique du remplissage de la "Caune de l'Arago" (Tautavel)*. D.E.A. thesis (Pédologie), Nancy.

Leroi-Gourhan, Arl. 1961. Analyse pollinique de niveaux acheuléens de la Cotte de Saint Brelade (Jersey). *Int. Kongr. für Vor- und Frühgeschichte* (Hamburg) V: 501–4.

1975. The flowers found with Shanidar IV, a Neanderthal burial in Iraq. *Science, N.Y.* 190: 562–4.

1983. Glaciaire ... ou pas encore Glaciaire? *Bull. Soc. Préhist. Fr.* 80(7): 203.

Levine, M.A. 1979. *Archaeo-zoological analysis of some Upper Pleistocene horse bone assemblages in Western Europe*. Unpublished Ph.D. thesis, University of Cambridge.

Libby, W.F. 1952. *Radiocarbon Dating*. University of Chicago Press, Chicago.

Loose, H. 1975. Pleistocene Rhinocerotidae of W. Europe with reference to the recent two-horned species of Africa and S.E. Asia. *Scr. Geol.* 33: 1–59.

Lumley, H. de, 1969. Une cabane acheuléenne dans la grotte du Lazaret (Nice). *Mem. Soc. Préhist. Fr.* 7.

— 1975. Cultural evolution in France in its paleoecological setting. In Butzer, K.W., and Isaac, G.Ll. (eds.) *After the Australopithecines: Stratigraphy, Ecology and Culture Change in the Middle Pleistocene*, pp. 745–808. Mouton, The Hague.

— 1976a. Baume Bonne (Quinson, Alpes de Haute-Provence). *UISPP* 9 (Nice). Livret-guide de l'excursion C2: 29–38.

— 1976b. Cadre chronologique absolu, paléomagnétisme, chronologie paléontologique et botanique, esquisse paléoclimatologique, séquences culturelles. In Lumley, H. de (ed.) *La Préhistoire Française*, pp. 5–23. CNRS, Paris.

— 1976c. Grotte du Lazaret. *UISPP* 9 (Nice). Livret-guide de l'excursion 1976d (ed.). *La Préhistoire Française*. CNRS, Paris.

Lumley, H. de, and Boone, Y. 1976. Les structures d'habitat au Paléolithique moyen. In Lumley, H. de (ed.) *La Préhistoire Française*, pp. 645–55. CNRS, Paris.

Lumley, H. de, *et al.* 1979. L'homme de Tautavel. *Dossiers d'Archaeol.* 36.

Lumley-Woodyear, H. de 1969. Le paléolithique inférieur et moyen du Midi méditerranéen dans son cadre géologique. I: Ligurie-Provence. *Gallia Préhistoire*, 5ème suppl. (1).

— 1971. Le Paléolithique inférieur et moyen du Midi méditerranéen dans son cadre géologique. II: Bas Languedoc-Roussillon-Catalogne. *Gallia Préhistoire*, 5ème suppl. (2).

Luttropp, A. and Bosinski, G. 1971. Der altsteinzeitliche Fundplatz Reutersruh bei Ziegenhain in Hessen. *Fundamenta* A/8.

McBurney, C.B.M. 1950. The geographical study of the older palaeolithic stages in Europe. *Proc. Prehist. Soc.* 26: 163–83.

— 1962a. Excavations at the Cotte de St. Brelade, Jersey 1961–2. *UISPP* 6 (Rome). Communications: 194–5.

— 1962b. Report on the Cambridge University excavations at La Cotte de St. Brelade. *Ann. Bull. Soc. Jersiaise* 18 (2): 225–6.

— 1963. Report on the Cambridge University excavations, 1962. *Ann. Bull. Soc. Jersiaise* 18 (3): 339–40.

— 1967a. *The Haua Fteah (Cyrenaica) and the Stone Age of the South-East Mediterranean.* Cambridge University Press, Cambridge.

— 1967b. Preliminary report on the current programme of research at La Cotte de St Brelade. *Ann. Bull. Soc. Jersiaise* 19 (3): 222–3.

— 1968. The cave of Ali Tappeh and the epipalaeolithic in North East Iran. *Proc. Prehist. Soc.* 34: 385–413.

— 1969. Second summary report of excavations at La Cotte de St. Brelade. *Ann. Bull. Soc. Jersiaise* 20 (1): 29–31.

McBurney, C.B.M., and Callow, P. 1971. The Cambridge excavations at La Cotte de St. Brelade, Jersey. A preliminary report. *Proc. Prehist. Soc.* 37: 167–207.

MacCalman, H.R. 1967. The zoo park elephant site, Windhoek. In van Zinderen Bakker, E.M. (ed.) *Palaeoecology of Africa, 1964–65*, pp. 102–3. Balkema, Cape Town.

McManus, I.C. 1981. Handedness and birth stress. *Psych. Med.* 11: 485–96.

Maglio, V.J. 1973. Evolution of the Elephantidae. *Trans. Am. Phil. Soc.* 63 (3): 1–149.

Mai, D.H., Mania, D., Nötzold, T., Toepfer, V., Vlček, E., and Heinrich, W.D. 1983. *Bilzingsleben II. Homo Erectus — Seine Kultur und seine Umwelt.* VEB Deutscher Verlag der Wissenschaften, Berlin.

Mania, D. 1976. Altpaläolithischer Rastplatz mit Hominidresten aus dem Mittelpleistozänen Travertin-Komplex von Bilzingsleben (D.D.R.). *UISPP* 9 (Nice). Colloque 9: 35–47.

Mania, D., Toepfer, V., and Vlček, E. 1980. *Bilzingsleben I. Homo Erectus — Seine Kultur und seine Umwelt.* VEB Deutscher Verlag der Wissenschaften, Berlin.

Marcus, L.F., and Newman, W.S. 1983. Hominid migrations and the eustatic sea-level paradigm: a critique. In Masters, P.M., and Flemming, N.C. (eds.) *Quaternary Coastlines and Marine Archaeology: Towards the Prehistory of Land Bridges and Continental Shelves*, pp. 63–85. Academic Press, New York.

Marett, R.R. 1912. Pleistocene man in Jersey. *Archaeologia, Lond.* 62 (2): 450–80.

— 1916. The site, fauna and industry of La Cotte de St. Brelade, Jersey. *Archaeologia, Lond.* 67: 75–118.

— 1918. Exploration of La Cotte de St. Brelade. Report on work done in 1917. *Bull. ann. Soc. Jersiaise* 43: 354–7.

— 1919. Exploration of La Cotte de St. Brelade. Report for the year 1918. *Bull. ann. Soc. Jersiaise* 44: 50–3.

Marett, R.R., de Gruchy, G.F.B., Keith, A., Andrews, C., Dunlop, A., Gardner Warton, R., and Balfour, H. 1916. Exploration of the Palaeolithic site known as La Cotte de St. Brelade, Jersey. *Rept. British Ass.* 1916: 292–4.

Marks, S.A. 1976. *Large Mammals and a Brave People.* University of Washington Press, Washington.

Marr, J.E., Moir, J.R., and Smith, R.A. 1921. Excavations at High Lodge, Mildenhall. *Proc. Prehist. Soc. East Anglia* 3: 353–79.

Marshall, L. 1976. Sharing, talking and giving: relief of social tensions among the !Kung. In Lee, R.B., and DeVore, I. (eds.) *Kalahari Hunter-Gatherers. Studies of the !Kung San and their Neighbours*, pp. 349–71. Harvard University Press, Cambridge, Mass.

Mary, G. 1982. Rôle probable de l'isostasie dans les modalités de la transgression holocène sur la côte atlantique de l'Europe et de l'Afrique. *Bull. Assoc. Fr. Et. Quat.* 19(1): 39–45.

Masset, C., and van Vliet, B. 1974. Observations sur les sédiments d'une sépulture collective. La Chaussée Tirancourt (Somme). *Bull. Soc. Préhist. Fr.* 71: 243–8.

Matthews, J.V., Jr. 1982. East Beringia during late Wisconsin time: a review of biotic evidence. In Hopkins, D.M., Matthews, J.V., Jr., Schweger, C.E., and Young, S.B. (eds.) *Paleoecology of Beringia*, pp. 127–50. Academic Press, New York.

Matthews, R.K. 1973. Relative elevation of Late Pleistocene high sea-level stands: Barbados uplift rates and their implications. *Quat. Res.* 3: 147–53.

Michel, D. 1971. Contribution à l'étude du Paléolithique de Montfarville (Manche). *Mém. Soc. Nationale Sci. Nat. et Math. de Cherbourg.* 55: 17–78.

Milankovitch, M.M. 1941. *Canon of Insolation and the Ice-Age Problem.* U.S. Dept. of Commerce and the National Science Foundation, Washington.

Miskovsky, J.C. 1976. Le Pléistocene du Midi méditerranéen (Provence et Languedoc) d'après les remplissages de grottes et abris sous roche. In Lumley, H. de (ed.) *La Préhistoire Française*, pp. 201–24. CRNS, Paris.

Mitchell, G.F. 1970. The Quaternary deposits between Fenit and Spa on the north shore of Tralee Bay, Co. Kerry. *Proc. R. Ir. Acad.* Series B, 70: 141–62.

— 1977. Raised beaches and sea-levels. In Shotton, F.W. (ed.), *British Quaternary Studies — Recent Advances*. Clarendon Press, Oxford.

Mitchell, G.F., Catt, J.A., Weir, A.H., McMillan, N.F., Margarel, J.P., and Whatley, R.C. 1973. The Late Pliocene marine formation at St. Erth, Cornwall. *Phil. Trans. R. Soc. Lond.* Series B, 266: 1–37.

Mitchell, G.F., Penny, L.F., Shotton, F.W., and West, R.G. 1973. A correlation of Quaternary deposits in the British Isles. *Geol. Soc. Lond., Spec. Rep.* 4: 1–99.

Monnier, J.L. 1980a. Le station paléolithique du Mont-Dol (Ille et Vilaine). *Dossiers du Centre Régional Archéologique d'Alet* C–1980: 3–19.

— 1980b. *La Paléolithique de Bretagne dans son cadre géologique.* Thèse Sc.,Univ. de Rennes, 1980. Travaux Lab. anthr. préhist. protohist., Rennes.

— 1982. Le Paléolithique inférieur et moyen en Bretagne. Habitats et économie des matières premières. *Bull. Assoc. Fr. Et. Quat.* 19(2–3): 93–104.

Morlan, R.E. 1983. Spiral fractures on limb bones: which ones are artificial? In LeMoine, G.M., and MacEachern, A.S. (eds.) *Carnivores, Human Scavengers and Predators: a Question of Bone Technology*, pp. 241–69. Archaeol. Assoc. Univ. Calgary, Calgary.

Morzadec-Kerfourn, M.T. 1974. Variations de la ligne de rivage armoricain au Quaternaire. *Mém. Soc. Géol. minér., Bretagne* 17: 1–201.

Mottershead, D.N. 1971. Coastal head deposits between Start Point and Hope Cove, Devon. *Field Stud.* 3: 433–53.

Mourant, A.E. 1933. The raised beaches and other terraces of the Channel Islands. *Geol. Mag.* 70: 58–66.

— 1935. The Pleistocene deposits of Jersey. *Bull. ann. Soc. Jersiaise* 12: 489–96.

Movius, H.L. 1950. A wooden spear of Third Interglacial age from

Lower Saxony. *South-western J. of Anthrop.* 6 (2): 139–42.

Murray, A.S. 1981. *Environmental Radioactivity Studies relevant to Thermoluminescence Dating.* Unpublished D.Phil. thesis, Oxford University.

Myers, T.P., Voorhies, M.R., and Corner, R.G. 1980. Spiral fractures and bone pseudotools at paleontological sites. *Am. Antiq.* 45 (3): 483–90.

Newcomer, M.H., and Hivernel-Guerre, F. 1974. Nucléus sur éclat: technologie et utilisation par differentes cultures préhistoriques. *Bull. Soc. Préhist. Fr.* 71(4): 119–28.

Nicolle, E.T., and Sinel, J. 1911. Report on the exploration of the palaeolithic cave-dwelling known as La Cotte, St. Brelade, Jersey. *Bull. ann. Soc. Jersiaise* 36: 69–74.

1912. Report on the resumed exploration of La Cotte de Saint Brelade by the Société Jersiaise. *Man* 12: 158–62.

Nie, H.N., Hull, C.H., Jenkins, J.G., Steinbrenner, K., and Bent, D.H. 1975. *Statistical Package for the Social Sciences* (2nd edn.). McGraw-Hill, New York.

Nilsson, T. 1983. *The Pleistocene: Geology and Life in the Quaternary Ice Age.* Reidel, Dordrecht.

Ninkovich, D., and Shackleton, N.J. 1975. Distribution, stratigraphic position and age of ash layer "L", in the Panama Basin region. *Earth and Planet. Sci. Lett.* 27: 20–34.

Oakley, K.P. 1955. Fire as a palaeolithic tool and weapon. *Proc. Prehist. Soc.* 21: 36–47.

1958. Use of fire by Neanderthal Man and his precursors. In von Koenigswald, G.H.R. (ed.) *Hundert Jahre Neanderthaler 1856–1956.* Wenner-Gren Foundation, New York.

1961. On Man's use of fire, with comments on tool-making and hunting. In Washburn, S. (ed.) *Social Life of Early Man.* Viking Fund Publications in Anthropology. 31: 176–93.

Oakley, K.P., Andrews, P., Keeley, L.H., and Clark, J.D. 1977. A reappraisal of the Clacton spearpoint. *Proc. Prehist. Soc.* 43: 13–30.

Oakley, K.P., Campbell, B.G., and Molleson, T.I. 1971. *Catalogue of Fossil Hominids. Part I. Europe.* British Museum (Natural History), London.

Oates, C.G. (ed.) 1881. *Matabele Land and the Victoria Falls. Letters and Journals of the late Frank Oates.* C. Kegan Paul, London.

Odell, G. 1977. *The Application of Microwear Analysis to the Lithic Component of an Entire Prehistoric Settlement: Methods, Problems and Functional Reconstructions.* Unpublished Ph.D. Dissertation, Harvard.

Odell, G., and Odell-Vereecken, F. 1980. Verifying the reliability of lithic use-wear assessments by 'blind tests': the low-power approach. *J. Field Archaeol.* 7: 87–120.

Ohel, M.Y. 1979a. The Clactonian: an independent complex or an integral part of the Acheulean? *Current Anthrop.* 20(4): 685–726.

1979b. The Clactonian notch reconsidered. *Quartär* 29–30: 167–8.

Olivier, R.C.D. 1982. Ecology and behavior of living elephants: bases for assumptions concerning the extinct woolly mammoths. In Hopkins, D.M., Matthews, J.V., Jr., Schweger, C.E., and Young, S.B. (eds.) *Paleoecology of Beringia,* pp. 291–305. Academic Press, New York.

Osborn, H.F. 1936, 1942. *Proboscidea* (2 vols). American Museum Press, New York.

Otte, M. 1980. Le 'Couteau de Kostienki'. *Helinium* 20: 54–8.

Ovey, C.D. (ed.) 1964. *The Swanscombe Skull. A Survey of Research on a Pleistocene Site.* R. Anthrop. Inst. Occasional Paper no. 20.

Paepe, R., and Sommé, J. 1970. Les loess et la stratigraphie du Pléistocène récent dans le Nord de la France et en Belgique. *Ann. Soc. Géol. Nord* 90(4): 191–201.

Passingham, R.E. 1982. *The Human Primate.* Freeman, Oxford.

Penck, A., and Bruckner, E. 1909. *Die Alpen im Eiszeitalter.* Tauchnitz, Leipzig.

Peneaud, P. 1978. *La paragénèse phosphatée de la grotte de la Caune de l'Arago (Pyrénées Orientales).* Thèse de Doctorat, 3ème cycle, Univ. de Paris VII.

Perlès, C. 1976. Le feu. In Lumley, H. de (ed.) *Le Préhistoire Française,* pp. 679–83. CNRS, Paris.

1977. *Préhistoire du Feu.* Masson, Paris.

1981. Hearth and home in the Old Stone Age. *Nat. Hist.* 90(10): 38–41.

Pfizenmayer, E.W. 1939. *Siberian Man and Mammoth.* Blackie, London.

Piningre, J.F. 1978. Quelques caractéristiques du débitage du site paléolithique de Biache-Saint-Vaast (Pas-de-Calais). *Bull. Assoc. Fr. Et. Quat.* 15(1–3): 56–9.

Pope, S.T. 1974. *Bows and Arrows* (reprint edition). University of California Press, Berkeley.

Reher, C.A. 1977. Adaptive processes on the shortgrass plains. In Binford, L.R. (ed.) *For Theory Building in Archaeology,* pp. 13–40. Academic Press, New York.

Renouf, J.T. 1971. Geological report for 1970. *Ann. Bull. Soc. Jersiaise* 20: 224–5.

Rey, C. 1980. Climatological report for 1979. *Ann. Bull. Soc. Jersiaise* 22(4): 472–3.

Rigaud, J.Ph., and Tixier, J.P. 1981. A propos des particularités techniques et typologiques du gisement des Tares, commune de Sourzac (Dordogne). *Bull. Soc. Préhist. Fr.* 78(4): 109–17.

Roe, D.A. 1968. British Lower and Middle Palaeolithic handaxe groups. *Proc. Prehist. Soc.*

1981. *The Lower and Middle Palaeolithic Periods in Britain.* Routledge and Kegan Paul, London.

Rolland, N. 1981. The interpretation of Middle Palaeolithic variability. *Man* (n.s.) 16: 14–52.

1986. Recent findings from La Micoque and other sites in southwestern and Mediterranean France: their bearing on the 'Tayacian' problem and Middle Palaeolithic emergence. In Bailey, G.N., and Callow, P. (eds.). *Stone Age Prehistory. Studies in Memory of Charles McBurney.* Cambridge University Press, Cambridge.

Ruddiman, W.F., and McIntyre, A. 1982. Severity and speed of Northern Hemisphere glaciation pulses: the limiting case? *Bull. Geol. Soc. Am.* 93: 1273–9.

Sainty, J., and Thévenin, A. 1978. Le sol 74. *Rech. Géogr. Strasbourg* 7: 99–112.

Sakanoue, M., and Yoshioka, M. 1974. Uranium-series dating of bone samples from the Amad Cave site. *Quat. Res., Tokyo* 13: 220–4.

Schenkel, R., and Schenkel-Hulliger, L. 1969. *Ecology and Behaviour of the Black Rhinoceros* (Diceros bicornis Linn.): a field study. Parey, Hamburg.

Schiffer, M.B. 1978. Methodological issues in ethnoarchaeology. In Gould, R.A. (ed.) *Explorations in Ethnoarchaeology,* pp. 229–47. Univ. of New Mexico Press, Albuquerque.

Schmid, E. 1969. Cave sediments and prehistory. In Brothwell, D., and Higgs, E. (eds.) *Science in Archaeology,* 2nd ed., pp. 151–66. Thames and Hudson, London.

Schwarcz, H.P. 1980. Absolute age determination of archaeological sites by Uranium series dating of travertines. *Archaeometry* 22: 3–24.

Schwarcz, H.P., and Blackwell, B. 1983. $^{230}$Th/$^{230}$U age of Mousterian site in France. *Nature, Lond.* 301: 236–7.

Schwarcz, H.P., and Latham, A.G. 1984. Uranium-series age determination of travertines from the site of Vértesszöllös, Hungary. *J. Archaeol. Sci.* 11: 327–336.

Scott, K. (in prep.). *Archaeological and Non-archaeological Faunal Remains of the Devensian (Last) Cold Stage in Britain.* Ph.D. thesis in prep., University of Cambridge.

1980. Two hunting episodes of Middle Palaeolithic age at La Cotte de Saint Brelade, Jersey. *World Archaeol.* 12(2): 137–52.

Scott, W.B. 1937. *A History of Land Mammals in the Western Hemisphere.* MacMillan, New York.

Semenov, S.A. 1964. *Prehistoric Technology.* Cory, Adams and Mackay, London.

Shackleton, N.J. 1969. The last interglacial in the marine and terrestrial records. *Proc. R. Soc. Lond.* B. 174: 135–54.

1975. The stratigraphic record of deep-sea cores and its implications for the assessment of glacials, interglacials, stadials and interstadials in the mid-Pleistocene. In Butzer, K.W., and Isaac, G.Ll. (eds.) *After the Australopithecines: Stratigraphy, Ecology and Culture Change in the Middle Pleistocene,* pp. 1–24. Mouton, The Hague.

Shackleton, N.J., and Opdyke, N.D. 1973. Oxygen isotope and palaeomagnetic stratigraphy of equatorial Pacific core V28–238; oxygen isotope temperatures and ice volumes on a $10^5$ and $10^6$ year scale. *Quat. Res.* 3: 39–55.

1976. Oxygen isotope and palaeomagnetic stratigraphy of Pacific core V28–239, Late Pliocene to latest Pleistocene. In Cline, R.M., and Hays, J.D. (eds.) *Investigation of Late Quaternary Paleoceanography and Paleoclimatology*, pp. 449–64. Mem. Geol. Soc. Am. 145.

Shafer, H.J. 1970. Notes on uniface retouch technology. *Am. Antiq.* 35: 480–7.

1971. Investigations into South Plains prehistory, West Central Texas. In *Papers of the Texas Archaeological Salvage Project No. 20*. University of Texas, Austin.

Shotton, F.W. *et al.* 1983. United Kingdom contribution to the International Geological Correlation Programme; Project 24, Quaternary glaciations of the northern hemisphere. *Quat. Newsl.* 39: 19–25.

Shotton, F.W., and Williams, R.E.G. 1971. Birmingham University radiocarbon dates. *Radiocarbon* 13: 141–56.

Sinel, J. 1912. The prehistoric cave dwelling "Cotte à la Chèvre", St. Ouen. *Bull. ann. Soc. Jersiaise* 37: 209–13.

1914. *Prehistoric Times and Men of the Channel Islands* Bigwood, St. Helier, Jersey.

Singer, R., and Wymer, J.J. 1982. *The Middle Stone Age at Klasies River Mouth in South Africa*. University of Chicago Press, Chicago.

Sirodot, S. 1873. Fouilles exécutées au Mont-Dol (Ille-et-Vilaine) en 1872. *Mém. Soc. Emul. Côtes-du-Nord* 11: 59–108.

Skarlato, O.A. (ed.) 1977. *Mammoth Fauna of the Russian Plain and Eastern Siberia*. Acad. Sci. USSR, Leningrad.

Smith, E.A. 1981. Optimal foraging theory and the analysis of hunter-gatherer group size. In Winterhalder, B., and Smith, E.A. (eds.) *Hunter-Gatherer Foraging Strategies. Ethnographic and Archeological Analysis*, pp. 36–65. University of Chicago Press, Chicago.

Smith, W.G. 1894. *Man the Primeval Savage: his Haunts and Relics from the Hill-tops of Bedfordshire to Blackwall*. Stanford, London.

Solecki, R.S. 1975. Shanidar IV, a Neanderthal burial in Northern Iraq. *Science, N.Y.* 190: 880–6.

Solomon, J.D. 1933. The implementiferous gravels of Warren Hill. *Jl. R. Anthrop. Inst.* 63: 101–10

Sommé, J., Paepe, R., and Lautridou, J.P. 1980. Principes, méthodes et système de la stratigraphie du Quaternaire dans le nord-ouest de la France et la Belgique. In Chaline, J. (ed.) *Problèmes de Stratigraphie Quaternaire en France et dans les Pays Limitrophes*, pp. 148–62. Bull. Assoc. Fr. Et. Quat. (supplément).

Sommé, J., and Tuffreau, A. 1976. Le gisement acheuléen supérieur de Vimy. *UISPP* 9 (Nice). Livret-guide de l'excursion A10: 191–4.

Sparks, B.W., and West, R.G. 1972. *The Ice Age in Britain*. Methuen, London.

Stearns, C.E. 1976. Estimates of the position of sea-level between 140,000 and 75,000 years ago. *Quat. Res.* 6: 445–9.

Stephens, N. 1970. The West Country and Southern Ireland. In Lewis, C.A. (ed.) *The Glaciations of Wales and Adjoining Regions*. Longmans, London.

Straus, L.G. 1977. Of deerslayers and mountain men: Paleolithic faunal exploitation in Cantabrian Spain. In Binford, L.R. (ed.) *For Theory Building in Archaeology*, pp. 41–76. Academic Press, New York.

Stuart, A.J. 1982. *Pleistocene Vertebrates in the British Isles*. Longman, London.

Sturdy, D.A. 1975. Some reindeer economics in prehistoric Europe. In Higgs, E.S. (ed.) *Palaeoeconomy*, pp. 55–95. Cambridge University Press, Cambridge.

Sutcliffe, A.J. 1976. The British Glacial-Interglacial sequence. *Quat. Newsl.* 18: 1–7.

1980. Progress report on excavations in Minchin Hole, Gower. *Quat. Newsl.* 33: 1–17.

Sutcliffe, A.J., and Kowalski, K. 1976. Pleistocene rodents of the British Isles. *Bull. Br. Mus. Nat. Hist. (Geol.)* 27(2): 133–147.

Sutcliffe, A.J., and Zeuner, F.E. 1962. Excavations in the Torbryan Caves, Devonshire. *Proc. Devon Archaeol. Explor. Soc.* 5: 127–45.

Szabo, B.J. 1980. Results and assessment of uranium-series dating of vertebrate fossils from Quaternary alluviums in Colorado. *Arct. and Alp. Res.* 12: 95–100.

Szabo, B.J., and Collins, D. 1975. Ages of fossil bones from British

interglacial sites. *Nature, Lond.* 254: 680–2.

Szabo, B.J., Malde, H.E., and Irwin-Williams, C. 1969. Dilemma posed by uranium-series dates on archaeologically significant bones from Valsequillo, Puebla, Mexico. *Earth and Planet. Sci. Lett.* 6: 237–44.

Tavoso, A. 1984. Réflexions sur l'économie des matières premières au Moustérien. *Bull. Soc. Préhist. Fr.* 81(3): 79–82.

Teilhard de Chardin, P. 1965. *The Making of a Mind; Letters from a Soldier-Priest 1914–19*. Collins, London.

Tester, P.J. 1965. An Acheulian site at Cuxton. *Archaeol. Cant.* 80: 30–60.

Texier, A. 1972. Le gisement paléolithique du lieu-dit 'Les Vignes', Commune de Voulème (Vienne). *L'Anthropologie* 76 (7–8): 727–40.

Thévenin, A. 1976. Les premières industries humaines en Alsace. In Lumley, H. de (ed.) *La Préhistoire Française*, pp. 810–816. CNRS, Paris.

Thieme, H., Brunnacker, K., and Juvigné, E. 1981. Petrographische und urgeschichtliche Untersuchungen im Lössprofil von Rheindalen/Niederrhenische Bucht. *Quartär* 31/32: 41–67.

Tite, M.S., and Waine, J. 1962. Thermoluminescent dating: a reappraisal. *Archaeometry* 5: 53–79.

Tixier, J. 1956. Le hachereau dans l'Acheuléen Nord-Africain. Notes typologiques. *Congrés Préhistorique de France* XV (Poitiers — Angoulême): 914–23.

Toepfer, V. 1957. Die Mammutfunde von Pfännerhall im Geiseltal. *Veröffentlichungen des Landesmuseums für Vorgeschichte in Halle* 16: 1–58.

Toth, N. 1985. Archaeological evidence for preferential right-handedness in the lower and middle Pleistocene. *J. Hum. Evol.* 14: 607.

Tuffreau, A. 1971. Quelques aspects du Paléolithique ancien et moyen dans le Nord de la France. *Bull. Soc. Préhist. Nord*, Special Number 8.

1976. Les civilisations du Paléolithique moyen dans la région parisienne et en Normandie. In Lumley, H. de (ed.) *La Préhistoire Française*, pp. 1098–1104 CNRS, Paris.

1978. Les fouilles du gisement paléolithique de Biache-Saint-Vaast (Pas-de-Calais): années 1976 et 1977 — premiers résultats. *Bull. Assoc. Fr. Et. Quat.* 15(1–3): 46–55.

1979. Recherches récentes sur le Paléolithique inférieur et moyen de la France septentrionale. *Bull. Soc. R. Belge Anth. et Préhist.* 90: 161–77.

1982. The transition Lower/Middle Palaeolithic in northern France. In Ronen, A. (ed.) *The Transition from Lower to Middle Palaeolithic and the Origin of Modern Man*, pp. 138–49. British Archaeological Reports, International Series 151.

1984. Le Paléolithique dans le Nord de la France. *Cahiers Géog. Phys., Lille* 5: 7–29.

In press. *Chronostratigraphie et Faciès Culturels du Palaéolithique Inferieur et Moyen dans l'Europe du Nord-Ouest*. Papers given at Congr. Préhist. France (Lille, 1984).

Tuffreau, A., Munaut, A.V., Puisségur, J.J., and Sommé, J. 1982a. Stratigraphie et environnement de la séquence archéologique de Biache-Saint-Vaast (Pas-de-Calais). *Bull. Assoc. Fr. Et. Quat.* 19(2–3): 57–62.

1982b. Stratigraphie et environnement des industries acheuléennes de la moyenne terrasse du bassin de la Somme (région d'Amiens). *Bull. Assoc. Fr. Et. Quat.* 19(2–3): 73–82.

Tuffreau, A., and Zuate y Zuber, J. 1975. La terrasse fluviatile de Bagarre (Etaples, Pas-de-Calais) et ses industries: note préliminaire. *Bull. Soc. Préhist. Fr.* 72(8): 229–35.

Turner, C., and West, R.G. 1968. The subdivision and zonation of interglacial periods. *Eiszeit. und Gegen.* 19: 93–101.

Valoch, K. 1976. Un groupe spécifique du Paléolithique ancien et moyen d'Europe Centrale. *UISPP* 9 (Nice). Colloque 10: 86–91.

1982. The Lower/Middle Palaeolithic transition in Czechoslovakia. In Ronen, A. (ed.) *The Transition from Lower to Middle Palaeolithic and the Origin of Modern Man*. British Archaeological Reports, International Series 151.

van der Hammen, T., Wijmstra, J.A., and Zagwijn, W.H. 1971. The floral record of the Late Cenozoic of Europe. In Turekian, K.K. (ed.) *The Late Cenozoic Ice Ages*, pp. 391–424. Yale University Press, New Haven.

# BIBLIOGRAPHY

van Montfrans, H.M. 1971. *Palaeomagnetic Dating in the North Sea Basin*. Princo N.V., Amsterdam.

van Vliet, B. 1975. *Bijdrage tot de paleopedologie van Boven Pleistoceen voornamelijk in het bekken van de Haine*. Thèse de Doctorat, Univ. de Gand.

van Vliet, B., Faivre, P., Andreux, F., Robin, A.M., and Portal, J.M. 1982. Comportement [sic] of some organic components in blue and U.V. light. Application to the micromorphology of podzols, andosols and planosols. In Bullock, P., and Murphy, P. (eds.) *Soil Micromorphology; Proceedings of 6th Inter. Work Meet. 1981*. London.

van Vliet-Lanoë, B. 1980. Approche des conditions physico-chimiques favorisant l'autofluorescence des minéraux argileux. *Pédologie* 30(2): pp. 369–90.

——— 1982. Structures et microstructures associées à la formation de glace de ségrégation: leurs conséquences. In French, H. (ed.) *The R. Brown Memorial Volume; Proceedings of 4th Canadian Permafrost Conference, Calgary, Alberta, March 2–6, 1981*, pp. 116–22. NRC, Canada.

——— 1983. *Etudes paléopédologiques*. Internal publication, Laboratoire de Géomorphologie du CNRS, Caen.

van Vliet-Lanoë, B., and Valadas, B. 1983. A propos des formations déplacées des versants crystallins des massifs anciens: la rôle de la glace de ségrégation dans la dynamique. *Bull. Assoc. Fr. Et. Quat.* 20(4): 153–60.

van Zeist, W., and Bottema, S. 1977. Palynological investigations in western Iran. *Palaeohistoria* 19: 19–95.

Vaughan, P. 1986. Wear analysis of a lower Magdalenian flint assemblage from S.W. France. In Newcomer, M., and Sieveking, G. de G. (eds.) *Human uses of flint and chert: papers from the Fourth International Flint Symposium*. Cambridge University Press, Cambridge.

Vayson de Pradenne, A. 1929. La station paléolithique du Mont-Dol. *L'Anthropologie* 39: 1–42.

Vértes, L. 1966. The Upper Palaeolithic site on Mt. Henye at Bodrogkereszur. *Acta Archaeol. Acad. Sci. Hung.* 18: 3–15

Vereschagin, N.K., and Baryshnikov, G.F. 1982. Paleoecology of the mammoth fauna in the Eurasian Arctic. In Hopkins, D.M., Matthews, J.V., Jr., Schweger, C.E., and Young, S.B. (eds.) *Paleoecology of Beringia*, pp. 267–279. Academic Press, New York.

Villa, P. 1981. Matières premières et provinces culturelles dans l'Acheuléen français. *Quaternaria* 23: 19–35.

Vita-Finzi, C., Higgs, E.S. with Sturdy, D.A., Harriss, J., Legge, A.J., and Tippet, H. 1970. Prehistoric economy in the Mount Carmel area of Palestine: site catchment analysis. *Proc. Prehist. Soc.* 36: 1–37.

Vogel, J.C., and Waterbolk, H.T. 1963. Groningen radiocarbon dates IV. *Radiocarbon* 5: 163–202.

Vrba, E.S. 1980. The significance of bovid remains as indicators of environment and predation patterns. In Behrensmeyer, A.K., and Hill, A.P. (eds.) *Fossils in the Making: Vertebrate Taphonomy and Paleoecology*, pp. 247–71. University of Chicago Press, Chicago and London.

Waechter, J. d'A. 1973. The late middle Acheulian industries in the Swanscombe area. In Strong, D.E. (ed.) *Archaeological Theory and Practice*. Seminar Press, London.

Walker, P.L. 1978. Butchery and stone tool function. *Am. Antiq.* 43: 710–715.

Watson, E., and Watson, S. 1970. The coastal periglacial slope deposits of the Cotentin peninsula. *Trans. Inst. Br. Geogr.* 49: 125–43.

Wernert, P. 1957. Stratigraphie paléontologique et préhistorique des sédiments quaternaires d'Alsace: Achenheim. *Mém. Serv. Carte Géol. Alsace et Lorraine* 14.

West, R.G. 1977. *Pleistocene Geology and Biology with especial reference to the British Isles* (2nd edn.). Longman, London.

West, R.G., and Sparks, B.W. 1960. Coastal interglacial deposits of the English Channel. *Phil. Trans. R. Soc. Lond.* Series B, 243: 95–133.

Wheat, J.B. 1972. The Olsen-Chubbock site — A Paleo-Indian bison kill. *Mem. Soc. Am. Archaeol.* No. 26.

Williams, R.B.G. 1975. The British Climate during the last glaciation: an interpretation based on periglacial phenomena. In Wright, A.E., and Moseley, F. (eds) *Ice Ages: Ancient and Modern*, pp. 95–120. Seel House Press, Liverpool.

Wilson, D. 1885. Palaeolithic dexterity. *Proc. Trans. R. Soc. Can.* 3(2): 119–133.

Wilmsen, E.N. 1968. Functional analysis of flaked stone artefacts. *Am. Antiq.* 33: 156–61.

——— 1891. *Left-handedness*. Macmillan, London.

Windels, F. 1948. *Lascaux*. Centre d'Etudes et de Documentation préhistoriques, Montignac.

Winterhalder, B. 1981. Foraging strategies in boreal forest: an analysis of Cree hunting and gathering. In Winterhalder, B., and Smith, E.A. (eds.) *Hunter-Gatherer Foraging Strategies: Ethnographic and Archeological Analysis*, pp. 66–98. University of Chicago Press, Chicago.

Wintle, A.G. 1973. Anomalous fading of thermoluminescence in mineral samples. *Nature, Lond.* 245: 143–4.

——— 1977. Detailed study of a thermoluminescent mineral exhibiting anomalous fading. *J. Lumin.* 15: 385–93.

——— 1982. Thermoluminescence properties of fine-grain minerals in loess. *Soil Sci.* 134: 164–70.

Wintle, A.G., and Huntley, D.J. 1982. Thermoluminescence dating of sediments. *Quat. Sci. Rev.* 1: 31–53.

Wintle, A.G., Shackleton, N.J., and Lautridou, J.P. In press. Thermoluminescence dating of periods of loess deposition and soil formation in Normandy. *Nature Lond.*

Wishart, D. 1978. *Clustan User Manual* (3rd edn.) Program Library Unit, Edinburgh University, Edinburgh.

Woillard, G. 1974. *Exposé des recherches palynologiques sur le Pléistocène dans l'Est de la Belgique et dans les Vosges Lorraines*. Trav. Lab. de Palynologie et de Phytosociologie, Univ. de Louvain.

——— 1975. Recherches palynologiques sur le Pléistocène dans l'Est de la Belgique et les Vosges lorraines. *Acta Geogr. Lovaniensia* 14.

——— 1978. Grand Pile peat bog: a continuous pollen record for the last 140,000 years. *Quat. Res.* 9: 1–21.

——— 1980. The pollen record of Grand Pile (N.E. France) and the climatic chronology through the last interglacial-glacial cycle. In Chaline, J. (ed.) *Problèmes de Stratigraphie Quaternaire en France et dans les Pays Limitrophes*, pp. 95–103. Bull. Assoc. Fr. Et. Quat., (supplément).

Woillard, G., and Mook, W.G. 1982. Carbon-14 dates and Grande Pile: correlation of land and sea chronologies. *Science* 215: 159–61

Woldstedt, P. 1954. *Das Eiszeitalter*, vol. 1. Ferdinand Enke Verlag, Stuttgart.

——— 1958. *Das Eiszeitalter*, vol. 2. Ferdinand Enke Verlag, Stuttgart.

Wolpoff, M.H. 1979. The Krapina dental remains. *Am. J. Phys. Anthrop.* 50: 67–113.

Wooldridge, S.W. 1927. The Pliocene history of the London Basin. *Proc. Geol. Ass., Lond.* 38: 49–132.

Wu, R., and Lin, S. 1983. Peking man. *Sci. Am.* 248(6): 78–86.

Wymer, J.J. 1968. *Lower Palaeolithic Archaeology in Britain*. John Baker, London.

——— 1974. Clactonian and Acheulean industries in Britain — their chronology and significance. *Proc. Geol. Ass.* 85(3): 391–421.

——— 1983. The Lower Palaeolithic site at Hoxne. *Proc. Suffolk Inst. Archaeol., and Hist.* 35: 169–89.

Yellen, J. 1977. *Archaeological Approaches to the Present*. Academic Press, London.

Zagwijn, W.H. 1975. Variations in climate as shown by pollen analysis, especially in the Lower Pleistocene of Europe. In Wright, A.E. and Mosely, F. (eds.) *Ice Ages, Ancient and Modern*, pp. 137–52. Geol. J. Spec. Issue 6. Seel House Press, Liverpool.

Zagwijn, W.H., van Montfrans, H.M., and Zandstra, J.G. 1971. Subdivision of the "Cromerian" in the Netherlands: pollen analysis, palaeomagnetism and sedimentary petrology. *Geol. en Mijn.* 50: 41–58.

Zeuner, F.E. 1940. *The Age of Neanderthal Man, with notes on the Cotte de St. Brelade, Jersey, C.I.* London Univ. Inst. of Archaeol. Occasional Papers, 3.

——— 1959. *The Pleistocene Period*. Hutchinson, London.

# INDEX